About the Author

Rodney Stark grew up in Jamestown, North Dakota, and received his Ph.D. from the University of California, Berkeley, where he held appointments as a Research Sociologist at the Survey Research Center and the Center for the Study of Law and Society. He left Berkeley to become Professor of Sociology and of Comparative Religion at the University of Washington. In 2004, he became University Professor of the Social Sciences at Baylor University. e has published twenty-seven books and more than 150 scholarly articles on subjects as diverse as prejudice, crime, suicide, and city life in ancient Rome. However, the greater part of his work has been on religion. He is past president of both the Association for the Sociology of Religion and the Society for the Scientific Study of Religion. He has won a number of national and international awards for distinguished scholarship and some of his books and articles have been translated and published in foreign languages, including: Bahasa Indonesian, Chinese, Dutch, French, German, Greek, Italian, Japanese, Polish, Portuguese, Romanian, Spanish, and Turkish. Among his recent books are *For the Glory of God: How Monotheism Led to Reformations, Science, Witch-Hunts, and the End of Slavery* (2003); *Exploring the Religious Life* (2004); *The Victory of Reason: How Christianity Led to Freedom, Capitalism, and Western Success* (2005), and *Cities of God: Christianizing the Urban Empire* (2006). He maintains a website at rodneystark.com

www.wadsworth.com

www.wadsworth.com is the World Wide Web site for Thomson Wadsworth and is your direct source to dozens of online resources.

At *www.wadsworth.com* you can find out about supplements, demonstration software, and student resources. You can also send e-mail to many of our authors and preview new publications and exciting new technologies.

www.wadsworth.com
Changing the way the world learns®

Sociology

TENTH EDITION

RODNEY STARK
Baylor University

THOMSON

WADSWORTH

Australia • Brazil • Canada • Mexico • Singapore • Spain • United Kingdom • United States

THOMSON
™
WADSWORTH

Sociology, **Tenth Edition**
Rodney Stark

Senior Sociology Editor: *Robert Jucha*
Assistant Editor: *Kristin Marrs*
Editorial Assistant: *Katia Krukowski*
Technology Project Manager: *Dee Dee Zobian*
Marketing Manager: *Michelle Williams*
Marketing Assistant: *Jaren Boland*
Marketing Communications Manager: *Linda Yip*
Project Manager, Editorial Production: *Cheri Palmer*
Creative Director: *Rob Hugel*
Print Buyer: *Rebecca Cross*

Permissions Editor: *Joohee Lee*
Production Service: *Dusty Friedman, The Book Company*
Text Designer: *Gopa & Ted2, Books & Design*
Photo Researcher: *Myrna Engler*
Copy Editor: *Frank Hubert*
Illustrator: *Impact Publications*
Cover Designer: *Yvo Riezebos, Yvo Riezebos Design*
Cover Image: *GJON MILI/Getty Images*
Compositor: *Lachina Publishing Services*
Printer: *Quebecor World/Versailles*

Library of Congress Control Number: 2005934429

ISBN 0-495-09344-0

Thomson Higher Education
10 Davis Drive
Belmont, CA 94002-3098
USA

For more information about our products, contact us at:
Thomson Learning Academic Resource Center
1-800-423-0563
For permission to use material from this text or product, submit a request online at **http://www.thomsonrights.com**. Any additional questions about permissions can be submitted by e-mail to **thomsonrights@thomson.com**.

Brief Contents

PART I   PRINCIPLES OF SOCIOLOGY

CHAPTER 1   Groups and Relationships: A Sociological Sampler   1

CHAPTER 2   Concepts for Social and Cultural Theories   31

CHAPTER 3   Microsociology: Testing Interaction Theories   67

CHAPTER 4   Macrosociology: Studying Larger Groups and Societies   95

PART II   INDIVIDUALS AND GROUPS

CHAPTER 5   Biology, Culture, and Society   125

CHAPTER 6   Socialization and Social Roles   147

CHAPTER 7   Crime and Deviance   175

CHAPTER 8   Social Control   209

PART III   INEQUALITY

CHAPTER 9   Concepts and Theories of Stratification   231

CHAPTER 10   Comparing Systems of Stratification   255

CHAPTER 11   Racial and Ethnic Inequality and Conflict   283

CHAPTER 12   Gender and Inequality   325

PART IV   INSTITUTIONS

CHAPTER 13   The Family   359

CHAPTER 14   Religion   387

CHAPTER 15   Politics and the State   415

CHAPTER 16   The Interplay between Education and Occupation   445

PART V   CHANGE

CHAPTER 17   Social Change: Development and Global Inequality   471

CHAPTER 18   Population Changes   499

CHAPTER 19   Urbanization   533

CHAPTER 20   The Organizational Age   563

CHAPTER 21   Social Change and Social Movements   587

EPILOGUE   Becoming a Sociologist   608

References   613

Credits   631

Name Index   633

Subject Index / Glossary   641

Contents

Part I Principles of Sociology

Chapter 1

GROUPS AND RELATIONSHIPS: A SOCIOLOGICAL SAMPLER   1

Science: Theory and Research   1
The Discovery of Social Facts   3
 Stability and Variation   4
 The Upward Trend   6
The Sociological Imagination   6
Sociology and the Social Sciences   7
 BOX 1-1 FOUNDERS OF AMERICAN SOCIOLOGY DEPARTMENTS   8
Units of Analysis   9
Micro- and Macrosociology   11
A Global Perspective   11
Scientific Concepts   13
Groups: The Sociological Subject   13
 Primary and Secondary Groups   14
Solidarity and Conflict: The Sociological Questions   15
Analyzing Social Networks   15
 Networks and Social Solidarity   15
 Networks and Social Conflict   16
 A CLOSER VIEW Samuel Sampson: Monastic Strife   16
Studying Self-Aware Subjects   17
 Unobtrusive Measures   17
 Validation   19
 Bias   20
The Social Scientific Process   21
Free Will and Social Science   23
Conclusion   24
Review Glossary   24
Suggested Readings   25
Sociology Online   26

Chapter 2

CONCEPTS FOR SOCIAL AND CULTURAL THEORIES   31

The Concept of Society   33
Social Structural Concepts   33
Stratification   35
Networks   35
    Networks and Power   35
    Local and Cosmopolitan Networks   37
    Societies as Networks   38
The Concept of Culture   39
Cultural Concepts   39
    Values and Norms   39
    Roles   41
    Multiculturalism and Subcultures   41
    Prejudice and Discrimination   42
    Assimilation and Accommodation   42
Modernization and Globalization   43
Jews and Italians in North America   43
    Prejudice and Discrimination   44
    Assimilation and Accommodation of Jews and Italians   47
Theorizing about Ethnic Mobility   48
The Cultural Theory   48
    A CLOSER VIEW   Zborowski and Herzog: Jewish Culture   48
    A CLOSER VIEW   Leonard Covello: Italian Culture   50
The Social Theory   51
    A CLOSER VIEW   Stephen Steinberg: The Jewish Head Start   52
    A CLOSER VIEW   Joel Perlmann: A New Synthesis   55
Reference Groups and Italian Traditionalism   57
    A CLOSER VIEW   Andrew Greeley: The Persistence of Italian Culture   61
Conclusion   62
Review Glossary   62
Suggested Readings   63
Sociology Online   65

Chapter 3

MICROSOCIOLOGY: TESTING INTERACTION THEORIES   67

The Rational Choice Proposition   69
Interaction Theories   71
Symbolic Interaction   72
    Social Construction of Meaning   73
    Social Construction of the Self   74
Exchange Theory   75
    Solidarity   76

Inequality   76
Agreement and Conformity   76
Theory Testing: Measurement and Research   77
Variables   77
Criteria of Causation   78
On Abstractions   78
BOX 3-1 CORRELATION   80
The Experiment: Studying Group Solidarity   81
A CLOSER VIEW   Solomon Asch: Solidarity and Conformity   81
Manipulating the Independent Variable   82
Randomization   83
Significance   83
Field Research: Studying Recruitment   83
A CLOSER VIEW   Lofland and Stark: Religious Conversion   84
Covert and Overt Observation   85
Conversion as Conformity   85
Four Principles of Recruitment   87
BOX 3-2 HEAVEN'S GATE   88
Conclusion   89
Review Glossary   90
Suggested Readings   92
Sociology Online   93

Chapter 4

MACROSOCIOLOGY: STUDYING LARGER GROUPS AND SOCIETIES   95

Describing Populations   96
Simple and Proportional Facts   96
BOX 4-1 EXPERIMENTAL EFFECTS OF SOCIAL STRUCTURE   97
Censuses and Samples   98
Why Sampling Works   98
Survey Research: Ideology and Conformity   100
A CLOSER VIEW   Hirschi and Stark: Hellfire and Delinquency   100
Spuriousness   101
Conflicting Results   102
Contextual Effects   102
Network Analysis: Group Properties   104
A CLOSER VIEW   Benjamin Zablocki: Networks of Love and Jealousy   104
The Sociology of Societies   107
Societies as Systems   108
Sociocultural Components   108
Interdependence Among Structures   109
Equilibrium and Change   109
Theories about Social Systems   110
Functionalism   110
Social Evolution   111
Conflict Theory   113

Comparative Research: Violence and Modernity   115
   Comparative Research   115
   Cross-Cultural "Samples"   115
     A CLOSER VIEW   Napoleon Chagnon: Life along the Amazon   117
     A CLOSER VIEW   Jeffery Paige: Conflict and Social Structures   118
Conclusion   120
Review Glossary   120
Suggested Readings   121
Sociology Online   122

Part II   Individuals and Groups
Chapter 5

BIOLOGY, CULTURE, AND SOCIETY   125

Heredity   127
   BOX 5-1 IDENTICAL TWINS   128
Behavioral Genetics   129
   BOX 5-2 MINORITIES AND STARDOM   130
The Growth Revolution   133
   Environmental Suppressors   134
Hormones and Behavior   135
     A CLOSER VIEW   The Vietnam Veterans Study   136
DNA and Culture   136
     A CLOSER VIEW   Origins of the Lemba   137
Humans and Other Animals   137
     A CLOSER VIEW   Jane Goodall's Great Adventure   138
   BOX 5-3 BIRD BRAIN   140
   Nonhuman Language   142
     A CLOSER VIEW   Washoe Learns to Sign   142
Conclusion   144
Review Glossary   144
Suggested Readings   144
Sociology Online   145

Chapter 6

SOCIALIZATION AND SOCIAL ROLES   147

Human Development   149
   Suppressing Development   149
   Accelerating Development   150
     A CLOSER VIEW   Studying the Mythical "Mozart Effect"   150
Cognitive Development   151
     A CLOSER VIEW   Jean Piaget: Theory of Cognitive Stages   151
   Language Acquisition   154

A CLOSER VIEW   Noam Chomsky: Universal Grammar   154

Evidence of a Language Instinct   154

A CLOSER VIEW   Judy Kegl: The Nicaraguan Sign Language Study   155

Emotional Development   156

Emergence of the Self   156

Personality Formation   157

Culture and Personality   158

Differential Socialization   161

A CLOSER VIEW   Melvin Kohn: Occupational Roles and Socialization   163

A CLOSER VIEW   Erving Goffman: Performing Social Roles   165

Sex-Role Socialization   166

A CLOSER VIEW   DeLoache, Cassidy, and Carpenter: Baby Bear Is a Boy   168

Women and Science   169

Conclusion   170

Review Glossary   171

Suggested Readings   172

Sociology Online   173

Chapter 7

CRIME AND DEVIANCE   175

Ordinary Crime   177

Robbery   178

Burglary   179

Homicide   180

Offender Versatility   180

The Criminal Act   181

Biological Theories of Deviance   182

"Born Criminals"   182

Behavioral Genetics   183

A CLOSER VIEW   Walter Gove: Age, Gender, Biology, and Deviance   184

Mental Illness   186

Personality Theories   187

Elements of Self-Control   187

Deviant Attachment Theories   187

Differential Association: Social Learning   188

Subcultural Deviance   189

Structural Strain Theories   190

"White-Collar" Crime   192

Control Theories   194

Attachments   195

Investments   197

Involvements   197

Beliefs   197

A CLOSER VIEW   Linden and Fillmore: A Comparative Study of Delinquency   198

Anomie and the Integration of Societies   199

Climate and Season   200

The Labeling Approach to Deviance   201
Drugs and Crime   202
Conclusion   203
Review Glossary   205
Suggested Readings   206
Sociology Online   207

Chapter 8

SOCIAL CONTROL   209

Informal Control   210
    A CLOSER VIEW   Hechter and Kanazawa: Solidarity and Conformity
       in Japan   211
    Dependence   211
    Visibility   212
    Extensiveness   212
Formal Control   212
Prevention   213
    A CLOSER VIEW   The Cambridge-Somerville Experiment   214
    Other Prevention Programs   216
Deterrence   216
    A CLOSER VIEW   Jack Gibbs: A Theory of Deterrence   218
    The Capital Punishment Controversy   220
The Wheels of Justice   221
    The Police   221
    A CLOSER VIEW   Corman and Mocan: Cops and Deterrence   222
    The Courts   223
Reform and Resocialization   224
    A CLOSER VIEW   The TARP Experiment: Salaries for Ex-Convicts   225
Conclusion   226
Review Glossary   228
Suggested Readings   228
Sociology Online   229

Part III   Inequality

Chapter 9

CONCEPTS AND THEORIES OF STRATIFICATION   231

Conceptions of Social Class   231
Marx's Concept of Class   232
    The Bourgeoisie and the Proletariat   232
    Class Consciousness and Conflict   233
    The Economic Dimension of Class   235
Weber's Three Dimensions of Stratification   236
    Property   236
    Prestige   236
    Power   237

Status Inconsistency   237
Social Mobility   238
   Rules of Status: Ascription and Achievement   238
   Structural and Exchange Mobility   239
   Class Cultures and Networks   241
Marx and the Classless Society   242
   Dahrendorf's Critique   244
   Mosca: Stratification Is Inevitable   245
The Functionalist Theory of Stratification   245
   Replaceability   246
   A Toy Society   247
The Social Evolution Theory of Stratification   248
The Conflict Theory of Stratification   249
   Exploitation   249
   The Politics of Replaceability   249
Conclusion   251
Review Glossary   251
Suggested Readings   252
Sociology Online   253

Chapter 10

COMPARING SYSTEMS OF STRATIFICATION   255

Simple Societies   256
Agrarian Societies   259
   Productivity   259
   Warfare   259
   Surplus and Stratification   260
   Military Domination   261
   Culture and Ascriptive Status   261
   BOX 10-1 STIRRUPS, SADDLES, AND FEUDAL DOMINATION   262
Industrial Societies   265
   Industrialization and Stratification   266
   Social Mobility in Industrialized Nations   268
     A CLOSER VIEW   Lipset and Bendix: Comparative Social Mobility   269
     A CLOSER VIEW   Blau and Duncan: The Status Attainment Model   271
     A CLOSER VIEW   Biblartz, Raferty, and Bucur: Mothers and Mobility   273
     A CLOSER VIEW   John Porter and Colleagues: Status Attainment in Canada   273
   Female Status Attainment   275
     A CLOSER VIEW   Michael Hout: Trends in Status Attainment   275
     A CLOSER VIEW   Bonnie H. Erickson: Networks and Power in Canada   276
     A CLOSER VIEW   Yanjie Bian: Networks and Success in China   277
Conclusion   278
Review Glossary   279
Suggested Readings   280
Sociology Online   281

Chapter 11

RACIAL AND ETHNIC INEQUALITY AND CONFLICT   283

Intergroup Conflict   285
Race   285
Ethnic Groups   286
Cultural Pluralism   286
Preoccupation with Prejudice   288
Allport's Theory of Contact   291
A CLOSER VIEW   The Sherif Studies   292
Slavery and the American Dilemma   293
Status Inequality and Prejudice   296
A CLOSER VIEW   Edna Bonacich: Economic Conflict and Prejudice   297
Exclusion   300
Caste Systems   301
Middleman Minorities   301
Identifiability   301
Equality and the Decline of Prejudice   302
The Japanese Experience in North America   302
Japanese Americans   303
Japanese Canadians   305
Mechanisms of Ethnic and Racial Mobility   306
Geographical Concentration   307
Internal Economic Development and Specialization   308
Development of a Middle Class   310
Hispanic Americans   310
Going North: African American "Immigration" in the United States   312
African American Progress   315
Integration   317
A CLOSER VIEW   Firebaugh and Davis: The Decline in Prejudice   317
Barriers to African American Progress   318
BOX 11-1 DO WHITE FANS REJECT "TOO BLACK" TEAMS?   318
Conclusion   321
Review Glossary   322
Suggested Readings   322
Sociology Online   323

Chapter 12

GENDER AND INEQUALITY   325

An Operatic Insight   327
Sex Ratios and Sex Roles: A Theory   332
Sex Ratios   332
Causes of Unbalanced Sex Ratios   333
BOX 12-1 THE GILDED CAGE   334
Ancient Athens: Too Few Women   336
Ancient Sparta: Too Few Men   336
Sex Ratios and Power Dependence   338

Testing the Sex-Ratios and Sex-Roles Theory   340
    Trends in North American Sex Ratios   340
    Gender and Social Movements   342
    Women in the Labor Force   345
    The Sexual "Revolution"   349
    Sex Ratios and the African American Family   351
    Sex Ratios and the Hispanic American Family   353
Conclusion   354
Review Glossary   356
Suggested Readings   356
Sociology Online   357

Part IV   Institutions
Chapter 13

The Family   359

Defining the Family   360
Family Functions: An International Perspective   363
    Sexual Gratification   363
    Economic Support   364
    Emotional Support   364
    A CLOSER VIEW   Life in the Traditional European Family   365
    Household Composition   366
    Crowding   368
    "Outsiders"   368
    Child Care   369
    Relations between Husbands and Wives   370
    Bonds between Parents and Children   371
    Bonds among Peer Group Members   371
Modernization and Romance   372
Modernization and Kinship   373
Modernization and Divorce   375
    How Much Divorce?   375
    When Romance Fades   377
Gender and Extramarital Sex   377
    A CLOSER VIEW   Trent and South: International Comparisons in Divorce   377
Living Together   378
The One-Parent Family   379
Remarriage   380
    A CLOSER VIEW   Jacobs and Furstenberg: Second Husbands   381
    A CLOSER VIEW   White and Booth: Stepchildren and Marital Happiness   381
Household Chores   383
Conclusion   383
Review Glossary   384
Suggested Readings   384
Sociology Online   385

Chapter 14

RELIGION   387

The Nature of Religion   388
The Gods   389
Gods and Morality   391
Gender and Religious Commitment   393

Religious Economies   395
Church-Sect Theory   395
Secularization and Revival   398
Innovation: Cult Formation   399
Charisma   400
The American Religious Economy   400
Secularization and Revival   401
Secularization and Innovation   402

The Canadian Religious Economy   405
Cult Movements in Europe   406
The Protestant Explosion in Latin America   407
Eastern Revivals   408
The Universal Appeal of Faith   409
Conclusion   410
Review Glossary   411
Suggested Readings   412
Sociology Online   413

Chapter 15

POLITICS AND THE STATE   415

A CLOSER VIEW   Messick and Wilke: The "Tragedy of the Commons"
in the Laboratory   417

Public Goods and the State   419
Functions of the State   420
Box 15-1 THE REWARDS FOR FREE RIDING   421
Rise of the Repressive State   423

Taming the State   424
The Evolution of Pluralism   424
Elitist and Pluralist States   426
Pluralism or Power Elites?   428

Democracy and the People   429
A CLOSER VIEW   George Gallup: The Rise of Opinion Polling   429

The American "Voter"   432
Female Candidates   432
A CLOSER VIEW   Hunter and Denton: Do Canadian Voters Reject Women
for Parliament?   433
A CLOSER VIEW   Jody Newman: American Women as Candidates   435

Ideology and Public Opinion   436

Elite and Mass Opinion   438
  A CLOSER VIEW   Herbert McClosky: Elites and Ideology   439
  A CLOSER VIEW   Philip E. Converse: Issue Publics   439
  Organizations and Ideologies   440
Conclusion   441
Review Glossary   442
Suggested Readings   442
Sociology Online   443

Chapter 16

THE INTERPLAY BETWEEN EDUCATION AND OCCUPATION   445

Occupational Prestige   446
The Transformation of Work   448
The Transformation of the Labor Force   450
  Women in the Labor Force   451
  Professional Women   451
  Unemployment   452
  Job Satisfaction   453
The Transformation of Education   453
  A CLOSER VIEW   "Educated" Americans   455
Do Schools Really Matter?   456
  A CLOSER VIEW   Barbara Heyns: The Effects of Summer   457
High School Today   458
Homeschooling   461
  A CLOSER VIEW   Heyneman and Loxley: School Effects Worldwide   462
Does Education Pay?   462
  A CLOSER VIEW   John W. Meyer: Theory of Educational Functions   465
Conclusion   466
Review Glossary   468
Suggested Readings   468
Sociology Online   469

Part V   Change
Chapter 17

SOCIAL CHANGE: DEVELOPMENT AND GLOBAL INEQUALITY   471

Internal Sources of Social Change   472

  BOX 17-1 MODEST MILESTONES OF MODERNIZATION   473
  Innovations   473
  Conflicts   474
  Growth   474
Change and Cultural Lag   475

External Sources of Change   476
    Diffusion   477
    Conflict   478
    Ecological Sources of Change   478
The Rise of the West   480
Marx on Capitalism   481
    Capitalism   482
    Precapitalist Command Economies   482
The Protestant Ethic   483
The State Theory of Modernization   485
Dependency and World System Theory   486
    Elements of a World System   486
    Dominance and Dependency   488
    Mechanisms of Dependency   488
    A CLOSER VIEW   Jacques Delacroix: Testing the Dependency Theory   488
Dimensions of Global Inequality   492
Globalization   493
Conclusion   494
Review Glossary   495
Suggested Readings   496
Sociology Online   497

Chapter 18

POPULATION CHANGES   499

Demographic Techniques   501
    Rates   501
    Cohorts   503
    Age and Sex Structures   504
Preindustrial Population Trends   506
    Famine   506
    Disease   507
    War   508
Malthusian Theory   508
Modernization and Population   511
The Demographic Transition   511
    A CLOSER VIEW   Kingsley Davis: Demographic Transition Theory   512
The Second Population Explosion   514
    BOX 18-1 THE LIFE CYCLE OF THE BABY BOOM   515
    The Sudden Decline in Mortality   518
    High Fertility and Cultural Lag   519
The Population Explosion Wanes   519
    Economic Development   521
    Numeracy about Children   522
    Contraception and Wanted Fertility   523
The Crisis of Depopulation   523

Gray Nations   525
Culture and Immigration   526
   Low Fertility and Gender Bias   526
Conclusion   527
Review Glossary   529
Suggested Readings   530
Sociology Online   531

Chapter 19

URBANIZATION   533

Preindustrial Cities   534
   Limits on City Style   535
   BOX 19-1 CAUSES OF DEATH IN THE CITY OF LONDON, 1632   538
   Why Live in Such Cities?   538
Industrialization and Urbanization   540
   The Agricultural Revolution   540
   Specialization and Urban Growth   543
Metropolis   544
   The Fixed-Rail Metropolis   546
   The Freeway Metropolis   547
   Preferring a Decentralized Metropolis   548
   Commuting   549
Suburbs   550
Urban Neighborhoods   550
   A CLOSER VIEW   Park and Burgess: Ethnic Succession   551
   A CLOSER VIEW   Guest and Weed: Economics and Integration   551
Segregation in World Perspective   553
Theories of Urban Impact   554
   Anomie Theories   554
   Effects of Crowding   556
   Macrostudies of Crowding   556
   A CLOSER VIEW   Gove, Hughes, and Galle: Microstudies of Crowding   557
Conclusion   558
Review Glossary   559
Suggested Readings   560
Sociology Online   561

Chapter 20

THE ORGANIZATIONAL AGE   563

The Crisis of Growth: Inventing Formal Organizations   565
   The Case of the Prussian General Staff   565
   The Cases of Daniel McCallum and Gustavus Swift   567
   The Case of Civil Service   569
Weber's Rational Bureaucracy   570
Rational Versus Natural Systems   572

Goal Displacement   572

Goal Conflict   573

Informal Relations   573

The Crisis of Diversification   575

Functional Divisions   575

Autonomous Divisions   576

A CLOSER VIEW   Peter M. Blau: Theory of Administrative Growth   577

Rational and Natural Factors in Decentralization   578

Bureaucracy and the Bottom Line   580

Conclusion   582

Review Glossary   583

Suggested Readings   584

Sociology Online   585

Chapter 21

SOCIAL CHANGE AND SOCIAL MOVEMENTS   587

Sociological Approaches to Social Movements   589

Shared Grievances   590

Economic Domination   590

Political Domination   591

Personal Domination   591

Hope   592

A Precipitating Event   593

Network Ties   594

Mobilizing People and Resources   595

Internal Factors: Building the Movement   595

External Factors: Opponents and Allies   596

The Proliferation of Civil Rights Organizations   599

Freedom Summer: Mississippi, 1964   599

A CLOSER VIEW   Doug McAdam: The Freedom Summer Study   601

Becoming a Volunteer   602

Biographical Availability   602

Attitudes and Values   603

Social Networks   603

Participants Versus No-Shows   604

The Volunteers Twenty Years Later   604

Conclusion   605

Review Glossary   606

Suggested Readings   606

Sociology Online   607

Epilogue

BECOMING A SOCIOLOGIST   608

    Majoring in Sociology   608

    Going to Graduate School   609

        Selection   609

        Applying   610

        Succeeding in Graduate School   610

    On Learning to Write   611

    Sociological Careers   612

REFERENCES   613

CREDITS   631

NAME INDEX   633

SUBJECT INDEX / GLOSSARY   641

Preface

This textbook does not settle for presenting sociology as a somewhat unattractive descriptive vocabulary. The sociology that I know and practice is a social *science* that seeks to *explain* social phenomena and to *test* these explanations with appropriate *data*. Through these ten editions, I have been trying to make my readers familiar with some of the many powerful theories that sociologists have constructed to help us understand how social life works and also to give them a basic understanding of research as the means to confront these theories with reality.

These two aims have somewhat different implications for the process of revision. Theories tend to change slowly and require only occasional updating or reconfiguration. New research fills the journals and necessitates many changes from one edition to the next—but the sources of this material are well-known and widely available. What changes most rapidly, and sometimes most dramatically, are the fundamental statistics of social science—such as crime, fertility, and divorce rates—and the availability of recent cross-cultural data. Unfortunately, it is not possible simply to consult an annual book or two of social statistics to update those in the text—there are no such volumes. Instead, an aggressive hunt for new statistics is required. In addition, many of the statistics used in this text—including many of the most informative—must be calculated by me from new data bases such as the most recent set of World Values Surveys, the latest General Social Survey, or various major studies. That takes up a great deal of time for each new edition.

Moreover, new data often prompt new substantive sections. Thus, a new section on Women and Science now appears in Chapter 6, partly because this topic has been the focus of so much recent conflict at Harvard, but mainly because new time series data became available showing the very rapid closing of the gender gap in the proportion of doctorates going to women in the major scientific fields. Or, many discussions of Islamic societies are now scattered through the chapters because the most recent World Values Surveys included many Muslim nations for the first time.

In addition, Chapter 11 now includes a new section on Hispanic Americans that was made possible by a new national survey of this major population segment. New cross-cultural data presented in Chapter 16 reveal that American students now lag even further behind in logic, science, and mathematics compared with students in other industrial nations. However, Chapter 7 shows that American crime rates are lower than those in many of these same European nations. Moreover, as revealed in Chapter 18, the fertility rate has dropped far below replacement levels in nearly every European nation and in China too. Consequently, their populations have begun to shrink (or soon will) and are rapidly getting elderly since there are progressively fewer young people. This, in turn, is prompting immigration from outside Europe which has generated considerable conflict and prejudice. And so it goes—as with life, so with each revision.

However, new statistics and new sections are not limited to the text. Since this is the third Internet Edition, students will have the opportunity to try their hands at real analysis of real data—all of which are new as well.

On-Line Student Research

As will be obvious from a glance at the complete step-by-step instructions provided at the end of Chapter 1, students will be able to test interesting and relevant hypotheses on three professional quality data sets—thousands of students already have done so with no trouble at all.

At the end of each chapter in the text (except for Chapter 20), students are given specific questions and directed to examine specific data bearing on major issues or principles developed in the chapter. Three professional-quality data sets are provided. One of these offers a selection of items from the 2000 General Social Survey (the most recent one available as the book goes to press). Whenever a student selects a particular variable, such as "political activism," the software within the site will cross-tabulate it automatically with eight background factors: age, education, sex, race, income, region, religion, and political party. The software also provides a written interpretation of each table, including whether or not it is significant. In this way, students rapidly learn to read cross-tabulations without really knowing that they are doing so.

A second data set is based on the fifty states. When a student selects a variable—suicide rates, for example—the software will create a color-coded map and also will provide a list showing the actual rate for each state, ranking them from highest to lowest. A third set is based on the 174 nations having populations of 200,000 or more. Here again, all students must do is select a variable, and the software will produce both a map and a ranked list.

The value of these assignments rests on the fact that everything is *real:* Students explore real questions using real data and get real results. For students who would like more substantial experience in doing sociology, I suggest my workbook *Doing Sociology: A Global Perspective*, also published by Wadsworth. This workbook includes seven data sets and provides an opportunity to use additional analysis techniques such as correlation—the actual analysis software is included.

InfoTrac College Edition

Just as students now can go online to do sociology, they also can search the periodical literature for articles that expand discussions in each chapter by using the well-known InfoTrac® College Edition system. Because InfoTrac College Edition charges a substantial fee per student user, Wadsworth can provide individual access codes only to those who purchase new copies of the textbook when their instructor chooses to order it with the main text. InfoTrac College Edition gives students access to all issues published since 1988 by thousands of periodicals. It enables them to search for articles on specific topics. Having identified a set of articles, students can then bring up each and select one or more to read—they can even print them out for study. Several useful search terms are provided at the end of each chapter. It also enables students to search for specific articles. To take advantage of this relatively new feature, I have added specific article citations to the end of many chapters, along with a variety of related information and questions on the companion Web site.

New to This Edition

In addition to these online applications, I have updated the book in the usual ways—supplying the latest statistics, substituting new studies for old, and dropping less essential materials to make room for new areas of research activity. Examples include an entirely new section on divorce and remarriage; a new discussion of the complex relationship between religion and morality; an examination of the rather poor showing by American 15-year-olds on achievement tests when compared with their peers in 25 other nations; a section on home-schooling; and a discussion of the "law of constant travel time," which shows that even in ancient societies people will move closer to their work if their commute time exceeds about thirty minutes each way. The many other changes are less dramatic, consisting of a new paragraph or pages inserted or substituted through the book, but together they make the book better than ever.

As noted in previous editions, I am overwhelmed by the generosity of my colleagues across North America, who have volunteered so many useful suggestions. I thank each of you for taking the time to write.

However, if I was surprised by the supportive and valuable mail I received from sociologists, I remain absolutely astonished at the number of students who write to me. It seems clear from their comments that the reason hundreds of them have written is because the "over-the-shoulder" style lets students recognize that sociology is a human activity and that by writing to me they can participate. Not only have I greatly enjoyed these letters, but several brought new material to my attention—one letter even caused me to write a new chapter on gender.

I think this level of student response justifies my initial decision to break some norms of college textbook writing. Most textbooks take pains to sound as if they had no authors but were composed during endless committee meetings. Moreover, human beings are equally indistinct within most texts; the books present a field as consisting mainly of printed matter—of papers and books, of principles and findings. This misleads students about the real nature of scholarly disciplines, which consist not of paper but of people. Moreover, a Nobel laureate once told me that if, after the first ten minutes of the first day of introductory physics, his students didn't know that people go into science primarily because it's fun, he would consider himself a failure as a teacher.

So this is a book with a voice, in which a sociologist addresses students directly and describes the activities of a bunch of living, breathing human beings who are busy being sociologists for the fun of it. Moreover, it attempts to show students that the single most important scientific act is not to propose answers but to ask questions— to wonder. As I let students look over the shoulders of sociologists, be they Émile Durkheim or Kingsley Davis, I want students to first see them wondering—asking why something is as it is. Then I want students to see how they searched for and formulated an answer. For, as an advertising copywriter might put it, I want students to realize that sociology can be a verb as well as a noun.

Point of View and Approach

Sociologists considering a textbook often ask what "kind" or "brand" of sociology it reflects. What are the author's theoretical and methodological commitments? I find some difficulty framing a satisfactory answer to such questions because I don't think I have a brand. First of all, my fundamental commitment is to sociology as a social science. Hence, I want to know how societies work and why, not to document a perspective. Moreover, in constructing sociological theories, I am a dedicated, even reckless, eclectic. Competing theoretical sociologies persist, in part, not only because they tend to talk past one another but also because each can explain some aspect of social life better than the others can. Therefore, in my own theoretical writing, I tend to take anything that seems to work from whatever school can provide it. This textbook does much the same, but with care to point out which elements are being drawn from which theoretical tradition.

I also have not written a book that favors either micro- or macrosociology. Both levels of analysis are essential to any adequate sociology. Where appropriate, the chapters are structured to work from the micro- to the macrolevel of analysis. And the book itself works from the most microtopics to the most macro.

Methodologically, the text is equally eclectic. In my own research, I have pursued virtually every known technique—participant observation, survey research, historical and comparative analysis, demography, human ecology, even experiments. My belief, made clear in the book, is that theories and hypotheses determine what methods are appropriate (within practical and moral limits). That is why there is not one chapter devoted to methods and only one devoted to theory. Instead, Chapter 3 first introduces basic elements of microtheories and then demonstrates how such theories are tested through experiments and participant observation. Chapter 4 introduces social structure within the context of survey research methods. The chapter then assesses basic elements of major macroschools of sociological theory and concludes with an extended example of testing macrotheories through comparative research using societies as units of analysis. Throughout the book, the interplay of theory and research is demonstrated rather than asserted. No sooner do readers meet a theory than they see it being tested.

Countless publishers have stressed to me that introductory sociology textbooks, unlike texts in other fields, must not have an integrated structure. Because sociologists, I am told, have idiosyncratic, fixed notions about the order of chapters, books must easily permit students to read them in any order. That would be a poor way to use this book. The fact is that later chapters build on earlier ones. To do otherwise would have forced me to eliminate some of sociology's major achievements or else to write a book that repeats itself each time basic material is elaborated or built upon. Clearly, some jumping around is possible—the institutions chapters work well enough in any order (and could even be omitted without harming subsequent chapters)—but the basic ordering of the major parts of the book is organic. Thus, for example, the chapter on socialization expands upon material already presented in the biology chapter. And the discussion of theories of intergroup relations in Chapter 11 is basic to the examination of models of urban segregation taken up in Chapter 19. In my judgment, textbooks can be highly flexible only at the risk of being superficial (imagine a chemistry book with chapters that could be read in any order).

Study Aids

To assist readers, each chapter ends with a complete review glossary that includes concepts and principles. For example, the glossary for the population chapter includes not only such concepts as "birth cohort" and "crude birthrate" but also a succinct restatement of "Malthusian theory" and of "demographic transition theory." The glossary is ordered in the same way as the chapters, so it serves to summarize and review the chapter.

Anyone who reads all of the books and articles recommended for further reading will know a lot of sociology. To choose them, I asked myself what I had read that was of broad interest and had helped me to write the chapter. Obviously, I did not think anyone would rush out and read them all. But students attracted by a particular topic may find useful follow-up reading provided in these suggestions. I also have found these works useful in composing lectures.

Supplements

Sociology, Tenth Edition, is accompanied by a wide array of supplements prepared to create the best learning environment inside as well as outside the classroom for both the instructor and the student. All the continuing supplements for *Sociology*, Tenth Edition, have been thoroughly revised and updated, and several are new to this edition. We invite you to take full advantage of the teaching and learning tools available to you.

For the Instructor

Instructor's Resource Manual. This supplement offers the instructor extended abstracts (chapter summaries), key learning objectives, chapter outlines, key research studies (with page references), InfoTrac College Edition exercises, and teaching suggestions, including lecture ideas, class activities, discussion questions suitable for WebTutor, additional readings, Sociology Online, and 4–5 essay questions. Also included is a Resource Integration Guide (RIG), a list of additional print, video, and online resources, and concise user guides for SociologyNow™, InfoTrac College Edition, and WebTutor™.

Test Bank. This test bank consists of 100 multiple-choice questions and 15–20 true-false questions for each chapter of the text, all with answer explanations and page references to the text. Also included are multiple choice and short-answer questions for each chapter. All questions are labeled as new, modified, or pickup so instructors know if the question is new to this edition of the test bank, modified but picked up from the previous edition of the test bank, or picked up straight from the previous edition of the test bank.

ExamView® Computerized Testing for Macintosh and Windows. Create, deliver, and customize printed and online tests and study guides in minutes with this easy-to-use assessment and tutorial system. ExamView includes a Quick Test Wizard and an Online Test Wizard to guide instructors step by step through the process of creating tests. The test appears on screen exactly as it will print or display online. Using ExamView's complete word processing capabilities, instructors can enter an unlimited number of new questions or edit questions included with ExamView.

Extension: Wadsworth's Sociology Readings Collection. Create your own customized reader for your sociology class, drawing from dozens of classic and contemporary articles found on the exclusive Thomson Wadsworth TextChoice database. Using the TextChoice website (**http://www .TextChoice.com**), you can preview articles, select your content, and add your own original material. TextChoice will then produce your materials as a printed supplementary reader for your class.

Classroom Presentation Tools for the Instructor

JoinIn™ on TurningPoint®. Transform your lecture into an interactive student experience with

JoinIn. Combined with your choice of keypad systems, JoinIn turns your Microsoft® Power-Point® application into audience response software. With a click on a hand-held device, students can respond to multiple-choice questions, short polls, interactive exercises, and peer-review questions. You can also take attendance, check student comprehension of concepts, collect student demographics to better assess student needs, and even administer quizzes. In addition, there are interactive text-specific slide sets that you can modify and merge with any of your own PowerPoint lecture slides. This tool is available to qualified adopters at **http://turningpoint .thomsonlearningconnections.com.**

Multimedia Manager Instructor Resource CD: A 2006 Microsoft® PowerPoint® Link Tool. With this one-stop digital library and presentation tool, instructors can assemble, edit, and present custom lectures with ease. The Multimedia Manager contains figures, tables, graphs, and maps from this text, pre-assembled Microsoft PowerPoint lecture slides, video clips from DALLAS TeleLearning, ShowCase presentational software, tips for teaching, the instructor's manual, and more.

Introduction to Sociology 2006 Transparency Masters. A set of black and white transparency masters consisting of tables and figures from Wadsworth's introductory sociology texts is available to help prepare lecture presentations. Free to qualified adopters.

Video. Adopters of *Sociology*, Tenth Edition, have several different video options available with the text. Please consult with your Thomson Learning sales representative to determine if you are a qualified adopter for a particular video.

Wadsworth's Lecture Launchers for Introductory Sociology. An exclusive offering jointly created by Thomson Wadsworth and DALLAS TeleLearning, this video contains a collection of video highlights taken from the *Exploring Society: An Introduction to Sociology Telecourse* (formerly *The Sociological Imagination*). Each 3–6 minute long video segment has been specially chosen to enhance and enliven class lectures and discussions of 20 key topics covered in the introduction to sociology course. Accompanying the video is a brief written description of each clip, along with suggested discussion questions to help effectively

incorporate the material into the classroom. Available on VHS or DVD.

Sociology: Core Concepts Video. Another exclusive offering jointly created by Thomson Wadsworth and DALLAS TeleLearning, this video contains a collection of video highlights taken from the *Exploring Society: An Introduction to Sociology Telecourse* (formerly *The Sociological Imagination*). Each 15–20 minute video segment will enhance student learning of the essential concepts in the introductory course and can be used to initiate class lectures, discussion, and review. The video covers topics such as the sociological imagination, stratification, race and ethnic relations, social change, and more. Available on VHS or DVD.

CNN® Today Sociology Video Series, Volumes V–VII. Illustrate the relevance of sociology to everyday life with this exclusive series of videos for the introduction to sociology course. Jointly created by Wadsworth and CNN, each video consists of approximately 45 minutes of footage originally broadcast on CNN and specifically selected to illustrate important sociological concepts.

Wadsworth Sociology Video Library. Bring sociological concepts to life with videos from Wadsworth's Sociology Video Library, which includes thought-provoking offerings from Films for Humanities, as well as other excellent educational video sources. This extensive collection illustrates important sociological concepts covered in many sociology courses.

Supplements for the Student

Study Guide with Practice Tests. This student study tool contains extended abstracts (chapter summaries), key learning objectives, chapter outlines, key terms and key research studies (with page references), InfoTrac College Edition exercises, Internet exercises, and practice tests consisting of 25–30 multiple-choice questions, 10 true-false questions, 10 short-answer questions, and 3–5 essay questions. All questions include answer explanations and page references to the text.

Internet-Based Supplements

InfoTrac College Edition with InfoMarks™. Available as a free option with newly purchased texts, InfoTrac College Edition gives instructors and students four months of free access to an extensive online database of reliable, full-length articles (not just abstracts) from thousands of scholarly and popular publications going back as much as 22 years. Among the journals available 24/7 are *American Journal of Sociology, Social Forces, Social Research*, and *Sociology*. InfoTrac College Edition now also comes with InfoMarks, a tool that allows you to save your search parameters, as well as save links to specific articles. (Available to North American college and university students only; journals are subject to change.)

WebTutor™ Advantage on WebCT and Blackboard. This web-based software for students and instructors takes a course beyond the classroom to an anywhere, anytime environment. Students gain access to a full array of study tools, including chapter outlines, chapter-specific quizzing material, interactive games and maps, and videos. With WebTutor Advantage, instructors can provide virtual office hours, post syllabi, track student progress with the quizzing material, and even customize the content to suit their needs.

Wadsworth's Sociology Home Page at http://www.thomsonedu.com/sociology. Combine this text with the exciting range of web resources on Wadsworth's Sociology Home Page, and you will have truly integrated technology into your learning system. Wadsworth's Sociology Home Page provides instructors and students with a wealth of information and resources, such as Sociology in Action; Census 2000: A Student Guide for Sociology; Research Online; a Sociology Timeline; a Spanish glossary of key sociological terms and concepts; and more.

Turnitin™ Online Originality Checker. This online "originality checker" is a simple solution for professors who want to put a strong deterrent against plagiarism into place and make sure their students are employing proper research techniques. Students upload their papers to their professor's personalized website, and within seconds, the paper is checked against three databases—a constantly updated archive of over 4.5 billion web pages; a collection of millions of published works, including a number of Thomson Higher Education texts; and the millions of student papers already submitted to Turnitin. For each paper submitted, the professor receives a customized report that documents any text matches found in Turnitin's databases. At a glance, the professor can see if the student has used proper research and citation skills, or if he or she has simply copied the material from a source and pasted it into the paper without giving credit where credit was due. Our exclusive deal with iParadigms, the producers of Turnitin, gives instructors the ability to package Turnitin with the *Sociology*, Tenth Edition, Thomson textbook. Please consult with your Thomson Learning sales representative to find out more!

Companion Website for *Sociology*, Tenth Edition, at http://www.sociology.wadsworth.com/stark10e. The book's companion site includes chapter-specific resources for instructors and students. For instructors, the site offers a password-protected instructor's manual, Microsoft PowerPoint presentation slides, and more. For students, there is a multitude of text-specific study aids, including the following:
· Tutorial practice quizzes that can be scored and emailed to the instructor
· Web links
· InfoTrac College Edition exercises
· Flash cards
· MicroCase Online data exercises
· Crossword puzzles
· Virtual Explorations
And much more!

Acknowledgments

I do not create this book by myself. For one thing, I usually have a cat or two dozing on my monitor, inspiring me by example to always take a relaxed approach. In addition, I am helped in a more direct way by some extraordinary people: Gopa of Gopa & Ted 2 updated the text design and created the cover for this edition, Frank Hubert and Martha Ghent combed the manuscript and proofed for errors, Dusty Friedman supervised the production, and Cheri Palmer made everything happen on schedule. I am grateful to them all.

I am especially indebted to all of my colleagues who devoted time and effort to assessing portions of the manuscript.

First Edition Reviewers

Mary Frances Antolini, Duquesne University
David M. Bass, University of Akron
H. Paul Chalfant, Texas Tech University
Gary A. Cretser, California Polytechnic University at Pomona
Stephen J. Cutler, University of Vermont
Kay Denton, University of Utah
Thomas Egan, University of Louisville

Avery M. Guest, University of Washington
Geoffrey Guest, State University of New York, College at Plattsburgh
Faye Johnson, Middle Tennessee State University
Frederick R. Lynch, University of California at Los Angeles
Shirley McCorkell, Saddleback Community College
Jerry L. L. Miller, University of Arizona
Carol Mosher, Jefferson County Community College
Barbara Ober, Shippensburg State University
Vicki Rose, Southern Methodist University
James F. Scott, Harvard University
David A. Snow, University of Arizona
Steven Stack, Wayne State University
Kendrick S. Thompson, Northern Michigan University
Susan B. Tiano, University of New Mexico

Second Edition Reviewers

Ben Aguirre, Texas A&M University
Brian Aldrich, Winona State University
William Sims Bainbridge, Towson State University
Mary Beth Collins, Central Piedmont Community College
Larry Crisler, Milliken University
Howard Daudistel, University of Texas, El Paso
William Findlay, University of Georgia
Gary Jensen, Vanderbilt University
Bruce Kuhre, Ohio University at Athens
Mary Ann Lamanna, University of Nebraska
Fred Mahar, St. Michaels College
Martin Marger, Michigan State University
Dan McMurray, Middle Tennessee State University
Robert Miller, University of Central Florida
Kay Mueller, Baylor University
Roger Nett, University of Houston, University Park
Fred Pampel, University of Colorado
Robert Silverman, University of Alberta
Joel Tate, Germanna Community College
Charles Tolbert, Baylor University
David Treybig, Baldwin-Wallace College

Third Edition Reviewers

Patricia Albaugh, Southwestern Oklahoma State University
Robert Alexander, Paterson Counseling Center
John D. Baldwin, University of California, Santa Barbara
Peter Bearman, University of North Carolina
John R. Brouilette, Colorado State University
Joseph J. Byrne, Ocean County College
Greg Carter, Bryant College
Mary Ruth Clowdsley, Tidewater Community College
Maury Wayne Curry, Belmont College
Susan Brown Eve, North Texas State University
Roger Finke, Loyola University of Chicago
Sharon Georgianna, Seattle Pacific University
Erich Goode, State University of New York at Stony Brook
Robert Herrick, Westmar College
Paul Higgins, University of South Carolina
Robert Hirzel, University of Maryland
Brenda Hoke, Michigan Technological University
Mary Beth Kelly, Genesee Community College
James Kluegel, University of Illinois, Urbana
Michael Kupersanin, Duquesne University
Clark Lacy, Laramie County Community College
Henry Landsberger, University of North Carolina
Martin Levin, Emory University
Janet Huber Lowry, Austin College
Sam Marullo, Georgetown University
Charles Maxson, Grand Canyon College

William F. McDonald, Georgetown University
Jerry L. L Miller, University of Arizona
David O'Brien, University of Missouri, Columbia
B. Mitchell Peck, Texas A&M University
Peter Venturelli, Valparaiso University
Ira M. Wasserman, Eastern Michigan University
Frank Whittington, George State University
Timothy P. Wickham-Crowley, Georgetown University
John Wildeman, Hofstra University
Robert Wilson, University of North Carolina
Paul Zelus, Idaho State University

Fourth Edition Reviewers

Deborah A. Abowitz, Bucknell University
William Sims Bainbridge, Towson State University
Jay Bass, University of Science an Arts of Oklahoma
Peter Bearman, University of North Carolina, Chapel Hill
Jo Dixon, New York University
Maurice Garnier, Indiana University
Elizabeth Higginbotham, Memphis State University
Debbie D. Hoffman, Belmont College
Peter Lehman, University of Southern Maine
Jerry L. L. Miller, University of Arizona
Ephraim H. Mizruchi, Syracuse University
Aldon Morris, Northwestern University
Pamela Oliver, University of Wisconsin, Madison
Jeffrey Riemer, Tennessee Technical University
Gabino Rendon, New Mexico Highland University
David Snow, University of Arizona
Charles Tittle, Washington State University
Christopher K. Vanderpool, Michigan State University
Shirley Varmette, Southern Connecticut State University
J. Dennis Willigan, University of Utah
Sara Wolfe, Dyersburg State College

Fifth Edition Reviewers

Margaret Abraham, Hofstra University
David Brown, Cornell University
John Cochran, University of Oklahoma
Robert L. Deverick, John Tyler Community College
Frances L. Hoffman, University of Missouri
Roland Liebert, University of Illinois
Alan Miller, Florida State University
Jim Ranger-Moore, University of Arizona
William Schwab, University of Arkansas
Wes Shrum, Louisiana State University
John Tinker, California State University, Fresno
J. Thomas Walker, Susquehanna University
Russell E. Willis, Iowa Wesleyan College
Richard Zeller, Bowling Green State University

Sixth Edition Reviewers

Don Albrecht, Texas A&M University
Ken Ferraro, Purdue University
Cynthia C. Glover, Temple University
Jerry Michel, Memphis State University
Mitch Miller, University of Tennessee, Knoxville
David Mitchell, University of North Carolina, Greensboro
Donald Ploch, University of Tennessee, Knoxville
Lance Roberts, University of Manitoba
Harry Rosenbaum, University of Winnipeg
R. Guy Sedlack, Towson State University
Thomas Shriver, University of Tennessee, Knoxville
Javier Trevino, Marquette University
William L. Zwerman, University of Calgary

Seventh Edition Reviewers

William Canak, Middle Tennessee State University
Michael Dalecki, University of Wisconsin, Platteville
Shelly K. Habel, University of Hawaii
Frances Hoffman, University of Missouri
Fred Pampel, University of Colorado
Virginia Paulsen, Highline Community College
Edward Rowe, Southwest Texas State University
Mark Schneider, Southern Illinois University
Joyce Tang, City University of New York, Queens College
Michael Webber, University of San Francisco
Robert Wood, Rutgers University

Eighth Edition Reviewers

David J. Ayers, Grove City College
Michael Dalecki, University of Wisconsin, Platteville
J. Ross Eshleman, Wayne State University
Marilyn Leichty, Iowa Wesleyan College
Michael Miller, University of Texas at San Antonio
George Primov, University of Miami
M. Therese Seibert, Keene State College
Robert E. Wood, Rutgers University, Camden Campus
Robert A. Wortham, North Carolina Central University

Ninth Edition Reviewers

Jon Alston, University of Texas
Charles M. Brown, Albright College
J. Meredith Martin, University of New Mexico
Robin Perrin, Pepperdine University
Maurice N. Richter, Jr., State University of New York at Albany
Lawrence Scott, Bunker Hill Community College
Kathryn M. Williams, University of Central Oklahoma
Song Yang, University of Minnesota-Twin Cities

Tenth Edition Reviewers

Richard Anderson, University of Colorado
Ray Brannon, Wayne Community College
Dennis Downey, University of Utah
Douglas Gurak, Cornell University
Timothy McLean, Herkimer Community College
Virginia Paulsen, University of Washington
Theron Quist, Baldwin-Wallace College
Georgie Weatherby, Gonzaga University
Andrea Williams, Purdue University
Robert Wortham, North Carolina Central University

TITANIC
DISASTER
GREAT LOSS
OF LIFE!
EVENING NEWS

CHAPTER

1

When these Londoners read the first newspaper reports of the sinking of the "unsinkable" British luxury liner *Titanic,* during its very first voyage (April 14, 1912), they were in search of facts: *Why* did it sink (it hit an iceberg), *where* (in the North Atlantic near Newfoundland), and *how many* died (1,513)? But a sociologist would have been far more interested to read that when the passengers assembled on deck and realized that there was room in the lifeboats for only about half of those on board, it was immediately agreed that the women and children would go in the boats and the men would remain behind in hopes that other ships would come to their rescue in time. Why was this decision made? Why did everyone accept it? Why didn't the men simply force their way into the lifeboats? And finally, as the people in the lifeboats watched the *Titanic* go down, why could they hear the band keep playing and the men keep singing until the end?

Groups and Relationships: A Sociological Sampler

IN A COMMONSENSE SORT OF WAY, EVERY normal human being is a social scientist. If we could not predict one another's behavior with a very high degree of accuracy, social relationships would be impossible— and human life would be "solitary, poor, nasty, brutish, and short," as Thomas Hobbes wrote back in 1651. Only because we usually can anticipate how others will respond to our actions is it possible for us to have families and friends, let alone customers, students, professors, clerks, and employers. Each of us would be forced to sleep alone in a secret hiding place if we couldn't accurately predict that our relatives and neighbors won't try to kill us while we sleep.

If everyone is something of a social scientist, it also is true in the same sense that, from the dawn of human existence, everyone has been a physical and natural scientist. Cave dwellers were fully aware of gravity as a fact of life and

therefore that people on top of a hill can throw stones down much farther and harder than their enemies can throw stones back up at them. And even people in the Stone Age knew a great deal about the habits of animals and about which plants were edible.

SCIENCE: THEORY AND RESEARCH

Nevertheless, there were no physicists, chemists, or biologists in the Stone Age. Nor were there any sociologists, psychologists, or economists. As we understand the term today, **science** is a sophisticated and precise method for describing and explaining why and how things work—including human "things." The scientific method developed slowly and came to full fruition only quite recently. It consists of two components: theory and research.

CHAPTER OUTLINE

SCIENCE: THEORY AND RESEARCH

THE DISCOVERY OF SOCIAL FACTS
Stability and Variation
The Upward Trend

THE SOCIOLOGICAL IMAGINATION

SOCIOLOGY AND THE SOCIAL SCIENCES

Box 1-1: Founders of
American Sociology Departments

UNITS OF ANALYSIS

MICRO- AND MACROSOCIOLOGY

A GLOBAL PERSPECTIVE

SCIENTIFIC CONCEPTS

GROUPS: THE SOCIOLOGICAL SUBJECT
Primary and Secondary Groups

SOLIDARITY AND CONFLICT:
THE SOCIOLOGICAL QUESTIONS

ANALYZING SOCIAL NETWORKS
Networks and Social Solidarity
Networks and Social Conflict

A CLOSER VIEW
Samuel Sampson:
Monastic Strife

STUDYING SELF-AWARE SUBJECTS
Unobtrusive Measures
Validation
Bias

THE SOCIAL SCIENTIFIC PROCESS

FREE WILL AND SOCIAL SCIENCE

CONCLUSION

REVIEW GLOSSARY

SUGGESTED READINGS

SOCIOLOGY ONLINE

A **theory** is an abstract statement that explains why and how certain things happen or are as they are, whether these things be eclipses of the moon, chemical reactions, or outbursts of racism. In addition, scientific theories must have *empirical implications* (the word *empirical* means "observable through the senses"). That is, theories make definite predictions and prohibitions; they say some things will happen under certain circumstances and that other things will not happen.

For example, the theory known as Homans' law of inequality predicts that within any group the emotional attachments among members will be stronger among persons of similar rank than among members of different ranks (see Chapter 3). That does not mean that every person will like every other person of the same rank or that no persons of different ranks will have close ties. But it does prohibit the finding that rank has no influence on who likes whom or the finding that people favor close ties with persons of higher or lower rank than themselves. The theory makes definite empirical predictions and prohibitions. The world ought to resemble Figure 1-1. Should it not, this would show that Homans' law is faulty.

Research involves making *appropriate empirical observations* or measurements—activities usually referred to as collecting data (or facts). The purpose of research is to *test theories* (do their predictions hold?) or to *gain sufficient knowledge* about some portion of reality so that it becomes possible to theorize about it. Whenever sociologists chart friendship ties within a group, they make observations appropriate for testing Homans' law. But long before he formulated his law (in about 1950), many other people had often observed that rank seemed to influence relationships. Thus, when Homans began his theorizing, he already was in possession of data on the phenomenon to be explained.

Because science ultimately rests upon data—upon empirical observations—the emergence of the various scientific specialties has been influenced by the ease of gathering the pertinent data or making the necessary observations. Thus it is that astronomy was probably the earliest true science, since the heavens are there for all to see (especially in the days before artificial illumination made the night sky nearly invisible in large cities). Granted that the invention of the telescope in 1608 made far better observations available to astronomers, but the ancients did very well in predicting eclipses, the phases of the moon, and the movements of the planets.

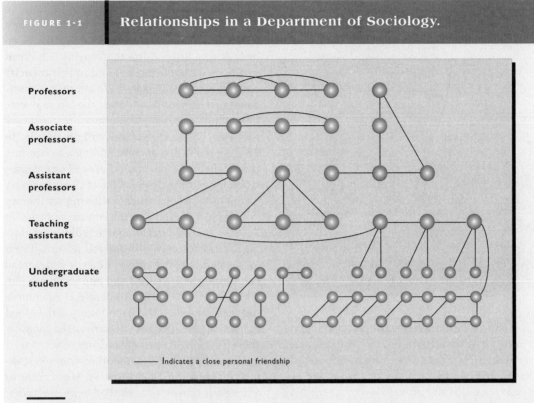

FIGURE 1-1 **Relationships in a Department of Sociology.**

Professors

Associate professors

Assistant professors

Teaching assistants

Undergraduate students

——— Indicates a close personal friendship

In keeping with Homans' law of inequality, friendships tend to be concentrated among people of the same rank. Notice that exceptions to the rule, members with close ties to those of another rank, tend to lack ties to others of their own rank—to be outsiders.

In similar fashion access to various physical results made it possible for the ancient Greeks to have a quite sophisticated physics as well as geometry. But the ancients never moved beyond speculation in the realm of social science. There simply weren't enough data. For example, Aristotle had no reliable information on something so basic as the size of the population of Athens or of any other ancient city, let alone on breakdowns of these populations by age, sex, income, or religion. Roman historians frequently complained about crime, but they had no knowledge of the actual rates at which particular crimes were occurring.

Thus, the social sciences languished for want of data while the physical sciences gained in sophistication and scope. The social scientific breakthrough did not begin until 1776, when Adam Smith published *The Wealth of Nations*, thereby founding economics as a social *science*. Smith's book consists of an elaborate theory in which the fundamental principles of supply and demand were introduced. But his achievements would have been impossible a century earlier, for the theory is firmly rooted in data on trade and economic activities that had only become available a few years earlier (Collins, 1994).

Fifty years later came the first scientific sociologist, a Frenchman named André Michel Guerry (pronounced "GAR-y"). And just as the new availability of data had made it possible for Adam Smith to found the science of economics, it was new data that made a science of sociology possible. Indeed, these data virtually *forced* Guerry and his successors to become sociologists!

THE DISCOVERY OF SOCIAL FACTS

In 1825 the French Ministry of Justice began to collect criminal justice statistics from the prosecutor's office of each of the nation's departments, the geographical units into which France is divided, then numbering eighty-six. Known as the *Compte general de l'administration de la justice criminelle en France* (General Account of the Administration of Criminal Justice in France), the

More than 3,500 years ago, people living in Britain, who probably had no written language and did not have metal tools, were able to construct this extraordinary monument at Stonehenge which allowed them to accurately predict many astronomical phenomena, such as the summer solstice.

Compte offered detailed statistics concerning criminal justice activities such as arrests and convictions. The data were submitted quarterly and published annually. The statistics are immensely detailed and broken down rather finely by age, sex, season, and the like. Once they began, the French soon were collecting data on a variety of things, including suicide, illegitimate births, military desertion, charitable contributions, literacy, and even per capita revenues raised by the royal lottery. These data soon became known as "moral statistics"—so-called because of the moral implications of most of the activities being recorded.

The first edition of the *Compte* was published in 1827. In the introduction the Minister of Justice, Comte (Count) de Peyronnet, wrote:

The exact knowledge of facts is one of the first needs of our form of government; it enlightens deliberations; it simplifies them; it gives them a solid foundation by substituting positive vision and the reliability of experience for the vagueness of theories. (in Beirne, 1993)

The *Compte* was initiated and supervised by Jacques Guerry de Champneuf (1788–1852), who was appointed Director of Criminal Affairs and Pardons in 1821 (he was dismissed from office immediately following the revolution of July 1830). However, Champneuf only collected and published the data; he did little or no analysis of them. Thus, it was left to a young attorney to seize the opportunity.

André Michel Guerry was born in Tours in 1802 and died in Paris in 1866. He set out to be a lawyer and was admitted to the bar in Paris. Soon he became a Royal Advocate (roughly equivalent to a prosecuting attorney) and in this capacity encountered the *Compte*.[1] He became so fascinated that he soon abandoned the law to devote himself to interpreting these amazing statistics—particularly those on crime and on suicide. In 1831 he published his first findings, attempting to see whether education influenced crime rates. Contrary to the widespread belief that crime rates were lower in departments with higher literacy rates, Guerry found that the reverse was true. In 1832 he analyzed data on mental illness, exploring the relationships with age and seasons. These initial publications earned Guerry widespread recognition as a creative and important scholar. Then, in 1833 came his masterpiece, *Essai sur la statistique morale de la France* (Essay on the Moral Statistics of France) published under the sponsorship of the French Royal Academy of Science.

Guerry produced many other volumes, capping his career in 1860 with *The Moral Statistics of England Compared with the Moral Statistics of France*. For this book Guerry was awarded the prize in statistics in 1861 by the French Academy of Sciences. Nevertheless, the 1833 *Essai* is the book that launched sociology. It did so because the data revealed two very pronounced patterns that people at this time found absolutely astonishing.

STABILITY AND VARIATION

First, the rates were *extremely stable* from year to year. Thus, in any given French city or department, year after year almost exactly the same number of people committed suicide, stole, murdered their parents, or gave birth out of wedlock. And the *kinds* of people who did such things also was incredibly stable. Table 1-1 shows the percentage of females and the percentage of persons age 16–25 accused of theft over a five-year period (1826–1830). The percentage of females never rose above 23 percent and never fell below 21 percent! The percentage of persons age 16–25 never fell below 35 percent and never rose above 38 percent!

Second, the rates often *varied greatly* from one place to another. For example, the number of suicides per 100,000 population—calculated by Guerry as an average for the years 1827–1830—varied from 34.7 in the Department of the Seine

1. At first the *Compte* was distributed only to the nobility, members of parliament, and state functionaries (Beirne, 1993).

(which includes Paris) down to fewer than 1 per 100,000 in Aveyron and the Hautes-Pyrenees. As for property crimes, Guerry's data showed a rate of 73.1 per 100,000 in Seine in contrast with a rate of 4.9 in Creuse. Violent crimes also varied immensely from department to department: from 45.5 in Corse (the island of Corsica) down to 2.7 in Creuse and in Ardennes.

These patterns forced Guerry to reassess the primary causes of human behavior. What could be more individualistic actions, more clearly motivated by private, personal, idiosyncratic motives, than committing suicide, property crime, or murder? But if they are indeed the ultimate individual acts, why didn't the rates fluctuate wildly from year to year? If individual motives alone were involved, how could it be that year after year the same number of people in Paris or in Marseille took their own lives or killed their spouses?

There was no alternative but to conclude that there are very powerful forces *outside* the individual that cause the incredible stability and the equally incredible variations from place to place that the *Compte* data revealed to Guerry. Faced with this conclusion, Guerry devoted much of his book to careful analysis of possible *social forces,* such as population density and economic development (measured as the percentage of the labor force employed in skilled trades), as causes of suicide and crime.

Almost immediately, others followed Guerry's lead and took up the study of moral statistics. Thus, Adolphe Quetelet (pronounced "ket-ah-LAY") came to Paris from Belgium to see how the *Compte* was conducted and in 1835 made a major contribution, *Sur l'homme* (On Man), in which he developed the idea of that now-familiar statistical being, the "average person" (*homme moyen*). Quetelet noted that "we can assess how perfected a science has become by how much or how little it is based on calculation." Referring to moral statistics as a form of "social physics," Quetelet identified the lack of empirical data about social phenomena as the chief obstacle faced by this new science (Beirne, 1993). Quetelet also stressed that the stability of crime and suicide rates in any particular place forces the admission that society plays an immense role in shaping individual behavior.

Unfortunately, Quetelet soon had a very heated quarrel with Guerry over who had been the first to recognize the stability of these rates, and they ended up as lifelong enemies.[2] But between them they inspired many others to join

TABLE 1-1	Percentage Female and Percentage of Persons Age 16–25 Accused of Theft in France	
	FEMALE (%)	AGE 16–25 (%)
1826	21	37
1827	21	35
1828	22	38
1829	23	37
1830	22	37
Average	**22**	**37**

Source: Guerry (1833)

in the new venture. One of the earliest was Alexandre Parent-Duchatelet, who in 1836 published a two-volume work on prostitution in Paris, which showed the number of prostitutes in each of the forty-eight sections of the city and included data on their age, religious sentiments, health, and backgrounds.

The moral statistics movement spread rapidly (Stark and Bainbridge, 1997). Almost at once it gained many followers in Britain. The Statistical Society of London was founded in 1834 by readers of Guerry's book, and by the end of the decade, the society began to publish a very influential journal. The British were not especially interested in suicide, being far more concerned with education (especially literacy), religion, and crime. In 1838 Edgell Wyatt-Edgell published a report in the society's journal on three neighborhoods in the city of Westminster, in which he reported the number of local churches, their seating capacities, Sunday school enrollments, the most popular books checked out from the local libraries (novels by Walter Scott led the list), newspaper circulation, and a very elaborate table reporting the number of persons arrested for crimes committed during the previous year. Among the statistics: 7 men had been arrested for dog stealing, 58 men and 35 women had been taken in for passing counterfeit money, and 2,367 men and 1,424 women were arrested for public drunkenness.

Lambert Adolphe Jacques Quetelet was born in 1796 in Ghent, Belgium. At age twenty-four he was elected to the Royal Academy of Science in Brussels. In addition to his famous work in astronomy and in probability theory, Quetelet wrote the libretto for an opera, a historical survey of romance, and a great deal of poetry, meanwhile playing a major role in founding the moral statistics movement.

© Bibliotheque Royale Albert, Bruxelles (Cabinet des Estampes) Jean-Baptiste Madou, artist.

---

2. Since the stability of the rates was obvious to any sophisticated person who examined the *Compte* closely, each probably had discovered it independently. But Guerry gets the credit because he published first.

FIGURE 1-2   **Suicides per 100,000 in Europe, 1870.**

1870: Suicides per 100,000.

Ireland	1.5	Belgium	6.0	Italy	9.0
Spain	1.7	United Kingdom	6.6	Germany	14.8
Finland	2.9	Austria	7.2	France	15.0
Netherlands	3.6	Norway	7.6	Switzerland	19.6
		Sweden	8.5	Denmark	25.8

The darker the nation, the higher its rate; no data were available for the uncolored nations. Here we can see the amazing international contrasts in suicide rates. The data for this map were adapted from Henry Morselli's *Suicide* (1882). However, data for seven small states were combined to create a rate for "Germany," which was then still in the process of uniting. This map is of special historical interest because it is based on the same data Émile Durkheim used at the turn of the century to write a classic sociological study, which he also called *Suicide*.

Moral statistics soon attracted participants in the United States, Germany, and Italy. Thus, in 1848 the American moral statistician Pliny Earle published the first of his many insightful works using statistical data on those afflicted with mental illness. In 1864 the German scholar Adolf Wagner produced a monumental study of suicide based on data from many parts of Europe. The Italian physician Henry Morselli fol-

lowed up Wagner's study in 1882 when he published *Suicide: An Essay on Comparative Moral Statistics.*

THE UPWARD TREND

By Morselli's time the moral statistics movement had become so widespread that he was able to compare the 1870 suicide rates for many European nations—as is shown in Figure 1-2. The differences are extremely large: In Ireland and Spain, there were fewer than 2 suicides per 100,000 population, while in Denmark there were more than 25. Moreover, sufficient data had accumulated so that a third amazing fact about moral statistics had been recognized. Throughout the century the rates for crime and for suicide had been *rising*, inching ever upward. Table 1-2, adapted from Morselli, shows that the rate of suicide rose from 6.3 to 6.7 per 100,000 population in England from 1830 through 1870, while in Sweden it rose from 6.8 to 8.5. Similar increases occurred in Paris and London.

Morselli addressed both the issue of variation from place to place and the issue of increases in the rates. He asked: Why do some places have higher rates than others, and why are the rates rising everywhere? He answered: because of the shift from societies based on small-town and rural life to modern, industrialized societies with their huge, impersonal, and disorderly cities. Or as Morselli put it, suicide reflects "that universal and complex influence to which we give the name of *civilization*." He argued that some nations had higher rates because they were more modernized ("civilized") and that the rates were rising because modernization was taking place throughout Europe.

THE SOCIOLOGICAL IMAGINATION

In 1897 a Frenchman named Émile Durkheim, who called himself a sociologist rather than a moral statistician, also published a book entitled *Suicide*, which he based on the data previously published by Wagner and by Morselli. He repeated Morselli's thesis that suicide rates are stimulated by modernization and argued at length that modern societies have high rates because they are deficient in the kinds of secure and warm *interpersonal relationships* typical of more traditional rural and village life. As Durkheim put it, the suicide "victim's acts which at first seem to express only his personal temperament are really [caused by] a social condition." That is, high suicide rates reflect *weaknesses in the web of relationships among members of a society*, not weaknesses of character or personality in the individual.

Thus did Durkheim display something that came to be called the **sociological imagination.** The term was coined by American sociologist C. Wright Mills (1959) to describe the ability to see the link between incidents in the lives of individuals and large social forces. Mills argued that no matter how private or personal our actions, we can understand ourselves and our intimates much better if we can place ourselves within a larger framework. Putting the same point another way, Peter Berger (1963) stressed that sociology is devoted to discovering the *general in the specific:* When sociologists see particular people do specific things, they attempt to visualize the general patterns that apply to these specific actions. This is not to say that our sense of personal freedom and of making choices is an illusion, but that the choices we face often are very limited and structured by factors beyond our control. For example, couples often decide when and whether to have a child. But they are far more likely to decide not to have one during economic depressions. Like suicide rates, fertility rates are extremely stable in the short run and rise and fall in response to other general trends.

Of course, Mills only named the sociological imagination. In one sense people have always had it—even when we lived in caves, we knew that "it's a jungle out there." And from Guerry through Durkheim, the data on moral statistics forced the early social scientists to fully exercise and develop sociological imaginations—to recognize the existence of, and devote their primary theoretical attention to, the *social world surrounding the individual.* It is in this world that the sources of variation in rates are to be found as well as the reasons for the stability of the rates. Consequently, the facts on which the emergence of the science of sociology was based were above all else *social facts.* That is, a suicide rate is far more than the sum of individual actions. It reflects profound aspects of the social world, or environment, surrounding the individual.

SOCIOLOGY AND THE SOCIAL SCIENCES

The progressive uncovering of *social causes of human behavior* produced the field now called **sociology,** which can be defined as the scientific study of the patterns and processes of human social relations. Sociology is one of several related fields known collectively as the **social sciences.**

TABLE 1-2	Number of Suicides per 100,000 Population	
	ENGLAND	**SWEDEN**
1830–1840	6.3	6.8
1845–1855	6.2	6.9
1856–1860	6.5	6.4
1861–1865	6.6	7.6
1866–1870	6.7	8.5
	PARIS	**LONDON**
1827–1830	34.7	—
1861–1870	35.7	8.1
1872–1876	42.6	8.6

Source: Adapted from Morselli (1882).
Note: — indicates data not available.

They share the same subject matter: human behavior. They are called social sciences because humans are not solitary beasts. Our daily lives intertwine with the lives of others—what we do, even much of what we hope, is influenced by those around us.

Despite their common subject matter, there are a number of different social sciences. Psychologists, economists, anthropologists, criminologists, political scientists, and even many historians, as well as sociologists, are social scientists. Divisions among these fields are often hazy—sometimes it is impossible to tell to which field a social scientist's work belongs. The field may be determined by something as trivial as the university department in which the person is trained or employed. Nevertheless, the following rules of thumb may help you distinguish sociologists from other social scientists.

Sociologists differ from psychologists because we are not concerned so exclusively with the individual, with what goes on inside people's heads; we are more interested in what goes on *between* people. Sociologists differ from economists by being less exclusively interested in commercial exchanges; we

Émile Durkheim.

© Bettmann/Corbis

BOX 1-1   FOUNDERS OF AMERICAN SOCIOLOGY DEPARTMENTS

Albion Small.

W. E. B. Du Bois.

THE SOCIAL SCIENCES are quite new. Economics, the oldest, can be dated from 1776 and the moral statistics movement began about fifty years later. The first experimental psychology laboratory was founded in Germany in 1879 by Wilhelm Wundt. In 1880 Karl Marx made one of the first attempts to conduct an opinion survey by distributing 25,000 questionnaires to English workers (almost none of whom responded).

The name *sociology* was first suggested in the 1830s by the French philosopher Auguste Comte. But for many years it remained only a suggestion; Comte urged others to do sociology, but he never got around to doing any himself.

It is not until late in the nineteenth century that we can identify people who called themselves sociologists and whose work contributed to the development of the field. Among these were Herbert Spencer in England, who published the first of his three-volume *Principles of Sociology* in 1876, and Ferdinand Tönnies in Germany, who published *Gemeinschaft und Gesellschaft* in 1887 (see Chapter 19). A decade later Émile Durkheim published *Suicide*.

The earliest sociologists in North America, as in Europe, studied moral statistics. Their work proved so popular that it led to the rapid expansion of census questions. However, sociology as an academic specialty was imported into North America from Germany and first appeared at the University of Chicago and at Atlanta University, then an all-black school.

The first department of sociology in North America was founded at the University of Chicago in 1892 by Albion Small (1854–1926). Small did his graduate study at Leipzig, Germany. Returning to the United States, he published *An Introduction to the Science of Sociology* in 1890, which led the newly opened University of Chicago to select him to create a department there. He also founded the *American Journal of Sociology* and edited it from 1895 to 1925 (Faris, 1967).

The second initial base for sociology in North America was created by W. E. B. Du Bois (1868–1963), an African American who grew up in Great Barrington, Massachusetts, and was educated at Fisk, Harvard, and then the University of Berlin. In Atlanta in 1897, Du Bois created a sociological laboratory and directed the Atlanta University Conferences—a major annual meeting devoted to sociological research and analysis of the circumstances of African Americans. From 1896 through 1914, Du Bois published a book based on his sociological research *every* year and wrote many articles and gave many speeches as well. As a sociologist, Du Bois was devoted to taking an objective, scientific approach. He believed that if the facts were known about social concerns, reasonable actions would follow. "We simply collect the facts: others may use them as they will," he once wrote. After a few years, however, Du Bois decided to shift his efforts from sociology to direct action. He was a founder of the National Association for the Advancement of Colored People (NAACP) and served as editor of its periodical, *The Crisis,* from 1910 through 1934. Although Du Bois was still alive when the Civil Rights Movement began to break the bonds of racism against which he had battled so long and so eloquently, as both sociologist and journalist, he had by then committed his hopes to revolutionary Marxism and gone to live in Ghana, where he is buried (Broderick, 1974; Green and Driver, 1978).

Despite these early beginnings in Chicago and Atlanta, sociology grew slowly in North America. The first sociology department in Canada opened in 1922 at McGill University in Montreal. Not until 1930 did Harvard have a sociology department, and the University of California at Berkeley and Johns Hopkins did not until the 1950s.

As a field barely 100 years old and pursued by a small number of people, no wonder sociology cannot yet match the achievements of much older and more widespread sciences such as physics and chemistry. Nevertheless, much has been achieved—as will become evident as you read this book.

are equally interested in the exchange of intangibles such as love and affection. We differ from anthropologists primarily because the latter specialize in the study of preliterate or primitive human groups, while we are primarily interested in modern industrial societies. And while most criminologists are trained in and

employed in sociology departments, they specialize in illegal behavior, whereas sociologists are interested in the whole range of human behavior. Similarly, political scientists focus on political organization and activity, while sociologists survey all social organizations. Finally, sociologists share with historians an interest in the past but are equally interested in the present and the future.

As these contrasts make evident, sociology is a broader discipline than the other social sciences. In a sense the specialty of sociologists is generalization: to find the connections that unite the various social sciences into a comprehensive, integrated science of society. When I had to decide which social science to pursue, I chose sociology precisely because of its greater scope and grander aspirations. It is the field I would like to introduce you to in this book. Although many sociologists continue to study crime, suicide, and the other "moral" concerns on which the field was founded, sociology abounds with less gloomy topics. But throughout, the central topic of sociology remains the link between individuals and their environment.

UNITS OF ANALYSIS

All science is based on analyzing data obtained through systematic observations. But there is immense variation across the sciences in the "things" that they observe. Some scientists base their observations on molecules, cells, atoms, genes, earthworms, galaxies, or ants. The things being observed by researchers are called the **units of analysis.** Because social scientists study people, *individuals* are often their units of analysis. However, social scientists—and especially sociologists—often base their observations on units made up of more than one person. These often are created by summing, or aggregating, information on individuals, and consequently, they also are called *aggregate* units of analysis. In the following examples, the data are real but the sociologists are imaginary.

Suppose that several sociologists had decided to study current conflicts over abortion. One of them wanted to know whether or not most Americans favor abortion. To answer questions like this, *individuals* are the appropriate *units* of analysis. So, this sociologist conducted a national survey of American adults, asking questions about approval of abortion under various circumstances. The answers would have been as shown in Table 1-3. Here we see that the overwhelming majority of Americans thought it ought to be possible for a woman to obtain a legal abortion if her health is endangered by the pregnancy (90 percent) or if there is a strong chance of a serious defect in the baby (76 percent). Were these the only questions asked, one might conclude that opposition to abortion is limited to a small,

TABLE 1-3	Using INDIVIDUALS as the Units of Analysis

"Please tell me whether or not you think it should be possible for a pregnant woman to obtain a *legal* abortion."

1. "If the woman's own health is seriously endangered by the pregnancy?"

Yes	90%
No	8%
Don't Know	2%
	100%

2. "If there is a strong chance of a serious defect in the baby?"

Yes	76%
No	21%
Don't Know	3%
	100%

3. "If she is not married and does not want to marry the man?"

Yes	41%
No	56%
Don't Know	3%
	100%

4. "If the woman wants it for any reason?"

Yes	41%
No	56%
Don't Know	3%
	100%

Source: Prepared by the author from the General Social Survey, 2002.

noisy minority. But the third and fourth questions show that isn't true. Only a minority of Americans think it ought to be possible to obtain a legal abortion if a woman is not married and doesn't want to marry the father (41 percent) or if the woman simply wishes to have an abortion "for any reason" (41 percent). Keep in mind, too, that some respondents no doubt were willing to allow others to have an abortion who would not themselves do so in these particular circumstances. Thus, the sociologist discovered there is no single answer to the question of whether or not most Americans approve of abortion.

A second sociologist decided to compare local "Pro-Life" and "Pro-Choice" groups. In this instance the *units* of analysis are *groups*. Having selected a number of groups of each kind, the sociologist sent a questionnaire to each member. However, instead of analyzing the responses of individuals, as was done in the example above, the sociologist *aggregated* the responses of individual respondents to create statistics characterizing each group. For example, by summing responses of each member to a question about how many children they have, the sociologist was able to calculate the average number for each group.

Or basing calculations on answers to the question "What do you think is the ideal number of children for a family to have?" the sociologist was able to calculate the average "ideal" number of children for each group. By using a question on the frequency of church attendance, the sociologist could determine what percentage of each group attended weekly. It also was easy to determine the percentage who voted for Bush in the 2000 presidential election.

The results would have been like those shown in Table 1-4. Notice that here we are not comparing individuals. These are the properties of two different groups, and groups are the units of analysis. For individuals, children only come in whole numbers—no real family has 2.1 children as the "Pro-Life" group does. We see that those in the "Pro-Life" group have an average of nearly one more child than do members of the "Pro-Choice" group. They also prefer to have more children than do those in the "Pro-Choice" group (2.9 compared to 1.9). We also can see that "Pro-Life" members are far more likely to attend church weekly (38 percent compared with 12 percent). It also is true that the "Pro-Life" group was considerably more likely to vote for George W. Bush for president in the 2000 election (against Gore or Nader), while Bush lost badly among the "Pro-Choice" group.

A third sociologist was more interested in actual abortion rates than in attitudes toward abortion and based a study on information about abortions published in the *Statistical Abstract of the United States*. Using the most recent edition, the sociologist noted the abortion rates for each of the fifty states. The *units* of analysis for this study were *states*. The rates were created by obtaining the total number of abortions performed in each state during the previous year, dividing this number by the total number of live births, and multiplying the result by 1,000. The result is the number of abortions per 1,000 live births. Whereas only an individual woman can have an abortion or give birth, only aggregates can have an abortion rate.

In this instance the aggregate data also have a geographical basis—the units of analysis not only are states, but they also are *places*. Sociologists often use places such as neighborhoods, cities, counties, states, and even nations as their units of analysis. One of the nice things about places is that they can be mapped. Thus, Figure 1-3 maps the abortion rate. A glance at the map shows that the geography of abortion is bicoastal—highest in the Far West and the Northeast. The actual rates, shown below the map, reveal that New York has the highest rate, 694 abortions per 1,000 live births, closely followed by Hawaii, Nevada, and California and then by Delaware, Massachusetts, Rhode Island, and New Jersey. Wyoming, with only 74 abortions per 1,000 live births, has the lowest rate in the nation; South Dakota, Idaho, and Utah also are very low. The next step for this sociologist would be to seek an explanation for this geographical pattern.

Still another sociologist wanted to compare attitudes toward abortion internationally. *Nations* were the *units* of analysis. The data for each nation were based on the results of a nationwide opinion poll conducted within each nation. In each of these eighty-one nations, respondents were asked (in their native language) to rank their attitude toward abortion on a scale of one to ten, with one meaning "never justified" and ten meaning "always justified." For each nation the percentage who answered one was calculated. The results are shown in Table 1-5. Notice the huge differences among nations: 92 percent of the population of Zimbabwe said abortion is never justified whereas only 5 percent of Swedes gave this answer. In Mexico 67 percent thought abortion is never justified as did 30 percent in both Canada and the United States (all three nations are shown in boldface). Interestingly, most Chinese (55%) said abortion is never justified despite the fact that until very recently forced abortion was government policy for many years, in support of laws allowing couples to have only one child.

TABLE 1-4	Using GROUPS as the Units of Analysis	
	"PRO-LIFE" GROUP	"PRO-CHOICE" GROUP
Average number of children	2.1	1.3
Average "ideal" number of children	2.9	1.9
Percentage who attend church weekly	38%	12%
Percentage who voted for Bush in 2000	61%	40%

Source: Although the example itself is hypothetical, these statistics were created from the General Social Survey of 2000, using those who think legal abortions should be available for "any reason the woman wants" as the "Pro-Choice" group and those who said no as the "Pro-Life" group, corrected for the overrepresentation of males in the subset of respondents asked the question.

MICRO- AND MACROSOCIOLOGY

Many sociologists work with individual units of analysis; others work with very small groups and focus on the patterns of face-to-face interaction between humans. This part of sociology is known as **microsociology**—*micro* means "small" as in *microscope*. Microsociologists use close-up lenses. Other sociologists do not. Instead, they concentrate on larger units of analysis. From their viewpoint the individual is simply one small dot among many dots that help form a larger picture, much like the dots on a TV screen. This wide-angle lens approach is known as **macrosociology** (*macro* means "large"). Macrosociologists pursue questions about aggregate units of analysis; for example, Why are suicide rates so much higher in some places than in others? or Why are some places more racially segregated than others?

A GLOBAL PERSPECTIVE

As exemplified by Table 1-5, macrosociologists often use nations as their units of analysis. Their primary reason is to test theories about why societies have certain features—a topic introduced at length in Chapter 4. But there are three other important reasons modern sociology stresses a global perspective and tables like this appear throughout the book.

The first is to provide a meaningful basis of *comparison*. Suppose I told you that a recent national opinion poll found that 47 percent of Americans agreed that "sharing household chores" is "very important" for a happy marriage. Is that high or low? Such a question can only be answered with another question: Compared with what? Compared with Nigerians, 74 percent of whom said "very important," Americans are quite low. Compared with Canadians (53 percent) Americans are a bit low. Compared with Germans (24 percent) they are quite high, and compared with the Japanese (10 percent) Americans are very high. Comparisons like this are much more informative than the simple poll result for one nation. As we shall often see, contrasts like these often prompt macrosociologists to ask what causes them and to formulate and test answers.

A second reason sociologists favor a global perspective has to do with the fact that for many purposes it is impossible to study a single society in isolation. Because all nations are increasingly *linked* by trade and communication, much

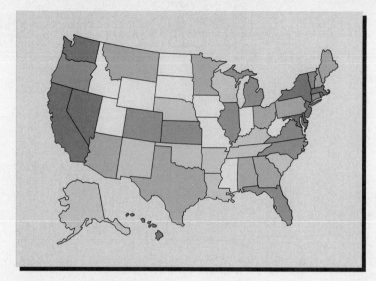

FIGURE 1-3	Using STATES as the Units of Analysis.

Number of Abortions per 1,000 Live Births

New York	694	Nebraska	246
Hawaii	617	Tennessee	243
Nevada	591	South Carolina	229
California	519	New Mexico	228
Delaware	502	Wisconsin	223
Massachusetts	472	Alaska	222
Rhode Island	461	Arkansas	213
New Jersey	460	Louisiana	195
Maryland	454	Oklahoma	193
Washington	447	Kentucky	191

Connecticut	444	Indiana	185
Florida	438	Iowa	185
Vermont	393	Mississippi	176
Michigan	393	Missouri	175
Virginia	373	North Dakota	149
Oregon	372	West Virginia	134
Colorado	362	Utah	104
Illinois	361	Idaho	97
North Carolina	357	South Dakota	92
Kansas	353	Wyoming	74

Georgia	350
Pennsylvania	302
Montana	298
Texas	297
Arizona	295
Ohio	294
Maine	282
Alabama	277
New Hampshire	269
Minnesota	251

TABLE 1·5	Using NATIONS as the Units of Analysis		
Percent who said abortion is "never justified."			
NATION	NEVER JUSTIFIED (%)	NATION	NEVER JUSTIFIED (%)
Zimbabwe	92	Bosnia	35
Tanzania	89	Latvia	35
Bangladesh	89	Portugal	34
Malta	89	Hungary	32
El Salvador	89	Lithuania	31
Indonesia	88	Belgium	30
Jordan	84	Ukraine	30
Morocco	84	**United States**	**30**
Algeria	78	**Canada**	**30**
Uganda	76	Montenegro	28
Brazil	75	Azerbaijan	28
Iran	74	Georgia	28
Colombia	74	Israel	27
Nigeria	74	Albania	26
Peru	73	Spain	26
Puerto Rico	73	Germany	25
Venezuela	71	Slovakia	24
Mexico	**67**	Australia	23
Chile	66	Austria	23
Argentina	63	Armenia	21
India	60	Switzerland	21
Vietnam	60	Serbia	19
Saudi Arabia	60	Bulgaria	19
Dominican Republic	57	Slovenia	19
Egypt	56	Estonia	18
China	55	Luxembourg	18
Philippines	55	Greece	18
Pakistan	54	New Zealand	18
South Africa	52	Turkey	17
Ireland	49	Belarus	17
Uruguay	46	Russia	16
Taiwan	45	Netherlands	15
Croatia	43	Norway	15
Northern Ireland	43	France	14
Poland	42	Czech Republic	13
Singapore	41	Denmark	13
Moldova	39	Japan	13
Macedonia	39	Finland	11
Romania	38	Iceland	11
South Korea	37	Sweden	5

Source: Prepared by the author from the World Values Survey, 2000–2001.

of what goes on in one society is influenced by what goes on in others. This is obvious in the economic sphere. For example, the size of harvests in one part of the world immediately influences commodity prices everywhere. Chapter 17 will pay extended attention to these linkages and to how they are tied to global inequality.

But perhaps the most important motive for cross-national comparisons is that science seeks *general theories*—a theory such as Homans' law of inequality must hold *everywhere* it is applied. It isn't enough for research in the United States to show that people are more likely to form friendships with others of the same rank rather than with persons whose rank is higher or lower than theirs. Nor is it sufficient to know that this holds true in Canada, too. Or even in all Western nations. To be general it must hold everywhere. This means that sociological research should be conducted in as wide a range of settings as possible. That way we can sort out things that are specific to a time and place (such as "people think tattoos are sexy") from those that apply across the board.

Of course, like all scientific generalizations, there may be rules governing when and where social scientific generalizations apply. The gravitational law that all bodies fall at the same rate is restricted to vacuums, for in an atmosphere bodies of different shapes will fall at different rates due to friction. In Chapter 4 we will discover contextual effects and see how generalizations about individual behavior may be modified because of aspects of the group or even the whole society. Specifically, we will see that the generalization that religious people are less likely to commit crimes holds *only* if the majority of those around them are religious. With this modification added, the principle about religion and crime remains general, but now we know an additional rule that limits its application, just as the rule about vacuums limits the law of gravity. We cannot discover limits such as this if we restrict our research to one time and place.

SCIENTIFIC CONCEPTS

The wisdom of the East notwithstanding, it is impossible to regard the world in a holistic fashion. Everywhere you look you see things that exist apart from other things—reality consists of parts and pieces. Moreover, at any given moment, you ignore most of what is within your field of vision and block out most sounds within hearing range. Only by selecting what to see and what to hear can we see or hear anything—to see and hear "everything" would overload our senses to

the point that we would hear and see nothing. This alerts us to the most fundamental scientific activity: to isolate and define those portions of reality to be studied and explained. Put another way, to explain the world, scientists must take it apart and classify the various pieces. Hence, chemists separate the material world into the elements shown in periodic tables, while biologists classify all living organisms into groups and subgroups such as genera and species. Such scientific classification is based on *concepts*.

Concepts are names used to identify some set or class of things that are said to be alike. Biologists, for example, define all living organisms that are warm-blooded and give birth to living offspring (as opposed to laying eggs) as mammals. Used this way *mammal* is a scientific concept. Sociologists define many concepts. For example, a **group** is defined as any set of two or more persons who maintain a stable pattern of social relations over a period of time. Notice that this definition says virtually nothing about the people making up a group. Just as the biological concept of mammal ignores all of the many differences among cats, dogs, mice, and elephants, treating each as a member of the class of living organisms defined as mammals, the concept of group ignores gender, race, age, and thousands of other traits of members when identifying which sets of persons qualify as groups.

Concepts are the building blocks of theories. That is, concepts are useful for taking the world apart and naming the pieces, but concepts don't explain anything. To classify certain kinds of behavior as robbery or as suicide says nothing about why robberies or suicides occur. Theories link a set of concepts with statements meant to explain why and how the concepts influence one another; theories attempt to say why and how certain things—suicide, for example—occur. When Guerry explored the relationship between literacy and crime, he was responding to a current theory linking the two.

There are a number of very basic sociological concepts, and you will encounter many of them in Chapter 2. Here the emphasis will be on the most fundamental sociological concept: the group.

GROUPS: THE SOCIOLOGICAL SUBJECT

Even when sociologists use individual humans as their units of analysis, their focus is not really on the person. Even if sociologists ask why Mary Johnson voted for a Pro-Life political candidate, they will not seek the answer within Mary's head but within her social situation: What church does she attend? What is her racial or ethnic group, her age group, her political affiliation? Where does she live? To understand Mary, sociologists want to know about the groups that may shape her opinions and encourage her behavior. In doing so they reveal that the fundamental subject matter of sociology is the group.

As defined previously a *group* consists of two or more persons who maintain a stable pattern of relations over a significant period of time. Some groups, such as a married couple, are tiny. Others groups, such as a sorority, may be quite large. However, not just any gathering of people qualifies as a group in the sociological sense.

In everyday speech we often refer to ten people waiting for the walk light as a "group." But sociologists would call them an **aggregate** of individuals: They have come together only briefly and accidentally. They are not acquainted with and may not even notice one another. For sociologists, people constitute a group only when they are united by social relations. If the ten people waiting for the walk light were all members of the same family or baseball team, then they would be a group in the sociological sense of the term.

PRIMARY AND SECONDARY GROUPS

Not all groups are of equal significance to their members. For example, we usually will be more willing to withdraw from a group of persons working in our office than one made up of family or close friends. The concepts of primary and secondary groups isolate this difference.

Primary groups are characterized by great intimacy among the members. People in these groups do not merely know one another and interact frequently, but they also have strong emotional ties. As a result people gain much of their self-esteem and sense of identity from primary groups.

Moreover, sociologists regard the relationships among primary group members as the essential glue holding social life together. When Morselli and Durkheim blamed high suicide rates on modernization, they were arguing that, compared with traditional societies, modern societies made it difficult to maintain primary groups. Hence, people were increasingly without the social support needed to sustain them during times of trouble and despair. By now there is an immense amount of evidence that people lacking primary group ties display many harmful symptoms, including poorer physical health (Litwak and Messeri, 1989). However, Morselli, Durkheim, and other early sociologists greatly exaggerated the impact of modernization on primary groups: Even in the midst of large, seemingly impersonal cities, primary groups still thrive, as can be seen in Table 1-6. The data are based on a national survey of Americans age eighteen and over. Respondents who live in major cities are compared with those living in small towns and on farms. If modernization were as incompatible with primary groups as Morselli and Durkheim feared, people in cities should spend far less time socializing with relatives and neighbors. But reading across the table you can see that this is not the case. There simply are no sizable or consistent differences.

The family is the most common primary group, but many other groups can also gain this level of member commitment, sports teams and small work groups often being examples. Charles H. Cooley, who coined the term *primary group*, said a group is primary if its members routinely refer to themselves as "we." Primary groups involve "the sort of sympathy and mutual identification for which 'we' is the natural expression" (Cooley, 1909).

Secondary groups consist of less intimate groups within which people pursue various collective goals but without the same consuming sense of belonging. Business organizations, social clubs, political parties, even hobby clubs typically are secondary groups. People find it relatively easy to switch from one secondary group to another and refer to themselves and the group members as "we" only casually. However, primary groups often form *within* secondary groups—a fact that often accounts for considerable conflict within secondary groups when primary group

TABLE 1-6	Social Relations by Place of Residence: National Sample of American Adults		
		PLACE OF RESIDENCE	
		MAJOR CITY	SMALL TOWN OR FARM
Percentage who spend a social evening *almost daily* with:			
Relatives		10	11
Neighbors		9	10
Friends		6	3

Source: Prepared by the author from the General Social Survey, 2002.

members treat other members as outsiders or several primary groups compete for control of the secondary group.

SOLIDARITY AND CONFLICT: THE SOCIOLOGICAL QUESTIONS

Because groups are *the* sociological subject, the two most fundamental sociological questions are about the two most obvious features of groups. The first question is, What binds people together? The second question asks, What separates us? Put another way, What causes *social solidarity* and what causes *social conflict*?

Social solidarity refers to the density and emotional intensity of attachments within a group. In other words social solidarity describes the capacity of group members to generate a sort of glue that enables them to stick together, to "belong," to be loyal. Moreover, as noted in the discussion of primary and secondary groups, some groups have far greater levels of solidarity than do others.

We will explore the causes and consequences of solidarity throughout the book, as it is the source of some of life's most intense satisfactions. But as solidarity makes groups strong, viewed in a larger context solidarity clearly sets groups off from one another. And therein lies an extraordinary dilemma of human life: The benefits of belonging to groups do not come free. To the degree that groups surround us with warmth, affection, and a deep sense of belonging, they also increase the potential for bitter conflicts with outsiders, especially with other groups that also have high levels of solidarity. It even could be said that a primary reason we stick together is so that we can better stick it to somebody else.

Social conflict refers to unfriendly interactions *between groups*, ranging in degree of seriousness from disagreements to violent encounters. Notice that conflicts between two individuals are excluded from this definition. Although such a conflict often leads to social conflicts as friends and associates of the two disputants are drawn into a quarrel, until that takes place an event involving two individuals is not a social conflict. That is, for a conflict to be social, at least three people must be involved.

Now, let's explore aspects of social solidarity and of social conflict using one of sociology's sharpest new tools: network analysis.

ANALYZING SOCIAL NETWORKS

One of the most important recent innovations in sociology has been to greatly improve our ability to examine networks of social relationships. The word **network** refers to a pattern of ties or connections among some set of units, as a computer network links many computers and a TV network links many local stations, allowing them to communicate and exchange. In the case of a **social network,** the connections consist of a *pattern* of social links or relationships among some set of social units—usually people but sometimes groups. A **social relationship** can be defined as repeated actions between social units or the persistence of stable, shared features among units. Thus, we might examine a network among individuals based on friendships, noting that the repeated actions between friends involve exchanges of emotions. Or we could examine a network based on hatred, on kinship, on sexual involvement, or on commercial exchanges. We also could construct a network based on shared features such as which members are linked by attending the same church or which share a racial or ethnic heritage.

A particular advantage of the social network approach is that it lends itself to measurement and depiction. If we limit the number of units, it is possible to draw accurate and easily interpretable diagrams of social networks. If you look back to Figure 1-1, you will see that it diagrams the network of social relationships among members of a sociology department. Consider how helpful this diagram would be to a newcomer, as it reveals not only aspects of social solidarity (who likes whom) but also the internal divisions which may produce social conflict.

NETWORKS AND SOCIAL SOLIDARITY

A glance at the network diagrams shown in Figure 1-4 lets us see differences in group solidarity. Each group consists of four persons (each represented as a circle), and the lines linking two circles represent a strong, positive relationship between them. In Group 1 everyone is strongly attached to everyone else. We can assume that each feels an intense tie to the group as well as a strong sense of obligation to live up to the expectations of other members. In Group 2, however, each member is strongly tied to one other member but lacks a strong relationship with the

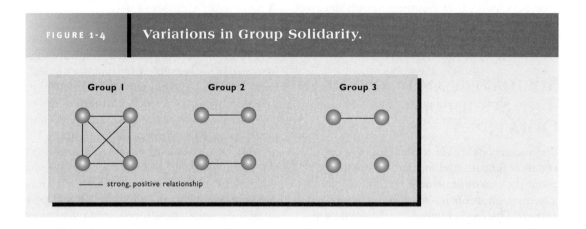

FIGURE 1-4   Variations in Group Solidarity.

other two. It is safe to assume that this group sustains a lower level of solidarity than does Group 1. Finally, in Group 3 two members have a close relationship to one another, but the other two do not. We don't know how they feel about their ties to the group, but it is entirely possible that the two without strong attachments might feel left out and might be quite lacking in loyalty.

NETWORKS AND SOCIAL CONFLICT

Since all groups consist of networks, conflicts between groups are conflicts between networks. Of greater interest, however, is the presence of relatively unconnected networks *within* groups, organizations, and even whole societies. These often generate and sustain the most intense conflicts. That is, when conflicts erupt, they usually don't require most people to choose sides; the sides already exist and may have done so for a long time. Indeed, it is the existence of sides that usually causes a conflict. Let's look over the shoulder of a sociologist as he studied a dramatic example of this.

A CLOSER VIEW
SAMUEL SAMPSON: MONASTIC STRIFE

Some years ago a graduate student at Cornell University decided to study life in a Catholic monastery for his Ph.D. dissertation. Samuel Sampson (1969), now professor of sociology at the University of Vermont, arranged to spend a year observing a Roman Catholic monastery in New England. This study was conducted in the late 1960s, which was a critical moment in the history of Catholic religious orders. In response to many changes initiated by Vatican II (a coun-

cil of all the bishops and cardinals of the church who met from October 1962 to December 1965), male and female religious orders were introducing many changes in how they lived and worshiped. These changes were welcomed by some and resented by others. Moreover, beginning the first year after the council adjourned, there was a very rapid decline in the numbers entering religious orders and a corresponding rise in the numbers who withdrew from their orders. So, Sampson thought it worthwhile to take a long and very close look at one group to see how it was dealing with this time of crisis.

Not content merely to watch and to listen, Sampson questioned each member closely about his relations with others, and from their answers he was able to map networks among eighteen men living in the monastery. Figure 1-5 was constructed from some of his data.

This diagram of the network of the monastery reveals the presence of three extremely isolated and dense internal cliques (pronounced "cleeks") or factions. Each clique had strong internal ties. To the extent that group solidarity existed in the monastery, it was limited to specific cliques. Despite the existence of these isolated internal networks, when Sampson began his study disagreements remained latent and things were quite peaceful. But not for long. Suddenly, an immense conflict erupted over whether to dismiss several young men in training to become monks. During the next several months, twelve of these eighteen men left the religious life. Predictably, five of the six who stayed belonged to the same clique: Numbers 4, 5, 6, 9, and 11. Number 12 was the only outsider to remain.

Even without any additional information, it is entirely predictable from the network diagram that eventually there would be a big fight in this monastery, and it is easy to see who would be on one side or another. In this specific case, the *Loyalists* were somewhat older men who were satisfied with things as they were. The *Rebels* were younger men who were eager to change many aspects of monastery life. Finally, three of the four *Outcasts* were isolated from the rest and formed a clique around Number 13, who tried to be friends with everybody. As will be seen in Chapter 3, in many situations monk Number 13 would have been extremely influential, able to talk to all sides—and possibly he might have been able to mediate the dispute. But when things really heat up, someone with friends in all factions gets on everyone's hate list.

STUDYING SELF-AWARE SUBJECTS

Suppose a chemistry teacher regularly demonstrated that when two harmless chemicals are mixed together, they explode. Then one day, just as the demonstration was about to begin, one of the chemicals said to the other, "I'm really getting sick of this. Today, let's not explode. Let's do something different. How about helping me make a huge stink instead?" If this could ever happen, chemistry would be a very different field. However, such things routinely happen in the social sciences. Unlike chemicals, people are able to choose among various possible actions. Moreover, they are self-aware. They often know when someone is looking at them, and unlike bacteria under a biologist's microscope, people do blush when they are looked at. Social scientists must overcome the problem of disturbing what they look at—of being misled because people sometimes act differently when they know they are being observed. While this makes the social sciences difficult, it also makes them fun. People are more interesting than bacteria or isotopes—at least to me. And part of the pleasure of doing sociology is finding ways to keep people from outsmarting your research procedures. A few examples may help demonstrate this point.

UNOBTRUSIVE MEASURES

Sociologists don't always need to observe what people do *at the time they are doing it* to obtain a good record of what they did. Much of human behavior leaves clear traces. When sociologists examine such traces, they need not worry that self-conscious subjects altered their normal behavior. For example, one sociologist who wanted to know

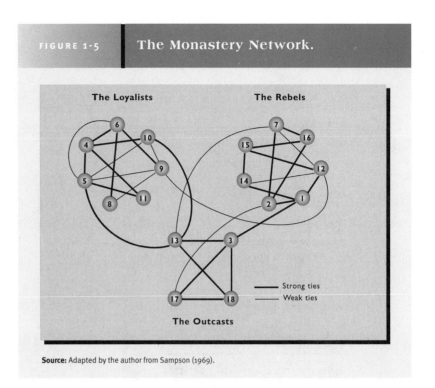

FIGURE 1-5   The Monastery Network.

The Loyalists   The Rebels

The Outcasts

Strong ties
Weak ties

Source: Adapted by the author from Sampson (1969).

what radio station people really listened to while driving arranged to have auto mechanics note the dial position of radios in the cars that they serviced.

Recently, my colleague Chris Bader and I began to discover the geography of mystical and occult groups and interests in America. Where is there more or less interest in astrology, UFOs, alien abductions, psychics, monsters, and the like? We began with the *Yellow Pages.* There we discovered listings for professional astrologers (Figure 1-6). After searching the *Yellow Pages* of the various telephone companies across the nation, we found a total of 1,513 astrologers were listed—if you think that's a lot, more than 40,000 people in France give astrologer as their full-time occupation on their income tax returns (Brierley, 1993). When we broke down the listings by state, California was highest with 255 astrologers, while there were only 12 in Delaware. This comparison is meaningless, however, because on the basis of population, Delaware should have far fewer astrologers than California. To compare states properly, it is necessary to eliminate variations in population size. We did this by dividing the total number of astrologers for each state by the state's population and then multiplying by 100,000. This results in a *rate*, the number of astrologers in each state per 100,000 population. Rates such as this in effect make each state the same size, and variations in the rates reflect real differences in the relative popularity

FIGURE 1-6 **Computing an Unobtrusive Measure.**

Astrologers

ADVANCED ASTROLOGICAL SERVICES
Counseling-Specializing In Horary & Comparison Charts-By Appointment Only
15346 Stone N ---------- **364-3198**
American Institute Of Astrology ---------- 362-2606
Astrological Counseling 16735 10th NE ---- 362-6517
ASTROLOGY CENTER OF THE NORTHWEST
Bookstore-Computer Calculated Horoscopes-Classes-Herbs
522 NE 165th ---------- **363-5313**
ASTROLOGY ET AL
BOOKS—CLASSES—HERBS
Professional Counseling
4728 University Wy NE ---------- 524-6365

5TH ELEMENT THE
Astrology & Tarot
Consultations By Appt.—Books
Open 10-5 Mon thru Fri
7527 Lake City Wy NE ---------- 522-4535

Green Jeff 4728 University Wy NE ---------- 524-6365
HUGHES DOROTHY B 2322 6th ---------- **622-7866**
INTERNATIONAL COLLEGE OF ASTROLOGY
Horoscopes By Appointment Or Mail
Astrology Home Courses-Books
4248 A Southeast Auburn ---------- **833-2361**
Jinni ---------- 524-3114
Meehan Rosalie M
Rosalie M Meehan-Room 819
1932 1st ---------- 622-4770
Moy Barbara Assoc 1331 3d ---------- 625-0575
MYRON & CORN
Bruce I Myron
Johanna E - Savannah Corn
By Appointment-Bank Cards Accepted
Smith Tower ---------- **622-2554**
Nalbandian Margaret
4728 University Wy NE ---------- 524-6365
REFLECTING POND
Books For Creative Living
Astrological Services By
David Vikaso Pond
6211 Roosevelt Wy NE ---------- 527-4333

ROBERTSON MARC & ASSOCIATES
---------- **324-3480**
Search Astrology And The Arts ---------- 523-1970
Wickenburg Joanne ---------- 523-1970

Here is the "Astrologers" section of the Seattle, Washington, *Yellow Pages* from several years ago. As you can see, there are eighteen separate listings for astrologers in Seattle. By itself, that total isn't very helpful. To compare the figure with those of other cities, population must be taken into account. By dividing the total number of astrologers by the population of the city in which they practice, you can obtain a *rate*. Notice, however, that one of these listings is not actually in Seattle. The International College of Astrology is in Auburn, so we must eliminate it. (Phone exchanges—the first three numbers—indicate those listings in the city being studied because prefixes have specific geographical boundaries.) You will also notice that several listings have the same phone number. For the sake of accuracy, calls were made to these listings to make sure that they represented several astrologers in a "group practice," not one astrologer using two listings.

By these means we can see that during this year, Seattle had seventeen persons or firms whose astrological practice was sufficiently active to warrant a business listing. Divided by the population of the city, this produces a rate of professional astrologers of 35 per 1 million residents, making Seattle one of the top U.S. cities in terms of astrological interest.

of astrology. The comparisons are shown in Figure 1-7. With population taken into account, California is far from being highest, while little Delaware is number two, just behind New Jersey. No astrologers are listed in the *Yellow Pages* for North Dakota and Mississippi.

When astrologers took out listings in the *Yellow Pages*, they were not worried that Bader and I were looking. Nor were we looking when people selected an astrologer by consulting the *Yellow Pages*, thereby ensuring that the listing would be continued. Such a measure of activity is called *unobtrusive*. An **unobtrusive measure** gains information without disturbing the objects of research.

Bader and I also planned to consider UFOs in our study. We thought it would be informative to see what role the news media might have in generating and sustaining such beliefs. One very economical way to study newspaper coverage on any particular topic is to employ a press clipping service. These are companies that subscribe to all of the daily newspapers published in the country. Staff members then read each paper every day looking for stories about particular topics or people. Whenever they find one, they "clip it" and send a copy to the client who is paying for press attention to that particular person or topic. Many celebrities employ clipping services to get copies of every story printed about them, and companies often use them to monitor their publicity. So, we hired a clipping service to look for stories on UFOs (we also had them look for stories on monsters, Big Foot, cattle mutilations, and similar topics). During the years 1994–1996, a total of 1,217 UFO stories ran in America's daily newspapers. Here, too, we needed to create rates to make states comparable. In this case, rather than differences in population size, our concern was with the fact that some states have far more newspapers than do others. To make states comparable, we divided the number of stories from papers in each state by that state's total number of daily newspapers. We then multiplied the result by 100 (to avoid rates requiring many decimal places). The results are shown in Figure 1-8. New Mexico has the highest rate of stories about UFOs—679 per 100 papers. No UFO stories appeared in newspapers in Maine and New Hampshire during the period. In this instance, too, our observations were unobtrusive in that no one involved in these stories knew we were looking.

Of course, sociologists do not rely only on traces of unobserved behavior. Most of the time when we examine what people believe and do, they know we are observing them. In fact, we often ask them to tell us what we want to know. We are able to trust such data because a great deal of effort has gone into checking to see if they are accurate and into finding ways to increase their accuracy.

VALIDATION

To ensure that they are getting accurate information, sociologists frequently conduct **validation research.** One way to assess validity is to test data against some independent standard of accuracy.

For example, we can often obtain data known to be accurate, but it can be expensive and difficult to do so. We could determine the exact ages of a sample of American adults by checking official documents (driver's licenses, birth records, and the like), but it is much cheaper simply to ask them their age. To make sure this information is accurate, we can periodically test the information so obtained by checking it against official records. If what people tell us is close enough to the information obtained when we check up, we feel confident in the information that is not checked. In this case, studies found that people gave more accurate answers when they were asked their year of birth than when they were asked their age. So that's how we often ask the question now.

In Chapter 7 we shall examine the results of many studies of delinquency. Most of these studies are based on *self-reports.* That is, samples of teenagers were interviewed or asked to fill out questionnaires, and some of the questions pertained to various illegal acts: "Have you ever stolen things from a store?" "How often have you done this?" "When was the most recent time you did this?"

The conclusions drawn from such studies obviously depend on the validity of people's answers. Because a lot of people may lie, the accuracy of self-reports on delinquency has repeatedly been subjected to stringent validity checks. Self-report data on crime and delinquency have been checked against official police and court records, against answers given while hooked to a lie detector, and even against reports given by each person's friends. The results show that self-report data are quite valid.

A second way to assess the validity of sociological data is to compare results when different measures are used. For example, for our study on the geography of interest in UFOs, Bader and I gathered many different measures in addition to the one shown in Figure 1-8. One of these was based on letters to the editor written by readers of *Flying Saucer Magazine.* These, too, were

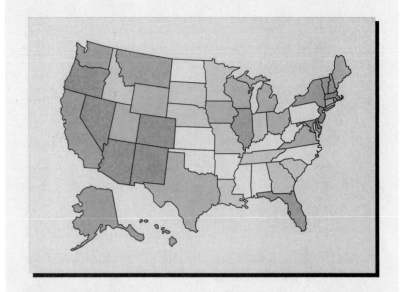

FIGURE 1-7   Number of Astrology Listings per 100,000 Population.

New Jersey	1739		Idaho	364
Delaware	1714		South Dakota	280
Alaska	1503		Georgia	275
New Mexico	1361		Tennessee	255
Oregon	1253		Louisiana	233
Hawaii	1195		South Carolina	220
Colorado	1150		Minnesota	199
Arizona	1067		Indiana	193
Massachusetts	965		Nebraska	187
Nevada	936		Missouri	172
Illinois	932		North Carolina	158
New Hampshire	889		Arkansas	124
Washington	875		Kansas	119
Maryland	866		Alabama	119
California	817		Kentucky	106
Florida	746		Oklahoma	93
Montana	715		Pennsylvania	75
Rhode Island	700		West Virginia	55
Vermont	692		North Dakota	0
New York	681		Mississippi	0
Ohio	676			
Connecticut	610			
Utah	538			
Michigan	517			
Virginia	493			
Maine	484			
Wisconsin	457			
Texas	427			
Wyoming	426			
Iowa	391			

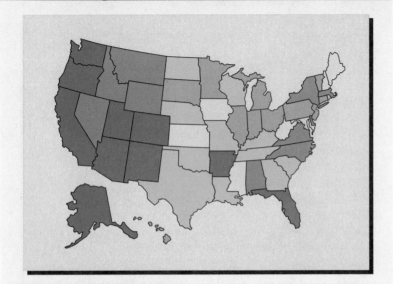

FIGURE 1-8   **Number of UFO-Related Newspaper Stories per 100 Daily Newspapers.**

New Mexico	679		Kentucky	30
Alaska	557		Rhode Island	29
Utah	433		Nebraska	26
Washington	381		South Carolina	24
Colorado	333		Wisconsin	22
Oregon	245		Georgia	22
Florida	240		Louisiana	21
Arkansas	175		North Dakota	20
Arizona	105		Missouri	18
California	102		Tennessee	18
			Oklahoma	18
			Texas	18
Delaware	100			
Virginia	88			
Idaho	83		West Virginia	13
Montana	82		Iowa	11
Wyoming	70		Mississippi	9
Massachusetts	67		Maryland	7
Nevada	63		Kansas	2
North Carolina	61		Maine	0
New York	61		New Hampshire	0
Alabama	59			
Ohio	55			
Pennsylvania	47			
Minnesota	44			
New Jersey	41			
Illinois	39			
Vermont	38			
Michigan	38			
Indiana	37			
Connecticut	35			
South Dakota	33			
Hawaii	33			

separated by state, and rates were calculated by dividing the number of letters by the state's population. The geography of these rates was extremely similar to that shown in Figure 1-8. Once again New Mexico had the highest rate while Maine again had a rate of zero. Another measure was based on circulation rates for *Flying Saucer Magazine.* Here, too, New Mexico had the highest rate and Maine was at the bottom. Any questions raised about the validity of any one of these measures lose their force when various measures collected in different ways all tell the same story.

BIAS

The essence of the scientific method is systematic skepticism: Take nothing for granted. Every step of logic in an argument and every set of facts produced by research are to be tested, checked, and retested. Moreover, the proper approach to research is to try to *disprove* those things that the researcher actually believes to be true.

Scientists must be willing to lay their opinions and beliefs on the line—to subject them to rigorous tests and accept the test results. Must scientists then free themselves from their personal biases, commitments, or hopes to function properly? The image of scientists as neutral, unemotional beings, like Mr. Spock in *Star Trek,* is romantic nonsense. All human beings are inescapably biased; we all have deep personal beliefs. The scientific method does not aim to strip scientists of their fundamental humanity or to make them into computers but rather to prevent our personal biases from distorting our work. The scientific method consists of rules that, if followed, lead us to the facts, regardless of what we might hope or believe the facts to be. Some of the most important moments in scientific progress have occurred when research turned up results very different from those expected by the researchers.

It would be naive to suggest that scientists never distort their findings or cheat in an effort to support their own convictions. But the public nature of science weakens this temptation and eventually exposes those who cannot overcome their biases. Scientists must report not only what they found but also how and where they found it. This procedure lets others check the results and even repeat the research to see if they obtain the same results.

Personal bias is possibly a more serious problem for social than for natural and physical scientists. Few of us grow up with deep convictions about what color a shark's liver ought to be or about the proper behavior for atomic particles.

But we all grow up with many firm beliefs about what people are like and how they ought to behave. Hence, social scientists have to try harder to lay their beliefs aside and look carefully at how things really are.

But this situation also offers some advantages. For if compared with chemicals or tree toads human subjects are more difficult objects of research, social scientists have the ability to relate to their subjects in ways that chemists and biologists never can. We can draw upon our own experience to help us anticipate research findings. For this reason many results produced by social research are quite obvious. For example, using *individuals* as the units of analysis (*micro*sociology):

■ Kids from poor families are more apt to be juvenile delinquents than are kids from wealthy homes.

■ Among married couples, those who often attend church have sex less frequently than those who don't attend church very often.

■ Women are more likely than men to read a newspaper daily.

Or using *aggregate* units of analysis (*macro*-sociology):

■ The higher the poverty rate in a community, the higher its rate of alcohol consumption.

■ The Jehovah's Witnesses are growing more rapidly in nations where the average person has very little education.

Whether or not these statements struck you as obvious, a great deal of research has consistently shown that *each of them is wrong!*

Family income is unrelated to juvenile delinquency, church attendance has no influence on the sex lives of married couples, and men are far more likely than are women to read the newspaper daily (mainly because of sports). The higher the poverty rate is in a community, the lower the rate of alcohol consumption. The Jehovah's Witnesses are growing more rapidly in nations with the most educated populations.

Too often, what everyone knows to be true isn't. That's why we do research—to find out what's true and what's false.

THE SOCIAL SCIENTIFIC PROCESS

At this point it will be useful to give you a preliminary overview of sociology as a *process* by which questions are raised and then answered.

The social scientific process consists of eight essential steps (as illustrated in Figure 1-9):

1. **Wonder.** Science always begins with someone wondering why. The founders of moral statistics wondered why suicide rates were so stable across time yet so variable from place to place. Others have wondered about gender bias in elections or about why people join unusual religious groups.

2. **Conceptualize.** To proceed, all scientists must be precise about what it is they are wondering about. Hence, they must *isolate* and *define* their key terms or *concepts*. To explain the world, we first must take it apart and identify the pieces: primary groups, networks, solidarity. However, concepts do not *explain* anything; they are merely the building blocks of theories. For example, to identify a set of behaviors as suicide is the essential first step in constructing an explanation of suicide. But simply to classify deaths as suicides tells us nothing about *why* people take their own lives.

3. **Theorize.** To explain something, we must say how and why some set of concepts are related—we must propose a theory. Recall that not just any set of statements containing concepts is a theory. A theory must be vulnerable to disproof. It must predict and prohibit certain things that can be checked. That is, theories must be *testable*.

4. **Operationalize.** The first step in making a theory testable is to identify indicators of each concept. Recall Homans' law of inequality from the start of the chapter. Figure 1-1 tests that theory. But to do so, it was necessary to apply the theory to something observable—in this case to a sociology department. In addition it was necessary to select an observable measure of inequality—in this instance ranks within the department. It also was necessary to measure emotional attachments—in this case the expressed degree of friendship. Having selected these measures of Homans' concepts, it became possible to observe whether or not things were as predicted. When social scientists select observable measures of concepts, it often is said that they **operationalize** each concept—make it possible to perform observational operations on it.

5. **Hypothesize.** The second step in making a theory testable is to formulate predictions

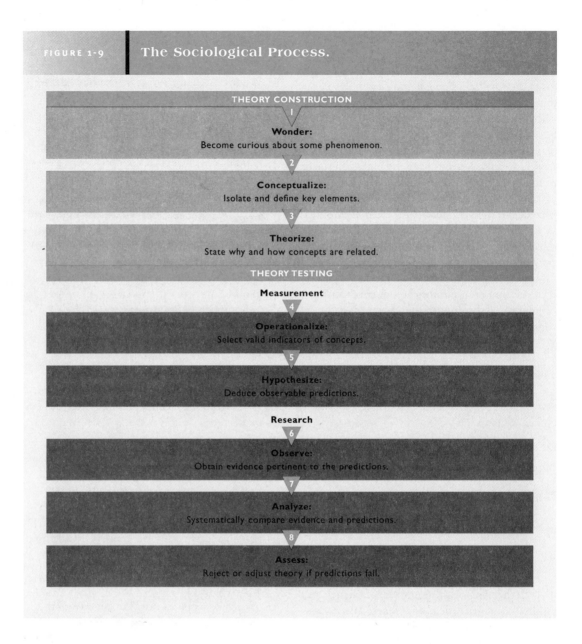

FIGURE 1-9   The Sociological Process.

about what will be observed in the connections (or relationships) among the indicators of the concepts. Put another way, theories make general statements about the world, and hypotheses make specific statements about observable things. That is, a **hypothesis** is a statement about the expected relationship between (or among) observable measures of concepts. In this case Homans' theory led to the hypothesis that members of a sociology department will tend to limit their close friendships to those of similar rank.

6. **Observe.** Once we know what to look at and where, the next step is to use the appropriate research design to gather the observations.

7. **Analyze.** Looking is not enough. We must *compare* what we observe with what the hypothesis said we would see. In this case we confirm that friendships do tend to be limited in the predicted way.

8. **Assess.** The final step in the sociological process is to make whatever adjustments are necessary in a theory on the basis of the analysis—we change theories to fit the evidence; we do not change evidence to fit theories. When the predicted outcomes keep occurring each time a theory is tested, fine. But when predictions fail, we must try to reformulate or adjust the theory so that it yields correct predictions, or we must abandon the theory and try to create a new one.

FREE WILL AND SOCIAL SCIENCE

For many centuries a major dispute in theology concerned individual responsibility. According to the doctrine of religious determinism (or fatalism), all human actions are preordained, determined by the gods or God; humans are helpless to alter their fates. But if this were so, how could humans be asked to observe moral codes? If our actions are not ours to choose, how can we be blamed for our evil deeds? Early Christian theologians dealt with this problem by declaring that each individual possesses **free will.** God did not make the world and create robots required to do His bidding. Instead, He gave humans the capacity to choose freely among alternatives, and He can therefore reward those who choose good and condemn those who choose evil.

This was a very powerful religious idea. God was no longer to be regarded as capricious, unjust, and terrible, as He must be if He is responsible for what we do, whether good or evil. Instead, it was possible to conceive of a God of mercy, justice, and absolute virtue—a God who gave humans life and choice and asked only virtue in return.

But what does the doctrine of free will have to do with social science? It mistakenly led to the conclusion that because humans possess free will, it is impossible to construct scientific theories to predict and explain their behavior! Thus, it was assumed that if it were possible to achieve social science, then humans must not possess free will; if their behavior was lawful and predictable, then it was inescapable and preordained. If we can predict who will commit crimes, then criminals can't choose good over evil and therefore bear no personal responsibility for their deeds.

While philosophers and theologians often used this argument to "prove" that social science was impossible, since it violated the principle of free will, ironically many social scientists have used much the same argument to "prove" that humans do not possess free will. Thus, Quetelet (1835) claimed that the stability of crime and suicide rates not only challenged the doctrine of free will; it also demonstrated that humans do not really choose among courses of action but are essentially puppets dancing to the beat of society. As he put it: "The greater the number of individuals observed, the more do individual peculiarities, whether physical or moral, become effaced, and allow the general forces to predominate."

Another version of this view underlies claims by some psychotherapists that there are no "bad" people in prison, only "sick" people. But as we

© Elizabeth Crews

shall see in Chapters 7 and 8, sociological theories of crime do not assume that criminals have no choices. Instead, they concentrate on how different people have a different basis for making choices and different alternatives from which to choose. This is contrary to an image of human robots programmed to steal and kill.

Similarly, the first assumption of virtually all social science theories is the same: Humans possess the ability to reason and therefore to select among different lines of action. Social science proceeds from this starting point by postulating that the choices people make can be predicted and explained by assuming that they will *attempt* to do the most reasonable thing, given their circumstances, information, and preferences. That is, *people will seek to maximize their rewards and minimize their costs.* In Chapter 3 we shall see how this simple assumption about human behavior quickly leads to explanations of why and how codes of morality come to exist and why these in turn shape individual calculations about what choices are rewarding. Here let us concentrate on the fact that it is *only* because people's choices *are* predictable that it is possible to claim that people have free will.

If people's behavior is not predictable, it must be random. That is, if knowledge of their past actions and their present situation tells us nothing about what people will do next, then the human mind does not reason but operates like a slot machine or a pair of dice. For only then can there be no consistent patterns between past and future

Choosing is the essential feature of human existence. Whether we are building with Tinkertoys or composing music, throughout our lives we attempt to make good choices and to learn from our mistakes.

behavior, no link between circumstance and action. Such people would indeed be unpredictable and would frustrate all attempts at social science. But such people would not be human. And surely they could not be judged for their acts or be said to possess free will, for they would have no capacity to choose and no reasoning power at all.

Let's approach this matter in another way. Suppose I set up a money store and offered to sell genuine $20 bills for 35¢ each. As soon as people made sure that there was no hidden gimmick, that they really could buy money from me at a huge discount, I'm sure I would have more customers than I could handle. When I predict that people will opt for a good deal, I am not reducing them to predetermined robots lacking the power of choice. I am merely saying that by assuming people can select reasonable choices, it is often easy to predict what they will choose to do.

Free will is the essential assumption of social science. Or as W. E. B. Du Bois (1904) put it, sociology "is the science of free will." We do not assume that humans are puppets but that they are reasoning, feeling organisms who learn from experience, who respond to the world around them, who have the power to love, hope, dream, and plan. The goal of social science is to understand why and how humans have these capacities. Moreover, it is because humans can draw on the findings of social science in making their choices that social science is worthwhile. Social science is not dehumanizing. Rather, it is in some ways the most humanizing of disciplines: It asks what the nature of humanity is and how human life can be enhanced.

CONCLUSION

The purpose of this book is to introduce sociology not simply as a subject but also as an activity. In letting you see what sociologists do, I hope to convey more effectively what we know, but I also want to show why and how we go about our trade. In this way I hope to enlist some of you to do sociology.

This chapter has attempted to lay an introductory basis for the chapters to come and to engage your interest and curiosity—to serve as an enticing invitation to the rest of the book. Rather than teaching you a first lesson in sociology, it offers sample previews of what is to come, just as TV previews are meant to encourage you to tune in a new show.

I know very well that not every part of every chapter or even some whole chapters will arouse your interest or stir your imagination. That's no surprise. Although every topic included in this book is of primary interest to some group of sociologists, probably no professional sociologist is especially interested in *all* of these topics. To be candid, I'm much less interested in some of them than in others. However, since this book is an introduction to the whole field of sociology, it must cover all active topics in the field. Although you should sample and explore all of these topics, it is not necessary that you find each topic of special interest. Any one of the chapters that follow contains enough important, unsolved problems to provide you with a life's work. Indeed, every chapter covers an area of sociology in which any major department of sociology offers at least one whole course and sometimes several advanced courses.

In a sense, then, this book is a sociological sampler that surveys what sociologists do. In this way you can discover for yourself what kind of sociology most appeals to you.

Finally, I have asked you, as a reader of this book, to explore sociology with me. Although we probably will never meet, we nevertheless will spend a good deal of time together over the next several months. I have made a serious effort to let our author-reader interaction be more intimate than is usual in college textbooks. I want you to share in the fun I have in being a sociologist and telling you about it. For this reason, from time to time, I let you look over my shoulder as I do some sociology. Hence, while many of my studies are reported in other introductory textbooks, my work will receive more space in this one. Because I am presuming to tell you what sociology is and how it is done, it seems fair to let you see for yourself what kind of sociologist I am. But I have also chosen to let you watch me work because it was by doing these studies that I truly became a sociologist. Letting my students see me at work is the most effective way I have found to reveal to them what being a sociologist means to me.

Review Glossary

Terms are listed in the order in which they appear in the chapter.

science A *method* for describing and explaining why and how things work (human as well as material "things"). It consists of two components: theory and research.

theory An abstract statement that *explains why and how certain things take place*, whether these things be eclipses of the moon, chemical reactions, or outbursts of racism. In addition scientific theories must have *empirical implications* (the word *empirical* means "observable through the senses"). That is, theories make definite predictions and prohibitions; they say some things will happen under certain circumstances and that other things will not happen.

research The process of making *appropriate empirical observations* or measurements—this usually is referred to as collecting data (or facts). The purpose of research is to *test theories* (do their predictions hold?) or to *gain sufficient knowledge* about some portion of reality so that it becomes possible to theorize about it.

sociological imagination A term coined by American sociologist C. Wright Mills to describe the ability to see the link between incidents in the lives of individuals and large social forces.

sociology The scientific study of the patterns and processes of human social relations.

social sciences Those scientific fields devoted to the study of human behavior, including sociology, psychology, economics, political science, anthropology, criminology, and some branches of history.

units of analysis The "things" on which a set of research observations are based. Sociologists use many different units of analysis; among them are individuals, small groups, large organizations, counties, cities, states, and nations.

microsociology The study of small groups and of face-to-face interaction among humans.

macrosociology The study of large groups and even of whole societies.

concepts Names used to identify some set or class of things that are said to be alike. Concepts are the building blocks of theories.

group Two or more persons who maintain a stable pattern of social relations over a significant period of time.

aggregate A collection of people lacking social relations; for example, pedestrians waiting for a walk light.

primary groups Groups whose members have close and intimate emotional attachments to one another.

secondary groups Groups whose members have only limited emotional attachments to one another.

social solidarity The density and emotional intensity of attachments within a group; put another way, the capacity of group members to generate a sort of glue that enables them to stick together, to "belong," to be loyal.

social conflict Unfriendly interactions *between groups*, ranging in degree of seriousness from disagreements to violent encounters. Notice that conflicts between two individuals are excluded from this definition. For a conflict to be social, at least three people must be involved.

network A pattern of ties or connections among some set of units, as a computer network links many computers and a TV network links many local stations, allowing them to communicate and exchange.

social network A *pattern* of social relationships or links among some set of social units—usually people, but sometimes groups.

social relationship Repeated actions between social units, or the persistence of stable, shared features among units.

unobtrusive measures Techniques used to measure behavior without disturbing the behavior of the subjects.

validation research Studies conducted to determine whether particular measures used in research are accurate.

operationalize To select measures of concepts to make it possible to perform observational operations on them.

hypothesis A statement about the expected relationship between (or among) observable measures of concepts.

free will The philosophical and theological doctrine that humans possess the capacity for choosing among alternatives and, therefore, can be held responsible for the choices they make.

Suggested Readings

Du Bois, W. E. B. [1899] 1973. *The Philadelphia Negro.* Millwood, N.Y.: Kraus-Thomson.

Fletcher, Joseph, Esq. 1849. "Moral and Educational Statistics of England and Wales." *Journal of the Statistical Society of London* 12:189–335.

Stark, Rodney, and William Sims Bainbridge. 1997. *Religion, Deviance, and Social Control.* New York: Routledge.

U.S. Bureau of the Census. *Statistical Abstract of the United States* (published annually). Washington, D.C.: U.S. Government Printing Office.

Sociology Online

www.socstark10.com

GO TO THE INTERNET AND TYPE: www.socstark10.com. This screen will appear.

Sociology Tenth Edition - Rodney Stark

 Wadsworth

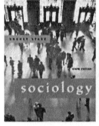

This textbook offers students an active introduction to sociology. They watch sociologists formulate, dispute, and revise theories. Better yet, students look over the shoulders of real researchers as they design and conduct important studies.

In keeping with the conception of sociology as an activity, this Web site offers readers of the Stark textbook a chance to do some sociology—to explore real data and reflect on real results.

The three data sets are widely used by professional researchers.

Choose a Data Set for Analysis
2000 General Social Survey
Nations of the Globe
The 50 States

2000 General Social Survey This is a national survey now conducted every other year by the National Opinion Research Center at the University of Chicago. This survey consists of interviews with 2,817 randomly selected Americans age eighteen and over. Many of the questions were asked of everyone. But some questions were asked only of a random subsample. When that is the case, many respondents will be designated as missing data.

Nations of the Globe This data set is based on the 174 nations having populations of 200,000 or more. The data come from many sources including the United Nations, the World Bank, and comparable surveys conducted in a subset of the nations. The data are the most recent available. When no value is shown for a nation on any specific variable, that is because no data were available.

The 50 States Here are data that permit comparisons among the American states. Most of the data come from government sources such as the Bureau of the Census and the Department of Justice.

⬈**CLICK ON:** 2000 General Social Survey

This screen will appear. (**Note:** Different Web browsers produce slight variations in the appearance of the graphics. The screens shown here have been accessed by Internet Explorer 4.0.)

For the moment, ignore the lower box. The entire list of variables can be examined one by one in the upper window, using the arrow to the right. When you click on the arrow, a window opens to show a selection of the available variables. As you scroll up and down the list, the variable on which the arrow is resting becomes highlighted. To select a variable, click on it when it is highlighted. To begin:

GSS2000 file: Select a variable for analysis

Select your variable by clicking on its name on the following list:

> SUICIDE OK? ⬍

Analyze Now

- OR -

Search for a variable with the following words or phrases:

Word/Phrase 1: _____

Word/Phrase 2: _____

Word/Phrase 3: _____

FIND VARIABLES with ANY of these words/phrases or ALL of these words/phrases? [Select Choice]

ANY ⬍

Start Search

✓**SELECT:** SUICIDE OK? Does a person have the right to end own life if this person is tired of living and ready to die?

⬈**CLICK ON:** Analyze Now

This screen will appear.

GSS2000
2000 General Social Survey (Selected and Modified Variables)

1) SUICIDE OK
Does a person have the right to end own life if this person is tired of living and ready to die?

Category	Freq.	%
YES	307	17.0
NO	1494	83.0

Among all respondents, 16.8% of the sample think a person who is tired of living has a right to commit suicide.

24) AGE

	<30	30-49	50 AND UP
YES	18.7%	15.5%	18.2%
NO	81.3%	84.5%	81.8%
Number of cases	331	822	643

The higher the Age, the less likely the individual is to think a person who is tired of living has a right to commit suicide. The most significant difference comes between those under 30 and those aged 30 to 49.

25) EDUCATION

	NO HS GRAD	HS GRAD	COLL EDUC
YES	13.3%	14.0%	19.8%
NO	86.7%	86.0%	80.2%
Number of cases	166	307	542

The higher the Education, the more likely the individual is to think a person who is tired of living has a right to commit suicide. The most significant difference comes between those who graduated high school and those with college education.

26) SEX

	MALE	FEMALE
YES	19.1%	15.5%
NO	80.9%	84.5%
Number of cases	769	1032

Among males, 20.3% think a person who is tired of living has a right to commit suicide. Among females, this percentage was only 14.1%. The difference is statistically significant.

27) RACE

	WHITE	AFRICAN AM
YES	17.9%	11.9%
NO	82.1%	88.1%
Number of cases	1405	303

Among whites, 17.7% think a person who is tired of living has a right to commit suicide. Among African-Americans, this percentage was only 9.2%. The difference is statistically significant.

↗**CLICK ON:** The 50 States

Does this data set include anything on suicide? There are two ways to find out. First, simply click on the down arrow beside the upper box and then scroll down through the variables to see if any of them concerns suicide. Better yet, use the lower box to search for the word *suicide.* Click on the Word/Phrase 1 box and then type the word *suicide* in the window, as shown.

STATES
The 50 States

1 variables matched your search criteria.
Click on any variable name for on-line analysis.

5) SUICIDE
Suicides per 100,000 population.

Here we see that Americans overwhelmingly reject the principle that people have a right to end their lives when they wish to: only 17.0 agreed. Although this result is interesting, it raises even more interesting questions. *Who* is more favorable toward suicide? By who, sociologists mean what *categories* of people. Whenever you analyze a variable using the General Social Survey, the program will automatically produce tables allowing comparisons by age, education, sex, race, income, region, party, and religion. The software also will interpret each table for you. By scrolling up and down, you can examine each table.

Many would suppose that people will be more favorable toward suicide if their own life circumstances are less rewarding. If so, then older people, the less educated, people with lower incomes, women, and African Americans should be more apt to answer "Yes." But is that so? Why do you think that is the case? Might networks be involved?

↗**CLICK ON:** Select New Data Set

Select a variable for analysis

Select your variable by clicking on its name on the following list:

1) STATE NAME--STATE NAME

Analyze Now

- OR -

Search for a variable with the following words or phrases:

Word/Phrase 1: suicide
Word/Phrase 2:
Word/Phrase 3:

FIND VARIABLES with ANY of these words/phrases or ALL of these words/phrases?
[Select Choice]
ANY

Start Search

Then:

↗**CLICK ON:** Start Search
This screen will appear.

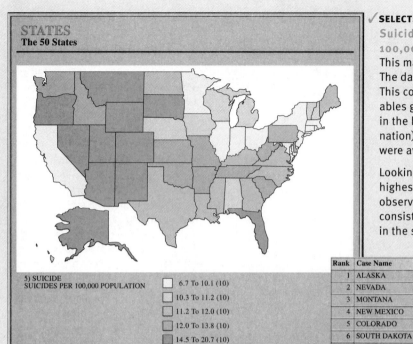

STATES
The 50 States

5) SUICIDE
SUICIDES PER 100,000 POPULATION

- ☐ 6.7 To 10.1 (10)
- ☐ 10.3 To 11.2 (10)
- ☐ 11.2 To 12.0 (10)
- ☐ 12.0 To 13.8 (10)
- ■ 14.5 To 20.7 (10)

✓ **SELECT:**

Suicide: Suicides per 100,000 population
This map of the United States will appear. The darker a state, the higher its suicide rate. This color key applies to all maps of all variables generated by this software, as is shown in the legend below the map. If a state (or nation) is uncolored, that means no data were available for it.

Looking at the map, or by comparing the five highest states with the five lowest, can you observe a regional pattern in suicide? Is this consistent with the regional pattern revealed in the survey?

Rank	Case Name	Value
1	ALASKA	20.7
2	NEVADA	19.8
3	MONTANA	18.6
4	NEW MEXICO	17.9
5	COLORADO	16.8
6	SOUTH DAKOTA	16.1
7	WYOMING	16.1
8	ARIZONA	15.6

Beneath the map, a list of the fifty states also appears, ranked from highest to lowest on their rates of suicides per 100,000 population.

⬏ **CLICK ON:** Select New Data Set

⬏ **CLICK ON:** Nations of the Globe
What about this data set? Does it include anything on suicide? To find out, use the lower box to search for the word *suicide*. First type in the word suicide. Then:

⬏ **CLICK ON:** Start Search

✓ **SELECT:** Suicide: Suicides per 100,000 population
This map of the world appears, with the highest nations shown in the darkest shade. Nations that are not colored are those for which no rate is available.

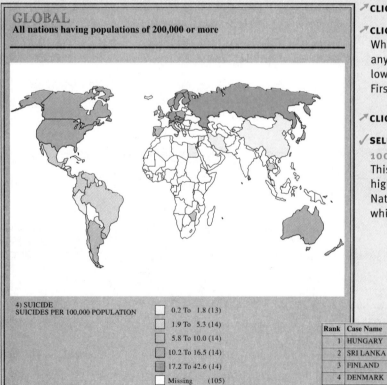

GLOBAL
All nations having populations of 200,000 or more

4) SUICIDE
SUICIDES PER 100,000 POPULATION

- ☐ 0.2 To 1.8 (13)
- ☐ 1.9 To 5.3 (14)
- ☐ 5.8 To 10.0 (14)
- ☐ 10.2 To 16.5 (14)
- ■ 17.2 To 42.6 (14)
- ☐ Missing (105)

Rank	Case Name	Value
1	HUNGARY	42.6
2	SRI LANKA	30.0
3	FINLAND	29.9
4	DENMARK	26.6
5	AUSTRIA	23.1
6	BELGIUM	22.3
7	SWITZERLAND	21.0
8	RUSSIA	20.8

Beneath the map, the list of nations appears, ranked from highest to lowest in terms of their number of suicides per 100,000 population.

Morselli believed that suicide rates were rising because of modernization. If he was correct, then the most modernized nations should tend to have high suicide rates and the least modernized should have low rates. Is that so?

USING INFOTRAC COLLEGE EDITION

www.infotrac-college.com

A passcode for gaining access to InfoTrac College Edition is provided with all new copies of this textbook. InfoTrac College Edition is a database containing nearly 1 million articles that have appeared in more than 600 periodicals during the past several years—it is updated constantly.

GO TO THE INTERNET AND TYPE:
www.InfoTrac-college.com

This screen will appear.

> ↗**CLICK ON** Register New Account
> You will be asked to enter your Access Code and to create and enter a User Name and Password. Having done so,

> ↗**CLICK ON** InfoTrac College Edition or Log On.
> Now you can enter search terms.
>
> Try **DU BOIS, W.E.B.**

> ↗**CLICK ON** Search.
> InfoTrac College Edition will tell you how many periodical articles are available on this very important early sociologist.

InfoTrac College Edition

Five years of full-text articles from The New York Times... 138,000 new reasons that InfoTrac College Edition is your best resource for online research.

We are delighted to announce that The New York Times has been added to our list of more than 5,000 scholarly and popular periodicals available through InfoTrac College Edition! No other source offers this depth and range of articles from The New York Times . . . and no other source offers anywhere close to the number of peer-reviewed, reliable articles that you'll find on InfoTrac College Edition. In fact, today, you can draw from more than 18 million articles from more than 5,000 sources when doing your online research. And, articles are added daily, so you will always have the most current information! Log on today and see how InfoTrac College Edition puts the world at your fingertips.

InfoTrac College Edition
• Available 24/7
• Over 15 million articles
• Over 5000 journals
• Updated daily
• Over 20 years of content

Now Introducing InfoWrite!
Learn how to...
• Write a successful research paper
• Evaluate and compare sources
• Improve grammar and word usage
...and much more!

> ↗**CLICK ON** View to see them listed.
> To read any of these articles, simply

> ↗**CLICK ON** View text and retrieval Choices.
> For articles on other famous early sociologists, try these terms:

SEARCH TERMS:
Durkheim
Quetelet

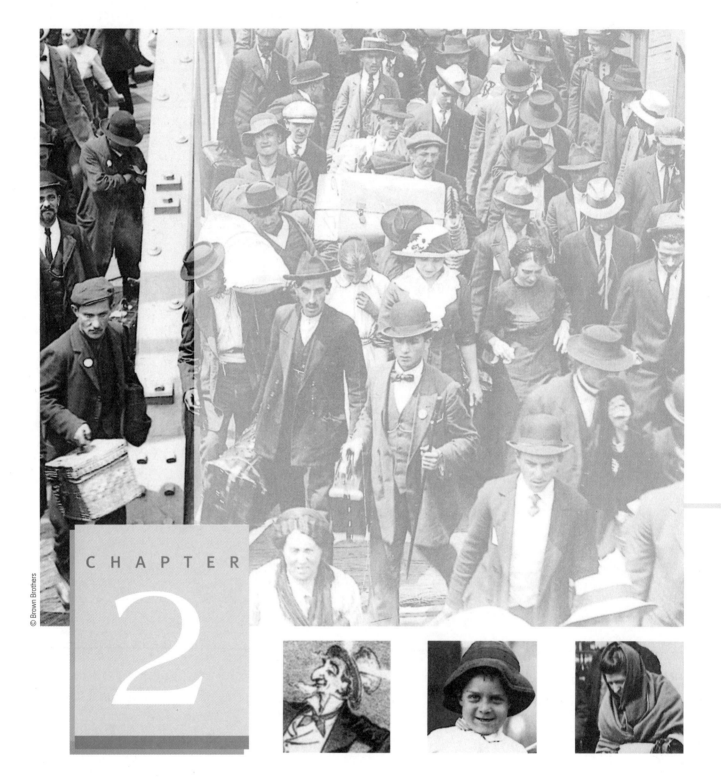

CHAPTER

2

immigrants stepping ashore at new york in 1910

These people were among the more than 1 million who came in that year. The lack of women suggests that many of the men in this group hoped to return to Europe with savings earned rapidly in the booming American economy. Notice, too, how well-dressed these immigrants are—nearly all of the men are wearing suits and ties, and the several women in view are wearing pretty hats and nice dresses. This helps us recognize that the image of immigrants as desperately poor peasants was not entirely accurate.

Concepts for Social and Cultural Theories

IN 1907 AN APPREHENSIVE U.S. CONGRESS appointed a commission to study whether steps should be taken to slow or halt the flood of immigrants pouring into the nation. That year a record 1,285,349 persons were admitted. Actually, it wasn't *how many* were coming in but *who they were* that really was causing concern. Of all the immigrants arriving in 1907, 81 percent came from southern and eastern Europe—principally from Italy, Greece, Russia, Poland, and what later became Czechoslovakia, Hungary, and Yugoslavia. Of all these people, the two largest groups of immigrants in 1907 were Italians, mainly from southern provinces, and Jews from Russia, Poland, and other parts of eastern Europe.

The commissioners identified the current flow as the "new immigration." They contrasted this with the "old immigration," the pattern that had prevailed from before the American

Revolution through the 1880s. The commission noted that in 1882 only 13 percent of immigrants were from southern and eastern Europe, while 87 percent came from Great Britain (including Ireland) and from northwestern Europe—Belgium, France, Germany, Scandinavia, and Switzerland.

The commission was formed primarily because of growing concerns about the impact of the new immigration. In its opening declaration of purpose, the commission suggested that the old immigration brought good citizens "from the most progressive sections of Europe" to America, where "they mingled freely . . . and were quickly assimilated." But the new immigrants were said to be "far less intelligent than the old" and unwilling and unable to assimilate. Many expert witnesses testified that Jews, Italians, Greeks, Poles, Slavs, Hungarians, and other new immigrants would never fit in, never become

CHAPTER OUTLINE

THE CONCEPT OF SOCIETY

SOCIAL STRUCTURAL CONCEPTS

STRATIFICATION

NETWORKS
Networks and Power
Local and Cosmopolitan Networks
Societies as Networks

THE CONCEPT OF CULTURE

CULTURAL CONCEPTS
Values and Norms
Roles
Multiculturalism and Subcultures
Prejudice and Discrimination
Assimilation and Accommodation

MODERNIZATION AND GLOBALIZATION

JEWS AND ITALIANS IN NORTH AMERICA
Prejudice and Discrimination
Assimilation and Accommodation
of Jews and Italians

THEORIZING ABOUT ETHNIC MOBILITY

THE CULTURAL THEORY

A CLOSER VIEW
Zborowski and Herzog: Jewish Culture

A CLOSER VIEW
Leonard Covello: Italian Culture

THE SOCIAL THEORY

A CLOSER VIEW
Stephen Steinberg:
The Jewish Head Start

A CLOSER VIEW
Joel Perlmann: A New Synthesis

REFERENCE GROUPS AND ITALIAN
TRADITIONALISM

A CLOSER VIEW
Andrew Greeley:
The Persistence of Italian Culture

CONCLUSION

REVIEW GLOSSARY

SUGGESTED READINGS

SOCIOLOGY ONLINE

educated, never contribute to the culture or to the economy.

However, in addition to hearing such testimony, the commissioners did something much more creative: They organized some massive research projects to discover just what was going on. What were the new immigrants *really* like? How were they fitting in? What were their lifestyles?

Published in many massive volumes of statistics, the Immigration Commission Report of 1911 offers a priceless glimpse into the nation at that time. And it includes many surprising findings.

The cleanliness of the new immigrants and their households was found to be "unexpectedly good." And the new immigrants did not form permanent ethnic enclaves in the poor districts of cities—enclaves that would insulate them from all outside efforts to make them into "real Americans." Instead, "as their economic status improves . . . they very generally move to better surroundings." The commission also could find no evidence of greater criminal activity among the new immigrants.

But perhaps the most surprising finding was that some of the new immigrants had already caught up economically. Table 2-1, based on data the commission gathered in 1908, shows that employed Jewish men born in eastern European nations other than Russia had an average weekly income *higher* than that of native-born white men, while Jews from Russia were just slightly behind native-born whites. In contrast, Italians earned only 69 percent as much as native-born whites, while men born in Mexico and Greece earned only 62 percent as much. In fact, these latter groups earned less than African Americans. (The data exclude agricultural workers, and most African Americans still worked on farms in this era.) Moreover, this pattern was not limited to the United States. Jews had also achieved very rapid economic success in Canada, and there, too, they stood out in sharp contrast to the much slower progress of Italians and other new immigrants.

As these facts came to the attention of early sociologists, many of them began to wonder why. Why had the Jews done so well, so rapidly? Or put in relative terms, why had the Jews achieved success so rapidly compared with the Italians? Both arrived here during the same period and brought little money with them. Both faced considerable hostility from the native-born population. But there were the Jews, new immigrants already on a par financially with the old immigrants. How had they done it?

This is exactly the sort of wondering that is the first step in the sociological process. In this chapter we are going to follow up on this question, looking on as social scientists formulate and test theories aimed at providing an answer. And we are going to give particular attention to efforts to fulfill the second step in the sociological process: to conceptualize.

I have chosen to pursue this particular question because it prompted two broad forms of theory that sociologists often use. Hence, the concepts these two forms of theory require are those most widely used by sociologists. By encountering these key concepts now, you will acquire a sociological vocabulary that will help you in later chapters. Perhaps I could just have written a chapter that defined these concepts and asked you to memorize them. But it's very boring to learn terminology for which you cannot yet see any use—even professors of sociology don't sit around reciting lists of key concepts. Concepts are not meant to be appreciated but to be used in theories. And the best way to learn them is by using them. So that's how you will encounter concepts in this chapter.

Keep in mind that in time Italians also achieved economic parity with the old immigrant groups in Canada and the United States, so our focus will be not on why only the Jews succeeded but on why they did this so much sooner. As you will see, this question has implications far beyond these particular groups in a certain time and place. For the underlying issue involves the general conditions under which ethnic and cultural minorities achieve economic parity. Moreover, we will be following a question that has attracted the attention of many first-rate sociologists. As you watch them work, you will gain greater insight into what sociologists actually do. And when sociologists wonder about the relative speed of economic progress of Italians and Jews, the next thing they need to do is *conceptualize*. Let's watch.

THE CONCEPT OF SOCIETY

Society refers to any relatively self-contained and self-sufficient group of human beings who are united by social relationships. A distinct social boundary (and often a geographical one) sets one society off from others, and members

TABLE 2-1	Income of Employed Males 18 and Over in the United States, 1908 (nonfarm only)
	AVERAGE WEEKLY INCOME
Native-Born	
White	$13.89
Black	10.66
Foreign-Born	
Canadian (French)	10.62
Canadian (English)	14.15
English	14.13
Irish	13.01
Greek	8.41
Mexican	8.57
Italian (south)	9.61
Jewish (Russia)	12.71
Jewish (other)	14.37

Source: U.S. Senate Committee on Immigration, *Abstract of Reports of the Immigration Commission,* vol. 1 (Washington, D.C.: U.S. Government Printing Office, 1911).

will know to which society they belong. Usually, a society occupies a definite physical location— even nomadic societies tend to travel a familiar route within a specific area. The terms *society* and *nation* often are used synonymously, but not all societies are nations (many small preliterate groups are societies but are not nations), and some nations include several societies. For example, the former Soviet Union split into many nations because it had consisted of a number of quite separate societies.

SOCIAL STRUCTURAL CONCEPTS

Sociologists see societies as composed of *social structures*. When Guerry attempted to explain variations in suicide and crime rates, he considered population density, the number of people living within a designated area. Population density is a **social structure** in that it is a characteristic of a group rather than of individuals. Individuals may have home addresses, but they do not have population densities. Such density exists only as a property of the group or neighborhood. If a family moved from Singapore, where the population density is 11,441 persons per square mile (see Table 2-2), to Mongolia, where there are only 4 people per square mile, the *family* wouldn't become less dense, but the lives of its members would change rather dramatically in response to such a huge shift in their social

TABLE 2-2	Population Density as a Social Structure		
NATION	**POPULATION PER SQUARE MILE**	**NATION**	**POPULATION PER SQUARE MILE**
Singapore	11,441	Austria	240
Bangladesh	2,255	Spain	204
Taiwan	1,659	Costa Rica	159
Netherlands	1,146	Egypt	142
South Korea	1,138	Ireland	131
Belgium	850	**Mexico**	**121**
Japan	757	Ecuador	101
India	697	Zimbabwe	72
United Kingdom	617	**United States**	**71**
Haiti	591	Venezuela	59
Germany	588	Sweden	54
Italy	509	Brazil	48
Switzerland	442	New Zealand	32
Nigeria	348	Russia	22
China	320	Congo	18
Denmark	314	**Canada**	**8**
France	269	Australia	6
Uganda	242	Mongolia	4

Source: Prepared by the author from *Nations of the Globe: An Electronic Data Base in MicroCase Format, 2002*, distributed by Wadsworth.

Note: Population density is an average and can be very misleading. Most Canadians and Australians, for example, live in cities where the population density is quite high, and large parts of each country are essentially unpopulated.

environment. In similar fashion the sex ratio is a social structure. If a person moved from Russia, where there are 114 females for every 100 males, to Saudi Arabia, where there now are 84 women for every 100 men, that person would notice, but her or his sex would not have changed. Technically speaking, all group characteristics are social structures. Usually, however, sociologists reserve the term for group characteristics that are relatively independent of culture—this distinction will become clear when culture has been defined.

The reason that such characteristics of groups are called "structures" is that they influence our behavior in much the same way that physical structures such as doors and stairways channel our movements. The age composition of a group is a social structure. Suppose a nineteen-year-old college student was placed within a group in which the youngest person was four and the oldest nine. Suppose that after a month that same student was moved to a group in which the youngest person was sixty-five. It is likely that this student's behavior would have changed dramatically when he or she shifted from the first group to the second,

having gone from being by far the oldest to being by far the youngest member of the group. So, sociologists think of the social world around us as consisting of structures, in that they shape our behavioral options to an almost physical extent.

Émile Durkheim, the early French sociologist, was among the first to equate social structures with physical and material forces in terms of their impact on the individual. In *Suicide* (1897) he wrote that characteristics of groups "have an existence of their own; they are forces as real as [physical] forces, though of another sort; they, likewise, affect the individual from without, though through other channels . . . which cause us to act from without, like the physico-chemical forces to which we react." He went on to argue that "these forces must be of the moral order and since . . . there is no other moral order in existence in the world but society, they must be social."

Ever since Durkheim, when sociologists use the word *society*, they usually are thinking about a set of social structures. In addition to those just mentioned, let's examine some social structures of greatest interest to sociologists.

STRATIFICATION

One of the most important social structures is stratification. **Stratification** refers to the unequal distribution of rewards (or things perceived to be valuable) among members of a society (or group). In Chapter 9 we will see that these rewards are separated into three major types: *property, power,* and *prestige* (or honor). In all known societies, these rewards are unequally distributed—some people have more and some have less. However, societies differ greatly in the shape and degree of their stratification structures. In some a tiny elite enjoys immense power, property, and prestige, while the huge majority has very little of each. In other societies the gap between the rich and poor is much smaller and power is widely shared. Put another way, some societies are far more *stratified* than are others.

Stratified literally means "to be layered"; the term *upper crust* reflects the idea that the stratification structure of any society is made up of layers. Sociologists identify these layers as **classes**— these are people who share a similar position or layer within the stratification structure. Societies differ in their number of classes as well as in the principles employed to place people within a given class.

Sometimes individuals or whole groups change their level or class position. Upward movement is known as **upward mobility;** the reverse is called **downward mobility.** For example, when an engineer's child becomes a factory worker, we say that she or he has been downwardly mobile. When a factory worker's child becomes a banker, we say that he or she has been upwardly mobile.

The term **status** refers to the position or rank of a person or group within the stratification structure—as in everyday speech we often say someone lacks status or has a high or a low status. Status can be determined in two general ways. When people are placed within the stratification structure on the basis of their individual merits or achievements, it is known as **achieved status.** When one's position is derived primarily through inheritance, we call this **ascribed status;** that is, a person's position in society is fixed (or ascribed to him or her by others) on the basis of family background or genetic inheritance. Racial, ethnic, and religious differences, as well as gender, often serve as the basis for ascribed status.

The caste system in India long has been an extreme example of a stratification structure based on ascribed status. Each level in the stratification structure is known as a caste. Everyone is born belonging to a specific caste. The caste of the parents thus generally determines the status of their children, regardless of ability or merit.

NETWORKS

Chapter 1 identified groups as the primary sociological subject, and groups also are the primary social structure. Groups have many important attributes and therefore come in an immense variety. For example, sociologists often compare groups on the basis of their relative solidarity, as was evident in the discussion of primary and secondary groups. Or groups can be compared on the basis of the extent to which they contain internal cliques or factions and the amount of conflict that takes place within the group. Both these aspects of groups come into sharp relief when we examine their most fundamental aspect: their *network structures.* So, let's pause and examine some of the most important things sociologists know about network structures.

NETWORKS AND POWER

During his classic study of how people get jobs, Mark Granovetter (1973) discovered that people seldom heard about a job from their close friends, but usually from a distant acquaintance. This led him to formulate a principle about networks to which he gave the ironic name **the strength of weak ties.** As used in network analysis, a **tie** is another word for a link or a relationship, and Granovetter operationalized the strength of a tie as "a combination of the amount of time, the emotional intensity, the intimacy (mutual confiding), and the reciprocal services which characterize the tie." It is obvious that because the stronger the tie, the more time and effort it involves, an individual can maintain far more weak ties than strong ones. It also follows that networks based on weak ties will be far larger than networks of strong ties. Finally, Granovetter was ready to propose that *for purposes of spreading information, weak ties are stronger, or more effective.*

Let's take a close look at this principle. Consider the problem of getting a job in light of your network position as depicted in Figure 2-1. You have two very strong ties (C and D), both of whom also have strong ties to people in the same group or clique within a larger network. You also have weak ties to A and to B, each of whom is in a different group or clique with which you have no other contacts. Between them, C and D have strong ties to six people, compared with thirteen for A and B. Other things being equal, you are more than twice as likely to hear of a job from A or B than from C or D. Your weak ties are more apt to land you a job because there are more of them. This explains why so many of Granovetter's respondents reported that they got their job because they accidentally ran into someone they knew only casually, not because their close friends put them onto an opening.

However, weak ties are only more effective in spreading information; for exerting influence strong ties are more effective. As will be seen in Chapter 10, in China hiring is based far more on *who* you know than on *what* you know. There, local networks have greater impact on getting a job (Bian, 1997).

But there is more to be learned from Figure 2-1. Following up on Granovetter's insights about the strength of weak ties, Ronald Burt (1995) pointed out that networks made up of strong ties also tend to have redundant ties in terms of influence and the flow of information. In network terms a **redundant tie** is one that duplicates links among members. For example, in Group 1 two members directly linked to C also are directly linked to D. In a fully redundant network, any member can send a person-to-person message to another member by several routes. That has several implications. First, information will get around rapidly within such a group. Second, everyone's information will be the same. In terms of a job search, each member will know of the same openings and will have heard about the same one many times. Third, the scope of the information reaching the group will be very limited. This reveals the flip side of Granovetter's principle: the weakness of strong ties.

Now reexamine your position in Figure 2-1. Your ties with A and B are *nonredundant*. You are the *only* person linking these three clusters, which, in terms of the flow of information, makes you indispensable. Consider gossip. You hear about everything that people in all three groups find out. So, you are the one who brings the latest dirt to your network of close friends. But you also are the only source that A has for the gossip from your group and from Group 3. And you are the only source that B has to the news from Groups 1 and 2. That means you can limit what any group knows and you can personally benefit accordingly. It also gives you a valuable resource for bargaining with A and B as with C and D. This is what is meant by the common expression that "knowledge is power."

Another way of looking at this is that there are *structural holes* in the networks (Burt, 1995). If every person shown in Figure 2-1 had ties of equal strength to every other person, there would be no clusters and no holes. **Structural holes** are unlinked pairs, and they show up on network diagrams as blank space. Two kinds of structural holes appear in Figure 2-1. One kind is a *hole between groups* (or cliques), such as the holes that separate most people in Group 1 from most people in Groups 2 and 3. Holes are relative, not absolute. There is only a relative lack of links among these three groups, for you are the one individual with across-group (or across-hole) links. Links across network holes are referred to as **bridge ties,** and a person having such ties is said to occupy a **bridge position.** Bridge positions, as noted, tend to confer power.

The second kind of structural hole is located *inside a local group*. Here the thing to notice is

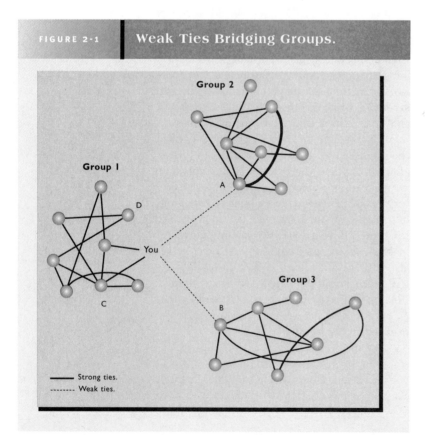

FIGURE 2-1 **Weak Ties Bridging Groups.**

Group 2

Group 1

D

You

A

Group 3

C

B

——— Strong ties.
- - - - - Weak ties.

that there aren't many structural holes locally as almost everyone is connected to everyone else. *Inside* Group 1, you have no network advantage over anyone else. You all get the local gossip more or less at the same time. Because of the lack of local structural holes, there is not much bargaining power inside this cluster. However, this too is variable, and some local networks do have holes, as distinctive cliques or factions exist. In fact, Figure 2-1 might depict the network within an organization, and thus, each group could be a clique or perhaps each group represents a separate department within the organization (accounting, marketing, and so on). We will examine networks within organizations in Chapter 20.

To sum up: Within any network persons with nonredundant ties have an advantage over those who have only redundant ties. The greater the number of network groups or cliques to which an individual maintains ties, the greater her or his power—but as we shall see, the greater his or her vulnerability during conflicts among clusters or cliques.

LOCAL AND COSMOPOLITAN NETWORKS

It isn't just placement in a network that determines power; the shape and density of the network matter, too. Networks tend to take one of two basic shapes: local and cosmopolitan. Both are depicted in Figure 2-2.

In the previous discussion, Group 1 was referred to as a local network. **Local networks** are dense with strong, redundant ties. They are called *local* because members often engage in the direct, person-to-person interaction that is necessary to form and sustain strong ties. Hence, members tend to be clustered geographically.

Cosmopolitan networks are relatively full of holes, consisting of weaker, nonredundant ties. The word *cosmopolitan* means worldly, at home throughout the world, or widely distributed. Hence, members of a cosmopolitan network seldom engage in face-to-face interaction and tend to be scattered geographically.

Local networks have many virtues. These are, after all, the intimate social groups given so much emphasis by early sociologists as the source of belonging, affection, and meaning. We could describe this as the strength of strong ties. As we shall see throughout the book, there is a mountain of research showing that people with strong ties are happier and even healthier because in such networks members provide one another with strong emotional and material support in times of grief or trouble and someone with whom to share life's joys and triumphs.

For example, Paul DiMaggio and Hugh Lough (1998) found that when people make large and risky purchases, such as buying a new car or house, they prefer to buy from someone with whom they have previous social ties (such as a friend or relative) because they feel that this will impose additional obligations on the seller to offer an attractive deal. Not surprisingly, DiMaggio and Lough also found that sellers prefer not to sell to members of their social networks.

The weakness of local networks lies in their self-containment, for they lack input as well as outreach. In a classic study of urban politics, Herbert Gans (1962) found that neighborhoods with the highest levels of solidarity often were unable to block unfavorable policies and programs for lack of ties to possible allies elsewhere in the city—their ties were too redundant. It was for this reason that Gans referred to them as "urban villagers."

As the opposite of local networks, cosmopolitan networks offer little solidarity and have little capacity to comfort and sustain members. But members benefit from a constant flow of new information and from the great reach of their influence, even if it tends to be somewhat lacking in strength. Local networks tend to be small. In contrast, cosmopolitan networks can

FIGURE 2-2 Cosmopolitan and Local Networks.

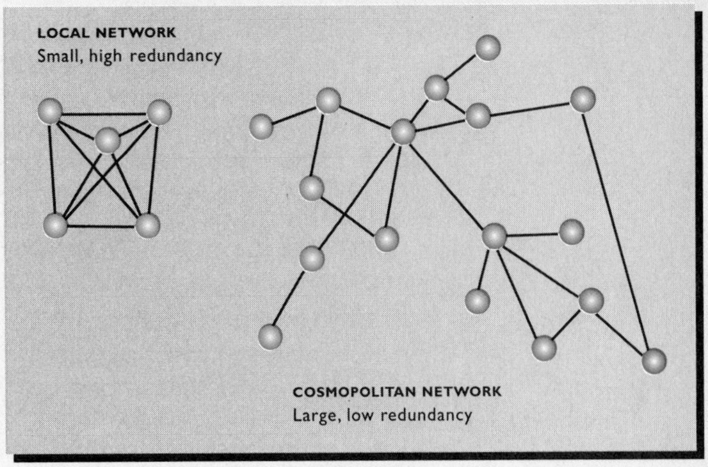

LOCAL NETWORK
Small, high redundancy

COSMOPOLITAN NETWORK
Large, low redundancy

be huge. Thus, while the "urban villagers" lacked ties even to their local city government, cosmopolitan network ties often lead into the White House.

SOCIETIES AS NETWORKS

Not only can societies differ in the overall inclusiveness and intensity of the social relations among members, but they also can differ in terms of having a single internal network or having two or more parallel networks. Thus, in principle, every member of a society could have an abundance of close ties to others, but these ties could be to networks of social relations having almost no overlaps. For example, half of the people in a society could be tied into Network A and the other half into Network B, with virtually no one having friends in both networks. When societies contain parallel networks, there is apt to be serious conflict between (or among) networks, and these may increase the isolation of the networks from one another. The many violent racial and ethnic conflicts in nations around the world reflect the existence of parallel social networks. In many African nations, for example, internal conflicts occur between isolated networks, each based on a different tribe. In Canada, as we have seen, there tend to be two quite separated networks based on language and on historical grievances between them.

Whether or not they contain several isolated, cosmopolitan networks, societies always include many local networks. A major difference across societies lies in the relative strength of their local networks. In some societies large numbers of people are so strongly attached to their local networks as to severely limit geographical mobility—people simply won't move regardless of economic opportunities. Table 2-3 allows us

TABLE 2-3	Nations and Local Networks	
"If you could improve your work or living conditions, how willing or unwilling would you be to:"		
NATION	"MOVE TO ANOTHER NEIGHBORHOOD (OR VILLAGE)?"	"MOVE TO ANOTHER TOWN OR CITY WITHIN THIS COUNTRY?"
	UNWILLING (%)	UNWILLING (%)
Russia	72	77
Latvia	70	77
Bulgaria	61	63
Hungary	59	68
Austria	58	69
Ireland	56	63
Japan	56	61
Slovenia	51	58
Poland	50	59
Philippines	45	46
Czech Republic	44	60
Spain	44	48
Slovakia	40	52
Italy	38	53
Germany	34	47
Sweden	31	46
Netherlands	30	38
New Zealand	29	41
Great Britain	28	38
Norway	26	38
Australia	25	36
Canada	**21**	**33**
United States	**19**	**29**

Source: Prepared by the author from the International Social Survey Program, 1995.

to compare twenty-three nations as to the willingness of citizens to move.

The first column shows responses to a question of whether people would move to another *neighborhood* in the same town or city if they could improve their "work or living conditions." In Russia nearly 75 percent say they would be unwilling to move to another neighborhood, and more than 50 percent of the Japanese would not leave their present neighborhood. But only 19 percent of Americans and 21 percent of Canadians are this locked into a local network.

The second column asks about willingness to move to another city (or town). Everywhere, more people are firmly attached to their town or city than to their neighborhood—77 percent of Russians would not move to another city, compared with 33 percent of Canadians and 29 percent of Americans. The strength of local ties may sustain people emotionally, but it often will cost them a lot materially. In fact, it can impose serious difficulties on a nation's economy if people living in an area of high unemployment will not move to an area where there is a shortage of workers.

Looking back on these discussions of social structures, it will be evident that they are *universal* but *devoid of content*. That is, all societies have a population density and a sex ratio. But they differ greatly in how they interpret and define these structures. All societies have stratification structures, but the concept of stratification ignores that societies differ greatly in how much emphasis is placed on class differences and in the rules governing how people gain their positions in the structure. And all societies have one or more networks of social relationships, but the basis for membership and for network differentiation remains to be discovered. There is nothing in the concept of networks, stratification, or society that tells us what it is that separates Canadians, for example. Thus, social structures identify only a portion of what is needed to understand social life. So, let's examine basic concepts needed to examine the *contents* of social life.

THE CONCEPT OF CULTURE

Culture is the sum total of human creations—intellectual, technical, artistic, physical, and moral. If societies consist of structures, culture consists of *content* that interprets these structures. Culture is the complex pattern of living that directs human social life, the things each new generation must learn and to which eventually they may add. Daniel Chirot (1994) characterized culture as "the codes or blueprints" of societies. "Cultures interpret our surroundings for us and give them meaning and allow us to express ourselves. Language, religions, science, art, notions of right and wrong, explanations of the meaning of life—these are all part of the cultural system of a society."

Thus, culture is everything that humans learn. It also includes all the *things* they learn *to use*. Technology, whether it be bows and arrows or computers, is as much a mental as a physical phenomenon. It is human knowledge that turns a stick and a string into a bow. Technology does not exist naturally; it must be created and it must be *understood*. To an ant, a statue is only another piece of rock.

CULTURAL CONCEPTS

As is obvious, culture isn't one "thing." It is an elaborate system of mental and physical "things." To help us take these things apart and keep track of them, sociologists have developed a basic set of cultural concepts.

VALUES AND NORMS

The **values** of a culture identify its ideals—its ultimate aims and the most general standards for assessing good and bad, desirable and undesirable. When we say that people need self-respect, dignity, and freedom or that people should always be truthful, we are invoking values. Values not only are lofty, but they are also quite general. Frequently, values are linked to religion, which gives them a sacred quality. The majority of people in many nations, including most North Americans, believe we should be kind to other people, for example, not only because it is the right thing to do but because not to do so is a sin.

Values differ across societies, both as to their content and the emphasis placed on a particular value. This is clear in Table 2-4 where nations are compared in terms of the emphasis their cultures place on telling the truth. In Bulgaria 78 percent of the adult population thinks it is "never justified" to tell a lie for your own benefit. In Belarus only 19 percent share this value.

TABLE 2·4	Cultural Contrasts: Truth as a Value		
NATION	"LYING IN YOUR OWN INTEREST IS NEVER JUSTIFIED" (%)	NATION	"LYING IN YOUR OWN INTEREST IS NEVER JUSTIFIED" (%)
Bulgaria	78	Russia	49
Iceland	69	**Canada**	**45**
South Korea	69	Ukraine	42
Turkey	66	Sweden	41
Argentina	65	Nigeria	41
Chile	62	Finland	40
South Africa	61	Luxembourg	39
Denmark	60	Hungary	37
Poland	59	Estonia	35
Brazil	59	France	34
India	57	Austria	33
Norway	56	Czech Republic	33
Slovenia	55	China	32
United States	**55**	Greece	32
Portugal	55	Belgium	31
Croatia	54	Slovakia	30
Japan	54	Germany	30
Ireland	53	**Mexico**	**29**
Switzerland	50	Netherlands	25
Romania	50	Spain	20
Italy	50	Belarus	19

Source: Prepared by the author from the World Values Survey, 1995–1996, and 2000–2001.

If values are very general, **norms** are quite specific. They are rules governing behavior. Norms define what behavior is required, acceptable, or prohibited in particular circumstances. They indicate that a person should or must act (or must not act) in certain ways at certain times. We all have been in situations where we were somewhat anxious about how we ought to act. Our anxiety reflects that we sometimes are not sure of what the norms are, and we also know that violation of the norms often will lead to disapproval or even punishment. Conversely, conforming to the norms often brings approval and other rewards. In Chapter 3 we shall begin discussing a theme that runs throughout the course: Why do people conform or fail to conform to the norms? We also will examine where norms come from.

Values and norms are related: Values justify the norms. For example, we can invoke the values of human dignity and self-respect to justify the norm against ridiculing people because of their appearance. When we break this norm, we are doing something that the values of our society define as morally *wrong*.

There is immense variation across groups and societies in terms of values and norms. In some cultures the values stress family obligations and thus sustain norms concerning how family members ought to behave toward one another, while in other cultures the stress is on individual freedom, and therefore, family obligations are not emphasized. But even among cultures that stress family obligations, there can be very great differences. For example, while some cultures teach that men should be very respectful toward their wives, other cultures make men responsible for controlling their wives' behavior and condone wife beating as a means for doing so (as we will see in Chapter 13). In some cultures the values stress group solidarity to such an extent that members are expected to be very hostile toward all outsiders. And so it goes.

ROLES

A **role** is a collection of norms associated with particular positions in a group or society. These norms describe how we can expect someone in a particular position to act or not to act. Put another way, important social roles have scripts that those who perform those roles are supposed to follow. Consider college class-rooms. Assume there are two roles: student and instructor. Persons in one of the roles are expected to act quite differently from persons in the other role. Students are expected to listen and take notes. If they wish to speak, they ought to raise their hands and wait to be called on. The instructor is expected to give a lecture and to lead discussions. Some students fail to perform their role properly because they day-dream, sleep, come in late, talk out of turn, or eat their lunch. Some instructors also fall short of fulfilling their roles because they don't pre-pare, talk too rapidly or too softly, arrive late, or take much too long to grade exams.

Just as people are rewarded for obeying the norms and punished for violating them, people tend to be rewarded or punished on the basis of their role performances. Faculty who live up to their role expectations tend to be popular, while students try to avoid those who fail to measure up. Students who sleep in class will get no breaks when final grades are assigned.

Cultures differ not only in the roles they pro-vide (there are no auto mechanics in the isolated tribes of the Brazilian rain forest), but they also differ in the value they place on a given role. For example, the role of professional soccer player is seen as far more important in Europe than in North America, while the role of chess master is considered far more important in Russia than in China. Differences in the rewards given various roles in a particular society largely determine what roles people will seek.

Chapter 6 will have much more to say about roles—about how people learn to perform roles, about conflicts among roles, and the like. For now, it is enough to understand that different cul-tures can evaluate a given role quite differently.

MULTICULTURALISM AND SUBCULTURES

Although all societies have cultures, it frequently is the case that one society may include several quite distinct cultures—which often is referred to as **multiculturalism.** Considerable multi-

© Brown Brothers

culturalism exists in both the United States and Canada, for example. In each society immigrants have created a cultural mosaic, for most ethnic and racial groups have retained some distinctive elements of their native cultures. Moreover, there is constant addition to the overall cultures of each nation as material from various groups comes to be shared by all. For example, almost everyone eats Chinese, Mexican, and Italian food. Yet, even as the larger culture has embraced ele-ments of ethnic culture, distinctively Chinese, Mexican, and Italian cultures still flourish in the United States and Canada—as do many other ethnic cultures.

To deal conceptually with multiculturalism, sociologists developed the concept of *subculture*. A **subculture** is a culture within a culture—a distinctive set of beliefs, morals, customs, and the like are developed or maintained by some set of persons within the larger society. Sometimes differences between the general culture and a particular subculture are rather minor, as

Women taking medical examinations at Ellis Island, the reception station in New York Har-bor through which most American immigrants passed at the turn of the twentieth century. Look-ing at these women, we see that they are not simply entering a new *society* but are bringing with them a way of life, a *culture,* that is quite different from that of the two American women examining them.

between the general American culture and the subculture of Norwegian Americans. Sometimes the differences are great, as between the general American culture and that of the rural Mennonite subcultures.

Usually, people are born and raised in a subculture. But often, too, they *join* them, as when people convert to unusual religious movements (which we will examine in Chapter 3).

Multiculturalism very often leads to violent conflicts as groups seek to impose their cultural standards—language, religion, values, and norms—on the other(s). Moreover, multiculturalism often serves as the basis for ascribed status. For a group to be ascribed a low status doesn't just happen; it requires actions by other members of the society to limit the opportunities of those ascribed a low status. These actions in turn usually are justified by negative beliefs about these people. That is, ascribed status is always rooted in *prejudice* and *discrimination*.

PREJUDICE AND DISCRIMINATION

Prejudice refers to negative or hostile beliefs or attitudes about some socially identified set of persons. People become the objects of hatred, contempt, suspicion, or condescension simply because of who they are, without regard for their individual qualities.

Discrimination refers to actions taken against some socially defined set of people to deny members, collectively, rights and privileges enjoyed freely by others. When members of a racial, ethnic, or religious minority are refused employment, promotion, or residence in a neighborhood, these actions constitute discrimination. Put another way, prejudice consists of thoughts and beliefs, while discrimination consists of actions.

But while multiculturalism often leads to conflict, prejudice, and discrimination, history provides examples of more pleasant solutions.

ASSIMILATION AND ACCOMMODATION

Assimilation refers to the process of exchanging one culture for another. Usually, this term is applied to people who adjust to new surroundings by adopting the prevailing culture as their own. Think of assimilation in terms of fitting into or *disappearing into* a new culture. For example, to become fully assimilated into Canadian culture, an immigrant from the United States would need to change those ways of speaking, acting, and thinking that distinguish the two cultures. The same would be required of an English-speaking Canadian who wished to become assimilated into the culture of French-speaking Canada (and vice versa). Or an American would be fully assimilated into Mexican culture when native-born Mexicans could not detect his or her American background.

Only rarely do individuals manage to become fully assimilated, especially if the cultures differ greatly. Usually, it is their children or their grandchildren who assimilate. Moreover, some groups are unwilling to fully assimilate. Many groups of immigrants to the United States and Canada made many adjustments to the local culture but

There are a number of somewhat similar North American subcultures deriving from the Anabaptist Movement that arose in Europe during the sixteenth century; among them are the Amish (often called the Pennsylvania Dutch), the Mennonites, and the Doukhobors. Often referring to themselves as the "Plain People," they resist many aspects of modernization, opting for horse-drawn transportation and observing modest standards of dress.

© Tim Fitzharris/Masterfile

continue to strongly identify with their native culture, as anyone who has been in the American Southwest during *Cinco de Mayo*, in Vancouver during Chinese New Year celebrations, or in St. Louis during *Oktoberfest* must be aware.

The failure of groups to assimilate sometimes has led to intense conflicts—but not always and not forever. A second outcome is **accommodation,** which describes the situation where two groups find they are able to ignore some important cultural differences between them and emphasize common interests instead. During the nineteenth and early twentieth centuries, Catholic-Protestant conflicts agitated American society. A rapid influx of Irish Catholics, fleeing the terrible famine of 1845–1846 in Ireland, caused many Protestants in the United States to fear that their religious culture was being threatened. Over the decades the cultural antagonisms between Protestants and Catholics waned until it became possible for them to emphasize their common Christianity rather than their historical theological disputes. At that point accommodation had occurred. Chapter 11 will pursue these matters in greater detail.

Finally, we must consider two of the most important cultural concepts, which refer to a culture's level of technical, economic, organizational, and scientific development—or what Morselli called "civilization" and sociologists now call *modernization*—and to the spread of modern culture around the world, or *globalization*.

MODERNIZATION AND GLOBALIZATION

The word *modern* implies that something is the latest, most up-to-date, and most recent. Thus, the term *modernization* refers to the process of becoming or remaining modern. Applied to cultures, **modernization** usually refers to the processes of industrialization, economic development, and technological innovation by which a culture sustains high standards of living and maximizes control over the physical environment. Contemporary cultures differ incredibly in terms of modernization—from isolated cultures that do not have a written language or know how to work metal to cultures able to voyage into space. In Chapter 17 we explore why modernization occurred in some societies but not others.

Implicit in the concept of modernization is the conclusion that the world is moving in the direction of one culture—that as all societies modernize they will increasingly become alike. Indeed, anyone who has seen African villagers gathered to watch tapes of American TV shows knows that the spread of mass communications is ending the isolation of less modern societies and nations. In similar fashion a world economy is emerging in which trade plays a vital role in the standard of living in all nations—something we will pursue at length in Chapter 17. The development of global communications, a global economy, and a global culture often are referred to as **globalization** (Beyer, 1994; Simpson, 1996; Bradshaw and Wallace, 1996). In Chapter 17 we will consider recent analyses of globalization, including arguments that the globalization of culture never will be more than superficial.

These, then, are basic concepts for analyzing society and culture—for exploring how social structures and the cultural contents defining the structures interact. Keep in mind that in real life, society and culture are inseparable. There are no societies without cultures, and culture exists only as it is sustained by societies. For this reason sociologists often refer to variations in the sociocultural environment. However, when we try to explain variations across societies—in terms of suicide rates, for example—it often is useful to keep in mind whether we are examining differences in social structures or differences in culture.

Having examined these sociological concepts, now let's put them to use.

JEWS AND ITALIANS IN NORTH AMERICA

As already noted, the Immigration Commission was appointed not because of concerns that too many foreigners were entering the country but because of anxieties about their being the "wrong kind" of people. And what made them "wrong" was their culture.

Those making up the new immigration seemed very alien to eyes conditioned by the cultural traditions of western Europe. They spoke strange-sounding languages and ate odd foods. Many went to unusual churches—the Jews even went to church on the "wrong" day. To make matters worse, most of them were desperately poor. Many Americans and Canadians were convinced that these newcomers would always be "foreigners" who would never fit in.

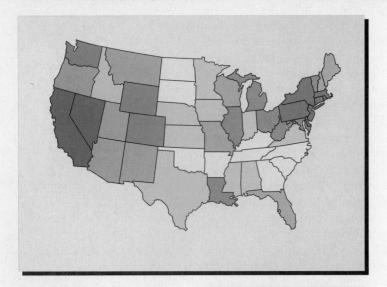

FIGURE 2-3

Persons Born in Italy per 100,000 U.S. Population, 1920.

North Carolina	39	Maryland	1,360
South Carolina	46	Wyoming	1,369
Georgia	48	Washington	1,382
North Dakota	52	Michigan	1,523
South Dakota	129	West Virginia	1,831
Arkansas	169	Ohio	2,075
Kentucky	172	Vermont	2,188
Tennessee	208	Louisiana	2,579
Oklahoma	234	Illinois	3,019
		Colorado	3,034

Virginia	238		
Mississippi	242	Delaware	3,730
Alabama	275	California	4,895
Kansas	397	Pennsylvania	5,394
Iowa	403	Nevada	5,823
Indiana	433	Massachusetts	6,183
Texas	435	New York	10,827
Nebraska	518	New Jersey	10,914
Idaho	577	Rhode Island	11,699
Minnesota	620	Connecticut	12,218

Arizona	698
Maine	727
Wisconsin	843
New Hampshire	885
Missouri	914
New Mexico	964
Oregon	964
Florida	1,064
Montana	1,190
Utah	1,341

Italian immigrants to the United States settled mainly on the East and West Coasts—a pattern typical of all immigrants at the turn of the century. The reason is simple: These were the areas of greatest economic opportunity at the time. The free land that once drew waves of immigrants to the Midwest had been claimed and settled by the start of the twentieth century. The persistent poverty of the American South held no attraction for immigrants.

Note: Hawaii and Alaska were not states in 1920.

Fears about admitting "inferior racial stocks" rose as the initial stream became a torrent. The 55,000 Italian immigrants to the United States during the 1870s were followed by more than 300,000 the next decade, more than 650,000 during the 1890s, and then more than 2 million in the first decade of the twentieth century. The pattern of Jewish immigration to the United States closely paralleled that of the Italians, both in numbers and timing. Meanwhile, Canada also was getting its first substantial immigration from eastern and southern Europe, albeit with smaller totals. Between 1900 and 1914 (when the outbreak of World War I cut off immigration from Europe), about 120,000 Italians entered Canada, many coming north after a brief stop in the United States (Spada, 1969). During this same period, about 150,000 Jews also arrived in Canada (Vigod, 1984).

You can see from Figures 2-3 through 2-6 that on both sides of the border Jewish and Italian immigrants tended to settle in the urban East, although a significant number of Jewish immigrants to Canada took up farming in Manitoba and Saskatchewan (Sack 1965; Rosenberg, 1970; Kage, 1981). And on both sides of the border, Jews and Italians often faced unfriendly receptions (Musmanno, 1965; Ages, 1981).

PREJUDICE AND DISCRIMINATION

In 1895 Henry Gannett, chief of the U.S. Geological Survey and geographer for the tenth and eleventh censuses, published a popular book, *The Building of a Nation*, that summed up developments in American society since the founding of the republic. Filled with lavish color maps and illustrations, the book was also regarded as a serious work of social history. In a section presenting statistics on immigration, Gannett noted the concentration of Italians and Jews in the largest cities and remarked:

Hence it appears that the most objectionable elements of the foreign-born population have flocked in the greatest proportion to our large cities, where they are in a position to do the most harm by corruption and violence.

Gannett scattered similar remarks about foreigners throughout the book. Moreover, his views reflected respectable opinion about Italians, Jews, and other eastern and southern European groups at that time.

FIGURE 2-4	Italians per 100,000 Canadian Population, 1921.

N. W. Territories	12
Pr. Edward Island	29
Saskatchewan	91

New Brunswick	95
Nova Scotia	309
Manitoba	317
Yukon	524

Quebec	684
Alberta	684
Ontario	1,137
British Columbia	1,637

The pattern of Italian settlement in Canada somewhat resembles that for the United States, although Canada at this time still had considerable open land. British Columbia had the highest rate, followed by Ontario. The rates for Canada are much lower than those for the United States despite the fact that the Canadian data refer to anyone claiming Italian descent, whereas data for the United States are based on place of birth.

Note: Newfoundland was not part of Canada at this time.

This newspaper cartoon, published at the turn of the twentieth century, shows Uncle Sam as a latter-day Moses parting the seas marked *oppression* and *intolerance* to let the children of Israel reach new Western homes. The cartoon thus sympathetically acknowledges the outbreak of violence against Jews in Russia that prompted millions to emigrate. But it does not label these immigrants as Jews—readers were expected to recognize Jews. Notice the use of harsh anti-Semitic stereotypes: Everyone (including Uncle Sam) has a huge, hooked nose and meaty lips. The women are fat. Despite showing some men with bundles hung from poles over their shoulders, most of the men wear top hats and fine coats—symbols of the "rich" Jewish banker and moneylender.

Library of Congress

FIGURE 2-5   **Jews per 100,000 U.S. Population, 1926.**

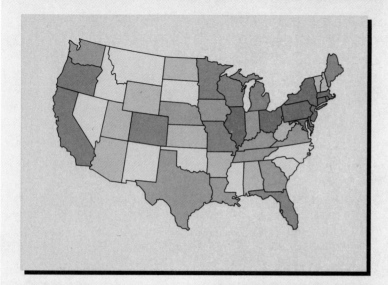

South Dakota	56	Florida	944
Idaho	71	Maine	965
New Mexico	92	Washington	1,130
Montana	123	Oregon	1,355
North Carolina	127	Minnesota	1,601
Mississippi	149	Colorado	1,899
Oklahoma	182	Michigan	1,901
Nevada	192	Delaware	2,155
South Carolina	230	Missouri	2,164
		California	2,569
Arizona	236		
North Dakota	243	Ohio	2,640
Kansas	270	Rhode Island	3,676
Arkansas	272	Pennsylvania	4,246
West Virginia	304	Maryland	4,487
Alabama	364	Illinois	4,736
Wyoming	391	Massachusetts	5,184
Vermont	401	Connecticut	5,945
New Hampshire	466	New Jersey	5,952
Utah	472	New York	16,226
Iowa	520		
Kentucky	613		
Georgia	632		
Louisiana	698		
Texas	729		
Indiana	758		
Tennessee	798		
Washington	881		
Nebraska	912		
Virginia	943		

Unlike the Canadian census, the individual U.S. census form does not ask about religion. However, beginning in 1890 the U.S. government conducted an enumeration of religious membership each decade through 1936 by sending questionnaires to all religious congregations. The data shown here for 1926 indicate that, like Italians, Jews tended to settle on the East Coast. But relative to Italians, Jews were more concentrated in Illinois and Ohio and less so in California and Nevada.

Note: Hawaii and Alaska were not states in 1926.

Look closely at the cartoon on page 45 showing Uncle Sam welcoming Jews. It was meant to be *sympathetic* to the Jews, who were then being persecuted in Russia. Yet it also depicts Jewish immigrants in terms of harsh anti-Semitic stereotypes. Or consider a book published in Canada in 1909 by the Methodist Church and intended to *improve* attitudes toward immigrants (Woodsworth, 1909). In a chapter devoted to "Hebrew" immigrants, we are told that Jews "may be miserly along some lines" but that they are generous with their own kind. We also are told that "the same keen business instincts are common to both" the Jewish peddler and the Jewish "money-barons who control the world's finances." This is classic **anti-Semitism** (prejudice and discrimination against Jews) written with the best intentions by one of the most respected churchmen in Canada at that time.

Italians fared no better (Musmanno, 1965). In 1929 the Italian World War Veterans' Association of Toronto published a book meant to refute public beliefs that Italian Canadians were criminal, illiterate "undesirables." The book reported, for example, that of all immigrants to Canada between 1901 and 1909, Italians had the lowest rate of deportation (Gualtieri, 1929). Ironically, the book is dedicated to "Giovanni Caboto, the first emigrant in Canada." At the time every English-speaking schoolchild in Canada was taught that the name of the man who "discovered Canada" was John Cabot, while French speakers learned of him as Jean Cabot (with a silent "t"). He was actually an Italian in service of the English Crown.

Known as "dagos," "wops," and "guineas," Italians were among those "foreigners" described as "human flotsam" by Madison Grant, chairman of the New York Zoological Society and a trustee of the American Museum of Natural History. He went on to say that "the whole tone of American life, social, moral, and political has been lowered and vulgarized by them." Grant's famous and influential book, *The Passing of the Great Race*, published in 1916, developed elaborate "scientific" and "biological" reasons for halting immigration by people from "the Mediterranean basin and the Balkans."

In 1921 the U.S. Congress heeded Grant's warnings and imposed strict quotas on immigration. Only 154,000 persons a year would be allowed in. Different nationalities were assigned their own quotas. Great Britain received a quota of 65,621 per year. Germany's quota was set at 25,927 and Italy's at a minute 5,802 (Allen, 1985). No Africans could enter, and most Asians were also excluded. These policies remained essentially unchanged until 1965, when the blatant discrimination was finally ended. Canada also adopted very restrictive immigration policies, easing them only in recent years.

ASSIMILATION AND ACCOMMODATION OF JEWS AND ITALIANS

The prejudice and discrimination that greeted Jews and Italians in North America were not universal or unrelenting. Among my early sociological research was a study of anti-Semitism in the United States. My colleagues and I found that in the middle and late 1960s, prejudice against Jews had declined greatly, and few people still harbored serious feelings against Jews (Glock and Stark, 1966; Stark et al., 1971). Moreover, throughout the United States during the post–World War II period, restrictive real estate covenants preventing Jews from buying homes in certain neighborhoods were overturned by the courts, and many other forms of discrimination subsided as well. For Italians, too, the worst is long past.

In Chapter 11 we will see what sociologists have learned about how prejudice arises and how it dies. Here we will note several factors that contributed to changes in attitudes toward Jews and Italians. The first is the substantial degree to which Jews and Italians have become assimilated. Most were born in North America, and so they talk, dress, and act much like others born here. A major aspect of assimilation is *intermarriage*—marrying someone of another ethnic background. More than half of the U.S. citizens who listed their ancestry in the 1990 census as Italian listed a second ancestry as well, which means that their parents or grandparents had intermarried. Alba (1977, 1985) estimated that by the third generation, 60 percent of Americans of Italian ancestry married non-Italians. The same pattern seems to hold in Canada. A. V. Spada (1969) noted that between 1900 and 1961, 439,714 Italians entered Canada. However, despite Italian Canadians having a high birthrate, only 450,351 Canadians gave their ancestry as Italian in the 1961 census. How could this be? Spada concluded, "The major reason

| FIGURE 2-6 | Jews per 100,000 Canadian Population, 1921. |

N. W. Territories	12
Pr. Edward Island	24
Yukon	190

Saskatchewan	710
Ontario	1,629
Quebec	2,032
Manitoba	2,732

New Brunswick	320
British Columbia	323
Nova Scotia	413
Alberta	551

Jews in Canada tended to settle not only in the urban East (Quebec and Ontario) but also in several prairie provinces: Jews made up a larger part of the population of Manitoba at this time than of any other province. The first Jews in Manitoba came as traders in the 1870s. Many arrived in the early 1880s as laborers building the Canadian Pacific railroad. Then, when Jews began to flee Russia and Poland in large numbers, Baron Maurice de Hirsch organized and financed a huge resettlement project that founded many Jewish farming communities on the prairies of Manitoba and Saskatchewan.

Note: Newfoundland was not part of Canada at this time.

for the disappearance of the Italian population is the high degree of assimilation."

Jewish intermarriage rates are lower than those for Italians, in both Canada and the United States. The primary reason is that Italians are able to marry people of many other ancestries and still marry *within their religion*. For Jews the boundaries of ethnicity and religion coincide, thus making intermarriage of greater significance. Nevertheless, Jewish intermarriage rates have risen steadily on both sides of the border and are especially high in the western parts of both nations where the Jewish population is relatively smaller (Goldstein, 1971; Cohn, 1976; Brym, Gillespie, and Gillis, 1985).

A second aspect of changed attitudes is accommodation. Despite substantial assimilation, large Jewish and Italian subcultures still exist in the United States and Canada. Jews have not changed their religion, nor have Italians. But these cultural differences no longer generate much conflict. Agreement has been reached within the societies that these differences are not important in comparison with the many common bonds linking all Canadians and all Americans.

Chapter 11 will argue that the underlying basis of such accommodation is economic equality—that prejudice and discrimination against a minority subculture seem to persist until the group has managed to achieve equal status. Put another way, cultural differences are not accommodated until a group has "made it." And this statement brings us back to the sociological question underlying this chapter. By now, both Jews and Italians have achieved economic success in North America. But the Jews were upwardly mobile much sooner than were the Italians. Why?

THEORIZING ABOUT ETHNIC MOBILITY

Now that we have clarified the needed concepts and identified the question we wish to answer, we can move to the third step in the sociological process. It's time to *theorize*. We now must say how some set of concepts fits together in a way that offers an answer to our question. We shall examine two such theories. The first was suggested long ago, soon after sociologists first began to wonder about why Jews had been upwardly mobile so much sooner than had Italians. Over many decades this theory was refined and tested, until by the 1970s social scientists had gained great confidence in it. This theory stresses *cultural causes*. It contrasts the values, norms, and roles dominant in Jewish and Italian cultures and concludes that one culture helped while the other hindered the economic progress of the two ethnic groups. The second theory we will examine arose recently to challenge the older theory. It stresses *social causes*, especially aspects of stratification systems, to account for the differences in the economic position of Jewish and Italian immigrants. We shall develop each theory and let them collide with each other as well as with appropriate research evidence.

THE CULTURAL THEORY

Human beings are shaped by their cultures, and most people obey the norms most of the time. If we want to predict how people will behave in any given circumstance, the best way to do that is to know their cultural background. In this way, for example, we can be confident that Italian men will be careful to remove their hats and caps when they enter a church, while Jewish men will be careful to cover their heads when entering a synagogue.

Following this logic, if we want to know why Jews achieved such rapid economic success compared with Italians, chances are that the culture of one group gave them advantages in adjusting to the circumstances they faced in North America—both economic and cultural circumstances.

To state the theory briefly: Sociologists proposed that Jewish values of learning, their norms of educational achievement, and the immense respect given to the role of scholar paved the Jewish road to success. Conversely, it was proposed that Italians valued not learning but family loyalty; their norms led them to drop out of school; and the immense importance placed on the role of father made their original culture slow to change.

Now let's see how this theory squares with appropriate research evidence. In doing so we will not work through specific steps in the sociological process; that is postponed to Chapter 3. Here we simply will watch as researchers attempt to discover whether Jewish and Italian cultures were different in the ways specified by the theory.

A CLOSER VIEW
ZBOROWSKI AND HERZOG: JEWISH CULTURE

Over the decades many prominent sociologists have attributed the rapid upward mobility of Jews to cultural advantages they brought with them from eastern Europe (Slater, 1969; Steinberg, 1974). However, it was two anthropologists who assembled the most detailed and compelling cultural explanation of Jewish success in the United States and Canada.

Mark Zborowski and Elizabeth Herzog painstakingly reconstructed shtetl life in Poland and western Russia, from which the great waves of Jewish immigrants came during the latter nineteenth and early twentieth centuries. *Shtetl* (rhymes with *kettle*) is the Yiddish word for "village" or "small town." During the centuries of shtetl life, the cultural traditions of ancient

Judaism were transformed into the way of life the Jews brought to North America.

Outside eastern Europe as well, Jews lived almost exclusively in towns and cities. From early medieval times, Jews were prohibited from farming or owning land in most parts of Europe. Thus, most Jews in western Europe were required by law to live in crowded *ghettos*[1]—neighborhoods reserved exclusively for Jews. These ghettos often were walled, and the gates were locked at curfew.

Since ancient times the Jewish religion has stressed literacy for men because each man is expected to read the scriptures and spend time studying their meaning. This tradition stimulated great respect for learning; indeed, learning became a value in Jewish communities. Because they lived in towns and cities, Jews could easily maintain schools and gather in study and discussion groups. Consequently, scholarship became "the dominant force in the Jewish culture" (Zborowski and Herzog, 1962).

These facts about Jewish culture were familiar. What Zborowski and Herzog wanted to do was to re-create the details of shtetl life so that they could see how this "cult of scholarship" worked on a day-to-day basis and understand how the roles and norms of Jewish life reflected and sustained the values about learning and scholarship. Accordingly, they conducted in-depth interviews with more than 100 elderly people who had been shtetl residents. (What remained of shtetl life was destroyed during World War II, when shtetl inhabitants were sent to Nazi death camps.) From these recollections by Jewish immigrants as well as a vast supply of letters, diaries, and life histories, Zborowski and Herzog (1962) produced a rich and compelling account of shtetl life in their book *Life Is with People*.

They discovered that the *norms* governing schooling, even in the early 1800s, were strict and demanding by modern standards. Children began school between the ages of three and five years old, and the school day began at 8 A.M. and lasted until 6 P.M. six days a week! Males who showed the greatest academic aptitude were expected to adopt the *role* of scholar and devote their lives to study and learning. In fact, scholars were so highly respected that most parents hoped their sons could become scholars. Wealthy merchants sought to gain scholars as sons-in-law;

© Alter Kacyzne/Raphael Abramovich Collection, Yivo Institute

indeed, the life of a scholar was made possible by his marrying a woman whose family could support him as he devoted himself to full-time study.

When Jews began to immigrate to North America in large numbers, both the United States and Canada were in a period of rapid economic development. This boom attracted millions to come to North America. In both nations, although more so in the United States, the greatest economic opportunities were in the large cities and in the skilled occupations. Consequently, the educational systems were also expanding rapidly—especially the higher educational systems. Enrollment at colleges and universities grew by almost 1,200 percent in the United States between 1870 and 1920. Thus, during the period when most Jewish immigrants arrived, the colleges and universities were making room for and actively seeking much larger enrollments. An amazing number of these new students were the sons and daughters of Jewish immigrants (Steinberg, 1974).

Jews were accustomed to sending their children to school, exulting in their academic achievements, demanding hard study, and making family sacrifices to educate their children. Indeed, an editorial in a Jewish newspaper published in New

A teacher in a shtetl school drills Jewish boys in the Hebrew alphabet. Understandably, they are more interested in the photographer than in their books. But most of the time, they were required to study long and hard.

1. The term *ghetto* originated in Venice, where the section of the city in which Jews were required to live was, in late medieval times, called the "borghetto." This word derived from the Italian word *borgo*, which meant "borough"— a major section of a city. *Borghetto* was the diminutive form meaning "little borough." Over time the word was shortened to *ghetto*, and its use spread to all European languages. Today, the term is often applied to any neighborhood occupied by an ethnic or racial minority.

© Brown Brothers

A Jewish boy delivering bundles of partly sewn men's suit coats in New York about 1910. Like other immigrants, Jewish children often had to help their families earn a living. But compared to most other immigrant groups, they were much less likely to drop out of school to take full-time jobs. Consequently, Jews rapidly entered technical and professional occupations.

York in 1902 boasted of the Jewish "love for education, for intellectual effort," and went on to say that "the Jew undergoes privation, spills blood, to educate his child" (quoted in Sanders, 1969).

It was not an idle boast. A 1922 study of high school students (Counts, 1922) found that for every 100 freshmen who were children of native-born, white Americans, there were 44 seniors. Thus, about 56 percent of those who began high school did not finish. Among the children of Italian immigrants, there were only 17 seniors for every 100 freshmen. But among children of immigrant Jews, there were 51 seniors for every 100 freshmen—a slight majority were graduating.

Thus, it is no surprise that by the turn of the century U.S. colleges and universities experienced waves of Jewish enrollment. Indeed, Jews soon formed a majority of the students enrolled in New York City College and constituted a very sizable minority in other eastern schools, such as Harvard and Columbia. In fact, by the 1920s many of these schools imposed quotas on the number of Jews admitted. Thus, Columbia reduced its Jewish enrollment from 40 percent to 22 percent within two years. A public furor erupted over formal limits Harvard placed on Jewish enrollment, which exceeded 20 percent by 1920. Consequently, Harvard adopted a policy of regional balance, seeking students from all forty-eight states. In effect this policy imposed a limit on

Jews by limiting enrollments from New York and other eastern states with large Jewish populations.

But despite quotas and simmering anti-Semitism, plenty of room remained in the educational system for Jewish students, and Jews rapidly made their way into high-prestige, high-paying occupations, especially professions requiring advanced degrees: medicine, law, dentistry, and education. As early as 1913, the proportions of doctors, lawyers, and college professors who were Jewish were substantially greater than the proportion of Jews in the general population (Steinberg, 1974).

Thus, Zborowski and Herzog attributed the rapid upward mobility of Jews to the favorable fit between their learning- and schooling-oriented culture and the opportunities existing in the United States and Canada at the time of the greatest Jewish immigration.

A CLOSER VIEW
LEONARD COVELLO:
ITALIAN CULTURE

Unlike the children of Jewish immigrants, the children of newly arrived Italians did not excel in school and did not seek higher education to pursue professional occupations. Italian children tended to quit school at an early age. In the mid-1930s Leonard Covello, a young teacher and school administrator in an Italian part of New York City, began a lengthy research project to try to understand the educational problems of Italian American children such as "truancy, absence, cutting classes, lateness, and disciplinary problems." His research lasted until 1944, and he finally published his findings in 1967. His book, *The Social Background of the Italo-American School Child*, is a superb counterpart to the work of Zborowski and Herzog. For by conducting his research both in Italy and among his fellow Italian Americans, Covello drew a portrait of the culture Italians brought to North America—a culture that did not fit well with life on this side of the Atlantic.

Covello, like Zborowski and Herzog, focused his attention on education because Italian Americans were failing to take advantage of their educational opportunities. Dropping out of school early, they typically were forced to settle for low-paying, unskilled jobs. By the time Covello published his book, this pattern had changed dramatically: Italian Americans were staying in school and many were going to college. But it had taken many decades longer for this pattern to emerge than it had among Jews. Why?

Covello concluded that Italians had arrived in North America suspicious of schools and accustomed to sending children to work at an early age. These cultural patterns were appropriate to conditions in Italy, particularly the region most Italian immigrants came from—southern Italy.

The government of Italy was controlled by the populous northern regions. Southern Italy was rural, impoverished, and exploited by northerners who regarded the south as backwater. Consequently, southern Italy had relatively few schools. Worse yet, these few schools represented the culture of the north; even the language used in them was quite different from the common southern dialects.

Southern Italian parents regarded schooling as a threat to their own values, especially the key value of loyalty to the family, for the schools reflected negative judgments of local life. Moreover, little of what was learned in school was of much importance to life in southern Italy. Whether they went to school or not, the children grew up to be peasants. As one old Italian father told Covello, "What good if a boy is bright and intelligent in school, and then does not know enough to respect his family? Such a boy would be worth nothing."

Furthermore, the absence of children from home while they attended school often threatened the family's well-being. As another father told Covello, "If our children don't go to school, no harm results. But if the sheep don't eat, they will die. The school can wait but not our sheep."

The belief that school was not important and was possibly harmful was appropriate to life in southern Italy. The things children really needed to learn had to be learned at home anyway, such as how to plant crops and tend sheep. Thus, academic learning was not a value in southern Italian culture. And sending children to school and fostering rigorous study habits were not norms.

Finally, the role of scholar was of little importance. The primary value was family loyalty. The most important norms concerned behavior within the family, and the father was the most important social role.

Transplanted to the United States and Canada, these cultural patterns proved inappropriate. Here the Italians did not become farmers. Instead, they found themselves in rapidly growing industrial cities where child labor was of little

Library of Congress

In 1903, when this picture was taken, being a newsboy was not something a boy did only after school. He had to be on his corner all day long or lose his spot to another boy. (The Newsboy Law in New York said boys could work only from 6 A.M. to 10 P.M.) Thus, it is almost certain that this Italian boy had already dropped out of school to help feed his family. At that time most Italian American children did quit school early.

value and large families were an economic burden rather than an asset. Furthermore, unskilled physical labor paid low wages and offered no opportunity for advancement. Covello concluded that these cultural patterns thwarted the social progress of Italians. As long as these patterns persisted, Italians could not achieve upward mobility.

THE SOCIAL THEORY

Because the patterns of Jewish and Italian cultures differ so much in the way proposed by the cultural theory, social scientists tended to treat the cultural theory as almost self-evident. Nevertheless, by the 1970s some sociologists were beginning to notice evidence that didn't fit and to discover predictions that didn't hold. One very troublesome fact was that there was nothing at all unusual about the time the Italians had taken to achieve economic parity; most other immigrant groups took just as long. So it seemed unlikely that the specific cultural barriers to achievement identified in Italian culture actually mattered. Groups lacking these values and norms did not rise any faster. From this evidence it followed that the theory really applied only to the Jews as an explanation of why they were different from other groups.

Library of Congress

Italian housewives buying crusty loaves of bread from bakery boys on Mulberry Street in the heart of New York's "Little Italy" in about 1900. Notice that one boy's shoes are worn out—he and his friend may have already dropped out of school. Many of the people shown here probably planned to return to Italy, and many of them probably did.

However, a closer look *within* the Jewish community produced an even more awkward fact. At the pertinent time, most Jews did not attend college—most didn't even finish high school. Still, when compared to other recently arrived immigrant groups, poorly educated Jews were far ahead in terms of pay. What do educational advantages have to do with explaining differential rates of economic success between people with the same amount of education? Clearly, nothing.

When a theory and the evidence don't agree, it is time to go back to the drawing board and try to adjust or replace the theory. And that's exactly what Stephen Steinberg did.

A CLOSER VIEW
STEPHEN STEINBERG: THE JEWISH HEAD START

In the early 1970s, a young graduate student at the University of California, Berkeley, began a study of the religious and ethnic origins of college and university faculty. Noting the substantial overrepresentation of Jews and wondering how this had come about, Stephen Steinberg paused to brush up on the history of Jews in America. He immediately encountered the cultural theory of rapid Jewish upward mobility. But he also noted the growing body of contradictory facts and faulty predictions. This led him to formulate a new theory that suggested the rapid upward mobility of Jews may have been largely an illusion—that it hadn't really happened. How could this be?

Steinberg admitted that the immigrant Jews were poor when they arrived. Table 2-5 shows that only 12 percent of Jewish immigrants from 1904 to 1910 had at least $50 upon their arrival in the United States, a figure that is close to that for most other immigrants.

But, Steinberg asked, is money really the issue? What if we meet a ditchdigger and an engineer as they arrive in America? What if each has only $50 in his pocket? Are they starting off on equal footing? Steinberg said no. Obviously, the engineer brings tremendous potential economic advantages with him compared with the ditchdigger, and we would be surprised if the engineer were not making much more money than the ditchdigger after a few years. Steinberg argued that the Jews were more like the engineer and Italians were more like the ditchdigger in terms of their social class origins.

Specifically, Steinberg attributed the superior economic position of Jewish immigrants in America to their superior economic and social positions in eastern Europe. While people from lower-class origins must acquire new skills to rise within a stratification system, he argued, people from more advantaged backgrounds obtain these skills as a matter of course. Upon arriving in new surroundings, people with higher-status backgrounds are likely to be able to *regain* higher-status positions.

Let's put this a bit more formally: *Among first-generation immigrants, status in their new society will be determined in large measure by their status in their former society.* Or applied to *groups,* their average status in the new society will reflect their average status in the old society.

Steinberg's theory predicts that Jewish status attainment in the United States and Canada reflected their higher status back in eastern Europe, that Jews arrived in North America with experience, training, and technical skills qualifying them for highly skilled jobs and enabling them to successfully pursue business and commercial opportunities.

As an initial test of his theory, Steinberg turned to two vital, but neglected, sources of information. The first was an analysis of the economic situation of Jews in Russia and Poland (then under Russian control) based on the Russian census of 1897 (Rubinow, 1907). The second was the massive forty-one-volume report of the Immigration Commission, quoted at the start of this chapter. And so, to work.

First, Steinberg examined the Russian census. At that time Jews were required to live in a restricted region of Poland and western Russia called the Pale of Settlement. Only a few Jews were permitted to live beyond the Pale. As a result Jews were a concentrated population. As we saw previously, these Jews lived in villages and small towns. Correspondingly, the Russian census found that only 3 percent of Jews were farmers, while 61 percent of non-Jews were (see Table 2-6). In contrast, the Jews were heavily concentrated in higher-status occupations. Nearly a third engaged in commerce, and an even higher percentage were in manufacturing or highly skilled trades. Five percent worked at professions, such as law and medicine. Within the Pale, a third of the factories were owned by Jews, who also dominated commerce.

Of course, many of these factories were tiny operations, and much of the commerce was nothing more than selling household items door to door. What is important, however, is not how much money Jews were earning in Russia and Poland but the training and skills involved. These occupational skills markedly set off the Jews from the non-Jewish, mainly peasant populations in eastern Europe. These skills could have equally distinguished them from most other immigrants to North America, who also were mainly peasants.

We must consider the possibility that only unsuccessful Jews emigrated from eastern Europe. However, since Jews in the late nineteenth century emigrated primarily to flee severe persecution under the Russian czar, that seems unlikely. In any event Steinberg found detailed records in the Immigration Commission reports that demonstrated the occupational advantages of Jewish immigrants.

Table 2-7 shows that the occupational backgrounds of Jews arriving in the United States differed from those of other immigrants.

TABLE 2-5	Immigrants Having at Least $50 upon Arrival in the United States, 1904–1910	
IMMIGRANT GROUP		**PERCENT**
Jews		12
Southern Italians		5
Irish		17
Germans		31
English		55
All other immigrants		14

Source: Reports of the Immigration Commission, vol. 3 (1911), reproduced in Steinberg (1974).

Two-thirds of Jewish immigrants worked in skilled crafts in the old country, in contrast with 15 percent of the southern Italians, 13 percent of the Irish, and 6 percent of the Poles. Even immigrants from England were less likely to have skilled occupations than were the Jews. In contrast, the Jews had rarely worked as laborers, farmers, or servants, whereas these were the most common occupations of the Italians, the Irish, and the Poles—indeed, of all immigrants as a group.

Canadian authorities did not begin to keep data on the occupations of immigrants until 1931. Joseph Kage (1981) points out, however, that since "the sources of Jewish immigration to the two countries were the same, we may assume the American information is valid for Canada."

Jews arrived in North America with highly skilled occupational backgrounds precisely when rapid economic development and industrial growth offered them immense opportunities (Kage, 1981). Consequently, the Jews rapidly reentered their old occupations as printers, jewelers, tailors, watchmakers, cigar makers,

TABLE 2-6	Occupations of Jewish and Non-Jewish Adults in Russia, 1897	
	PERCENTAGE	
OCCUPATION	**JEWS**	**NON-JEWS**
Professions	5	3
Commerce	32	3
Manufacturing and skilled trades	38	15
Service	19	16
Transportation	3	2
Agriculture	3	61
TOTAL	**100**	**100**

Source: Compiled from the Russian census of 1897 and presented in Rubinow (1907).

TABLE 2-7	Occupations of Immigrants Entering the United States, 1899–1910					
	PERCENTAGE					
PREVIOUS OCCUPATION*	**JEWS**	**SOUTHERN ITALIANS**	**IRISH**	**POLES**	**ENGLISH**	**ALL IMMIGRANTS†**
Higher Status						
Professions	1	0	1	0	9	1
Commerce	5	1	1	0	5	2
Skilled labor	67	15	13	6	49	20
Lower Status						
Labor	12	42	31	45	18	36
Farming	2	35	7	31	4	25
Service	11	6	46	17	5	14
Other	2	1	1	1	10	2

Source: Adapted from the Reports of the Immigration Commission (1911), as reproduced in Steinberg (1974).

*Excludes immigrants with no previous occupation, including most women and children.

†Also includes other groups not separately listed in the table.

Jewish Canadian tailors at work for the Eaton Company in Toronto in about 1904. Because so many Jewish immigrants to North America arrived with qualifications for skilled occupations such as this, they rapidly achieved economic success. In contrast, many other immigrant groups in Canada and the United States had to depend on unskilled laboring jobs to sustain them, and hence, their climb to economic equality took longer and was more difficult.

Eaton's of Canada Archives

tinsmiths, furriers, and the like. Such jobs paid much better wages than did the laboring jobs available to most other new immigrants. Although the Jews arrived poor, they came with marketable skills that permitted them to escape poverty rapidly. In contrast, most other groups had to develop such skills after they arrived. Indeed, the majority of them arrived illiterate, whereas the overwhelming majority of Jewish immigrants could read and write.

Several other immigrant groups have repeated the Jewish pattern of rapid success and for similar reasons. In the late 1950s and early 1960s, for example, a large number of Cuban refugees came to the United States to escape Castro's revolutionary government. Most of them had held middle- and upper-class positions in Cuba before the revolution. Although they came with little more than the clothes on their backs, they rapidly regained their class positions here. More recently, middle-class refugees from Vietnam have shown a similar tendency toward rapid economic advancement.

Steinberg's reassessment of the actual circumstances of Jewish immigrants does not mean that Jewish culture played no role in their success. Rather, he argued that these aspects of Jewish culture reflected the position of the shtetl Jews. Compared with the surrounding non-Jewish populations, the Jews of eastern Europe were middle class. Many studies have shown that middle-class people everywhere in the world are very concerned about education and push their children to do well in school (Steinberg, 1974). They are especially likely to do so if they have momentarily lost their positions in society (as had Jews, Cubans, and Vietnamese when they fled to the United States). As we shall see in Chapter 11, education is the cheapest, most rapid, and most reliable path to economic advancement under present conditions.

Keep in mind that Steinberg's theory is not limited to predictions about the relative social class positions of first-generation Jews and Italians in North America. His concept of a first-generation immigrant group applies to *all* such groups, and when he links status in the society of origin with status in the society of residence, he is generalizing to *all* societies. Thus, many specific *hypotheses* can be deduced from his theory. Recall that a hypothesis is a statement of what we ought to be able to observe according to predictions from the theory.

So why don't we formulate a new hypothesis from Steinberg's theory and check it out? Back when the Immigration Commission was gathering data, a substantial number of Canadians were coming into the United States. The vast majority of these people were French speaking, while the others were English speakers. At the turn of the century, the stratification system in Canada was highly skewed on the basis of language. The social status of the average French Canadian was considerably below that of the average English Canadian (which was why so many French Canadians were moving south). Applying Steinberg's theory, we can formulate the following hypothesis: Among first-generation Canadian immigrants in the United States, the English speakers will have higher incomes than will the French speakers. Is this true? Turn back to Table 2-1 and see for yourself.

Once again the theory survives an effort to prove it false. Keep in mind, however, that no amount of research can prove a theory to be true. Moreover, in science it is very rare for a publication, no matter how brilliant, to be the last word on the subject. Thus, even though most sociologists agreed that the social theory was far more adequate than the cultural theory, research continued on this subject. In 1981 Steinberg published a new book in which he spelled out the social theory more fully and applied it far more generally: *The Ethnic Myth: Race, Ethnicity, and Class in America.*

But as so often happens in science, once again some sociologists began to have second thoughts on the matter. Do cultural differences count for nothing? Did it really make no difference *at all* that Jews valued education while Italians seemed not to?

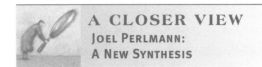

A CLOSER VIEW
JOEL PERLMANN: A NEW SYNTHESIS

In 1988 a young professor at Harvard's Graduate School of Education opened a new chapter in this debate when he published the results of a ten-year study of schooling and occupational achievement among ethnic groups in Providence, Rhode Island, from 1880 through 1935.

The impact of Joel Perlmann's book came not so much from its new ideas as from the statistical findings based on new data of extraordinary quality and detail that he had painstakingly assembled. He started with the manuscript schedules for the 1880 and the 1900 U.S. Census. These are the actual handwritten forms filled out by the census-takers when they called at each household (now available on microfilm). For each household the census reported the name of each resident, their address, age, sex, race, place of birth, relation to the head of the household, occupation, school attended (if any) in the past year, and many other important facts. In addition to the U.S. Census, Perlmann used the Rhode Island State Census manuscript schedules for 1915 and 1925 (until recently, many northeastern states conducted a census of their own, usually in the middle of each decade). From these census documents, Perlmann drew random samples from all adolescents living in Providence (see Chapter 3).

Then Perlmann used the information on the names and addresses of these adolescents to locate their school records, including the actual academic record of courses taken, grades earned in each—even absences and tardiness.

So now Perlmann knew a lot about the home environment and schooling of each young person in his sample. But Perlmann also wanted to know what these young people ended up doing for a living later in life. He therefore consulted city directories from the period; these list all residents and, in those days, the current occupation of each adult male.

TABLE 2-8	Ethnicity, Education, and Occupation of Sons in Providence, Rhode Island, 1915		
	FATHER'S ETHNICITY		
	YANKEE*	ITALIAN†	JEWISH‡
Percentage of sons who:			
Enrolled in high school	49.6	17.2	54.3
Graduated from high school	13.8	5.9	21.9
Entered a white-collar occupation	45.7	33.2	77.2

Source: Perlmann (1988).

*Whites whose parents and grandparents were born in the United States.

†Father born in Italy or Sicily.

‡Father born in Russia, including the part that today is Poland.

With these data in hand, Perlmann set out to reconstruct a portrait of ethnic patterns of achievement in Providence. The first thing he discovered was that by 1880 "virtually all children spent at least a few years in school—roughly between the ages of seven and eleven." From there, however, enrollment dropped rapidly. By age thirteen fewer than three-fourths of Providence children were still in school, by age fifteen this had fallen to only 42 percent, and by seventeen fewer than 20 percent were still in school. Moreover, most of those over age fifteen who were still in school *were not in high school!* Instead, they were still in the lower grades—many immigrant children found it necessary to repeat grades. In fact, in 1880 only about 12 percent of adolescents in Providence ever enrolled in high school. By 1900 this had risen to 18 percent, and in 1915, 35 percent enrolled.

This average completion rate, however, hides immense variation among groups, as can be seen in Table 2-8. Slightly less than half

TABLE 2-9	Occupations of Fathers of Providence Schoolchildren, 1915		
	PERCENTAGE		
OCCUPATION	YANKEE*	JEWISH†	OTHER IMMIGRANTS
Self-employed	16.1	70.2	19.1
Engaged in trade	19.3	54.9	18.5

Source: Perlmann (1988).

*Whites whose parents and grandparents were native born.

†Born in Russia, including the part that today is Poland.

(49.6 percent) of the sons of Yankee fathers enrolled in high school in 1915 compared with 17.2 percent of the offspring of fathers born in Italy. In sharp contrast more than half (54.3 percent) of the sons of Jewish men born in Russia (or Poland) went to high school in that year. When high school graduation rates are examined, the differences are even greater: Jewish sons were nearly twice as likely as Yankees and nearly four times as likely as Italian sons to graduate. The apparent payoff of their added education can be seen in the proportion of Jewish sons who ended up in higher-status, white-collar occupations: more than three-fourths of the Jews compared with only a third of the Italians.

Thus far, Perlmann's data confirmed what everyone else had found: that Jews gained rapid success in the American stratification system, while Italians took far longer. And Table 2-9 suggests that Jewish success was in fact based on their superior starting position, just as proponents of the social theory have claimed. Jewish fathers in Providence in 1915 were incredibly concentrated in the ranks of the self-employed, most of them engaged in trade. That is, the majority of Jewish fathers were involved in business at some level. To pursue these activities, they had to be able to read, do arithmetic, and make a presentable personal appearance. Why would it be surprising that their children stayed in school and then got good jobs? Which is, of course, Steinberg's primary point.

However, Perlmann wasn't willing to let matters rest there. His data permitted him to take account of the background advantages Jewish adolescents had over their Italian classmates and over other ethnic groups as well. The way he did this was to in effect calculate the odds that persons of different ethnic backgrounds would enter high school, get good grades, graduate, and take good jobs, given the advantages or disadvantages provided them by their backgrounds. What he discovered was that Jewish children far surpassed their odds. Thus, while the Jewish head start did account for a significant amount of their more rapid success, compared with other immigrant groups, "these explanatory factors left large unexplained ethnic differences" in schooling and occupational success. For example, Italian children from middle-class homes tended to leave school earlier than Jewish children from working-class homes.

Put another way, when Perlmann used statistical techniques to give Jewish and Italian children an even start, the Jewish children were still much more likely to enter high school, graduate, and take good jobs. These findings revealed that the social theory was inadequate to explain the full range of Italian-Jewish differences. Thus, Perlmann concluded that *both* theories are needed to account for what actually took place.

Recall that early in this chapter I pointed out that it is somewhat artificial to separate society and culture—that they are opposite sides of the same coin and never truly occur in isolation of each other. Here we see further evidence of this point. For the fact is that the social positions occupied by Jews in eastern Europe were conducive to developing and sustaining the key cultural elements thought to have been so valuable to them in America. No one has expressed this more forcefully than Nathan Glazer (1955):

We think that the explanation of the Jewish success in America is that Jews, far more than any other immigrant group, were engaged for generations in the middle-class occupations. . . . The special occupations of the middle class—trade and professions—are associated with a whole complex of habits [or culture]. Primarily these are the habits of care and foresight. The middle-class person is trained to save his money, because he has been taught that the world is open to him, and with the proper intelligence and ability, and with resources well used, he may advance himself. He is also careful—in the sense of being conscious—about his personality, his time, his education, his way of life.

Thus, Glazer drew a direct link between society and culture and hence between the social and cultural theories, showing that each generates and necessitates the other. To the question of whether Jewish success was due to their social positions in Russia and Poland or to their cultural values of hard work and study, Glazer answered "Yes!"

Before we can leave this topic, one more aspect of Jewish and Italian differences must be addressed, which also will allow us to consider one final sociological concept.

REFERENCE GROUPS AND ITALIAN TRADITIONALISM

At the turn of the century, government studies revealed that, compared with Jews and many other immigrant groups, Italians seemed slow to learn English. For example, in 1911 two-thirds of the Jews who had been in the United States less than five years could speak English. The figure for Italians was only one-fourth. Even Jewish and Italian immigrants who had been in the country ten years or longer showed marked differences in their knowledge of English.

Obviously, it is a considerable difficulty to be unable to speak the language of the country in which one is trying to earn a living. Furthermore, the children of parents who do not speak the language are hindered in learning to speak it and are thus at a disadvantage in school. In fact, Covello blamed many school problems experienced by Italian Americans in the 1930s on their poor English skills. The reason the Italians were slow to learn English was a reason for many of their economic problems as well.

Italians were slow not only to learn English but also to adapt to occupational conditions in the United States because most of them did not plan to become Americans. Instead, most Italian immigrants came to America to take advantage of the relatively high wages available to unskilled laborers (compared with wages back in southern Italy) and then returned to Italy with their savings to resume their old ways of life in greater comfort.

It is easy to demonstrate this temporary migration. Immigration Commission statistics reveal that of Italian immigrants between 1899 and 1910, only 21 percent were females (in contrast with 43 percent of Jews). Similarly, only 12 percent of the Italian immigrants were under age fourteen (compared with 25 percent of Jews). Therefore, few Italian families were coming to America; instead, it was mostly young, single men. Moreover, most of them did go back. Between 1908 and 1910, for every 100 Italians who entered the United States, 55 others left to go back to Italy. Among Jews, only 8 left for every 100 arrivals during this period. A great many Italians who ended up staying in America had probably not planned to do so. Many of them delayed their returns because of the outbreak of World War I in 1914. By the time the war ended in 1918, economic

This Italian mother, with a bundle on her head, and her children passed through Ellis Island in 1915, on their way to join the father who had come over several years before, planning to return to Italy. But the plans changed, perhaps because of the outbreak of World War I in 1914. They probably still were not certain they would stay in America forever.

Library of Congress

conditions in Italy were very bad, leading many Italians to stay on and wait for a better time to go back. In the end large numbers never did return. However, as long as they thought they would go back, they were likely to cling to their Italian culture.

Patterns in Canada were similar. There, too, the overwhelming proportion of Italian immigrants in this period were young males who had come to earn a nest egg and then return home (Woodsworth, 1909; Spada, 1969).

To describe this situation in a more sociological way, immigrants from Italy continued to regard the folks back home as their **reference group.** This concept refers to the groups that individuals identify with, the groups whose norms and values serve as the basis for self-judgment. In an important sense, our reference groups are the audiences before whom we lead our lives—the people whose approval counts most with us.

A reference group need not actually be present to influence a person's behavior. Even if no member of our reference group can actually see what we are doing, we can still act on the basis of how that group *would* react. This aspect of reference groups often allows sociologists to make sense of behavior that seems out of place.

Consider a nineteenth-century British gentleman exploring the upper reaches of the Nile. He sets out by himself, accompanied by forty men from an isolated African tribe who are acting as guides and carrying the supplies. There is no other British gentleman, perhaps not even any other white man, within 1,000 miles. Yet every night this explorer wears a tuxedo to dinner. His native companions think this is very strange, and if we could see him doing it, we

Jewish immigrants crowd the decks of the SS *Westernland* in about 1890, trying to escape the foul air of the jam-packed hold. Notice the many women and children, indicating that these are family groups.

might think it strange, too. But he doesn't care, since we are not the relevant audience. His reference group is other British gentlemen, and a gentleman *always* dresses for dinner. Once we know this man's reference group, we can explain and even predict much of his behavior. For example, we can be sure that he will always rise from his chair if a woman enters the room and that he will think it vulgar to mention the price paid for a possession.

The reference group for large numbers of Italian Americans was the inhabitants of rural villages in southern Italy. Once we know that, much of their behavior is understandable. Why reject traditional norms and values of rural Italian life if you are planning to return to it? Indeed, an observer reported hearing a mother in New York say the following to her son, in forceful Italian, "You *shall* speak Italian, and nothing else, if I must kill you; for what shall your grandmother say when you go back to the old country, if you talk this pigs' English?" (in Woodsworth, 1909).

As time passed, however, the ties to the old country began to fade. Young Italians found it hard to use as a reference group people they had never met, who lived in a place they had never been. Their reference groups began to change, and they began to adapt to the culture and conditions around them. In a few years, Italians achieved rapid upward mobility.

The process and results of changing from Italian to North American reference groups are pointedly illustrated by the story of Amadeo Giannini. Born in 1870 to immigrant parents, he founded a small bank in 1904 in the North

Bank of America

Following the San Francisco earthquake of 1906, the Bank of Italy occupied these temporary quarters on Montgomery Street. Today, this bank is known as the Bank of America.

A. P. Giannini in about 1906.

Bank of America

Beach district of San Francisco, an Italian neighborhood. Giannini's bank made loans to the small businessmen of the neighborhood, who found it very difficult to get credit at the city's other banks. Giannini named his institution the Bank of Italy. The bank survived the 1906 earthquake and the fire that destroyed much of the city. Under Giannini's brilliant management, the policy of lending to small businessmen brought the bank considerable success.

By 1928 Giannini had become uncomfortable about his bank's name, which sounded too foreign. Besides, its customers were no longer mainly Italians. So Giannini changed the name from the Bank of Italy to the Bank of America. When he died in 1949, his little neighborhood bank had grown into the largest privately owned bank in the world.

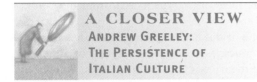

A CLOSER VIEW
ANDREW GREELEY: THE PERSISTENCE OF ITALIAN CULTURE

Our examination of Italian culture thus far has stressed assimilation, implying that as time passed and their reference group shifted, Italian Americans became just like everybody else. Moreover, logic suggests that if Italian culture once played a role in limiting the social mobility of Italians in America and Canada, these cultural differences must have vanished by now because Italians no longer are disadvantaged. In fact, most writers on ethnicity in America, and particularly on Italians (Alba, 1985, 1990), stress assimilation and describe the ethnic identification of the typical American of European descent as superficial (Waters, 1990). But if most sociologists take this view, a very distinguished scholar has vigorously dissented—Andrew Greeley (1991), the best-selling novelist, priest, and prolific sociological researcher.

In Greeley's judgment research by Covello and others describing the Italian family as unusually close-knit was not merely a description of the past. Instead, Greeley believes that these aspects of Italian culture have changed very little over the years: Italian families still are highly likely to share a household or to live close together and to interact daily, and Italian families display intense solidarity and loyalty.

However, Greeley was not content merely to depict Italian families this way in his novels. He did research to see if he was right. Since he was interested in the *persistence* of Italian family culture, he formulated these three hypotheses:

1. Italian Americans will differ significantly from other Americans in terms of living arrangements, amount of interaction among family members, and reliance on family members for help and financial aid.

2. Italians living in Italy will differ from persons in other European nations in the same way that Italian Americans differ from other Americans.

3. Italian Americans will closely resemble people in Italy in terms of family culture.

To test these hypotheses, Greeley used a set of surveys conducted by the International Social Survey Program (ISSP). These surveys have been conducted annually since 1983 in a number of nations, most of them in Europe. Although the survey interviews are conducted in many different languages, great effort is put into making the questions as similar as possible. In 1988 the ISSP focused on the family.

One item in the survey asked respondents if their father and/or mother shared their home. The results strongly supported Greeley's hypotheses. Italians were more than twice as likely as were other Europeans to be living with their fathers and mothers. Italian Americans were almost three times as likely as other Americans to share a home with their parents. In similar fashion Italians and Italian Americans were far more likely to visit family members, to live near them, and to talk with them frequently on the phone.

Two other items examined family solidarity. For the item "Suppose you had the flu and you had to stay in bed for a few days and needed help around the house with shopping and so on. Who would you turn to first for help?" respondents were offered sixteen response categories including "no one," "neighbors," "friends," and "family." Italians differed greatly from other Europeans in the proportion who selected "family." And Italian Americans differed greatly from other Americans in making the same choice.

For the item "Suppose you needed to borrow a large sum of money. Who would you turn to first for help?" in addition to other people, "financial institutions" was offered as a choice. The same patterns turned up again: Italians, both in America and Italy, were more inclined to turn to their families for financial aid.

Thus, Greeley concluded that "the unique Italian family style does not seem to have changed very much so far."

Are Greeley's conclusions compatible with the cultural theory of ethnic mobility? That is, how can Italian culture have ceased to hinder Italian occupational achievement if that culture hasn't changed very much? The answer appears to be that Italian family *values* have not changed, but the *norms* by which Italians fulfill their values have changed. For example, a century ago the children of Italian immigrants tended to drop out of school at an early age to contribute to the support of their families. Thus, in those days family solidarity may have delayed the upward mobility of the group. Today, Italian Americans still honor the values concerning financial obligations to other family members, but they regard education as a valuable means to meet those obligations. In this normative context, family solidarity ceases to impede success. In fact, it may make it easier for Italian Americans to go to college or to start their own businesses.

CONCLUSION

Let's look back over this chapter and see how it illustrates the sociological process. We began with a question, with sociologists wondering why Jews had achieved such rapid economic advancement after emigrating to North America. Over the years sociologists have constructed two major theories to answer this question, so the first thing we needed to do was grasp the concepts these theories require. Having separated society from culture, we examined a series of concepts used in social theories, and then we considered concepts often used in cultural theories. Then we looked at some theorizing.

Cultural theories proposed that the Jewish advantage lay in unique features of their culture—in such things as their love for learning and their strict norms of schooling and study. Initially, when sociologists began to test cultural theories of Jewish mobility, they found the predicted differences between Jewish and Italian culture. And the question of why seemed to be answered. But then problems began to turn up. For example, less educated Jews also found rapid economic success. What could account for them? What cultural advantages did

they have over Italians and other new immigrants?

So, it was back to step three in the sociological process—to more theorizing. This time a social theory was constructed. Italian immigrants came from the bottom of the stratification system in Italy and were qualified only for unskilled labor upon arriving in the United States. But Jews were not from the lower classes in Russia, Poland, and other parts of eastern Europe. They had been not peasants but merchants and craftsmen. This was their real edge over other immigrants. The Jews did not have to gain higher positions in the stratification system; they had merely to *regain* them. Peasants from Italy had to acquire skills that would allow them to rise economically—and typically, only their children or their grandchildren achieved higher status. But tailors, diamond cutters, or watchmakers from Europe simply had to find a chance to use their skills.

Thus, it appeared that social scientists had been wondering about things that either hadn't really happened or at least were rather ordinary: that the astounding rate of Jewish economic success was mainly an illusion based on misreading their initial starting point. And for students who read the first three editions of this textbook, that's where the story ended. But as students who read this edition will discover, science doesn't sit still. All scientific knowledge comes with a label that reads "Good until further notice." In this case further notice was given when Perlmann discovered that cultural factors play an important part in the upward mobility of immigrant groups after all.

Review Glossary

Terms are listed in the order in which they appear in the chapter.

society A relatively self-sufficient and self-sustaining group of people who are united by social relationships and who live in a particular territory.

social structure A characteristic of a group rather than of an individual.

stratification The unequal distribution of rewards (or of things perceived as valuable) among members of a society; the class structure.

classes Groups of people who share a similar position in the stratification system.

mobility, upward and **downward** A change of position within the stratification system.

status The position or rank of a person or group within the stratification structure.

achieved status A position gained on the basis of merit (in other words by achievement).

ascribed status A position assigned to individuals or groups without regard for merit but because of certain traits beyond their control, such as race, sex, or parental social standing.

the strength of weak ties Mark Granovetter's proposition that, for purposes of spreading information, weak ties are stronger, or more effective, than strong ones.

tie Another word for a link or a relationship among persons or groups.

redundant tie One that duplicates links among members of the same network.

structural holes Unlinked pairs that show up on network diagrams as blank space.

bridge ties Links across holes between groups.

bridge position One having bridge ties.

local networks Dense networks with strong, redundant ties. They are called *local* because members often engage in the direct, person-to-person interaction that is necessary to form and sustain strong ties. Hence, members tend to be clustered geographically.

cosmopolitan networks Networks that are relatively full of holes, consisting of weaker, nonredundant ties. The word *cosmopolitan* means worldly, at home throughout the world, or widely distributed. Hence, members of a cosmopolitan network seldom engage in face-to-face interaction and tend to be scattered geographically.

culture The sum total of human creations—intellectual, technical, artistic, physical, and moral. Culture is the complex pattern of living that directs human social life, the things each new generation must learn and to which they eventually may add.

values Ideals or ultimate aims; general evaluative standards about what is desirable.

norms Rules that define the behavior that is expected, required, or acceptable in particular circumstances.

role A set of expectations governing the behavior of persons holding a particular position in society; a set of norms that defines how persons in a particular position should behave.

multiculturalism The presence of several significant cultures.

subculture A culture within a culture; a group that maintains or develops its own set of beliefs, morals, values, and norms, which usually are at variance with those of the dominant culture.

prejudice Negative or hostile attitudes toward and beliefs about a group.

discrimination Actions taken against a group to deny its members rights and privileges available to others.

assimilation The process by which an individual or a group reacts to a new social environment by adopting the culture prevalent in that environment.

accommodation A situation where two groups ignore some cultural differences between them and emphasize common interests instead.

modernization The processes of industrialization, economic development, and technological innovation by which a culture sustains high standards of living and maximizes control over the physical environment.

globalization The development of global communications, a global economy, and a global culture.

anti-Semitism Prejudice and discrimination against Jews.

reference group A group that a person uses as a standard for self-evaluation.

Suggested Readings

Alba, Richard. 1985. *Italian-Americans: Into the Twilight of Ethnicity.* Englewood Cliffs, N.J.: Prentice Hall.

Covello, Leonard. 1967. *The Social Background of the Italo-American School Child.* Leiden, the Netherlands: E. J. Brill.

Greeley, Andrew. 1974. *Ethnicity in the United States.* New York: Wiley.

Howe, Irving. 1976. *World of Our Fathers.* New York: Harcourt Brace Jovanovich.

Perlmann, Joel. 1988. *Ethnic Differences: Schooling and Social Structure among the Irish, Italians, Jews, and Blacks in an American City, 1880–1935.* Cambridge: Cambridge University Press.

Porter, John. 1975. "Ethnic Pluralism in Canadian Perspective." In *Ethnicity: Theory and Experience,* edited by Nathan Glazer and Daniel P. Moynihan. Cambridge, Mass.: Harvard University Press.

Steinberg, Stephen. 1974. *The Academic Melting Pot.* New York: McGraw-Hill.

Steinberg, Stephen. 1981. *The Ethnic Myth: Race, Ethnicity, and Class in America.* Boston: Beacon Press.

Zborowski, Mark, and Elizabeth Herzog. 1962. *Life Is with People: The Culture of the Shtetl.* New York: Schocken Books.

Sociology Online

www.socstark10.com

GO TO THE INTERNET AND TYPE: www.socstark10.com.

↗**CLICK ON:** 2000 General Social Survey

✓ **SELECT:** SOCFREND: How often do you spend an evening with friends who live outside the neighborhood?

↗**CLICK ON:** Analyze Now
Assume that people who never socialize with friends from outside their neighborhood are limited entirely to *local* networks. Examine the tables for education and for income. Are these findings consistent with the discussion of local and cosmopolitan networks in the text?

↗**CLICK ON:** Select New Data Set

↗**CLICK ON:** The 50 States

✓ **SELECT:** % COLLEGE: Percentage of those over 25 who have completed a college degree.

↗**CLICK ON:** Analyze Now
On the basis of what you found out about education and local networks, in which state would you expect the highest percentage of people to be members of cosmopolitan networks? In which state should the highest percentage be members of local networks?

↗**CLICK ON:** Select New Data Set

↗**CLICK ON:** Nations of the Globe

✓ **SELECT:** COLLEGE: College and university students per 1,000 population.

↗**CLICK ON:** Analyze Now
In similar fashion, what five nations should excel in cosmopolitan networks compared with what five nations where local networks should predominate?

Can you suggest any ways in which states lacking college graduates are like nations that also are low in terms of graduates?

USING INFOTRAC COLLEGE EDITION

GO TO THE INTERNET AND TYPE:
www.InfoTrac-college.com

↗**CLICK ON:** Register New Account
You will be asked to enter your Access Code and to create and enter a User Name and Password. Having done so,

↗**CLICK ON:** InfoTrac College Edition or Log On.
Using the following search terms, find articles about the American experience of Jews and Italians that expand on the discussion in the chapter.

SEARCH TERMS:

Anti-Semitism

Italian America

CHAPTER

3

Microsociology: Testing Interaction Theories

SCIENTISTS OFTEN GET CARRIED AWAY by a new discovery. For decades following the discovery of x-rays, physicians misused them to treat just about everything, from colds and acne to baldness and infertility. Social scientists did much the same in response to the discovery of social facts. Having become aware that social and cultural forces could greatly shape what previously had been regarded as intimately personal actions, many social scientists (especially in Europe) entirely dismissed consciously made choices as a basis of behavior. So, for several generations students taking social science courses were introduced to humans as empty robots whose every significant action was caused by outside forces.

Thus, in 1895 the famous French sociologist Émile Durkheim wrote that social facts "consist of ways of acting, thinking, and feeling, *exterior* to the individual, that possess a power of *coercion*

by virtue of which they *impose themselves* upon [the individual]" (italics added). He went on to claim that since the "essential characteristic" of social facts "consists of the power they possess of exercising, from the *outside*, a pressure on individual minds, it must be that they do not derive from individual minds." Therefore, he announced triumphantly, it is to sociology, not to psychology, that we must turn to explain human behavior.

For the next several generations, many sociologists not only tended to deny that internal, or psychological, phenomena were primary *sources* of human behavior, but they also came to regard internal mental states as almost entirely the *consequences* of external forces. Thus, the very influential American sociologist Talcott Parsons (1937) noted that the norms "do not . . . merely regulate 'externally' . . . they enter directly into the [human consciousness]." It was widely believed

CHAPTER OUTLINE

THE RATIONAL CHOICE PROPOSITION

INTERACTION THEORIES

SYMBOLIC INTERACTION
  Social Construction of Meaning
  Social Construction of the Self

EXCHANGE THEORY
  Solidarity
  Inequality
  Agreement and Conformity

THEORY TESTING: MEASUREMENT
AND RESEARCH
  Variables
  Criteria of Causation
  On Abstractions
  Box 3-1: Correlation

THE EXPERIMENT: STUDYING GROUP
SOLIDARITY

A CLOSER VIEW
Solomon Asch: Solidarity and Conformity
  Manipulating the Independent Variable

RANDOMIZATION

SIGNIFICANCE

FIELD RESEARCH: STUDYING RECRUITMENT

A CLOSER VIEW
  Lofland and Stark: Religious
  Conversion
  Covert and Overt Observation
  Conversion as Conformity
  Four Principles of Recruitment
  Box 3-2: Heaven's Gate

CONCLUSION

REVIEW GLOSSARY

SUGGESTED READINGS

SOCIOLOGY ONLINE

that the basic framework of our internal mental states is mainly the result of **socialization,** the process by which culture is learned and *internalized* by each normal member of a society—much of which occurs during childhood.

Culture has been **internalized** when people accept the norms, values, roles, beliefs, and other primary aspects of their culture to the point that these are the taken-for-granted, fundamental basis for all of their decision making. No one disputes that people do tend to internalize their culture, but for a long time, many sociologists—especially Europeans—took a more extreme view, claiming that people who have internalized their culture will act in conformity to it, pretty much without considering, or even without recognizing the existence of, other possible options. As Dennis Wrong (1961) put it:

Internalization has . . . been equated with "learning," or even with "habit formation" in the simplest sense. Thus when a norm is said to have been "internalized" by an individual, what is frequently meant is that he habitually affirms it and conforms to it.

The image of humans presented by sociology in this period was so entirely at odds with the emphasis given individual sources of behavior, not only by psychologists but also by economists, that it prompted the quip that "economics is all about how people make choices. Sociology is all about why they don't have any choices to make" (Duesenberry, 1960). For as James S. Coleman (1990) explained:

The proposed causes of action are not persons' goals or purposes or intents, but some forces outside them or unconscious impulses within them. As a consequence, these theories can do nothing other than describe an inexorable fate. . . . At the mercy of these uncontrolled external or internal forces, persons are unable to purposefully shape their destiny.

The belief that socialization is the basis for nearly all behavior—leaving room to attribute only quirky, impulsive, and irrational actions to internal sources—was, of course, entirely at odds with notions of free will, as discussed in Chapter 1. People do not really choose; they merely act as their socialization has "programmed" them to behave. Eventually, these notions fell of their own obvious shortcomings. It's not that anyone began to doubt that socialization plays a major role in shaping human beings or that people do internalize culture. At issue was the balance between social and cultural forces, on the one hand, and the ability of human beings to make self-conscious choices, on the other hand. And clearly, the claim that the sociocultural environment explains everything was far out of balance. Thus, in 1961 Dennis Wrong, a very prominent sociologist, wrote an influential essay in which he condemned sociology for its "oversocialized conception" of humans.

Wrong asked the question, if socialization explains most of what people do, then why is it that so often we are unable to attribute behavior to socialization until after the fact? For example, when people donate money to charities, sociologists were quick to say they did so because they had been socialized to donate and that little or no individual reflection or weighing of options had gone on. But these same sociologists, fully informed about each person's socialization, could not very accurately predict who would give and who would not. Why not? Because within any given culture and even among people very similarly socialized, there remains an immense amount of latitude for individual choices, even for those whose behavior remains within the norms. Thus, during the latter part of the twentieth century, efforts to understand internal sources of behavior and the

basis for individual choice making regained major importance on the general sociological agenda (Collins, 1994). Of course, all along, a considerable number of microsociologists had retained their interest in understanding why and how people make choices.

This chapter will introduce you to the basic theoretical approaches of microsociology. We will see how modern sociologists understand the *interplay* between the individual and the group. Against that background we then watch sociologists conduct research to test important portions of microtheories and especially how they try to establish cause-and-effect relationships.

⚹ THE RATIONAL CHOICE PROPOSITION

Today, the dominant approaches to social theory share a common first assumption or proposition. This proposition has been stated in a great many ways, but each variant asserts the same insight: When faced with choices, humans try to select the most rational (or reasonable) option, defined as the one that will yield them the maximum benefit. This first assumption often is referred to as the **rational choice proposition,** and while it has been stated in many ways, I prefer this formulation: *Within the limits of their information*

If it were not for differences in tastes and preferences, there would only be one TV channel since everyone would always watch the same shows.

and available choices, guided by their preferences and tastes, humans will tend to maximize.

Let's analyze this sentence to see precisely what it does and does not mean. The first part of the sentence—*within the limits of their information*—recognizes that we can't select choices if we don't know about them, nor can we select the most beneficial choice if we have incorrect knowledge about the relative benefits of choices. The second part of the phrase—*within the limits of their . . . available choices*—notes that we cannot choose that which is unavailable. Recognizing these limits on choices is to acknowledge the extent to which external factors can impinge upon the individual and how these can differentially affect people within the same society. For example, the stratification system can influence both information and available options. Suppose the choice involves selecting an occupation. A college graduate usually will have many more choices than will a high school dropout; moreover, even when both are qualified for a job, the college graduate also may be far more likely to find out that the job is available and know how to get it—mainly because the college graduate will be more apt to be linked to a cosmopolitan network (the link between class and networks is discussed at length in Chapter 9). Nevertheless, even within these constraints, most people will have a substantial amount of discretion most of the time—that is, usually they will face multiple options.

However, if humans all seek to maximize, why is it that they don't always act alike? Why don't people reared in the same culture all seek the same rewards? Because their choices are *guided by their preferences and tastes.* Preferences and tastes define what the individual finds rewarding or unrewarding. Consequently, people may differ in what they want and how much they want it. This helps us understand not only why people do not all act alike but also why it is possible for them to engage in exchanges: to swap one reward for another. For example, John likes candy more than Jane likes it, while Jane likes fruit more than John does, so they often exchange candy and fruit.

Clearly, not all preferences and tastes are variable—there are some things that virtually everyone values, regardless of their culture: food, shelter, security, and affection being among them (Aberele et al., 1950). Obviously, too, culture in general and socialization in particular will have a substantial impact on preferences and tastes. It is neither random nor a matter of purely personal taste whether someone prays to Allah or Shiva or, indeed, whether one prays at all. Still, the fact remains that even within any culture there is substantial variation across individuals in their preferences and tastes. Some of this variation is also at least partly the result of socialization differences—for example, many people probably got interested in sports mainly because their parents were. But a great deal of variation is so idiosyncratic that people have no idea how they came to like certain things. It's as the old adage says, "There's no accounting for taste." As noted above, that people differ greatly in terms of tastes and preferences facilitates exchanges. But it also explains what often are rather remarkable differences in behavior, as we shall see.

Finally, as already mentioned, the phrase that *humans will tend to maximize* means that they will try to get the most of what they want at the least cost to themselves. The word *tend* is included to note that people don't *always* act in entirely rational ways. Sometimes we act impulsively—in haste, passion, boredom, or anger ("I really didn't stop to think about what I was doing"). But most of the time, normal human beings will choose what they *perceive* to be the more rewarding option, and whenever they do so, their behavior is rational even if they are mistaken about which is the better option. For example, people buy stocks hoping to profit. If their stocks decline in value, that doesn't mean they acted irrationally, only that they were wrong about which stocks to buy or when.

Many people are upset by the claim that normal humans seek to maximize, interpreting that to mean that social scientists think people are always selfish. What about **altruism,** they ask, defining this term to mean unselfish actions done entirely for the benefit of others? What about parents who rush into burning buildings to try to save a child? Or what about people like Mother Teresa who forgo a comfortable life to aid the sick and the poor? How are such people acting to maximize their personal rewards and minimize their personal costs? Or as British sociologist Anthony Heath (1976) put it: "The people who act out of a sense of duty or friendship, it is said, cannot be accounted rational and cannot be brought within the scope of [the] rational choice [proposition]." But Heath went on:

Of all the fallacies [about rational choice], this is the least excusable. Rationality has nothing to do with the *goals* which [people] pursue but only with the *means* they use to achieve them. When we ask whether someone is behaving rationally we are asking, for example, whether he [or she] is choosing the most efficient means to the goal. We are not asking whether he [or she] is choosing the "right" goal.

As Heath implies, social scientists are fully aware of people such as Mother Teresa. But we also recognize that their behavior violates the rational choice proposition *only* if we adopt a very narrow, materialistic, and entirely egocentric definition of rewards and ignore the immense variety of preferences and tastes. Human life and culture are so rich because of the incredible variety of our preferences and tastes, of things we perceive as rewarding. There is no need to suggest that a parent has acted against his or her self-interest by rushing into a burning building. Rather, let us recognize that the ability of humans to regard the survival of a child as more rewarding than their own survival is a credit to the human spirit and to our capacity to love. To call that altruism and place it in opposition to the rational choice premise is to reduce noble behavior to crazy and irrational action. In fact, the "selfish" rational choice proposition is humanistic in the fullest sense. It acknowledges our capacity to find rewards in our dreams, hopes, love, and ideals.

INTERACTION THEORIES

Although all microsocial theories begin with the rational choice proposition, they soon go their separate ways as more propositions are added. Microtheories become truly *sociological* when they incorporate two key insights.

First, compared with economists, *sociologists greatly expand the concepts of rewards and costs*, as was noted in the discussion of altruism. Although sociologists do not slight the importance of material rewards, they place at least as much (if not more) emphasis on intangible rewards, especially emotional satisfactions. Chief among these are affection and self-esteem. We will go to great lengths to be loved, liked, respected, and admired.

Second, compared with psychologists, *sociologists stress that much of what we want can be obtained only from other people*. To be loved or respected, we must induce others to love us or respect us. However, it costs other people to give us things, even intangible things such as love and respect. They invest time and energy in communicating their respect and love, and they risk the costs of wasting their love or respect on someone who, if the truth were known, does not deserve it or who does not reciprocate. So, to get others to reward

© Mike Valeri/Taxi/Getty Images

Firefighters do not rush into burning buildings because they have lost touch with reality. It's what they are paid to do and trained to do, and they go about it as carefully as they can, making full use of safety equipment. That doesn't mean they aren't "heroes" for risking their lives for others—it is the pursuit of generous goals that makes people "heroes."

We can identify this girl as a Christian because of the cross she is wearing around her neck—the cross is a *symbol* of Christianity. That is, the cross itself is not a Christian but merely stands for (symbolizes) being a Christian. In fact, the word *Christian* also is merely a symbol. The word itself is not a Christian; it conveys instead the idea of what it means for someone to be a Christian.

us, we must reward them. Perhaps nothing could be more obvious, yet throughout this course we will trace the many profound implications of this simple point.

Because humans seek rewards from one another, they are inevitably forced into *exchange relations*. That is, humans engage in **social interaction,** the process by which humans influence one another. The basis of social interaction is the exchange of rewards—a conclusion that is implicit in the rational choice proposition as soon as we shift our focus from the individual in

pursuit of rewards to two or more persons, each seeking rewards. Therefore, microsociology is concerned primarily with understanding social interaction and the exchanges it entails. Furthermore, microsociological theories are designed to explain the *regularities or patterns that arise out of interactions and exchanges*, especially those involving face-to-face relationships.

There are two basic sociological approaches to microtheory. The first is known as *symbolic interaction theory*, and the second is *exchange theory* (sometimes called rational choice theory). Of the two, symbolic interaction theory pays particular attention to the most microaspects of social behavior, asking how individuals learn to interpret and relate to the social world around them—how we develop a "self." Exchange theory is more concerned with social solidarity—how we learn to like one another and get along. Exchange theory usually is stated in more formal language than is symbolic interaction theory and places somewhat more stress on structures than on culture, while symbolic interaction reverses this emphasis.

If we were to sketch the networks of social relations of sociologists who identify themselves as exchange theorists and those who are symbolic interactionists, we would discover two quite dense and quite isolated clusters or cliques. Thus, it is hardly surprising that heated conflicts occur between the two. Having made substantial use of *both* theoretical approaches in my own work, I occupy a bridge position from which I find the two approaches very compatible, each offering its most powerful insights on precisely the issues which the other neglects.

SYMBOLIC INTERACTION

As the name suggests, the theoretical approach known as **symbolic interaction** also regards interaction among human beings as the fundamental social process. From this perspective people are endlessly tangled up in interactions, in influencing and being influenced by other people around them. We act and our action affects others. They respond and their responses affect our next action. This in turn affects theirs. And so it goes, as we constantly adjust and readjust our activities according to feedback from others.

It is easy to see that human interaction consists primarily of communication. Through language, gestures, and actions, we communicate

Herbert Blumer was not only a very productive and famous American sociologist but also a star athlete. In 1921, during his senior year at the University of Missouri, Blumer was selected to the All-American Team as a tackle (playing on both offense and defense, as everyone did in those days). Then he played for the Chicago Cardinals (later the St. Louis and now the Arizona Cardinals) of the National Football League while earning his Ph.D. at the University of Chicago. He continued to play for several years even after he joined the Chicago faculty. In 1958 he became the first chair of the newly founded sociology department at the University of California, Berkeley.

with others and they communicate with us. However, unlike the grunts and hoots of animals, human communication consists primarily of **symbols,** things that stand for or indicate another thing. The word *fish* is not a fish; it is intended to convey to a listener the *idea* of a fish. If humans could not use symbols, then you could know that Lake Washington is full of salmon only if someone led you to the lake and caught some salmon and showed them to you. That is precisely what a mother bear must do to teach her cubs about fish.

The use of symbolic communication can be extremely efficient, but it depends on the ability of others to interpret or *decode* the symbols used. When we tell people we are happy, they cannot directly perceive how we feel. They must have learned to interpret the meaning of the symbol *happy.*

Randall Collins (1994) has called symbolic interaction theory "the distinctively American tradition," and it long dominated microsociology in the United States. Those who use this theoretical approach often are called *symbolic interactionists.* The fundamental premise of this theoretical approach is that interaction through the use of symbols makes and keeps people human. Consequently, a primary focus of research by symbolic interactionists is on the personal, subjective meanings we attach to various symbols. How do people engaged in interaction perceive their situations and the intentions and meanings of those around them? Is some behavior the result of misunderstandings of the meanings of symbolic communication? Indeed, do symbols mean the same things to different people? If so, how does that happen? Such questions often have led symbolic interactionists to conduct detailed studies of people engaged in face-to-face interactions. And symbolic interactionists often have paid particular attention to children in an effort to learn how humans develop the capacity to understand one another's use of symbols.

SOCIAL CONSTRUCTION OF MEANING

Although the theoretical approach dates back to the late nineteenth century, the term *symbolic interaction* was coined "in an offhand way in an article written in 1937" by Herbert Blumer (1969). In addition to naming this approach, Blumer was one of its most influential proponents and was the first to identify its three basic premises. We shall examine each.

Charles Horton Cooley.

Bentley Historical Library/University of Michigan

According to Blumer, the first premise is: *Human beings act towards things* [including people] *on the basis of the meanings that the things have for them.* This premise clearly set symbolic interactionists in opposition to those social scientists who dismissed internal mental states and regarded the sociocultural environment as determinative of behavior. Not so, said the symbolic interactionists. The outer world is not a given; people must perceive it and interpret it before it is possible for them to respond to it.

In reaction to the sociocultural determinists, some of the most radical proponents of symbolic interaction, such as Charles Horton Cooley (1864–1929), virtually dismissed the external world, arguing that society really exists only in the mind. Cooley (1922) taught:

There is no separation between real and imaginary persons; indeed, to be imagined is to become real, in a social sense . . . an invisible person may easily be more real to an imaginative mind than a visible one. . . . A person can be real to us only in the degree in which we imagine an inner life which exists in us. . . . All real persons are imaginary in this sense. . . . My association with you evidently consists in the relation between my idea of you and the rest of my mind. . . . *The immediate social reality is the personal idea;* nothing, it would seem, could be more obvious than this. . . . Society, then, *is a relation among personal ideas.*

Blumer rejected such extreme "mentalism," recognizing that Cooley's claim was as overstated as the opposing claim that the mind is merely the product of external forces. Nevertheless, Blumer was skeptical of the capacity of externalities to impose meanings directly on the mind. Thus, his second premise of symbolic interaction: *The meaning of things is derived from, or arises out of, social interaction.* According to Blumer, this premise denies that anything, including physical objects, has a meaning other than those given them by humans and that these meanings are created collectively. He wrote:

The meaning of a thing for a person grows out of the ways in which other persons act toward the person with regard to the thing. Their actions operate to define the thing for the person. Thus, symbolic interactionism sees meanings as social products, as creations that are formed in and through the defining activities of people as they interact.

The third premise is: *The meanings of things are handled in, and modified through, an interpretative process used by persons in dealing with things they encounter.* That is, Blumer postulated that meanings are additionally the result of internal, mental processes which take the form of *internal interaction.* He explained:

This process [of interpretation] has two distinct steps. First, the actor[1] indicates to himself the things toward which he is acting; he has to point out to himself the things that have meaning. The making of such indications is an internalized social process in that the actor is interacting with himself . . . engaging in the process of communication with himself. Second, by virtue of this process of communicating with himself, interpretation becomes a matter of handling meanings. The actor selects, checks, suspends, regroups, and transforms the meanings in light of the situation in which he is placed and the direction of his action.

These premises have yielded some interesting insights. One of the earliest and most important involved the way in which human infants gain a sense of self—a concept of who, and what, they are. The pioneers of symbolic interaction proposed that our *conception of self is socially constructed.*

SOCIAL CONSTRUCTION OF THE SELF

Cooley (1922) introduced the term **looking glass self** to describe the process by which our sense of self develops. *Looking glass* is an old term for mirror, and Cooley argued that during interaction humans serve as mirrors for one another. Whether we come to hold a good or poor opinion of ourselves depends on the reflection others communicate to us. That is, we can feel good about ourselves only if other people give us reason to do so. In a classic study, Frank S. Miyamoto and Sanford M. Dornbusch (1956) found that college students' self-conceptions were remarkably similar to ratings of them made by their fellow students. Moreover, the greater the number and strength of a person's attachments, the stronger and more positive his or her self-conception.

George Herbert Mead (1863–1931) carried Cooley's ideas about the self considerably further. He distinguished two aspects of human identity: the *mind* and the *self* (Mead, 1934). To

Library of Congress

George Herbert Mead.

---

1. Social scientists often use the word *actor* to refer to any individual engaged in action.

participate in interaction, humans must learn to use and interpret symbols. Mead used the concept of **mind** to identify our *understanding of symbols*, arguing that the mind arises entirely through repeated interaction with others—a learning process that in infants is largely a matter of trial and error.

The self also arises through social interaction. As Mead defined the term, the **self** is our learned understanding of the responses of others to our conduct. Through long experience in seeing others react to what we do, we not only gain a general notion of who we are but also are able to put ourselves in another's place. Mead called this *taking the role of the other*. From doing this repeatedly, we form a generalized notion of others—of what they want and expect and how they are likely to react to us. Out of this tension between ourselves and others, the self is formed. And we can only truly know ourselves through the eyes of others.

Mead (1925) pointed out that until children develop a self, until they can take the role of the other, they cannot take an effective part in most games. In baseball, for example, children have to anticipate what others will do in order to play. It is not enough that a shortstop knows to cover second when balls are hit to the right side of the infield; the other players must also know how shortstops play. The next time you pass a soccer field where very young kids are playing, notice that they tend to be bunched closely around the ball. Each kid tries to get the ball, and no one runs to an open spot in anticipation of receiving a pass. It takes time for children to learn to take the role of the other.

Implicit in the idea of taking the role of the other is a more general premise emphasized by symbolic interactionists: *To understand why people act a certain way in a certain situation, it is necessary to understand how they define their situations.* It is this insight that places symbolic interactionists firmly within the "rational choice" theoretical tradition and in opposition to those who are quick to offer "irrationalist" accounts of common human action. If we fail to discover how people define their situation, it is often impossible to distinguish rational from irrational behavior. As James S. Coleman (1990) put it: "Much of what is ordinarily described as nonrational or irrational is merely so because observers have not discovered the point of view of the actor, from which the action *is* rational." Put yourself in the place of those left aboard the *Titanic*. Was it irrational for them to join in singing "Nearer My God to Thee" as the ship went down?

According to basic principles of exchange theory, it was a quite rational thing to have done.

EXCHANGE THEORY

The term **exchange theory** reflects the central concern to explain and reveal the key processes by which people (in accord with the rational choice premise) *seek to maximize by exchanging rewards with one another*. When we expand our scope from one person to two persons, the rational choice proposition lets us see that self-interest limits the conditions under which exchanges will take place. This was first recognized by Adam Smith more than 225 years ago when he included this proposition in his theory of economics: *If an exchange between two persons is voluntary* (if neither partner to the exchange is being forced to yield rewards to the other), *then an exchange will not take place unless both partners believe they will benefit from it.*

When we give something to someone else, we expect something of at least comparable value in return. For example, if we give someone a gift, we expect them to reciprocate, perhaps with a gift of equal value, and if not, then with emotional rewards such as gratitude and/or affection. Here the qualifications about preference and taste enter into calculations of value. Many exchanges do not occur because one person places greater value on his or her offering than on that which is offered in return. This is known

These children are too young to play games that would require them to take the role of the other. While I watched them, they showed the ability to play catch, sort of. The youngest one always tried to catch the ball with his face, and no one went out for a pass—the ones without the ball moved closer instead. Finally, several minutes after the picture was taken, the boy in the rubber boots took the ball and hid behind a tree. The other two did not look for him.

Courtesy of Lynne Roberts

as a win/lose situation, and when such exchanges do occur, they seldom are repeated. Usually, when exchanges occur, it is because each partner values what the other offers more than he or she values what must be offered in return. This is known as a win/win situation in that both persons come out ahead. At the very least, for an exchange to occur, people must regard the values of what is offered and what is asked as equivalent.

Beyond the question of value lies the issue of *risk*. Whenever we engage in exchanges, we risk loss because the others may fail to live up to their side of the bargain. They may cheat us by giving nothing or by giving us less than they promised or by misrepresenting the value of their offer. Thus, the following proposition: *In the absence of any restraints, cheating is the expected behavior of exchange partners since, if successful, cheaters maximize.* An essential question is, Why doesn't everyone cheat? Note the first six words of the proposition. For most of us, most of the time, there are *restraints* on our behavior. Let's see why.

SOLIDARITY

When only two persons are involved in an exchange, issues of risk are less severe and the restraints are relatively informal, so let's consider that situation first. The old saying, "Cheat me once, shame on you; cheat me twice, shame on me," captures the way we respond when we feel we have taken part in an inequitable exchange. Following Homans (1974), we may put this in the form of a theoretical proposition: *When one exchange partner fails to receive the expected level of reward, he or she will be angry.* And the way we often respond is to avoid future exchanges with that person. Conversely, when we exchange with someone and are satisfied with the outcome of the exchange, we are inclined to exchange with that person again should an appropriate occasion arise. If that exchange proves satisfactory, too, we are even more likely to trade with this person a third time.

This line of reasoning lets us see that over time people tend to develop stable patterns of exchange with regular exchange partners (Homans, 1974). Keep in mind that these exchanges are not limited to material goods or services; one of the most important exchange commodities is emotion. In fact, even when people give a material gift (jewelry, for example) to someone, often the last thing they want in return is a material gift of equal or greater value.

For them an equitable exchange would involve a response of gratitude and affection. Thus, we can recognize that stable exchange relationships tend to generate emotions and sentiments such as trust, admiration, affection—even love (Blau, 1986; Hechter, 1987; Coleman, 1990). This discussion was summed up by George Homans as the **law of liking**: *Participation in common activities (exchanges) causes people to like one another.* Symbolic interactionists might express the same thing by saying that persistent patterns of interaction result in positive evaluations of others.

Sociologists often use the term **attachment** to identify bonds of liking or affection between two people. Stated in terms of exchange theory, an attachment is a stable and persistent pattern of exchange between two people when positive sentiments are among the "commodities" exchanged. *Social solidarity*, then, reflects the degree to which members of a group or network are attached to one another. In this way we can recognize the far greater solidarity of local as opposed to cosmopolitan networks.

INEQUALITY

There is an important limit to Homans' law of liking. Common activities lead to liking, *provided that the participants are equal in rank.* Bringing persons together to carry out common activities does not lead to friendships if some of them have power over the others. Bosses and workers may do many activities together, but if there is a constant reminder that one gives the orders and the others carry them out, the result is usually not much informal solidarity. That is, subordinates may greatly respect their leaders, possibly even love them, but they will seldom or never become intimates—coaches rarely even try to become personal friends with their players in the same way that players can be friends of one another. The more formal deference one rank gives to another rank, the more difficult it is for them to become personal friends. As was discussed in Chapter 1, Homans identified this as his **law of inequality**: *Emotional attachments among members of a group will be weaker among members of different ranks than among those of similar rank.*

AGREEMENT AND CONFORMITY

When people of equal rank interact with and like one another, we can see an additional phenomenon that Homans stated as the **law of agreement**: *The more the members of a group like* *one another, the more apt they are to agree with each other.* If the group is closely knit, manifesting a lot of solidarity, tied together by informal feelings of friendship among its members, it also tends to have a high degree of uniformity of opinion. Conversely, if people dislike one another, the more apt they are to disagree.

Individuals believe what their friends believe, especially about those things that are most important to the group. This has both a voluntary and a coercive aspect. Each group member, if she or he is emotionally committed to the group, wants to believe what the group believes. Members feel it is important to realize that these are good ideas and that it is valuable to believe in them. In a religious group, an individual will feel that the most important thing is to believe in the group's doctrines; in a group of social movement activists, members will feel that protecting the environment or stopping abortions or freeing laboratory animals is the most valuable thing anyone can do.

The flip side of this personal commitment to the central beliefs of the group is that you expect other members of the group to agree. A member who doesn't agree with some central belief becomes the target of group pressure. Others will try to change his or her mind. At first these efforts may be low key, perhaps done in a joking way, but if disagreement persists, dissenting members will increasingly be ignored or attacked. Eventually, those who fail to conform will be dropped from the group. Later in the chapter, we will examine the classic Asch experiment in which group pressure is sufficiently potent to get people to falsify their own judgments about very simple and obvious visual comparisons.

Group pressure, as with group solidarity, is variable. Not all groups are equally intense in their conformity to group beliefs. But a mark of a very strongly integrated group is that it has this kind of cultural conformity. We can formulate this as the **law of conformity:** *The more intense the group solidarity, the more intense the demand for conformity.*

Which comes first, the agreement in beliefs or the solidarity? Do we like people because we agree with them, or do we agree with them because we like them? The process can work both ways, but there is a good deal of evidence that the main line of causality runs from solidarity to conformity—that we shift our beliefs to agree with our friends. We will see a dramatic example of this when we discuss religious conversion, but it occurs in all groups on all topics. We also will see the process in reverse. If people are prevented from meeting regularly with their friends (perhaps because of external circumstances such as moving to another place), their liking for one another declines, and their beliefs start to diverge from one another as well.

In any event this analysis of group pressures for agreement and conformity explains the origins of *norms.* As defined in Chapter 1, norms are rules governing behavior. They specify how people are supposed to act in various circumstances. The existence of norms is implicit in the risks involved in exchanges and in the basis of attachments. As noted in the very first paragraph of Chapter 1, if we can't predict how someone will behave, it is too risky to exchange with that person and impossible to establish or maintain an attachment. Since people *must* exchange, they evolve norms to reduce the risks by making one another's behavior sufficiently predictable. Furthermore, norms not only make attachments possible; *attachments provide the force behind norms.* When we violate norms, we risk our attachments. Attachments are the glue of social life. They represent emotional and material bonds among individuals, bonds that often define membership in a group, and they serve as a major potential cost for norm violations. As we will pursue at length in Chapter 7, people lacking attachments are relatively free to break the norms. It also is the case that people lacking attachments are at risk of poor mental health—depression, loneliness, and low self-esteem—as was clear in the discussion of symbolic interaction.

To sum up: *Although individual choices play a major role in human behavior, what we choose to do is greatly influenced by what those to whom we are attached want us to do.*

THEORY TESTING: MEASUREMENT AND RESEARCH

Theories attempt to explain how the world works. Research attempts to determine whether or not theories work. Suppose we wanted to test Homans' law that solidarity causes conformity. Before we can proceed, there are several things we need to understand.

VARIABLES

The first of these is that the world can be separated into constants and variables. A **constant** is something that never changes (or varies). The sun always comes up in the east. The speed of light in a vacuum is always 186,282 miles per second. A **variable** is anything that varies or changes. The time of sunrise changes daily. The brightness of a light increases as it is closer. Generally speaking, we are far more interested in variables than in constants. Why do days grow longer and shorter? Why are some crimes reported and not others? What *causes* people to commit burglaries? Does group solidarity really *cause* people to conform?

When we say that something is the cause of something, we are saying that something makes something else happen. A **cause** is anything producing a result, an effect, or a consequence. Put another way, in causal relationships variations in one variable cause variations in another variable.

It will be helpful to distinguish between **independent** and **dependent variables.** If we think something might be the cause of something else, we say that *the cause is the independent variable* and that *the consequence (or the thing that is being caused) is the dependent variable.* To help you remember the difference, think that variables being caused are dependent on the causal variable, while causal variables are not dependent but are independent.

CRITERIA OF CAUSATION

How can we demonstrate that some variable is the cause of another variable? To demonstrate causation, three tests or criteria must be met. When *any one* of these is not met, no cause-and-effect relationship can exist between variables.

TIME ORDER The first test involves the very simple principle of **time order:** A cause must occur *before* its effect. Put another way, the principle of cause and effect makes no sense backward. Suppose you claim that eating green apples made you sick. That's entirely reasonable, as long as you ate the apples *before* you got sick. If you already were sick when you ate the apples, the apples could not be the cause of your illness. Or to claim that the divorce of their parents often causes kids to misbehave, it is necessary to show that they didn't begin misbehaving until *after* the divorce. Or to argue that loss of friends causes mental illness, it is necessary to show that people first lost their friends and then became mentally ill. The claim of causation is refuted if it turns out that people first became mentally ill and then lost their friends.

CORRELATION If something is the cause of something else, then it *must* be the case that they *vary in unison.* Changes in the cause must produce changes in the proposed effect. When variables vary or change in unison, they are **correlated.** Suppose you turn a knob on the dashboard of your car and your music plays louder or softer depending on which direction you turn the knob. The position of the knob and the volume of your sound are correlated. But if you turn the knob and the volume does not change, you probably are turning the wrong knob because there is no correlation.

Correlations can be *either* positive or negative. The higher people's incomes, the more TV they watch is a positive correlation—as one variable increases, the other increases, too. But the higher people's incomes, the less alcohol they drink is a negative correlation—as one variable increases, the other decreases. Either negative or positive correlations can reflect causation.

NONSPURIOUSNESS Two variables often appear to be in a cause-and-effect relationship when in fact they are correlated *only* because each is correlated with some third, unobserved or unnoticed variable. If you went to any grade school in the world and measured the height of each student and then gave each of them a reading comprehension test, you would find a very strong positive correlation between height and test scores: The taller children would have the higher reading scores. Cause and effect? Hardly. The taller kids are *older* and older kids read better! Variations in age cause variations in both height and reading ability and thus produce a correlation between the two.

Correlations such as between height and reading are called **spurious relationships.** They appear to reflect causation, but they don't. Thus, to conclude that a cause-and-effect relationship exists, social scientists must make sure that they are not examining a spurious relationship.

Let's look at a second example. The first day of class, a sociologist often administers a questionnaire to all of the students in his very large introductory sociology course. Among the questions is one asking students if they have shoplifted and, if so, how often and how recently. After they have answered all of the questions, students are asked to trace their hand on the last page. Invariably, the results show that the larger their hand, the more often students will have shoplifted and the more recently they have done so. Do big hands cause shoplifting (perhaps by making it easier to palm things)? Hand size has nothing to do with it. Males tend to have larger hands, and males are far more likely than females to shoplift, as will be discussed at length in Chapter 7. But within each gender there is no correlation between hand size and shoplifting. Thus, although hand size and shoplifting are correlated, this is a spurious, not a causal, correlation. This is depicted in Figure 3-1.

ON ABSTRACTIONS

All scientific concepts are **abstractions;** that is, they are ideas, not things. We can see many animals that belong to that class of living organisms called mammals, but we can't see or touch the concept of mammal itself—it exists only in our minds. Nor can we see or touch the concept of a group, although we can observe countless sets of persons to whom it applies. The concept is always more general than any actual set of objects to which it applies. Hence, the concept of a group includes all sets of persons having the characteristics of stable and durable social relations, and it applies not only to all such groups that exist but also to all that have ever existed, will exist, or could exist.

Because theories are made up of concepts, theories are also abstractions. Consequently, no matter how often the empirical predictions and prohibitions of a theory are found to be accurate, we can never, ever, say for sure that the theory is true. This is because a theory, like concepts, applies not only to the past and the present but also to the future. Unless we run out of future, we never exhaust the opportunities for the theory to fail—it always remains possible that tomorrow or next week or ten centuries from now empirical predictions from the theory will be wrong. That limit applies to *all* scientific theories, including such basics as the "law" of gravity. That law predicts that if you throw this book up into the air, it will come down again. Should the book stay suspended in the air or just fly away, we could be sure that the law of gravity needed fixing. That has not happened yet (as far as we know), but it always *could*. The influential philosopher of science Rudolf Carnap (1953) explained that no "universal sentence" such as a "law of physics or biology" ever can be proved, for "the number of instances to which the law refers . . . is infinite and therefore can never be exhausted." However, when a theory has successfully survived many tests of its predictions, we gain so

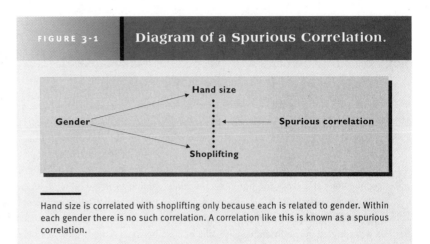

FIGURE 3-1   Diagram of a Spurious Correlation.

Hand size is correlated with shoplifting only because each is related to gender. Within each gender there is no such correlation. A correlation like this is known as a spurious correlation.

much confidence in it that we regard it as true. But we always must be aware that scientific truth is provisional—true only until further notice, never certainly true.

Since a theory is an abstraction, we can't see or touch it. How, then, can we connect it to the empirical world in ways that make it subject to disproof? That is, how do we connect the abstract to the concrete? Through hypotheses, as explained in Chapter 1. That is, we apply theories to specific instances and then see if predictions from the theory are supported. If the results are contrary to the theory, then questions arise as to its accuracy or adequacy, and we may be forced to discard or amend the theory. But even when many hypotheses from a theory have been supported, the theory is not proven. On the other hand, theories that survive the most frequent and most severe efforts to prove them false are those in which we place greatest confidence. Therefore, the duty of all scientists is to do their best to falsify a theory, to find instances when its predictions and prohibitions do not come true. In this way they contribute to a selection process whereby erroneous and incomplete theories are weeded out.

Hypotheses not only tell us where to look and what to expect to see; within moral and practical limits, they tell us *how* to look. "How to look" refers to how to do the necessary research, how to collect the data or facts, and how to analyze them. In the remainder of the chapter, we will examine the fundamentals involved in two research methods used to test microtheories. In subsequent chapters you often will encounter each as we let you look over the shoulders of good social scientists while they work.

BOX 3-1	CORRELATION

A CAUSE AND ITS EFFECT must be correlated—they must vary in unison. Put another way, as the independent variable takes different values, the values of the dependent variable must change in response. For example, because economic development brings improved medical care and public health programs, it should cause an increase in life expectancy. If that is true, then we ought to find a correlation between economic development and life expectancy—that the higher a nation's level of economic development is, the higher will be its life expectancy. But how can we discover whether development and life expectancy are correlated? We could start by selecting as our cases the 172 nations in the world having populations of at least 200,000. Then we could examine each and tabulate its level of economic development and its life expectancy. But then what? Social scientists use a very simple device called a **correlation coefficient** to tell them whether and how strongly any two variables are correlated—in this case whether economic development and life expectancy vary in unison.

A correlation coefficient consists of a number between 0.0 and 1.0. There is no need for you to know how correlation coefficients are calculated—there is software that does it automatically. But it will be helpful to know how to interpret the results. If variations in the independent variable are unrelated to variations in the dependent variable, the value of the coefficient is 0.0. If every change in the independent variable is matched by an equal change in the dependent variable, the coefficient is 1.0. Since the world isn't perfect, correlations of precisely 1.0 or 0.0 are rare. So, that leaves the problem of assessing the meaning of correlations above 0.0 and below 1.0. To do this, sociologists rely on tests of significance, which calculate the odds that a correlation of a given size (and based on a given number of cases) could have occurred randomly (by a fluke). Some actual examples will show you how easy it is to interpret correlations.

Table B-1 shows two sets of correlations. The upper set shows correlations between economic development and life expectancy as well as two other variables. As predicted, there is a very strong correlation of .80 (or 0.80) between development and life expectancy. This coefficient is quite close to 1.0, and if we calculated a test of significance, we would discover that the odds are less than 1 in 1,000 that this result is a random fluke. So, the first test of causation is met: Economic development *is* correlated with life expectancy.

Now look at the second correlation. Notice that it is preceded by a minus sign (−.79). That alerts us to the fact that correlations can be either positive or negative and still reflect causation. Economic development causes life expectancy to rise, but it also causes fertility to decrease. Why this happens will be explained in Chapter 18; here it is sufficient to see that variables can be correlated not only because as one rises the other rises but also because as one decreases the other rises. The third correlation coefficient in the table shows that economic development and the area of nations in square miles are not correlated—the odds are very high that the coefficient of .12 is nothing but a random fluke and should be interpreted as if it were 0.0.

Now, let's shift our focus to life expectancy and ask: What factors cause it to increase or decrease? The second set of correlations in the table offers some very interesting health advice. Life expectancy is positively correlated (.52) with the consumption of alcoholic beverages, with cigarette smoking (.72), and with the number of television sets per 1,000 population (.65). This suggests that if you want to live longer you should drink, smoke, and watch TV. But these are *spurious* correlations. Each of these variables is quite strongly correlated with economic development and therefore each is correlated with life expectancy, but this is not cause and effect. To the contrary, among nations having equal levels of economic development, smoking and drinking are negatively correlated with life expectancy.

TABLE B-1	Correlation Coefficients Based on 172 Nations
ECONOMIC DEVELOPMENT	
Life expectancy	.80
Fertility rate	−.79
Area	.12
LIFE EXPECTANCY	
Alcohol consumption	.52
Cigarette consumption	.72
TVs per 1,000	.65

Source: Prepared by the author from *Nations of the Globe*, MicroCase Archive, 2003.

In the remainder of the book, you often will see correlation coefficients. Those shown will all be significant. And none of them will be spurious (so far as is known).

THE EXPERIMENT: STUDYING GROUP SOLIDARITY

Both symbolic interactionists and exchange theorists stress the power of groups to shape the individual. As we learn to take the role of the other and to see ourselves as others see us, we learn to "fit in"—to behave in ways that earn us the affection and respect of our associates. To the extent that everyone in a group likes everyone else, the group sustains a high level of solidarity. The cost of high solidarity is intense pressure for conformity. So, when people disagree with other members of the group or act in other than the approved ways, they risk the affection of others and, ultimately, their own self-conceptions.

As competent human beings, we all know that these are the facts of social life, but it is the business of social science to make sure that what "everybody knows" is correct. Moreover, these general principles aren't very subtle. To what *extent* will people yield to group solidarity? How do *variations* in solidarity influence conformity? Will people even respond to solidarity in aggregates, as opposed to groups—or how fast can solidarity arise?

These are but a few of the considerations that have motivated an immense variety of studies of solidarity and conformity. The most famous and influential of these was an experiment conducted on college students more than fifty years ago by Solomon Asch (1952). Since then the "Asch experiment" has been repeated many times with many variations, all confirming his original results. Perhaps the best way to appreciate this study is to pretend that you were one of the students who took part.

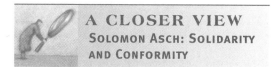

A CLOSER VIEW
SOLOMON ASCH: SOLIDARITY AND CONFORMITY

You have agreed to take part in an experiment on visual perception. Upon arriving at the laboratory, you are given the seventh in a line of eight chairs. Other students taking part in the experiment sit in each of the other chairs. At the front of the room, the experimenter stands by a covered easel. He explains that he wants you to judge the length of lines in a series of

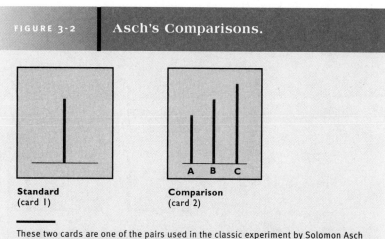

FIGURE 3-2 | Asch's Comparisons.

Standard (card 1)   Comparison (card 2)

These two cards are one of the pairs used in the classic experiment by Solomon Asch to measure conformity to group pressure. Subjects were asked to pick which of the three lines on the second card matched the line on the first card. The correct choice is very easy to see. But when everyone else said the line on card 1 matched line A on card 2, many subjects would agree.

comparisons. He will place two decks of large cards on the easel. One card will display a single vertical line. The other card will display three vertical lines, each of a different length. He wants each of you to decide which of the three lines on one card is the same length as the single line on the other card. To prepare you for the task, he displays a practice card. You see the correct line easily, for the other lines are noticeably different from the comparison line.

The experiment begins. The first comparison is just as easy as the practice comparison. One of the three lines is obviously the same length as the comparison line, while the other two are very different. Each of the eight persons answers in turn, with you answering seventh. Everyone answers correctly. On the second pair of cards, the right answer is just as easy to spot, and again all eight subjects are correct. You begin to suspect that the experiment is going to be a big bore.

Then comes the third pair (Figure 3-2). The judgment is just as easy as before. But the first person somehow picks a line that is obviously wrong. You smile. Then the second person also picks the same obviously wrong line. What's going on? Then the third, fourth, fifth, and sixth subjects answer the same way. It's your turn. You know without doubt that you are right, yet six people have confidently given the wrong answer. You are no longer bored. Instead, you are a bit confused, but you go ahead and choose the line you are sure is right. Then the last person picks the same wrong line everyone else has chosen.

FIGURE 3-3   **The Asch Experiment.**

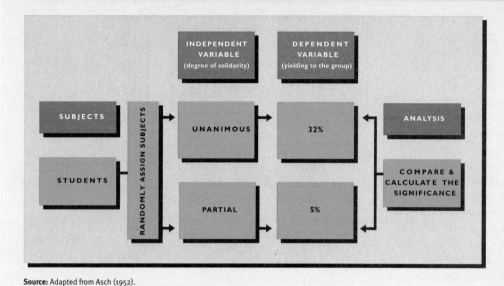

The precision of an experiment rests on the ability of researchers to control what happens, when, and to whom. Here we see the experimenter can randomly determine which situation a subject confronts. Subjects in one situation faced a unanimous group, whereas those in the other had support for the correct choice from one group member. Since subjects were randomized as to which group they confronted, the two groups of subjects should be alike in all ways. Therefore, any differences on the dependent variable ought to be caused by the independent variable. A test of significance is used to determine whether an observed difference is large enough to be more than a meaningless, random fluctuation. In this example the large difference observed here is highly significant.

Source: Adapted from Asch (1952).

A new pair is unveiled, and the same thing happens again. All the others pick an obviously wrong line. The experimenter remains matter-of-fact, not commenting on right or wrong answers but just marking down what people pick. Should you stick it out? Should you go along? Maybe something's wrong with the light or with your angle of vision. Your difficulty lasts for eighteen pairs of cards. On twelve of them, all the others picked a line you knew was incorrect.

When the experiment is over, the experimenter turns to you with a smile and begins to explain. You were the only subject in the experiment. The other seven people were stooges paid by Professor Asch to answer exactly the way they did. The aim of the experiment was to see if social pressure could cause you to reject the evidence of your own eyes and conform.

However, in other performances of the experiment, not all of the other subjects faced the same situation you did. Half of the time, one other person also picked the correct line; hence, subjects in this situation had some support for disagreeing with the group. Put another way, *solidarity* was the *independent variable* in Asch's experiment, taking two values: unanimous and partial. Applied to this situation, Homans' theory gives rise to the hypothesis that a higher proportion of people will yield and support the incorrect judgment of the group to the extent that the group is unanimous—when solidarity is complete. And that's exactly what Asch discovered.

Figure 3-3 depicts the Asch experiment. It reveals that **experiments** have two fundamental features: The researchers are able to (1) *manipulate* the independent variable (make it vary as much as they wish, whenever they wish) and (2) *randomly assign* persons to groups exposed to different levels of the independent variable. People who take part in an experiment often are referred to as the **subjects** because they are subjected to different values of the independent variable.

MANIPULATING THE INDEPENDENT VARIABLE

To test his hypothesis, Asch first recruited seven students to pretend to be subjects in the experiment and trained them how to behave during the experiment. Signals were arranged so that the student confederates would know which line to select in each comparison, and one person was selected to pick the correct line when signaled to do so. Then Asch recruited other students to be the actual subjects. As each showed up to take part in the experiment, Asch used random means (he flipped a coin) to see whether he or she would face a unanimous group (1) or the group having only partial solidarity (2). If group solidarity does cause conformity, then people in Group 1 ought to be more likely than people in Group 2 to yield. And that's exactly what happened. When faced with a unanimous group, a third of the subjects yielded to the group more than half of the time (and 75 percent yielded at least twice out of the twelve incorrect comparisons). But when one person broke ranks and the group was not unanimous in support of an incorrect choice, only 5 percent of subjects yielded half of the time or more.

The experiment is the most powerful research design because it so elegantly meets all three criteria of causation. *Time order* is certain—people knew how six of seven members of the group answered *before* they were asked for their judgment. *Correlation* is easily established: Either yielding shifted as solidarity shifted or it did not. But the real power of the experimental method comes from its capacity to rule out sources of *spuriousness*.

RANDOMIZATION

The key to preventing spurious experimental findings lies in the random assignment of subjects. **Randomization** involves the use of chance procedures to determine who will be exposed to which value of the independent variable. Sometimes experimenters do what Asch did and just flip a coin to assign subjects to a group; more typically, they use a table of random numbers. As is true of any set of human beings, the subjects taking part in experiments differ in many ways. The purpose of randomization is to make the *groups* seeing each combination *alike* so that they will include the same mix of individual characteristics. Thus, each group should include the same percentage of tall people, of "nonconformists," and so on. Of course, randomization does not guarantee that the groups will be *exactly* alike. Simply by chance there will be some variations between groups, and these variations will be smaller as the groups grow larger. However, when groups are assigned randomly, it is possible to compute the probability that any difference observed on the dependent variable was caused by variation in the independent variable as opposed to only a meaningless, random difference. Such a computation is called a test of significance.

SIGNIFICANCE

A **test of significance** specifies the odds that a given correlation (or a difference between groups in an experiment) was caused by random variations. Significance tells us when to take an observed result as real and when to reject it as a fluke. Two factors enter into calculations of significance. First is the *number* of subjects in each group—the larger the groups, the less likely that results are random flukes. Second is the size of the observed difference or correlation—the larger it is, the less likely it is to have been caused by a fluke. Usually, social scientists require that the odds be at least 20 to 1 against a random fluke before they take a result seriously, and they often will want the odds to be at least 100 to 1.

To sum up: The essential elements of all experiments are the same. The experimenters can **manipulate** the independent variable—that is, make the independent variable take whatever value they wish while holding everything else constant. *Randomization* holds everything else constant (the age, race, education, and other aspects of individuals) by making the composition of the groups the same. For this reason social scientists ought to use the experimental design whenever possible.

Unfortunately, it is impossible to use experiments to study much that social scientists must, or wish to, study. It is impossible to manipulate such important independent variables as sex, race, age, and religion. We can't randomly assign people to be male, white, thirty-seven, and Catholic or to be female, Asian, sixty-two, and Baptist. We must take these variables as they occur in the world. Moreover, there are many things we could manipulate but must not. We must not test theories of child-rearing practices by randomly deciding that some children will be neglected or abused and then comparing them with children who were not. So, social scientists are forced to use a number of nonexperimental research designs.

FIELD RESEARCH: STUDYING RECRUITMENT

Social movements are groups of people who organize to cause or prevent social change. Aside from everything else, the fate of social movements depends on their ability to attract participants. As we will see in Chapter 21, the Montgomery bus boycott that ignited the Civil

Rights Movement in the 1950s not only needed leaders and financial supporters; it also needed to recruit thousands of African American commuters to stay off the city buses. The Feminist Movement of the 1970s and 1980s had so much impact because it enlisted so many women (and some men, too). Once people join a social movement, they can be located within the network of participants, and as we have seen, such a network can be examined for potential factions, the density of ties, and the scope of the network. But such an examination fails to reveal why and how people join a social movement. How do successful movements recruit members?

Most people would answer this question on the basis of ideological attractions. An **ideology** is a connected set of strongly held beliefs based on a few very abstract ideas or principles, such as constitute religions or philosophies. Once people have adopted an ideology, usually through socialization, it then serves as a filter to screen out beliefs and proposed actions that do not fit and to accept opinions and proposed actions that are consistent with the ideology. It is widely believed that people join movements whose views they find to be consistent with their ideology—that people join the Feminist Movement, the Civil Rights Movement, or indeed, anti-Civil Rights organizations such as the Ku Klux Klan because the beliefs and purposes of the group appeal to them. Thus, some people became feminists because of books they read or ads they see or feminist speakers they hear. Perhaps some people do join movements in this way, but most don't. As we will see in Chapter 15, most people do not have an ideology and most who do *learned* it after joining an ideologically based movement. But if this is so, why *do* people join movements?

Most of what sociologists now know about how and why people join social movements has been learned from field studies of conversion to unusual religious groups. But subsequent studies show that these findings apply to nonreligious groups, too, as will be discussed in Chapter 21. So let's see how movements really recruit members.

A CLOSER VIEW
LOFLAND AND STARK:
RELIGIOUS CONVERSION

Why do Americans, many of them from conventional religious backgrounds, suddenly become Hare Krishnas or members of the Unification Church or go to live in an ashram and study yoga with a guru from India?

When I began graduate school, that was something I wondered about. During several years as a newspaper reporter, I had been assigned to cover stories about a number of exotic new religions, including several based on revelations said to have been brought to earth by friendly aliens in flying saucers. During my first week of classes at the University of California, Berkeley, I met John Lofland, who was also interested in conversion to new religions. Much had been written on this topic over the years, but we found this work unconvincing. For one thing nearly all of it seemed to have been researched in libraries because there was very little to suggest that the authors had ever met any members of the groups they claimed to explain.

There was widespread agreement among these authors, however, that people joined a new religion because of the correspondence between the tenets of the religion and the problems suffered by those who joined. The research procedure each scholar seemed to have followed was to study the ideology of a group to answer the question: What does this faith promise to do for people? Having determined this, they next asked: To whom do such promises most appeal? Thus, for example, if a new religious movement, such as Christian Science, claimed the ability to cure illness, it seemed likely that it would most appeal to those with chronic illnesses or physical disabilities. Having deduced who ought to join a particular religious movement, these scholars seemed content to conclude that in fact those were indeed the people who actually did join.

But was that true? Lofland and I were not so sure. Moreover, we suspected that ideological appeals were emphasized as the cause of conversion primarily because that's about all you can study about religious groups in the library. In the library you can't watch anyone join a religious movement or even observe members' activities. About all you can find are books and articles written by the group to explain their ideology and works written by others to attack or criticize that ideology. But what really happens, we wondered? The only way to find out was to go into the world and look, applying the methods of field observation research.

Field research involves going out to observe people as they engage in the activities that the social scientist wants to understand. It is called field research because the research is conducted in the field—in the natural settings in which the people and activities of interest are normally found.

So Lofland and I began looking around the San Francisco Bay Area for a group to observe. We wanted one that was new and growing but still small enough so that the two of us could closely observe most of what went on. After much hunting we found exactly what we wanted. We discovered a group of about a dozen people who had just come to San Francisco from Eugene, Oregon, where they had been recruited to be the first American members of a new religion that had begun in Korea.

The group was led by Dr. Young Oon Kim, a Korean woman who had once been a professor of religion at Ewha University in Seoul. She had been sent to America to seek converts for a religious movement founded by Sun M. Moon. Moon was trained as an electrical engineer, and one day he became convinced that God had chosen him to start a new form of Christianity that would unite all of the many competing denominations. He was quite successful in attracting converts in Korea, and so after a few years, he dispatched some missionaries to other nations. Dr. Kim and her young followers were the very first American members of the Unification Church. Although this group was tiny and unimpressive when Lofland and I found it, and was still very small when our study was finished, it subsequently has received a great deal of press coverage, most of it quite negative, in which members are referred to as "Moonies," after their founder. However, members prefer to be called "Unificationists."

Lofland and I began attending the group's religious services, which were held in the evening in the apartment building owned and occupied by the group. After a few weeks, we were sure that this group was exactly what we needed, and so we settled in to watch as new people came in contact with the group. What we hoped to do was identify factors that determined who among these people did or did not convert. At this point we shifted from *covert* to *overt* observation.

COVERT AND OVERT OBSERVATION

Our initial contact with the group was the result of accepting an invitation to come and hear the new religious message. While we made no secret about being sociologists, Lofland and I said nothing about studying the group. Hence, we were **covert observers** in that the Unificationists did not know we were studying them. For example, we waited until we were away from the group before recording our observations. The advantage of covert observation is that it minimizes **observer effects**—changes people make in their behavior when they know they are being studied. The Unificationists treated us the same way they treated other people whom they wanted to convert because that's how they viewed us. Consequently, our presence did not disturb normal patterns of behavior. The disadvantage of covert observation is that the observers must act as normal participants, too, and this can interfere with the research. For example, covert observers are very limited in how inquisitive they can be. In this instance, as potential converts to a new religious group, we couldn't conduct interviews or openly take notes about what was happening (we went into the john to write down things we didn't want to forget). **Overt observers** face none of these limitations. Because their role is known, they can conduct interviews, and they need not act like potential members.

Consequently, we sat down with Dr. Kim and explained that we wanted to do a study of the group to understand the process of conversion. We were careful to be sure she understood what that would involve: We would want to interview people at length and to attend meetings, and we would eventually publish our results. Dr. Kim told us to go ahead. If nothing else, she explained, the study might serve as a detailed historical record of the early years of the movement in America. Because members already knew us quite well, the shift to overt observation went smoothly. Moreover, since Lofland and I had observed a lot of normal behavior as covert observers, we were confident that our presence as known observers wasn't causing people to act differently.

CONVERSION AS CONFORMITY

As time passed Lofland and I reached the surprising conclusion that people did not join the Unification Church as the result of a religious search—most converts had not even been all that interested in religion before their conversions. Instead, we discovered that all of the current members were united by close ties of friendship predating their contact with Dr. Kim. Indeed, Dr. Kim's first converts constituted an intense local network, as shown in Figure 3-4.

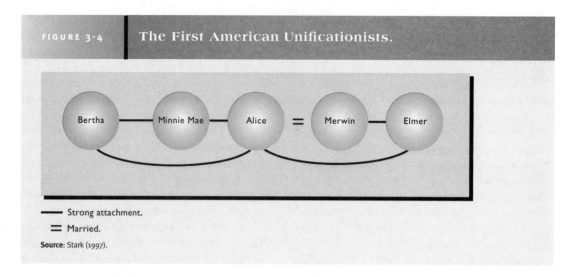

FIGURE 3-4   **The First American Unificationists.**

——— Strong attachment.

═══ Married.

Source: Stark (1997).

Bertha, Minnie Mae, and Alice (not their real names) were young housewives who were next-door neighbors. Merwin was Alice's husband and Elmer was his buddy from work. This network encountered Dr. Kim after she rented a basement room in the home of Alice and Merwin. Eventually, several other people with ties to this network joined, but by the time Lofland and I arrived to study them, the group had never succeeded in attracting a stranger.

We also found it interesting that although all the converts were quick to describe how their spiritual lives had been empty and desolate prior to their conversion, many claimed they had not been particularly interested in religion before. As Merwin told me, "If anybody had said I was going to join up and become a missionary I would have laughed my head off. I had no use for church at all."

It is instructive that during most of her first year in America, Dr. Kim had tried to spread her message directly by talks to various groups and by sending out many press releases. Later, in San Francisco the group also tried to attract followers through radio spots and by renting a hall in which to hold public meetings. But these methods yielded nothing. As time passed Lofland and I were able to observe people actually convert. The first several converts were old friends or relatives of members who came from Oregon for a visit. Subsequent converts were people who formed close friendships with one or more members of the group.

Eventually, Lofland and I realized that of all the people the Unificationists encountered in their efforts to spread their faith, the only ones who joined were those *whose interpersonal ties to members overbalanced their ties to nonmembers.* In

effect conversion is not primarily about seeking or embracing an ideology; it is about bringing one's religious behavior into alignment with that of one's friends and family members. *Conversion is an act of conformity.* But then, *so is nonconversion.* In the end it is a matter of network ties.

Becoming a Unificationist violates general norms defining legitimate religious affiliations and identities, and to join could cost someone his or her friends. Lofland and I saw many people who spent some time with the Unificationists and expressed considerable interest in their doctrines but never joined. In every instance these people had many strong ties to nonmembers who did not approve of the group. Their failure to join was their failure to shift networks. Of persons who did join, many were newcomers to San Francisco whose attachments were all to people far away. As they formed strong friendships with group members, these were not counterbalanced because distant friends and families had no knowledge of the conversion in process. In several instances a parent or sibling came to San Francisco intending to intervene after having learned of the conversion. Those who lingered eventually joined up, too. Keep in mind that becoming a Unificationist may have been regarded as deviant by outsiders, but it was an act of conformity for those whose most significant attachments were to other members.

Since Lofland and I first published our conclusion that attachments lie at the heart of conversion and therefore that conversion tends to proceed along social networks, our study has been replicated many times. In science **replication** means to repeat earlier research to guard against incorrect findings or findings that occurred merely by chance. Replications also extend the range of the findings, since they are done in different times and places than the original research. In this instance many others have found that conversion is a network process, basing their research on an immense variety of religious groups all around the world. A study based on Dutch data (Kox, Meeus, and 't Hart, 1991) cited twenty-five additional empirical studies, all of which supported the initial finding. And that list was far from complete (Stark and Finke, 2000).

A second aspect of conversion is that people who are deeply committed to any particular faith do not go out and join some other faith. Thus, Mormon missionaries who called upon the Unificationists were immune, despite forming warm relationships with several members. Indeed, the Unificationist who previously had "no use for church at all" was more typical. Converts were not former atheists, but they were essentially unchurched, and many had not paid any particular attention to religious questions. Thus, the Unificationists quickly learned that they were wasting their time at church socials or denominational student centers. They did far better in places where they came in contact with the uncommitted. This finding has also received substantial support from subsequent research. Converts to new religious movements are overwhelmingly from relatively irreligious backgrounds. The majority of converts to modern American cult movements report that their parents had no religious affiliation (Stark and Bainbridge, 1985). This may be stated as a theoretical proposition: *New religious movements mainly draw their converts from the ranks of the religiously inactive and discontented and those affiliated with the most accommodated (worldly) religious organizations.* Expressed in network language, people are more apt to join a religion if people in their network are neutral about or uninterested in religion. This also is to acknowledge that *if* people have strong ideological commitments, it will limit the degree to which they are open to other ideologies.

Had social scientists not gone out and watched people as they converted, this point might well have been missed entirely because when people retrospectively describe their conversions, they tend to put the stress on theology (or ideology). When asked why they converted, Unificationists invariably noted the irresistible appeal of the Divine Principles (the group's scripture), suggesting that only the blind could reject such obvious and powerful truths. In making these claims,

converts implied (and often stated) that their path to conversion was the end product of a search for faith. But Lofland and I knew better because we had met them well before they had learned to appreciate the doctrines, before they had learned how to testify to their faith, back when they were not seeking faith at all—when most of them regarded the religious beliefs of their new set of friends as quite odd.

You will notice that Lofland and I could not be nearly so precise about meeting the criteria of causation in our study as we could have through an experiment. Of course, there was no way we could randomly assign people to have or not have Unificationist friends and then sit back to await conversion. Nevertheless, we could meet causation criteria to some extent. First, by keeping careful records on each person who came into contact with the Unificationists, we established a correlation between attachment and conversion. No one ever joined without having close attachments. Second, because we observed the group over a considerable period of time, we were able to demonstrate time order. The friendships occurred before the conversions. Indeed, in many cases the friendships had already existed for years.

To demonstrate nonspuriousness is a much more difficult problem in nonexperimental research. We could do little to ensure that people who had friendships with Unificationists weren't different in other significant ways from those without such friendships. We didn't notice such differences, but they might have existed. Hence, there is always an element of risk in accepting nonexperimental research findings. In Chapter 4 we shall examine some statistical procedures that help to decrease the possibility of spuriousness in nonexperimental studies, but these usually do not apply to the observational variety of nonexperimental research.

FOUR PRINCIPLES OF RECRUITMENT

From the findings of our study of the Unificationists, four principles may be formulated about recruitment by social movements, whether or not they are religious.

The first can be called the **open-minded principle:** *It will be very hard to recruit someone who is strongly tied to a network wherein most people are opposed to the ideology of the recruiting movement, but recruitment does not depend on people*

| BOX 3-2 | HEAVEN'S GATE |

BEGINNING IN 1975 Herff Applewhite and Bonnie Nettles, who then called themselves Bo and Peep (as in Little Bopeep), began to seek followers to a new religion based on an exotic mixture of Christianity and belief in UFOs: Heaven exists in outer space and UFOs travel back and forth; Jesus came to Earth on a UFO and "ascended" to heaven on another one. Bo and Peep's primary message was that soon a small group of persons who accepted their leadership would be able to make the journey to heaven, too, as a UFO would be sent by God to pick them up. In pursuit of followers, Bo and Peep held a series of public meetings in various towns and cities, mostly in the Far West. At each meeting they offered members of the audience the opportunity to join—but new recruits had to make up their minds immediately and be ready to leave town within six hours. Those who took up the invitation found themselves on the road, moving daily and camping out at night. During the two years that Bo and Peep devoted to these recruitment meetings, about 200 people accepted their invitation. But their followers numbered only thirty to forty at any given moment, as most who joined didn't stay long.

© AP/Wide World Photos

Within the first year of their missionary efforts, Bo and Peep had been spotted by Robert Balch, a sociologist at the University of Montana, who began to do field research on them. During the first several years, he was assisted by David Taylor, an undergraduate student, also at the University of Montana. Eventually, Balch's study stretched over twenty years and produced many publications (Balch and Taylor, 1976, 1977; Balch, 1980, 1982, 1985, 1994).

As the years passed, the group changed in many ways. They ceased traveling and formed a commune, settling first in Denver, then in Texas, and eventually in a rural compound near Manzano, New Mexico. Bo and Peep changed their names to Do and Ti (pronounced as in music: "doe" and "tea"). They ceased seeking members. Attempting to overcome their sexual urges (which were believed to interfere with the ability to achieve sufficient perfection), some of the men, including Do, had themselves castrated. In 1985 Ti died of cancer. Their numbers slowly declined until there were only twenty-four in 1994.

Suddenly, in 1994 Do regained the urge to recruit. He bought a full-page ad in *USA Today* explaining the group's beliefs and embarked on a twenty-two-state recruitment tour. Having relocated in Rancho Sante Fe, California—an expensive community north of San Diego—the group went

---

having any prior attitudes favorable to the ideology of the movement.

Second is the **network recruitment principle**: *Books, articles, advertisements, and speeches may serve many important functions, such as boosting the solidarity and confidence of members, but recruitment is primarily a person-to-person phenomenon.*

A third principle follows from the second and can be called the **cosmopolitan growth principle**: *The opportunities for the growth of a movement are maximized when movements can establish ties to cosmopolitan networks having many nonredundant ties.* This is another example of the strength of weak ties.

But a very difficult paradox about the growth of social movements is revealed by a final principle, the **principle of dense origins**: *Social movements tend to originate within dense local networks.* The reason is that it is far easier to gather a founding nucleus of a movement on the basis of very strong ties. During her first year in America, Dr. Kim made many acquaintances, most of whom found her a pleasant, likable person. But she failed to gather any followers until she built very strong ties to several people who already were part of a local network. That is, it is much easier to found a movement when a faction or clique already exists and can be collectively enlisted in the cause. This means, of course, that to succeed, movements must find a way to escape from the limits imposed by a local network to link up with cosmopolitan networks. It is here that people in bridge positions become vital to the future of social movements.

into business creating websites for various organizations and businesses. In search of new members, they created their own home page where they explained why they soon would be departing aboard a "spacecraft from the Level Above Human." By now they referred to their group as Heaven's Gate. On March 26, 1997, San Diego sheriff's deputy Robert Brunk discovered that Do and thirty-eight of his followers had indeed "gone," leaving their bodies behind. Do was sixty-six. Most of the followers were well past forty (having joined during their twenties). There were eighteen men and twenty-one women.

Isn't this group an exception to the principles of recruitment? They did not recruit on the basis of networks or through the formation of interpersonal attachments. Instead, they used public meetings and required an almost instant response from strangers. That their tactics were entirely contrary to what sociology now knows about recruitment is precisely why the group was such a relative failure. In the end only thirty-nine people "left" because Bo and Peep had incorrect notions about how to recruit. Instead of trying to build bonds to others, and especially to a network of others, they asked potential recruits to in effect run away with strangers (as Bo and Peep were themselves runaways from their families and professions). Consequently, the only people who accepted their offer were loners and runaways—people without attachments or who wished to be free of their attachments, as did several women who abandoned husbands and children to go on the road with Bo and Peep. Moreover, the use of public meetings that promised to reveal "the real story behind UFOs" limited their audiences to people who already shared some of Bo and Peep's basic ideas. That is, unless they already were interested in UFOs, people were very unlikely to attend. Thus, the pool of potential converts was very small—there just aren't a lot of loners and potential runaways who are into UFOs. And even they require direct, interpersonal contact, however brief, as the lack of results from the national newspaper ad and the website demonstrated.

© Corbis/Sygma

Herff Applewhite.

CONCLUSION

This chapter has introduced the fundamental building blocks of microsociological theory. We have seen that humans must interact and exchange with one another and that they therefore tend to form attachments and conform to one another's expectations. In this way norms arise to guide and structure our behavior. Throughout the rest of the book, we shall see how this elementary theoretical proposal serves as the basis for elaborate theories from which a great variety of hypotheses can be derived.

This chapter also more closely looked at portions of the sociological process, especially those steps devoted to research. We saw that the experimental method excels in satisfying the criteria of causation, which is why it is the common research method in fields that do not require human subjects: Not only do bacteria not blush when they are being observed, but they also have no legal rights and elicit no sympathy. However, because sociologists study people, we often cannot use the experimental method. In this chapter we examined an observational field study and saw how it attempted to approximate the precision of experiments. In the next chapter, we shall encounter more sophisticated nonexperimental forms of research. We shall also encounter macrosociological theories.

Review Glossary

Terms are listed in the order in which they appear in the chapter.

socialization The process by which culture is learned and *internalized* by each normal member of a society—much of which occurs during childhood.

internalization The process by which people accept the norms, values, roles, beliefs, and other primary aspects of their culture to the point that these are a fundamental basis for all of their decision making.

rational choice proposition The proposition that within the limits of their information and available choices, guided by their preferences and tastes, humans will tend to maximize.

altruism Unselfish actions done entirely for the benefit of others.

social interaction The process by which humans seek to influence one another.

symbolic interaction A microsociological theoretical approach that, like exchange theory, regards interaction among human beings as the fundamental social process but places far more emphasis on how people influence one another and communicate than it does on the exchange process as such. In addition, symbolic interaction stresses the social construction of meaning. The theory rests on three essential propositions:

1. Human beings act toward things (including people) on the basis of the meanings that the things have for them.

2. The meaning of things is derived from, or arises out of, social interaction.

3. The meanings of things are handled in, and modified through, an interpretive process used by persons in dealing with things they encounter.

symbol Something that stands for or indicates something else. A cross worn on a chain around a person's neck is often a symbol indicating that person is a Christian. We know that the cross is not a Christian but merely stands for (symbolizes) being a Christian. Notice, too, that the word *Christian* also is but a symbol. The word itself is not a Christian; it conveys instead the idea of what it means for someone to be a Christian.

looking glass self The process by which our sense of self develops through interaction as we come to see ourselves as others see us—humans serve as mirrors for one another.

mind Our *understanding of symbols.* According to Mead, the mind arises entirely through repeated interaction with others—a learning process that in infants is largely a matter of trial and error.

self Our learned understanding of the responses of others to our conduct. Through long experience in seeing others react to what we do, we not only gain a central notion of who we are but also are able to put ourselves in another's place. Mead called this *taking the role of the other.* From doing this repeatedly, we form a generalized notion of others—of what they want and expect and how they are likely to react to us.

exchange theory A microsociological theoretical perspective having as its central concern to explain and reveal the key processes by which people seek to maximize by exchanging rewards with one another. Some basic propositions of the theory are:

1. If an exchange between two persons is voluntary, an exchange will not take place unless both partners believe they will benefit from it.

2. In the absence of any restraints, cheating is the expected behavior of exchange partners since, if they are successful, cheaters maximize.

3. When one exchange partner fails to receive the expected level of rewards, he or she will be angry.

4. Over time people tend to develop stable patterns of exchange with regular exchange partners.

5. Stable patterns of exchange relationships tend to generate emotions and sentiments such as trust, admiration, affection—even love.

law of liking Participation in common activities (exchanges) causes people to like one another.

attachment A stable and persistent pattern of exchange between two people when positive sentiments are among the "commodities" exchanged. That is, attachments involve bonds of liking or affection between two people.

law of inequality Emotional attachments among members of a group will be weaker among members of different ranks than among those of similar rank.

law of agreement The more the members of a group like one another, the more apt they are to agree with each other.

law of conformity The more intense the group solidarity is, the more intense will be the demand for conformity.

constant Something that never changes (or varies).

variable Something that changes (varies) in that it can take more than one value.

cause Something that makes something else happen. A cause is anything producing a result, an effect, or a consequence. To demonstrate causation, three criteria, or tests, must be met: **time order, correlation,** and **spurious relationships.** Failure to satisfy *any one* of these tests means that no cause-and-effect relationship can exist between the variables in question.

independent variable Something we think might be the cause of something else.

dependent variable The consequence (or the thing that is being caused). To help you remember the difference, think that variables being caused are dependent on the causal variable, while causal variables are not dependent but are independent.

time order The sequence in which variables occur and a vital test of causation. A cause must occur *before* its effect. This is simply to recognize that the process of cause and effect cannot work backward in time.

correlation Variation in unison, another test of causation. For something to be the cause of something else, the two must be correlated, or vary in unison. Causes must produce changes in their supposed effects.

spurious relationship A false correlation between a proposed cause and its effect—for instance, both the "cause" and its "effect" may be the result of some third factor.

abstractions Ideas or mental formulations that are apart from the concrete, material world. "Beauty" is an abstract mental formulation that exists only in our mental responses to the world rather than as a quality of the things observed, a fact that is acknowledged in the common saying "Beauty is in the eye of the beholder."

correlation coefficient A measure of the degree of correlation between two variables; it varies from 0.0 (no correlation) to 1.0 (perfect correlation).

experiment A method wherein researchers are able to (1) *manipulate* the independent variable (make it vary as much as they wish, whenever they wish) and (2) *randomly assign* persons to groups exposed to different levels of the independent variable.

subjects Persons on whom an experiment is based; they are subjected to the experiment.

randomization The use of chance procedures to determine who will be exposed to which value of the independent variable. Sometimes experimenters do what Asch did and just a flip a coin to assign subjects to a group; more typically, they use a table of random numbers.

test of significance A calculation of the odds that a given correlation (or a difference between groups in an experiment) was caused by random variations (or flukes).

manipulate In experimental designs controlling the value or level of the independent variable to which subjects are exposed.

social movements Organizations created to cause or prevent social change (see Chapter 21).

ideology A connected set of strongly held beliefs based on a few very abstract ideas or principles, such as constitute religions or philosophies. Once people have adopted an ideology, usually through socialization, it then serves as a filter to screen out beliefs and proposed actions that do not fit and to accept opinions and proposed actions that are consistent with the ideology.

field research Going out to observe people as they engage in the activities that the social scientist wants to understand; so called because the research is conducted in the field—in the natural settings in which the people and activities of interest are normally found.

covert observers Field researchers who are operating without the knowledge or permission of those who are being studied.

observer effects Changes that people make in their behavior when they know they are being studied.

overt observers Field researchers who are operating with the knowledge and permission of those who are being studied.

replication Repetition of earlier research to guard against incorrect findings or findings that occurred merely by chance.

open-minded principle (of social movement growth) It will be very hard to recruit someone who is strongly tied to a network wherein most people are opposed to the ideology of the recruiting movement, but recruitment does not depend on people having any prior attitudes favorable to the ideology of the movement.

network recruitment principle Books, articles, advertisements, and speeches may serve many important functions, such as boosting the solidarity and confidence of members, but recruitment is primarily a person-to-person phenomenon.

cosmopolitan growth principle The opportunities for the growth of a movement are maximized when movements can establish ties to cosmopolitan networks having many nonredundant ties. This is another example of **the strength of weak ties.**

principle of dense origins Social movements tend to originate within dense local networks.

Suggested Readings

Barker, Eileen. 1984. *The Making of a Moonie—Brainwashing or Choice?* Oxford and New York: Basil Blackwell.

Blau, Peter M. 1964. *Exchange and Power in Social Life.* New York: Wiley.

Hechter, Michael, ed. 1983. *Microfoundations of Macrosociology.* Philadelphia: Temple University Press.

Homans, George C. 1974. *Social Behavior: Its Elementary Forms.* New York: Harcourt Brace Jovanovich.

Phelps, Edmund S., ed. 1975. *Altruism, Morality, and Economic Theory.* New York: Russell Sage Foundation.

Popper, Karl R. 1959. *The Logic of Scientific Discovery.* New York: Basic Books.

Robbins, Thomas. 1988. *Cults, Converts and Charisma.* Beverly Hills, Calif.: Sage.

Stark, Rodney, and Lynne Roberts. 2002. *Contemporary Social Research Methods.* 3rd ed. Belmont, Calif.: Wadsworth.

Sociology Online

www.socstark10.com

GO TO THE INTERNET AND TYPE: www.socstark10.com.

↗**CLICK ON:** 2000 General Social Survey

✓**SELECT:** ATTEND: How often do you attend religious services?

↗**CLICK ON:** Analyze Now
The open-minded principle of recruitment suggests that a new religious movement will best be able to attract members from among those who are religiously less active. If that is so, then should such movements be more or less able to attract:

___Older people or ___Younger people

___High school dropouts or ___College educated

___Whites or ___African Americans

___Higher income people or ___Lower income people

___Southerners or ___Westerners

↗**CLICK ON:** Select New Data Set

↗**CLICK ON:** The 50 States

✓**SELECT:** %NO RELIGION: Percentage of the population who say they have no religion.

↗**CLICK ON:** Analyze Now
Does there seem to be a substantial regional pattern?

Suppose you wanted to start a new religious movement. Based on the open-minded principle, what five states would you consider as the most promising places to locate your group? What five would be the least promising (do not consider Alaska and Hawaii, as no data are available for them on this measure).

↗**CLICK ON:** Select New Data Set

↗**CLICK ON:** Nations of the Globe

✓**SELECT:** CH. ATTEND Percentage who attend religious services once a month or more.

↗**CLICK ON:** Analyze Now
Based on the open-minded principle, in what nation should people be the most receptive to new religious movements? Has there been anything in the news during the past year or two to suggest that is the case?

USING INFOTRAC COLLEGE EDITION

GO TO THE INTERNET AND TYPE:
www.InfoTrac-college.com

↗**CLICK ON:** Register New Account
You will be asked to enter your Access Code and to create and enter a User Name and Password. Having done so,

↗**CLICK ON:** InfoTrac College Edition or Log On.

These search terms will let you evaluate some popular articles about conversion and cult groups based on the discussions in this chapter. You may wish to consult discussions in Chapter 14 as well.

SEARCH TERMS:
Conversion
Cults

CHAPTER

4

Macrosociologists are not concerned with the details of human interaction but rather seek to discover and explain larger social structures. From the vantage point of macrosociology, human beings recede to small dots that form larger patterns in the same way that the dots on a TV screen create a complete picture. This metaphor is brought to life as 12,500 U.S. soldiers and army nurses form an American eagle for the camera of Arthur S. Mole, who perched atop a seventy-five-foot ladder. It was December 1918. World War I had ended a few weeks before, and everyone shown here was waiting for his or her discharge papers.

Macrosociology: Studying Larger Groups and Societies

Because there were very few American Unificationists at the time Lofland and I studied them, it was easy for us to make accurate generalizations about them. For example, while the first members had all been neighbors in Eugene, Oregon, each had been born and raised in another state and all were relative newcomers to Oregon. To make that statement, all Lofland and I had to do was run a dozen people through our memories (or look them up in our field notes). Twenty years later, when Eileen Barker, a sociologist at the London School of Economics, began her study of the Unification Church in Great Britain, she had to be able to describe more than 2,000 British Unificationists. To make generalizations about them, such as that most had attended college, she couldn't just consult her memory or even her field notes (Barker, 1984). Instead, Barker relied on questionnaires filled out by all of the members.

Whenever sociologists study larger populations, they can't depend on the informal techniques common in field research. Since this chapter is about the sociology of larger units of analysis, we will first examine why and how research is based on samples. Then we will explore the most common method used to study data collected from samples, *survey research*. As we do so, you will see how sociologists attempt to eliminate spurious correlations when they are unable to use the experimental method. Next the chapter explores the borderline between micro- and macrosociology as revealed by *contextual effects*, wherein surveys produce different results depending on when or where they are conducted. The remainder of the chapter is devoted to *comparative sociology*—the name applied to the sociology of larger groups and whole societies. We start by watching a sociologist test hypotheses using "communes" as the units of analysis. This sets the stage for the *sociology of societies*. We examine three major approaches to macrosociological theories—theories about how societies operate and why they contain certain features and arrangements. The chapter concludes by

CHAPTER OUTLINE

DESCRIBING POPULATIONS
Simple and Proportional Facts
Box 4-1: Experimental Effects
of Social Structure
Censuses and Samples
Why Sampling Works

**SURVEY RESEARCH: IDEOLOGY
AND CONFORMITY**

A CLOSER VIEW
Hirschi and Stark: Hellfire and
Delinquency

Spuriousness
Conflicting Results

CONTEXTUAL EFFECTS

NETWORK ANALYSIS: GROUP PROPERTIES

A CLOSER VIEW
Benjamin Zablocki: Networks of Love
and Jealousy

THE SOCIOLOGY OF SOCIETIES

SOCIETIES AS SYSTEMS
Sociocultural Components
Interdependence among Structures
Equilibrium and Change

THEORIES ABOUT SOCIAL SYSTEMS
Functionalism
Social Evolution
Conflict Theory

**COMPARATIVE RESEARCH:
VIOLENCE AND MODERNITY**
Comparative Research
Cross-Cultural "Samples"

A CLOSER VIEW
Napoleon Chagnon:
Life along the Amazon

A CLOSER VIEW
Jeffery Paige: Conflict
and Social Structures

CONCLUSION

REVIEW GLOSSARY

SUGGESTED READINGS

SOCIOLOGY ONLINE

watching two social scientists study the basis for violent conflict in preliterate societies.

DESCRIBING POPULATIONS

Suppose you were a sociologist who wanted to study burglary. To begin, you would want to know as many facts as you could about this crime, about those who commit it, and about its victims. Fact gathering often is referred to as **data collection** in that researchers use various means to collect or gather the facts or data needed to answer specific questions about the world. In social science there are essentially two ways to collect data: direct observation and reports from others. That is, we can go out and look for ourselves and try to discover what's going on or we can ask people to tell us what's going on (or read their reports about what went on). In either case we must recognize the distinction between simple and proportional facts.

SIMPLE AND PROPORTIONAL FACTS

A **simple fact** is an assertion about a concrete, quite specific, and limited state of affairs, often merely the claim that something happened or exists and usually having to do with only one or very few cases. Some examples will help:

■ Some homes in Los Angeles were burglarized last year.

■ Laurie Olson's house was burglarized, but she did not report it to the police.

■ Kevin Wilson confessed to thirty burglaries, each of which took place within a mile of his own home.

It usually is quite easy and inexpensive to establish simple facts. Almost anyone in Los Angeles knows of a recent burglary. A casual conversation with Laurie Olson would be sufficient to learn of her burglary and that she hadn't bothered to report it. Reading Kevin Wilson's confession would reveal that he worked close to home.

A **proportional fact** asserts the *distribution* of something, or even the joint distribution of several things, among a number of cases. Questions requiring proportional facts include:

■ How *many* burglaries occurred in Los Angeles last year and what *proportion* of all homes were burglarized?

BOX 4-1   EXPERIMENTAL EFFECTS OF SOCIAL STRUCTURE

SUPPOSE THAT YOU HAVE VOLUNTEERED to take part in a discussion group organized by two of your professors. When you arrive, they show you to a booth equipped with a microphone. They explain that to let each participant remain anonymous, the discussion will take place over an intercom system. When it is your turn to talk, a light will come on and your mike will be open. Five other people will take part in the discussion, which will be about problems in adjusting to college. After explaining this, the professors leave.

The discussion begins, and each person takes a turn. Suddenly, one of the others, who had already mentioned that he had been a bit scared about going to college because he sometimes has epileptic seizures, begins to stammer and breathe hard into the microphone. He gasps out, "Help!" Then there is a crash, like a body falling to the floor. You can't speak to the others because your mike is not on. You don't know where the professors have gone, but they promised not to listen in. What do you do?

What you are most likely to do depends on *how many other people you believe are part of the discussion group*. In the actual experiment, no one else was really present. All the other voices were simply tape recordings, including that of the young man who had the seizure. When John Darley and Bibb Latané (1968) did this experiment, some subjects were told that they would be one of two people in the discussion. Others were told there were three, and still others were told there were six (Figure B-1). When subjects thought that they were part of a two-person group—therefore, they alone knew the young man was having a seizure—*all* of them quickly left the booth to seek help. But when subjects thought it was a three-person group—and thus one other person was also aware of the emergency—only 80 percent left the booth to seek help, and they also took longer to respond. Subjects in the six-person situation were even slower to seek help, and only 60 percent attempted to do so.

What is going on here? Clearly, these differences in the apparent willingness of people to help someone else are not due to differences of individual character. Because people were randomly assigned to groups of different sizes, the three groups should have contained the same mix of personality types and other individual characteristics. What is going on here is *social structure*. Quite beyond individual characteristics are characteristics of the groups—such as size—in which we exist, and these group features operate as structures that influence our behavior in the same way that physical structures—such as the position of doorways—do. All characteristics of groups are social structures.

In this instance Darley and Latané were conducting research on *group size* as a social structure. They had hypothesized that the larger the group believed to be present, the less an individual will feel personal responsibility to act in an emergency. And that is precisely what the results of their experiment showed.

FIGURE B-1   The Experiment.

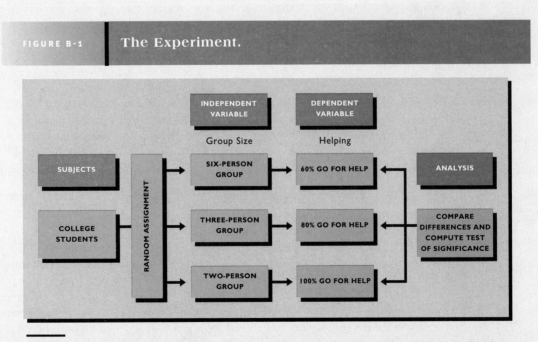

Random assignment of subjects ensures that, although each subject differs from all others, the three groups of subjects will be alike (except for random variation). For this reason perceptions of group size, not something else, must have caused the differences among these three groups on the dependent variable—going for help. As the perceived size of the group declined, the proportion who went for help increased.

Simple fact: John has a pet cat.

- What *percentage* of burglaries are reported to the police?
- Do *most* burglars strike close to home?

Informal conversations and observations *cannot* provide proportional facts. The only way sociologists can accurately describe social life in terms of proportional facts is to observe or otherwise rate or classify *every* case or a representative *sample* of all cases.

CENSUSES AND SAMPLES

As defined in most dictionaries, a census is an official count of the population and the recording of certain information about each person. But in social science the term is used more broadly: A **census** is a collection of data from *all cases* (or units of analysis) in the relevant set. Thus, a census of Fargo, North Dakota, would involve obtaining information on every single inhabitant of that city. A census of high schools in Texas would involve obtaining information (such as enrollment, graduation rate, grade average, and so on) from every high school in the state. Or if one obtained a copy of every book ever published on the lives of every person who had a speaking part in an American movie between 1930 and 1940, that too would be a census.

Put another way, it's a census as long as "everybody" is included, whether that be "everybody"

in the smallest village in Mexico, "everybody" taking chemistry at the University of Mississippi, or "everybody" in Canada. But it's also a census if "every*thing*" is included—all college football teams, all chapters of Delta Gamma, or all prisons.

If we wanted to know how many people in Los Angeles were burglarized last year, we could find out by conducting a census to determine whether or not each household in L.A. had been burglarized. If in addition we wished to know what percentage of burglaries had been reported to the police, we could include that in our census study. Whenever we are able to collect accurate data from each case, then our description *must* be accurate. That is, if we were able to interview everyone in Los Angeles and if they all gave us correct answers to our questions about being burglarized, then our calculation of the percentage who had been burglarized would be precise.

The trouble with censuses is that they are so expensive. It costs billions of dollars to conduct a census of the United States once every ten years. If a census were the only reliable way to obtain proportional facts, very little sociological research would get done. Fortunately, it is not necessary to collect data about everybody or everything to achieve an accurate description of a set of cases. Instead, we can select relatively few cases from the whole set (or population), obtain data on each of these cases, and assume that these results apply to the entire set. When the Gallup Poll wants to know how many Americans drink tea, they don't have to ask everyone—it's enough to ask only 1,000 or so. This is called *sampling*. The logic involved is that if the people included in the sample are like those who are not included, then the results from the sample will be like the results we would get by including everyone.

WHY SAMPLING WORKS

A **sample** consists of a set of cases (or units) *randomly selected* from the entire set of cases or units to be described. **Random selection** means that *all* cases have an equal (or at least a known) probability of being included in the sample. When people call a TV station or a newspaper to participate in an "opinion poll" on some issue, they are not a random sample—many people have no probability of being included. In fact, such studies are referred to as SLOPS[1] (self-selected listener opinion polls). No matter how many people take

---

1. The acronym is attributed to Norman Bradburn, longtime director of the National Opinion Research Center, at the University of Chicago.

part in a SLOPS study, the results have no credibility. One of the largest SLOPS ever done was based on more than 5 million Americans who sent back postcards in 1936 indicating their choice in the upcoming presidential election. The results of this "poll" by *The Literary Digest* showed that the Republican challenger, Alf Landon, would receive over 60 percent of the vote while the incumbent Democrat, Franklin D. Roosevelt, would lose with less than 40 percent. As it turned out, it was Roosevelt who received more than 60 percent of the vote. In contrast, that same year the Gallup Poll very accurately predicted the election on the basis of interviews with about 1,400 Americans.

Let's select a random sample of Detroit. First, put the name of each resident of Detroit on an individual metal tag and then put all the tags in a big barrel. Then spin the barrel until the tags are well mixed up. Now, draw out 1,000 tags. You have a random sample of Detroit. Everyone's tag was in the barrel so everyone had an equal chance of being selected. In practice, survey researchers don't actually draw names to obtain a random sample, but the basic logic always is the same—that everyone has an equal chance to be picked. For example, people who conduct surveys by telephone know that many people have unlisted numbers. So they use automatic dialing machines that randomly select telephone numbers from all possible numbers within a particular area code (the area codes also are selected at random from the list of all area codes). This ensures that listed and unlisted numbers have an equal chance of being called.

Sampling works *only* because the laws of probability can be applied to random selections of cases. This is exactly the same application of randomization as used in experiments. In experiments the reason for randomization is to make sure that people in various treatment groups (levels of the independent variable) are the same. In sampling the object is to make sure that those included in the sample are the same as those left out. Even so, samples are never *exactly* like the entire population from which they are selected. However, the *larger the sample, the more accurate it will be*. Suppose we selected a sample of the United States and stopped after we had interviewed only two people. Chances are that they both would be white, and there is a good chance that both would be female. However, if we continue to randomly select Americans, as the total grows the odds will begin to even out and our sample will increasingly begin to resemble the total population as to race and gender and other characteristics.

© Jeff Zaruba/Stone/Getty Images

What's important is the absolute size of the sample, the total number of cases included, *not* the ratio of the sample to the population. What that means is that a sample including 1,000 cases will give as accurate a picture of the population of California as it will of the population of Green Bay, Wisconsin. However, a sample including 2,000 cases will be more accurate than one including 1,000 cases, and a sample having 25,000 cases will be extremely accurate. Researchers prefer samples with at least 1,000 cases, and 1,500 cases is typical of good survey studies. When you encounter reports of opinion polls, they often will mention that the results are accurate plus or minus some number of percentage points. For example, a poll may indicate that 31 percent of Americans favor some political action, plus or minus 3 percentage points. What that means is that 95 percent of the time the true percentage in favor of this action will not be higher than 34 percent (31 plus 3) or lower than 28 percent. This calculation is based entirely on the size of the sample—the larger it is, the smaller the misrepresentation of the true value.

Proportional fact: There is a pet cat in 48 percent of the homes in this community.

SURVEY RESEARCH: IDEOLOGY AND CONFORMITY

In Chapter 3 we discovered that most people do not join social movements out of ideological motives. Instead, in the usual case, a person's ideology—a system of general beliefs—has only a passive effect on recruitment, serving as a filter that prevents some people from joining particular movements but not often playing an active role in attracting recruits. Does this mean that ideology is not a major factor in human behavior more generally? It certainly would seem likely that religious ideologies, for example, often must enter into the moral choices made by individuals. Keeping in mind that even obvious "facts" need to be checked out, let's watch.

A CLOSER VIEW
HIRSCHI AND STARK: HELLFIRE AND DELINQUENCY

For several years Travis Hirschi and I were on the staff of the Survey Research Center at the University of California, Berkeley. One afternoon, sitting on the lawn, Hirschi mentioned that there was virtually no published research on the effects of religion on delinquency. People had explored an incredible range of possible causes of delinquency, but no one had demonstrated that kids who believed in doctrines of sin and hellfire were less likely to commit crimes. "Actually," he added, "there isn't even research showing whether church or Sunday school attendance matter." This seemed particularly odd since many theories of crime and delinquency assume that there is a strong link between religious ideology and conformity. Reflecting on this, I suggested that maybe lots of research had been done, but journal editors had thought the results too obvious to be worth publishing.

One thing led to another, and we decided to go back inside and use his huge survey of high school students to see if hellfire does deter delinquency. As will be seen, it was a rather fateful decision.

The method Hirschi and I used is called **survey research.** It has two identifying features. First, it is based on a sample of the population to be studied. Second, the data are collected by *personal interviews* or by having each individual complete a *questionnaire*. Thus, if you want to

know how religious people are, you ask them a series of questions about what they believe and do. If you want to know about their delinquency, then you ask them questions about violations they have committed.

Obviously, problems can arise about whether people answer truthfully and whether they are even able to give accurate answers. For example, research has demonstrated that people do not report their TV viewing habits very accurately. To get good information about this subject, it is necessary to have people keep daily logs on what they watch. However, an immense amount of effort has gone into perfecting survey research techniques, and the data so obtained can be quite reliable when researchers employ proper techniques and safeguards.

Hirschi and I proceeded to investigate the effects of religion on delinquency on the basis of questionnaires filled out by a sample of students enrolled in junior high schools and high schools in Richmond, California, a city of about 100,000 people across the bay from San Francisco.

Our first concern was to see if there was a correlation between religion and delinquency. To do this, we first separated the data according to two categories of students: those who attended church frequently (at least once a month) and those who did not. Then we sorted each of these groups into delinquents (those who had recently committed two or more delinquent acts) and nondelinquents (those who had recently committed no more than one delinquent act). The results are shown in Table 4-1, which lists the percentages of delinquents and nondelinquents within each religious category. The findings clearly seem to support our hypothesis. Frequent churchgoers are much less likely to be delinquent (22%) than are infrequent churchgoers (38%). Thus, our results fulfilled the correlation criterion of causation (see Chapter 3). But what about the other two criteria?

It is much more difficult to establish *time order* for questionnaire data than for data from experiments. Perhaps people become delinquents and then become infrequent churchgoers. However, other research shows that patterns of church attendance among teenagers primarily reflect family religious patterns, which are usually established well before a child was born. So, it is reasonable to assume that this study met the criterion of time order.

But what about the criterion of nonspuriousness? Obviously, delinquent and nondelinquent teenagers are likely to differ in many ways other than church attendance. Could one of these other uncontrolled factors cause a spurious correlation between church attendance and delinquency? The answer turned out to be yes.

SPURIOUSNESS

Although there had been no significant prior research on the connections between religion and delinquency, much had been devoted to each of these topics. Therefore, it was well known that boys are much more likely than girls to be delinquent. It was equally well known that girls are more likely than boys to attend church (see Chapter 14). So, we knew we had to examine the possibility that sex differences were the real cause of the correlation in Table 4-1. We used a simple technique to check this out. First, we divided the sample into males and females. Then, we examined the relationship between church attendance and delinquency separately for males and females.

The results are shown in Table 4-2. There we can see that boys who attend church are no less likely to be delinquent than boys who do not (50% of both groups are delinquents). The same holds among females: Ten percent of the girls who attend church frequently are delinquents, as are 10 percent of the girls who do not.

Thus, we must conclude that the correlation found in Table 4-1 is spurious: Religion does not cause people to refrain from delinquent behavior. We know a relationship is spurious if it disappears when some third variable is controlled. As a check, we tried many other measures of religiousness, including belief in heaven and hell, Sunday school attendance, and even parents' church attendance. None of these was correlated with delinquency either when sex differences were controlled.

Hirschi and I were astonished at these results. After all, we had set out to test something that everyone knew to be true. We had not even been sure the study was worth the time and trouble. In fact, had the data not already been available to us, we probably would not have bothered with it. However, what we found turned a lot of what everyone had believed about the world upside down . . . for a while.

After our findings were published (Hirschi and Stark, 1969), most sociologists accepted our results, and the paper was frequently cited and often reprinted. Within several years the "knowledge" that religion fails to guide teenagers along the straight and narrow was enshrined in

TABLE 4-1	Church Attendance and Delinquency	
CHURCH ATTENDANCE	**FREQUENT (%)**	**INFREQUENT (%)**
Delinquent	22	38
Not delinquent	78	62
	100%	100%

Source: Adapted from Hirschi and Stark (1969). The results shown here have been modified and simplified for clarity. However, they accurately reflect patterns in the actual data.

TABLE 4-2	Controls for Sex Reveal a Spurious Relationship	
CHURCH ATTENDANCE	**FREQUENT (%)**	**INFREQUENT (%)**
Boys		
Delinquent	50	50
Not delinquent	50	50
	100%	100%
Girls		
Delinquent	10	10
Not delinquent	90	90
	100%	100%

Source: Adapted from Hirschi and Stark (1969). The results shown here have been modified and simplified for clarity. However, they accurately reflect patterns in the actual data.

© Ellis Herwig/Stock Boston

Will these children be less likely to commit juvenile crimes because they attend church? The answer to that question depends on *where* this picture was taken. If it was taken in a city where the majority of citizens belong to a church, which is the case in most North American cities, then going to church greatly reduces the probability of juvenile delinquency. But if it is in a city like those along the West Coast, where the majority do not belong to a church, then church attendance seems to have no influence on delinquency.

undergraduate textbooks. But then problems began to turn up as other researchers tried to *replicate* our research.

CONFLICTING RESULTS

Several years after our study was published, two other scholars replicated it with a sample of teenagers from several cities in the Pacific Northwest (Burkett and White, 1974). They, too, could find no religious effects on delinquency. While no one could explain why religion did not influence delinquency, it still seemed that it did not.

But then four more studies yielded very different results. The first, based on a sample of teenagers in Atlanta (Higgins and Albrecht, 1977), found a very strong negative correlation between church attendance and delinquency—exactly what Hirschi and I had expected to find. The second, based on teenagers living in six wards (congregations) of the Mormon Church (Albrecht, Chadwick, and Alcorn, 1977), found the same thing. At this point a study that had gone unnoticed came to light (Rhodes and Reiss, 1970). Based on a sample of Nashville students, it also reported a substantial negative correlation between church attendance and delinquency. Meanwhile, two of my students and I analyzed data from a huge study based on Seattle, and we could find no religion effect (Stark, Kent, and Doyle, 1982).

Does religion inhibit delinquency or not? The research score stood at three to three—three studies said yes, three said no. To make the confusion worse, the conflicting findings could not be blamed on poor research. Each study was well done. To just shrug and say that sometimes religion inhibits delinquency and sometimes it doesn't is not an adequate scientific response. So I wondered why, and wondered some more.

CONTEXTUAL EFFECTS

Some years later I realized that these findings had a consistent pattern and were a striking example of a contextual effect. A **contextual effect** exists when a relationship found among individuals is conditional upon social contexts, when different results occur in different social surroundings. Contextual effects mark the borderline between micro- and macrosociology.

Table 4-3 displays a clear example of contextual effects. It has long been "known" that married people tend to be happier than unmarried people. This difference has shown up consistently in surveys done in the United States during the past fifty years. As can be seen by reading across the first line of the table, 46 percent of married Americans say they are "very happy," while only 33 percent of the unmarried gave this response. The same is true for all ten of the Western nations shown in the table. These differences all have a very high level of statistical significance (recall Chapter 3). In all ten nations, the odds are greater than 1,000 to 1 against these being random results.

Much has been written to explain the link between marriage and happiness. Among the many reasons cited are the fact that married people have greater emotional security, higher incomes, and more frequent sex, while unmarried people more often suffer from loneliness and an uncertain future. Sounds convincing, even obvious, right?

But to everyone's surprise, as data from non-Western nations became available, it was discovered that in many places married people are not happier than unmarried people.

In each of the nine non-Western nations shown in the table, unmarried people are as happy or happier than married people. So, now we know that the effect of marriage on happiness *depends* on *context*. Results vary depending on where one looks. Of course, the next question is, Why? Why aren't married people happier than unmarried in the nine non-Western nations? What's different about married life in the United States and Canada, for example, from married life in India and Pakistan? Sociologists don't yet have an answer.

Meanwhile, the relationship between religion and delinquency also varies by context. As shown in Table 4-4, the three studies that failed to find a relationship were based on data from communities along the West Coast, while all three studies done further east found a strong effect. Here, too, the question is, Why? What's different about the sociocultural context of the West Coast that prevents religion from deterring delinquency? The answer is that *religion sustains conformity to the moral order* only as religion is a *vital aspect of group solidarity*. That is, for religion to deter delinquent behavior, not only must individuals accept religious beliefs but these beliefs must also be sustained and reinforced by those with whom they interact. Let me spell this out and then apply it to this particular contextual effect.

Teenagers form and sustain their interpretations of norms in day-to-day interaction with their friends. If most of a young person's friends are not actively religious, then religious considerations will rarely enter into the process by which norms are accepted or justified. Even if the religious teenager does bring up religious considerations, these will not strike a responsive chord in most of the others. This is not to suggest that nonreligious teenagers don't believe in the norms or discuss right and wrong but that they will do so without recourse to religious justifications. In such a situation, the effect of the religiousness of some individuals will be smothered by group indifference to religion, and religion will tend to become a very compartmentalized component of the individual's life— something that surfaces only in specific situations such as at Sunday school and in church. In contrast, when the majority of a teenager's friends are religious, then religion enters freely into everyday interactions and becomes a valid part of the normative system.

So, why doesn't religion affect delinquency along the West Coast? Because the most striking feature of American religion is an "unchurched belt" running along the shores of the Pacific from California through Washington (also including Alaska and Hawaii). In these states only about a third belong to a local church, while in the rest of the nation church membership is around 60 percent (in Utah, Rhode Island, and North Dakota, three-fourths belong). Studies that found no religious effect on delinquency were done in states having very low church membership rates, while those showing an effect were done in states with relatively high rates. Table 4-4 also shows the church membership rate for each state during the decade the study was conducted. The pattern is perfect.

Recently, I was able to replicate this contextual pattern using a national sample of high school students (Stark, 1996b). Because the sample was so large (11,995), it was possible to examine the effect of religion on delinquency in *each* of the five major census regions. In the East, Midwest, South, and Mountain regions, religion has a strong, negative impact on delinquent behavior—the more often young people attend church, the less likely they are to get in trouble with the law. But in the Pacific region, there is no effect! Students who go to church every week

are no less likely to get in trouble than are those who never attend.

So, that's what a contextual effect looks like and how it operates. And the existence of contextual effects shows that if we want to understand human behavior, it is vital that we understand groups.

TABLE 4-3	A Contextual Effect	
"Taking all things together, would you say you are very happy, quite happy, not very happy, or not happy at all?"		
	VERY HAPPY (%)	
	MARRIED	NOT MARRIED
United States	**46**	**33**
Canada	**50**	**37**
Great Britain	40	27
Australia	51	34
Belgium	50	31
France	37	27
Ireland	50	32
Sweden	44	25
Denmark	53	36
Finland	32	13
Non-Western Nations		
India	25	27
Pakistan	16	27
Bangladesh	14	16
Iran	23	23
Indonesia	15	21
Philippines	38	38
Tanzania	53	59
Nigeria	67	67
Zimbabwe	20	20

Source: Prepared by the author from the World Values Survey, 2000–2001.

TABLE 4-4	Studies of Religion and Delinquency: Results, Location, Church Membership		
	RELIGIOUS EFFECT?	LOCATION OF SAMPLE	CHURCH MEMBERS (%)
Hirschi and Stark	No	California	37
Burkett and White	No	Oregon	36
Stark, Kent, and Doyle	No	Washington	32
Higgins and Albrecht	Yes	Georgia	60
Albrecht, Chadwick, and Alcorn	Yes	Utah	76
Rhodes and Reiss	Yes	Tennessee	61

NETWORK ANALYSIS: GROUP PROPERTIES

Sociologists are fond of saying that "a group is more than the sum of its parts." By that they mean that groups have properties of their own that cannot be reduced to the characteristics of individuals. Put another way, if you examine each group member in terms of purely individual characteristics, you will learn little or nothing about, for example, the lines of communication and influence within the group. That requires knowledge of *relationships* among group members. And that's why many sociologists are so interested in social networks, for it is here that the properties of groups are most clearly revealed.

Network diagrams have been used in previous chapters to describe several group properties, including solidarity (Figure 1-4), inequality (Figure 1-1), conflict (Figure 1-5), and local and cosmopolitan structures (Figure 2-2). But there was little or no effort to make or test causal hypotheses linking one group property with another. **Network analysis** is a research method that has been developed specifically to test hypotheses about properties of social networks. It involves constructing measures of network variables and using networks as the units of analysis. The same criteria of causation apply as in all scientific methods. And as with any method, the best way to learn about it is to watch someone use it.

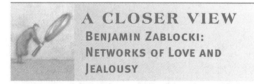

A CLOSER VIEW
BENJAMIN ZABLOCKI: NETWORKS OF LOVE AND JEALOUSY

From time to time, people decide to experiment with group living in pursuit of a more satisfying or exciting lifestyle, so they form a commune. A **commune**[2] is a group of people who organize to live together, often choosing to equally share duties, resources, and finances (Kanter, 1972). Typically, communes also attempt to live a distinctive lifestyle in accord with an ideology that sets them apart from the surrounding society. Therefore, communes are subcultures. In a sense communes reflect efforts by members to break with society and to create a new one in a nook or cranny of the host society.

People have been forming communes for at least 2,000 years (Zablocki, 1980). For example, cloistered religious orders are communes and have a long history in all major religions. As currently used, however, the word *commune* does not apply to religious orders affiliated with *conventional* denominations, but is applied to groups that observe unconventional religious teachings (at least unconventional in the society in which the group exists), such as groups that form around founders of new religions. Many such communes formed in the United States during the nineteenth century, as did many others based on radical political ideologies. But perhaps the most prolific period of commune formation in American history occurred during the late 1960s and early 1970s (Zablocki, 1980). Communes were an intrinsic part of the hippie drug culture of the time, as summed up in the popular slogan: "Turn on, tune in, and drop out." And in that era a lot of young people did just that. They dropped out of mainstream American society and dropped into a commune.

Some communes were in the remote countryside, where the group tried to be self-sufficient, sometimes even growing their own food. Others were urban communes, often located in a rented house, and everyone chipped in to pay expenses. Some communes were made up of hippies, who symbolized their distinctness from conventional manners by their long hair, beads, exotic clothes, and casual use of marijuana, peyote, LSD, and other psychedelic drugs. Other communes consisted of religious groups—in this era these often were followers of a guru from India (or someone pretending to be from India); some were formed by political radicals; some were followers of rock musicians. Despite the variation, all communes were attempts to get away from what members considered the materialist, aggressive values of "straight" society. Each in its own way was attempting to create an egalitarian new culture, so much so that they began to describe themselves as "countercultures" and to talk endlessly about breaking down the old norms (Zablocki, 1980; Berger, 1981; Bainbridge, 1997).

Among the old norms that had to go were those against drug use. But of even greater importance were norms restricting sexual relations to stable pairs and to being conducted in private. Why couldn't anyone and everyone have sex whenever and wherever they wanted? Hence, many communes attempted to maintain norms of "free love," or of "indiscriminate promiscuity" as Bennett Berger (1981) put it. As we will

---

2. The word *commune* comes from the Latin *communis*, meaning "common," and thus implies having or holding things in common. *Communis* also is the root for the words *communal*, *community*, and *communism*.

see, these communal sexual experiments showed that there probably are valid reasons for norms putting at least some limits on sexual options. And in fact, some communes chose norms limiting all sexual activity, opting for celibacy. However, whether committed to unlimited sexuality or to celibacy, nearly all communes in this era emphasized the ideal of love. Indeed, LOVE (or LUV, as the British rock groups put it) became the identifying cliché of the times.

Since northern California was the hotbed of hippiedom, of psychedelic rock, and of communes, it is no surprise that sociologists in that area led the way in studies of communes. The most influential study was directed by Benjamin Zablocki, then at the University of California, Berkeley, and now at Rutgers University.

Zablocki (1980) and a crew of his graduate students went out to see for themselves what really happens in communes. They tracked down many communes and spent time in each of them observing and asking questions—they even had each member fill out a brief questionnaire. Eventually, Zablocki obtained data on 120 communes and conducted in-depth studies of 60 of these.

One of his first discoveries was that communes differed quite a lot in how fully they achieved the ideal of mutual love. The ideology of the commune movement was that everyone should love one another, whether or not the loving relationship involved sexuality. But Zablocki observed that not only were some members of a commune loved more than others but that in some communes a higher *proportion* of members were linked by loving sentiments than in some other communes. These were not simply Zablocki's impressions. He was able to say *who loved whom* and just *how many members* loved others and were loved in return. And he did this by diagraming the network of love relationships in each commune.

Figure 4-1 depicts the loving networks of each of four communes. Each member of each commune was asked to rate her or his relationship with each other person in the group on a number of characteristics, including "tense," "loving," "jealous," "sexual," "hateful," and so on. Zablocki classified a relationship as loving only when *both* members of the pair so classified it. Although none of these communes was very large, I selected four of the smaller ones to keep the diagrams simple. I made up a name[3] for each commune consistent with the intensity of its

loving network, which is the percentage of all possible relationships in the group that are classified as loving.

Look first at Faction Flat. Here there are twenty-eight possible relationships, and of these only three, or 10.7 percent, are loving. This calculation enables us to say that love is rather lacking in this commune, and therefore, its network can be classified as a *cool* one. While this calculation is quite informative, it is not as informative as the diagram, which lets us see that the pattern of loving relationships reveals potential factions within the group. Person 1 has a loving relationship with both 5 and 6, but 5 and 6 do not love one another. Persons 3 and 8 have a loving relationship, but it is exclusive. Persons 2, 4, and 7 are loners when it comes to love. One might suppose that this group could be lacking in solidarity and have a relatively high potential for breaking up.

Things are a bit better in Line House, where 30 percent of all possible relationships are loving, earning a rating as a *warm* network. Moreover, the diagram reveals that four of the five members have a loving relationship, and these unite, rather than divide, the group (although Person 4 must feel rather left out).

The ideology of "love one another" is much more fully met by the Guru Group, where 57.5 percent of all possible relationships are loving, making this a *hot* network. But the network diagram is even more revealing since it displays the utter lack of factions within the group. Everyone has many loving relationships, except for Member 6 who has such a relationship only with Member 9. Member 9 has a loving relationship with everyone else except Member 10. Not sur-

Nothing is more characteristic of the hippie culture of the 1960s than a Volkswagen bus with an elaborate psychedelic paint job emphasizing LOVE, bearing the peace symbol, and indicating opposition to nuclear arms (ecological slogans came a bit later). The typical occupants were under age thirty, wore their hair long, and often expressed their "love" by flipping the finger to everyone they passed.

© James Marvy / The Stock Market

---

3. To protect the privacy of participants, Zablocki has never revealed the names of these communes.

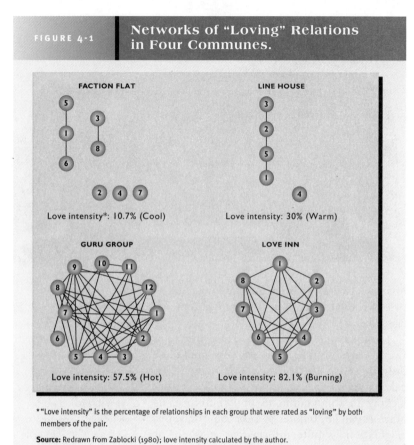

FIGURE 4-1   Networks of "Loving" Relations in Four Communes.

FACTION FLAT

Love intensity*: 10.7% (Cool)

LINE HOUSE

Love intensity: 30% (Warm)

GURU GROUP

Love intensity: 57.5% (Hot)

LOVE INN

Love intensity: 82.1% (Burning)

* "Love intensity" is the percentage of relationships in each group that were rated as "loving" by both members of the pair.

Source: Redrawn from Zablocki (1980); love intensity calculated by the author.

prisingly, Member 9 is the guru of this group—its founder and unquestioned leader. You probably would not be surprised to learn that Person 6 is a newcomer.

Love Inn is so named to reflect that 82.1 percent of all relationships in this commune are loving—a *burning* level of intensity. Unlike Guru Group, relationships in Love Inn are not centered on one member: Four members have seven love ties and none has fewer than four.

Based on these networks and what you have learned about group solidarity, it would seem probable that Love Inn would be the most durable commune and Faction Flat would be the least durable. That's what Zablocki expected, too. Imagine his surprise when he examined the data presented in Figure 4-2. Here the *independent variable* is love intensity, and each commune is placed in a category based on its love intensity score, from burning to cool. His *dependent variable* is stability, and Zablocki calculated two measures: the average membership turnover rate and the disintegration rate (based on a two-year period). To see the effect of intensity on stability, Zablocki read across each column, comparing

the turnover rates and the disintegration rates of each intensity level. To his astonishment, the *higher* a commune's level of love intensity was, the *less stable* the commune. Most burning groups were very short lived, and all experienced very high rates of member turnover (the percentage of members who left during the period of the study). In contrast, all of the cool communes survived, and the majority of their members stayed.

How can we explain this? Where there is love there is apt to be jealousy, and where there is a lot of love, there is apt to be a lot of jealousy. This is particularly the case when many of the loving relationships within a group also are sexual liaisons. Thus, what happens to burning communes is that in effect they get so hot they melt.

A number of studies have revealed the internal conflict produced in communes that attempt to follow the rule of "free love" (Jaffe, 1975; Berger, 1981). In principle everyone in the commune could have sex whenever and with whomever he or she wanted. But in reality some persons were more sexually popular than others. Typically, women who wanted to practice free love could have large numbers of partners, while men who wanted free love tended to find they weren't in as much demand as they hoped. There were more men than women who wanted to practice free love, so some guys were always leftovers. Despite the ideology of "love one another," in practice there wasn't enough time, enough energy, or the inclination to actually love everyone equally. Thus, although communes were based on the ideal that everyone would be equal, and although members tried to share all material things in common, they overlooked the fact that love, too, is a valuable "good" and that it is far harder to parcel it out equally than it is to give everyone the same clothing allowance.

Thus, many communes were so "full of love" that they burst, often in a spectacular fashion, leaving many bitter ex-members. In contrast, the groups that were the most durable tended to be those that minimized jealousy and emotional entanglements. Some of these began with strict limits on sexual relationships, and others turned to such norms on the basis of experience. In fact, some successful communes adopted norms of celibacy (Zablocki, 1980; Berger, 1981; Stark and Bainbridge, 1997). In regulating or prohibiting sex, of course, these modern communes followed the pattern of successful religious communes throughout history. And that finding sheds light on another question about the survival of communes.

A number of studies have shown that communes founded on a religious ideology are far more successful than those founded on a political or other secular ideology (Stark and Bainbridge, 1997). For example, Karen H. and G. Edward Stephan (1973) found that of 143 American communes founded between 1776 and 1900, nearly all of those based on political or secular ideologies died out within their first ten years of existence, while almost two-thirds of communes based on religious ideologies survived. In fact, no secular commune lasted forty years, while a third of the religious communes did so. For a long time, sociologists argued about *why* religious communes lasted longer. It now is recognized that a major factor has to do with the ability of religious communes to protect relationships within the group from the strains of jealousy and competition for love. As one former member of a California commune put it, "You don't know what grief is until you have to spend half of each day hassling about who you are going to sleep with that night and then maybe it doesn't happen." Or as a sociologist might put it, in hot and burning communes, the network diagram is different every day.

Zablocki continued to keep in touch with many of the respondents to his initial questionnaire and to track the fate of each commune. As of 1999, only 3 of the 120 communes still existed—2 of them religious.

THE SOCIOLOGY OF SOCIETIES

What many regard as the "big" sociological questions are about the largest units of analysis: *societies.* If microsociologists ask about the causes of conflicts within small groups, many macrosociologists ask about the causes of the largest and most brutal conflicts: wars. If microsociologists ask about group solidarity, macrosociologists may ask about solidarity written in large print: national pride (Table 4-5). But the sociology of whole societies is not a subject unto itself. All macrosociology rests upon microsociology. For example, all macrosociologists assume, even if only implicitly, that human beings operate on the basis of rational choice, or self-interest.

In defining the most basic elements of societies, American macrosociologist Talcott Parsons (1951) included a set of individuals

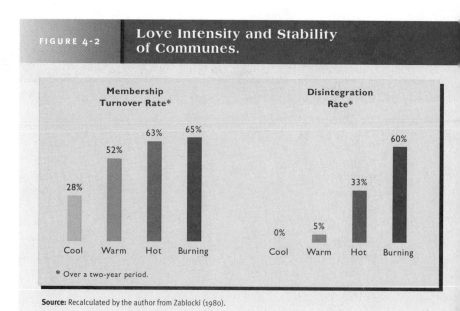

FIGURE 4-2 Love Intensity and Stability of Communes.

Source: Recalculated by the author from Zablocki (1980).

"interacting with each other" within a definite physical environment, noting that these individuals "are motivated in terms of a tendency to the optimization of gratification" and whose relationship to one another and to their surroundings "is defined and mediated" by "culturally structured and shared symbols." That could have been written by any symbolic interactionist and would be acceptable to any exchange theorist. In similar fashion, when Karl Marx constructed theories about societies, he placed major emphasis on the concepts of class consciousness and conflict, arguing that members of a society all seek their self-interests (the rational choice premise), and given their differential placements in the stratification system, the interests of the classes often collide—hence, conflict. Or when macrotheories address the consequences of sudden migration creating communities filled with newcomers and strangers, they predict that crime rates will rise on the basis of the underlying micropremise that, in the absence of attachments to restrain their behavior, people will be more likely to cheat their exchange partners (Chapter 7).

Not only do the primary approaches to macrosociological theory make very similar assumptions about microsociology; they also make some common assumptions about the fundamental features of societies. In particular, all assume societies are not happenstance collections of people, culture, and social structures; they assume instead that *things fit together* and that societies are *systems.*

TABLE 4-5	National Pride		
NATION	"VERY PROUD" OF THEIR NATIONALITY (%)	NATION	"VERY PROUD" OF THEIR NATIONALITY (%)
Puerto Rico	93	Slovenia	55
Venezuela	91	Israel	54
Iran	90	Norway	52
Morocco	89	Spain	51
Philippines	87	Austria	50
Colombia	85	Hungary	48
El Salvador	85	Indonesia	48
Egypt	82	Romania	47
Pakistan	81	Great Britain	47
Tanzania	81	Denmark	45
Mexico	**79**	Singapore	44
Zimbabwe	78	Armenia	42
Vietnam	77	Croatia	41
Peru	76	Sweden	39
Portugal	76	Italy	38
South Africa	75	Serbia	38
Dominican Republic	74	France	37
Malta	74	Bosnia	35
Bangladesh	73	Bulgaria	32
Algeria	73	Luxembourg	31
Chile	72	Russia	30
Albania	72	Latvia	29
Nigeria	72	China	25
Uruguay	72	Czech Republic	25
United States	**71**	Belarus	24
Ireland	70	Slovakia	24
Australia	70	Montenegro	24
Poland	70	Switzerland	24
India	67	Ukraine	23
Jordan	66	Moldova	22
Canada	**65**	Japan	21
Iceland	65	Lithuania	20
Argentina	65	Belgium	20
Georgia	64	Netherlands	20
Azerbaijan	63	Northern Ireland	19
Turkey	62	Estonia	18
New Zealand	61	South Korea	17
Macedonia	58	Taiwan	14
Finland	55		

Source: Prepared by the author from the World Values Surveys, 2001–2002.

SOCIETIES AS SYSTEMS

The idea of a **system,** whether it be the solar system or a social system, has three important features (von Bertalanffy, 1967): (1) a set of components (or parts) that are (2) interdependent and (3) maintain some degree of stability or equilibrium.

SOCIOCULTURAL COMPONENTS

One: Any system consists of a number of separate components or parts.

In science concepts are used to isolate and identify the key parts of the phenomena to be studied. Chapter 2 was devoted to identifying parts of societies and cultures, each of which exists in every society. Some of the most important components of social systems have both social structural and cultural aspects. Class is an example. Limited to a structural definition, the concept of class merely refers to a group of people having a similar position in the stratification system. But any actual class in any real society is equally a cultural phenomenon. Rules governing class membership and mobility, as well as the specific issues of class conflicts and consciousness, must be understood before anything else involving specific classes can be understood.

One of the most important sociocultural components of societies as systems is the *institution*. Sociologists recognize that social roles, groups, and activities are not randomly arranged within societies but tend to be clustered. Moreover, each of these clusters makes fairly specific contributions to the overall welfare of a society by satisfying basic needs the society requires to exist. For example, children must be born and prepared to replace adults, or else the group will die out. Arrangements must also exist to produce and distribute goods and services among members of a society; without food, shelter, and clothing, humans cannot survive.

Relatively permanent patterns, or clusters, of specialized roles, groups, organizations, customs, and activities devoted to meeting fundamental social needs are called **social institutions.** From examining many societies, both primitive and modern, sociologists have concluded that at least five basic social institutions exist in all societies: the family, the economy, religion, the political order, and education. Chapters 13–16 discuss each of these institutions in detail.

INTERDEPENDENCE AMONG STRUCTURES

Two: The components of systems are *interdependent*, or connected, so that changes in one part produce changes in at least one other part.

If you have a number of parts spread out on a workbench—say, small gear wheels, springs, and screws—moving one part will have no effect on the remaining parts. But if these parts are assembled into a watch and thus become *parts of a system*, then the movement of one part affects at least one other part (and usually several other parts), and the overall state of the system changes. When this is the case, the parts demonstrate **interdependence.**

Macrosociologists work from the premise that societies resemble watches more than an array of disassembled watch parts. Hence, they seek to explain why and how various parts are interdependent (Figure 4-3). They do not,

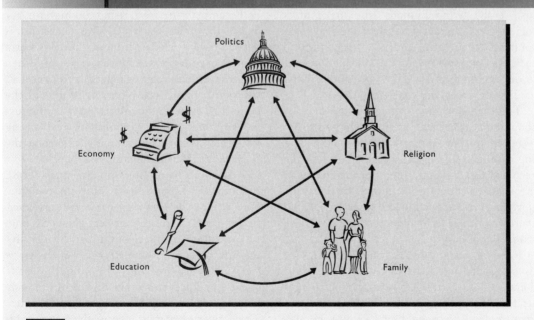

FIGURE 4-3   Society as a System of Institutions.

To discover why and how one social institution influences another is a major activity among macrosociologists. For example, will changes in family structure prompt religious or economic changes? Part IV of this book is devoted to the study of institutions.

however, assume that every component is related to every other component, nor do they assume that the same degree of interdependence among structures exists in all societies. In fact, as we will see in Chapter 17, sometimes such weak connections exist among sociocultural components that a society falls apart.

EQUILIBRIUM AND CHANGE

Three: Because the components of a system are interdependent, they tend to fall into some kind of **equilibrium,** or balance, or steady state, in their interrelations.

The interdependence among parts of a social system necessarily limits the possible variation of any given part. To continue the watch analogy, a given part can be moved only in certain ways and to a certain extent before its connections to other parts begin to limit further movement. Such mutually limiting connections of the parts making up societies mean that the overall system tends toward a state of balance, or *equilibrium.* That is, the freedom of movement of parts is limited by their connections to other parts. However, societies are *never entirely static.* Even in societies with the greatest degree of equilibrium, the components fluctuate constantly (as do the parts of a watch) as each shifts to remain in alignment with other parts. This occurs if for no other reason than that societies are **open systems,** systems open to external influences. Just as the living humans who make up a society must interact with their physical environment (to breathe and eat, for example), societies also must respond to outside forces—to their topography, weather, seismic activity, natural resources, and other societies.

There is nothing magical or mysterious about the notion that systems, including social systems, tend toward equilibrium. If we bend a cogwheel in a watch, the watch will no longer keep time. So, too, not just any arrangements of the parts of a society will suffice to keep that society going. Societies can and often have fallen apart when their fundamental components got too far out of balance.

Keeping in mind how characteristics of systems shape and delimit what goes on within them, let's now examine how macrosociologists attempt to explain how basic social arrangements come into being, how they fit together, and how they change.

THEORIES ABOUT SOCIAL SYSTEMS

Macrosociological theories come in three major forms: functionalist, social evolutionary, and conflict. Although some sociologists regard the disputes among these three as so basic that one must choose sides, I tend to regard them more as complementary than contradictory. In my own work, I often have blended the three because each approach helped me answer some questions better than the others.

FUNCTIONALISM

Because societies are open systems—systems that exchange with their environments—it proves useful to explain their structures and culture on the basis of their *consequences* for other parts of the system, especially as these parts come under pressure from the environment. Theories taking this form are called **functionalist theories** (or **functionalism** for short), and macrosociologists who favor this approach are called *functionalists.*

Functionalist theories are common throughout science. Sociologists adopted functionalism from biology, so let's start with a simple biological example to clarify the logical structure of functionalism. Why do we have sweat glands? Biologists answer that humans have sweat glands (a part of the system called the human body) because of their function, namely, to prevent the body temperature from rising too high for our organs to survive. Such a cooling mechanism is needed because the environment often is warm enough to endanger the human organism. When the environment causes a person to overheat, the sweat glands release water stored in the body. The evaporation of this water on the surface of the body causes cooling. By explaining sweat glands in this way, biologists display the basic elements of all functionalist theories.

According to Arthur Stinchcombe (1968), functionalist theories have three components. First, there is *the part of the system to be explained.* In our example this is the sweat gland. Second, we explain the existence of this part by identifying how it *preserves some other part of the system from disruption or overload.* In our example the sweat gland prevents other organs of the body from being damaged by high temperatures. Third, the theory must identify the *source* of this *potential disruption* or overload. In our example this is identified as high temperatures in the environment.

Now, let's apply these principles to a sociological example. When we examine less developed societies, we frequently find that the family unit is defined differently from the family in North America and Europe. While most North Americans and Europeans grow up in a **nuclear family** (one adult couple and their children), less developed societies often have **extended families,** which include several adult couples and their children. For example, in many societies the sons remain at home; when they marry, they bring their wives home rather than set up a new household.

Suppose we want to explain why the extended family is common in less developed societies. According to functionalism, we must see what contribution the extended family makes to some other part of these societies—what function the extended family has. Looking closely, we can see that in such societies the family *serves to support dependents*, whether they are young, disabled, ill, or elderly. To provide this support, the family must always include an adequate number of able-bodied adults. That is, there must be enough family members to support dependents.

However, under the conditions of life in many less developed societies, the death rate is so high that it constantly threatens the capacity of the family to support dependents. So many adults die while still young that if other adults are not on hand to assume their responsibilities, many dependents will go unsupported. By clustering adult couples into a single unit, the extended family minimizes the impact of high death rates as a source of social disruption (Figure 4-4). When parents die, leaving young children, other adults remain to care for them. And when people grow too old to work, younger adults remain to support them.

Although functionalist theories attempt to explain parts of societies by describing their contributions to the system, they do not assume that only one particular part can fulfill any given function. For example, while all mammals die if they become too hot, many mammals, including dogs, do not have sweat glands. Instead of sweating when they get hot, dogs hang out their tongues and pant. The evaporation of water from their large tongues cools the air as they inhale to keep their body temperature within tolerable limits. This is a **functional alternative** to the sweat gland—a different part or process

FIGURE 4-4   Diagram of a Functional Theory.

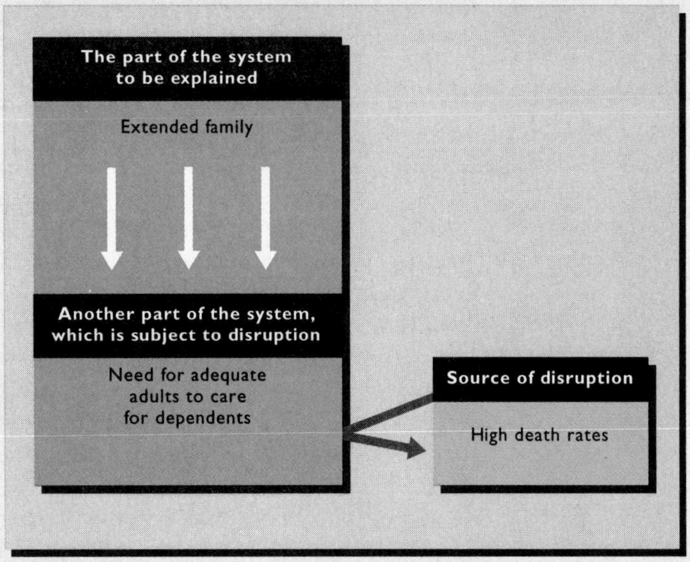

Functional theories explain why a given part of any system exists by showing how it prevents some other system part from being disrupted. In this example the extended family is explained as a social structure designed to provide an adequate number of able-bodied adults within families to support dependent family members. Here the source of disruption is a high death rate in preindustrial societies where extended families are common. High death rates threaten the capacity of smaller families to provide for their dependents.

Source: Adapted from Stinchcombe (1968).

by which the same function can be accomplished. Hence, developed societies have released the family from having sole responsibility for supporting dependents by creating insurance policies, retirement plans, and welfare programs—these are functional alternatives to the extended family.

Functionalists do not assume that all aspects of societies necessarily have beneficial effects. In fact, functionalists use the term **dysfunctions** to identify arrangements that harm or distort the social system—racial and ethnic conflict being an obvious example. Nor do functionalists assume that all social systems *ought* to survive. Mention of survival brings us to the fundamental subject matter of the second approach to macrotheories: social evolution.

SOCIAL EVOLUTION

Functional theories imply evolution. Thus, they do not directly answer the question of how various parts or processes of a system developed.

For example, explaining how the sweat glands or the extended family make vital contributions to other parts of the system does not tell us how they got there in the first place. Biologists use the principle of natural selection to explain the origins of the sweat glands. They argue that animals with more efficient cooling mechanisms are more likely to survive and reproduce than are animals with less efficient cooling mechanisms. Thus, biologists rely on evolutionary theories to explain the origins of the sweat glands while relying on functionalist theories to explain why sweat glands improve the chances of survival.

Functionalist theories in sociology imply a social evolutionary mechanism to explain the origins of parts of societies just as functionalist theories in biology do. **Social evolutionary theories** are based on the premise that societies having characteristics that enable them to better adapt to their physical and social environments have a better chance for survival than do societies that fail to develop such characteristics; those who propose such theories are referred to as *social evolutionists*. Thus, through the process of selection, certain highly adaptive structures and cultural traits tend to exist in societies. Perhaps many early human societies failed to develop the extended family or a functional alternative. If so, the lack of such societies reflects their failure to survive.

The implicit evolutionary assumptions of functionalism have led a number of contemporary macrosociologists to construct theories about the evolution of human societies (Lenski, 1976; Stark and Bainbridge, 1987; Chirot, 1994). As noted, the essence of all of these social evolutionary theories is that societies adapt or evolve in the direction of substantially increasing their capacity to survive. Consequently, over the centuries societies have tended to become *larger*, to accumulate more *effective culture*, especially *technology* (thereby becoming more able to modify the environment, more efficient at producing food, and more formidable in making war), to become more *complex* (in the sense of having more specialized occupations and organizations), and to become more *urban*. These are, of course, recognized to be tendencies, not invariable shifts. As Gerhard Lenski (1976) put it, "Not every society has grown in size and complexity. . . . Some have remained hunters and gatherers or [simple gardeners] down to the present day." But if not all societies have evolved, most have.

And as noted, the bases for these evolutionary trends are the pressures the environment puts on human societies; these pressures favor societies that are larger, more complex, more urban, and more technologically advanced. These pressures can be natural. For example, if the climate turns colder (as it did during the Ice Age), societies having shelter, fire, and clothing will be far more likely to survive. However, the primary pressure on societies comes from other societies. A society with a larger number of warriors equipped with better weapons will be able to seize the resources of less numerous and less sophisticated neighbors. The competition for survival has long resulted in winners and losers, and hence, there is a cruel survival bias in favor of aggressive cultures with superior numbers and firepower.

Despite acknowledging the dominant evolutionary trends in human history, modern theorists do not make the mistake made by social evolutionists in the late nineteenth century who assumed that societies always change for the better, that change is both inevitable and progressive. Indeed, Daniel Chirot (1994) has introduced the term **involution,** which in biology means a retrograde or degenerative change, to characterize how societies can decrease their survival capacities in response to various conditions. He noted:

This is most obvious when one looks at the highly successful classical agrarian empires [such as ancient Egypt, Greece, China, and Rome]. All of them became increasingly involuted with time. Their administrative structures became larger but less efficient, they became drains on the economies that supported them, and they lost the flexibility to adapt to various emergencies. This was a matter of slowly drifting toward inefficiency. In the long run, such involution could make a big difference in decreasing adaptability. Rome was able to handle a number of barbarian invasions rather easily, but by the fifth century, it had lost its capacity to do this.

In addition modern theorists have been careful not simply to apply biological theories to human evolution directly to explain social phenomena. That is, today sociologists try to explain the evolution of culture and of social structures as these have survival value for societies. They do not fall into the simplistic and racist position, common among late nineteenth century social scientists, that some societies were culturally more complex and sophisticated than others because they were populated with more highly evolved human beings. Human evolution is not well understood, but this much is clear: It is extremely slow, taking tens and even hundreds of thousands of years for small changes to occur. Consequently, human biological evolution could not possibly account for the social and cultural evolution of societies, because as the archeological and historical records demonstrate, it often has been extremely rapid. Indeed, as Daniel Chirot (1994) pointed out, "No language spoken two thousand years ago or even one thousand years ago would be understood by anyone today (except a few specialized scholars)."

The reason social evolution can be so rapid is that it so often is the product of conscious plans, unlike biological evolution, which occurs through random genetic variations. To quote Chirot once again, "Humans can learn from their experience, but genes cannot. We are conscious of our cultural memories, of our knowledge, and we can choose to use or discard what we know." And in accord with the rational choice proposition, faced with these contingencies, humans will attempt to make the most beneficial selections.

CONFLICT THEORY

A third approach to macrosociological theory stresses that the *conflicts that occur within a society* engender its particular cultural patterns and social structures, hence the name **conflict theory.** Where functionalists ask how particular aspects of society and culture serve other aspects of the social system, and where social evolutionists ask how a particular aspect improves a society's chances of survival, *conflict theorists* ask, Who benefits from this state of affairs?

Karl Marx (1818–1883) was an early conflict theorist. He argued that culture and social structures are created by the most powerful members of a society, the ruling class. He wrote that "the ideas of any age are the ideas of its ruling class." He further argued that the ruling class constructs society and culture to best serve its own interests and, conversely, that these arrangements in fact

Karl Marx (1818–1883) did his most important writing in England and is buried in London's Highgate Cemetery. Some years after his death, the English Communist party erected this elaborate headstone, and Marx's grave is the frequent site of tributes and ceremonies drawing communists from around the world. Here we see a ceremony marking the 100th anniversary of Marx's death, featuring a tribute read by the general secretary of the British Communist party.

determine who will be the ruling class. Thus, Marx traced all aspects of social life to class interests and conflicts.

Marx also was an early social evolutionist who believed that class conflicts *within* societies, rather than external pressures, are the primary causes of evolution. Thus, he predicted that communist societies would inevitably emerge from the final moments of class conflict, the inevitable goal of social evolution being societies that have but a single class wherein class conflicts no longer are possible. We shall pursue this further in Chapter 9.

Many non-Marxist macrosociologists also seek to explain social and cultural phenomena on the basis of internal conflicts, but most do not limit their attention to class conflicts. Taking a very broad view of the potential bases for conflict, they follow the lead of the great German sociologist Max Weber (1864–1920). Weber argued that while class conflicts are an important social influence, there are many other causes of group conflict within societies. For example, people often band together to pursue common aims on the basis of a great variety of social and cultural interests or identities: race, ethnicity, religion, region, occupation, gender, sexual preference, age, and the like. Weber ([1921] 1946) called these *status groups*, noting that they "are normally communities . . . and above all else a specific *style of life* can be expected of all those who wish to belong." The mention of a specific lifestyle lets us recognize that status groups are a particular kind of *subculture*. What distinguishes status groups from other subcultures is that a **status group** is a subculture having a rather specific rank (or status) within the stratification system. That is, societies tend to include a hierarchy of status groups, some enjoying high rankings and some low.

Racial and ethnic groups often are status groups—each having a distinctive culture (sometimes even a separate language)—and each group can be ranked in terms of its prestige and power, and sometimes in terms of property as well. Gender also has been the basis of status groups, as indicated by such terms as *men's* or *women's culture*. Regions are another common basis for status groups—think of the different cultural assumptions we make about people depending on whether they are from the East, Midwest, South, or West.

When we think of status groups based on region or religion, for example, it also will be evident that status groups often are themselves quite internally stratified. As Weber put it, "Both propertied and propertyless people can belong to the same status group." That is, rich and poor African Americans belong to a status group consisting of African Americans, as do rich and poor white Protestants and rich and poor Hispanic Catholics. This means that self-interests based on class may be at odds with self-interests based on status group. For example, wealthy African Americans sometimes must choose between siding with wealthy whites or with poor African Americans. In Canada similar choices confront members of the English-speaking and French-speaking status groups.

As Marx noted, classes may or may not be very self-conscious and therefore may or may not engender solidarity. In contrast, status groups tend to be very self-conscious and to generate strong solidarity, reflected in powerful cultural activities and symbols. In the United States, there is no holiday called "Rich Day" or "Poor Day" or even "Celebrity Day." But there are elaborate celebrations and symbols for Christmas, Passover, Kwanzaa, Cinco de Mayo, St. Patrick's Day, Columbus Day, and so on.

Conflict theories, then, are concerned with how power is distributed in societies and how various interest groups (including classes) seek and gain power and then use their power to reshape the society in their favor. From this perspective any given society at any given moment exists in a form resulting from past compromises and power struggles (Habermas, 1975).

To sum up: *Conflicts among classes and/or status groups supply the energy and the motivation for constructing and maintaining patterns of culture and particular social structures.*

Chapter 1 identified attempts to understand social solidarity and conflict as the primary sociological questions. The three approaches to macrosociology can be distinguished in terms of which of these questions they find more important and interesting. Functionalist and social evolutionary approaches stress social solidarity, emphasizing the contribution any social or cultural component makes to the *overall* welfare of the society and, therefore, how everyone benefits. For example, the evolutionary and functionalist perspectives emphasize that effective military capacities evolve and exist in societies because they protect *all* members of a society from the ravages of conquest by enemies. In this fashion social solidarity in support of military capacity is rewarded. But from the conflict perspective, it is argued that military capacity benefits the ruling class far more than the lower classes, as it is used primarily by the elite to dominate everyone else. In this sense conflict theorists often condemn general social solidarity as the enemy of class solidarity. Or while functionalist and social evolutionary theorists often stress how religion helps to sustain social solidarity (see Chapter 14), conflict theorists have long attacked religion as a primary source of false class consciousness. As Marx put it, "Religion is the sigh of the oppressed creature, the feeling of a heartless world, and the soul of soulless circumstances. It is the opium of the people."

COMPARATIVE RESEARCH: VIOLENCE AND MODERNITY

It is widely supposed that modern life is the primary cause of human distress. Talking heads on the "serious" television shows often nod knowingly when someone suggests that we probably would be better off if we could somehow return to simpler times when humans may have lacked conveniences such as flush toilets and freeways, but when we lived closer to nature and had time to know our neighbors. In support of these views, countless documentaries explain how simple societies lived at peace with one another before the coming of Europeans. Sound familiar? But is it true? To find out, we need to do research using societies as the units of analysis, which is called *comparative research*.

COMPARATIVE RESEARCH

Of course, all research is based on comparisons. To see the results of an experiment, we compare the results for groups subjected to different values of the independent variable, and in surveys we also compare across categories or levels of the independent variable. However, the term **comparative research** usually identifies comparisons of large social units, typically whole nations or societies, but sometimes states, cities, or counties (Ragin, 1987).

This use of the term *comparative research* began in the nineteenth century, long before social scientists did surveys or even many experiments, back when some social scientists attempted to compare several societies in an effort to account for differences between them. These studies typically compared no more than three or four

European nations, and the research topics ranged from why some nations had higher suicide rates than did others to why some had proportionately more beggars. While these studies often presented much quantitative data, there was little attempt to analyze the data. Few statistical techniques were yet known, and most such techniques require far more than three or four cases to produce meaningful results.

Today, journals are filled with quantitative and statistically sophisticated studies based on scores of nations and societies. In part this is because so many nations now publish so much comparable data on so many important matters. For example, next time you are in a bookstore or library look in any almanac and you will find about a half page devoted to each nation in the world (except for a few tiny ones). Each of these sections is crammed with statistics: population size; density; ethnic, religious, and language composition; area; life expectancy; infant mortality rate; percentage of the labor force employed in agriculture; rates of TV, radio, and telephone ownership; number of physicians and hospital beds; literacy rate; and much more. And these data represent but a fragment of what's easily available. Clearly, a great deal of comparative research can be, and is, based on such data, using nations as the units of analysis. But there is another major source of data on societies.

CROSS-CULTURAL "SAMPLES"

Anthropologists pioneered field research methods, and for nearly two centuries, they have been conducting ethnographic (or descriptive) field studies of human societies (Evans-Pritchard, 1981). Usually, the emphasis has been on recording detailed accounts of the culture of preliterate societies—an effort meant to preserve this information before the group was destroyed or its original culture was changed beyond recognition through contact with more developed societies. Besides preserving knowledge of cultural patterns, these ethnographies have served as the basis for comparative studies that seek to demonstrate how various aspects of culture "fit together"—for example, to show that in societies where women are more important in providing subsistence, there is less gender inequality.

For a long time, the method most often used in such studies was to compare two or a very few societies. However, as large numbers of ethnographic accounts of different societies

TABLE 4-6	A Cross-Cultural Data Sampler

FROM THE STANDARD CROSS-CULTURAL SAMPLE (186 SOCIETIES)

Percentage of Subsistence Labor Supplied by Females	Frequency	Percent
Under 10%	15	8
10%–25%	42	23
26%–50%	104	57
Over 50%	22	12

FROM THE ATLAS OF WORLD CULTURES (563 SOCIETIES)

Degree of Dependence on Hunting for Subsistence	Frequency	Percent
None	171	30
Some	179	32
Much	213	38

FROM THE NORTH AMERICAN SUBSET OF THE ATLAS OF WORLD CULTURES (124 SOCIETIES)

Importance of Slavery in This Society	Frequency	Percent
Absent	82	68
Some	25	21
Much*	14	11

No data for three cases.

Source: Prepared by the author from the Cross-Cultural Data Files.

* Most people are unaware that slavery was common among Northwest Coast tribes, who enslaved members of other tribes captured during raids.

began to pile up, some researchers began to propose that such comparative studies be based on many more cases and that statistical methods be used to analyze the data (Kluckhohn, 1939). To do so, however, required that the field reports written by many different ethnographers about many different societies be systematically transformed or coded to create a uniform set of variables in numerical form. For example, lengthy accounts of subsistence activities had to be converted into a set of scores, including such things as how dependent each culture is on hunting and gathering for subsistence (Table 4-6).

After World War II a group of anthropologists at Yale, led by George P. Murdock, began a vast project to code a data set based on ethnographic accounts in which simple societies were the units of analysis. By 1971 data on 1,264 cultures (societies) had been published. At that point Murdock and his associates became

concerned that many of the cases included in the data set were poorly documented. Worse yet, it seemed that many of them were "duplicates" in the sense that many cultures appeared several times because ethnographers had studied different villages or subgroups belonging to the same society and having essentially identical cultures. Thus, by appearing multiple times in the data set, one case might greatly influence the results. So, in 1969 Murdock and Douglas R. White published an initial set of variables for a subset of cases that they identified as the Standard Cross-Cultural Sample; tables based on these data appear often throughout this book.

It might better have been called the Standard Sample of Preindustrial Societies, but anthropologists typically use the word *culture* to refer to groups other social scientists call societies. Perhaps it shouldn't have been called a "sample," since technically these cases are not a sample of a well-defined universe. However, because Murdock invested so much effort in selecting a "representative" set of societies from the ethnographic literature, social scientists treat them as if they were a true sample. In any event the completed Sample includes 186 cases worldwide. Although the cultures included come from different periods in human history (the ancient Hebrews are one of the cases, for example), data on each case are from the same point in time. That is, all codes for the ancient Hebrews date from Old Testament times, the data for the Aleut of North America date from 1840, and those for the Pawnee Indians apply to the late 1860s.

But Murdock still was not satisfied. So, in 1981 he decided to carve out a new "sample." His primary purpose was to have enough cases to permit analysis within the major geographical regions of the world. This time he selected 563 societies to make up what he called the *Atlas of World Cultures*, and there are no fewer than 65 cases (and usually many more) in any major region of the globe.

Murdock's work has been of immense value for comparative research, but we mustn't overlook the fact that none of these data sets could have been created (and no societies will be added) had people not gone out and conducted field studies of simple societies. To appreciate the interdependence of comparative and field researchers, let's watch as two of them grapple with the issue of why some societies are so violent.

A CLOSER VIEW
NAPOLEON CHAGNON:
LIFE ALONG THE AMAZON

Napoleon Chagnon is one of the world's foremost anthropologists and is a member of the faculty at the University of California, Santa Barbara. He has spent a substantial part of his adult life among the Yanomamö, a small tribe living in the remote tropical rain forests of the Amazon Basin, along the border between Brazil and Venezuela (1988, 1992). Chagnon has been conducting field studies among the Yanomamö for nearly thirty years and has lived among them during thirteen long visits.

The Yanomamö are among the least modern people still in existence, and when he first arrived among them, Chagnon was impressed by their simple lifestyle and their carefully tended gardens—they seemed like the peaceful primitives so widely admired for their gentle ways. But as his field observation continued, Chagnon began to hear about murders. As these reports began to add up, he began to catch on that the Yanomamö were much less peaceful than they appeared. He learned that, in fact, revenge killings are very frequent among the Yanomamö. However, Chagnon never saw a murder. Most of them occur in raids by a male kin group from one village during which they ambush and kill one or two males from another village and then run for home. Apparently, the Yanomamö postponed attacks on whatever village was hosting Chagnon, not striking until after he had moved elsewhere. These raids, Chagnon was told, usually were in retaliation for an earlier killing—chains of revenge often spanning many years.

But even after Chagnon knew murders were common among the Yanomamö, he really didn't know how violent they were. He never would have known had he not decided to construct a complete kinship chart of all twelve Yanomamö villages. In doing so he expanded his efforts to include a recent genealogy as well, leading him to ask not only about dead relatives but how they had died. The results were so astonishing that they were quickly published in *Science*, a publication mainly limited to new discoveries in the natural and physical sciences. Here are some major statistical results:

■ Nearly 70 percent of adults over age forty have had at least one close relative—

parent, spouse, sibling, or child—killed by another member of the tribe.

■ More than 30 percent of all male deaths were due to violence.

■ Nearly half of the males over age twenty-five have taken part in a killing.

What events initiated a series of revenge killings? Almost always a conflict over the most prized possession in Yanomamö society: women. Murder is caused by infidelity, seduction, kidnapping brides, and the like. For the Yanomamö, as do many societies, have a severe shortage of women; in some villages there are 130 men for every 100 women. This occurs because the Yanomamö and many premodern societies practice female infanticide. In Chapter 12 we shall explore the whole syndrome of high male sex ratios caused by killing many females at birth. Here our concern is with violent conflicts.

By spending years among the Yanomamö, Chagnon produced sufficient material so that this case could be included in the Standard Cross-Cultural Sample. His findings have also been widely publicized and have added to a growing awareness among social scientists that extraordinary violence often lurks beneath the peaceful appearance of primitive groups. However, *some* primitive groups are as peaceful and nonviolent as they appear. In fact, social scientists now tend to classify primitive societies into two basic types: *factional* and *communal* (Swanson, 1968, 1969). The basis for this classification rests on patterns of decision making within the group. In *factional* societies such as the Yanomamö, groups or factions within the society pursue their own interests at one another's expense. As a consequence levels of conflict and competition within the society are high and frequently erupt in violence. In *communal* societies such as the Navajo of the American Southwest, factional disputes and conflicts are regarded as improper, and elaborate processes of deliberation are used to delay decisions until general agreement can be reached. Having avoided conflicts, these societies also minimize violence within the group.

Another way of describing these two kinds of societies is that in communal societies solidarity is based on the whole community, while in factional societies solidarity is based on subgroups within the community.

The Film Study Center, Harvard University

This rare photograph taken from above a valley in New Guinea shows a battle among primitive warriors armed with spears. Sometimes such battles occur between societies competing for resources, but often they erupt between factions within a single society. Paige's research showed that the degree of internal conflict in primitive societies was highly correlated with their rules of residence.

Social scientists were aware of these contrasts for a long time, but little progress had been made toward understanding why they existed. Most thought the answer would be found in culture. But to say that the culture of one society admires males who are prone to violence while another society's culture encourages men to be tolerant is only another way of saying what we already knew: that these two societies differ in terms of their levels of internal violence. We would expect them to have different norms and values about violence, given how people act. But we still don't know why.

A CLOSER VIEW
JEFFERY PAIGE: CONFLICT AND SOCIAL STRUCTURES

Wondering why is usually how social science begins. And so, one day a young graduate student at the University of Michigan was wondering

why some simple societies are factional and some are communal when he had a sudden inspiration. Maybe too much attention has been paid to cultural differences when we should have been looking more closely at social structures.

Jeffery Paige (1974) began with the observation that conflicts within simple societies tend to occur between factions or cliques—subgroups having a high level of attachments with the subgroup and few and weak attachments across subgroups (see Chapter 1). So the critical question is, What causes factions to form? And why do some societies seem relatively immune to factionalism, whereas in others factions flourish? Paige believed the answer must lie in differences in their social structures. Moreover, he thought he knew *which* social structures were involved.

Paige began with two key concepts that identify the two primary bases for factionalism in primitive societies. The first is *kinship*, ties based on family relationships. The second is *residence*,

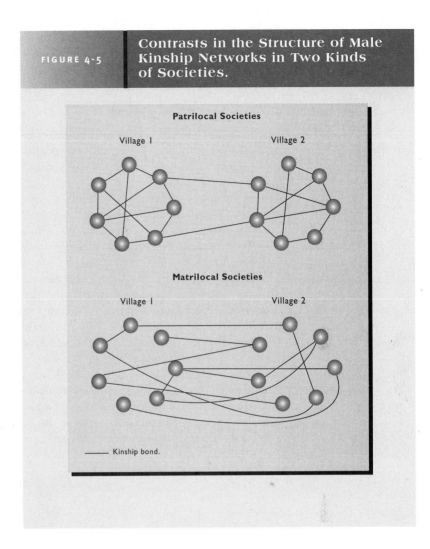

FIGURE 4-5

Contrasts in the Structure of Male Kinship Networks in Two Kinds of Societies.

the composition of households. The first two steps in his theory link each of these concepts to the concept of faction: (1) *Factions in primitive societies will form along divisions based on kinship,* and (2) *factions in primitive societies will form along divisions based on residence.* (See Figure 4-5.)

The reasoning behind these steps in Paige's theory is simple. Family members will interact more often with one another than with outsiders; hence, attachments within families will be stronger than attachments between families. Residents of the same household will interact more often with one another than with residents of other households; hence, attachments will be stronger within households than between them.

The third step in Paige's theory identifies males as the primary participants in violence: (3) *In primitive societies violence primarily occurs between male factions.* Finally, he was ready to lay out the truly significant step in his theory: (4) *Primitive societies will tend to lack factions and*

to have little violence when kinsmen do not share common residence.

Paige's fourth step reflects variations in rules of residence—the norms about who lives with whom. In some primitive societies, the bride leaves home after marriage, and the couple take up residence with or close to the husband's family. This is called the **patrilocal rule of residence.** In such societies male kin live in close proximity. A man's male neighbors are primarily his father, uncles, brothers, and sons. This maximizes interaction among males who are already united by ties of kinship. It also minimizes a man's interaction with males who are not his relatives. In such societies kinship and residence structures coincide and are mutually supportive.

However, some primitive societies observe a **matrilocal rule of residence,** and newlyweds reside with or near the bride's family. Consequently, male kin in matrilocal societies are scattered and lack day-to-day interaction. After marriage, males interact mostly with males who are not their relatives. Therefore, their tendencies to form factions with their kinsmen would bring them into conflict with the men with whom they live, while residential factions would force men to oppose their closest relatives.

Put another way, where there is matrilocal residence, men develop crosscutting loyalties: Commitments to kin crosscut those to neighbors. Each of these usual bases for conflict cancels the effects of the other.

Finally, Paige was ready to formulate a hypothesis: *Matrilocal societies tend to be communal and have low levels of internal violence, while patrilocal societies will be factional, with high levels of internal violence.*

His next steps were to examine appropriate data to see if they supported or rejected the hypothesis. What he needed to do was compare matrilocal and patrilocal societies as to their levels of violence. Turning to the *Atlas of World Cultures*, Paige selected a sample of simple societies. When he compared them, he discovered that nearly all matrilocal societies are communal and had little internal violence, whereas patrilocal societies (like the Yanomamö) are overwhelmingly factional and violent. Since most simple societies have patrilocal rules of residence, few actually resemble the peaceful groups imagined by the critics of modernity. It is far more dangerous to walk around in the wrong neighborhood in most simple societies at night than it is to walk around in the most crime-ridden urban neighborhoods.

Conclusion

This chapter examined three major methods of sociological research and three primary approaches to theorizing about societies. This ends the preliminaries. Now you have the basic tools—conceptual, theoretical, and methodological—used by sociologists. In the remainder of the book, you will see the tools and the sociologists at work as we take a guided tour of the principal sociological topics. Along the way I not only want to share with you the many important things sociologists know about social life but also show you what we are trying to find out and the many inviting challenges ahead.

Review Glossary

Terms are listed in the order in which they appear in the chapter.

data collection Systematic fact gathering.

simple fact An assertion about a concrete, quite specific, and limited state of affairs, often merely the claim that something happened or exists and usually having to do with only one or very few cases.

proportional fact The *distribution* of something, or even the joint distribution of several things, among a number of cases.

census Data from *all cases* (or units of analysis) in the relevant set. It's a census as long as "everybody" is included, and it's also a census if "everything" is included.

sample A set of cases (or units) *randomly selected* from the entire set of cases or units to be described.

random selection A method of selection whereby *all* cases have an equal (or at least a known) probability of being included in the sample.

survey research (sometimes called public opinion polling) A method of sociological research that has two identifying elements. First, it is based on personal interviews or by having each individual fill out a questionnaire. Second, it is based on a random sample of the relevant population. All studies having these two features are correctly identified as survey research.

contextual effect The dependence of a relationship found among individuals on social contexts, when different results occur in different social surroundings. Contextual effects mark the borderline between micro- and macrosociology.

network analysis A research method that has been developed specifically to test hypotheses about properties of social networks. It involves constructing measures of network variables and using networks as the units of analysis.

commune A group of people who organize to live together, often choosing to equally share duties, resources, and finances. Typically, communes also attempt to live a distinctive lifestyle in accord with an ideology that sets them apart from the surrounding society.

system Anything with these three features: (1) a set of components (or parts) that are (2) interdependent and (3) maintain some degree of stability or equilibrium.

social institutions Relatively permanent patterns or clusters of specialized roles, groups, organizations, customs, and activities devoted to meeting fundamental social needs. Five major social institutions are the family, economy, religion, political order, and education.

interdependence A relationship among parts of a system such that if one part changes, at least one other part is affected.

equilibrium A state of balance among interdependent parts of a system.

open system A system that is open to external influences. A society is open to such forces as weather, topography, seismic activity, natural resources, and other societies.

functionalist theories, functionalism Theories that attempt to explain some part of a system by showing its consequences for some other part of the system. These consequences are called *functions;* for example, the function of the sweat gland is to keep organisms from overheating.

nuclear family A family group containing one adult couple and their children.

extended families Families containing more than one adult couple.

functional alternative The existence of more than one system structure that satisfies the same system need.

dysfunctions Social arrangements that harm or distort a social system.

social evolutionary theories Theories that account for the existence of a social structure on the basis of its survival benefits for societies. For example, technologically superior societies will be better able to withstand environmental challenges; hence, societies will evolve toward increased technological capacity.

involution A retrograde or degenerative change in the survival capacity of societies.

conflict theory An explanation of social structures and cultural patterns on the basis of conflicts between classes and status groups, each seeking to gain the most benefits.

status group A subculture having a rather specific rank (or status) within the stratification system.

comparative research Comparisons of large social units, typically whole nations or societies but sometimes states, cities, or counties.

patrilocal rule of residence A situation in which married couples live with or near the man's family.

matrilocal rule of residence A situation in which married couples live with or near the woman's family.

Suggested Readings

Blau, Peter M. 1977. *Inequality and Heterogeneity: A Primitive Theory of Social Structure.* New York: Free Press.

Boulding, Kenneth E. 1970. *A Primer on Social Dynamics.* New York: Free Press.

Chagnon, Napoleon A. 1992. *Yanomamö: The Last Days of Eden.* San Diego, Calif.: Harcourt Brace Jovanovich.

Keeley, Lawrence H. 1996. *War before Civilization.* New York: Oxford University Press.

Stinchcombe, Arthur L. 1968. *Constructing Social Theories.* New York: Harcourt Brace Jovanovich.

Sociology Online

www.socstark10.com

GO TO THE INTERNET AND TYPE: www.socstark10.com

↗ **CLICK ON:** 2000 General Social Survey

✓ **SELECT: HAPPY?** Taken all together, how would you say things are these days—would you say you are very happy, pretty happy, or not too happy?

↗ **CLICK ON:** Analyze Now

TRUE OR FALSE: Younger people are more apt to be very happy than are older people.

How would you explain this?

TRUE OR FALSE: Women are less apt to be very happy than are men.

How would you explain this?

TRUE OR FALSE: African Americans are less apt to be very happy than are whites.

How would you explain this?

TRUE OR FALSE: People who say they have no religion are more apt to be very happy than those who have a religious preference.

How would you explain this?

↗ **CLICK ON:** Select New Data Set

↗ **CLICK ON:** The 50 States

✓ **SELECT: AV. AGE:** Average age of the population.

↗ **CLICK ON:** Analyze Now

The average age of the residents of each state is calculated by adding up the ages of everyone and then dividing by the total number of residents. When a measurement is based on everyone, is this a census or a sample?

Is there a substantial regional pattern? Based on the findings about age and happiness, in which state would you expect people to be happiest and in which state would you expect them to be least happy?

↗ **CLICK ON:** Select New Data Set

↗ **CLICK ON:** Nations of the Globe

✓ **SELECT: VERY HAPPY:** Percentage who say that are very happy.

↗ **CLICK ON:** Analyze Now

These percentages are based on surveys of a random sample of the adult population of each nation. (Because the question was worded slightly differently in this international study than in the General Social Survey used above, and perhaps because the international surveys were conducted earlier in the decade, the percentage of Americans who are very happy differs between the two.) Look at the bottom eleven nations. Can you notice something they have in common that might help explain why so few of their citizens say they are very happy?

TRUE OR FALSE: As you answer the question above, you are doing microsociology.

↗ **CLICK ON:** Select New Variable

✓ **SELECT: INTERESTED:** Percentage of respondents who were rated by the interviewer as very interested during the interview.

↗ **CLICK ON:** Analyze Now

In light of the discussion of survey research in the chapter, do these findings suggest anything about possible differences across nations of survey results?

USING INFOTRAC COLLEGE EDITION

GO TO THE INTERNET AND TYPE:
www.InfoTrac-college.com

↗**CLICK ON:** Register New Account
You will be asked to enter your Access Code and to create and enter a User Name and Password. Having done so,

↗**CLICK ON:** InfoTrac College Edition or Log On.

Many articles discuss surveys and sampling techniques and expand on the discussions in this chapter.

SEARCH TERMS:
Surveys
Surveys, Demographic
Surveys, Social

In addition, I have selected a specific article that will usefully supplement the chapter: Seeing around Corners. You can find it by searching for this title. You may read the article on the screen or print it using the usual print commands. If you also go to www.socstark10.com and click on the InfoTrac College Edition icon, you can read an explanation of why I selected this article and find several questions that will help you connect the article to material in this chapter.

CHAPTER

5

Eugene Sandow became an overnight celebrity when the legendary showman Flo Ziegfeld featured him as the "Mighty Monarch of Muscle" at the Chicago World's Fair in 1893. As part of his publicity campaign, Ziegfeld announced that any woman who donated $300 to charity would be permitted to squeeze Sandow's biceps. Sandow's act included lifting pianos and letting horses walk on him—he even wrestled a lion. Sandow went on to be a star on the vaudeville stage. He was a short man, and that may have been largely a result of biological factors—he probably had short parents. But much more than biology was involved in Sandow's big shoulders and upper arm development. He got those from lifting weights. Sandow's physique demonstrates especially well something that is true of all human beings—we are the product of an intricate interplay between our biology and our culture.

Biology, Culture, and Society

A CENTURY AGO MANY SOCIAL SCIENTISTS were really biologists in disguise. They argued that most human behavior was caused by inborn, or inherited, features of our biology.

Major figures in psychology at the turn of the twentieth century, including Sigmund Freud, E. L. Thorndike, and John Dewey, embraced instinctual theories of behavior (Allport, 1968). An **instinct** is a form of behavior that occurs in all normal members of a species *without having been learned*. For example, a spider hatched in isolation from all other spiders will still spin webs identical to webs made by others of the same species. Thus, psychologists who proposed instinct theories of human behavior discounted the impact of the *environment*—of cultural and social influences—on human development. To them, what was given to humans by heredity was final.

The major proponent of instinctual theories was the social psychologist William McDougall, who was on the faculty of Harvard before moving to Duke. In a very influential textbook of its time, *An Introduction to Social Psychology*, McDougall (1908) listed a number of human instincts and attempted to show how even elaborate forms of behavior are produced by "compounds" of several instincts. Religiousness, he explained, is caused by a blend of four instincts: curiosity, self-abasement, fear, and the protective, or parental, instinct. Other psychologists suggested other instincts, and in his 1932 book, McDougall revised his list and proposed that human behavior could be explained by eighteen different instincts.

Not to be outdone by psychologists, many sociologists of the period proposed biological and hereditary explanations for cultural differences among societies. For instance, less

CHAPTER OUTLINE

HEREDITY

Box 5-1 Identical Twins

BEHAVIORAL GENETICS

Box 5-2 Minorities and Stardom

THE GROWTH REVOLUTION

Environmental Suppressors

HORMONES AND BEHAVIOR

A CLOSER VIEW
The Vietnam Veterans Study

DNA AND CULTURE

A CLOSER VIEW
Origins of the Lemba

HUMANS AND OTHER ANIMALS

A CLOSER VIEW
Jane Goodall's Great Adventure

Box 5-3 Bird Brain
Nonhuman Language

A CLOSER VIEW
Washoe Learns to Sign

CONCLUSION

REVIEW GLOSSARY

SUGGESTED READINGS

SOCIOLOGY ONLINE

technologically advanced societies were believed to be made up of people with inferior intellects. Others claimed that the Swedes were stubborn, the Italians were excitable, the Spanish cruel, and the Dutch obsessed with cleanliness because of their heredity. Recall from Chapter 2 that biological arguments were used to justify the exclusion of "genetically inferior" groups such as southern and eastern Europeans, Africans, and Asians from entry to the United States in the early twentieth century.

All of these biological theories were simpleminded and obviously inadequate. Studies of children subjected to extreme neglect showed that they were not capable of speech, let alone sophisticated thought, and thus proved that most human behavior is not instinctual (see Chapter 6). As for the "backward races," some members soon were going to Harvard. Moreover, no one could demonstrate that humans exhibit any instinctual behavior, with the possible exceptions of a sucking response in infants and an infant's tendency to imitate facial expressions (Figure 5-1). Soon social science books ceased to mention instincts, and even some of the major proponents of the instinct approach eventually discarded it. In fact, social science textbooks soon made no mention of human biology at all.

By the 1930s the social sciences were dominated by purely environmental theories. Heredity was assigned no role. Everything humans do was said to be entirely the result of cultural and social influences (White, 1949). Some researchers grudgingly admitted that societies and cultural patterns would be different if human biology had produced but one sex, if infants grew to adulthood in only several months, or if the average person lived for 10,000 years. But these were regarded as nothing more than silly hypothetical possibilities. The accepted view was that our biology may set some limits, but within these human nature is essentially plastic and can be shaped into virtually any form.

As we shall see in Chapter 6, anthropologists such as Margaret Mead studied remote tribes for the express purpose of "proving" how plastic human nature really is. We shall also see that at least some of these reports probably contained as much fantasy as fact. In addition more careful research began to show that heredity does sometimes overcome environmental influences. For example, while environmental factors can make people short, they cannot make them taller than their hereditary potential.

Today, the absolute environmentalist position is judged to be as extreme as the absolute hereditarian position it was reacting against. Few social scientists accept that humans have more than very limited instincts, but most believe that human beings are the result of the interplay between their biology and their social and cultural environment. To illustrate this, let us consider symbolic interaction, which was defined in Chapter 3 as the essential human capacity. Clearly, we are not born with this capacity. If no one talks to us, we never learn speech on our own. To learn the meaning of symbols takes much time and immense stimulation from the environment. However, it is equally true that our biology makes symbolic interaction possible. Only because the human brain is of sufficient size and complexity are we able to learn symbolic interaction skills; persons whose brains are too damaged or deficient cannot learn these skills.

In this chapter we shall focus on how biology interacts with culture and society to shape human behavior. First, we shall examine several basic concepts and principles of heredity. Next, we shall explore the developing field of behavioral genetics. Here we shall find out how geneticists demonstrate whether a human trait is hereditary. For example, geneticists have recently found evidence for a hereditary element in mental illness and in intelligence.

Next, we shall examine how social and cultural factors modify the fulfillment of our genetic potentials, using the recent "revolution" in human growth as our focus. Then, we will examine evidence that human hormones have a substantial impact on many kinds of behavior—from wife beating and other acts of violence to educational and occupational achievement. Finally, we shall examine recent studies that compare humans and other primates to see what similarities and differences exist. Are humans truly unique? Do only humans possess culture and language? Can animal studies provide us with useful insights?

HEREDITY

We know that cats never give birth to pups and that the offspring of humans will always be human. The reason is that tiny parts of the male's sperm and the female's ovum contain an amazing amount of information that determines the kind of organism which results from the mating. Since offspring grow up to be whatever their

FIGURE 5-1   "Same to You, Fella!"

For a long time, it was believed that human infants did not learn to imitate facial expressions until they were eight to ten months old. However, research has shown that infants are born with this ability. Andrew N. Meltzoff and M. Keith Moore (1977, 1983a,b) found that newborns, some only sixty minutes old, would imitate facial expressions. These sample photographs from video recordings show two-week-old infants imitating tongue protrusion, mouth opening, and lip protrusion. More recently, Meltzoff and Moore (1994) found that six-week-old infants retained the ability to imitate a facial expression for twenty-four hours. That is, having imitated an expression, they did so again upon seeing this same person the next day.

Source: A. Meltzoff & M. K. Moore, "Imitation of facial and manual gestures by human neonates." *Science,* 1977, 198, 75–78. American Association for the Advancement of Science and A. Meltzoff.

parents' cells specify, an individual biological organism inherits its particular makeup; that is, the physical aspects of an organism are inherited.

In the case of humans, a male sperm contains twenty-three **chromosomes,** as does the female ovum. These combine to form twenty-three pairs (Figure 5-2). The specific instructions on each pair of chromosomes combine to determine various traits, such as eye color. These instructions are encoded in complex chemical chains called DNA, which are in tiny structures called **genes.** Each chromosome contains many genes, and humans are estimated to have more than 1,000 genes. No two sperm and no two ova, even from the same male or female, are likely to have the same array of genes. That is why the same parents can have one child with red hair and another with brown. However, a person's hair color is determined by the

BOX 5-1   IDENTICAL TWINS

BECAUSE IDENTICAL TWINS are genetic duplicates, any physical or mental differences they display can be caused only by environmental factors. Thus, identical twins offer social scientists a priceless natural experiment for attempting to isolate and untangle biological, cultural, and social factors in human development. Of course, since identical twins are always of the same sex, look exactly alike, have the same parents, and usually grow up in the same home, it is more difficult than it first appears to tell whether their intellectual and psychological similarities are genetic or environmental in origin.

The two little boys shown here are Gordon and Gary Shepherd (Gary is the one on the right). The Shepherds were born and raised in Salt Lake City, and after high school they attended the University of Utah, where they both majored in sociology. Upon graduation they entered graduate school—Gary at Michigan State and Gordon at State University of New York, Stony Brook—and each earned a Ph.D. in sociology. Today, Gordon Shepherd (the one on the right in this recent photograph) is a professor of sociology at Central Arkansas State University, while Gary is a professor of sociology at Oakland University. Each has done studies in the sociology of religion.

No one thinks that the Shepherds both chose careers in sociology because of their identical genetic inheritance. But there is considerable evidence that their common heredity caused them to be very similar in terms of some more basic traits.

(left and right) Courtesy of Gordon and Gary Shepherd

particular genes he or she has received. Geneticists have worked out some precise rules of heredity, and in many cases they can specify the odds that a given trait will show up in a child.

Two other genetic concepts will be useful in this chapter: *genotype* and *phenotype*. The sum total of the genetic instructions that an organism receives from its parents is called the **genotype.** However, the physical development of organisms does not always exactly follow these genetic blueprints, or genotypes. Environmental forces sometimes deflect or prevent the fulfillment of the genotype. For this reason we refer to any specific organism as a phenotype to take into account the interplay between the genotype and the environment in physical development. In other words the **phenotype** is what we see when

we look at any organism. As you will learn later in this chapter, for example, throughout nearly the whole period of human existence, the genotype of the average person "planned" for him or her to become much taller than the environment actually permitted. We can view the genotype in part as a genetic potential, whereas the phenotype is the actual outcome of the interplay between the genotype and the environment.

A second reason to distinguish between genotype and phenotype is that much of a person's genetic inheritance does not show up in his or her phenotype, but it can show up in the phenotypes of that person's children. For example, brown-eyed parents can have blue-eyed children, thus showing that both parents had a blue-eyed gene in their genotypes. The importance of heredity

FIGURE 5-2   **Human Chromosomes.**

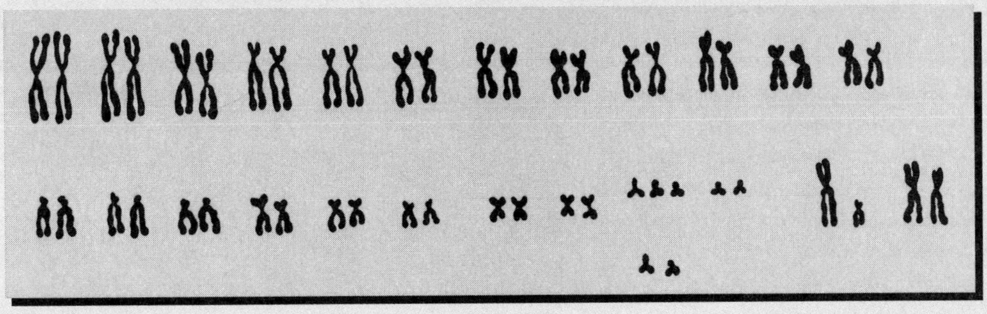

A photomicrograph of the twenty-three pairs of chromosomes in a human cell. The actual hereditary information (the genes) carried by each chromosome pair is contained in DNA molecules.

to sociologists does not depend on whether some human characteristics are wholly determined by genetics, for without question some are. Blue-eyed parents, for example, can produce only blue-eyed children (although brown-eyed parents also can produce blue-eyed children). The important question is, What traits are determined to what degree by genetic inheritance? Do any of these traits influence human activities of interest to social science? Sociologists don't care much about eye color, but they do care very much about variations in humans that determine what people can and cannot do and how they do or do not act.

BEHAVIORAL GENETICS

Just because instinctual theories were silly does not mean that heredity plays no role in human behavior. In 1960 the publication of the first textbook in **behavioral genetics** marked the rapid rise of a new scientific field (Fuller and Thompson, 1960). In recent years behavioral geneticists have claimed considerable success in isolating human characteristics and behavior that are influenced to a substantial degree by genetic inheritance (Sherman et al., 1997). Among these are such characteristics as intelligence, major forms of mental illness, alcoholism, and a tendency toward impulsive and aggressive behavior.

To show that any particular trait is inherited, one must first demonstrate that blood relatives are more alike in terms of this trait than are randomly selected, unrelated individuals. For example, David Rosenthal (1970) summarized dozens of studies reporting that schizophrenia, a mental illness, clusters in families; that is, relatives of a schizophrenic are considerably more likely to become schizophrenic than are people without schizophrenic relatives.

However, this approach has encountered major criticism. Not only do two brothers who develop schizophrenia have a similar genetic inheritance, but they also grew up in the same home with the same parents and were exposed to similar social circumstances outside the home. Opponents of behavioral genetics argue that the similar environment in which relatives are raised only makes it seem as if heredity plays a part.

For this reason studies of twins, especially identical twins, are central to research in behavioral genetics. We have seen that children of the same parents each have a unique genotype, although their genotypes are more alike than are those of children with different parents or children with only one parent in common. Most twins are not identical and are the result of their mother producing two ova at the same time, each of which was impregnated by a different sperm. Such twins are no more genetically similar than are children born at different times to the same parents.

However, once in a while, after an ovum has been impregnated, it splits in half and develops into two babies (Figure 5-3). When such splits take place, each half has the same genetic content. This is how identical twins occur—twins who not only look alike but also are *exactly alike* in terms of their genetic makeup. Any hereditary trait that shows up in one identical twin must

BOX 5-2 | MINORITIES AND STARDOM

THE MAJORITY OF PLAYERS on every team in the National Basketball Association are African American. White boxing champions are rare. A far greater proportion of professional football players are African American than would be expected based on the size of the African American population. Furthermore, African Americans began to excel in sports long before the Civil Rights Movement broke down barriers excluding them from many other occupations. This has led many people, both African American and white, to conclude that African Americans are born with a natural talent for athletics. How else could they have come to dominate the ranks of superstars?

The trouble with this biological explanation of African Americans in sports is that it ignores an obvious historical fact: It is typical for minorities in North America to make their first substantial progress in sports (and for similar reasons, in entertainment). Who today would suggest that Jews have a biological advantage in athletics? Yet at the turn of the twentieth century, the number of Jews who excelled in sports far exceeded their proportion in the population. And late in the nineteenth century, the Irish dominated sports to almost the same extent as African Americans have done in recent decades.

By examining an encyclopedia of boxing, for example, we can draw accurate conclusions about patterns of immigration and periods at which ethnic groups were on the bottom of the stratification system. The Irish domination of boxing in the latter half of the nineteenth century is obvious from the names of heavyweight champions, beginning with bare-knuckle champ Ned O'Baldwin in 1867 and including Mike McCoole in 1869, Paddy Ryan in 1880, John L. Sullivan in 1889, and Jim Corbett in 1892. The list of champions in lower-weight divisions during the same era is dominated by fighters named Ryan, Murphy, Delaney, Lynch, and O'Brien.

Early in the twentieth century, Irish names became much less common among boxing champions, even though many fighters who were not Irish took Irish ring names. Suddenly, champions had names like Battling Levinsky, Maxie Rosenbloom, Benny Leonard, Abe Goldstein, Kid Kaplan, and Izzy Schwartz. This was the Jewish era in boxing.

Then Jewish names dropped out of the lists, and Italian and eastern European names came to the fore: Canzoneri, Battalino, LaMotta, Graziano, and Basilio; Yarosz, Lesnevich, Zale, Risko, Hostak, and Servo. By the 1940s fighters were disproportionately African American. Today, African American domination has already peaked, and Hispanic names have begun to prevail.

Carnegie Museum of Art, Pittsburgh

Jackie Robinson, the first African American to play in the major leagues.

show up in the other, unless outside forces intrude. This fact has caused scientists around the world to seek out identical twins and to subject them to an immense number of tests and measurements. Sweden and Finland have been the center of much of this research because their governments keep such extensive records on all citizens that it is easy to identify and then locate pairs of twins. Thus, Swedish researchers have assembled data on 12,798 twin pairs, while researchers in Finland are at work on 7,144 twin pairs. Of the Finnish twin sample, 2,320 are *monozygotic* (MZ), or identical, twin pairs.

The logic of twin studies is simple. Suppose that the mental illness known as schizophrenia has a genetic basis. If that is so, then if one of a pair of identical twins develops this illness, the other should do so as well. By the same token, if

The current overrepresentation of African Americans in sports reflects two things: first, a *lack of other avenues to wealth and fame*, and second, the fact that minority groups can overcome discrimination most easily in occupations where *the quality of individual performance is most easily and accurately assessed* (Blalock, 1967). These same factors led to the overrepresentation of other ethnic groups in sports earlier in history.

It is often difficult to know which applicants to a law school or a pilot training school are the most capable. But we can see who can box or hit a baseball. The demonstration of talent, especially in sports and entertainment, tends to break down barriers of discrimination. As these fall, opportunities in these areas for wealth and fame open up, while other opportunities remain closed. Thus, minority groups will aspire to those areas in which the opportunities are open and will tend to overachieve in these areas.

In an important theoretical contribution to racial and ethnic relations, Hubert M. Blalock (1967) was one of the first to explain why minorities more rapidly overcome discrimination in sports. Let's consider several of his propositions.

First, Blalock argued, work groups differ in the extent to which an outstanding individual can bring success to the whole group. A worker on an assembly line, for example, does not increase the earnings of other workers by working faster. But a great quarterback or a great hitter can transform an average team into champions. Thus, Blalock theorized, the more an individual can increase the benefits of all work-group members, the less that group will discriminate against minority members. This will be particularly so when it is easy to judge how much a person could add to the group's success.

To illustrate Blalock's point, consider an all-white baseball team, many of whose players are prejudiced against African Americans. However, they also want to win the pennant and the World Series, but they need a better power hitter to do so. Such a team will be inclined to ignore their prejudice against African Americans if they have a chance to get a star hitter who is African American.

Blalock also suggested that when employers compete intensely for talented people, they will be much less likely to discriminate. Because such competition is the essence of management in sports, highly talented minority players will be an irresistible temptation for owners and managers. Discrimination should cease in sports long before it does in most other high-status occupations. Blalock's proposition also implies that less successful teams would take the lead in ending discrimination, whereas the most successful teams would resist it. This is reflected in the fact that during the many years when they routinely won the pennant and the World Series, the New York Yankees were among the least integrated teams in baseball.

Thus, the overrepresentation of an ethnic or racial minority in sports often signals that group's early progress in struggling up from the bottom of society. However, the real signal that a group is making it comes when their overrepresentation in sports begins to decline, for it means that young people of this racial or ethnic background have other possible roads to success. This is not to suggest that it is better for people to become lawyers or dentists than to become linebackers. (I much prefer to watch a linebacker fill a hole than a dentist.) But no group should face such limited opportunities that playing sports is their only escape from poverty and prejudice. The overrepresentation of a racial or ethnic minority in sports does not reflect inborn athletic talent any more than their underrepresentation in science reflects an inborn lack of academic talent. Instead, both reflect limited opportunities.

These same principles apply to overrepresentation in the entertainment world. The early success of African Americans in music, for example, led to the belief that they were born with a "natural sense of rhythm." Again, when opportunities are few, people will concentrate their efforts. African Americans who could play musical instruments, dance, sing, or write music dedicated themselves to perfecting their skills, as did other ethnic groups when their opportunities were limited. Like athletic talent, entertainment skills are very visible and easily demonstrated. Bill "Bojangles" Robinson could have become a star just by dancing on a street corner (which he often did even after he was world famous). Louis Armstrong's trumpet playing was as obviously inspired as Michael Jordan's dunk shots. To claim that Fats Waller couldn't play the piano would have been as silly as to say Joe Louis couldn't punch. As in sports, barriers of discrimination tend to fall early in the entertainment industry.

intelligence is genetic, then the IQ score of one twin should be virtually the same as the score of the other (allowing for slight discrepancies in the accuracy of the test). And for a long time now, that's what the studies have found. Indeed, a rather amazing array of personality and intellectual characteristics seem to have a very substantial genetic basis, given the very high concordances on these traits between pairs of identical twins (Rose et al., 1988; Tellegen et al., 1988; Bouchard et al., 1990).

However, even these studies based on identical twins have been subject to the same criticism mentioned earlier. Since these twins look exactly alike and are the same age, they will have been treated in almost identical ways. Once again, similarity of environment offers a very plausible counterexplanation.

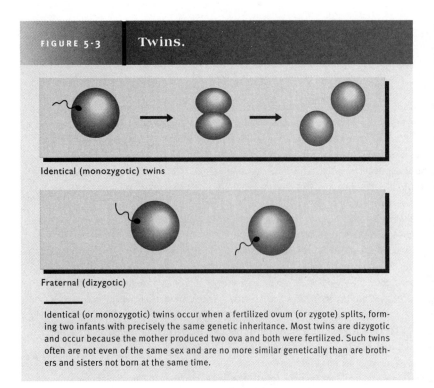

FIGURE 5-3   **Twins.**

Identical (monozygotic) twins

Fraternal (dizygotic)

Identical (or monozygotic) twins occur when a fertilized ovum (or zygote) splits, forming two infants with precisely the same genetic inheritance. Most twins are dizygotic and occur because the mother produced two ova and both were fertilized. Such twins often are not even of the same sex and are no more similar genetically than are brothers and sisters not born at the same time.

However, not all identical twins grow up in the same house. Sometimes they are adopted by two different families and grow up without even knowing they have a twin. For some decades now, behavioral geneticists have focused on locating such sets of identical twins because they permit a more stringent test of hereditary and environmental explanations.

For example, if intelligence has no hereditary basis, the IQs of identical twins raised separately should not be more similar than IQs of randomly selected persons. In fact, a great many studies, done in different countries by many different geneticists, show that the IQs of identical twins raised apart are extremely similar, although not identical. Environmental factors, therefore, can and do influence intelligence. But the IQs of identical twins reared apart are far too much alike to allow us to reject heredity as a major factor determining intelligence. Indeed, the most recent results, based on the Minnesota Study of Twins Reared Apart, suggest that about 70 percent of the variation in intelligence is caused by genetic variation (Bouchard et al., 1990). Studies based on identical twins also have produced strong evidence in favor of a substantial genetic factor in a wide range of personality traits (Rose et al., 1988; Tellegen et al., 1988; Bouchard et al., 1990). Similar studies also have supported the role of heredity in many forms of mental illness, including schizophrenia and depressive disorders. The distinguished studies by Irving Gottesman (1991, 1993) have demonstrated that the siblings of a person with schizophrenia are ten times more likely to develop this very serious form of mental illness than are persons selected at random and that 13 percent of the children of a parent with schizophrenia acquire the disease, as opposed to 1 percent of the general population. In contrast, adoptive siblings or adoptive children of schizophrenics are no more likely to be afflicted than are persons in the general population, which suggests there is no environmental effect (Sherman et al., 1997).

Aside from studies of identical twins, behavioral geneticists also rely on studies based on persons who were adopted as infants. They seek to separate heredity from environment by testing whether adoptees more closely follow a behavioral pattern of the adoptive parents who raised them or that of their biological parents. Here the work by Marc Schuckit and his colleagues on alcoholism is an excellent example. He found that adoptees with an alcoholic biological parent had a nearly 50 percent rate of alcoholism regardless of whether one of their adoptive parents was an alcoholic. This finding suggests a strong genetic influence. Among adoptees whose biological parents were not alcoholics, the risk of alcoholism was far lower. But here environmental factors were felt, for *the odds of becoming an alcoholic were twice as high if an adoptive parent was an alcoholic* (Schuckit et al., 1972a,b, 1979). These results have been replicated in Denmark and Sweden using the same technique of comparing adoptees with their real and adoptive parents. Moreover, studies based on twins, including identical twins, have yielded similar support for a genetic component in alcohol abuse (Hay, 1985).

Some social scientists resent behavioral geneticists and cling to the belief that the environment causes all or nearly all variation among humans. But it seems to me unwise to ignore a field conducted in a scientific manner and guided by theories that have amply explained less controversial human traits. Of course, social scientists did not feel threatened when geneticists unraveled the hereditary transmission of such diseases as sickle-cell anemia or diabetes. But when geneticists discuss the heredity of mental illness or intelligence, they tread on ground social scientists were accustomed to having all to themselves. Yet social scientists have as much to tell geneticists as geneticists have to tell us. This will be obvious in the next two sections

Special Collections Division, University of Washington Libraries

Here we see the University of Washington's "super varsity" basketball team of 1929. They went undefeated until losing the Coast Conference championship game to the University of California's Golden Bears. The tallest man in the picture is six feet two inches. The previous year Washington had been a real power-house because they had a center who was six feet six inches. Today, these men would be too short to have much chance of making the Washington team—or even a Seattle high school team.

of this chapter, where we shall see the immense power of the social and cultural environment to modify fulfillment of human potentials. Here the combined resources of sociologists and geneticists help to illuminate many mysteries.

THE GROWTH REVOLUTION

Over the past several decades, people have been getting bigger. The change has been especially obvious in sports. For example, the line of the Chicago Bears' famous "Monsters of the Midway" averaged less than 200 pounds in their National Football League championship season of 1933. Good high school football players are much bigger than that today. The offensive line of the *USA Today* 2001 All-USA High School Team averaged 302 pounds.

Purdue University won the Big Ten basketball championship for three straight years (1928–1930) because their All-American center, Charles "Stretch" Murphy, was so tall no one could stop him. Stretch was six feet six inches. And in 1944 the University of Utah won the NCAA basketball tournament, beating Dartmouth in overtime. The Utes' tallest player was six feet four inches.

Similar changes have gone on in professional basketball. In 1964–1965 only three players in the NBA were seven feet tall. In 2001–2002

forty-eight players were at least seven feet tall, and the number one draft choice was Yao Ming from China who is seven feet five inches tall.

To some degree, athletes are larger because they are now recruited from a much larger population; not only has the U.S. population grown, but the population enrolled in high school and college has grown even more. Thus, if a person seven feet tall occurs only once in a million births, there will be more seven-footers when the population is over 200 million than when it is 100 million.

However, not only sports stars have become bigger. As Table 5-1 shows, dramatic changes have taken place in the height of the American population. The data come from actual measurements made under the supervision of the U.S. National Center for Health Statistics during the years 1976–1980. Here we can see that men born in 1956–1962 were three times as likely as men born in 1906–1915 to reach an adult height of six feet or taller. Moreover, nearly a third of those born in 1906–1915 were under five feet seven inches, while fewer than 10 percent of those born in 1956–1962 were this short. The same comparisons hold for women. Women born in 1956–1962 were five times as likely to be taller than five feet five inches as were women born near the turn of the century.

Nor is this rapid increase in height a uniquely American phenomenon. People in all of the advanced industrial nations are bigger than their

TABLE 5-1	Trends in the Height of Americans		
	YEAR OF BIRTH		
HEIGHT	1906–1915	1926–1935	1956–1962
Males			
Percent over 6'	10.4	14.2	31.8
Percent under 5'7"	29.4	11.4	8.2
Females			
Percent over 5'5"	5.4	15.4	25.2
Percent under 5'3"	44.3	27.7	17.8

Source: *Statistical Abstract of the United States, 1990.*

parents and grandparents. For example, the average height of Japanese men rose from five feet two inches in 1950 to five feet seven inches in 1996. And of course, this average still includes many men from the short birth cohorts born prior to 1950. This dramatic increase in average height has required major changes: A few years ago, most of the desks in Japan's elementary schools had to be replaced by larger ones.

In addition a second change has recently been noticed. Not only are people growing larger than

before, but they are also growing faster, achieving their full size at a much younger age. Data from various nations reveal that a century ago most people still grew a lot in their late teens and usually did not reach their full adult height until age twenty-five or even later (Tanner, 1970). Today, most people stop growing by eighteen or so, and most achieve nearly their full growth much sooner than that.

Clearly, there has been a dramatic shift in the human phenotype in modern times. Suddenly, we are much bigger than our ancestors. How could this have happened?

ENVIRONMENTAL SUPPRESSORS

No trained biologist could believe that these rapid changes in human physiology were caused by genetic changes. No such change could spread so far so fast. Thus, the answer had to be sought elsewhere. In fact, the answer lies in the potent capacity of environmental factors to modify genetic potential. Indeed, to ask why humans suddenly began to get so large so fast is to raise the wrong question. The right question is, What kept humans so small and delayed their maturation for so long? What we are examining is a rapid change in the phenotype of human beings that reveals the previously *unfulfilled potential* of the human genotype.

Much research still needs to be done to explain how the environment suppressed our natural growth patterns, but the major factors are easy to determine: inadequate nutrition (especially shortages of vitamins and proteins) and chronic poor health.

As we shall see many times in this book, one of the major impacts of the Industrial Revolution, of modernization, has been healthier people living longer lives. Until modern times, most people, even farmers, ate meager diets that were deficient in vitamins and proteins. Most people ate meat, eggs, or dairy products very rarely, hardly ever ate fruit, and ate vegetables only in summer (Braudel, 1981). They lived almost exclusively on bread and on mush and soup made from grain. As a result their growth was stunted and their maturation delayed. Poor diet also made people more susceptible to illness and contagious diseases, the latter being most common among children. A lack of sanitation and poor personal hygiene (see Chapters 13 and 19) further contributed to poor health. For these reasons large numbers of children, often as many as half, did not reach adulthood. Nor were those who survived strong and

One aspect of this portrait of a Chinese American man posing with his parents would be found in the huge majority of similar families—the son, born and raised in the United States, is much taller than his father. While people in industrial nations generally have been getting larger, the shift has been especially marked for Asian immigrants to Europe and North America.

© Charles Kennard / Stock Boston

healthy. The average person died by age thirty-five.

Under such privation, we are hardly surprised that human development was stunted and maturation long delayed. Then the Industrial Revolution suddenly began to change these conditions. People began to eat much more—and much more nutritious—food (see Chapter 18). Sanitation and immunization practices eliminated many common diseases. Infant mortality declined rapidly as the average life expectancy doubled. And suddenly, people got much bigger and began to mature much younger.

A century ago anyone who said that the average person soon would grow to be much larger would have been laughed at. Moreover, anyone who had even hinted that diet, not heredity, was the primary basis for the differences in stature between Europeans and Asians would have been dismissed as a crank. Nevertheless, we now know that not only does heredity set limits on human potentials, but the environment—society and culture as well as the physical ecology—can suppress these potentials very severely.

Having used a noncontroversial example to explore the capacity of the environment to suppress full expression of the genotype in the phenotype, we may now use our knowledge to examine how the interplay of biology, culture, and society shapes something that determines the whole of human behavior: the ability to think and to learn.

HORMONES AND BEHAVIOR

The human endocrine system consists of glands—among them the adrenal, thyroid, pancreas, pituitary, hypothalamus, thymus, pineal, testes, and ovaries—each of which secretes hormones into the bloodstream. **Hormones** serve as "chemical messengers" that control or regulate other organs and bodily functions. For example, the thyroid hormones stimulate oxygen consumption in the cells, influence the metabolism, affect cholesterol levels, and regulate normal growth.

For decades scientists have suspected that hormones not only have impact on our cells and on particular organs but that they have some direct influences on human behavior. Particular interest has focused on the sex hormones—**testosterone** (produced by the testes) and **estrogen** (produced by the ovaries)—because these play such a profound role in producing secondary sexual

characteristics. However, until very recently connections between various hormones and behavior were little studied, and what research was done suffered from very limited resources. In part this is due to the fact that it has been difficult and expensive to measure variations in hormonal levels. But another limitation involved the difficulties of interdisciplinary research: Those best qualified to study hormones were not particularly interested in or trained to study behavior, and vice versa.

Meanwhile, beginning in the 1970s, the widespread abuse of anabolic steroids (synthetic testosterone) by athletes and body-builders encouraged the view that hormones can greatly influence behavior. As steroid use spread, observers concluded that steroids have marked behavioral side effects. Athletes taking large doses seemed unusually subject to rapid and unpredictable mood swings, to uncontrolled fits of temper, and to outbursts of violent behavior. They also seemed to experience greatly increased sexual appetites. The implications seemed obvious: Men having high natural levels of testosterone might be prone to violent behavior.

Slowly, a body of findings began to build up. Reid Daitzman and Marvin Zuckerman (1980) found that men with higher levels of testosterone scored higher on tests of impulsiveness and sensation seeking. Typical of the time, however, their results were based on only forty male college students. James M. Dabbs et al. (1987) studied eighty-nine men in prison and found those convicted of violent offenses had higher levels of testosterone than those convicted of nonviolent crimes. The next year Dabbs et al. (1988) found similar results based on eighty-four female prisoners—women do have testosterone, although much less than do men. Teresa Julian and Patrick C. McKenry (1989) found that men with high levels of testosterone had poor marital and parental relationships—based on thirty-seven men aged thirty-nine to fifty having professional occupations. J. Richard Udry (1988) found that testosterone level had a huge impact on adolescent sexual activity, among both boys and girls, even when the most important social variables, such as parental supervision, were controlled. But again, the data were from 102 males and 99 females.

Then, almost by accident a huge, comprehensive database became available to scientists wishing to study the link between testosterone and behavior.

A CLOSER VIEW
THE VIETNAM VETERANS STUDY

In 1985 researchers at the U.S. Centers for Disease Control in Atlanta were assigned to study the "health status" of Vietnam veterans. To do so, they first selected a random sample of men (selected from military records) who had served in Vietnam during the period 1965–1971—officers were not included. Efforts were made to locate each man drawn in the sample and to recruit him to participate in a very comprehensive study that included elaborate interviews covering an immense array of topics, such as educational and work histories, family and marital histories, drug and alcohol abuse, childhood delinquency, adult criminality, and sexual behavior. Each vet also was given many standardized psychological and neuropsychological tests, and in addition each underwent a comprehensive medical examination at the Lovelace Medical Foundation in Albuquerque, New Mexico. The extensiveness of the data collection is all the more remarkable because 4,462 men took part.

In 1988 the centers' report on the veterans was released. It was long, highly technical, and not very stimulating. But amid the blizzard of facts about veterans was the amazing revelation that the data included a measure of testosterone level for every one of these veterans. Suddenly, researchers who had struggled along with tiny numbers of cases could base their research on this huge sample, and because data collected by the government are in the public domain, any scholar could obtain the complete database free.

Subsequent to the release of this huge study, the research has been pouring out. And the results are far clearer and more extensive and powerful than had been anticipated (Dabbs and Morris, 1990; Dabbs, 1992; Booth and Dabbs, 1993). The higher their level of testosterone, the *more* likely men are:

- to get divorced.
- to physically abuse their wives.
- to engage in extramarital sex.
- to have many sexual partners.
- to have problems with alcoholism.
- to use drugs.
- to have been punished while in military service.

- to have been a juvenile delinquent.
- to get in trouble with the law as an adult.
- to be unemployed.

But the higher their level of testosterone, the *less* likely men are:

- to get married.
- to become highly educated.
- to obtain a high-status occupation.

None of these relationships turned out to be spurious when multivariate analysis techniques were used.

So, now we know that hormones strongly intrude into domains social scientists have been accustomed to regard as theirs alone. Of perhaps even greater significance for future progress, the leading studies were not relegated to obscure journals, nor were they published only in medical journals unread by social scientists. Instead, they have appeared in the major social science journals.

DNA AND CULTURE

In 1953, when James D. Watson and Francis Crick discovered the structure of deoxyribonucleic acid, now known as DNA, they provided the key to unlock the secrets of human genetics. But back then no one dreamed their work would ever have implications for the social sciences. In the past several years, however, studies of the distribution of DNA characteristics in human populations around the world have yielded many insights into cultural patterns, some of them extending far back into prehistory. For example, we now have firm evidence that women have been far more geographically mobile than men. This is evident in the fact that in all parts of the world, the DNA of women is far more diverse than that of men, indicating that men have tended to stay put, while women often have migrated from their birthplaces. Social scientists already knew, of course, that this was implied by the prevalence of patrilocal societies (see Chapter 4) and that women often have been bought and kidnapped. But no one suspected the magnitude of female mobility—that women are more closely related than are men (Seielstad et al., 1998).

This ability to trace human relatedness through DNA recently resulted in a discovery so surprising and exciting that it was reported on the front page of *The New York Times* (Wade, 1999). Let's watch.

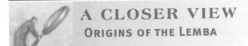

A CLOSER VIEW
ORIGINS OF THE LEMBA

For more than a century, anthropologists have known that the Lemba, a tribe of black Bantu-speaking people living in southern Africa, claim to be Jews. They practice circumcision, do not eat pork, and observe the Sabbath. Their oral traditions tell of them having come south from a city called Senna in Ethiopia following a flood that destroyed their homes. However, given the long history of misguided, false, or occult claims concerning "Lost Tribes of Israel," these Lemba traditions were dismissed as mythical (Brough, 1992). That began to change when, ten years ago, Tudor Parfitt, director of the Center for Jewish Studies at the School of Oriental and African Studies in London, discovered that there is a ruined city in Ethiopia that once was called Senna, which was destroyed by a flood about 1,000 years ago (Kaplan, Parfitt, and Semi, 1995; Parfitt and Semi, 1999). Even so, the idea that perhaps the Lemba really are of Jewish descent was given little credence. Most social scientists suspected that the "Jewishness" of the Lemba was the result of conversion, perhaps due to contact with Jewish traders, albeit that conversion may have occurred more than 1,000 years ago.

Meanwhile, scientists in Israel and America made the remarkable discovery that there is a characteristic genetic "signature" found only in the DNA sequences of the Y chromosomes of some Jewish men. Around 5 percent of all Jewish men (both Ashkenazi and Sephardic) have this signature, and about 50 percent of Jewish men claiming to be *cohanim*—members of the priestly "class" descended from Aaron—have it (Skorecki et al., 1997; Thomas et al., 1998). Thus, the stage was set for the sensational announcement by Oxford's David Goldstein in 1999 that Lemba men possess the Jewish genetic signature as frequently as do other Jewish populations! Moreover, more than 50 percent of the members of the Lemba's hereditary priestly clan have it, indicating that they, too, are *cohanim*. This "scientific" confirmation that the Lemba actually are *descended* from Jews may cause scholars to place greater faith in purely social scientific evidence, such as Parfitt's, and perhaps to place greater confidence in oral "traditions."

Courtesy of Dr. Tudor Parfitt

Matshaya Mthivha is a spiritual leader of the Lemba. For centuries no one believed their traditions that they were Jews.

HUMANS AND OTHER ANIMALS

Social scientists have always been very ambivalent about classifying humans as animals. None denies the obvious biological similarities. But when it comes to saying just what *sort* of animals we are, social scientists have tended to see little resemblance between humans and other species. At odds are two long traditions among social scientists: One stresses the immense *intellectual superiority* of humans over other primates; the other stresses the *moral and ethical inferiority* of humans when compared with the rest of nature—or even in comparison to our biologically more primitive ancestors.

The first tradition stresses the unique human capacity to *create culture* and to pass it on to younger generations. Some social scientists argue that only humans need years of education and training to be able to fulfill normal adult roles—for only humans can create and acquire knowledge and technology. Indeed, social scientists long took it for granted that only humans have *sufficient intelligence* to acquire a *language* and thus to have the means to share culture.

Louis Leakey painstakingly probes for fossils with a dental pick. Here in Olduvai Gorge, located in Tanzania's Serengeti Plain, Leakey and his wife and sons uncovered bones of creatures who lived millions of years ago and who many paleontologists believe were early human ancestors. Because Leakey thought knowledge of the behavior of primates might shed light on the behavior of the hominids he was unearthing in Olduvai, he promoted field observational studies of primates. After a stint at Olduvai, Jane Goodall agreed to Leakey's suggestion that she study chimpanzees in a nearby game reserve.

© Ian Berry/Magnum Photos

The second of these traditions suggests that the gift of intelligence is, itself, the curse of Cain—that humans are, in the words of Jean-Jacques Rousseau (1712–1778), *"un animal dépravé"* (a depraved animal). Back when humans lacked true consciousness, they, like other animals, were happy and good, according to Rousseau. But consciousness permits humans to suffer from envy, to remember grievances and seek revenge, to recognize that "crime" can pay. Since Rousseau's day, the idea that humans represent nature gone wrong—that we are a species of depraved apes who have lost their natural virtues—has been influential. Indeed, Konrad Lorenz (1966) argued that murder is uniquely human, that other animals rarely attack members of their own species. Lorenz blamed this on humans having become meat eaters only late in their evolution, far too late to have developed instinctual inhibition mechanisms that curb intraspecies aggression among carnivores. To make matters worse, according to Lorenz, humans have developed a murderous arsenal.

In 1973 Lorenz won the Nobel prize, partly for his work on human aggressive instincts. Among others who have sustained the image of humans as uniquely carnivorous killer apes is Robert Ardrey in his three immensely influential bestsellers, *African Genesis* (1961), *The Territorial Imperative* (1966), and *The Social Contract* (1970).

But is this image of humanity correct? Indeed, is it true that human intelligence is unique, giving us alone the capacity to create culture and master language? To assess these views, let's travel to Africa and watch Jane Goodall, one of the most talented and celebrated field observers of our time, studying a band of chimpanzees in the wild.

A CLOSER VIEW
JANE GOODALL'S GREAT ADVENTURE

Few scholars in modern times have come to their careers in a less likely fashion than did Jane Goodall. She grew up in England and at eighteen left school, took a secretarial course, and then spent four years working in London. In 1957 a friend invited her to visit Kenya, so Goodall left London and returned home to Bournemouth, where she spent the summer working as a waitress to save money for her trip. After spending a month at her friend's farm, Goodall decided to stay on in Africa for a while and took "a somewhat dreary office job" in Nairobi (Goodall, 1971). One day someone who knew of her great interest in animals suggested that she ought to meet Louis Leakey (1903–1972), the most famous anthropologist and paleontologist in the world and curator of the National Museum of Natural History in Nairobi.

Leakey, his wife and sons, and his many students and assistants spent much of each year at their immensely important digs for early hominid fossils in Olduvai Gorge, across the border in Tanzania (Cole, 1975). When Jane Goodall turned up in his office and explained that she wanted to "get closer to animals," Leakey promptly hired her as a secretary. Soon she accompanied the Leakeys back to Olduvai. She found the digging fascinating, and she was even more fascinated to sit in camp in the evenings as Louis Leakey conducted a freewheeling seminar for everyone present.

After a few months, Leakey decided that the quiet, carefully organized, young Jane Goodall was precisely the right person to go off to study a band of several hundred mountain chimpanzees—a study he had been suggesting for twenty years. Never mind that she lacked a Ph.D. and hadn't even gone to college. Never mind that she had no training in fieldwork or in biology. Never mind that she was only twenty-three. And never mind that she had no money. All that mattered to Leakey was that these chimps be studied properly. It would take a long time to do it right, he pointed out; two years would probably just be a good start. Leakey told Goodall that he was

particularly interested in the behavior of a group of chimpanzees living on the shores of a lake—for the remains of prehistoric man were often found on a lakeshore and it was possible that an understanding of chimpanzee behavior today might shed light on the behavior of our Stone Age ancestors. (Goodall, 1971)

Goodall could "hardly believe that he spoke seriously" when he asked if she would undertake this work. When she agreed, Leakey dismissed all other objections as minor matters easily dealt with.

Within a short time, Leakey sent Goodall back to London to study primates for a year. Next, he talked a small foundation in Des Plaines, Illinois, into putting up $3,000 to buy Goodall a tent, some supplies, and a plane ticket back from London. Then, Leakey secured a government permit to conduct a study on the Gombe Stream Chimpanzee Reserve. However, when officials at the Game Department learned that the actual study was to be done by a young woman, they refused to permit her to live alone in the bush. Goodall solved this problem without Leakey's aid—she simply took her mother along!

So, in 1960 Jane Goodall and her mother headed into the bush to observe the behavior of chimpanzees in the wild. Each day Goodall climbed into the mountains to seek the chimps while her mother passed the day in camp, sometimes providing minor first aid to local fishermen. And every evening Goodall returned disappointed. The chimps wouldn't let her near them. Once in a while, she spotted them on a distant slope, but she could never get close enough even for the most superficial observations. This went on for six months. Then, one day Goodall walked into a clearing, and there, less than twenty yards away, were the two largest male chimps calmly grooming themselves. Soon a female and a youngster peeked at Goodall from the tall grass nearby. Goodall sat motionless. After about ten minutes, the two big chimps arose, looked at her, and sauntered away. She was ecstatic:

The depression and despair that had so often visited me during the preceding months were as nothing compared with the exultation I felt when the group had finally moved away and I was hastening down the darkening mountainside to my tent on the shores of Lake Tanganyika. (Goodall, 1971)

Baron Hugo van Lawick © National Geographic Society

Thus began a decade during which Goodall virtually became a member of this pack of chimpanzees. She came to know each one and learned to recognize and simulate the facial expressions, gestures, and hoots by which they communicated. And once the chimps began to let her come close, Goodall soon made discoveries that brought her international fame.

First of all, Jane Goodall saw chimps make and use tools. As she looked on, several chimps plucked long blades of grass, stripped off the leaves, licked one end to make it sticky, and poked the stem into a termite nest, pulling it out covered with termites. These they licked off as a child would a lollipop. Was this tool use? A **tool** is an object that has been *modified* to suit a particular purpose. A long blade of grass is not a tool. But when it has been stripped so it will slip down the hole in a termite nest, and when it has been made sticky so the termites will adhere to it, and when it is used to pull out termites, then it is a tool. Granted it isn't a very fancy or sophisticated tool. But then neither were the stone tools the

Jane Goodall accepts a banana from a wild chimpanzee on the Gombe Reserve. But the chimps were not always this friendly. It took Goodall six frustrating months to get them even to let her come within sight. But she kept trying until finally the two largest males, apparently as curious as she, let her come quite close. Over the next several years, Goodall established close relationships with the chimps, learning to imitate the hoots and gestures they used to express themselves, and eventually, she got to be one of the troop.

BOX 5-3   BIRD BRAIN

CHIMPANZEES aren't the only animals that can make and use tools. The scientific world recently was stunned when a careful study published in *Science* by three Oxford zoologists demonstrated that crows can do so, too (Weir et al., 2002). Without any training, this female crow—captured as a juvenile bird in New Caledonia—bent a wire to form a hook and used it to lift food from a vertical plastic pipe. The crow has repeated this performance many times. As the authors concluded, "Our finding, in a species so distantly related to humans and lacking symbolic languages, raises numerous questions about the kind of understanding of 'folk physics' and causality available to nonhumans, the conditions for the abilities to evolve, and their associated neural adaptations."

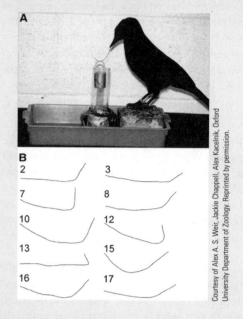

Courtesy of Alex A. S. Weir, Jackie Chappell, Alex Kacelnik, Oxford University Department of Zoology. Reprinted by permission.

Leakeys were discovering in Olduvai Gorge. In any event this one observation was sufficient for Louis Leakey to tour the international lecture circuit proclaiming that Goodall's work meant we must revise our definition of humans as "the tool-making animal." Chimps could do it, too.

Goodall's second sensational discovery directly challenged the image of humans as depraved killer apes. Maybe humans did not take up meat eating until late in their evolution, as Lorenz and others have claimed, but chimpanzees, who clearly are less evolved primates than are the hominids at Olduvai, eat meat. Regularly. Goodall frequently saw them catch, kill, and devour bushbucks and bush pigs. She also saw them frequently kill and eat monkeys and young baboons. Indeed, Goodall saw her chimps throw rocks and use sticks as clubbing weapons.

These early discoveries ensured Goodall's fame and ended all problems of financing further research—the National Geographic Society gave her annual grants. But what has ensured her immense scholarly reputation is the cumulative impact of her work. Goodall has now spent more than thirty years, off and on, with her chimps. Over the years she has been able to modify and clarify many of her early conclusions on the basis of her increased knowledge. Indeed, Goodall (1990) eventually discovered that, like

their human cousins, chimps sometimes commit murder; that is, they kill other chimps. This observation by Goodall destroyed the last element in the charge that humans are a unique, unnatural, depraved species of killers.

Moreover, despite her lack of formal training, Goodall has earned the uniform respect of her scientific colleagues for her superb mastery of observational field methods. Always evident in her work are these primary rules:

■ Record everything you notice, always, and file it so you can find it again.

■ After you have drawn a conclusion, search for evidence that *you are wrong*.

■ Never, ever think you understand everything that's going on.

Goodall was only the first person to do field research on chimpanzees. Soon many others followed her example, and in time it became possible to compare the behavior of chimps in various troops in various parts of Africa. The results leave no room to doubt that chimps have cultures. In 1999 nine field researchers (including Jane Goodall) published a joint article in *Nature*, one of the world's most respected scientific periodicals. In it they listed thirty-nine distinct behavior patterns and use of tools found in some troops and not in others, for which ecological explanations can be eliminated (Whiten et al., 1999). The tools include hammers, anvils, sponges made of chewed-up leaves, combs made by shredding a stem, stepping sticks used to walk over thorns, digging sticks, and clubs. The fact that these examples of technology are not universal among chimpanzees, but that only some troops have any given example, means that the use of such tools is learned and passed down from parents to children.

Of course, chimps have only very simple technology. But if we are to dismiss these tools as not adding up to culture, then we must doubt whether human groups in the Stone Age and earlier had cultures. To keep our ancestors within the human race, we must accept the very simple definition of tools already given, as purposeful alterations of objects. And if we do, then we must acknowledge the existence of technology among chimpanzees. Although recent human culture is immeasurably more complex and powerful than animal culture, we are not unique as toolmakers or as teachers of the young.

Studies of chimpanzees are not the only work to undermine the notion that animal behavior is mainly unlearned and instinctive. A lot of animal behavior is instinctive, of course. That is, a pattern of behavior is somehow programmed into the biological heritage of a species so that all normal members of the species exhibit this behavior automatically and without any learning. For example, squirrels raised in an isolation cage and fed only a liquid diet will, in their first trial, bury nuts in exactly the same way as do adults that have observed and practiced nut burying from infancy (Eibl-Eibesfeldt, 1970). We can cite such examples almost indefinitely.

However, the list of animal behaviors that are *not* instinctual but that must be learned also has been growing in recent decades. What could be more catlike than bathing? Yet kittens separated from their mothers at a young age grow up to be cats that do not bathe. Moreover, many animals raised in a zoo are unable to mate. Thus, in 1975 the Sacramento Zoo in California finally resorted to showing a movie of two adult gorillas mating to their prized young gorilla pair, who seemed to want to mate but didn't know how. The film successfully provided them with the needed sex education. Clearly, then, for these species these are learned patterns of behavior that adults must pass on to the young just as humans must pass along their culture.

Harlow Primate Laboratory, University of Wisconsin

For many scientists some of the most overpowering evidence of animal culture came from Harry and Margaret Harlow's years of experimentation with rhesus monkeys at the University of Wisconsin. In one major experiment, the Harlows (1965) raised three groups of infant monkeys. Monkeys in the first group were raised in total isolation and fed mechanically. Those in the second group were raised in isolation cages, each with its mother. Infants in a third group were raised together but isolated from adults. Later, all of the monkeys were placed in a normal monkey colony, where their behavior was carefully recorded and assessed.

The behavior of monkeys raised in total isolation was extremely abnormal. They avoided all contact, cowered in corners, and never learned to engage in sex. Perhaps surprisingly, the monkeys who had been isolated with their mothers were nearly as abnormal as the first group; they also failed to learn sexual behavior or to adjust to social relations. Those raised in isolated groups without mothers showed the least abnormality and adjusted best to life in

In another famous experiment, the Harlows isolated infant monkeys with two artificial "mothers." The mother at left is made of wire and has a nipple at which the infant can nurse. The second mother has no nipple but is made of cuddly terry cloth. When the Harlows frightened an infant, it fled not to the wire mother (as Freudian theorists predicted) but to the terry cloth mother. This was evidence that the primary tie between infants and their mothers is not based on nursing but on cuddling.

the colony, but they never became completely normal either.

These results indicate two things. First, monkeys must be exposed to monkey society from an early age, or they will fail to become normal monkeys. Second, many of the effects of isolation seem irreversible. None of the monkeys ever fully made up the deficits of infant isolation. As we shall see in Chapter 6, the effects of severe deprivation on humans bear a striking resemblance to the effects on the Harlows' monkeys.

NONHUMAN LANGUAGE

One reason nonhuman animals cannot develop more elaborate cultures is their lack of efficient communication. We recognize that animals do have some capacity for communication. Cries and calls meaning "Danger!" or "I'm looking for a mate!" are widespread among birds and other animals. Also, animals often communicate with gestures and body movements. Male baboons challenge others to a fight by displaying their very light-colored eyelids, for example. But animals can communicate little information through such means compared with the immense amount that human language can communicate quickly and accurately. Lacking language, other animals have been stymied in sharing their experience and thus in accumulating and passing on knowledge. Consider how little a human parent without speech could teach a child. Teaching by showing is slow, and it seriously limits what can be taught. Without doubt, language is the human trait most responsible for our great superiority over other animals. If we are the kings of beasts, we talked our way into the title.

But why do only humans have language? No one is quite sure. For centuries people assumed that only humans were sufficiently intelligent to develop speech. But in recent decades many careful observers of the higher apes, such as chimpanzees, have been uneasy with this assumption. Through long observation of ape behavior, they became convinced that apes display considerable intelligence. Their ability to solve problems, for example, is sufficient to suggest that they are smart enough to learn a language. And this has led to a series of efforts to teach chimpanzees to speak.

For example, several couples have taken an infant chimpanzee into their home and raised it with their own children, hoping that the chimp would learn speech as the human child did. The Kelloggs (1933) may have been the first to try. Their chimp did not learn to speak, but their child learned to scrape paint off the wall with his teeth and to give

the characteristic chimp "food bark." Perhaps the most celebrated of these attempts involved a chimp named Viki, reared by two psychologists in the late 1940s. In six years of intense effort, they managed to teach Viki to utter four sounds which some listeners thought crudely approximated English words (Hayes and Hayes, 1951).

Speaking is impossible for chimps because they lack the vocal apparatus needed to produce speech sounds. In the wild they are usually silent and produce vocal sounds only during moments of extreme excitement. To try to teach them to talk was a little like trying to teach a human to swing by the tail. And blaming their failure to talk on a lack of intelligence was a little like blaming the human inability to swing through trees on stupidity instead of an unsuitable physique.

A CLOSER VIEW
WASHOE LEARNS TO SIGN

The breakthrough in determining the language-learning capacity of higher primates came in the late 1960s, when two daring psychologists at the University of Nevada at Reno realized that language does not require speech. In the wild, chimps seem to communicate mainly through gestures. Although these gestures do not constitute anything like a language, Beatrice and Allen Gardiner struck upon the idea of exploiting this natural tendency of chimps by attempting to teach a chimp American Sign Language (used for communication by the deaf).

In June 1966 the Gardiners obtained a year-old female chimpanzee from Africa, whom they named Washoe. Verbal speech was never used in Washoe's presence; only American Sign Language was used. The Gardiners also undertook teaching sign language to Washoe. They were soon successful. The chimp began to acquire a vocabulary, much as any human toddler begins to pick up words. In time Washoe began to form the simple word combinations that pass for sentences among young humans (Gardiner and Gardiner, 1969).

As word began to leak out in the scientific community of these efforts, the Gardiners were much ridiculed, and their claims were quickly dismissed. But the Gardiners kept on working, and Washoe continued to develop a bigger vocabulary and to fashion more sentences. Scientists were impressed when the Gardiners invited some deaf people without any previous contact with the project to come and communicate with Washoe. The scientists were sign-deaf, but the deaf people

Courtesy of Lynne Roberts

had no difficulty understanding Washoe. And Washoe understood them.

By October 1970 five-year-old Washoe had a vocabulary of 160 signs (Fleming, 1974). Since chimps mature at much the same rate as humans, this was no minor accomplishment for her age. However, she was also getting big. A full-grown chimp weighs more than 120 pounds and is much stronger than any human male. So the Gardiners decided it was time to find her a different environment. Roger Fouts, who had studied with the Gardiners and worked with Washoe, took her with him when he joined the Institute for Primate Studies at the University of Oklahoma.

In Oklahoma Washoe joined a number of other chimps and monkeys gathered for research on communication. A number of them were taught sign language, and Fouts and his colleagues were then able to study communication among chimps (in contrast with chimp-human communication). They were especially eager to see what would happen when chimps who had learned sign language had infants. Would they teach their infants to sign?

When Washoe was sexually mature, she was allowed to mate, and the researchers waited for her to give birth. Unfortunately, she miscarried. Two subsequent pregnancies ended in miscarriage and in the early death of the infant. So Fouts obtained an infant male chimp named Loulis from the Primate Research Center at Emory University and gave him to Washoe to raise. Within eight days of being adopted by Washoe, Loulis began to imitate her signs. By

the age of seventeen months, Loulis knew ten signs, including "hug," "drink," "food," "fruit," and "give me." Thus, chimps can transmit at least some language to their young.

What do chimps have to say when they learn sign language? Obviously, they do not begin to recount the history and traditions of chimpanzees. Lack of language has prevented such a culture from ever developing. If chimps do develop their own history, it can begin only now. Perhaps they will not prove to be sufficiently bright to develop and pass on a complex language or to contemplate their own existence. But before you dismiss the idea of chimps developing an elaborate culture as something that could happen only in *Planet of the Apes,* ponder the following account based on Washoe's first few days in her new Oklahoma home.

When she arrived in Oklahoma, Washoe had never seen any monkeys. Fouts therefore taught her a new sign: "monkey."

She was happy to use it for the squirrel monkeys and for the siamangs, but she concocted a different name for a rhesus macaque who had threatened her. She called him *dirty monkey.* When Fouts asked her the sign for the squirrel monkeys again she quickly went back to just plain monkey. But when they returned to the macaque, it was *dirty monkey.* Before this incident, Washoe had used *dirty* to describe only soiled objects or feces. Since her meeting with the aggressive macaque, she has applied this sign to various teachers when they refuse to grant her wishes. (Fleming, 1974)

Washoe had invented an appropriate invective for what she didn't like. And as with any human five-year-old, her first "bad word" was the equivalent of "poop."

Despite all the excitement stirred up by the ability of Washoe and other primates to master sign language, research on primate communication did not develop into a major area of social scientific research. Once social scientists had established that primates *could* do it, there seemed little to be learned from training more of them to do it. Instead, the focus has shifted from nonhuman communication to the genetic basis of language among humans. In the next chapter, we will examine some truly exciting breakthroughs that recently have been achieved.

CONCLUSION

This chapter has traced the radical turnabouts in how sociologists and other social scientists have dealt with the implications of biology for society and culture. At the start of the twentieth century, many social scientists attributed most individual behavior and many cross-cultural differences to biology. Then biology was excluded from all consideration. More recently, scientists have recognized the fascinating interactions among biological, cultural, and social elements. Moreover, the study of animal behavior, once the sovereign province of biology, has now attracted the interest of many social scientists. As a result, in about eighty years, we have moved from the view that nearly all human behavior is produced by instincts governed by heredity to the view that much animal behavior is learned rather than instinctive.

The next chapter builds on the interplay of biology, culture, and society as we ask how the potential humanity of newborn infants is fulfilled. We shall see that the process of socialization does not simply mean growing up or how children are raised. In a significant sense, we never finish growing up. Socialization is a lifelong process.

Review Glossary

Terms are listed in the order in which they appear in the chapter.

instinct Any behavior that occurs in all normal members of a species without having been learned.

chromosomes Complex genetic structures inside the nucleus of a cell, each containing some of the basic genetic units (genes) of the cell. Chromosomes combine in pairs; thus, in humans twenty-three chromosomes from the father combine with twenty-three from the mother.

genes The basic units of heredity within which specific genetic instructions are encoded in complex chemical chains.

genotype The sum total of genetic instructions contained in an organism's genes.

phenotype The observable organism as it has developed out of the interplay between the genotype and the environment.

behavioral genetics A scientific field that attempts to link behavior, especially human behavior, with genetics.

hormones "Chemical messengers" excreted into the bloodstream by the glands comprising the endocrine system.

testosterone The male sex hormone.

estrogen The female sex hormone.

tool An object altered or adapted from natural materials that increases the ability of an organism to achieve some goal.

Suggested Readings

Booth, Alan, and James M. Dabbs Jr. 1993. "Testosterone and Men's Marriages." *Social Forces* 72:463–477.

Bouchard, Thomas J., Jr., David T. Lykken, Matthew McGue, Nancy L. Segal, and Auke Tellegen. 1990. "Sources of Human Psychological Differences: The Minnesota Study of Twins Reared Apart." *Science* 250:223–228.

Gardiner, R. Allen, and Beatrice T. Gardiner. 1969. "Teaching Sign Language to a Chimpanzee." *Science* 165:664–672.

Goodall, Jane. 1971. *In the Shadow of Man.* Boston: Houghton Mifflin.

Goodall, Jane. 1990. *Through a Window: My Thirty Years with the Chimpanzees of Gombe.* Boston: Houghton Mifflin.

Meltzoff, Andrew N., and M. Keith Moore. 1994. "Imitation, Memory and the Representation of Persons." *Infant Behavior and Development* 17:83–99.

Sociology Online

www.socstark10.com

GO TO THE INTERNET AND TYPE: www.socstark10.com.

↗**CLICK ON:** 2000 General Social Survey

✓**SELECT:** PAIN: During the past four weeks, how much did pain interfere with your normal work (including both work outside the home and housework)?

↗**CLICK ON:** Analyze Now

Pain is a blend of the biological and the cultural. People usually experience pain because of something affecting their bodies, but how much pain they feel, and especially how much they are willing to admit feeling pain, are shaped by culture.

TRUE OR FALSE: Older people are significantly more likely to report having pain.

TRUE OR FALSE: Women are more likely than men to report pain.

TRUE OR FALSE: People with high family incomes are less likely to report pain.

Can you think of a way in which all three of these findings might be related?

↗**CLICK ON:** Select New Data Set

↗**CLICK ON:** The 50 States

✓**SELECT:** PLASTIC DR. Plastic surgeons per 100,000 population.

↗**CLICK ON:** Analyze Now

Is there a substantial regional pattern? Which state is highest in plastic surgeons? Which one is lowest? Can you suggest something about the cultures of these two states that might account for this difference?

↗**CLICK ON:** Select New Variable

✓**SELECT:** % FAT: Percent of population 18 and over who are "overweight."

↗**CLICK ON:** Analyze Now

Is there a substantial regional pattern?

↗**CLICK ON:** Select New Data Set

↗**CLICK ON:** Nations of the Globe

✓**SELECT:** MEAT EAT: Annual meat consumption (excluding fowl and fish) per person in pounds.

↗**CLICK ON:** Analyze Now

What might the top five nations have in common? Examine the nations where people eat less than five pounds of meat per year. What might they have in common?

USING INFOTRAC COLLEGE EDITION

GO TO THE INTERNET AND TYPE:
www.InfoTrac-college.com

↗**CLICK ON:** Register New Account

You will be asked to enter your Access Code and to create and enter a User Name and Password. Having done so,

↗**CLICK ON:** InfoTrac College Edition or Log On.

If you found the sections on chimpanzees and on Jane Goodall interesting, the following search terms will turn up more on these topics. And you can find a lot more on identical twins, too.

SEARCH TERMS:

Chimpanzees

Goodall, Jane

Twins, Identical

CHAPTER

6

Socialization and Social Roles

SEVERAL YEARS AGO a Seattle judge gave permanent custody of a six-year-old boy to the county welfare department. He asked that efforts be made to have the child adopted, despite protests from the child's father and mother. The judge based his decision on the fact that the child could not speak intelligibly, crawled rather than walked, and barked like a dog when people approached him. When found in his home by social workers, who had been alerted by a neighbor, the child was "filthy, smelled of urine, his teeth were rotten, he had sparse and brittle hair and a pale, pasty complexion." After placement in a foster home, the boy was soon toilet trained and began to walk and talk.

This little boy's tragic condition was the result of isolation and almost total neglect. From infancy he had spent most of his days and nights alone in a filthy one-room house. His father was in prison, and his mother was rarely home, stopping by now and then only to feed him—it's a wonder that he didn't die from eating spoiled food. Aside from his mother, he rarely saw a human being, and it's not clear he even knew he was a little boy.

There have been countless cases like this, of children whose parents so neglected them and so isolated them from all human contact that when they were discovered, they acted more like wild animals than human beings. In fact, children like this are often called **feral children** (the word *feral* means "untamed"), and some people have mistakenly assumed that such children had been reared in the wild by an animal (Malson, 1972; McClean, 1978; Shattuck, 1980).

Human children cannot be raised by mother wolves, dogs, or other animals—such stories are fantasies. Unfortunately, demented or cruel human parents can raise their children like ani-

CHAPTER OUTLINE

HUMAN DEVELOPMENT

Suppressing Development

Accelerating Development

A CLOSER VIEW

Studying the Mythical "Mozart Effect"

COGNITIVE DEVELOPMENT

A CLOSER VIEW

Jean Piaget: Theory of Cognitive Stages

Language Acquisition

A CLOSER VIEW

Noam Chomsky: Universal Grammar

Evidence of a Language Instinct

A CLOSER VIEW

Judy Kegl: The Nicaraguan Sign Language Study

EMOTIONAL DEVELOPMENT

Emergence of the Self

Personality Formation

CULTURE AND PERSONALITY

DIFFERENTIAL SOCIALIZATION

A CLOSER VIEW

Melvin Kohn: Occupational Roles and Socialization

A CLOSER VIEW

Erving Goffman: Performing Social Roles

SEX-ROLE SOCIALIZATION

A CLOSER VIEW

DeLoache, Cassidy, and Carpenter: Baby Bear Is a Boy

Women and Science

CONCLUSION

REVIEW GLOSSARY

SUGGESTED READINGS

SOCIOLOGY ONLINE

mals. Sometimes children thought to be feral are actually victims of mental retardation or severe mental illness, not neglect. But others, like the little boy in Seattle, are born with normal capacities, for they make rapid progress once they are rescued from isolation (Davis, 1940, 1947, 1949).

Feral children demonstrate an important principle: Our biological heritage alone cannot make us into adequate human beings. Only through social relations—constant intimate interaction—can the rich cultural legacy that sets humans apart from other animals be transmitted to new humans. An infant is born without culture. Reared in isolation a human being will not even learn to talk or walk, let alone to sing or read. Consider that the little boy in Seattle often was exposed to television—his mother frequently left the set on while she was away—but these electronic images apparently conveyed nothing to him. No matter how much talk he heard on TV, he did not learn to speak until he was surrounded by live human beings.

The learning process by which infants are made into normal human beings, possessed of culture and able to participate in social relations, is called **socialization.** This process, which literally means to be "made social," begins at birth and continues until death; we never cease to be shaped by our interactions with others. As we pass through life, what others expect of us and what roles we are expected to fill all change. And we change, too.

In this chapter we shall examine how humans are socialized. First, we will see how infants acquire basic skills such as speech and the ability to reason. Next, we examine how we discover that we are human—that we have a unique and independent identity separate from the external world. That is, how do humans develop a *self?* We then consider the concept of *personality.* What is its place in social science? How do people develop personalities? This sets the stage for an extended look at the link between culture and personality. How do societies shape people so they fit into particular cultural patterns? Here we will assess the rise and fall of cultural determinism.

Midway in the chapter, we reach the central sociological contribution to the study of socialization: Even in the same family, not all children are socialized in the same way. Instead, from infancy we are sorted out on the basis of the roles our parents expect us to perform as adults. Moreover, each time we enter a new role, we undergo additional socialization as we learn how to perform it adequately. For example, much socialization is involved in performing such roles as parent, spouse, teacher, lawyer, or priest. Because of the crucial link between roles and socialization, the rest of the chapter will explore *differential socialization*—socialization related to roles and to role expectations. Next, we will connect roles and differential socialization with the development of *self-conceptions*—our perceptions of ourselves. Finally, we apply the principles of socialization to a major focus of current sociological interest: *gender.* How are we socialized on the basis of roles rooted in male-female differences? That is, how are infants raised to be boys and girls, men and women?

HUMAN DEVELOPMENT

Physicians and psychologists have done much of the research on socialization, aiming to discover the normal processes involved in the biological and physical development of infants and children. Their ultimate aim is to further discover how normal development can be modified by the environment.

Years of testing infants at various ages have made it possible to determine patterns of normal development. For example, with reasonable parental care, children usually learn to stand up by about the age of eight months and to walk by about fourteen months.

Statistical norms like these help parents and physicians be alert for signs of slow development, in which case treatment or special training may be sought. But the norms also serve as a basis for assessing whether development is impeded by particular environmental circumstances or even whether enrichment of the environment can accelerate development.

SUPPRESSING DEVELOPMENT

Chapter 5 mentioned the work by Harry and Margaret Harlow (1965) with infant monkeys raised in isolation. As with feral human children, monkeys raised in isolation failed to develop normal monkey skills and displayed symptoms of acute maladjustment when placed in contact with normal monkeys. Similar studies cannot be conducted on human infants, yet cases of child neglect have supplied comparable data. For example, researchers have studied infants raised in orphanages for signs of impaired development.

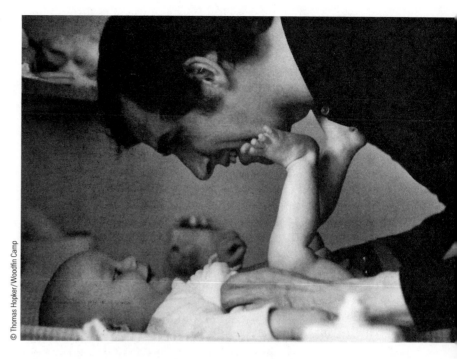

© Thomas Hopker/Woodfin Camp

In a classic study, Skeels and Dye (1939) conducted research in an orphanage on infants whose development at nineteen months was judged so retarded that they were considered unfit for adoption. These infants were transferred to an institution for the mentally retarded, where Skeels and Dye arranged for each to be put under the personal care of an older mildly retarded girl. Four years later the infants showed dramatic improvement. Their estimated IQs had risen by an average of thirty-two points, while similar infants who had remained in the orphanage had lost an additional twenty-one points over the same period.

Twenty-seven years later, Skeels (1966) did a follow-up study of these same subjects. Most of the orphans who had been mothered by retarded girls had graduated from high school, and a third had gone to college. Nearly all of them were self-supporting and rated as normal. However, most of the people who had been in the control group and remained in the orphanage had not progressed beyond the third grade and either remained in institutions or were not self-supporting.

Clearly, infants who lie unattended and unstimulated in crowded orphanages are deprived in much the same way as the Harlows' monkeys and as feral children. As a result, despite a normal biological heritage, they fail to develop normally. Skeels and Dye's research prompted major reforms in the treatment of orphaned infants to prevent this sort of retardation. Today,

While special efforts to speed up infant development have little effect, neglect can cause development to be tragically retarded. Children need a great deal of daily interaction like this to fulfill their potential. This is not surprising if we remind ourselves how much each infant must master in a few short years to be normal.

most such infants are placed in foster homes rather than in institutions and thus receive more normal levels of attention. Even when infants are raised in institutions, great effort is made to hold and cuddle each one often.

ACCELERATING DEVELOPMENT

If normal development can be suppressed, is the reverse also possible? Can normal development be speeded up? Lots of people seem to think so. Thus, each year millions of parents accept claims about new technologies that can accelerate their children's development. One of the most famous of the recent accelerating technologies is the "Mozart Effect" which promises that the IQs of infants and young children can be raised substantially if they are exposed to classical music—especially that of Wolfgang Amadeus Mozart (1756–1796). Consequently, in South Dakota, Georgia, and Tennessee, every new mother is provided with a free CD of Mozart's music. Florida now requires that, to receive state aid, all child-care facilities must play classical music for at least half an hour a day. Concerned parents made a bestseller out of Don Campbell's *The Mozart Effect: Tapping the Power of Music to Heal the Body, Strengthen the Mind, and Unlock the Creative Spirit* (Avon Books, 1997), and several Mozart Effect CDs have been high on the *Billboard* charts. The size and speed of the response are due to the fact that the Mozart Effect is not some unsubstantiated claim but is firmly based on scientific research. Or is it? Let's watch.

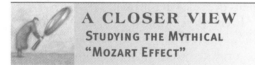

A CLOSER VIEW
STUDYING THE MYTHICAL "MOZART EFFECT"

The October 14, 1993, issue of *Nature*, one of the oldest and most prestigious scientific journals in the world, published a study by Frances Rauscher, Gordon Shaw, and Katherine Ky, researchers at the University of California, Irvine. Their experiment was based on thirty-six undergraduate students. Having arrived to participate in an experiment, each student was given a paper folding test selected from a section of the Stanford-Binet IQ Test. After being tested, the students were randomly divided into three groups. The twelve students in the first group of subjects spent ten minutes listening to a tape of Mozart's Sonata in D Major for Two Pianos. The second group heard ten minutes of

a "relaxation tape" instead of Mozart. Those in the third group were directed to sit in silence for ten minutes. Then all subjects were given a second paper folding task from the Stanford-Binet. Those who had heard Mozart scored about eight points higher on the second test than on their first, while the scores of those who sat in silence or who heard the relaxation tape did not improve. The difference was statistically significant. Ten minutes after taking the second test, all students were given an additional paper folding test. This time all groups scored at their original level—the IQ increase of those who heard Mozart had lasted only a few minutes.

And that's it. That is the "scientific" evidence on which the claims about the Mozart Effect were based. Notice that the subjects were not infants or young children. Notice, too, that the effect was small and very short lived. Moreover, the researchers did not use the term *Mozart Effect*, nor did they interpret their results in support of accelerating mental development or of raising IQs. Instead, the importance given to their study, which caused it to be published in a distinguished journal, was entirely concerned with implications for a complex model of neural functions. Unfortunately, these aspects of the research were soon lost as the findings were picked up by the mass media and repeated by writers who did not consult the *Nature* article but who added new speculative interpretations of their own to previous media accounts. Had any of them bothered to read the original article (it took up less than one page), they would have realized at once that it offered no justification for playing Mozart to infants—not even if the effect reported in *Nature* really occurs. But does it?

Recall from Chapter 3 that statistical significance does not rule out the possibility that an observed effect is nothing but a random fluke. All a significance test does is specify the odds against a result being random. In the case of the original Mozart Effect experiment, the odds were calculated to be at least twenty to one that the result was not merely random, thus satisfying the minimum standard observed by researchers. Of course, since thousands of experiments are conducted each year, chances are that many of them will meet the minimum standard of significance even though they are nothing but a random result. How can we find out which ones these are? By *replication*, by doing the research over again. The odds are stacked against getting two random results in a row, each of which meets the twenty to one standard.

In 1999 Christopher F. Chabris, a Harvard psychologist, published an article in *Nature* which summarized the results of the sixteen studies that had attempted *unsuccessfully* to replicate the Mozart Effect. So, now we know that it was just a fluke. It surely will do infants and children no harm to listen to Mozart, but it won't make them smarter—not even for a few minutes.

There are two reasons to doubt *all* claims about accelerating the development of infants and young children. First, at least in most societies, nearly all infants receive adequate stimulation to achieve normal development. Second, children cannot develop faster than their physiological development will permit. For example, certain fundamental neural developments must occur before certain mental activities can take place (Bruer, 1999). Coaching in advance of these stages of maturation is to no avail, as we now shall see.

COGNITIVE DEVELOPMENT

For a long time, it was believed that cognitive abilities such as reasoning and speaking were entirely the result of learning. Consequently, learning has always been the central aspect of social scientific approaches to socialization. And for several generations, **stimulus-response (S-R) theory** dominated social scientific explanations of how humans learn. The S-R theory is relentlessly social. It proposes a simple model of learning in which humans play only a passive role. The theory maintains that behavior is merely a response to external stimuli (hence its name), that what humans do is to repeat whatever behavior has been reinforced by their environment in the past. According to this view, the human brain is little more than a computer hard drive, capable of storing and retrieving past reinforcements but not able to play an active role even in the memory process. Thus, the brain does not condense past experience to formulate general rules which then guide future behavior. For example, the S-R approach dismisses the idea that people acquire grammatical rules as part of learning to talk. Instead, it postulates that we acquire language word by word and then sentence by sentence as we repeat what

we have heard, *if* we are reinforced for repeating it. In a later section, we will see why this view of language acquisition is not now tenable (if it ever was).

S-R theory did not limit itself to speech but went so far as to propose that the same process of reinforcement provides all of our higher mental capacities as well—that in effect people don't really reason. Of course, some social scientists such as the symbolic interactionists wouldn't go quite that far and left some room for individuals to think and to shape their behavior to some extent. But even most of these dissenters stressed the social sources of learning and dismissed studies of brain physiology as merely a branch of medicine. While admitting that brain damage may prevent people from learning, they argued that neural anatomy is largely irrelevant to understanding how people learn or think. Nevertheless, even in the heyday of S-R theory, a pioneering Swiss scientist was discovering otherwise.

A CLOSER VIEW
JEAN PIAGET: THEORY OF COGNITIVE STAGES

Jean Piaget (1896–1980) never ceased to be amazed by the fact that a strict interpretation of S-R theory leads to the conclusion that no such theory can be formulated. If nothing more than a process of responses to external stimuli is taking place inside the brain, invention is impossible. How can anyone ever say a new sentence, for example, if we only acquire language by repeating what we are taught? *Any behavior that goes beyond what is present in our environment cannot possibly be a mere copy of the environment.* And Piaget pointed out that we invent all the time. In fact, much of modern mathematics cannot possibly be regarded as a reflection of external reality because it has no counterpart in reality. It is a human mental creation. Thus, Piaget (1970) wrote, "To present an adequate notion of learning one first must explain how the [person] manages to construct and invent, not merely how he repeats and copies."

Early in his career, Piaget became convinced that the human mind develops and functions on the basis of **cognitive structures,** or general rules that govern reasoning. His initial insight into this matter came from administering IQ tests to youngsters. What struck him were the consistent patterns of *wrong answers* to

The **sensorimotor stage** begins at birth and lasts until around the age of two years. During this period, infants discover and develop their senses and their motor skills. A major cognitive discovery during this stage is the *rule of object permanence*—the principle that objects continue to exist even when they are out of sight. Young infants immediately lose all interest in an object as soon as it is blocked from view; they do not search for it. But by about ten months, children will search for an object when it is suddenly covered with a cloth or otherwise removed from sight: They know the rule of object permanence. However, rather than searching for an object where it was last seen, young children often will search where they last found the object.

The **preoperational stage** begins at about age two and ends at about seven. The earliest years of this period are devoted to language learning. However, the other major task during this period is to overcome egocentrism, or to use a symbolic interaction concept discussed in Chapter 3, to learn to take the role of the other.

During the preoperational stage, children cannot solve problems that require them to put themselves in someone else's place. In a classic experiment, Piaget and Edith Meyer-Taylor constructed a model mountain range out of clay. Children were placed by the model and asked to describe it. Then they were asked to describe how it would look from where another child was standing (from a quite different perspective). Children in the preoperational stage were unable to comprehend the request and continued to describe the model from where they stood. As George Herbert Mead pointed out, this kind of limitation prevents younger children from participating adequately in team games.

The **concrete operational stage** begins at about seven and ends at about twelve (although many people never progress beyond this stage). In this stage children develop a number of logical principles that permit them to deal with the concrete, or observable, world. One such principle is the *rule of conservation*, which states that a given amount of material does not increase or decrease when its shape is changed. Children still in the preoperational stage do not yet understand this. For example, if you present them with identical clay balls and then flatten one of the balls, they will say that the flattened ball is smaller and contains less clay. Similarly, when such children are presented with two identical rows of checkers and then one row is clustered into a smaller space, they will say

The elderly man sitting on the bench, smoking his pipe and watching the children play, is Jean Piaget, the famous Swiss psychologist. Piaget's work on stages of cognitive development redirected the field.

open-ended questions. That is, time and again kids gave the same wrong answer to a given question. Why? Piaget concluded that the children were applying the same, but incorrect, rule to a problem.

This led Piaget to suspect that cognitive development involves coming to comprehend a set of basic principles or rules of reasoning; normal development consists of acquiring these rules by particular ages. Thus, the difference between children who correctly answered a particular IQ test question and those who chose the same incorrect answer was that the first group had acquired the rule needed for giving a correct response, whereas the other group was still applying an inadequate or incorrect rule.

Thus, Piaget set out to discover basic rules of reasoning and the ages at which normal children acquire them. First, he carefully observed his own children as they grew up; then he conducted a long series of experiments with large numbers of children. In the end he proposed that cognitive development passes through four fundamental stages: (1) sensorimotor, (2) preoperational, (3) concrete operational, and (4) formal operational. In a number of experiments, he showed that humans at one stage of cognitive development cannot solve problems requiring understanding at a higher stage and cannot be taught to solve these problems before they reach that stage of development. Let us briefly examine these stages and some of the pertinent experimental evidence.

the cluster contains fewer checkers than the row contains.

The **formal operational stage,** the final stage in Piaget's theory of cognitive development, generally begins at about age twelve. In previous stages children learn mainly by trying things out to see what happens. Eventually, however, some people learn to think abstractly and to impose logical tests on their ideas.

Recall the distinction between theories and hypotheses. Theories are abstract, general statements of a principle, whereas hypotheses are specific statements about the concrete world. In the concrete operational stage, people deal with the world only at the level of hypotheses, and they test hypotheses by examining concrete evidence. But when they reach the formal operational stage, people can formulate and manipulate theories and logically deduce from these theories that certain things are likely to be true or false. With this comes the ability to think hypothetically—to say "What if?" and then trace the logical implications of this supposition.

People still in the concrete operational stage miss the point of hypothetical assertions and often find them distressing. Suppose someone says, "Let's assume, for the moment, that humans were of only one sex and that every individual could bear infants. Would humans still form family units in order to make child rearing easier?" People in the formal operational stage are able to pursue such a hypothetical line of reasoning, add to it, and evaluate it. In so doing they may more fully understand why the family is a universal human institution (see Chapter 13). But people in the concrete operational stage will immediately object: "But people don't come in only one sex. Why are you saying something so stupid? How can I suppose that a lie like that is true? That's all just idle nonsense. Let's go study families if we want to really know what goes on."

Obviously, the formal operational stage never ends. Some humans can continually refine their ability to think logically and abstractly. Unfortunately, not everyone can do so. After testing large numbers of people, researchers have concluded that perhaps half of all adults do not reach the formal operational stage of cognitive development and thus are limited to literal interpretations of the world around them (Kohlberg and Gilligan, 1971). Several factors may prevent people from achieving the ability for formal operations. Some people may lack the necessary intelligence. Others may not have the opportunity or the need to develop powers of abstract reasoning at this level. Although all normal humans have constant practical experience with the concrete world, many have little occasion for abstract thought.

Because of Piaget's work, few social scientists now propose that all learning is purely the result of S-R mechanisms. Even many who devote most of their own research to improving the S-R model acknowledge that cognitive structures influence what is learned and when (Bandura, 1974).

Piaget's theory does not reject the importance of S-R mechanisms for a great deal of what we learn. Rather, it adds to this model. In particular it adds an active human consciousness that is capable both of formulating rules of reason to interpret the constant flow of environmental stimuli and of generalizing from a few instances. For example, by falling from several objects (such as trees and bicycles), we learn not only the specific lesson that it hurts to fall from them but also the general principle that falling from anything can hurt. That knowledge saves a lot of bruises.

Because Piaget discovered that it was impossible to accelerate passage from one cognitive stage to another, he always believed that the cognitive stages reflected underlying stages in the maturation of the brain. During his lifetime, research on the brain could offer no confirmation of this linkage. Today, however, many scientists believe such links can be demonstrated. Perhaps even more important has been the rapidly growing awareness among linguists that a fundamental ability to acquire language seems to be "hardwired" into the human brain. S-R theory contributes to our understanding of *what* language a child acquires and how well she or he is able to speak it—children learn these things from exposure to external stimuli and improve as they are corrected and rewarded. Nevertheless, language acquisition also seems to have a crucial genetic component—each normal infant is born having what linguists now refer to as a "language instinct" (Pinker, 1994). The **language instinct** was not discovered by scientists studying the brain. Here, too, social scientists led the way by observing children as they learned language and people in very special circumstances as they *invented* languages.

LANGUAGE ACQUISITION

Chapter 5 pointed out that at the start of the twentieth century most social scientists attributed much human behavior to instincts, but by the 1930s things had gone to the opposite extreme and most denied that humans had any instincts at all. As applied to language, the conventional S-R view claimed that everything we say is an imitation of things we have heard said and have been reinforced for repeating. We have seen that this approach broke down when Piaget tried to apply it to reasoning, since it precludes invention and originality. For quite similar reasons, the strictly S-R approach eventually broke down when it was applied to language, for the fact is that children often say things they could not possibly ever have heard said. For example, in a classic study of speech in early childhood, Roger Brown and Ursula Bellugi (1964) recorded many sentences such as:

"Cowboy did fighting me."

"Don't giggle me."

"A this truck."

"Write Cromer shoe."

They concluded that sentences such as these demonstrate that children experiment with speech in ways that appear to involve a search for grammatical rules—trial-and-error attempts to discover rules determining what words are allowed in what positions in sentences in their native language. Brown and Bellugi recognized that this was in accord with Paiget's ideas about cognitive structures underlying learning. These and many similar observations of children led the way to a revolution in linguistic theory.

A CLOSER VIEW
NOAM CHOMSKY:
UNIVERSAL GRAMMAR

Noam Chomsky, professor of linguistics at the Massachusetts Institute of Technology, identified three key aspects of language. First, language cannot consist of a set of learned words and sentences stored in a mental "hard drive" and trotted out as the need arises. Too much of what is said is an original combination of words that the person has never heard—often these are combinations that have never even been said before by anyone. Therefore, each child must possess a "program" that can construct an unlimited set of sentences out of the vocabulary the child has acquired. Chomsky called these programs mental grammars.

Second, children develop complex programs or grammars very rapidly and at a very young age. These enable them not only to construct and utter novel sentences but also to correctly interpret novel sentences and constructions they have never heard before.

Third, infants learn one language as easily as another: Chinese and French babies acquire speech at the same age. Hence, the underlying grammar must be similar in all infants rather than differ across cultures.

From these three points, Chomsky (1975) deduced that there must exist a universal grammar that is *inborn*. Chomsky pointed out that no one would accept claims that the "human organism learns through experience to have arms rather than wings." He then asked, Why do we "study the acquisition of a cognitive structure" as if it has no physical reality and can take any form imposed on it by external forces? He dismissed the seemingly immense variety in human languages as "superficial," noting that to a Martian, all humans speak a single language in that the same underlying principles are observed in each (Pinker, 1994).

Unfortunately, Chomsky's descriptions of the principles that constitute the universal grammar are extremely complex and impossible to explain adequately either briefly or in nontechnical language. Since this is not an advanced course in linguistics, and with full knowledge that Chomsky would disapprove of this summary of his work, it can be said that the universal grammar exists as an instinctive awareness of nouns and verbs and of how they can be combined. That is, as infants begin to observe the world and hear people use language, they instinctively distinguish between *things* and *actions* relating to things. In any event Chomsky was able to revolutionize the field of language acquisition because his theory stimulated new lines of research. As a result there is a growing body of evidence to support his predictions.

EVIDENCE OF A LANGUAGE INSTINCT

There are thousands of languages spoken on earth, and each is highly sophisticated. As Steven Pinker (1994) put it, we have encountered many "Stone Age societies, but there is no such thing as a Stone Age language." Rather, according to Edward Sapir (1921), "When it comes to linguistic form," the language of "the head-hunting savage of Assam" is equal to that spoken by Confucius or Plato. Efforts to find a "cradle" of language from which languages spread around the world have been to no avail, which is what would be expected if there is a language instinct, just as efforts to discover where humans first learned to breathe would be futile. However, the truly compelling evidence in favor of a language instinct comes not from normal human situations but from aspects of language development and acquisition in very unusual circumstances.

PIDGIN AND CREOLE From time to time, people speaking different languages have been forced to work together without an opportunity to learn one another's languages. This often happened as a result of the Atlantic slave trade that brought millions of people from Africa to the Western Hemisphere (see Figure 11-2). Frequently, a group of slaves spoke many different African languages and none spoke the language of their masters and overseers—English, Spanish, Portuguese, French, or Dutch. In situations such as these, in order to communicate, people develop a very simple form of speech called pidgin. **Pidgin** is a jargon of made-up nouns and verbs, often including some borrowed from the language of the dominant group (such as plantation owners or colonizers), highly variable in terms of word order, and with little in the way of grammar. Over time, pidgin can develop into a real language, entirely distinct from any other language, and having the full range of grammatical rules—such a language is called a **creole**.

Sometimes a pidgin is transformed into a creole very rapidly in a fashion that strongly supports the existence of a universal grammar. If infants are exposed primarily to pidgin—as in the case of a group separated from parents during the day and cared for by adults who speak to them in pidgin—at the age when they would normally acquire language, they do not simply learn to talk in pidgin. Rather, they begin to speak a creole that had not previously existed. That is, the children inject grammatical complexity into pidgin. They could not have learned this new creole because it was unknown to the adults around them; hence, they must have created it. Derek Bickerton (1981, 1999) discovered such an invention of creole among the children

of workers from China, Japan, Korea, Portugal, the Philippines, and Puerto Rico who had been brought to work on the Hawaiian sugar plantations early in the twentieth century. Even in the 1970s, the first generation could only communicate with one another in pidgin. But that's not the way their children spoke to one another: They spoke a unique creole using many terms from their parents' pidgin, but constituting a fully developed language that their parents could not understand.

Bickerton argued that when a creole is created by children, unprompted by hearing complex language, it reveals the nature of the inborn linguistic mechanisms of the brain. Thus, Bickerton went on to demonstrate the striking similarities among creoles made up of quite unrelated language mixtures. Bickerton's work has been criticized because it depended on reconstructions of past events—he did not actually observe children create a creole in Hawaii. This objection does not apply to an extraordinary study of the evolution of a new creole among deaf children in Nicaragua.

A CLOSER VIEW
JUDY KEGL: THE NICARAGUAN SIGN LANGUAGE STUDY

Many less developed nations make no provisions for educating deaf children. Consequently, those born deaf remain at home and are raised by parents or relatives who have no knowledge of a sign language. Usually, these children develop a very limited ability to communicate through pantomimes and gestures. For example, when they are hungry, they may point to their mouths and chew. Such systems of communication constitute a very simple pidgin, limited to a particular family.

This was the situation of deaf children in Nicaragua until 1979 when the government first instituted schools for the deaf. Deaf children from all over the country were brought to these schools where the staff attempted to teach them to lip-read and speak Spanish. These efforts were not very successful. One reason for the failure was that as soon as deaf students were assembled and enabled to play with one another, they began to pool the makeshift pantomimes they had used to communicate at home into a common pidgin. Very soon these pantomimes were streamlined into more efficient signs—increasingly simplified and abstract, thus making communication far more rapid. For instance,

Courtesy of Judy Shepard-Kegl, Nicaraguan Sign Language Project

Two girls in Nicaragua comment in sign language.

having an effective grammar strongly attests to Chomsky's claims about the existence of an innate human language capacity.

But if we are born with a substantial capacity for language, why didn't these deaf children each develop a unique and complex language of her or his own? Or consider the neglected little boy described at the start of the chapter. While his mother was hardly ever around, she did leave the television on. Why didn't he learn to talk from the TV? The answer is *because language requires interaction.* The TV never talks *with* you. Parents with hearing did not try to develop a sign language with their deaf children. It was only when deaf children interacted with others having the same need to communicate that their capacities for language were activated.

a slap to the stomach replaced the pantomime of eating to indicate hunger.

Very early on Judy Kegl, now on the faculty of the University of Southern Maine, discovered what was happening among deaf children in Nicaragua and initiated a field study of the birth of this new language (Kegl, Senghas, and Coppola, 1999). As she watched the students, their sign language rapidly developed a complex grammar—a pidgin was transformed into a creole. This occurred despite the fact that these students did not know any existing language. Nor was their sign language influenced by teachers or others with hearing because until the sign language was quite well developed, no one but the deaf children could understand it. The continuing arrival of younger children seemed to spur the linguistic development of Nicaraguan Sign Language, as it now is known. In fact, the younger children are far more fluent in the sign language than are older children who began its development.

As Kegl noted, the ideal test of the existence of a language instinct would involve separating a number of normal infants from their parents at birth, placing them together in isolation, and waiting to see if they developed a language. To actually do such a thing is, of course, unthinkable. But the deaf children in Nicaragua approximate such a test in that they had no significant language input prior to being brought together. They knew nothing of grammar, for example. In this instance the rapid development of a language

EMOTIONAL DEVELOPMENT

Humans not only think and speak; they also have feelings about things—feelings called emotions. Even newborn infants display strong emotions: Deny an infant something it wants and watch the anger boil up. Part of socialization is learning self-control so that we have appropriate emotional reactions to the world around us. But to do so, we must first develop what George Herbert Mead called the *self.* Recall from Chapter 3 that Mead identified the self as our learned understanding of the responses of others to our conduct. By seeing how others react, we gain understanding not only of them but also of ourselves. Once a self is formed, it begins to take on particular patterns of preference and response— patterns that can be described as a *personality.*

Let's briefly examine the origins of the self and the formation of personality.

EMERGENCE OF THE SELF

If you asked a baby, "Who are you?" you could not expect an answer. We have to discover who we are. In fact, we first must discover that we exist. Just as infants must discover their hands, they also must discover they are separate from the world around them. And just as Mead argued, they must learn that from others through interaction.

In a classic sociological study, Reed Bain (1936) charted the vocabulary development of infants. Unlike language researchers, Bain didn't want to know *how* kids learned to talk. Instead, he wanted to know *when* they acquired words to indicate *others* and to indicate *self.* He discovered that we acquire other-related words sooner than self-related words. That is, we know "they" are out there before we learn that "we" are in here. And we learn about "us" from "them."

In Mead's analysis the self emerges but gradually, and the essential self-discovery occurs as we gain the ability to "take the role of the other"—to put ourselves in another person's place and gear our behavior to external perspectives. Until we can do that, we cannot play team sports or perform any actions requiring coordination with others.

In addition to Bain's research on self-related and other-related words, the emergence of the self has been charted carefully in many studies of cognitive and linguistic development. For example, young children's speech is very egocentric in the sense that it often is uttered with little or no regard for being heard or understood. In fact, at the same time Mead was writing and teaching (in English) about taking the role of the other, Piaget (1926) was teaching and writing (in French) that children speak egocentrically, without regard for the hearer:

The talk is egocentric, partly because the child speaks only about himself, but chiefly because he does not attempt to place himself at the point of view of his hearer.

This point was well documented in a study by Flavell and his colleagues (1968). They asked a group of eight-year-olds and a group of fourteen-year-olds to explain the rules of a board game successively to two other students their age. One of the students in each pair was blindfolded. The eight-year-olds gave the same instructions to each listener—instructions completely inadequate for the blindfolded listener. The fourteen-year-olds gave much more elaborate instructions when addressing the blindfolded student than when addressing the one who could see. They could put themselves in the place of the blind, whereas the eight-year-olds could not yet do so. Many other studies confirm this aspect of the emergence of the self (Schmidt and Paris, 1984).

PERSONALITY FORMATION

In Chapter 1 we saw that because humans have the capacity of choice—of free will—their behavior is predictable. In Chapter 3 we saw that most behavior is governed by norms that arise out of interaction. Norms are rules governing behavior and serve the function of making behavior predictable, for only if we can predict one another's behavior with considerable reliability is it possible for us to interact. If I have no idea of how you will react, I will not feel sufficiently safe to risk interaction. When we meet strangers, we rely on our general knowledge of the norms and of human nature to anticipate how they will act. But through repeated interactions with others, we gain much greater precision in anticipating their actions. Such anticipations are made possible because individuals typically display a *consistent pattern* of thoughts, feelings, and actions. We refer to such a pattern as their **personality.**

In terms of their personalities, all humans are alike in certain ways; in other ways every human is like *some* other humans; and in some ways each human is unique, like no one else (Kluckhohn and Murray, 1956). By seeing how all three claims can be true, we can examine elements of personality.

All humans are alike because some factors that form our personalities are universal. We all share the biological endowments of humanity. We all are subject to the same general physical environment governed by the same laws of nature. All normal humans acquire language. All of us reason. All of us seek to make more rewarding choices from among the options we perceive. All of us must learn the culture of our group and society.

All humans are like only *some* other humans because the contents of culture vary from group to group. Norms do not simply limit behavior but frequently are incorporated into the personality as well. That is, usually people not only know the rules but also *believe in them.* Because norms differ from place to place, so, too, will the belief systems of typical members of different societies. This also will be true of their emotional reactions and, ultimately, of their behavior. Thus, for example, a normal person in one society will be provoked into a towering rage by something that will be passed off as trivial in another culture.

Finally, no two people ever have identical biographies, and therefore, no two people ever have precisely the same personality. Each of us has a unique historical pattern of interaction with our social and physical environment, and this pattern provides us with individuality. While we are interesting to one another because of our unique qualities, these are of limited interest to social scientists (except for those engaged in psychotherapy). The business of science is to seek *generalizations*, and that turns our attention from the unique to the common. No social scientist cares that your friend Jack hates cats. But we would be interested in a cat-hating society.

Thus far in this chapter, we have attended to the universal aspects of socialization. There is every reason to suppose that children deep in the Amazon jungle acquire language in much the same way as do children in Boston. We expect children everywhere to develop a self and learn to take the role of the other, much as Mead and Piaget proposed. However, for the remainder of the chapter, we shall explore how particular cultural and social structural patterns and the socialization process interact. That is, how are people socialized to be like *only some other people?*

CULTURE AND PERSONALITY

Chapter 5 discussed the reaction of social scientists in the 1920s and 1930s against extreme biological determinism: theories that attributed most of human behavior to such things as instincts rather than learning. This reaction involved the equally extreme claim that biology was insignificant and that environment alone determined human behavior.

A leader of the environmentalists was Franz Boas (1858–1942), the first person to be appointed professor of anthropology at Columbia University. Boas wrote about and taught the principle of **cultural determinism.** According to this principle, regardless of how a given culture came into being, the culture wholly determined the behavior of all persons who were socialized within it. Moreover, Boas argued that human nature is infinitely plastic—that cultural forces can create virtually any kind of personality type and any patterns of roles and behaviors. For example, he argued that sex roles have no biological basis but simply reflect cultural forces that blind us to other possibilities. In addition Boas took the view that culture makes its deep and permanent impression on people during

Neg. #2A 5161, American Museum of Natural History

Franz Boas.

early childhood. That is, a particular culture determines how its members will think and act as a result of specific patterns of child rearing, which shape the immature personality into the desired form.

By postulating this powerful link between child rearing and personality, Boas was in effect *equating culture and personality.* Or as his famous student Ruth Benedict (1934) put it, culture is "personality writ large." That is, the individual personalities of the members of a society are tiny replicas of their overall culture, while their culture is the summation of their personalities.

The major problem facing Boas and his supporters was to demonstrate exceptions to all generalizations that had been made about human behavior and personality so they could claim that existing patterns are only common, not necessary—that an almost infinite array of cultural and social forms is possible.

Cultural determinism reached its peak of influence during the long career of Margaret Mead (1901–1978), Boas's most famous student. In 1931 Mead traveled to New Guinea in search of tribes that would prove Boas to be correct in his claims that patterns of socialization in early childhood strictly determine the adult personality—or what he and Mead often called "temperament." More specifically, Mead intended to prove that sex roles can take virtually any form. To do this, she hoped to locate a tribe in which

© Barbara Kirk / Peter Arnold

Among New Guinea tribes such as this one, Margaret Mead claimed to find proof that early child-rearing practices determine the adult personality and that human nature is essentially plastic, making virtually any cultural pattern possible. For sociologists, however, the most striking thing seen in this picture is differential socialization. The little boy in the foreground already is becoming a very different kind of person from those the young girls in the background are becoming. This process is not limited to childhood. The adult males in this picture lead daily lives far different from those of their wives, and thus, basic male-female personality differences are re-created and constantly reinforced.

both male and female members develop "feminine" temperaments and another in which both genders have "masculine" temperaments. And she planned to demonstrate that these patterns of gender temperament were the result of early childhood socialization practices.

Once in New Guinea, Mead quickly discovered tribes she believed exhibited precisely these patterns, and she found them "conveniently within a hundred mile area" (Mead, 1935). In her famous book *Sex and Temperament in Three Primitive Societies* (1935), Mead claimed that among the Arapesh of New Guinea, the ideal personality type for both men and women is gentle, unaggressive, responsive, cooperative, and passive; that is, both male and female Arapesh have "feminine" temperaments. Mead argued that this was the result of extremely gentle child-rearing practices in which both parents played equal roles. For example, Arapesh children were not weaned or toilet trained until they were relatively old. This avoided repression and shaped their personalities into a gentle, feminine disposition.

In contrast, among the cruel Mundugumor, Mead found the temperament of both men and women to be "masculine"—that is, unrelentingly aggressive, cruel, suspicious, and violent. The Mundugumor were (and are) cannibals and treated one another almost as savagely as they did enemy tribes. Mead explained this on the basis of their brutal child-rearing practices. Mothers resented nursing their infants and used slaps and shoves to wean them at an early age. Children were cuffed and kicked whenever they displeased their parents. This, according to Mead, soon turned them into little monsters destined to become adult monsters like their parents.

From these findings Mead concluded that male and female temperaments have no biological basis whatever. She further concluded that culture is the source of personality, imposing its marks indelibly during infancy.

Mead was often criticized for working too fast and for invariably finding the world as she expected it to be. Thus, when asked to write a preface for a new edition of her book, she took

TABLE 6-1	External Pressures and Socialization in Premodern Societies	
	SOCIETIES SOMEWHAT ISOLATED FROM OTHER SOCIETIES (%)	SOCIETIES WITH OTHER SOCIETIES CLOSE BY (%)
Rarely engage in external wars	44	5
Much prefer male infants	17	44
Parents often hit kids	31	55
Much emphasis on fortitude, aggression, and competitiveness in socialization of *males*	35	70
Much emphasis on fortitude, aggression, and competitiveness in socialization of *females*	12	50
Stress the virtues of inflicting violence on outsiders	39	78

Source: Prepared by the author from the Standard Cross-Cultural Sample.

the opportunity to scorn unnamed critics who suggested that her findings were "too good to be true" and that she must have just "found what I was looking for" (Mead, 1950). Since her death, a raging battle has erupted between Mead's supporters and critics of her scientific objectivity and methods (Freeman, 1983).

But let's not worry more about the accuracy of Mead's observations in New Guinea. Let's take her at her word. In doing so we can spot some of the omissions and logical shortcomings of extreme cultural determinism.

The first shortcoming is insensitivity to physical realities. Mead tells us that the gentle Arapesh live in an almost inaccessible part of New Guinea, unsuited to growing crops. The Mundugumor, on the other hand, live on a rare tract of excellent growing land—high, well drained, and clear of the jungle. Let's pretend we know nothing about the Mundugumor except the desirability of their location and the fact that dozens of other tribes would like to live there as well. Might we not suspect the Mundugumor to be the meanest, toughest bunch of warriors on the island? If they weren't, why wouldn't tougher folks have driven them off and taken their land?

This reminds us that culture isn't just the accidental result of mental processes. There is always a real world to be reckoned with that imposes its tests of fitness on any culture (Harris, 1979). Indeed, the Arapesh's gentleness might explain why they ended up with the worst piece of real estate on the whole island.

This point is strongly supported by Table 6-1, which shows how greatly childhood socializa-

tion practices are shaped by external factors. The data are based on the 186 premodern societies selected in 1969 by George Peter Murdock and Douglas R. White from the *Ethnographic Atlas*. This selection has come to be known as the Standard Cross-Cultural Sample. By now various researchers have coded more than 500 variables for these societies.

One of these variables characterizes societies on the basis of the extent to which they face external threats: Are they somewhat isolated from other societies, or do they have other societies close by? In Table 6-1 societies are separated on the basis of their isolation. When we compare the two groups, the first thing we see is that societies with close neighbors tend to face chronic warfare! Only 5 percent of them rarely go to war, while nearly half of the somewhat isolated societies rarely engage in war. When cultural determinists are presented with a correlation between parents hitting their children and the tendency for the society to engage in warfare, they suggest that the one causes the other—that rough child rearing creates people who seek war. But when the environment is taken into account, as it is here, it seems more reasonable to suppose that isolated societies are less warlike for lack of the need to withstand external threats than it is to suppose that societies are isolated because they are not warlike. That being the case, the great differences in socialization patterns between more isolated and less isolated societies make good sense.

Societies faced with chronic warfare will have reasons to place exceptional value on male infants—to believe that their survival depends on having enough warriors. Moreover, it will be in such a society's interest to make sure that these boys grow up to be brave, aggressive, competitive—even nasty—men, ready to inflict violence. And to make sure of raising tough males, it may help to raise tough females as well. The Mundugumor are surrounded by dangerous and jealous neighbors, and like most other such societies, they raise their kids to be tough. The Arapesh, on the other hand, are so utterly isolated that Mead had a very difficult time reaching their village. This isolation gives them the opportunity to raise gentle kids.

A second, even more compelling problem with strict cultural determinism is the assumption that a direct and everlasting link exists between early childhood socialization and adult personality. Consider this question: At what age would a child taken from the Arapesh and given to the Mundugumor still be able to grow up with an Arapesh personality? Would it be too late to change at two, or five, or fifteen? At what age is it impossible to transform an individual? There is no clear answer because *socialization is not something that happens to infants and then stops.* It is a lifelong process. A lot happens between the time children are weaned and the time they become adults, just as a lot can happen to change people remarkably between the ages of thirty and forty or sixty and seventy.

The whole structure of Mundugumor and Arapesh life, not just early childhood experience, supported their particular styles of behavior. A Mundugumor warrior who began to mellow might end up in a neighbor's cooking pot. Being fully aware of that, he would not need to draw on his early childhood training to remember to stay tough.

Despite these problems, however, we must recognize that the major shortcoming of the cultural determinists was that they pushed their position too far. No one would deny that cultures can differ greatly and that among these differences are great variations in how people are expected to act. For example, the average Mundugumor, male or female, is obviously very different from the average German. If we acknowledge such differences, we must also acknowledge that they are produced by differences in socialization—that a Mundugumor infant sent off to be raised in Germany would become a German. Of course, early childhood socialization is not irrelevant; it simply is the start of a lifelong pattern by which, for example, Germans become and remain Germans.

The failures of the cultural determinism advocated by Boas and Mead are those of exaggeration and omission. Boas and Mead were correct in arguing that culture is extremely important in shaping humans, but they were excessive in saying that it is the *only* thing that matters. They were correct in arguing that human nature can be shaped into a great variety of expressions, but they were excessive in saying that virtually any cultural and personality pattern is possible. Finally, they were correct in arguing that child-rearing practices constitute a major aspect of the socialization process, but they were excessive in saying that only childhood socialization matters.

Suppose an adult Mundugumor warrior moved to Texas. He probably would never become just like a native Texan, but he would soon be very different from his friends back in New Guinea. Moreover, the ways in which he would change would depend a lot on the kinds of roles he assumed in Texas. Suppose he became a linebacker for the Dallas Cowboys or a cowboy on a ranch. The socialization involved in these two roles would produce rather different outcomes.

We will now leave the South Seas and examine processes of **differential socialization**: how roles determine how people are socialized and how this process prolongs socialization throughout an individual's lifetime.

DIFFERENTIAL SOCIALIZATION

Whatever the shortcomings of Margaret Mead's fieldwork, one thing is certain: In no society she visited were all children being raised in the same way. For example, not all Mundugumor children were expected to be equally fierce, to develop the same basic personality, or to learn the same physical and mental skills. Instead, from infancy children are sorted out in a variety of ways and socialized in different directions because they are expected to lead different lives. Put another way, they are being groomed to fill quite different social roles.

Recall from Chapter 2 that a social role is a set of shared expectations about the behavior of a person occupying a particular position in a society. A role consists of a set of norms applying to a particular position, and these norms serve as a script to be followed by those people filling that position.

Every society can be conceived of as a collection of related roles. Even simple societies have a number of different positions—son, daughter, father, mother, aunt, uncle, cousin, warrior, hunter, cook, gardener, grandfather, grandmother, chief, and priest, to name but a few. Each of these positions has a unique role associated with it; the role of priest, for example, is quite different from that of hunter, although the same person may alternate between these roles.

Many social roles make major demands on those who fulfill them, and thus, people must often undergo long and rigorous training before taking on such roles. Moreover, not just anyone is thought to be eligible for any role; for example, males are not qualified to be daughters. In addition people often aspire only to some roles and not to others. In short, people are socialized to hold certain positions and fulfill those roles.

All societies have differentially socialized people on the basis of sex: Males and females fulfill quite different sets of roles, and from infancy they are prepared to lead different lives. Although gender is the most universal and dramatic instance of differential socialization, it is but one of innumerable bases determining how people are socialized.

To illustrate this point, let's examine the differential socialization of two brothers born a year apart in England at the end of the nineteenth century. From birth the eldest son was the legal heir to his father's title as Baron of Buncombe and to the family estate: 2,000 acres of land and Buncombe Hall, an eighteen-room Gothic manor house. The younger son would inherit no title and no estate. At most he could hope for a modest cash inheritance. From the day of their births, these boys faced very different futures, and they knew they would play completely different roles as adults. Almost from the day of their births, they were treated differently by parents and servants.

The elder son was taught to ride and hunt and was instructed in the management of the estate. While still a lad, he rode with his father to call on the tenants who farmed various portions of the estate. On these trips he was constantly instructed in how to act like a proper lord of the manor, and he often encountered sons of tenants who were being prepared to replace their fathers and one day be his tenants.

Meanwhile, the second son was left out of most of these socialization activities. As long as he could remember, he had been aware that he would have to prepare for an occupational career—not just any career, of course, but one fit for the younger son of a baron. Usually, this meant a career in the army or in the church. Since his mother preferred that he go into the church, his father agreed that he should be prepared to become a priest of the Anglican Church. As a result he was left in the company of his mother more often than of his father, and his tutor made him study hard. The elder brother could get away with little study and with modest school achievements, but the younger brother was drilled hard in Latin, Greek, history, classical literature, and music, although, unlike his brother, he was not made to learn accounting.

At the appropriate age, each boy was sent off to a famous boarding school, but the elder was given more money to spend and was not expected to do better than passing work. The younger was pushed relentlessly in school because he would need to attend the university.

As an adult, the elder brother was self-confident, bold, and extremely well mannered; many regarded him as a born leader. He was also a terrific rider and sportsman. In contrast, the younger brother grew up to be somewhat shy. He often seemed ill at ease in large gatherings and preferred not to ride, hunt, or fish. While the elder brother spent many years as a popular and eligible bachelor, and did not marry and settle down until he was past forty, the younger brother married at twenty-five and soon had a large family. As a father the priest was mild and undemanding and treated all of his sons much the same. In contrast, the baron was soon taking his eldest son on visits to the tenants. His younger son, however, was meant for the army, so both boys did an equal amount of riding and hunting.

This is a classic case of differential socialization based on differing expectations of roles that the children would assume. Notice, too, that each brother was socialized somewhat differently to perform some of the same roles. The role of father carried a different script when it was combined with the role of Anglican priest instead of with the role of baron. Indeed, the role of husband was also scripted somewhat differently.

Differential socialization has long been the object of considerable sociological research. Of particular concern has been the way in which parents' expectations about their children's future influence patterns of socialization. Perhaps no one has pursued this question longer and more effectively than Melvin Kohn.

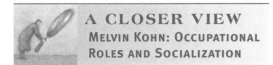

A CLOSER VIEW
MELVIN KOHN: OCCUPATIONAL ROLES AND SOCIALIZATION

Melvin L. Kohn was for many years chief of the Laboratory of Socioenvironmental Studies of the National Institute of Mental Health in Washington, D.C. In 1956 he conducted a study of the values guiding child-rearing practices among American parents. The results revealed some notable class differences. For example, working-class parents (manual laborers and blue-collar workers) placed greater stress on such values as obedience, neatness, and cleanliness than did middle-class parents. The latter thought such values as curiosity, happiness, consideration for others, and especially self-control were more important. Both groups gave equally high importance to honesty (Kohn, 1959).

Kohn argued that these findings reflected two underlying value clusters: (1) Working-class parents were more concerned about their children *conforming* to the expectations of others, especially expectations concerning good behavior, and (2) middle-class parents were more concerned about their children being capable of *self-expression* and *independence*. Put another way, working-class parents placed more importance on values involving external judgments (on pleasing others), while middle-class parents stressed values involving internal judgments (on pleasing oneself).

Kohn was careful to point out that these class differences were only tendencies. Parents of all classes regarded all of these values as important in raising children. But the emphasis given to these two clusters of values differed in the two classes.

Kohn then asked whether these differences in how parents socialized their children affected how they actually treated their children. His research led him to identify two primary differences in child-rearing practices. The first pertained to why parents punished their children, and the second to who punished them.

Kohn found that working-class families tended to punish children on the basis of what a child did. That is, if a child was prohibited from jumping up and down on the couch and then did so, the child was punished. In contrast, middle-class parents tended to be more concerned about the motives behind behavior than about rule violations as such. Thus, if a child broke a rule against yelling in the house, the parents would

Courtesy of Lynne Roberts

punish on the basis of why the child yelled. Yelling done out of anger or a loss of self-control would tend to be punished. But if it was done out of enthusiasm or happiness, middle-class parents would ignore it. In effect working-class parents reinforce conformity to external authority, while middle-class parents reinforce self-control and self-expression.

Middle-class parents feel an obligation to be *supportive* of children to encourage self-expressiveness. Moreover, they feel both parents should be supportive, just as they feel both should share responsibilities for disciplining children. Working-class parents, on the other hand, tend to have a division of responsibility in child rearing. Mothers take primary responsibility for the children and tend to be the supportive parent, while fathers are delegated the responsibility of enforcing the rules. Thus, when a child is caught jumping on the bed in a middle-class family, a family conference is likely to be held. In a working-class family, the child is told, "Well, when your dad gets home, he's not going to be happy about this." Again, Kohn

These two Seattle fishermen do not hold high-status, high-income jobs. But they are their own bosses. Not only does no one supervise their work, but usually, no one else is even in sight when they are pulling in their nets or their crab traps. In contrast, many high-status professionals work under close supervision and constant observation. Melvin Kohn has found that the degree to which people work under supervision is more important than their income or job status in determining how they raise their children. Degree of supervision even influences the basic personality of many workers.

was careful to point out that these are only tendencies; in many middle-class families, husband and wife divide the discipline and support responsibilities, while in many working-class families, parents take equal roles in child rearing.

The next question Kohn faced, of course, was, Why do these different tendencies exist in working- and middle-class families? Kohn's initial answer was that these tendencies reflect a parent's own experience with how the world works. He especially emphasized work experience. Working-class parents are more successful in jobs when they observe the rules: when they are prompt, do what they are told, and show up for work looking neat and clean. Middle-class parents find they are more successful in jobs when they can work without supervision, take individual initiative, and get along well with coworkers. Parents draw on this personal experience in deciding how to raise their own children: They raise them to succeed in the adult roles they expect their children to adopt.

This interpretation led Kohn to decide that social class as such was not the significant independent variable. Instead, he realized that the actual work conditions the parents experienced should be the real focus of research. Although class differences are correlated with the nature of working conditions, the correlation is far from perfect. Thus, some working-class people have jobs, such as repairing home appliances, that require a great deal of self-supervision and individual initiative. Conversely, many middle-class people work in highly supervised and controlled office jobs. So, Kohn set out to refine his results by examining occupational, and not simply class, effects.

The results were as predicted: Parents' work conditions, not their social class, influenced whether they stressed conformity or self-expression in raising their children (Pearlin and Kohn, 1966; Kohn and Schooler, 1969).

Kohn also demonstrated that experience with adult occupational roles not only influenced child-rearing practices but also reflected basic personality characteristics of the parents. That is, parents whose work experience placed a premium on conformity stressed this in raising children and had more conformist and less flexible and expressive personalities themselves (Kohn and Schooler, 1969). This raised a new question: Did correspondence between job experience and personality reflect *selection?* Were people with flexible, self-expressive personalities more likely to obtain jobs requiring little supervision and rewarding initiative and innovation? Or was

this evidence of **adult socialization?** Were people changed by socialization into adult roles and thus taught to be more flexible or conformist depending on the conditions of their occupation?

Such a question could be answered only by a **longitudinal study,** in which observations are made of the same people at several different times. In this way one can see what people are like before they enter a role and after they have performed it for a while. Since Kohn had already based one of his studies on a national sample of more than 3,000 employed American men, ten years later (in 1974) he arranged to reinterview them. Once again the results were striking.

Both selection and adult socialization account for the correlation between work conditions and personality. People with more flexible, self-directed personalities are more likely to obtain jobs with little supervision and much opportunity for individual initiative. Conversely, less flexible, less self-directed people gravitate toward more structured and more supervised occupations. However, people's personalities also tend to suit their jobs better as time goes by. Thus, people in highly structured jobs become less flexible and less self-directed, while people in less-structured jobs tend to become more flexible and self-directed (Kohn and Schooler, 1982).

The two primary lessons from Kohn's pioneering research are:

1. Children are socialized differentially on the basis of parental expectations about the roles the children will assume as adults; understandably, parents base these expectations on their own experiences.

2. Socialization is a lifelong process. In Kohn's sample even men age fifty and older showed shifts in basic personality traits in response to their occupational roles.

Kohn's findings have been replicated in many other nations, including Taiwan, Italy, Japan, Ireland, Germany, and Poland (Kohn and Schooler, 1983). In each society the nature of the parent's work experience greatly influences his or her values governing child-rearing practices.

Thus far we have seen that socialization prepares people for roles and that roles in turn shape socialization. At this point we need to consider more closely how people go about performing social roles.

A CLOSER VIEW
ERVING GOFFMAN: PERFORMING SOCIAL ROLES

Elizabethan playwright William Shakespeare (1564–1616) wrote in *As You Like It:*

All the world's a stage,
And all the men and women merely players:
They have their exits and their entrances,
And one man in his time plays many parts.

But it was sociologist Erving Goffman (1922–1982) who most effectively analyzed social interaction from the point of view that life really is a stage and that much of the time we are putting on performances for one another.

Goffman (1961) made an important distinction between role and role performance. *Role* refers to how a person would act if he or she did only what the norms attached to a particular position directed. **Role performance,** in contrast, is "the actual conduct of a particular individual while on duty in [a] position."

Goffman pointed out that while roles do greatly shape our behavior, we rarely act only according to the script. "Perhaps there are times," he wrote, "when an individual does march up and down like a wooden soldier, tightly rolled up in a particular role." But most of the time, we are not wholly confined by a role. Instead, we constantly display glimpses of ourselves, or the individual "behind" or "inside" the role. Moreover, we sometimes give an unconvincing, discreditable, incompetent, resentful, or even defiant performance.

Like a drama teacher, Goffman identified the basic techniques for giving adequate role performances (Goffman, 1959, 1961, 1963, 1971).

First of all, Goffman examined the costumes and props we use in role performances. Simply by wearing the appropriate clothing or carrying the right props, we can often make a convincing appearance in a role. As a young newspaper reporter, I once gained access to the medical files of a celebrated mass murderer by putting on a white coat and hanging a stethoscope around my neck. In a huge county hospital, that was sufficient "proof" that I was a resident physician with the right to wander onto wards and look at charts. (Such a deception would violate the role of sociologist, but not the role of reporter.)

Of course, if I had not also had the right kind of haircut, been of the appropriate age, and been able to exude the impression of bored self-confidence, someone might have asked me who I was. This illustrates Goffman's insight that a major aspect of the way we present ourselves in roles involves **impression management:** the conscious manipulation of scenery, props, costumes, and our behavior to convey a particular role image to others.

Goffman also pointed out that roles in life, like those in the theater, have both a *stage* and a *backstage.* Waiters in a restaurant, for example, give a stage performance to customers and a more relaxed backstage performance in the kitchen, where only the cooks and other waiters see them. Professors act out their roles somewhat differently when they chat with students in the hall from when they are in front of a class.

In addition *teamwork* is often involved in an adequate role performance. For example, a husband and wife holding a party will often prompt one another in their roles. Moreover, an effective role performance often requires that others play along. A hostess cannot play her part if the host begins to swear at her, throws his plate on the floor, takes off his clothes, or otherwise falls out of his proper role. Similarly, a parent cannot convincingly scold a child if the other parent is laughing at the misbehavior.

Another kind of teamwork plays a vital, but often little noticed, role in facilitating adequate role performances. Goffman called this **studied nonobservance.** For Goffman this is a powerful and civilizing norm based on the acceptance of our common humanity. It is summed up

A week ago these young men were like most other high school seniors. They had the role of teenager down cold—they laughed at the right times, listened to the right music, wore the right clothes, and said the right words. Then they joined the U.S. Marines. That first day, no matter what they did, their D.I. (drill instructor) yelled at them for doing it wrong. Having taken away their clothes, cut off their hair, and stood them on top of their footlockers wearing only towels, the Marine Corps had begun the process of socializing them into a new role. Eight weeks from now, the D.I. will smile as often as he snarls, and these recruits will be U.S. Marines.

© David Wells/The Image Works

by the expression, "After all, we're only human." Through nonobservance, we come to one another's aid by covering up miscues in role performance. For example, when we are embracing a lover and his or her stomach grumbles, we pretend not to hear it. When someone unconsciously scratches inappropriately, we pretend not to see it. Day in and day out, we are careful not to notice the little slips that mar role performances. In fact, a major aspect of good manners is to not notice bad manners.

Of course, we don't ignore all failures in role performance. While actors on the stage may only risk bad reviews for a poor performance, in real life people sometimes lose their friends, their jobs, their families, and even their liberty because they have violated their roles.

Moreover, people sometimes suffer these consequences even when they have given a superb role performance—because they were acting out a **deviant role.** Some positions in society and their roles are against the law. Thus, while the bank officer who embezzles risks jail for an inadequate role performance, the robber who withdraws funds from the same bank risks jail for performing a deviant role. In Chapters 7 and 8, we shall examine deviant behavior and deviant roles.

SEX-ROLE SOCIALIZATION

Most human societies have drawn sharp distinctions between male and female roles. And most of them have thought that these differences simply reflected biological facts of life. Obviously, males and females are different anatomically. Equally obviously, these differences have played *some* part in structuring a division of responsibilities. For example, only women can bear children, and only women can nurse an infant. In premodern times these two facts were quite significant. With high rates of fertility, women often had restricted mobility, and they also tended to suffer from chronic infections caused by childbirth. Prior to the invention of modern baby formulas, infants could only be nourished by breast-feeding. That meant a mother had to stay sufficiently near her infant to be able to feed it frequently. In addition to reproductive differences, most women also are smaller than men, which had social significance when much labor depended on muscle power.

But while these biological facts are self-evident, how important are they? Most human societies have thought them to be so important that they justified the subordination of women to the control of men. The data in Table 6-2 are based on the Standard Cross-Cultural Sample (as was Table 6-1). These are mostly very small societies having little technology, and many of them are nomadic.

The first set of statistics shows that it is rare for women to have equal political rights in such societies; in only 10 percent are women's rights equal to those of men. In fact, in 60 percent of these societies, women have *no* political rights. Despite the prevalence of gender bias revealed by these data, there is variation—in some societies women do have equal political rights, and in about a third, they have some rights, if fewer than men. It is variations like these that reveal that most of the differences we can see between the sexes in most societies are cultural, not biological. Hence, Chapter 12 will be devoted entirely to exploring and explaining cross-cultural variations in gender inequality: why gender inequality is so much greater in some societies than in others. Here our interests are in how gender inequalities and differences are *maintained within cultures:* how children are raised to accept and perform gender roles.

Of course, no matter what view they take of proper sex roles, to most members of any society it is seldom obvious that such basic features of their culture are optional and arbitrary. As we have seen, that's not how socialization works. Instead, to the extent that a culture defines gender roles as distinctly different, parents will raise boys and girls so they *will be different*. In addition these boys and girls will grow up *wanting* to be different, believing that these differences in sex roles are not only normal but necessary. For example, if a society conceives of sex roles so that men should be aggressive and women should be gentle, then most men and women in that society will try to live up to these standards.

Looking at the second set of statistics in Table 6-2, we see that premodern societies do tend to emphasize toughness and aggression when raising boys but are much less likely to do so when raising young women—only 10 percent of these societies raise "tough" girls. The table also shows that most societies begin socializing male and female children for their adult roles at the same age—but many (24%) begin training girls sooner than boys.

In the majority of premodern societies, women control child rearing for children under age four. But then things change, and the final child-rearing authority shifts to men—women continue to be the final authority in only 13 percent of these societies.

Finally, patterns of child rearing are greatly influenced by adult sex roles. In most of these societies, both hunting and gathering play a major role in the food supply, and these tasks involve a very sharp gender division of labor—overwhelmingly, men hunt and women gather.

These data demonstrate that there is a reality behind socialization. If boys and girls are raised quite differently, it is because they will lead quite different adult lives. When this is so, both men and women believe that gender differences are meant to be. Put another way, sex-role socialization is symmetrical: Both men and women agree about how men and women ought to act and how boys and girls should be raised.

This is obvious in Table 6-3. National samples of adults in eight Islamic nations were asked whether they thought it is important that women wear veils when they go out in public. The percentage who thought it is important varied from 95 percent in Egypt to 70 percent in Indonesia. Now look at the comparison between men and women in each nation. There are no meaningful differences. Both genders believe women should wear veils.

With this in mind, let's look at patterns of differential sex-role socialization in North America. From the earliest days of life, an infant is not simply a child but a boy or a girl. Boys get blue blankets, while girls get pink ones; little boys are nicknamed Buck and Butch, while little girls are called Honey and Sweetie. Not surprisingly, one of the first things a child learns is whether he is a "he" or she is a "she." And from a very early age, children show marked gender-specific preferences, such as for sex-typed toys and activities.

How does this happen? Do little boys arrive already programmed by their biology to prefer playing with toy guns and dump trucks while little girls have a biological propensity for tea parties? Of course not. These are things they must learn. Hence, our interest must focus on *who* is socializing young children and *how*.

Although it is fashionable to blame many of our current child-rearing problems on such impersonal forces as "society" or the "media," the fact is that parents, and especially mothers,

still do most of the socializing of youngsters. If we want to know how children in North America so quickly and so firmly adopt traditional conceptions of sex roles, the place to look is in parent-child interactions. And that's exactly what three sociologists at the University of Illinois did. So let's watch them while they observed mothers reading children's books to their offspring.

TABLE 6-2	Gender Roles and Gender Socialization in 186 Premodern Societies

Degree to which women have political rights, compared with men

Equal	10
Less	30
None	60
	100%

Degree to which socialization in late childhood stresses toughness and aggression

	Boys	Girls
Low	32	51
Medium	44	39
High	24	10
	100%	100%

Does training for adult duties begin sooner for boys or for girls?

Boys	1
Same	75
Girls	24
	100%

Do men or women have the final authority over the care and discipline of children?

	Children under Age 4	Children over Age 4
Men	18	37
Equal	17	50
Women	65	13
	100%	100%

A gender division of labor

	Amount of Hunting Done by Women	Amount of Gathering Done by Women
Under 10 percent	92	10
10 to 50 percent	8	28
Over 50 percent	0	62
	100%	100%

Source: Prepared by the author from the Standard Cross-Cultural Sample.

Two young doctors at a military hospital in Saudi Arabia obey the norm that women should wear veils when out in public. Contrary to what many people in the West assume, they probably would not rather go without veils (see Table 6-3).

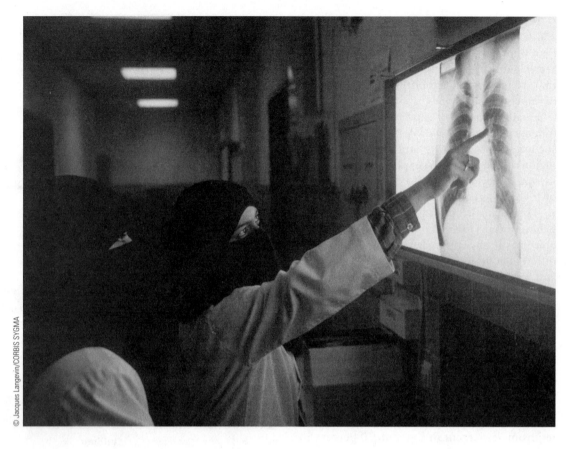

© Jacques Langevin/CORBIS SYGMA

TABLE 6-3	How Important Is It for Women to Wear a Veil in Public Places?		
	"IMPORTANT" (%)		
	All	Men	Women
Egypt	95	95	95
Saudi Arabia	91	92	91
Iran	87	88	85
Jordan	83	83	83
Nigeria	77	76	77
Bangladesh	75	76	75
Algeria	75	75	74
Indonesia	70	70	71

Source: Prepared by the author from the World Values Surveys, 2000–2001.

A CLOSER VIEW
DeLoache, Cassidy, and Carpenter: Baby Bear Is a Boy

One of the most common forms of interaction between parent and child takes the form of reading to the child. With very young children, the reading usually involves picture books with a minimum of text, and the child frequently sits on the parent's lap and follows along. Often, the gender of characters is unknown: We know the gender of Goldilocks and of Papa and Mama Bear, but not that of little Baby Bear. And in many other stories, there are no clues as to the gender of any characters.

How do parents deal with gender-neutral characters? Do they assign them a gender? If so, which? And how do they decide?

To find out, Judy S. DeLoache, Deborah J. Cassidy, and C. Jan Carpenter at the University of Illinois recruited mothers with young children to participate in an observational study. Each mother and child were scheduled for a session in a playroom designed to permit all interactions to be taped. The women were told that the researchers were interested in mother-child interaction in two situations, playing with toys and looking at books together. The mother and child were left alone to play with some toys for about ten minutes and then were seated in a large upholstered chair and given three books. In the first phase of the study, these were widely known books. The researchers reported that the mother was asked to look at, talk about, or read the books with her child however they would normally do at home. The mothers thus had no idea we were interested in their assignment of gender to the characters in the books. (DeLoache, Cassidy, and Carpenter, 1987)

Each of the three books included a number of gender-neutral characters. They did not have male or female names, did not wear any gender-specific clothing (such as hair ribbons), and were not engaged in any sex-typed activities. So, how did mothers refer to these characters? As *males:*

In all, the mothers used a total of 104 gender-specific labels, of which 102 (98%) were masculine references. . . . One of the only two feminine labels was one mother's reference to a large duck pictured with two smaller ducks as "Mrs. Duck."

The researchers conducted a second version of their study, this time using books specially prepared to present gender-neutral characters engaged in highly sex-typed activities. Thus, for example, an animal without any gender identity was shown standing in the front of a classroom filled with small animals. The researchers expected that this "teacher" would be identified as female. Another picture showed two bears in the front seat of a car, one of them driving. They expected the driver to be identified as male and the passenger as female.

Once again a group of mothers and children were recruited, and the instructions given were as before. And so were the results, or nearly so. Sixty-two percent of the characters were identified as male, 22 percent as neither, and only 16 percent as female. Variations in gender-typed behavior had only the weakest effects. For example, the "teacher" was more likely identified as female than as male, but not by much.

The importance of these findings is related to prior studies that have revealed that characters depicted in children's books are overwhelmingly shown to be male. Many have claimed that this enforces harmful perceptions of sex roles: that children learn it is a man's world in which women are on the sideline. Consequently, many people have demanded that children's books be revised to present gender-neutral characters. But what DeLoache, Cassidy, and Carpenter's work shows is that such books do not reduce perceptions that characters are males. Mothers rarely identified

animal characters as female unless they had female names, wore female clothing, or were obviously female. Otherwise, they referred to them as male. Hence, the researchers concluded, "The only strategy that is likely to succeed in achieving more egalitarian sex roles in young children's picture books is to portray more overtly female characters in a wide variety of nonstereotyped roles" (DeLoache, Cassidy, and Carpenter, 1987).

But why do we care? Why does it matter if boys and girls are differentially socialized? When people had very large families and women devoted most of their adult lives to having and raising children, it may have been beneficial to raise boys and girls differently to prepare them for their very different adult lives. But that's not true today when most women expect to spend nearly all of their adult lives employed outside the home. Hence, a major concern has been with whether and how women were diverted from successful careers by inappropriate, differential socialization. And a major focus has been on gender and scientific careers.

WOMEN AND SCIENCE

The central question has always been posed this way: Why do so few women pursue careers in science and engineering? Many answers are proposed. A common one has been that these are socially defined as male occupations, and women are discouraged from entering them, either by negative attitudes and/or by actual discrimination. Even more popular is the idea that childhood socialization gives women negative feelings about these careers (Lee, 2002). Some even have proposed that biological differences are involved, that women excel at relational thinking while men excel at linear thinking, the latter better suited for doing science. Ironically, some feminists have, perhaps unwittingly, encouraged this biological view by claiming that women do think differently and therefore that a feminist approach would lead to superior science (for a summary see Kohlstedt and Longino, 1997).

Meanwhile, as the hot debate continues among social scientists, something rather remarkable has gone on. If well-established trends continue, the dispute about why women don't become scientists may become of historical interest only. As can be seen in Table 6-4, there has been

TABLE 6-4	Percentage of Doctoral Degrees in Scientific Fields Awarded to Women	
	1970–1971 (%)	2001–2002 (%)
Engineering	0.6%	17.3%
Computer Science	2.3%	22.8%
Physics	2.9%	15.5%
Geology	3.4%	28.5%
Mathematics	7.6%	29.0%
Chemistry	8.0%	33.9%
Biological Sciences	16.3%	44.3%
All Doctorates	14.3%	44.9%

Source: National Center for Education Statistics, *Earned Degrees Conferred*, 2003.

a large, rapid increase in female scientists. A generation ago, in 1970–1971 only 1 of each 167 new doctorates in engineering was awarded to a woman. In 2001–2002 about 1 of 5 was earned by a woman. In computer science a generation ago, only 2.3 of each 100 new doctors were female. By 2001–2002 it was nearly 1 of 4. And in biological sciences, where once only a sixth of the doctoral degrees went to women, now they earn half. In fact, women now earn about half of all doctoral degrees awarded in the United States.

Will these trends continue? One reason to think so is that current female undergraduates are almost as likely as males to major in chemistry, mathematics, and geology, and they are more likely to major in biological sciences. It will be a few years before these undergraduate trends can be reflected in doctoral degrees, just as the recent increases of women in science are not yet reflected in senior faculty appointments; these, of course, reflect gender differences back in the 1970s when senior faculty earned their degrees. But the trend data very strongly suggest that the gender and science issue is becoming moot.

CONCLUSION

This chapter initially focused on early childhood socialization but then noted that socialization is a lifelong process. Because socialization is the transmission of culture and the shaping of character, it does not cease as long as we continue to interact and have new experiences. Moreover, because so much socialization is aimed at preparing people for, and comes as a result of, playing new roles, socialization continues as we pass through life's successive roles.

Socialization during infancy and early childhood has received the greatest attention from social scientists because serious failures in this period can have profound, long-lasting effects. A child who never learns to take the role of the other, for example, will never be able to play team sports; he or she may even grow up to be an egomaniac. A child who grows up in a modern society without learning to read is condemned to a life of marginal employment and ignorance.

However, even if early socialization is crucial, not all problems or patterns of adult behavior are rooted in childhood. Socialization failures can occur at any age and be wholly independent of anything that came before. In the next chapter, we shall see that people who fall into "bad company" during their teens, or even as adults, may become socialized into new and deviant patterns of behavior. People who have successfully performed a whole series of adult roles suddenly may fail to be adequately socialized into a new one—as grandparent or person who is retired, for example.

Just as socialization occurs throughout our lives, so does the topic of socialization occur throughout this book. For example, when we examine social institutions such as the family, religion, schools, or the economy, we also examine major sources of socialization. In fact, this textbook is part of your present socialization.

Review Glossary

Terms are listed in the order in which they appear in the chapter.

feral children The name often applied to children who, because of severe neglect, act as if they were raised in the wild (*feral* means "untamed").

socialization The process by which culture is learned and internalized by each normal member of society—much of which occurs during childhood.

stimulus-response (S-R) learning theory The theory in which behavior (responses) of organisms is said to be the result of external stimuli; that is, organisms only repeat behavior that has been reinforced by the environment.

cognitive structures General rules or principles that govern reasoning.

sensorimotor stage According to Piaget, the period from birth to about age two years during which the infant develops perceptual abilities and body control and discovers the rule of object permanence: that things still exist even when they are out of sight.

preoperational stage According to Piaget, the period from age two until about seven during which a child learns to take the role of the other.

concrete operational stage According to Piaget, the period from seven until about twelve during which humans develop a number of cognitive structures, including the rule of conservation.

formal operational stage According to Piaget, the time after about age twelve when some humans develop the capacity for abstract thought—that is, for using theories rather than only empirical observations.

language instinct An inborn, elementary universal grammar (as Chomsky calls it) that enables normal infants to rapidly acquire complex languages and even to create languages.

pidgin A jargon of made-up nouns and verbs, often including some borrowed from the language of the dominant group (such as plantation owners or colonizers), highly variable in terms of word order, and with little in the way of grammar.

creole A complex and grammatical language evolved from a pidgin.

personality Consistent patterns of thoughts, feelings, and actions.

cultural determinism The claim that an almost infinite array of cultural and social patterns is possible and that human nature can be shaped into almost any form by cultural forces.

differential socialization The process by which different members of the same society or even the same family are raised differently because of varying expectations about the roles that each will need to fill as an adult.

adult socialization Processes by which adults are enabled to perform new roles.

longitudinal study Research in which observations are made of the same people at different times.

role performance The actual behavior of people in a particular role, in contrast to how they are supposed to behave.

impression management Conscious manipulation of role performance.

studied nonobservance The way in which people pretend not to notice minor lapses in one another's role performance.

deviant role A set of norms attached to a position that in turn violates the norms adhered to by the larger society. For example, a proper performance of the role of burglar will deviate from other people's norms.

Suggested Readings

Freeman, Derek. 1983. *Margaret Mead and Samoa: The Making and Unmaking of an Anthropological Myth.* Cambridge, Mass.: Harvard University Press.

Goffman, Erving. 1959. *The Presentation of Self in Everyday Life.* New York: Doubleday.

Goffman, Erving. 1971. *Relations in Public.* New York: Basic Books.

Kohn, Melvin L., and Carmi Schooler. 1982. "Job Conditions and Personality: A Longitudinal Assessment of Reciprocal Effects." *American Journal of Sociology* 87:1257–1286.

Piaget, Jean, and Barbel Inhelder. 1969. *The Psychology of the Child.* New York: Basic Books.

Pinker, Steven. 1994. *The Language Instinct.* New York: William Morrow.

Sociology Online

www.socstark10.com

GO TO THE INTERNET AND TYPE: www.socstark10.com.

↗ **CLICK ON:** 2000 General Social Survey

✓ **SELECT:** WATCH TV On the average day, about how many hours do you personally watch TV?

↗ **CLICK ON:** Analyze Now
Television plays a very significant role in the socialization of many Americans. Can you identify three groups of people who are exposed to more TV than others?

↗ **CLICK ON:** Select New Data Set

↗ **CLICK ON:** The 50 States

✓ **SELECT:** NUC FAM: Percentage of households that are nuclear families—married couples with own children.

↗ **CLICK ON:** Analyze Now
Which state is highest? Which is lowest? How might this be explained? (Hint: Compare them on average age.)

↗ **CLICK ON:** Select New Variable

✓ **SELECT:** FEM. HEADED: Percentage of households that are female headed, no spouse present.

↗ **CLICK ON:** Analyze Now
There is a regional pattern to having few female-headed households; what is special about this region?

↗ **CLICK ON:** Select New Data Set

↗ **CLICK ON:** Nations of the Globe

✓ **SELECT:** SINGLE MOM: Percentage who approve of a woman choosing to be a single parent.

↗ **CLICK ON:** Analyze Now
Compare the top two nations with the bottom four: In what ways do they differ?

USING INFOTRAC COLLEGE EDITION

GO TO THE INTERNET AND TYPE:
www.InfoTrac-college.com

↗ **CLICK ON:** Register New Account
You will be asked to enter your Access Code and to create and enter a User Name and Password. Having done so,

↗ **CLICK ON:** InfoTrac College Edition or Log On.

A lot has been written about sex-role socialization as well as on socialization in general, and these search terms will let you explore this literature.

SEARCH TERMS:
Sex Role
Socialization

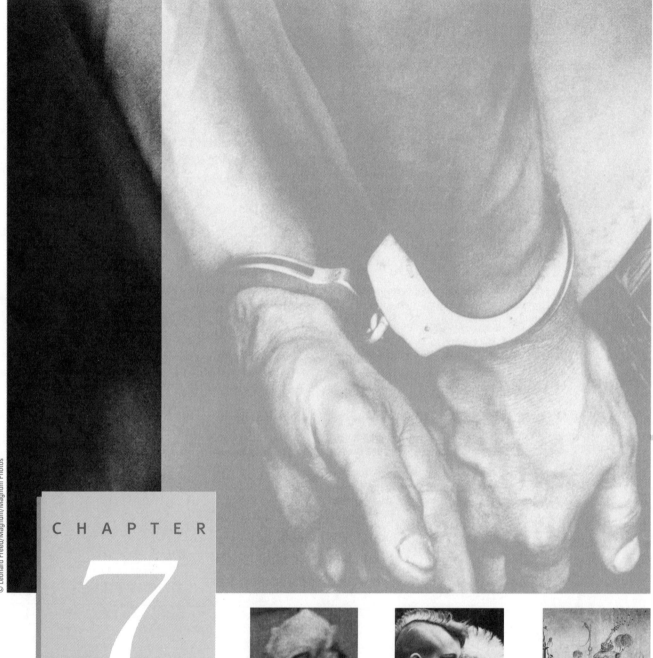

CHAPTER

7

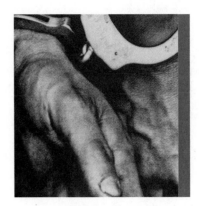

Suppose you had driven past a police car and had seen this young man handcuffed in the backseat. You might have wondered what he had done. You won't find out in this chapter what this young man did *this time*. But you will learn that it is very likely he had been arrested before and that he will be arrested again and that he probably doesn't specialize in a particular offense but is apt to commit many different crimes. Finally, when this picture was taken, he probably was within a mile of his home.

Crime and Deviance

OST OF US PLAY BY THE RULES. As we shift from role to role and from situation to situation, we usually conform to the norms that define how we are supposed to act—but not always. Each of us breaks the rules some of the time (sometimes we may be unaware of particular norms), and some of us break them a lot of the time. When we violate norms, our behavior is called **deviance:** We deviate from, or fail to conform to, the norms. The concept of deviance includes all norm violations, from the trivial to the tragic—from sleeping in class or wearing weird clothes to drowning one's children and blaming it on a carjacker. And although some social scientists specialize in forms of deviance that are no real threat to others, most social scientists who are concerned with norm violations are interested in only the most serious kind of deviance, crime. In fact, most of

these social scientists describe themselves as criminologists.[1] Unlike police investigators who ask *who* did it, criminologists try to explain *why* they did it. Unfortunately, crime often has been defined in a way that makes it difficult to explain why it occurs.

The problem of definition arises because some actions are criminal only in some times and places. For example, prostitution is legal in Denmark, whereas adultery is a capital crime in Saudi Arabia. In an effort to achieve a general definition of crime applicable everywhere, social scientists often define crime as "actions that violate the law." Thus, crime is whatever local legal codes say it is, and the task of criminology can be identified as the need to explain why people break laws. But this approach to defining crime entails several problems.

First, explanatory attention tends to be diverted from criminal actions to the processes by

1. The majority of criminologists are trained by and employed in sociology departments.

CHAPTER OUTLINE

ORDINARY CRIME
Robbery
Burglary
Homicide
Offender Versatility

THE CRIMINAL ACT

BIOLOGICAL THEORIES OF DEVIANCE
"Born Criminals"
Behavioral Genetics

A CLOSER VIEW
Walter Gove: Age, Gender, Biology,
and Deviance

MENTAL ILLNESS

PERSONALITY THEORIES

ELEMENTS OF SELF-CONTROL

DEVIANT ATTACHMENT THEORIES
Differential Association: Social Learning
Subcultural Deviance

STRUCTURAL STRAIN THEORIES

WHITE-COLLAR CRIME

CONTROL THEORIES
Attachments
Investments
Involvements
Beliefs

A CLOSER VIEW
Linden and Fillmore:
A Comparative Study of Delinquency

ANOMIE AND THE INTEGRATION OF SOCIETIES

CLIMATE AND SEASON

THE LABELING APPROACH TO DEVIANCE

DRUGS AND CRIME

CONCLUSION

REVIEW GLOSSARY

SUGGESTED READINGS

SOCIOLOGY ONLINE

which various acts are legally prohibited—to studies of the "criminalization" of certain behaviors (Turk, 1969). For example, many sociologists have argued that what we need to understand is not why people use illegal drugs, but why some drugs are illegal in some places (Becker, 1963; Lindesmith, 1965). Indeed, this line of analysis has led to many proposals to solve the drug problem by decriminalizing drugs, thus shifting the focus of attention from drug users to legislative behavior (Vallance, 1993).

Explorations of the working of political and legal systems are both interesting and valuable, but this approach skirts the fundamental issue that makes criminology one of the most highly funded subspecialties in the social sciences. Although it may be problematic whether or not various drugs should be illegal, no one is going to propose decriminalization as the solution to murder or armed robbery, and hence, these acts force consideration of the means by which criminal *behavior* can be controlled. This in turn leads to questions about causation: Why do they do it? But it is very difficult to discuss the causes of crime when the definition of crime includes diverse illegal activities that may have little in common and might stem from entirely different motives.

Thus, the second major problem is that, if crime is defined as violations of law, officials and politicians are permitted to dictate the parameters of an essential social scientific concept. For example, some nations outlaw actions such as criticizing the government, holding religious services, or organizing labor unions. Are social scientists therefore required to agree that such behavior is the moral equivalent of burglary, rape, or auto theft? Of course not. But by defining crime as whatever is illegal, social scientists have been driven to identify many "kinds" of crimes: violent crimes, property crimes, white-collar crimes, "victimless" crimes, political crimes, and so on. Faced with this variety, social scientists often have been forced to suggest an equally confused variety of causes. The need for a clear definition of crime, *independent of legality*, was long apparent.

Michael R. Gottfredson and Travis Hirschi (1990) proposed such a definition. **Crime** refers to "acts of force or fraud undertaken in pursuit of self-interest." The legal standing of these acts is ignored, and thus, this definition is entirely free of political contingencies. Protest demonstrations by Chinese students, for example, are not crimes whatever the Chinese government proclaims. Wife beating qualifies as a crime even if there is no law against it in many societies. Notice, too, that the use of fraud or force on behalf of the public interest (by the police, the military, or even ordinary citizens) is not defined as crime.

An additional virtue of this definition is that it permits us to focus on criminal *acts* rather than primarily on criminal *actors*—that is, we are able to closely examine crime before we try to identify why some people, but not others, commit crimes. Indeed, much can be learned about *why they do it* from seeing more clearly *what it is they do.*

Although many of the theories we will examine attempt to explain deviance in general, the emphasis of this chapter will be on forms of deviance that constitute crimes.

Despite the immense coverage given to crime stories by the news media, the fact is that very few Americans, other than law enforcement professionals, have a realistic portrait of crime. Therefore, the first section of this chapter sets the stage for theories of why they do it by offering careful depictions of what they do. Against this background we shall assess the primary theories of crime and deviance. As will be clear, these theories are more complementary than competitive. That is, many are directed toward slightly different aspects of the problem and try to answer somewhat different questions. Of course, the major theories are not entirely compatible, and the chapter will explain these theoretical disagreements and relate them to the latest and best research findings.

ORDINARY CRIME[2]

The typical crimes reported in the media are extremely unusual. Some of them are unusual because they are so horrible. But most of the remainder are unusual because they are carefully planned and skillfully carried out in pursuit of substantial gains. Thus, burglars make off with jewels, furs, and fine art; embezzlers steal huge sums from their unsuspecting employers; executives loot their companies; con artists trick the elderly out of their life savings; people kill their parents to inherit fortunes; robbers take thousands in bank stickups; car thieves specialize in luxury cars, stealing them on demand.

While all of these things do happen, they are extremely rare and completely at variance with ordinary crime. The overwhelming majority of

FIGURE 7-1	Burglaries per 100,000 Population (2003).

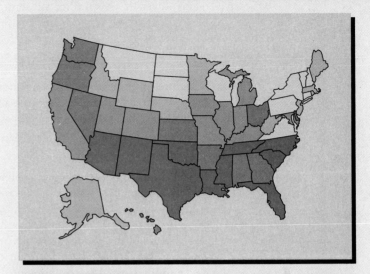

North Carolina	1197.6
Tennessee	1082.0
South Carolina	1050.9
Arizona	1050.3
Mississippi	1035.6
New Mexico	1025.2
Florida	1002.7
Louisiana	998.1
Texas	993.7
Oklahoma	992.3

Nevada	980.6
Alabama	960.8
Washington	950.3
Arkansas	913.6
Georgia	909.2
Hawaii	907.2
Ohio	830.1
Oregon	804.2
Kansas	803.6
Delaware	729.8

Missouri	717.1
Utah	713.1
Colorado	711.3
Maryland	701.4
California	682.8
Michigan	677.2
Kentucky	671.6
Indiana	671.1
Illinois	618.7
Iowa	596.0

Alaska	594.2
Nebraska	579.1
Idaho	570.2
West Virginia	562.2
Minnesota	547.4
Massachusetts	539.7
Wyoming	520.9
Rhode Island	513.3
Maine	503.9
New Jersey	503.0

Wisconsin	485.4
Vermont	477.8
Connecticut	448.1
Pennsylvania	436.0
Montana	405.6
New York	393.4
Virginia	391.5
South Dakota	375.9
New Hampshire	353.5
North Dakota	306.2

The media create the image that the United States has extremely high crime rates and that the rates are especially high in the urban East. It frequently is reported that people in eastern cities have been driven to install three and even four locks on their outer doors to attempt to keep out burglars. Maybe there are more locks and more fearful people in the East, but apparently, burglars prefer the Sunbelt. No eastern state is among the top fifteen in terms of its burglary rate.

Source: U.S. Department of Justice, 2004.

---

2. This section was prompted by the excellent chapter on the subject in Gottfredson and Hirschi (1990).

criminal acts lack planning, are performed incompetently (as many as half of criminal attempts are entirely bungled), and result in trivial gains—even if they inflict substantial losses on victims. To see the fundamental features of ordinary crime, it will be useful to look closely at several crimes. As we examine robbery, burglary, and homicide, you will note that the essential features of these seemingly rather different kinds of crimes are in fact the same and that these features typify crimes in general.

ROBBERY

The Uniform Crime Reporting Program defines robbery as "the taking or attempting to take anything of value from the care, custody, or control of a person or persons by force or threat of force or violence and/or by putting the victim in fear." In 2003 there were 413,402 robberies reported to the police in the United States, for a rate of 142.2 per 100,000 population. It may surprise you to know that not only is this a much lower rate than ten years earlier (256.0 per 100,000), but it is lower than in Spain (233.8), France

(186.5), or England and Wales (180.7). The robbery rate also varies a lot within the United States, as can be seen in Table 7-1, which reports the data for selected American cities: from 380.1 in Memphis down to 13.5 in Fargo.

More than half of all robberies occur on the street. Usually, the robber displays a gun or knife (a third of robbers rely on strong-arm tactics) and demands money and other valuables. Most victims have little money, as most of them are low-income people, and the usual take is much less than $50. Given that the majority of street robberies involve multiple offenders, the take per robber isn't much. However, few robbers attempt more prosperous targets—fewer than 10 percent of robberies target service stations or convenience stores, and fewer than 2 percent are bank robberies. The most recent data show that the average robbery of a service station produced only $546. If someone robbed a station every week, the take would add up to only $28,392 for the year—and he or she would have been caught long before the year was out. Bank robberies yield the highest average payoff ($4,516), but the risk of arrest is far greater.

Most robberies are unplanned. Based on his interviews with convicted robbers, Floyd Feeney (1986) found that they usually acted on the spur of the moment, selecting their targets more as a matter of convenience than anything else. Asked why they robbed a particular target, robbers gave replies such as, "Just where we happened to be, I guess," "Nothing else was open at 2 A.M.," and "We thought it would be quickest, you know, it's a small donut shop." Feeney summarized:

The impulsive, spur-of-the-moment nature of many of these robberies is well illustrated by two adult robbers who said they had passengers in their cars who had no idea that they planned a robbery. One passenger, who thought his friend was buying root beer and cigarets, found out the hard way what had happened. A clerk chased his robber-friend out the door and fired a shotgun blast through the windshield. . . .

Not only do robbers seldom do much planning, but most select targets in their own neighborhoods, typically within a mile of their residence (Feeney, 1986; Gottfredson and Hirschi, 1990).

The most recent data show that 91 percent of those arrested for robbery were males under the age of twenty-five. The average robber has previous arrests for a variety of offenses and will be arrested again for various offenses.

TABLE 7·1	Robberies per 100,000 Population for Selected Metropolitan Areas
	ROBBERY RATE
Memphis	380.1
Miami	360.3
Detroit	332.3
Houston	285.4
Baltimore	276.3
New York City	270.0
Dallas	265.3
Philadelphia	246.2
Los Angeles	245.3
New Orleans	238.9
Albuquerque	164.8
Seattle	131.4
Boston	124.4
Anchorage	122.6
Denver	120.1
Minneapolis	116.0
San Diego	114.9
Honolulu	109.2
Salt Lake City	97.6
Des Moines	68.4
Cheyenne	34.8
Fargo	13.5

Source: U.S. Department of Justice, 2004.

BURGLARY

Burglary is defined by the Uniform Crime Reporting Program as "the unlawful entry of a structure to commit a felony or theft." In 2003 there were slightly more than 2 million burglaries reported to the nation's police, a rate of 740.5 per 100,000 population. The average loss of residential burglaries (which make up two-thirds of the total) was $1,299. However, this figure is very misleading. Only about half of all burglaries are reported to the police. As a result middle- and upper-income homes are very overrepresented in burglary statistics because it is mainly people with insurance who report their burglaries. Thus, the average burglary actually involves a loss of only several hundred dollars. But whatever the average loss, this represents far more money than the average burglar nets for two reasons. First, damage to the house—mainly to doors and windows—is included in the loss but is of no value to burglars. Second, burglars can sell their stolen goods for only a tiny fraction of the actual value.

In the typical instance, the burglar knocks on the door or rings the bell. If no one answers, he (more than 90 percent of those arrested for burglary are male) tries the door. If it is unlocked, the burglar enters; fewer than a third of burglaries involve forced doors or windows. But if someone is home, if the door is locked and substantial, if a dog barks inside, if entryways are visible to neighbors or passersby, the burglar moves on to another house or apartment (Mayhew, 1984). The probability that any particular home or apartment will attract a burglar is very dependent on how close it is to the home of a burglar or to "personal activity space" of potential burglars, the routes they travel during the course of an ordinary day. Like robbers, burglars stick close to home (Brantingham and Brantingham, 1984).

The average person arrested for burglary is a white male under twenty-five. He will have previous arrests for a variety of offenses and will be arrested again for serious, but unpredictable, offenses.

For many years everyone went on and on about how high America's crime rates are compared with more "civilized" nations such as those in Europe. As it turns out, that was mostly

© Corbis/Sygma

Years ago Willie Sutton was up before a judge on charges of robbery after having served many prior sentences for the same offense. "Willie," the judge asked, "Why do you keep robbing banks?" "Your Honor," he replied, "'Cause that's where the money is." Here we see someone else doing what Willie did, going where the money is. And like Willie, he will probably spend a lot of time in prison because the FBI investigates bank robberies. So, most robbers don't try banks. Instead, they pick on much safer targets—even though that's where the money *isn't*.

TABLE 7-2	Burglaries per 100,000 Population for Selected Nations
NATION	**BURGLARY RATE**
Netherlands	3227.3
Denmark	1866.5
England and Wales	1587.3
Norway	1560.8
Sweden	1470.7
Austria	1077.4
Poland	943.9
Scotland	935.5
Finland	881.1
Switzerland	831.5
United States	**728.8**
France	629.9
Spain	561.9
Germany	564.8

Sources: Foreign data, *European Sourcebook of Crime and Criminal Justice Statistics*, 2003; American data, U.S. Department of Justice, 2004.

because Europeans did not collect and publish their crime statistics. In recent years they have begun to do so, and it seems that they weren't so civilized as had been thought. Notice in Table 7-2 that many nations of Europe have a far higher burglary rate than does the United States—the

TABLE 7-3	Homicides per 100,000 Population for Selected Cities	
CITY		**HOMICIDE RATE**
Pretoria, South Africa		43.0
New Orleans, U.S.		**25.5**
San Juan, Puerto Rico		24.1
Moscow, Russia		18.9
Tallin, Estonia		14.5
Detroit, U.S.		**10.0**
Dallas, U.S.		**6.9**
Amsterdam, Netherlands		6.8
Belfast, Northern Ireland		6.3
Warsaw, Poland		5.5
Bern, Switzerland		5.4
Prague, Czech Republic		5.3
New York City, U.S.		**5.2**
San Diego, U.S.		**4.4**
Seattle, U.S.		**3.5**
Madrid, Spain		3.3
Stockholm, Sweden		3.0
London, England		2.7
Boston, U.S.		**1.9**

Sources: American data, U.S. Department of Justice, 2004; foreign data, International Comparisons of Criminal Justice Statistics, 2001.

rate in the Netherlands is more than four times as high.

HOMICIDE

Homicide (or murder) is defined as "the willful (non-negligent) killing of one human being by another." In 2003, 16,503 Americans were murdered, a rate of 5.7 per 100,000. Is this high? Well, ten years earlier in 1993, the rate was almost twice as high—9.5 per 100,000.

For many years most murder victims had a prior relationship with their killer—about a third of homicides involved family members. This is no longer true. These days many victims are killed by strangers. Consequently, murders are far less frequently solved than before. When people are killed by a friend, workmate, or family member, the killer typically is still at the scene when police arrive, others saw the homicide occur, and the solution is not in doubt. Hence, back in 1965 when most homicide victims knew their killer, 91 percent of homicides were closed by arrest. In 2003 only 62.4 percent resulted in an arrest.

About 10 percent of murders now occur during a robbery and 5 percent involve drug

disputes. Press coverage to the contrary, rape-related homicides are rare—forty-three cases nationally in 2003.

Those arrested for homicide are very similar to their victims. Overwhelmingly, killer and victim are of the same sex, race, and age, and they also tend to have similar past criminal involvements—most have been arrested for other offenses.

Few homicides involve planning. According to Michael Gottfredson and Travis Hirschi (1990), in about half of homicide incidents:

People who are known to one another argue over some trivial matter, as they have argued frequently in the past. In fact, in the past their argument has on occasion led to physical violence, sometimes on the part of the offender, sometimes on the part of the victim. In the present instance, one of them decides he has had enough, and he hits a little harder or with what turns out to be a lethal instrument. Often, of course, the offender simply ends the dispute with a gun.

Homicides that occur during robberies or other crimes usually are as unplanned as the crimes during which they occur:

The standard robbery . . . becomes a homicide when for some reason (sometimes because the victim resists, sometimes for no apparent reason at all) the offender fires his gun at the clerk or store owner. (Gottfredson and Hirschi, 1990)

Because homicide is a very urban offense, the best international comparsions are based on cities. As can be seen in Table 7-3, New Orleans has an extremely high homicide rate (25.5 per 100,000), exceeded only by Pretoria, South Africa, among the cities for which statistics are available. But New Orleans also is far higher than any other American city (more than twice as high as Detroit, the American city with the second highest rate). Looking down the table you will see that many European cities have homicide rates substantially higher than does New York City and that Boston's rate is lower than any of them.

OFFENDER VERSATILITY

In the preceding discussions of specific crimes, words such as *robbers*, *burglars*, and *murderers* appear. This implies that offenders specialize,

TABLE 7-4	Violent Crimes Per 100,000 Students for Selected* Universities		
UNIVERSITY	**VIOLENT CRIME RATE**	**UNIVERSITY**	**VIOLENT CRIME RATE**
Vanderbilt University	299.9	University of Michigan	47.1
University of Delaware	124.1	University of Alaska	46.5
Massachusetts Institute of Technology	117.7	Harvard University	40.8
University of New Mexico	113.7	Ohio State University	39.2
Duke University	109.0	Indiana University	34.2
Boston University	104.5	University of North Carolina	31.4
University of California, Berkeley	99.6	University of Iowa	31.3
University of California, Los Angeles	85.3	University of Oklahoma	27.9
Arizona State University	83.2	University of Georgia	27.8
Tulane University	76.1	University of Washington	21.4
Michigan State University	70.1	Baylor University	21.1
Rutgers University	67.3	Pennsylvania State University	19.6
Florida State University	65.7	University of Minnesota	19.3
Brown University	64.3	University of Mississippi	15.8
University of Colorado	63.2	Texas A&M University	13.4
University of Missouri	59.2	University of Nebraska	13.2
Southern Methodist University	58.4	North Dakota State University	9.5
University of Alabama	57.5	Brigham Young University	9.2
University of Cincinnati	47.6	University of Texas	7.9
University of Florida	47.3	West Virginia University	4.4
		University of Wyoming	0.0

Source: U.S. Department of Justice, 2004.

*Many schools do not report their campus crimes. The data refer to the main campus unless otherwise indicated.

that some people prefer to rob, others to burglarize, and some people become murderers mainly because they have terrible tempers. But it isn't so. Very few offenders specialize even to a limited extent. To know who committed a burglary today is of little help in predicting their next crime; it is about equally likely to be a robbery, a car prowl, an assault, a murder, or a drug violation as to be another burglary (allowing for the different frequencies of these offenses). In fact, only a rather small proportion of rapes are committed by men who are specialized "sex offenders." Most rapists have committed a great variety of other crimes, and will again. Moreover, all offenders seem to "specialize" in traffic violations and in having auto accidents.

These patterns are referred to as *offender versatility*: "Offenders commit a wide variety of criminal acts, with no strong inclination to pursue a specific criminal act or a pattern of criminal acts to the exclusion of others" (Gottfredson and Hirschi, 1990).

THE CRIMINAL ACT

By now you will have drawn some conclusions about the typical criminal act. First of all, you may have noticed that people seldom *work* at crime. Few burglars have developed any special skills for forcing doors or windows or wiring around alarm systems—they simply break things or move on to an easier target. Few embezzlers maintain false accounts to shield a long-term pattern of theft—most just steal from the cash register (usually on impulse); dip into petty cash; pad their expenses; or take home tools, supplies, or products belonging to their employers. Few robbers invest time studying a target and planning how to maximize their take.

This leads to the observation that criminal acts involve *short-range choices*, that they tend to occur on the spur of the moment. Crimes are committed so close to the offender's home because offenders will not invest the time and effort to go far afield. And they frequently occur in response to perceived opportunities of the moment: a rape in response to observing a woman asleep with her window open; a robbery in response to entering a convenience store for cigarettes and seeing that the clerk is alone. It also is the case that criminal attempts very frequently are frustrated by momentary impediments; many robberies fail to take place because of a sudden increase in the number of people on the street or because police arrive for a coffee break at an all-night restaurant.

This also lets us see that most criminal actions are *brief* in duration. Moreover, the *rewards are small and fleeting*. The monetary rewards are almost always small and soon spent; robbers frequently spend all of their take within a few hours. However, it is important to see that whatever rewards are produced by a crime, they usually are *immediate*. The murderer silences an immediate source of irritation, be it a crying infant, an insolent buddy, or an uncooperative robbery victim. The rapist gains immediate sexual gratification. The burglar gains a DVD player, a piggy bank full of coins, or some liquor—now.

Most crimes are *easy* to commit and very *simple* in design. Nearly anyone willing to commit them can do so with little or no preparation. But even more important, most criminal acts are *exciting*. They involve the thrill of risk and danger as well as the rush that some gain from having domination over victims.

These aspects of criminal actions will help us to understand criminal actors: who they are and why they do it. Let us now take up this task.

BIOLOGICAL THEORIES OF DEVIANCE

For centuries humans have wondered why some people are chronic deviants—why some people cannot be trusted to conform to important norms. Virtually every facet of life has been blamed by someone as a cause of crime and deviance. But perhaps the oldest claim about deviance is that some people are just "born bad": Some people have an inborn personality flaw that stimulates misbehavior or prevents them from controlling their deviant urges. This view became very influential in the 1870s, when an Italian physician, Cesare Lombroso (1836–1909), began to gather systematic data on prison

© Bettmann/Corbis

Cesare Lombroso.

and jail inmates and to develop a biological theory of criminal behavior.

"BORN CRIMINALS"

Lombroso believed he had found the key to criminal behavior in human evolution. His years of careful observation and measurement of prison inmates convinced him that the most serious, vicious, and persistent criminals (who he believed made up about one-third of all persons who commit crimes) were "born criminals" (Lombroso-Ferrero, 1911). **Born criminals** were less evolved humans who were biological "throwbacks" to our primitive ancestors, according to Lombroso. The born criminal is "an atavistic being who reproduces in his person the ferocious instincts of primitive humanity and inferior animals."

Lombroso believed that, because of their genetic makeup, born criminals could not restrain their violent and animalistic urges. Because the trouble was biological, he argued, little or nothing could be done to cure born criminals; society could be protected only by locking them up. However, because their criminality was not their fault, born criminals ought to be treated as kindly as possible in dignified, decent prisons.

In Lombroso's day statistical techniques were virtually unknown, and his data were based on

832 Italian prison inmates, and especially on 390 of "the most notorious and depraved." Therefore, when he compared his measurements of various features of criminals with those based on 868 Italian soldiers, he did so in a rather impressionistic fashion. Among the differences reported by Lombroso were a preponderance of low foreheads and overlarge jaws among criminals—traits associated with primitive human types. In addition Lombroso reported that criminals lacked strength and weight and that they were of low intelligence.

Lombroso was the most influential criminologist of the late nineteenth century, although his views were vigorously attacked in French sociological circles, especially by Gabriel Tarde and Émile Durkheim. The latter believed that crimes were committed by entirely normal persons for reasons beyond their control[3] (Beirne, 1993).

Then in 1913 an immensely influential study appeared: Charles Goring's *The English Convict: A Statistical Study*. Goring's work was based on observations of 2,348 male convicts whom he compared with a variety of nonconvict populations. With the help of Karl Pearson, one of the founders of modern statistics, Goring used proper methods to analyze his data.

Goring devoted his first two chapters to refuting Lombroso, showing that skull shapes and sizes and other traits of the physiognomy of the born criminal did not distinguish convicts from others, nor did they distinguish the worst offenders from those convicted of less serious crimes. Goring's results caused a rapid rejection of Lombroso's evolutionary theory of the "born criminal." However, the findings based on English convicts reaffirmed some of Lombroso's most interesting findings. In particular Goring found that English convicts also were of "inferior stature and weight." Moreover, he observed a marked lack of intelligence among convicts and concluded that this was "the principal constitutional determinant of crime."

Today, the lack of stature found by Lombroso and by Goring is thought to have been caused by the dietary deficiencies suffered by the poor in that era. In contrast, studies show that today people with more muscular body types are more likely to commit crimes (Cortés and Gatti, 1972; Cortés, 1982). But the findings about lack of intelligence have stood the test of time and subsequent research. By now there is an immense body of evidence, much of it recent, that IQ is negatively associated with committing crimes (Wilson and Herrnstein, 1985). Considering the lack of judgment involved in most ordinary criminal actions, that is hardly surprising.

BEHAVIORAL GENETICS

In Chapter 5 we examined the field of behavioral genetics, which has attempted to assess the role of heredity in various forms of human behavior. As we saw, studies of twins have been a primary research method for behavioral geneticists. One study done in Denmark examined 3,586 twin pairs (Christiansen, 1977). The researcher checked each twin through the criminal record files of the Danish police, recording only serious offenses. The results were highly suggestive. For identical (or monozygotic) twins, if one twin had a serious criminal record, the odds were fifty-fifty that the other twin did, too. But for fraternal (dizygotic) twins (using male sets only to eliminate gender differences within pairs), if one twin was a criminal, the odds were only about one in five that the other twin also was a criminal. Because each set of twins grew up in the same home, their environment was held constant, and thus, these differences suggest that the more genetically similar, the more similar the pattern of deviance or conformity. Adoption studies also have sustained interest in a hereditary component in criminal behavior. In terms of criminal records, adoptees much more closely resemble their biological than adoptive parents (Mednick et al., 1984).

Keep in mind that even if there is a genetic "predisposition" to break the law, much more is involved in criminal acts. For one thing such actions, like all human behavior, must be learned. Spiders may be genetically programmed to spin webs, but no human is born with instincts to break into houses or to write bad checks. Moreover, geneticists have yet to discover just *what* people inherit that can predispose them to deviance. Wilson and Herrnstein (1985) have suggested that much deviant behavior results because some people seem unable to control their impulses or to consider long-term costs versus short-term gains.

The most recent research suggests that physiological factors may be involved in impulsive, deviant behavior. For example, it recently was

---

3. These views reflected Quetelet's 1842 dictum: "Society prepares crime, and the guilty are only the instruments by which it is executed" (in Beirne, 1993).

learned that a group of men who engaged in "antisocial" behavior had significantly less prefrontal gray matter than did a control group (Raine et al., 2000). Even more compelling is a study that found that boys identified by their peers and by their records as guilty of persistent antisocial behavior were quite deficient in the cortisol level found in their saliva after they had been subjected to stress (McBurnett et al., 2000). Cortisol is produced by the body in response to fear and anxiety. What the data show is that there exist boys who do not have normal fear reactions. In effect they cannot be "scared" into conforming.

So, here we are. It appears that scientists are closing in on substantial physiological sources of criminality. Nevertheless, most sociologists are uncomfortable with the idea that a tendency to commit crimes might be partly rooted in physiology and genetics. On the other hand, they long have been puzzled by the marked gender and age differences in patterns of deviant behavior, traits having obvious physical as well as social aspects. Then Walter Gove suggested a new synthesis of biology and sociology.

A CLOSER VIEW
WALTER GOVE: AGE, GENDER, BIOLOGY, AND DEVIANCE

Walter Gove (1985) began with three well-known, but little understood, facts about crime and deviance. First of all, no other variables influence criminal activity as much as do gender and age. As illustrated in Table 7-5, in all societies for which data exist, males are far more likely than are females to commit crimes (South and Messner, 1987). Moreover, as the data in Table 7-6 show, male arrest rates for violent crimes decline very steeply with age—a pattern that holds in all societies for which data are available.

However, Table 7-7 reveals something that too often is overlooked: Gender differences vary greatly by *type* of offense. Men are far more likely than women to be charged with violent crimes or, as in the case of burglary, crimes that have the potential for a violent confrontation. But gender differences are far smaller when "sit down" crimes such as larceny, forgery, and fraud are involved, and women are even slightly more likely than men to be arrested for embezzlement.

Many social scientists have attributed gender differences in crime rates to socialization, claiming that females are socialized in ways that make them more law abiding (Bowker, 1981). This may well be true, but it doesn't help explain why gender differences are so much smaller for some crimes than for others. Presumably, a well-socialized person should be as unwilling to write bad checks as to rob stores. In similar fashion attachments often have been invoked to explain the decline in criminality as people get older. Thus, as people marry and begin to have children, they have more to lose by being detected in deviant behavior (Sampson and Laub, 1990). While this explanation has con-

TABLE 7-5	Percentage Female among Those Arrested		
NATION	FEMALE (%)	NATION	FEMALE (%)
China	2.2	Hungary	12.5
England and Wales	4.0	Finland	13.5
Peru	6.0	Scotland	13.5
Romania	7.9	Soviet Union	13.8
Poland	8.2	**Canada**	**14.1**
Paraguay	9.2	Chile	14.5
South Korea	9.6	Swaziland	16.5
Czechoslovakia	9.7	Switzerland	16.7
Botswana	9.9	France	17.3
Netherlands	10.0	**United States**	**18.4**
Bahamas	10.1	Austria	19.1
Greece	10.3	Japan	20.7
Northern Ireland	11.5	West Germany	23.5

Source: Interpol (1992).

TABLE 7-6	Arrest of Males for Violent Crimes by Age per 100,000 Population (United States)
AGE	ARREST RATE
15–19	655.4
20–24	719.8
25–29	479.2
30–34	378.7
35–39	311.3
40–44	271.6
45–49	190.4
50–54	114.7
55–59	73.4
60–64	44.5
65 and over	23.1

Source: U.S. Department of Justice, *Crime in the United States*, 2004.

siderable merit, it does not address the question of why age has so much greater impact on some offenses—homicide and robbery, for example—than on others, such as larceny-theft.

Pondering these issues Gove concluded that the common factor linking these patterns is that *crimes differ in the extent to which they involve aggressive and physically demanding behavior.*

At this point Gove assessed a growing literature that links aggressive or assertive forms of deviance to physique. Beginning with the work of W. H. Sheldon (1940) in the 1930s through studies by J. B. Cortés and F. M. Gatti (1972) and Cortés (1982), research has found that an athletic (or mesomorphic) body build is conducive to these forms of behavior. This is not to suggest that being muscular causes people to commit assault or robberies but that some degree of strength and self-confidence is required to act in these ways: The proverbial ninety-eight-pound weakling does not make a successful mugger. Moreover, the lifestyle of the drunk or the addict is often very physically demanding as well. Consider the testimony of this heroin addict:

[It was] the worst period of my life. I found myself wandering around the streets of New York filthy all the time. I had no place to stay. I slept on rooftops, in hallways, in damp cellars, any available place and always with one eye open. . . . I was really low then, not eating . . . and cold all the time. (Tardola, 1970)

Finally, Gove was ready to put these pieces together.

Why do these forms of deviance rapidly decline at around age thirty? First, because *physical strength* peaks in the early twenties and then declines. Second, *physical energy* also peaks in the twenties. Gove postulates two aspects of energy. The first is endurance, or conditioning. As we age, we remain strong enough to perform various actions but lose the endurance or conditioning to do them for a long time. For example, an aging boxer may still be a dangerous opponent for a few rounds but may be forced to try for an early knockout, knowing he will tire badly in later rounds. A second aspect of energy, according to Gove, is the ability to *rebound*, or recover, from injury and exertion. For example, an older alcoholic will recover more slowly from a drunken spree. Third, Gove argues that *psychological drives*, especially those sustained by the production of such hormones as testosterone and adrenaline, decline suddenly, too.

TABLE 7-7	Percentage Female among Those Arrested for Various Crimes (United States)	
CRIME		**FEMALE (%)**
Homicide		10.6
Robbery		10.7
Burglary		14.3
Larceny-theft		37.6
Forgery		40.3
Fraud		45.9
Embezzlement		50.2

Source: U.S. Department of Justice, *Crime in the United States*, 2004.

This line of theorizing led Gove to suggest that females are much less likely than males to commit certain acts simply because they are weaker and smaller. He notes that percentages such as those in Table 7-7 indicate the need to explain not simply why women are less likely than men to commit crimes but also why the difference is so much greater for the high-risk, physically demanding crimes. Moreover, among both males and females, those with more muscular builds are more prone to these forms of deviance, and both genders show a notable drop in these behaviors as they begin to pass their physical prime. Gove concluded with the observation that, at the age when some people are winning Olympic medals, others the same age are busy committing assaults, robberies, rapes, and burglaries and that both groups consist of "young adults who withdraw from the field as they age."

Gove's work is entirely consistent with the findings about testosterone. While testosterone level is positively related to all forms of criminal and delinquent activity, the correlations are stronger with violent forms of crime than with property crime (Dabbs et al., 1987; Dabbs et al., 1988). Indeed, testosterone level is related to thrill-seeking behavior generally (Daitzman and Zuckerman, 1980). The well-known decline in violent crimes as people get older is very similar to the decline in testosterone levels also associated with age (Dabbs, 1990). In addition variations in testosterone levels are associated with violent behavior among *women*, despite the fact that women have very low levels of testosterone compared with men. Finally, that women have far less testosterone than men would appear to account for a substantial proportion of gender differences in violent behavior.

Although Gove's new approach may help explain certain aspects of deviance, it fails to address many others. Most people do not stop committing risky and physical crimes as they pass their prime because *most people never commit these offenses at any age.* What distinguishes those who do from those who don't? Let's turn to other theories of crime and deviance in search of answers.

MENTAL ILLNESS

Mental illness has long been regarded by the public, the courts, and the media as a major factor in criminal behavior, especially when acts of extreme violence and cruelty are involved. But until quite recently, many sociologists disputed this claim, arguing that the mentally ill are no more dangerous than the general public (Scheff, 1984). In 1987 the National Mental Health Association distributed a pamphlet claiming "people with mental illness pose no more of a crime threat than do members of the general population" (in Link, Andrews, and Cullen, 1992). And for a time research seemed to support these views. Studies of discharged mental patients showed them to be *less likely* to be arrested than the general public (Rabkin, 1979). But then the findings suddenly changed. Research has found that, since about 1965, discharged mental patients have been arrested more often than the general public (Rabkin, 1979; Holcomb and Ahr, 1988; Harry and Steadman, 1988; McFarland et al., 1989; Link, Andrews, and Cullen, 1992). At present discharged mental patients are three times as likely as others to be arrested, and the ratio is especially high when only violent crimes are included.

What happened? Why did researchers find discharged patients more law abiding than the general public before 1965 but far more of a threat thereafter? Radical changes in commitment laws and treatment policies caused the release of persons who previously would have been retained in mental hospitals. In 1960 more than 630,000 persons were being treated in mental hospitals, but by 1990 the number had declined to 119,062. The decline is even greater than it appears since the U.S. population grew by almost 70 million during this period. Expressed as rates, in 1960 there were 351 mental hospital patients per 100,000 population; in 1990 there were only 48. Part of the decline is due to the discovery of drugs able to control some forms of mental illness. But these drugs work only if people continue to take them, and without supervision and aftercare, many discharged patients fail to do so. In addition many others for whom drugs are not effective are nevertheless released back into the community when their symptoms momentarily are mild. When their condition worsens again, many of these victims of mental illness end up on the streets, where they make up an estimated third of the homeless population. And homeless or not, many of these people explode into episodes of violence.

In his in-depth study of the homeless, Fuller Torrey (1988) noted that "mentally ill individuals, dumped from mental hospitals and left to wander in society without treatment or aftercare, are responsible for an increasing number of violent acts." Because prior to 1965 persons committed to mental hospitals who had a history of violent behavior tended not to be released, researchers who compared *ex-patients* with the general public not surprisingly found the former to be less prone to violent offenses. Today, when even people with long histories of violent behavior often are quickly released back into the community, the importance of mental illness as a cause of violent crime is reflected in their very high offense rates. In fact, studies of arrest data on former mental patients likely *under*estimate their true offense rates. In many instances, when a former mental patient commits new offenses, he or she is simply returned to the hospital rather than arrested; thus, no record of these offenses turns up when social scientists check police files (Krakowski, Volavka, and Brizer, 1986).

When research is based on samples of *all* persons under treatment for mental illness, not just those who have been discharged, the correlation between mental illness and crimes of violence also shows up very clearly (Link, Andrews, and Cullen, 1992). Moreover, these studies show that most offenses by the mentally ill occur during episodes of acute illness, when they are experiencing hallucinations and delusions (Taylor and Gunn, 1984; Link, Andrews, and Cullen, 1992). That is, the mentally ill are not more dangerous than the general public when they are not ill. Does this mean that, aside from those who are in the grip of a psychotic episode, people who commit serious crimes are just like everybody else in terms of their personalities?

PERSONALITY THEORIES

Despite an immense amount of research, efforts to link various forms of criminal and deviant behavior to abnormal features of the personality have been disappointing (Liska, 1987). Thus, an assessment of ninety-four studies conducted between 1950 and 1960, meant to distinguish between criminals and noncriminals by using various personality tests, found the overall results to be weak and contradictory (Waldo and Dinitz, 1967). More recent reviews of the literature have also yielded very mixed findings (Wilson and Herrnstein, 1985). Many studies have failed to find the expected correlations between various psychological tests and criminal behavior, and too many of those that found such links were circular—the personality trait was itself a measure of criminal behavior. For example, the Q test requires subjects to use a pencil to trace their way through various mazes (Riddle and Roberts, 1977). These tasks were constructed so that it is necessary to cheat in order to perform them. Hence, the higher the score, the more a person cheated. The finding that Q scores are good predictors of delinquency may be useful diagnostically, but as an explanation it really amounts to nothing more than saying that people who will use fraud to their benefit on one occasion will do so on other occasions.

Nevertheless, the hunt for personality traits conducive to criminality has continued because it seems so obvious that they must exist. When you read the characteristics of the typical crime, I am sure you must have wondered something like, What could these people be thinking of? What they do is stupid and cruel. Reflecting on the same facts, Michael Gottfredson and Travis Hirschi (1990) proposed that the fundamental psychological feature characterizing those who engage in all varieties of criminal actions is *weak self-control*. They do not regard this as an inborn psychological characteristic and in fact propose that the primary task of criminology is to identify the sources of weak self-control. They also have given extended discussion to ways in which many well-known theories of crime and deviance can be utilized to do precisely that. Before turning to these theories, however, it will be useful to sketch the elements of self-control identified by Gottfredson and Hirschi.

ELEMENTS OF SELF-CONTROL

Low self-control involves the *unwillingness or inability to defer gratification*. Given the choice between receiving $5 today or $15 after a two-month wait, people with little self-control take the $5 today. As a corollary, people with low self-control also lack "diligence, tenacity, and persistence in a course of action" (Gottfredson and Hirschi, 1990). They prefer actions that are *simple and easy* such as getting money without working or obtaining sex without establishing a relationship. Indeed, people with little self-control will tend to have poor work records (high rates of absenteeism when employed) and to have unstable marital and family relationships—for they won't "work" at life. Consequently, they also will tend to learn little while in school, to quit school, and to lack all skills that require practice and training—they won't know how to fix a car or play a trumpet.

Because criminal acts are exciting, thrilling, or risky, people who commit them will be *thrill seekers*, and this also is consistent with weak self-control in the form of preferring immediate gratification: The thrill of risk is immediate, while the potential costs of risky behavior are uncertain and in the indefinite future (Gottfredson and Hirschi, 1990). Thus, people who engage in criminal acts also pursue immediate gratification and thrills through behavior that is not illegal—they smoke, drink, gamble, and engage in unprotected sex with strangers.

Finally, given that they impose loss and suffering on others, it is no surprise to find that people with a lack of self-control "tend to be *self-centered, indifferent, or insensitive* to the suffering and needs of others" (Gottfredson and Hirschi, 1990).

These traits of criminal actors are not merely deduced from what they do. Rather, these traits can be identified independently through testing and interviewing, and they are not transitory. Instead, they seem to become a permanent part of an individual's personality. So, now let's explore theories of deviant behavior with a special concern to see how they might help account for the development of weak self-control.

DEVIANT ATTACHMENT THEORIES

We have already encountered the fundamental proposition on which most sociological theories of deviance are based: Our behavior is shaped through interaction with others, especially those to whom we have formed strong attachments. These attachments to others cause us to try to live up to their expectations about how we should act. In Chapter 3 we examined research showing the power of attachments to influence our behavior: People will even adopt a new religion to please their friends.

However, although attachments produce conformity to the expectations of others, that does not necessarily mean we will conform to the norms of our society. What happens if our attachments are to people who themselves break norms? What if our friends are burglars or auto thieves? What if they encourage us to seize the moment, to live for now? Questions such as these have led to the formulation of several sociological theories seeking to explain deviance on the basis of attachments to *deviant others*. These theories elaborate on the sentiments of many proverbs about falling in with "bad company."

DIFFERENTIAL ASSOCIATION: SOCIAL LEARNING

Edwin H. Sutherland (1883–1950) is remembered as the most influential early American sociologist of deviant behavior. In 1924 he first proposed a theory based on deviant attachments, a theory known as the **differential association theory.** Sutherland argued that all behavior is the result of socialization through interaction. That is, how we act depends on how those around us desire us to act. How much we deviate from or conform to the norms depends on differences (or differentials) in whom we associate with.

Thus, Sutherland argued, boys become delinquent because too many of their attachments are to others who engage in and approve of delinquent acts. In this view the causes of deviance lie not in the individual but rather in the normal processes of social influence. However, Sutherland's differential association theory did not precisely explain how our friends influence our behavior and teach us to conform or deviate. Therefore, in 1966 Robert Burgess and Ronald Akers reformulated Sutherland's theory into a set of specific propositions based on learning theory.

Burgess and Akers argued that Sutherland had correctly identified the source of deviant behavior: attachments to people who supported deviance. Such friends teach us to deviate by rewarding us for deviant behavior and not rewarding (or reinforcing) nondeviant (conformist) behavior. Thus, a boy may receive a great deal of attention and respect from his friends when he steals a bike or breaks a window, but be ignored when he is "good."

Consider this more elaborate example: A person may have companions in the local tavern who reinforce heavy drinking and nightly visits but do not praise him or her for staying home or abstaining. Through such a pattern of *selective reinforcement*, a person may become a habitual heavy drinker. As time goes by, such a person may become increasingly dependent on "the regulars" at the tavern for social relations and become a problem drinker. Indeed, this may be why so many persons treated for alcohol or drug abuse resume such behavior even after treatment has cured their physical addiction. The patients return to the same social settings and the same sets of attachments through which they originally formed their problem behavior and quickly relearn it.

OFFENDING FAMILIES The strongest support for the differential association–social learning theory comes from the discovery of "offending families." While it has long been recognized that sometimes several members of a family will engage in deviant behavior and become involved in criminal activity, only recently have researchers realized the magnitude of this phenomenon. A national survey of delinquents confined in state institutions revealed that the majority had an immediate family member who had served (or was serving) time in jail or prison (Bureau of Justice Statistics, 1988). Twenty-five percent reported that their fathers had served time, 9 percent said their mother had been incarcerated, 25 percent had a brother or sister go to jail, and 13 percent listed uncles, aunts, or cousins. A substantial number also reported that several relatives had done time. Altogether, 52 percent reported that at least one member of their immediate family had been (or was currently) in jail or prison. A subsequent study of adult prisoners found that more than a third had a family member who had served time as an adult (Bureau of Justice Statistics, 1991).

But perhaps more important, children who are raised by parents who lack self-control and who commit crimes are likely to be socialized in ways that fail to develop substantial self-control. Recently, the Seattle newspapers reported on a teenage boy who hit a pedestrian while out driving his father's pickup truck. Although the victim clearly was seriously injured, the boy drove away. When he reached home, he told his parents what had happened. Their response was to drive the truck across the state line, abandon it, and then report it as stolen. What is noteworthy here is that both parents had arrest records and had allowed the boy to take the truck despite the fact that his driver's license had been suspended.

This anecdote illustrates the conclusion drawn by Donald West and David Farrington (1977) from their study of English families "that delinquency is transmitted from one generation to the next is indisputable." In their study 5 percent of families accounted for nearly 50 percent of the criminal convictions for the entire sample. This finding is congruent with American data (Wolfgang, Figlio, and Sellin, 1972).

In Chapter 13 more attention will be paid to how offending families are re-created from one generation to the next through faulty socialization.

DEVIANT FRIENDS Deviant attachments outside the family can also provide models of low self-control as well as involvement in crime. Having assembled the findings of many studies done over a forty-year period, Maynard L. Erickson (1971) showed that the great majority of delinquent acts occur in cooperation with other youth. Vandalism, for example, is rarely done by a lone juvenile. Most juveniles who steal a car or burglarize a house do so in groups. And research consistently finds that delinquents tend to have delinquent friends (Liska, 1987). Despite this supportive evidence, the theory fails to explain many aspects of deviance and does not jibe with many research findings.

A major problem is research suggesting that delinquent kids end up together because others reject them (Hirschi, 1969; Jensen, 1972; Liska, 1987), not because these friends encouraged one another's delinquency. Diana Gray (1973) found that teenage prostitutes did not begin to associate with pimps or other prostitutes until *after* they had already begun to sell sexual favors. Travis Hirschi (1969) found that only some boys with delinquent friends engaged in delinquency—those who did well in school and were attached to their parents were not more likely to engage in delinquency if they had delinquent friends than if they did not.

However, some teenagers are immersed in a *dense social network*, most or all of whose members are delinquent. In fact, some of these networks form around the commission of deviant acts. Obviously, this affects the continuing misbehavior of each member, as it is rewarded and encouraged by their peers (Haynie, 2001).

Membership in a deviant group may explain the deviance of any given member, but it raises a second compelling question: How do we explain the existence of such groups? That is, if attachments to deviant friends cause deviance, what caused the deviance of the friends? At some point someone must have begun to act in a deviant way without supportive friends. In addition why do some people but not others have deviant friends or family members? And why are only some people susceptible to the influence of deviant attachments, as research shows?

Finally, what explains the many crimes and other deviant acts that could not possibly be responses to the expectations of others (deviant or otherwise) because they are done in secret in the hope that no one finds out? The man who kills his wife and hides the body, the embezzler, or the secret drug user. Here we must be reminded of the nature of criminal acts—*other things being equal, they are directly rewarding,* even if these rewards sometimes seem petty compared to the potential costs. Murder satisfies the desire for revenge or to escape further annoyance; embezzlement yields money or goods; drugs produce a high. Granted, our friends and relatives can lead us astray, but people are perfectly capable of getting into mischief on their own (Wilson and Herrnstein, 1985).

The problems of the differential association–social learning approach are but sins of omission. That is, problems do not arise because of what the theory explains but because of what it does not explain. Other theories of deviance attempt to fill these gaps.

SUBCULTURAL DEVIANCE

One way to supply the deviant others needed by differential association–social learning theory is to recognize the existence of subcultures within a society. In Chapter 2 we discussed how a subculture is a culture within a culture: a group of people who maintain or develop a set of values, norms, and roles that are different from those of the surrounding society. As we saw in Chapter 3, sometimes subcultures arise around a new religion; those who become attached to persons in

To conform to the norms of one group often is to seriously break the norms of another group. These punk rock fans have gone to great pains to be "normal" in the eyes of other punks. Their dress and deportment break norms of society in general and especially those of older professionals **like the two men on the next page.** Of course, these men also make a considerable effort to be "normal." Some forms of deviance, such as these having to do with norms of appearance, reflect subculture conflicts. Which of these styles is seen as more acceptable often will be determined by which subculture is the more powerful.

St. Duroy/Rapho

such a subculture can subsequently become members themselves. In a similar fashion, a subculture can arise around drug use; in fact, during the 1960s many people were recruited to be drug users just as others were recruited to be Unificationists or Hare Krishnas.

Recognition of subcultures lets us understand that deviance is often a matter of definition. An outside observer noting the behavior of a member of a subculture may regard that person as deviant. But that same person is conforming to the norms of his or her group. For example, Jehovah's Witnesses are members of a religious group who, among other things, accept a norm against saluting flags (they regard it as idolatry). Many Americans observing a Jehovah's Witness refusing to salute the American flag might define the behavior as deviant, but to other Jehovah's Witnesses, it would be an act of conformity.

Particularly in complex modern societies, many subcultures exist, and thus, some deviant behavior can be explained as conflicts over norms, or **subcultural deviance.** Public controversies over pornography, marijuana, abortion, and sexual behavior are conflicts over whose norms will be represented in legal codes and public policies and whose will be judged deviant.

Some subcultural views of deviance can be summed up as "different strokes for different folks." Thus, the norms in any society are often determined by who has the power to pass laws and set policies. Because Jehovah's Witnesses are a tiny minority, they are unable to ban flag ceremonies. Because the majority of Americans oppose the use of narcotics, narcotics are illegal.

Subcultural theories also help explain how some forms of deviance that do not begin as conformity to the norms of a subculture can become stabilized as an individual enters into a subculture. We have seen that Diana Gray (1973) found that girls became prostitutes on their own. However, in time some of them did begin to associate with pimps and prostitutes and become part of this deviant subculture. Many of the other girls soon gave up prostitution.

Although subcultural theories add to our understanding of deviance, they fail to provide a full explanation and they seem especially inadequate for explaining crime. Aside from the rare gangland slaying, very few homicides seem to be the result of conforming to the norms of a subculture (Parker, 1989). In fact, although people in prison have more lenient attitudes about crime, they overwhelmingly identify criminal actions as "wrong" and are so opposed to certain crimes (sex crimes against children, for example) that precautions must be taken to protect those offenders from other prisoners. Moreover, prisoners resent being the victims of crime—having possessions stolen, for example—as much as does anyone else. And that reveals the major omission of subcultural theory: *its inability to account for crime and deviance within any subcultural group.*

All human groups, regardless of their norms, include some members who fail to conform. Some members of juvenile gangs steal from other members. Some gangsters squeal to the authorities. Some ministers commit adultery. Some scientists fake their results. And many people feel deeply ashamed and guilty after committing some act, which makes no sense if what they were doing was "right" according to their group norms. Subcultural explanations cannot account for deviance of this kind. Again, the solution is not to discard a theory but to add to it.

STRUCTURAL STRAIN THEORIES

We now encounter a closely related group of theories that attempt to explain deviance on the basis of **structural strain,** or frustration caused by a person's position in the social structure, especially the stratification system.

The ideas underlying structural strain theories are very old. Whenever someone says that people commit crimes because of poverty or some other disadvantage, that person is invoking strain theory. Back in 1938 Robert K. Merton formulated a theory of how disadvantage can lead to deviance, and theories based on his work are called **structural strain theories.**

Merton began with the assumption that humans have a natural tendency to observe norms, a tendency instilled in us by regular processes of socialization. Indeed, this tendency is reflected by that part of our personality often called the conscience. Because of our conscience, breaking the norms causes us to feel some degree of guilt and remorse. Yet people often act against their conscience. Why? Merton thought it was because of terrible strains on them.

These strains arise because people are socialized to have certain desires, or goals, and to regard certain means as proper ways to achieve these goals. However, the proper means don't work as well for some people as for others. People who are poorly placed in the stratification system will find themselves unable to achieve their goals or at least unable to achieve them as easily as people better placed in the system, if they use only legitimate means (that is, if they obey their conscience). As long as disadvantaged people stick to the rules and obey the norms, they will experience frustration because they will fail to achieve wealth, happiness, fame, comfort, influence, and all the other things socialization has taught them to value. The resulting strain forces people to use deviant or illegitimate means to achieve goals.

Merton argued that deviance is a built-in consequence of stratification. *Strain theory portrays a deviant as a person torn between guilt and desire, with desire gaining the upper hand* (Hirschi, 1969). In effect Merton argued that poverty causes the poor to turn to crime, to alcohol, to drugs, and even to killing friends and relatives to escape their unfulfilled desires for a better life.

By itself the structural strain theory of deviance runs into serious problems. First, the theory would seem to predict far more crime and deviance than actually occurs. Of those poorly placed in the stratification system, the great majority do not commit acts of significant deviance. What distinguishes them from those who do?

© Richard Kalvar/Magnum Photos

Second, most of the deviant behavior committed by persons under structural strain cannot alleviate their frustrations. Most crime pays very little—even menial jobs would provide more luxury. That is, illegal behavior offers scant hope for achieving goals that cannot be met by legitimate means.

Third, the theory offers no explanation for deviant acts committed by people in the privileged social positions, such as middle-class and upper-class teenage delinquents. Nor does it explain why wealthy people shoplift or why bankers embezzle. Indeed, if people commit crimes only when driven by intense frustrations caused by societal deprivation, crimes committed by privileged people are inexplicable.

Perhaps the worst problem strain theory faces is that a person's social class is barely, if at all, related to committing crimes. Table 7-8 is based on two national samples of American adults. It shows that 14 percent of those who reported that their family income when they were sixteen was "far below average" have been picked up by the police for something other than a traffic offense. Of those whose family income was "above average," 14 percent have been picked up by the police. This is as close to a perfect *zero correlation* as one could expect. Moreover, scores of studies have failed to show that poor kids are more likely to commit delinquent acts than are kids from privileged homes (Tittle et al., 1978; Tittle and Meier, 1990). On the other hand, when we focus on very serious offenses such as arson, robbery, and burglary

TABLE 7-8	Family Income of Americans at Age 16 and Being "Picked Up" by the Police			
"Thinking about the time when you were 16 years old, compared with American families in general then, would you say your family income was":				
	"FAR BELOW AVERAGE"	"BELOW AVERAGE"	"AVERAGE"	"ABOVE AVERAGE"
Percent who have been picked up by the police	14	13	12	14

Source: Prepared by the author from the General Social Surveys, 1983 and 1984.

of commercial firms, we find people from the very lowest stratum of society are substantially overrepresented (Hindelang, Hirschi, and Weis, 1981). But even if this behavior does represent a response to strain, it applies to only a tiny segment of the population and accounts for only a small proportion of the crime and deviance that occurs.

Whenever the unemployment rate rises, the news media begin to cite that rise as a "root cause" in their stories about crime or drug abuse. Yet although social scientists frequently have predicted a positive correlation between unemployment and property crime rates, research has failed to detect it (Cantor and Land, 1985).

Within cities, crime and delinquency *are* concentrated in the poorer neighborhoods (Shaw and McKay, 1942; Bursik and Webb, 1982; Simcha-Fagan and Schwartz, 1986; Taylor and Covington, 1988; Sherman, Gartin, and Buerger, 1989), but numerous reasons for this do not depend on strain theory. Poor neighborhoods tend to *attract* people who *already* engage in crime and deviance because social control is weakest in these parts of town. Control is weak here because of high rates of population turnover (residents constantly moving in and out), poor policing, and lack of community organization. Hence, chronic offenders tend to drift into the most run-down, poverty-stricken areas of cities because they are the safest places for them and because their crimes pay so poorly that they can't afford better surroundings. Indeed, this is where people tend to go following their release from jails and prisons. Ironically, this is also where they commit their crimes. Burglary rates, for example, are highest not in wealthy neighborhoods where homes contain many valuables but rather in the poorest neighborhoods. Most bur-

glars pick on homes and stores within a few blocks of their residence.

Another factor elevating crime rates in the poorest parts of cities is that these neighborhoods also tend to be mixed-use neighborhoods. Where homes, apartments, retail stores, and even light industries are all mixed together, the *opportunities* for crime and delinquency are higher. Kids living in many residential suburbs would have to ask mom or dad for a ride to shoplift or break into a liquor store. Many kids in mixed-use neighborhoods can do these things without even crossing a street (Bursik, 1988).

WHITE-COLLAR CRIME

Nothing is more embarrassing for strain theory than the frequency with which high-status people commit crimes. Merton's famous essay, "Social Structure and Anomie," essentially ignores the banker who embezzles or the physician who sells prescriptions to addicts, for only in this way could Merton sustain his thesis that crime and deviance are caused by the pains of poverty and want. While a graduate student, I asked a famous strain theorist how he explained upper-class criminals and was astonished when he said that was a job for psychotherapists because such people must be crazy. Eventually, however, social scientists accepted that many middle- and upper-class people commit crimes. And from this recognition arose a new concept: white-collar crime.

White-collar crime is distinguished from other crime on the basis of the *social status of the offender*. The term *white-collar* indicates that a person wears a suit (with a white shirt or blouse) to work as opposed to blue-collar workers, who wear "work clothes" and do manual labor. Hence, white-collar crimes are those committed by "a person of respectability and high social status in the course of his occupation" (Sutherland, 1983). A good example of white-collar crime would be a person in a position of power or authority who accepts bribes to perform his or her duties. Table 7-9 shows that while the majority of citizens of all nations included in the World Values Surveys say that it is never justified to take a bribe, substantial minorities in some nations think it is justified at least sometimes. People in Turkey are the most willing to justify bribes; people in Bangladesh are the least willing.

TABLE 7-9	"Is It Ever Justified for Someone to Accept a Bribe in the Course of Their Duties?"		
NATION	**NEVER JUSTIFIED (%)**	**NATION**	**NEVER JUSTIFIED (%)**
Bangladesh	99	Venezuela	75
Morocco	97	Taiwan	73
Jordan	95	Netherlands	73
Egypt	94	Dominican Republic	73
Denmark	92	Peru	72
Iran	91	Austria	72
Argentina	90	Uganda	72
Puerto Rico	89	Chile	69
Algeria	88	Russia	69
Iceland	86	**Mexico**	**68**
El Salvador	86	Sweden	68
Israel	86	Belgium	67
Bosnia	85	Great Britian	67
Australia	85	France	66
Norway	84	Germany	65
Indonesia	82	Estonia	65
New Zealand	81	Lithuania	65
Colombia	81	Nigeria	63
China	81	Armenia	62
India	81	Greece	62
Ireland	81	South Africa	61
United States	**80**	Albania	53
Canada	**80**	Czech Republic	51
Japan	80	Hungary	51
South Korea	80	Azerbaijan	50
Finland	79	Brazil	46
Croatia	78	Moldova	45
Italy	78	Slovakia	39
Switzerland	77	Belarus	38
Poland	76	Philippines	38
Singapore	76	Turkey	24
Bulgaria	75		

Source: Prepared by the author from the World Values Surveys, 2000–2001.

Defining a crime on the basis of *who* commits it has caused no end of difficulties. Paramount is the necessity of classifying identical actions as different if done by persons of different status. For example, performing unnecessary automobile repairs is a white-collar crime only if it is initiated by an executive in an auto repair firm. When it is done by a gas station attendant, it is something else. Or when a sales representative cheats on his income taxes by treating automobile mileage from his holiday trips as business-related, it is a white-collar offense, but when a janitor does likewise, it is not. Thus, Susan Shapiro (1990) suggested

that the status of offenders be dropped from the concept entirely and that a new concept of crimes involving violations of trust be substituted. She noted that upper-status persons are more often in positions of trust, and thus, social class can still be of importance in explaining these sorts of violations. Moreover, the concept of trust liberates social scientists from being unable to deal with research that shows that it is not the rich, but rather young, lower-income males, who are most likely to cheat on their taxes (Mason and Calvin, 1978).

For many social scientists, however, the problems inherent in the concept are outweighed by the capacity of the term *white-collar crime* to call attention to the fact that crime does not occur only among the lower classes—that when brokers dip into their clients' stock accounts, they are doing something as fully criminal as when high school dropouts steal cars and sell their parts. All social classes have thieves. However, for other sociologists white-collar crime serves as a conceptual device to protect strain theory from empirical falsification. That is, they argue that crime has different causes depending on the social class of the criminal. Hence, they reason that strain theory should be applied *only* to crimes committed by lower-class people. Crimes committed by middle- and upper-class people do not, therefore, challenge strain theory because the causes of their criminality must be sought elsewhere.

Indeed, there have been many efforts to formulate specific theories to explain white-collar crime (Sutherland, 1983; Coleman, 1987). In the end, however, each of these attempts is forced to argue that quite different motivations impel poor people and rich people to break the law. But is this true? Must we suppose that a gas station attendant stole from the cash register because he wanted more money but accountants have another motive when they embezzle? Is it useful to suggest that different causes lead doctors, on the one hand, and patients, on the other, to defraud Medicaid? Travis Hirschi and Michael Gottfredson (1987) have suggested that it is not. White-collar crime, they argued, causes problems only for some theories. Other theories have no difficulty with the fact that princes as well as paupers can be overcome by greed and temptation. Let's examine such a theory.

CONTROL THEORIES

To formulate a more comprehensive sociological theory of deviance, the famous French sociologist Émile Durkheim (1858–1917) proposed in effect that we dismiss the question, Why do they do it? and ask instead, Why *don't* they do it? Since Durkheim's time, his advice has been heeded by many leading sociological and criminological theorists; this approach to deviance is known as **control theory.**

The initial assumption all control theories make is that life is a vast cafeteria of temptation. By themselves *deviant acts tend to be attractive, providing rewards to those who engage in them.* To some, theft produces desired goods, and alcohol and drugs supply enjoyment. Indeed, control theorists argue that norms arise to prohibit various kinds of behavior because without these norms such behavior would be frequent.

Put another way, when we consider what things people should not be allowed to do, we don't bother to prohibit behavior that people find unpleasant or unappealing. We assume that people won't do these things anyway. So we concern ourselves with things that people find rewarding and therefore might be tempted to do.

Thus, control theorists take deviance for granted and concentrate instead on explaining why people conform. Their answer is that people vary greatly in the degree of control their groups have over them. In any group some people are rewarded more for conformity and punished more for deviance than other people are. Control theorists argue that conformity occurs only when people have more to gain by it than they have to gain by deviance.

In a classic paper, Jackson Toby (1957) described teenagers as differing in terms of their **stake in conformity.** This phrase refers to *what a person risks losing* by being detected in deviant behavior. Toby suggested that all of us are tempted but that we resist to the extent that we feel we have much to lose by deviant behavior; for instance, a boy with a low stake in conformity has little "incentive to resist the temptation to do what he wants when he wants to do it."

Therefore, like strain theorists, control theorists accept that access to desired rewards is unequal among members of any society: Some people succeed; some get left out. But while strain theory argues that inequality pushes the have-nots to deviate, control theory stresses how the have-nots are free to deviate. In the words of the golden oldie rock song, "Freedom's just another word for nothing left to lose." Some people are free to deviate because they risk very little if their deviant behavior is detected. But for others the costs of detection far exceed the rewards of deviance.

For control theory the causes of conformity are the **social bonds** between an individual and the group. When these bonds are strong, the individual conforms; when these bonds are weak, the individual deviates. Because the strength of these bonds can fluctuate over time, control theory can explain shifts from deviance to conformity (and vice versa) over a person's lifetime. Because many bonds are not related to social class, control theory can explain both the conformity of the poor and the deviance of the wealthy. But what are these bonds? There are

This famous painting by Hieronymus Bosch (1450–1516), entitled *Garden of Earthly Delights*, is an allegory on temptation prompted by the painter's deep religious concerns. Bosch's vision of the human condition as an unrelenting opportunity to sin foreshadowed the insights of modern control theory, which asks not why people deviate but why they ever conform.

Scala/Art Resource

four kinds between the individual and the group: *attachments, investments, involvements,* and *beliefs* (Hirschi, 1969; Stark and Bainbridge, 1997).

ATTACHMENTS

In accordance with Homans' law of liking, the more often people interact, the more they will come to like one another—they will become attached. Here group solidarity enters the picture. The degree to which an individual is attached to a group (has a social bond) is a function of the number and closeness of her or his

attachments—how much that individual cares about others and is cared about in return. When we are embedded in an intense local network, we are under intense pressure to live up to the standards and expectations of the group. In accordance with Homans' law of conformity, it is necessary for us to live up to group norms to retain the affection and respect of our intimates—we risk isolation from our primary group(s) if we are detected in serious norm violations. We shall explore this more extensively in the next chapter.

When we are alone, we often break norms—we pick our noses, belch, and otherwise act grossly. We usually do not break these norms so freely in company. Moreover, if we knew for certain that our friends would never find out, we might even commit serious norm violations. Those who lack significant attachments are in effect *always alone*, and their friends never know about the norms they break. They do not put relationships with others at risk because they have none to risk. For them, the costs of deviance are low.

By focusing on bonds of attachment, control theory is able to deal with a great many research findings about deviant behavior. The more that young people care about others—parents, friends, and teachers—the less likely they are to commit acts of delinquency (Hirschi, 1969; Laub et al., 1998; Giordano et al., 2002). Conversely, delinquents are very weakly attached, even to their delinquent friends (Hirschi, 1969; Kornhauser, 1978).

As Gove noted, sociological theories have trouble explaining why deviance declines with age. Delinquency rates rise rapidly from age twelve through about age sixteen and then begin to fall rapidly. Gove's theory of physical fitness would not apply here. In addition deviant behavior is much higher in the late teens and early twenties than it is after age thirty. This is the phenomenon Gove addressed. But Gove's theory does not address the fact that all forms of deviance, not just those requiring physical prowess, decline with age. Gove's theory is pertinent because it may tell us what causes some deviance rates to fall so rapidly with age. But we also must know why "sit down" crimes such as embezzlement also decline with age—albeit not so dramatically. Control theory fills this gap. With the onset of the teens, deviance rates rise rapidly not only because people become stronger but also because attachments weaken for many. Adolescence is a time when parent-child relations often become stressful, and teenagers often feel alienated from other family members. As this occurs, they have reduced stakes in conformity.

Then, in their later teens, many young people form strong new attachments, often to persons of the opposite sex. Hence, their stakes in conformity rise. Young adults are frequently very deficient in attachments. Often, they leave family and childhood friends behind and go out on their own, relatively unattached. With marriage, the birth of children, and steady employment, they form new attachments; thus, the tendency to commit even sedentary crimes lessens as people get older.

Perhaps the most powerful aspect of control theory is its ability to account for weak or missing correlations between social class and most forms of deviance. Close attachments are not confined to the middle and upper classes. Most poor kids love their parents, too, and most poor adults love their families and friends. Thus, most poor people have a strong stake in conformity. By the same token, some middle- and upper-class kids don't love their parents, and not all privileged people love their families and friends. Thus, their stake in conformity is low.

Durkheim answered the question, Why don't they do it? on the basis of attachments. As he put it, "We are moral beings to the extent that we are social beings."

The family serves as the primary source of the strong attachments that most effectively make us "social beings." Table 7-10 shows the impact of marital attachments on the percentage of American adults who have been picked up by the police. Among both men and women, those who were currently married were much less likely to have been picked up by the police than were persons who had never married or persons who were currently divorced or separated. When they are detected in deviant behavior, married people have more to lose: their families.

But that doesn't hold if the spouse is an offender, too! Recent research by Peggy Giordano and her colleagues (2002) found that serious female offenders (who had served sentences) were not "reformed" by marriage. That is, those who married were as likely to reoffend as those who remained single. The apparent reason is that although most male offenders who marry acquire a nonoffending spouse, serious female offenders tend to marry male offenders; hence, they do not risk spousal disapproval if they commit new offenses.

TABLE 7-10	Current Marital Status and Arrest among American Males and Females (Not Including Widowed)		
	MARRIED	SINGLE	DIVORCED AND SEPARATED
Percentage who have been picked up by the police			
Males	17	32	32
Females	5	9	10

Source: Prepared by the author from the General Social Surveys, 1982 and 1984.

INVESTMENTS

The idea of **investments** is simple. We are tied to conformity not only through our attachments to others but also through the stakes we have built up in life—the costs we have expended in constructing a satisfactory life and the rewards we expect. The more we have expended in getting an education, building a career, and acquiring possessions, the greater the risks of deviance. That is, we could lose our investments if we were detected in deviant behavior. An unemployed derelict may have very little to lose if caught sticking up a liquor store and a considerable amount to gain if he or she gets away with it. But it would be crazy for a successful lawyer to risk so much for so little, and most people rarely make really irrational decisions. When successful lawyers and bankers fail to resist the temptation to steal, they usually steal huge sums that seem to them to make the risks worthwhile. But the underlying processes of choice are the same for rich and poor.

Variations in investments also help account for the tendency of delinquents to reform as they reach adulthood. Teenagers have little investment at stake when they deviate. However, after people have begun to build normal adult lives, their investments mount rapidly. This has been demonstrated rather dramatically in new research by Christopher Uggen (2000), which showed that employment did not reduce the probability that individuals would commit a new crime until they reached their late twenties. That is, among offenders in their teens or early twenties, those who took a job were as likely to commit a new offense as those who remained idle. But after age twenty-seven, those who took a job were significantly more likely to cease offending.

INVOLVEMENTS

The **involvement** aspect of control theory takes into account that time and energy are limited. The more time a person spends on activities that conform to the norms, the less time and energy that person has to devote to deviant activities.

To a considerable extent, involvements are a consequence of investments and attachments. People who have families or play football after school or are engrossed in hobbies or are busy with careers have much less time and energy left for violating norms than do people with few attachments and investments. Popular wisdom has it that "idle hands are the devil's workshop."

Many studies have reported that the more time young people spend "hanging around" or riding in cars, the more likely they are to commit delinquent acts (Hirschi, 1969). And the more time they spend doing schoolwork or even talking with friends, the less likely they are to get into trouble. That is, people neglect to do all sorts of things for lack of time. College couples even delay breaking up until between terms or until the summer holiday, when they have more time (Hill, Rubin, and Peplau, 1976). People also tend not to do deviant things when they are pressed for time.

BELIEFS

Control theory stresses human rationality: Whether people tend to deviate or to conform depends on their calculations of the costs and benefits of deviance or conformity. But control theorists also recognize that through socialization we form **beliefs** about how the world works and how it *ought* to work. That is, we develop beliefs about how people, including ourselves, should behave. Sociologists often describe this as the **internalization of norms,** instead of using the word *conscience*.

We accept norms not only because our friends expect us to but also because we risk our self-respect if we deviate. The phrase "I'm not that kind of person" indicates that we hold certain beliefs about proper behavior. When a friend suggests a deviant act and assures us that nobody will know, we display internalized norms if we respond, "Yes, but *I* will know."

By themselves our beliefs may or may not cause us to conform. As we saw in Chapter 4, individual religious beliefs will prevent delinquent behavior only when the person is part of a community in which the majority belong to religious organizations. However, because most Americans live in such a community, a national sample of adults ought to display a negative correlation between attending church and having been picked up by the police. The data in Table 7-11 confirm this hypothesis: People who attend church less than once a year are more than four times as likely as weekly attenders to have been picked up by the police.

Put another way, religion gains the power to alter behavior when it is supported by attachments to others who accept the authority of the moral beliefs that religion teaches. This fact helps us to recognize that all four elements of control theory are interconnected. Attachments are also investments—much time and energy go into building close relations with others. Attachments and investments both act as involvements: Time spent with friends or at work is time not available for deviance. In addition our beliefs will also determine with whom we choose to become attached and what investments we decide to make.

As with other explanations of criminal and deviant behavior, control theory also cannot stand alone. It seeks to explain conformity to the norms of a social group, but it doesn't identify which group or note that conformity to the norms of one group may represent deviance from the norms of another group. In combination with subcultural theory, we have a more complete explanation of deviance. And control theory clearly implies and therefore requires elements of differential association–social learning theory to specify mechanisms by which attachments generate conformity.

To help pull these arguments together, let's watch as two sociologists combine aspects of control theory and differential association while replicating Travis Hirschi's classic study.

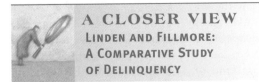

A CLOSER VIEW
LINDEN AND FILLMORE: A COMPARATIVE STUDY OF DELINQUENCY

Travis Hirschi's *Causes of Delinquency* (1969), based on a sample of American students in the public schools of Richmond, California, is one of the landmarks of delinquency research. In it he reformulated control theory and then deduced and tested many crucial hypotheses. We already have seen some of his important findings, including that success in school serves as a powerful investment and that such students

were impervious to the effects of deviant attachments.

Subsequently, Rick Linden and Cathy Fillmore (1981) undertook a parallel study based on a sample of Canadian students in Edmonton, Alberta. One of their aims was to demonstrate that studies of delinquency done in one modern society would generalize to other modern societies—that it would not be necessary to have different theories of delinquency for Canada, the United States, France, and Japan. A second aim was to more adequately combine elements of control and differential association theories of delinquency.

As mentioned earlier, differential association theory fails to explain why some teenagers have delinquent friends and others don't. The theory also has suffered from research evidence that teenagers often begin their delinquent activities *before* they begin substantial association with other delinquents. Moreover, Hirschi had found that having delinquent friends had no impact on some teenagers. Linden and Fillmore sought to repair these shortcomings of differential association theory by combining it with control theory. They proposed that teenagers with low stakes in conformity have little to lose by associating with other delinquents and that such association will further amplify their delinquency as they learn new criminal techniques and are reinforced for new acts of deviance. Figure 7-2 illustrates Linden and Fillmore's model.

Table 7-12 compares the delinquency of Richmond and Edmonton teenagers based on self-report data. More serious property crime was a bit more common in Richmond than in Edmonton at this time, but overall the data are very similar. Notice that here, too, we see huge gender differences.

Linden and Fillmore measured stakes in conformity on the basis of attachments to parents and liking school. They found that in both Edmonton and Richmond these were negatively correlated with being delinquent and with having delinquent friends. Finally, in both cities those with delinquent friends were more apt to be delinquent.

But the more important comparison involves applying the reformulated theoretical model to both samples. Linden and Fillmore found that in both nations their model fit the data well. In both Canada and the United States, low stakes in conformity seem to lead to the formation of attachments to delinquent peers, and these

TABLE 7-11	Church Attendance and Arrest among American Adults			
	ATTEND CHURCH			
	WEEKLY	MONTHLY	YEARLY	LESS THAN YEARLY
Percentage who have been picked up by the police	5	13	15	22

Source: Prepared by the author from the General Social Surveys, 1982 and 1984.

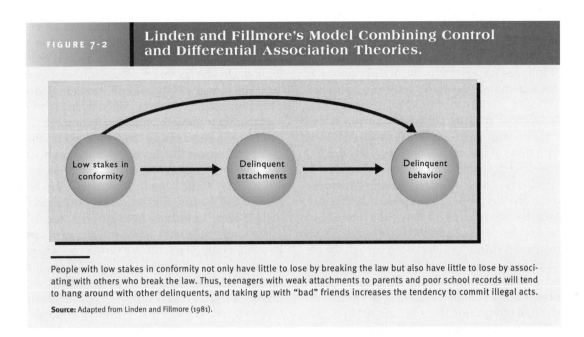

FIGURE 7-2

Linden and Fillmore's Model Combining Control and Differential Association Theories.

People with low stakes in conformity not only have little to lose by breaking the law but also have little to lose by associating with others who break the law. Thus, teenagers with weak attachments to parents and poor school records will tend to hang around with other delinquents, and taking up with "bad" friends increases the tendency to commit illegal acts.

Source: Adapted from Linden and Fillmore (1981).

attachments greatly increase the level of delinquency. That is, teenagers with little to lose by giving in to temptations are very likely to do so. But they are even more likely to do so if in addition they have friends who support them in their decisions to skip school, shoplift, stick up a store, steal a car, vandalize property, or beat somebody up.

Sociology is not a description of what goes on in a particular time or place. That is the job of journalism. Sociology is an attempt to formulate explanations that apply to human social life—here, there, and everywhere. Thus, Linden and Fillmore were not surprised to find that their model applied on either side of the U.S.-Canadian border.

TABLE 7-12	Self-Reported Delinquency for Richmond and Edmonton Youth			
	RICHMOND, CALIFORNIA		EDMONTON, ALBERTA	
	BOYS (%)	GIRLS (%)	BOYS (%)	GIRLS (%)
Have stolen something worth $2–50	19.1	7.7	14.1	5.6
Have stolen something worth over $50	6.6	2.0	2.4	0.2
Have stolen a car	10.8	3.6	5.9	0.7
Have damaged property	25.5	8.6	40.3	16.0
Have beaten someone up	41.7	15.6	40.9	16.0

Source: Adapted from Linden and Fillmore (1981).

While the theories we have examined thus far are meant to apply in any society or in different regions within any society, they are not designed to explain *differences* from one society or region to another. To do this, we now must examine a more macrotheory of deviance.

ANOMIE AND THE INTEGRATION OF SOCIETIES

In Chapter 1 we saw that the rapid growth of large cities during the nineteenth century led many early sociologists to predict a breakdown in human relations—that people would come to live in a world of strangers, lost and alone. Émile Durkheim, one of the founders of modern sociology, feared that as people lost their links to one another through longstanding attachments, society would suffer from **anomie:** a condition of *normlessness*. People would not know what the norms were, and attachments would not motivate them to obey the norms even if they knew them, for humans would lack the moral direction provided by others. Indeed, Durkheim argued that in modern urban societies individuals are morally in "empty space." Without social ties "no force restrains them" (Durkheim, 1897), and in such circumstances people literally lose the ability to tell right from wrong. Ephraim Mizruchi (1964) has suggested the term *deregulation* as the best translation of anomie into English. When social relations break down, society loses its ability to regulate behavior.

Durkheim contrasted conditions of city life with those of more traditional rural villages and identified the latter as **moral communities.** He stressed two components of the moral community. The first of these is **social integration.** Here he referred to the number and intimacy of attachments enjoyed by the average person. In traditional village life, Durkheim argued, the average person was firmly anchored in an extensive network of close attachments. The second aspect of moral communities, according to Durkheim, is **moral integration.** Here he referred to shared beliefs, especially religious beliefs, that provide members of a community with a common moral conception—mutual beliefs about what the norms are and why they are correct.

Both social and moral integration are undermined by life in large cities, according to Durkheim. People lose their attachments, and thus, social integration erodes. While a single religion unites traditional communities, many competing religions flourish in cities, and each weakens the others. Thus, Durkheim concluded that cities would destroy the power of societies to control their members and therefore become locations for excessive serious deviance.

Durkheim's theory describes how macrochanges in social structures can have devastating effects at the microlevel. Indeed, the correspondence between his anomie theory and control theory is obvious: The sheer size and disorganization of big cities reduce the average person's attachments and beliefs. Thus, free to deviate, people will do so.

Durkheim attempted to demonstrate his thesis by showing that suicide rates were much higher in urban areas and in areas where a variety of Protestant faiths existed than in rural areas and places that remained solidly Catholic (Durkheim, 1897).

As we saw in Chapter 1, research does not support the thesis that city life inevitably produces "mass societies" in which people lack attachments and thus fall victim to anomie. Thus, Durkheim's predictions about the future were overly pessimistic. However, that does not invalidate his theory. Durkheim's assertions about the consequences of weak moral and social integration can be applied to the extent that societies (or different areas of societies or even different areas in a city) vary in their degree of social and moral integration. Durkheim's theory predicts that deviance rates will change as integration varies; for example, crime and suicide rates will be higher wherever moral and social integration are lower.

Over the past seventy-five years, many studies have supported Durkheim's notions about social and moral integration (Shaw and McKay, 1929, 1931, 1942; Angell, 1942, 1947, 1949). And recent studies continue to support Durkheim. Beginning with the assumption that social integration will be weak to the extent that places have high rates of population turnover—where people are always moving in, moving out, or moving around—studies find crime rates to be highly correlated with turnover (Crutchfield, Geerken, and Gove, 1983). Research also confirms Durkheim's predictions about moral integration. Where rates of church membership and church attendance are higher, crime rates are lower (Stark and Bainbridge, 1997).

CLIMATE AND SEASON

Back in 1833, in his pioneering study of crime rates, André Michel Guerry (see Chapter 1) noted that French crime rates were substantially higher in summer than in winter, except in the most southerly departments, such as Corsica, where there really is no winter. Later in the century, Lombroso, Morselli, and Tarde each reported similar findings and suggested reasons climate could influence rates of crime and deviance—including the claim that hot climates produced hot tempers. But then, after Émile Durkheim (1897) ridiculed the notion that climate could have social consequences, the topic was ignored for most of the twentieth century. Recently, the stubborn fact that crime rates *are quite seasonal* prompted renewed interest.

Year after year, the *Uniform Crime Reports* show that the crime rate is highest in the warmest summer months and lowest in the coldest winter months. Seasonal effects are far greater in the colder states and are relatively small in the Sunbelt states. These variations are known to be caused mainly by weather—especially by variations in temperature. Thus, Richard Rosenfeld, Matthew Perkins, and Eric Baumer (1994) reported very strong correlations between variations in crime rates and variations in the daily temperature.

In addition to weather, there is a fascinating seasonal effect. Except for rape, crime rates hit a second high point in December. In fact, some years homicide, robbery, and burglary rates peak in December, although these rates are otherwise lower in winter than summer.

One interpretation of these findings seems rather obvious if we recall that a major factor in criminal actions is the requirement that they be easy and convenient. People who seldom are willing to travel over a mile from home (and usually less) to commit a crime also will be less likely to be active in bad weather. In January it gets cold waiting around for a pedestrian to rob (and there are fewer pedestrians) and potential burglars are not attracted by an open window or patio door. In winter people do not hang out on street corners or in other outdoor settings where situations frequently develop that produce assaults and homicides.

During December, however, the deterrent effects of cold weather are offset by a seasonal increase in opportunities and motives. Holiday shoppers are targets of opportunity for robbery. Gifts under the tree or in automobiles are an extra incentive for burglary and car break-ins. In fact, offenders often acknowledge that they are especially active in December because this is how they do their Christmas "shopping." As for homicide and assault, during the holidays liquor consumption rises sharply as does the incidence of family and neighborhood gatherings and office parties—occasions that often produce a homicide or an assault among friends or acquaintances. That rape does not have a second peak in December probably reflects the fact that the holidays produce no extra opportunities or special motivation, while the cold weather greatly reduces the opportunities.

THE LABELING APPROACH TO DEVIANCE

So far in this chapter, we have assumed that the members of any society or group know its norms. Consequently, when people seriously violate norms, it is reasonable to ask why they did so. Now we must consider a sociological approach to deviance suggesting that this question is often inappropriate. Instead, we should ask why we label people as deviant when they break norms. Sociologists taking this approach claim that most deviance results from some persons having been identified, or labeled, as

deviants. Not surprisingly, this approach is called **labeling theory.**

Labeling theorists distinguish between primary and secondary deviance. **Primary deviance** involves whatever behavior a person engaged in that caused others to identify or label him or her as deviant. **Secondary deviance** is behavior that is a reaction to having been labeled a deviant. Most labeling theorists suggest that primary deviance involves relatively transient, insignificant, quirky behavior that most people engage in from time to time. Usually, others ignore such behavior (Lemert, 1951, 1967; Schur, 1971; Scheff, 1984). Sometimes, however, others react strongly and negatively to the primary deviance of some people, who are then publicly labeled as deviant.

Having been stigmatized as deviants, many people are driven to fulfill our expectations of them. Because they have a "bad name," they come to see themselves as bad, and so they do bad things. The labeling approach, therefore, concentrates on identifying and criticizing the process by which norms arise and are enforced and by which some people are labeled for behaving a certain way while others are not. The labeling approach also specifies how a label forces people to adopt a career of secondary deviance.

Allen E. Liska (1987) has identified three major ways in which labels incline people to deviate. First, a deviant label, such as burglar, alcoholic, or prostitute, limits legitimate economic and occupational opportunities. Many employers will not hire ex-convicts, for example, and some reject applicants who have been arrested even if they were acquitted (Schwartz and Skolnick, 1962). At the same time, being labeled deviant may increase illegitimate economic opportunities (serving a jail term may make a person more proficient in crime).

Second, a deviant label limits a person's interpersonal relations. Thus, an ex-convict may find few chances for attachments with conventional people and therefore be limited to attachments with others who also have been labeled as deviant.

Third, being labeled a deviant can affect self-conceptions. Sociological theories of interaction have long held that we see ourselves as others see us. If others see us as deviants, we may come to accept their judgments. Then when we act as we are "supposed" to act, we will be acting in deviant ways.

Labeling theories have made some important contributions to understanding deviance. As with subcultural theories, they have sensitized us to the fact that norms are not absolute but are created by humans. Therefore, it is useful to examine how a particular norm was established, by whom, for what reasons, and perhaps, *against* whom. Many norms are born out of conflict, and many laws are passed by close votes. According to labeling theorists, this demonstrates that norms are often arbitrary. For example, thirty years ago, a doctor who performed abortions risked a prison sentence and expulsion from the medical profession. Today, abortions are routine medical procedures. Labeling theorists therefore suggest that the question to ask is not why deviants behave as they do but why we decide to prohibit certain actions.

Labeling theories also closely examine the process by which some people are labeled as deviant while others who do the same thing go unlabeled. Labeling theorists argue that the higher a person's status, the less chance he or she will be labeled for deviant behavior.

Finally, labeling theory calls attention to how the reactions by society to primary deviance can produce continuing patterns of secondary deviance—that by attempting to stop deviant behavior, we may cause it.

However, over the past thirty years, many hypotheses drawn from labeling theory have not been confirmed by research. For example, numerous studies have failed to show that people labeled as delinquents as a result of arrest and conviction subsequently increase their level of illegal activity (Gibbons and Blake, 1976; Klein, 1976; Liska, 1987). In fact, some studies show that juveniles who are labeled and punished for an offense are less likely to commit subsequent offenses than are those who escaped being labeled (McEachern, 1968; Thornberry, 1973). We shall consider this further in Chapter 8.

Nor has the view that primary deviance is usually insignificant and harmless been sustained. To the contrary, quirky, transient, and trivial norm violations are almost always ignored (Gove, 1975); people usually get labeled as a deviant after committing a serious act. People are labeled as rapists after they have raped someone, as robbers after they have robbed, as murderers after they have murdered. Thus, Liska asks, "Are these actions unimportant when committed by primary deviants?" That is, are they to be considered not serious if the person does not already have a deviant label?

This brings us to the major shortcoming of labeling theory. While it may help us to explain why ex-convicts often commit new offenses and go back to prison, it cannot explain the initial act of deviance by which people get labeled as deviants. Thus, explaining why an ex-convict commits new offenses does not explain why he or she became a convict in the first place. Why they *start* to do it lies beyond the scope of labeling theories.

DRUGS AND CRIME

Since drug offenses do not inherently involve fraud or force, they are not classified as crimes under the Gottfredson and Hirschi definition. That doesn't mean the use or sale of various drugs *shouldn't* be against the law. It merely means that this is a political decision. Even so, social scientists continue to have a great deal of interest in drug offenses because they postulate two primary links between drugs and crime. The first proposes a direct link between drugs and crime—that psychological effects of drugs and alcohol often stimulate violent actions. The second proposes an indirect link—that the high cost of illegal drugs causes many addicts to commit crimes to support their habits.

The evidence for a direct link between alcohol and violence is overwhelming. More than a dozen studies of homicide have found that in at least 60 percent of the cases studied either the killer or the victim, and usually both, were drinking at the time. Many studies report that at least 40 percent of rapists were under the influence of alcohol when they committed the crime (Wilson and Herrnstein, 1985). However, studies attempting to discover a direct link between narcotics and crime were, for many years, quite unsuccessful. The primary drug investigated in older studies was heroin, which is a strong sedative and tends to calm people rather than make them aggressive (Wilson and Herrnstein, 1985). When cocaine became the drug of choice, these patterns changed. Cocaine (and especially the form known as "crack") can arouse hyperactivity and hostile feelings. The primary evidence of its involvement in homicide comes from medical examiner reports showing that as many as 40 percent of homicide victims had used cocaine shortly before their deaths (Bureau of Justice Statistics, 1992). The assumption is that this applies to their killers as well, as it does in the case of alcohol.

There is substantial evidence that people sometimes do commit crimes to obtain funds to buy drugs. This seems particularly true as a motive for prostitution. However, the weight of evidence is that it is not so much that people rob and steal to support their need for illegal drugs as it is that drug and alcohol use are simply a part of the ordinary offender's lifestyle—another source of immediate, short-term pleasure (Gottfredson and Hirschi, 1990; Bureau of Justice Statistics, 1992). If many ordinary offenders sometimes steal or rob to buy crack, they also sometimes steal or rob to buy cigarettes, pizza, or movie tickets.

Whatever the link between drugs and crime, until recently no one was even sure how many people arrested for other felonies used drugs. The only data available were self-reports of drug use by persons arrested for other offenses, and it was thought (correctly) that many offenders who used drugs would deny it. In search of more trustworthy data, the National Institute of Justice launched an ambitious study. Initially, twenty-four major cities were selected as research sites, and arrangements were made to interview and obtain urine specimens from a sample of all persons arrested during a two-week period.

As arrestees were processed into jail, each was asked whether he or she had used any drugs during the previous forty-eight hours. Most said they had not. However, when their urine samples were tested, most were found to contain traces of drugs. The drugs tested for included cocaine, opiates, PCP, marijuana, methadone, methaqualone, benzodiazepines, barbiturates, and propoxyphene. In the general population, marijuana is the most widely used drug, but among arrestees the overwhelming majority of positive tests found traces of cocaine.

Table 7-13 shows the most recent results of drug tests of persons following their arrest. Everywhere, the great majority tested positive— from a high of 86 percent in Chicago down to 60 percent in Salt Lake City. Had the police continued to rely on self-report data, they would have thought only about 20 percent had used drugs within two days of their arrest.

The enormous extent to which drug use currently is involved in crime surprised even the experts. Moreover, such widespread use suggests that drugs are not so much causes of crime, either directly or indirectly, as they are a part of offender lifestyles. While some people undoubtedly commit crimes while their judgment is impaired and others may steal to buy drugs, most offenders were introduced to drugs

TABLE 7-13	Percentage of Male Arrestees Who Tested Positively for Drugs in Selected Cities
Chicago	86%
Boston	83
New Orleans	80
Phoenix	77
Cleveland	75
Atlanta	74
New York City	73
Denver	73
San Diego	71
Los Angeles	69
Philadelphia	69
Dallas	64
Houston	62
Tampa	62
Salt Lake City	60

Source: National Institute of Justice, 2004.

after they had begun to offend (Gottfredson and Hirschi, 1990; Bureau of Justice Statistics, 1992).

CONCLUSION

This chapter has reviewed a number of partial theories of deviance. I mentioned at the start that these theories were much more complementary than conflicting and that the next step for sociologists was to fit them into a general theory of deviance. An introductory textbook is hardly the place to undertake such a task. Nevertheless, as an effective summary of the chapter, some of the components of such a general theory can be described.

First, any general theory of crime and deviance must include elements of learning theory. Nearly all human behavior is learned, including deviant behavior. Next, a general theory must specify the sources from which conformity and deviance are learned. Here we need to apply theories of early childhood socialization as well as theories of differential and adult socialization (see Chapter 6). For example, because males are much more prone than females to commit criminal and delinquent acts, aspects of *sex-role* socialization must be part of a general theory. Furthermore, elements of differential association theory should prove useful. In learning to deviate or conform, what behavior a person's friends reward will matter.

We know, too, that some people seem to be more easily influenced by their friends than others are. Thus, a general theory of deviance may need to draw on personality theories dealing with traits such as low self-control.

To account for the existence of deviant groups that can provide differential association, elements of subcultural theory must be included. Although much deviance is not supported by subcultures, certainly some forms are, and some forms of deviance are stabilized by association with deviant subcultures.

All of the preceding elements help to clarify and extend control theories. Consider attachments—the bonds of affection between people. Clearly, learning provides the mechanism by which attachments form and influence our behavior. Differential association and subcultural theories help explain to whom we become attached. If we are attached to persons who reward conformity, then we tend to conform. If we are attached to persons who reward deviance, then we tend to deviate. And if we are unattached, then we tend to do as we please. Other things being equal, it will probably please us to commit deviant acts because most of them are inherently self-reinforcing. Here, too, psychological theories can help explain the inability of some people to form and maintain strong attachments (Wilson and Herrnstein, 1985).

The investment aspect of control theory incorporates the basic tenets of structural strain theory. If our investments in life are low, we are in a deprived condition: Life has rewarded us little and promises never to reward us very much.

This does not drive us to break the norms as much as it frees us to do so. Moreover, because control theory examines all crime and deviance in terms of relative gains and losses, it can explain the behavior of not only the unemployed burglar but also the wealthy embezzler.

Labeling theory can also contribute important insights to a general theory of deviance. To the extent that past acts of deviance stigmatize a person with a deviant label, his or her attachments to others may be severely limited. Conventional people may tend to avoid ex-convicts, prostitutes, or problem drinkers. This in turn frees the deviants of attachments that might cause them to conform. Moreover, lack of conventional attachments may prompt those who share a deviant label to associate with one another and thereby form a deviant subculture. Indeed, to the extent that deviant labels prevent people from building investments in conventional activities, they will also have greater freedom to deviate.

Finally, theories of social and moral integration place a general theory of deviance within the framework of large-scale social structures. That is, the overall condition of societies as whole systems can affect rates of internal deviance. For example, high rates of population instability hinder attachments, prompting deviance to rise. If there is a great decline in religious commitment, then deviance ought to rise as moral integration weakens.

This sketch of obvious connections among current theories is meant to show you that despite the proliferation of partial theories, social scientists have been making progress in understanding crime and deviance.

Review Glossary

Terms are listed in the order in which they appear in the chapter.

deviance Behavior that violates norms.

crime Acts of force and fraud undertaken in pursuit of self-interest.

born criminals Lombroso's term for people whose deviance he attributed to their more primitive biology.

differential association theory A theory that traces deviant behavior to association with other persons who also engage in this behavior.

subcultural deviance Behavior through which a person deviates from the norms of the surrounding society by conforming to the norms of a subculture.

structural strain Frustration or discontent caused by being in a disadvantaged position in the social structure.

structural strain theories Theories that blame deviance on the stress of structural strain; for example, one such theory claims that people commit crimes because of their poverty.

white-collar crime According to Sutherland (1983), crimes committed by "a person of respectability and high social status in the course of his [her] occupation."

control theory A theory that stresses how weak bonds between the individual and society free people to deviate, whereas strong bonds make deviance costly.

stake in conformity Those things a person risks losing by being detected committing deviant behavior; what a person protects by conforming to the norms. See **social bonds.**

social bonds Bonds that, as used in control theory, consist of the following:

  1. **attachments** Ties to other people.

  2. **investments** The costs expended to construct a satisfactory life and the current and potential flow of rewards expected.

  3. **involvements** The amount of time and energy expended in nondeviant activities.

  4. **beliefs** Our notions about how we ought to act.

internalization of norms The sociological synonym for *conscience;* refers to the tendency of people not simply to learn what the norms are but also to come to believe the norms are right.

anomie A condition of normlessness in a group or even a whole society when people either no longer know what the norms are or have lost their belief in them.

moral communities Groups within which there is very high agreement on the norms and strong bonds of attachment among members.

social integration The degree to which persons in a group have many strong attachments to one another.

moral integration The degree to which members of a group are united by shared beliefs.

labeling theory A theory that explains deviant behavior as a reaction to having been socially identified as a deviant.

primary deviance In labeling theory actions that cause others to label an individual deviant. More generally, any deviant acts that result in the commission of other deviant acts.

secondary deviance In labeling theory actions carried out in response to having been labeled as deviant. More generally, any deviant acts committed as a result of committing other deviant acts—for example, burglaries committed to support a drug habit.

Suggested Readings

Beirne, Piers. 1993. *Inventing Criminology.* Albany: State University of New York Press.

Criminology. 1987. Special issue on theory. 25:783–989.

Gottfredson, Michael R., and Travis Hirschi. 1990. *A General Theory of Crime.* Stanford, Calif.: Stanford University Press.

U.S. Department of Justice. *Sourcebook of Criminal Justice Statistics.* Washington, D.C.: U.S. Government Printing Office (issued annually).

Sociology Online

www.socstark10.com

GO TO THE INTERNET AND TYPE: www.socstark10.com.

➚ **CLICK ON:** 2000 General Social Survey

✓ **SELECT:** TRUST: Generally speaking, would you say that most people can be trusted or that you can't be too careful in dealing with people?

➚ **CLICK ON:** Analyze Now

TRUE OR FALSE: Most Americans are trusting.

TRUE OR FALSE: As they get older, people learn to be less trusting.

TRUE OR FALSE: Women are significantly more trusting than men.

TRUE OR FALSE: Democrats are significantly more trusting than Republicans.

➚ **CLICK ON:** Select New Data Set

➚ **CLICK ON:** The 50 States

✓ **SELECT:** COKE USERS: Estimated number of cocaine addicts per 1,000 population.

➚ **CLICK ON:** Analyze Now

Which two states have the most cocaine addiction? Which two have the least? Can you suggest something about how the states with the most cocaine addiction differ from those with the least that might account for their differences in cocaine use?

➚ **CLICK ON:** Select New Data Set

➚ **CLICK ON:** Nations of the Globe

✓ **SELECT:** DRUG USE: Estimated daily consumption of narcotic drugs, in doses per million population.

➚ **CLICK ON:** Analyze Now

Which nations are higher than the United States? How do nations with high levels of drug use differ from those with low levels?

USING INFOTRAC COLLEGE EDITION

GO TO THE INTERNET AND TYPE:
www.InfoTrac-college.com

➚ **CLICK ON:** Register New Account
You will be asked to enter your Access Code and to create and enter a User Name and Password. Having done so,

➚ **CLICK ON:** InfoTrac College Edition or Log On.

Based on what you have learned about crime and criminals in this chapter, you may find it interesting to assess (maybe even grade) recent articles in popular periodicals.

SEARCH TERMS:
Burglary
Criminals
Murderers

CHAPTER

8

The term *highwaymen* once referred to bandits who waylaid and robbed travelers. But early in the twentieth century, it took on a new meaning as prison inmates were assigned to build highways. The men seen in this photograph were serving prison terms in California and were engaged in building the scenic Yosemite Highway. In those days prison industries and labor contracts made the prison system self-supporting. Recently, there have been efforts to once again make prisoners perform productive tasks and thereby earn their keep. In some states the highwaymen are back at work—many crews are cleaning up litter along the roads. However, prisons were not instituted as highway projects or as a source of labor. Prisons exist because of the failure of group pressures to prevent some people from misbehaving, from acting in ways others regard as harmful and antisocial. In fact, that's why in some cities the agency administering the police force is known as the Department of Public Safety.

Social Control

W E HAVE EXAMINED WHY people deviate. In this chapter we take a different point of view: How do groups attempt to prevent deviance? All collective efforts to ensure conformity to the norms are forms of **social control.**

The most common mode of social control is **informal,** consisting of direct interpersonal pressures. Most of us conform most of the time to avoid offending those around us. Our responsiveness to social control is evident when we feel embarrassed for having committed mild norm violations in front of others. Even when we are alone, we often limit our deviance for fear that someone might find out.

Nevertheless, not everyone is responsive to informal social control. We all know people who just don't seem to care about others or what they think. All languages include a huge and colorful set of terms for such people, and their prevalence shows up among the societies making up the Standard Cross-Cultural Sample. Even among those societies having fewer than 100 members, there is a substantial amount of deviance. Consequently, 70 percent of all societies with more than 100 members have **formal** means of social control, including specialists in enforcement. In modern societies formal means of social control include the police, courts, psychiatrists, prisons, mental hospitals, even collection agencies. Hence, when informal methods of control fail, formal methods are activated.

As will be obvious, formal methods also fail to achieve full conformity—in all societies significant amounts of deviance occur. Nevertheless, *the more intensely any group or society applies informal and formal efforts to suppressing deviance,*

CHAPTER OUTLINE

INFORMAL CONTROL

A CLOSER VIEW
Hechter and Kanazawa: Solidarity and
Conformity in Japan

Dependence
Visibility
Extensiveness

FORMAL CONTROL

PREVENTION

A CLOSER VIEW
The Cambridge-Somerville Experiment

Other Prevention Programs

DETERRENCE

A CLOSER VIEW
Jack Gibbs: A Theory of Deterrence

The Capital Punishment Controversy

THE WHEELS OF JUSTICE
The Police

A CLOSER VIEW
Corman and Mocan: Cops and
Deterrence

The Courts

REFORM AND RESOCIALIZATION

A CLOSER VIEW
The TARP Experiment: Salaries
for Ex-Convicts

CONCLUSION

REVIEW GLOSSARY

SUGGESTED READINGS

SOCIOLOGY ONLINE

the less deviance there will be. This is simply to recognize that most people will adjust their behavior to the potential costs of misbehavior—for example, there will be less drunk driving where the law against it is enforced and the penalty is one year in jail than where it carries a $10 fine and the police ignore all but the most obvious offenders. This matter will be considered at length in the section on deterrence. Here I want to draw your attention to the fact that while deviance is costly, so is social control. These costs are not limited to the funds for formal methods such as police salaries or the cost of prisons but also include a substantial price in terms of restrictions on personal choices and freedom. To clarify this matter, the first part of the chapter will examine the informal mechanisms that result in Japan's very low rates of deviance. You will be invited to weigh the costs versus the benefits of the Japanese situation.

The bulk of the chapter is devoted to *formal methods* of social control: activities by organizations created to ensure conformity to the norms. First, we assess programs and policies designed to prevent deviance. We shall watch an amazing program designed to prevent young boys from becoming delinquents and see the results of the experiment used to measure its effectiveness. Next, we examine whether punishment can deter deviance: Can punishment cause some people to cease being deviant and prevent some others from ever starting to deviate? Finally, we look at the criminal justice system—the police, courts, and prisons. How well does this system operate? Can it serve to reform and resocialize criminals? Be prepared for some surprises. Much of the conventional wisdom about how to prevent crime and deviance is wrong or only partly true.

INFORMAL CONTROL

Nearly all Americans are concerned about our high crime rates. And when we look at comparative crime statistics such as those shown in Chapter 7, a natural response is to wonder why some comparable nations have such very low rates. In particular why couldn't we have low rates like the Japanese? What would we have to change? The best way to find out is to look over the shoulders of two sociologists as they discovered the answer.

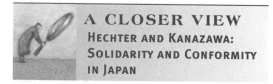

A CLOSER VIEW
HECHTER AND KANAZAWA: SOLIDARITY AND CONFORMITY IN JAPAN

For more than sixty years, social scientists have tried to explain Japan's high level of **social order,** defined as the extent to which citizens comply with important norms. Why do the Japanese exhibit such a high level of conformity to the norms? Most have credited the religious ideals associated with Confucianism for Japan's social orderliness (Smith, 1983; Befu, 1990). But this explanation fails when it is observed that other Confucian societies such as South Korea and China have relatively high levels of social disorder. Recently, two sociologists, Michael Hechter of the University of Washington and Satoshi Kanazawa of the University of Canterbury, New Zealand, offered a far more persuasive answer: *Japanese society is organized in ways that maximize the impact of local group solidarity on each individual.*

Hechter and Kanazawa (1993) proposed that the overall high level of social order in Japan was the result of high levels of conformity within the groups making up Japanese society. In this way they shifted attention from the society as a whole to its internal networks. Then they explained the high levels of conformity within local Japanese networks on the basis of three primary factors determining the *degree of social control* that groups can exert over their members. They formulated three theoretical principles.

First is the **principle of dependence:** *The more dependent members are on a group, the more they conform to group norms.* By dependent is meant the extent to which a group is the only available source of important rewards.

Second is the **principle of visibility:** *To the extent that the behavior of group members is easily observed (or otherwise monitored) by other members, their degree of conformity to group norms will be greater.*

Third is the **principle of extensiveness:** *The greater the scope and extent of norms upheld by the group, the greater will be the contribution to overall social order.*

The image of the orderly society that emerges from these theoretical principles is one in which the basic social groups hold immense and exclusive power over their members, demand conformity to a general set of norms, and are enabled to easily detect nonconformity. In the re-

mainder of their study, Hechter and Kanazawa examined how Japanese society fulfills these principles.

DEPENDENCE

"Throughout their lives," Hechter and Kanazawa wrote, "the Japanese are more dependent on groups than their counterparts in comparably developed societies. This pattern of dependence begins early in the life-cycle and continues throughout it."

Consider the extent to which Japanese students are quite literally at the mercy of their teachers and schools. Junior high schools determine if and where their students attend high school, and high schools play the primary role in determining their students' futures. For example, high schools determine if and where their students attend college, and those who do not go on to college are placed in jobs with specific firms by their high schools. "Employers simply tell a high school how many new recruits they want, and the school 'nominates' its students to them," Hechter and Kanazawa reported. The relationship between firms and schools is so tight that students cannot even apply for jobs without a nomination from their school. Moreover, once in a school, students do not have the option to transfer. Even if they can obtain permission to transfer (which is rarely granted), they must begin all over again. That is, a junior-year transfer would enter a new school as a freshman. Nor do Japanese who leave school have the option of going back later—adult students or programs are essentially unknown. Like getting a job, getting into college also depends entirely on recommendations by one's high school teachers. No wonder that Japanese students are obedient.

Once they have a job, the Japanese are as dependent on their employer and their supervisors as they were on their teachers. They very rarely change jobs. The positive side of this is that Japanese firms have offered lifetime job security to all who perform adequately, and nearly all promotions are from within the firm. The negative side is that those who leave one employer have difficulty finding a new job and will not be able to gain a better position by changing firms. Consequently, employers can impose on the private lives of their employees to an extraordinary degree. They not only can demand that employees lead exemplary lives on threat of being fired, but they can see if they are obeyed.

VISIBILITY

From an American perspective, it appears as if the average Japanese person has no private life. This situation starts at home. The Japanese family has very little living space, and their "rooms" are separated primarily by screens or by partitions less than one-tenth the thickness of the walls in American homes. At school Japanese students are under constant and close supervision. There are no individual activities, no study halls, no free periods, no library time. During class periods students may not leave the room for any purpose, not even to go to the bathroom. Teachers must give their approval for students' part-time and vacation jobs. Locker and body searches are common. When they get home from school, Japanese students remain under close observation. Local neighborhood patrols keep watch for students who should be home, paying "particular attention to dark, secluded patches of shrubbery, back alleys, and [other places] where adolescents might hope to hang out unnoticed" (Bestor, 1989).

High visibility continues for the Japanese worker on the job. Consider office workers. Most Americans in such jobs have a private office or at least a work station enclosed by partitions. Most Japanese work in an open office space without any partitions and with desks quite close together—in full sight and hearing not only of other workers but also of their supervisors. Moreover, single workers usually are required to live in dormitories provided and supervised by their employer. Here they are tightly controlled by a dorm director according to strict rules. For example, they need permission to stay out late or overnight. Visitors of the opposite sex are not allowed. Dorm directors have the right to enter anyone's room at any time.

Japanese workers do not escape from company control even when they play. They are expected to drink, party, go to movies, or play golf with co-workers—and this remains true even for married workers.

When they marry, some Japanese couples move into housing furnished by their employers, where supervision remains nearly as constant as in the dorms. But even those who move into private housing remain extremely visible to others. The neighborhood associations that patrol to prevent teenagers from "misbehaving" take a close interest in all other aspects of family life as well. Anyone who applies for a bank loan can be sure that the bank will interview the neighbors about all aspects of the applicant's life, including cleanliness. If that weren't enough, most Japanese couples remain under close observation by relatives and in-laws. Japanese wives take it for granted that their mother-in-law will drop by frequently and will be quick to criticize. Should the mother-in-law become a widow, she will live with the couple—this in housing where "privacy from other members of the family is almost nonexistent" (Shioji, 1980).

EXTENSIVENESS

It will be clear that the norms sustained by family, school, neighborhood, and work groups not only are strictly enforced but are of very general scope. Teachers do not limit their demands to how one behaves in school but encourage students to inform on one another about out-of-school activities. Employers require a proper lifestyle. Bankers can and do concern themselves with whether loan applicants show proper respect for their parents, and in-laws may even comment on the frequency of overheard marital sexual activity.

This is why the crime rates are so low in Japan, and these are the costs that Japanese people pay for social order. Do the Japanese think it's worth it? Well, in Chapter 13 you will discover how the Japanese feel about their family life. In Chapter 16 you will see how well they like their jobs. Meanwhile, even in Japan there are burglars, robbers, rapists, and murderers, so Japan has police, courts, and prisons.

FORMAL CONTROL

When you invite friends to dinner, you can assume that they will use silverware and not eat with their fingers, that they will use their napkins and not wipe their chins on the tablecloth, and that they will not throw plates of food against the wall. You can also assume that when they leave, they won't take your stereo system or your car. Your friends will not act this way because they *are* your friends; rather, they respond to informal pressures to conform. But not everyone does. When people do not, we often call upon formal agencies of social control to act on our behalf.

Formal social control is not applied to just any norm violation. The police won't come just because you happen to have friends who *do* eat with their fingers or wipe their chins on the tablecloth. They might come if plates thrown against the wall bother your neighbors. And they definitely will come if people steal your belongings. Formal means of social control are used only for acts of deviance that are also illegal or otherwise defined as providing legal grounds for intervention. Thus, even private collection agencies are limited to pursuing the payment of *legal* debts, rather than debts from illegal gambling or claims having no standing in common law.

Formal social control is attempted in three principal ways. The first is to *prevent* crime and deviance by removing opportunities for such acts to occur or by eliminating their causes. The second is to *deter* deviance by the threat of punishment: to make people afraid to do certain things or at least to do them again. The third is to *reform* or *resocialize* people so that they cease wanting to deviate. Let us see how each of these methods of social control functions in our society.

PREVENTION

For people to commit a criminal act, they must have the *opportunity* to do so. Thus, many approaches to formal social control are based on **prevention,** or the attempt to reduce opportunities. Campaigns encouraging people to lock their cars are meant to reduce the opportunity for auto theft, for example. The prohibition of alcoholic beverages in the United States during the 1920s was intended to prevent alcohol abuse, as laws today against the sale of certain drugs are intended to prevent drug abuse.

Many recent prevention programs are guided by an approach to deviance called *opportunity theory*. Lawrence Cohen and Marcus Felson (1979) recognized that the occurrence of a crime requires not only people motivated to commit an offense but also *suitable targets* (property or individual victims) and the *absence of effective guardians*. Opportunity theory traces a substantial part of the rise in household burglaries since World War II to the great increase in female employment. Female employment has greatly increased affluence and hence the proportion of households that contain possessions worth stealing. Moreover, female employment has substantially decreased the proportion of homes where someone is home during the day. Empty homes full of valuables create an opportunity for those motivated to commit burglaries.

Other proponents of opportunity theory have demonstrated the increased opportunities for crime caused by changes in the structure of urban neighborhoods. Many new house designs seek to increase privacy and inadvertently greatly decrease the ability of neighbors to note suspicious persons or circumstances and thus protect one another from becoming victims of crimes (Taylor et al., 1980). David Cantor and Kenneth C. Land (1985) have suggested that the reason burglary, larceny-theft, and auto theft decline as unemployment rises is an increase in the number of adults who stay home and who, therefore, can protect their own and their neighbors' property.

Anything that draws teenagers and young adults to a neighborhood increases the crime rate because people this age are the ones most apt to commit crimes. Hence, neighborhoods near public high schools have higher crime rates than do neighborhoods farther away (Roncek and Lobosco, 1983); crime rates also fall as one gets farther away from a fast-food restaurant (Brantingham and Brantingham, 1982).

A frequent and relatively successful product of the insights of opportunity theory is efforts such as block-watch programs designed to increase effective guardianship. The same aims are reflected in the booming sales of home security systems.

During the past few years, criminologists even have proposed substantial changes in the design of cities as a means of reducing crime. Marcus Felson (2002) noted how the physical setting of shopping malls produces a very low crime rate in comparison with retail areas laid out along regular streets. Similarly, by greatly restricting access by strangers and providing secure parking, condominiums provide much better security against crime than do single-family dwellings.

However, the major effort in preventing crime and deviance has been directed toward removing its causes. Here the emphasis has been placed on childhood *socialization*. A longstanding article of faith, among social scientists and nearly everyone else, is that adult misbehavior results from a long journey down the "wrong road," a journey that begins in youth. The most persistently addressed question has been, What can be done to intervene soon enough? How can we get kids headed on the right road before it's too late?

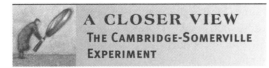

A CLOSER VIEW
THE CAMBRIDGE-SOMERVILLE EXPERIMENT

In the middle 1930s, a New England physician, Richard Clarke Cabot (a member of one of the oldest upper-class families in the East), decided that it was time to launch a major effort to help young boys grow up and not become delinquent. More important, he was determined to test the results of this effort rigorously. He thus initiated and funded a ten-year social experiment that even today is regarded as a model attempt at a delinquency prevention program and as a model sociological experiment.

In essence Cabot decided to give poor boys many of the same opportunities as boys from well-to-do homes. First, he tried to provide for their physical well-being. Second, he sought to enrich their experiences and broaden their horizons. Finally, he tried to provide close relationships with adults who could give the boys good advice and serve as models of well-adjusted adults.

In 1937 Cabot launched his project by selecting two economically distressed industrial communities near Boston: Cambridge and Somerville. He hired a staff of counselors and researchers, who selected 650 boys (with an average age of about eleven) as subjects for the delinquency prevention program. To know whether his program really did prevent delinquency, Cabot could not place all the boys in the program, for then there would have been no control group for comparing the delinquency rates. So, he *randomly* divided the group into two groups of 325.

The *experimental group* consisted of boys who would participate in the experimental prevention program. Another name for an experimental group is a *treatment group*—they get the treatment. In a medical experiment, for example, the experimental group would receive a new treatment, perhaps a new drug. In this case the experimental or treatment group was recruited to take part in Cabot's delinquency prevention program. The parents of each boy were contacted, and arrangements were made for their son to take part. Cabot had no trouble getting all the boys in the experimental group to take part. Little wonder, for during the depths of the Great Depression, he offered to furnish these youth with free health care, tutoring, vacations at summer camps, field trips, an elaborate recreational program, and individual counseling.

To determine whether his program worked, however, Cabot needed a *basis of comparison*. This was provided by the other 325 boys who randomly were excluded from the program. They made up what is called the *control group*. By comparing an experimental group with a control group, one can assess the effectiveness of some treatment. Suppose 650 boys with flu were located and 325 were randomly selected to receive a new drug. By comparing their speed of recovery with the 325 who were left untreated, we could see whether the new drug made any difference. In this case the boys left out of the program provided a basis for comparison to see if the program made a difference: Were boys in the program less likely to become delinquents than boys left on their own?

In May 1939 the program got under way. Unfortunately, Dr. Cabot died just as his ambitious undertaking began, but in his will he provided the needed funds, and his family supervised the project. Some aspects of the program had to be reduced during World War II—gas rationing cut into field trips, for example. Nevertheless, the project was carried through to its projected end (when the boys all reached eighteen).

What were the results of this massive project? To find out, the researchers compared the criminal records of boys in the control group with those of the boys in the experimental group. They found that 40 percent of the boys in the control group had been convicted of a crime. They also found that 40 percent of the boys in the program had been convicted of a crime! Moreover, boys in the experimental group had committed as many crimes as those in the control group had committed (Powers and Witmer, 1951). Obviously, the program had no effect on delinquency. Boys who were left on their own were no more likely to get into trouble than were boys who were part of the elaborate program (Figure 8-1).

Those associated with the study were astounded by these results and in the end refused to accept them. This is understandable. You will spend a number of years trying to help young people stay out of trouble only if you believe your efforts can make a difference. Sure, you can't help them all, but daily you see some kids going straight and you take comfort in that fact. Of course, you lack similar intimate observations of those outside the program, and so you cannot notice that they are doing just as well (or as badly) without your care. Having built up years of personal "experience" with the efforts

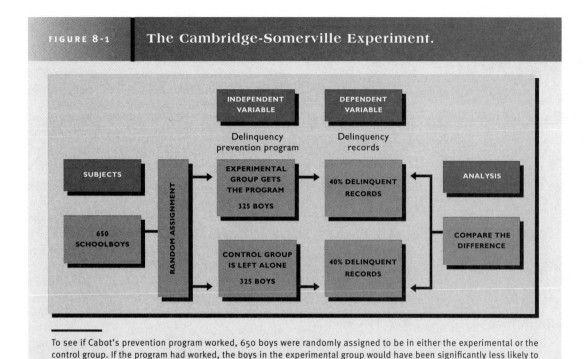

FIGURE 8-1 | **The Cambridge-Somerville Experiment.**

To see if Cabot's prevention program worked, 650 boys were randomly assigned to be in either the experimental or the control group. If the program had worked, the boys in the experimental group would have been significantly less likely to acquire delinquency records than would the boys in the control group. But as you can see, the program made no difference. Control and experimental groups were equally likely to have delinquency records—40 percent of each.

of your program, you will not readily accept research findings that your perceptions are biased and your effects negligible. Indeed, in their effort to show that the results of the experiment were wrong, the directors of the Cambridge-Somerville study simply discarded the control group altogether.

Instead, they engaged in a minute examination of the case records of each boy in the experimental group. For only these data could

satisfy the clinical worker, who, by the nature of his task, has little interest in averages. What such a person wants to know is whether there were any cases—even if only a few—in which the introduction of the [program] tipped the balance in a boy's favor. (Powers and Witmer, 1951)

From this private reading of the files, Edwin Powers and Helen Witmer concluded that there was clear evidence that fifty-one of the boys in the project had been helped. But if this were true, why did the experimental group not have a smaller proportion who were arrested? Did the program also push some boys into crime who otherwise would have gone straight? That seems unlikely. But it leads us to ask what was helping the boys in the control group. As Cabot understood, the experimental results were the only

reliable basis for judging the worth of the program. Because random assignment made the two groups equally prone to commit crimes, the program could be termed successful only if the experimental group had fewer crimes. There is no better evaluation method available.

This can be underscored by a poignant example. In an effort to show that the experiment was insensitive to the real results of the prevention program, the directors wrote to former participants and asked if the participants believed they had benefited from taking part. Many said they had, including one young man who wrote this testimonial about how much his counselor had helped him:

He told me what to do—what's right and wrong—*not to fool around with girls* [emphasis added]. It gives a boy a good feeling to have an older person outside the family to tell you what's right and wrong—someone to take an interest in you. (Powers and Witmer, 1951)

Several years later, this same young man began serving a five-year sentence for a serious sex offense (McCord and McCord, 1959).

Later, social scientists reopened this classic study, hoping that some evidence in favor of the program could still be found. William and Joan

control their children's behavior. We shall examine this project in depth in Chapter 13. But it now appears that many kids simply never outgrow the selfish, aggressive, antisocial behavior that is normal for two- and three-year-olds. In fact, children don't cease such behavior unless their parents punish them for it. Studies of really misbehaving children show that their parents do not consistently punish them. Although the Oregon group has had some success in training parents to cope with their children, it remains to be seen how applicable the results are. For one thing, many parents (especially those of the worst-behaved kids) are unwilling or unable to improve their parental performance. For another, intervention must occur at a very early age to be effective. Finally, such intense levels of training are needed that it may be economically impossible to make this training widely available.

Why did none of the other programs work? Because they failed to truly change the circumstances of people. Training adults to be better parents may strengthen weak attachments within a family; simply providing a kid with a counselor will seldom have such an effect. A counselor is not an adequate substitute for parents or close friends, and a delinquency program fails to give kids a stake in conformity—it may even do the reverse. Kids who mess up do not risk being kicked out of a delinquency program.

However, even if we can't set up programs to make kids want to conform, couldn't we make them afraid to be deviant?

Immense effort has been and continues to be invested in various therapeutic techniques aimed at correcting patterns of deviant behavior or preventing its recurrence. This group of young people is taking part in an encounter group led by an adult therapist. No research evidence suggests that such therapy succeeds.

McCord offered the useful hypothesis that perhaps the research had ended too soon, that possibly the effects of the program would not stabilize until later in life. So, they reexamined the police records of the members of the experimental and control groups for ten years after the original study had ended. But this hypothesis also turned out to be false; no difference could be found (McCord and McCord, 1959).

Thus ended the first experimentally evaluated attempt to prevent delinquency by trying to influence socialization of the young. But it did not end efforts to achieve this goal.

OTHER PREVENTION PROGRAMS

In the decades following the Cambridge-Somerville study, scores of delinquency prevention programs have been tried and evaluated. Until recently, the result was an unbroken string of failures (Weis, 1977). During the 1950s and 1960s, immense effort was made to utilize forms of psychotherapy to reform delinquents. None succeeded (Toby, 1965). Prevention efforts were also taken into the streets by social workers attempting to reform juvenile gangs. These efforts were also judged failures (Miller, 1962). Toward the end of the 1970s, just as sociologists became convinced that nothing would work, the first evidence appeared that something might. A group of psychologists in Oregon began to have at least modest success with their efforts to train the parents of problem kids to more effectively

DETERRENCE

As far back as written records go, humans have constructed legal codes that specify not only which acts are prohibited by law but also which punishments are to be given to offenders. The Code of Hammurabi, written about 3,700 years ago, tells us that if a man destroys another man's eye, the offender's eye should be taken out; if a son strikes his father, the son shall have his hand cut off. Early legal codes tried to achieve symmetry—to provide justice by matching the punishment with the offense. This approach is repeated in many places in the Old Testament. In the Book of Deuteronomy, for example, we read, "Life shall go for life, eye for eye, tooth for tooth, hand for hand, foot for foot."

Library of Congress

However, even in these early philosophies of justice, punishment was not meant merely to serve as revenge for victims or their relatives. Instead, punishment has long been intended as a means of making life and property more secure by reducing the likelihood of a person committing a crime or a second offense.

This aspect of social control is called **deterrence:** the use of punishment to deter people from deviance. As Plato put it 2,300 years ago, "Punishment brings wisdom; it is the healing act of wickedness." This occurs, Plato explained, because the point of punishment is not to "retaliate for a past wrong" but rather to make sure that "the man who is punished, *and he who sees him punished* [emphasis added], may be deterred from doing wrong again."

As human societies evolved into complex states, governments increasingly sought to deter crime; hence, punishments became increasingly severe as crimes continued to occur. **Capital punishment** (execution) became common; in England during the eighteenth century, more than 200 different crimes carried the death penalty. Additionally, executions typically were conducted in public, often drawing large crowds, in an effort to deter those who witnessed the punishment from committing similar acts.

However, many began to speak out, calling capital punishment cruel and barbaric: In time

the opponents of capital punishment succeeded in restricting the death penalty to fewer and fewer offenses in most of Europe and North America. By the twentieth century, some nations dispensed with it altogether. Canada suspended the use of capital punishment in 1967 for a five-year trial period. In 1973 another trial period was adopted. Then, in 1976 the Canadian Parliament abolished the death sentence (except under provisions of the National Defense Act). The last execution in Canada occurred in 1962. In 1972 the U.S. Supreme Court prohibited capital punishment in the United States on the grounds that it was applied in a discriminatory fashion—that African Americans and poor people stood in greater jeopardy. Since then, however, many states have redrafted their capital punishment statutes in a way acceptable to the Court, and in 1977 the first execution since the late 1960s took place in the United States. Since 1977, more than 350 persons have been executed and more than 3,500 are in U.S. prisons under sentence of death.

While much of the debate over capital punishment has centered on moral issues, a major element has been the argument that it does not deter crimes for which it has been used (mainly homicide, but also rape). Indeed, by the 1950s the accepted view among social scientists, presented in most introductory textbooks, was that the threat of punishment does not prevent

With his assistant pulling on the criminal's long pigtail to extend his neck, the official executioner is about to carry out a sentence of death by beheading. The photograph was taken in China around 1860. Most of these Chinese villagers probably had seen many similar executions, but few if any of them had ever seen a camera. That's probably why most seem more interested in looking at the photographer than at the execution. For centuries it was believed that public executions served to deter the onlookers from committing similar crimes. But for most of the twentieth century, social scientists rejected this view. Only recently has support for deterrence theory reappeared in social science journals.

deviant behavior and that the experience of having been punished does not cause people to cease their deviant behavior. Deterrence was dismissed as an obsolete notion, exposed as such by scientific research. In fact, very little research had ever been done. As we shall see, especially with regard to capital punishment, what little research had been done was flawed by poor methods and the great difficulty in obtaining appropriate data.

By the early 1970s, however, a growing number of social scientists began to reconsider the deterrent effects of punishment. For many it was simply a response to their realization that they attuned their own behavior on the basis of potential punishment. For example, on trips across the country, they drove faster or slower depending on a state's reputation for enforcing speed limits. Are these self-observations merely illusions? Can it really be true that if penalties for crimes were repealed, the crime rate would not go up? The case for deterrence was reopened.

This is "Old Sparky," New York State's electric chair. It was the world's first, going into service in 1890. A total of 695 persons were put to death in this chair from 1890 through 1963.

© Eddie Adams/TimePix/Getty Images

Jack Gibbs, a sociologist, led the way by reformulating the theoretical issues concerning crime. Isaac Ehrlich, an economist, reopened research on the effects of capital punishment on homicide. Both men provoked angry reactions from many of their colleagues. Together, however, they have stimulated an immense amount of theoretical and research effort; deterrence is no longer a discarded concept.

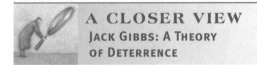

A CLOSER VIEW
JACK GIBBS: A THEORY OF DETERRENCE

As Jack Gibbs examined the grounds on which deterrence had been dismissed from social science, he recognized that much of the reasoning and the evidence cited missed the mark. The case against deterrence came down to this: Many people are punished for committing crimes, including some who are executed for homicide; nevertheless, the crime rate, including the homicide rate, remains high. Moreover, most people who serve a prison or jail sentence for a crime turn right around and commit new crimes when they are released. Clearly, then, punishment fails to deter.

Gibbs concluded that this argument is irrelevant to the fundamental issue. It claims in effect that if *some* people seem not to be deterred, *no one* is. When the issue is posed this way, it is impossible to demonstrate any deterrent effects of punishment unless it is 100 percent effective. This is the same as saying that aspirin has no effect on headaches unless it cures every headache for everyone. To see if punishments influence people to conform, Gibbs pointed out, we cannot look at just those who were not deterred; rather, we must look at everyone. That is, we need to know more than the fact that some people do risk punishment; we need to know whether the fear of punishment influences those who do not commit offenses. This is, of course, a much more difficult research problem.

Although Gibbs was very concerned with improving research on deterrence, he recognized that he first had to formulate an adequate theory. Why and how should punishment produce conformity? Until such an explanation was proposed, we could not say what empirical observations would be predicted or prohibited and therefore what research ought to be undertaken.

Gibbs (1975) set out to formulate a clear theoretical statement of the effects of deterrence.

Drawing on social learning and control theories of deviance, he proposed that it is not only the severity of punishment that matters but also the rapidity and certainty of punishment. Gibbs's **deterrence theory** can be summarized as follows: *The more rapid, the more certain, and the more severe the punishment for a crime, the lower the rate at which such crimes will occur.*

Thus, Gibbs predicted that severe sentences will not effectively deter crimes if people realize that they have little chance of being caught or that their punishment will be long delayed if they are caught. On the other hand, not even quick and certain punishment will deter crime if the punishment is very mild, for then the costs of detection often will be outweighed by the rewards of the crime. Here Gibbs's theory links with control theory to explain why the same punishment might be much less severe for some than for others. For example, two years in jail is a much greater cost to someone with a happy family and a good job than to an unemployed drifter. Deterrence thus fits into control theory as another of the costs to be considered by the potential deviant. The higher that cost and the swifter and more certain it is to be imposed, the greater the conformity.

Even though what Gibbs's theory predicts and prohibits is very clear, proper testing has been difficult. As we shall see later in this chapter, punishment for crimes in the United States today is often far from certain or swift and often not severe. This makes it difficult to test the theory against available statistics.

Gibbs was able to show, however, that people are less likely to be convicted for an offense a second time when the police have a high rate of success in solving such cases. For example, the police can solve rape cases far more often than burglaries. Thus, the certainty of arrest is greater for rape. Examining prison statistics, Gibbs found that far fewer people are arrested a second time for rape than are arrested a second time for burglary: Only 16 percent of those in prison for rape were second offenders compared with 51 percent for burglary.

Once Gibbs's theory was published, many social scientists began to design studies to test it more rigorously. The most important breakthrough was recognizing that what matters most is not the actual certainty, swiftness, or severity of punishment but rather the *perceptions* of these aspects of punishment. Suppose the actual odds of being caught for some offense were very high. A person who mistakenly thought that they were

very low would act on that perception. Indeed, Gibbs's theory assumes a link between reality and perception—that the true conditions governing punishment will act as a deterrent because people will perceive these conditions. Thus, Gibbs's theory can be tested without encountering problems concerning the present operations of the criminal justice system. Instead, we can simply find out how people perceive the situation and see if that perception influences their rates of deviance and conformity.

Later in this chapter, we shall see that people who have actual experience with police, courts, and the prison system often perceive lower risks in crime; this perception can be a major factor in sustaining chronic criminal behavior. This has led many deterrence researchers to discuss the importance of naiveté in social control. Indeed, many have suggested that a primary way in which differential association (see Chapter 7) can influence deviance is by exposing a naive individual to a perception of much lower risks (Jensen, 1969; Parker and Grasmick, 1979; Minor and Harry, 1982).

Despite the growing body of research supporting deterrence theory, vigorous opposition continues; some sociologists even claim that it doesn't matter whether punishments deter crime or not because, they argue, punishment is immoral. Closer examination of these views suggests that the real issue is not deterrence in general but rather concern that the renewed interest in deterrence will reopen a much more sensitive question: Does capital punishment deter homicide? For many people the idea of the legal system putting people to death is so repugnant that there can be no valid scientific purpose in researching the deterrent effects of capital punishment. For many others, including most American citizens (Figure 8-2), some acts of murder are so repugnant that the killer has no right to life.

Keep in mind, too, that the wording of the question about capital punishment tends to *minimize* public support. Respondents were asked: "Do you favor or oppose the death penalty for persons convicted of murder?" That wording seems to imply that most or all murders should get the death penalty. What if they had been asked: "Have some people committed murders so cruel and terrible that you could approve of them being given the death penalty?"

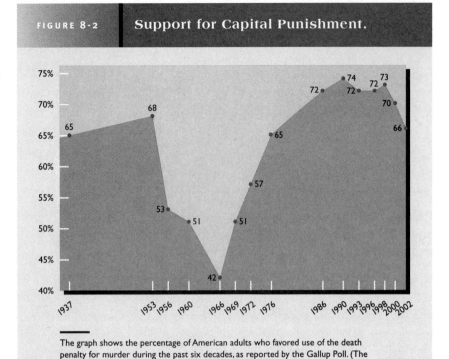

FIGURE 8-2   Support for Capital Punishment.

The graph shows the percentage of American adults who favored use of the death penalty for murder during the past six decades, as reported by the Gallup Poll. (The 1990, 1993, 1996, 1998, 2000, and 2002 data are from the General Social Survey.)

Here we can see the pressures that inevitably come to bear on social scientists. On the one hand, we are active participants in the world we study, and like all human beings, we must deal with moral issues. On the other hand, we have a responsibility to provide facts pertinent to many moral controversies. In this instance a most pertinent fact is whether capital punishment deters homicide. If it does not, then there may be no reason to debate the morality of the matter. Of course, we cannot know the facts until adequate research has been done. Often, sociologists have difficulty switching back and forth between the role of scientist and that of concerned citizen— to insulate our research procedures from our moral convictions. Keeping this in mind, let's observe the interplay of moral commitments and scientific research in this controversial area.

THE CAPITAL PUNISHMENT CONTROVERSY

People have opposed capital punishment for many reasons. Some have opposed it for religious reasons. In fact, we shall see later in this chapter that the Quakers were the first to institute prison sentences as punishment in an effort to eliminate torture and execution. Some have opposed it on grounds of racial discrimination because African

Americans were executed more often than whites (53.5% of the persons executed in the United States between 1930 and 1967 were African Americans). Others have even argued that executions actually increase homicides because capital punishment brutalizes public perceptions of the value of life (Bowers and Pierce, 1980).

However, a major aspect of the debate over capital punishment has been social scientific research that seemed to show that capital punishment does not deter homicides. If this is so, many have argued, then we should not risk executing an innocent person or risk other possible negative consequences of capital punishment, since nothing is gained. Unfortunately, this conclusion was based on uncritical acceptance of highly deficient research.

The deficiencies in the research were due partly to the limited statistical techniques social scientists used in the 1930s and 1940s, when the research was done. They also were due partly to problems in the data available for analysis. In effect these early studies simply compared the homicide rates of states that used capital punishment with the rates of states that did not. These comparisons showed no correlation: States without capital punishment did not have higher homicide rates than did states with capital punishment.

Unfortunately, little or no attention was given to the problem of spuriousness, even though states differed in many ways besides the use of capital punishment. For example, New York, which used capital punishment, had a homicide rate as high as or higher than North Dakota, which did not use capital punishment. However, to conclude that capital punishment did not deter homicide was to assume that these two states were identical in other ways. A much better test would have been to see what would happen to the homicide rate in New York if capital punishment were repealed and what would happen to the homicide rate in North Dakota if capital punishment were reinstated. But of course, no such data exist.

Nevertheless, most social scientists in the 1950s and 1960s, few of whom had read the pertinent research studies, assumed that capital punishment had no deterrent effect. But then Isaac Ehrlich challenged the accepted belief.

In an article in the *American Economic Review* of June 1975, Ehrlich used sophisticated statistical techniques to see if the homicide rate and execution rate for the United States from 1933 through 1969 were correlated. Ehrlich concluded that there is a negative correlation between the two and that the deterrent effects of capital punishment were huge. In fact, he concluded that each execution prevented eight additional homicides. That is, each time someone was executed for committing a murder, eight other people escaped becoming murder victims. Ehrlich concluded that, by putting to death those who have taken a life, we save many additional lives.

Ehrlich's article provoked a storm of protest and moral condemnation. Many attempted to dismiss his findings as the result of improper statistical procedures (McGahey, 1980). Other scholars rose to Ehrlich's defense, including statistical experts who claimed his methods were superior to those used by his critics (Yunker, 1982). Moreover, when the data used by Ehrlich were extended to the 1980s, when the first executions since the Supreme Court decision of 1972 occurred, the results were even stronger (Yunker, 1982).

The homicide rate tended to fluctuate with the execution rate from 1930 through the late 1950s. Then, when executions became very rare and eventually ceased entirely, the homicide rate began a rapid ascent. However, as we saw in Chapter 3, correlation alone is not sufficient proof of causation. And it is difficult to conclude from only these data whether executions really do deter homicide. A major problem is that the data are crude and apply to the United States as a whole. However, many states did not have capital punishment during this period, and states differed greatly in their homicide rates. Perhaps these are meaningless averages that are accidentally correlated.

Subsequently, there have been many efforts to obtain better data and to more adequately analyze them (Phillips, 1980; Kobbervig, Inverarity, and Lauderdale, 1982; Zeisel, 1982a; Stack, 1987; Bailey and Peterson, 1989). Unfortunately, the results have been quite mixed. The best that can be said at this point is that no one knows whether capital punishment deters homicide.

But we must also recognize that the question of whether capital punishment deters homicide is entirely distinct from the moral debate over the use of capital punishment. A society could choose not to execute murderers even if it could be demonstrated that many people are prevented from killing someone because they fear the death penalty. Similarly, a society could choose to impose the death penalty on killers as a matter of justice despite the knowledge that fear of the death penalty does not keep people from committing murders.

Even if we set aside the question of deterrence and capital punishment, many difficulties still surround the subject of deterrence. Common sense tells us that we often conform from fear of punishment. Yet as we have seen, it is difficult to show that crimes that do not happen were deterred. And it is easy to show that some people are not deterred—the prisons are full of them. Thus, we must return to the argument against deterrence that Gibbs dismissed as irrelevant: If punishment deters crime, why are the prisons filled with people, most of whom have served previous sentences? To answer this question, we must examine how the American and Canadian systems of criminal justice operate.

THE WHEELS OF JUSTICE

Suppose you are contemplating a career of crime. You will need to take into account three vital facts: It's far from certain that you will be punished for committing a crime; if you are punished, it will likely be long after the crime; and it is quite likely that the punishment will not be very severe, at least not the first few times you are caught. To see why this is so, we must look at the police and the courts.

THE POLICE

Except for traffic officers, Canadian and American police officers spend little time looking for violations of the law. Instead, the police respond primarily when offenses are reported to them; only then can they attempt to solve a case. A major reason the odds of getting caught for a crime are so low is that so many crimes are never reported to the police.

Table 8-1 is based on a huge national survey designed to measure the true incidence of major crimes. Periodically, the U.S. Department of Justice sends interviewers to 60,000 randomly selected American households to ask detailed questions about crime victimization. For each crime that respondents say has been committed against them during the past year, the question is asked, "Was this incident reported to the police?"

The data in Table 8-1 show that about half of the time the answer was no. Theft is the least frequently reported crime (29%), while auto theft is the most frequently reported (91%). The latter reflects the fact that victims may not collect their

TABLE 8-1	Percentage of Crimes Reported to the Police
OFFENSE	**REPORTED (%)**
Motor vehicle theft	91
Robbery	62
Rape	58
Assault	57
Burglary	53
Purse snatching/pocket picking	35
Theft	29
All crimes	39

Source: Criminal Victimization in the United States, 2000.

auto insurance unless the theft is reported; in addition people realize that there is a very good chance that the police will find their car. Overall, six of every ten crimes are unreported.

A reason that some people don't report crimes is that they lack confidence in the police. Table 8-2 compares nations on the basis of the percentage of their citizens who expressed a lot of confidence in the police. The overwhelming majority of people in most Western European nations as well as in Canada and the United States have a lot of confidence in their police. There is considerably less confidence in Asia, Eastern Europe, and Latin America. In Russia and Mexico, less than a third have confidence in the police. Despite these large international variations, the fact remains that in each nation many people lack confidence in the police and may therefore fail to report crimes. When the police don't even know about a crime, they are obviously unlikely to catch the criminal. Thus, about 60 percent of the time, one can expect to get away with a crime simply because the victim fails to report it.

Of course, even when a crime is reported, the police are often unable to find the criminal. Table 8-3 shows the proportion of various offenses that led to an arrest in 2004 (the results change little from year to year). Murder has the highest clearance rate—almost two-thirds are solved—and assaults also led to an arrest in more than half of the reported cases. Only 44 percent of all rapes result in an arrest, and after that, the clearance rates plummet. Only about one of every ten vehicle thefts or burglaries ends in an arrest.

Clearly, the odds of being caught seem low. Yet Table 8-3 overstates the chances of being

caught because it gives the percentages of *reported* cases that led to an arrest. About twice as many crimes are committed as are reported, so the proportion of cases that are solved by arrest is only about half of what is shown in the table. For example, 3.2 million burglaries were reported to the American police in 1988, 14 percent of which led to an arrest. However, based on the victimization survey for that year, an estimated 5.6 million burglaries actually occurred. If we base the computation on the actual number of burglaries, then slightly less than 9 percent were solved by the police. This means that the odds against getting caught for committing an act of burglary are about ten to one.

Of course, that does not mean that only one burglar in ten gets caught. Burglars tend to commit a series of crimes, so eventually, the odds catch up with them. Yet they have such a good chance of avoiding arrest that they tend to explain getting caught as simply "bad luck" or a "silly mistake" (Irwin, 1970). Similarly, studies find that juveniles who have had contact with the police by being arrested or stopped and questioned have a lower, not a higher, expectation of being caught for an offense (Piliavin and Briar, 1964; Jensen, 1969). Of course, this expectation is further depressed for such crimes as burglary, for which the actual odds of being caught are low anyway. Clearly, then, the threat of jail fails to deter burglars in part because they do not see punishment as certain; in fact, they see it as unlikely. Other offenders have similar feelings. Can anything be done to make burglars and other offenders more concerned about getting caught?

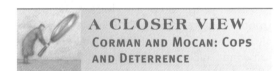

A CLOSER VIEW
CORMAN AND MOCAN: COPS AND DETERRENCE

A major reason that good studies of deterrence are rare is because they require a significant time dimension, and also the independent variable(s) must vary over that time period. For example, to discover whether increased policing reduces crime, it is necessary to have data on the amount of policing over time, for that amount to increase or decrease by a siginificant amount, and also to have data on the frequency of crime over that same period. Moreover, such a study would need to extend over a considerable period so that changes in policing had time to register with those apt to commit crimes. Consequently, if social scientists wished to conduct such a study, they would need to devote years to collecting data and hope that the amount of policing rose and fell significantly during that period. Or they would need to be lucky.

Recently, Hope Corman, an economist at the National Bureau of Economic Research and at Rider University, and H. Naci Mocan, an economist at the University of Colorado in Denver, got lucky (Corman and Mocan, 2000). They discovered that the New York City police department possessed reliable month-by-month data on crime rates, arrests, and police staffing covering a thirty-year period during which the number of police officers had fallen dramatically from about 32,000 in 1970 down to fewer than 22,000 during the early 1980s and then slowly climbed back to about 32,000 officers again by the late 1990s.

When they analyzed these data, they found very strong findings in favor of deterrence theory. As the number of police declined, so did arrests, while the number of major crimes rose substantially—at a lag of a few months. Then, as the number of police increased, the number of arrests rose and the number of major crimes declined. The changes were substantial and support the causal chain that when there are more police they make more arrests, and as the word spreads, many who might otherwise have committed an offense decide not to do so. There is, of course, an additional factor at work. When more offenders are arrested, not only do some potential offenders become apprehensive, but the number who would have been able to continue offending is reduced, too. That is, in addition to deterrence, there is the factor of **incapacitation:** *making people unable to commit crimes by locking them up.* According to Corman and Mocan, both factors were at work in New York City.

THE COURTS

What about people who are arrested for a crime? It is still not certain that they will be punished, and even if they are, punishment is likely to be long delayed.

In both Canada and the United States, many people arrested for a crime are never punished for it. First of all, prosecutors drop many charges because of flaws in the arrest procedures—officers didn't follow the rules with sufficient care or file their paperwork properly.

TABLE 8-2	"How much confidence do you have in the police?"		
NATION	HAVE CONFIDENCE IN THE POLICE (%)	NATION	HAVE CONFIDENCE IN THE POLICE (%)
Denmark	91	South Africa	56
Jordan	91	Chile	55
Finland	90	Poland	55
Egypt	87	Uganda	55
Norway	86	Bangladesh	53
Ireland	84	Morocco	53
Iceland	83	Croatia	53
New Zealand	81	Indonesia	52
Canada	**79**	Colombia	50
Argentina	76	Japan	50
Sweden	76	South Korea	50
Austria	76	El Salvador	49
China	73	Serbia	47
Luxembourg	72	Hungary	45
United States	**71**	Brazil	45
Turkey	71	Romania	45
Great Britain	70	Venezuela	41
Switzerland	69	Belarus	40
Italy	67	Latvia	40
Algeria	67	India	38
Tanzania	67	Nigeria	35
Albania	66	Estonia	34
France	66	Moldova	34
Portugal	66	Czech Republic	33
Zimbabwe	64	Ukraine	33
Bosnia	64	Armenia	32
Netherlands	64	**Mexico**	**30**
Philippines	61	Pakistan	29
Iran	61	Russia	29
Azerbaijan	59	Greece	28
Taiwan	59	Lithuania	26
Spain	59	Peru	16
Puerto Rico	57	Dominican Republic	13
Belgium	56		

Source: Prepared by the author from the World Values Surveys, 2000–2001.

TABLE 8-3	Reported Crimes Cleared by Arrest
Homicide	62.4%
Assault	55.9
Rape	44.0
Robbery	26.3
Larceny-theft	18.0
Vehicle theft	13.1
Burglary	13.1

Source: U.S. Department of Justice, *Crime in the United States,* 2004.

In many other cases, the charges are dismissed at preliminary hearings because of problems of evidence; often, key witnesses fail to appear. Of cases surviving these barriers, the majority are resolved by a plea bargain. That is, the charges against the person are reduced to less serious ones in exchange for a plea of guilty. For example, a frequent plea bargain known in the United States as "swallowing the gun" involves reducing charges of armed assault to simple assault, an offense carrying a far shorter maximum sentence, if the accused person will plead guilty. This spares the government the effort and expense of a trial, but it also makes punishment much less severe. In fact, the majority of persons convicted of even quite serious offenses do not go to prison but instead receive suspended sentences or are placed on probation. And of those who do go to prison, very few will serve their full sentence: Most will be out on parole long before their time is up. Moreover, time runs on a unique calendar in prison. In American prisons, for example, three days equal four in the outside world. That is, unless one gets in repeated trouble with prison officials, time off for good behavior equals 25 percent of one's sentence (Cole, 1983).

Zeisel (1982c) offered the following summary of the American criminal justice system. Of every 1,000 felonies committed in the United States, 540 are reported to the police. Of these, 65 result in an arrest. Of all those arrested, 36 are prosecuted and convicted; many of those arrested are juveniles, who usually are not taken to court. Of those convicted, 17 are sentenced to serve time in jail or prison. Of these, 3 are sentenced to serve more than one year. In sum, for those who commit crimes in the United States, punishment is very uncertain, not often swift, and rarely severe. Consequently, crime is not highly deterred, and those convicted of crimes are very likely to commit subsequent offenses.

The most recent data, released by the Bureau of Justice Statistics in 1999, showed a **recidivism rate** (the proportion of persons who commit a new crime after having served a prior sentence in jail or on probation) of more than 70 percent. This means that of those convicted of an offense, the overwhelming majority are repeat offenders. Indeed, not only have most people in our jails and prisons been there before, but many have been there many times before.

REFORM AND RESOCIALIZATION

Few people today realize that prisons as places where people serve sentences for crimes are quite new. For centuries prisons and jails were merely places where people were held while awaiting trial or until they received their sentences. Punishment did not involve spending time in prison but instead took the form of execution, mutilation, branding, or flogging. Authorities regarded it as unthinkably expensive to confine and feed able-bodied offenders for an extended period. Not until the late eighteenth century was confinement in prison used as an alternative to physical punishment.

The first serious experiments with prisons began in Pennsylvania under the direction of William Penn, whose Quaker beliefs caused him to oppose physical punishment. Penn directed that prisoners spend their sentences at hard labor to pay for their own upkeep and for damages to their victims. Penn's prisons became notorious for vice. Men, women, and juveniles were locked together, and the guards profited from the sale of liquor. So, in 1790 the Quakers of Pennsylvania tried a new approach: the penitentiary.

The new name reflected a whole new philosophy for dealing with criminal offenders. They were placed in circumstances much like those of monks in monasteries, forced to contemplate their sins and become penitent (hence the name). Each prisoner was isolated in an eight-by-six-foot cell and forced to live in silence. The Quaker example was copied by other states. Soon prisoners were required to work to pay the costs of their support, although the rule of silence prevailed.

Problems continued, however. Convicts learned to get around the rule of silence, continued to cause problems of disorder, and frequently escaped. Then, in 1816 a new prison was built in Auburn, New York. This prison was novel both architecturally and operationally. Cells were built in tiers five floors high, making it possible for a few guards to observe the interiors of cells from strategically placed viewing points. Prisoners were divided into three groups. Dangerous troublemakers were placed in solitary confinement. A second group was allowed out of their cells for work only. The third and by far the largest group worked and ate together during the day and went into seclusion in their cells only at night. Internal discipline was maintained by adopting military procedures, with prisoners marched from their cells to their places of work or the mess hall.

Modern prisons are typically modeled on the Auburn design. However, a major change occurred during the 1930s and 1940s, when finding work to keep most inmates busy was no longer possible. In state after state, political opposition, especially from labor unions, to the sale of prison-made goods and the use of prisoner labor to construct roads and bridges resulted in the termination of prison work projects. This left wardens with few jobs to occupy inmate time, and of course, it immensely increased the cost of prisons, which were once virtually self-supporting. At the same time, efforts to reform the prisons by making them more humane and turning them into therapeutic institutions also made them much more expensive to staff. At present it costs more than $50,000 a year to keep a person in prison in the United States.

Encouraged by the social sciences, especially psychology, many prisons have discarded their punishment philosophy in recent years and adopted a therapeutic philosophy of reform and resocialization. The modern therapeutic prison has sought to use psychological therapists and social workers to help inmates gain the necessary insights and to make the needed personality adjustments so they do not return to a life of crime. But just as counselors proved ineffective with juveniles in the Cambridge-Somerville project, so too have they failed to find a method for making adult prison inmates conform to patterns of nondeviant behavior. The recidivism rate is as high among the new therapeutic prisons as among the older punitive prisons: about 60 percent.

While the threat of going to prison probably causes many people to obey the law, clearly it does not deter some, even after they have served a prison sentence. Nor do the prisons seem able to achieve the **resocialization** of inmates so they will not want to continue in crime.

If we take seriously the theories of deviance discussed in Chapter 7, we should not be surprised that prisons fail to create conformity. If a lack of attachments tends to lead to deviance, then taking people who already have inadequate attachments out of the community will weaken these attachments even more.

Moreover, labeling theory suggests that, upon release from prison, the stigma of being an ex-convict will further limit the ability of offenders to form attachments with conventional people. Indeed, prison may well foster new attachments

© Alex Webb/Magnum Photos

to other deviants. In this fashion tiny subcultures of deviants form in prison and survive as members are released.

Furthermore, whatever investments the person had prior to prison (which failed to be a sufficient reason to conform) must diminish while a person is in prison. Most people leave prison with little money, no job, and no promising career. Ex-convicts thus have very little to lose by subsequent deviance. And in prison they may gain a distorted view of how much can be gained by deviance and learn how low the odds of being caught are. In this way the prison system contributes to career deviance, especially as inmates return again and again.

Many efforts have been made to break this vicious cycle and prevent ex-convicts from committing new crimes. One of the most ambitious of these programs, and one that was subjected to the most careful evaluation by research, involved paying convicts salaries during their first months back in society. By watching carefully as this study was conducted, we can gain a better understanding of the problems the criminal justice system faces.

Early in the twentieth century, inmates in American prisons spent little time in their cells during the day. Most were employed full time in prison industries; many worked on labor gangs building roads and bridges far from the prison. During the 1930s reformers stopped the use of work gangs and brought an end to prison industries in most states. As often happens, the reforms may not have improved prison conditions. Today, many prisoners have little or nothing to do with their time and spend long, tedious hours in their cells.

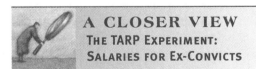

A CLOSER VIEW
THE TARP EXPERIMENT: SALARIES FOR EX-CONVICTS

Critics of the criminal justice system have stressed the economic plight of persons just released from prison. Lacking jobs, ex-convicts may be forced to commit new crimes to get money to live. Thus, these critics argued that recidivism could be reduced substantially if the government provided economic support to tide people over during their transition from prison back into society.

In 1976 Kenneth J. Lenihan persuaded the U.S. Department of Labor to support an experimental program to test this hypothesis. Lenihan also recruited his colleagues Peter H. Rossi and Richard A. Berk to direct research to study the effects of the program.

The study was designed as an experiment, thus allowing the most accurate evaluation possible. Arrangements were made with authorities in Texas and Georgia to identify all persons serving time in state prisons who were scheduled to be released soon; approximately 2,000 adult men and women were listed. These inmates were randomly assigned to two groups. Members of one group were given a weekly paycheck for about six months after their release. Members of the other group received no checks and left prison in the usual manner and condition.

The study, referred to as the TARP study (for Transitional Aid Research Project), was carried out with the strictest experimental controls. Because prisoners were randomly assigned to receive or not receive payments, and because a large number of subjects were involved, the group that was paid should not have been different from the group that was not paid.

The dependent variable was also clearly defined. The goal was to see if payments eased the way back to a conventional life, and people were judged successful in doing so if they managed to avoid arrest for one year after release from prison. Thus, the experimenters kept track of each person in the experimental and control groups and recorded when anyone was arrested.

What were the results? Among those who had not received TARP checks, 49 percent had been rearrested within a year of their release—a pretty dismal picture. But then the sociologists found that 49 percent of those who had been given financial aid had also been rearrested during the first year. The payments had made no difference at all (Figure 8-3). Keep in mind that this was only the proportion who got into trouble within the first year. Others undoubtedly got arrested in later years.

Like delinquency prevention programs, the attempt to prevent recidivism by giving released convicts financial aid totally failed. But just like

those who conducted the Cambridge-Somerville delinquency prevention study, the social scientists who conducted the TARP study also refused to accept the clear results of their own research. They, too, have ransacked the files on those in the experimental group and offered arguments that the program really did help some of them (Berk, Lenihan, and Rossi, 1980; Rossi, Berk, and Lenihan, 1980, 1982). Other scholars with no emotional stake in the study, however, regard the results as clear— the program failed (Zeisel, 1982b).

It would be comforting to argue that the failures of the prison system to reform convicts are caused by a lack of effort, investment, and public enlightenment—and people untrained in criminology often make such charges. But the facts are that legions of dedicated men and women have worked hard to provide effective therapies and programs to reform convicts, and immense energy and many resources have gone into the prison system. There has been no shortage of new ideas, and every few years new programs and styles of prisons are introduced. The problem is that nothing has worked. The most modern therapeutic treatment centers have achieved no better results than did the Quaker penitentiaries: The majority of those who leave the prisons commit new offenses.

But if prisons are unable to reform, they do serve to isolate from society for the duration of their stay persons who have been judged dangerous to people and property. The policy of locking up serious offenders is more acceptable in our culture than is physical torture or frequent resort to execution. That is, while we attempt to restructure prisons so that inmates are reformed or to discover other means of reformation, it is worthwhile to remember why prisons were invented in the first place: as a more humane form of punishment.

CONCLUSION

It is easy to draw overly pessimistic conclusions from this chapter and to assume that social control doesn't work. Granted, social control doesn't prevent some people from committing acts of serious deviance: More than 13 million crimes are reported to the police in North America each year, and the actual number of crimes committed is probably about twice that

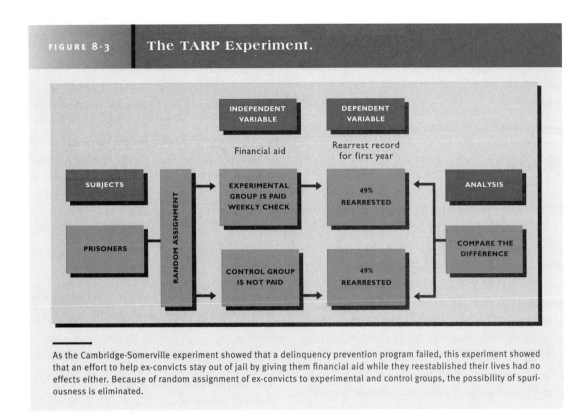

FIGURE 8-3   The TARP Experiment.

As the Cambridge-Somerville experiment showed that a delinquency prevention program failed, this experiment showed that an effort to help ex-convicts stay out of jail by giving them financial aid while they reestablished their lives had no effects either. Because of random assignment of ex-convicts to experimental and control groups, the possibility of spuriousness is eliminated.

number. Keep in mind, however, that even if each of these crimes was committed by a different person, then only about one person in ten engaged in a criminal act each year. And because we know that some people commit many crimes, only a tiny fraction of North Americans are responsible for the crime rate.

Viewed this way, social control usually seems to work—most of us usually conform to the law. And to a great extent, our conformity is rooted in informal social control. Long before we wonder if the police will catch us or the courts will jail us for committing a crime, we restrain such impulses because we know how our families and friends would react. Thus, as noted at the beginning of the chapter, formal methods of social control are activated only when informal methods fail. In an important sense, the police, the courts, and the prisons must assume responsibility for the people whom we—all of us in society—have failed to bind to the moral order. If the justice system typically fails to reform these people, this failure must be evaluated in light of the kinds of people it confronts. Keeping even 40 percent of such people from committing new offenses might be judged a substantial achievement.

Clearly, however, formal social control could be more effective. Improving the reporting of

crimes should be possible. In fact, recent police efforts to educate the public and to treat victims more sensitively have greatly increased the reports of rape. Better coordination among the police, prosecutors, and the courts could probably prevent the dismissal of so many cases on technical grounds. And many proposals exist for speeding up court procedures and preventing serious offenders from getting off lightly.

But if crime can be reduced somewhat by overhauling the criminal justice system, it would be unrealistic to expect truly dramatic changes. For the fact remains that in every human group known, some people break the norms—not just in small ways but also in very serious ways. Even in tiny, isolated, utopian religious communities, where each person has voluntarily chosen to join the group in its retreat from a "wicked" world, some people commit grave crimes. Surely, then, we shouldn't be surprised that some people commit crimes in a much less socially and morally integrated society such as ours.

As we have seen in the past two chapters, sociologists can explain much about why people deviate and conform. But to know why deviance occurs often does not provide the knowledge for preventing it. Moreover, social scientists do not know how to reform many people who exhibit chronic patterns of serious deviance.

Review Glossary

Terms are listed in the order in which they appear in the chapter.

social control All collective efforts to ensure conformity to the norms.

informal social control Direct social pressure from those around us.

formal social control Activities by specialists and specialized organizations devoted to ensuring conformity to the norms.

social order The extent to which citizens comply with important norms.

principle of dependence The more dependent that members are upon a group, the more they conform to group norms. By dependence is meant the extent to which a group is the only available source of important rewards.

principle of visibility To the extent that the behavior of group members is easily observed (or otherwise monitored) by other members, their degree of conformity to group norms will be greater.

principle of extensiveness The greater the scope and extent of norms upheld by the group are, the greater the contribution to overall social order will be.

prevention As a form of social control, all efforts to remove the opportunity for deviance or to deactivate its causes.

deterrence The use of punishment (or the threat of punishment) to make people unwilling to risk deviance.

capital punishment The death penalty.

deterrence theory The proposition that the more rapid, the more certain, and the more severe the punishment for a crime, the lower the rate at which that crime will occur.

incapacitation The inability of offenders to commit new offenses while they are in jail or prison.

recidivism rate The proportion of persons convicted for a criminal offense who are later convicted for committing another crime. Sometimes this rate is computed as the proportion of those freed from prison who are sentenced to prison again.

resocialization Efforts to change a person's socialization; that is, to socialize a person over again in hopes of getting him or her to conform to the norms.

Suggested Readings

Geerken, Michael R., and Walter R. Gove. 1977. "Deterrence, Overload and Incapacitation: An Empirical Evaluation." *Social Forces* 56:424–447.

Gibbs, Jack. 1975. *Crime, Punishment, and Deterrence.* New York: Elsevier.

Sociology Online

www.socstark10.com

GO TO THE INTERNET AND TYPE: www.socstark10.com.

➚**CLICK ON:** 2000 General Social Survey

✓**SELECT:** EXECUTE? Do you favor or oppose the death penalty for persons convicted of murder?

➚**CLICK ON:** Analyze Now
The media portray supporters of capital punishment as older, white, uneducated, male, Republicans. What's correct or incorrect about this portrait?

➚**CLICK ON:** Select New Data Set

➚**CLICK ON:** The 50 States

✓**SELECT:** % IN PRISON: Percentage living in correctional facilities.

➚**CLICK ON:** Analyze Now
What state has the largest percentage? Where does this state rank on cocaine use? Does Chapter 7 suggest a connection?

➚**CLICK ON:** Select New Data Set

➚**CLICK ON:** Nations of the Globe

✓**SELECT:** IN PRISON: Number of prison inmates per 100,000 population.

➚**CLICK ON:** Analyze Now
How does Japan's rate compare with that of the United States? Is this consistent with the discussion in the chapter?

USING INFOTRAC COLLEGE EDITION

GO TO THE INTERNET AND TYPE:
www.InfoTrac-college.com

➚**CLICK ON:** Register New Account
You will be asked to enter your Access Code and to create and enter a User Name and Password. Having done so,

➚**CLICK ON:** InfoTrac College Edition or Log On.

In addition to many new proposals for crime prevention, the periodical literature reports on the introduction of privately owned and operated prisons that accept prisoners on a contract basis.

SEARCH TERMS:
Crime Prevention
Private Prisons

CHAPTER

9

limousine look

A woman stares from the window of her limousine as she is driven through a poor neighborhood and perhaps is distressed by what she sees. This picture, taken in 1938 during the Great Depression by the famous photographer Dorothea Lange, captures a moment of awareness that has occurred countless times in human history, as people recognize the gulfs of wealth and privilege that separate them from one another.

Concepts and Theories of Stratification

STRATIFICATION—THE ORGANIZATION OF SOCIETY resulting in some members having more and others having less—has been a constant theme in moral, political, and philosophical writing through the ages. Millions of words have been written to denounce inequalities in wealth and power, as well as to justify these inequalities. From the point of view of modern sociology, this is an irresolvable conflict based on two inescapable facts. First, stratification has many undesirable consequences: People at the bottom of stratification systems often suffer greatly, both physically and emotionally. Second, some degree of stratification seems to be an unavoidable feature of social structure.

This chapter examines the basic concepts and theories sociologists use to describe and explain stratification. It begins by examining various concepts of social class. In Chapter 2 social classes were defined as groups of people who share a similar position, or level, within a stratification system. Now we shall pursue this definition in greater depth and explore differences in how leading sociologists have conceived of classes. We will also analyze the phenomenon of *social mobility:* upward or downward movement by individuals or groups within a stratification system. Armed with these conceptual tools, we shall then explore theories of stratification. Why are societies stratified? To what extent can stratification be minimized? In Chapter 10 we shall apply these principles by examining stratification and mobility in different kinds of societies.

CONCEPTIONS OF SOCIAL CLASS

People have used many different ideas to identify social classes, or divisions of rank and

CHAPTER OUTLINE

CONCEPTIONS OF SOCIAL CLASS

MARX'S CONCEPT OF CLASS
The Bourgeoisie and the Proletariat
Class Consciousness and Conflict
The Economic Dimension of Class

WEBER'S THREE DIMENSIONS OF STRATIFICATION
Property
Prestige
Power

STATUS INCONSISTENCY

SOCIAL MOBILITY
Rules of Status: Ascription and Achievement
Structural and Exchange Mobility
Class Cultures and Networks

MARX AND THE CLASSLESS SOCIETY
Dahrendorf's Critique
Mosca: Stratification Is Inevitable

THE FUNCTIONALIST THEORY OF STRATIFICATION
Replaceability
A Toy Society

THE SOCIAL EVOLUTION THEORY OF STRATIFICATION

THE CONFLICT THEORY OF STRATIFICATION
Exploitation
The Politics of Replaceability

CONCLUSION

REVIEW GLOSSARY

SUGGESTED READINGS

SOCIOLOGY ONLINE

wealth, within societies. Some have used broad distinctions while others have used narrow ones in deciding which people occupy similar positions in a stratification system. Are classes large and few in number, or are they small and numerous?

Plato saw only two classes in ancient Greek society, the rich and the poor, and he believed them to be locked in eternal conflict. Aristotle divided Greek society into three broad classes: a rapacious upper class, a servile lower class, and a worthy middle class that, having all virtues and all failings in moderation, could be trusted to see after the common good of all. The word *class* comes to us from the Romans, who used the term *classis* to divide the population into a number of groups for the purpose of taxation. At the top were the *assidui* (from which the word *assiduous* comes), who were the richest Romans. On the bottom were the *proletarii*, who possessed nothing but children.

However, not until the mid-nineteenth century was the concept of class given significant meaning for modern social theorists. The person who first did this was Karl Marx (1818–1883).

MARX'S CONCEPT OF CLASS

Marx aimed to explain social change and produce a theory of history: Why and how do societies change, and what will they be like in the future? He believed that the answers lay in conflicts among social classes. The whole of human history, Marx and Friedrich Engels wrote in *The Communist Manifesto* in 1848, has been "the history of class struggles." These struggles are the engines that pull societies into new forms; the history of human societies is a history of one ruling class being overthrown by a new one.

Marx saw that there is no single answer to the question of how many classes to identify in societies. Instead, the answer depends on which society and when. Thus, he identified four classes in ancient Rome—patricians, knights, plebeians, and slaves—and a larger number in Europe during the Middle Ages. But Marx expected modern capitalist societies to consist of only two classes.

THE BOURGEOISIE AND THE PROLETARIAT

A capitalist society, according to Marx, is one having a free-market economy based on private ownership of property. Chapter 17 develops a fuller definition of capitalism. In this chapter Marx's definition suffices. By the middle of the nineteenth century, when Marx wrote his major works, all of the nations of western Europe, as well as the United States and Canada, fit his definition of capitalism. Therefore, he predicted that each of them soon would undergo a great simplification of their stratification systems into two fundamental classes. As he wrote in *The Communist Manifesto*, "Society as a whole is more and more splitting up into two great hostile camps, into two great classes directly facing each other: Bourgeoisie and Proletariat."

Marx defined these two classes in terms of their different relationship to the **means of production**: everything besides human labor that goes into producing wealth. Chief among these are land (on which crops grow, cattle feed, and buildings stand), machines and tools, and investment capital. One class, the **bourgeoisie,** owns these means of production. The other class, according to Marx, contains everyone who does not own such means and therefore must sell his or her labor to the bourgeoisie. Marx called this class the **proletariat,** employing the name the Romans used to identify the poor. These terms essentially refer to owners (or employers) and workers (or employees).

Marx realized that all capitalist societies in his time had many people who did not fit into his two-class system, but he believed that these groups would not significantly affect history. One such group was the middle class, which included small merchants and self-employed professionals such as doctors and lawyers. Marx believed that as the capitalist system evolved, the middle class would eventually be crushed and forced into the proletariat.

Marx also dismissed many people who were marginal to the economy—vagrants, migrant workers, beggars, criminals, Gypsies, and the like. He classified such persons as **lumpenproletariat** (literally, the "ragamuffin proletariat"). These people had so little social purpose and self-respect, Marx believed, that they would have no effect on the impending revolutionary struggle.

Finally, Marx excluded farmers and peasants from his conception of class because he believed that the drama of historical change would occur in the urban industrial sector of capitalist societies; rural people would play lit-

Karl Marx.

Library of Congress

tle or no part in shaping social change. He wrote that the

> peasants form a vast mass, the members of which live in similar conditions, but without entering into manifold relations with one another. Their mode of production isolates them from one another . . . and the identity of their interests begets no unity, no national union, and no political organization; [therefore,] they do not form a class.

Ironically, the great communist revolutions Marx predicted never did occur in the urban, industrialized, capitalist nations. However, revolutions claiming to be Marxist have occurred in a number of less developed nations and have found their primary support among peasants. This suggests that, were he alive today, Marx would rethink his concepts to include rural populations.

CLASS CONSCIOUSNESS AND CONFLICT

In addition to *material* position in society, Marx also included an important *psychological* component in his notion of class. To be considered a real class, people must be similarly placed in

Library of Congress

A young radical, helping to organize demonstrations in support of the Industrial Workers of the World in 1913, wears his slogan on his hat.

society and share comparable prospects, but they must also be aware of their circumstances, their mutual interests, and their common class enemy. Marx called this awareness **class consciousness.** Much of his theory about the coming of the communist revolution concerns how the proletariat will achieve class consciousness, at which point their superior numbers will ensure their success. Marx also worried about the tendency for workers to believe they had common interests with the ruling class and called this **false consciousness.**

By incorporating assumptions about class consciousness into his definition of social class,

Marx inserted portions of his *theory* of revolution into his *concept* of class. This made key portions of his theory true by definition and thus untestable.

By Marx's definition, if people with a common economic position in society do not recognize their common interests and organize to pursue them, they are not a class. Indeed, Marx asserted that classes could not exist without class struggle. He wrote, "Individuals form a class only in so far as they are engaged in a common struggle with another class." Hence, when Marx said class struggle is inevitable, he had already made that statement necessarily true by his definition of the word *class.* To say that classes will be self-conscious and organized, then, is to predict nothing about the course of history; it simply states a definition.

Of course, Marx didn't coin the word class— people had been referring to upper and lower classes for centuries. But Marx and his political followers did make the notion of class membership so popular that most people around the globe have some idea of their **subjective class,** the class to which they think they belong. This

Class conflicts within industrial societies often take the form of strikes. Indeed, the right to form unions and to strike were major concessions employees won in the late nineteenth and early twentieth centuries. This young textile worker was arrested in Lawrence, Massachusetts, in 1912. She and hundreds of other strikers had demonstrated in the streets in violation of a court injunction.

Archives of Labor & Urban Affairs, University Archives, Wayne State University

can be seen in Table 9-1. Few people in any nation think of themselves as members of the upper class, which is as it should be since this label identifies a small social elite. Nations where the largest percentage say they are upper class tend to be less developed nations, and this also reflects reality. Another reality is that most people in most nations think they are middle class, and only in some less developed nations do most people think of themselves as in the lower or working class.

THE ECONOMIC DIMENSION OF CLASS

The most important feature of Marx's definition of class is that it is determined only by the *economic* dimension. Property ownership is the sole factor for ranking people, and then they are divided into only two groups: those who own the means of production and those who do not.

When he distinguished between the bourgeoisie and the proletariat based entirely on this single criterion, Marx necessarily implied that *all other differences* in position among people in society are *wholly* the result of property ownership. Thus, if some people are more powerful than others, or if some people are more admired or respected than others, it is due solely to the underlying economic differences between them. For, Marx claimed, the relationship to the means of production is "the final secret, the hidden basis for the whole construction of society." The rest of a society's culture results from the underlying economic arrangements. Indeed, for Marx culture was a "superstructure of various and peculiarly formed sentiments, illusions, modes of thought, and conceptions of life" that arises from economic relations.

Marx nowhere gave empirical evidence that economic differences were the sole basis of other social differences such as power and respect (Dahrendorf, 1959). His claim, however, was an empirical one that other social scientists could test. If power or prestige can be shown to *vary independently of property*, then Marx's statement is at least excessive and at worst false. Indeed, the likelihood that the economic dimension of property does not govern all aspects of stratification made many sociologists who came after Marx very uneasy with his single-factor conception of class. Among them was Max Weber.

TABLE 9-1	Subjective Social Class		
NATION	UPPER CLASS (%)	MIDDLE CLASS (%)	LOWER OR WORKING CLASS (%)
Azerbaijan	5.7	74	20
Nigeria	5.4	43	52
Dominican Republic	3.9	61	35
Moldova	3.4	63	34
South Africa	3.4	41	55
Switzerland	3.2	67	31
Armenia	3.2	66	31
Iran	3.1	70	27
Georgia	3.0	74	23
Philippines	2.6	44	54
Turkey	2.3	57	41
El Salvador	2.3	45	53
Tanzania	2.2	45	53
India	2.2	57	41
United States	**2.2**	**68**	**30**
Bangladesh	2.1	70	28
Peru	1.9	46	52
Pakistan	1.7	64	34
Indonesia	1.6	82	17
Albania	1.4	47	52
Puerto Rico	1.3	71	28
Brazil	1.3	63	36
Taiwan	1.3	64	35
Sweden	1.2	67	32
Mexico	**1.2**	**57**	**42**
Singapore	1.1	62	37
Bosnia	1.0	61	38
Colombia	1.0	59	40
Norway	1.0	66	33
Algeria	1.0	48	51
Canada	**0.9**	**64**	**35**
Egypt	0.9	65	35
Jordan	0.7	71	29
Morocco	0.6	43	56
Uganda	0.6	24	75
Zimbabwe	0.6	33	66
Spain	0.6	52	47
Australia	0.6	59	40
New Zealand	0.5	59	41
Macedonia	0.5	50	50
China	0.4	57	42
Japan	0.4	65	35
Chile	0.2	67	33
Vietnam	0.0	18	82
Argentina	0.0	39	61

Source: Prepared by the author from the World Values Surveys, 2001–2002.

WEBER'S THREE DIMENSIONS OF STRATIFICATION

Max Weber (1864–1920) is one of the great names in the history of sociology. We shall assess his work on religion and social change in Chapter 17 and his work on bureaucracy in Chapter 20. In Weber's lifetime, as in ours, the influence of Marx on social theory was immense. And some of Weber's major works were attempts to modify Marxist positions.

Weber believed that Marx's wholly economic view of stratification could not capture primary features of modern industrial stratification systems. Looking around in Germany, Weber noticed that social position did not always seem to be simply a matter of property ownership. Many Germans who belonged to the nobility lacked wealth yet possessed immense political power; for example, only they could be officers in the army. On the other hand, Weber noted that some wealthy German families, despite owning factories or large companies, lacked political power and social standing because they were Jewish.

In a strictly Marxist conception of class, these Jewish families would belong to the bourgeoisie, while many powerful *Junkers* (aristocrats) would belong to the proletariat. Thus, Weber came to view Marx's position as too simple. He proposed

Max Weber.

© German Information Center

that stratification is also based on other independent factors. He suggested three such factors: *class*, *status*, and *party*.

Modern social scientists have found several of Weber's terms somewhat confusing; therefore, they have renamed them to constitute "three Ps" of stratification: *property* (what Weber called "class"), *prestige* (what Weber called "status"), and *power* (what Weber sometimes called "party").

PROPERTY

By *class* Weber meant groups of people with similar "life chances" as determined by their economic position in society—their material possessions and their opportunities for income. This is what modern social scientists refer to as **property.** Weber stressed class membership based on *objective* economic position. Unlike Marx, he did not reserve the word *class* only for groups that had developed class consciousness and had organized for class conflict. Instead, Weber regarded the banding together by persons with the same economic position as merely one possibility. Thus, a key question for Weber was when and why class conflicts occur. Making class conflict part of the definition of class would not answer the question.

Furthermore, Weber did not stress ownership of property, but he realized that in some circumstances *control* of property might be independent of ownership. If a person can control property to his or her personal benefit, then it matters little whether the person legally owns the property. Thus, Weber was able to recognize the high class positions of managers (whether of capitalist corporations or socialized industries) who control firms they do not own. Marx had placed such persons in the proletariat.

PRESTIGE

Weber recognized that economic position can rest on control without ownership because he saw that prestige (or "status" in his terms) and power were not wholly the *consequence* of property relations. Instead, they could be the *source* of property relations. To use a trivial example, when famous athletes or military heroes endorse a commercial product, they are exchanging their **prestige,** or social honor, for economic advantage. Indeed, people often enjoy high prestige in a society while having little or no property. For example, poets and saints may have immense

All of these senior officers of the German General Staff in 1871 could claim membership in the nobility. Yet some of them had no private wealth and had to depend mainly on their army pay. Weber conceptualized stratification as multidimensional and therefore could take into account the power and prestige of these commanders while also recognizing their relative lack of property.

influence in a society while remaining virtually penniless.

POWER

The case for power as being independent of wealth is even more obvious. Weber defined **power** as the ability to get one's way despite the resistance of others. People may be very powerful without acquiring much property. For example, a corporation president may wield great power within the corporation and even in the political process of a society without personally owning any substantial part of the corporation. The same is often true of senior civil servants who run such powerful agencies as the CIA, the FBI, the RCMP, or the Federal Reserve while receiving relatively modest salaries. Additionally, power is often traded for economic advancement. Many politicians manage to retire rich even though they received only modest salaries while in office. The whole notion of influence peddling assumes the sale of power, while Marx seemed to believe that power can only be bought.

STATUS INCONSISTENCY

In Chapter 2 *status* was defined as any position within the stratification system. (This is not always what Weber meant by the term, and that is one reason sociologists often use the term *prestige* for Weber's concept.) This definition says nothing about the *basis* for status in a stratification system. Thus, a particular status or position can be high or low on the basis of the property, prestige, or power (or all three) associated with that position. We can also refer to *status characteristics:* certain individual or group traits that determine status. For example, various ethnic or racial minorities may be confined to a low status in society. Therefore, ethnicity and race are status characteristics: Variations in them influence position in society.

Because the term *status* denotes any particular position in the stratification system, it is a more general concept than class, which is just one measure of status. As Weber pointed out, there is more to stratification than simple economic differences.

If there are at least three basic dimensions of stratification in society, and if these can vary independently of one another, then individuals or groups can hold different ranks (or different status levels) on each of the three dimensions. For example, a person could be rich but have low prestige and little power. This state of affairs is called **status inconsistency.**

Status inconsistency theories predict that people whose status is inconsistent, or higher on one dimension than on another, will be more frustrated and dissatisfied than people with consistent statuses will be. A major proponent of status inconsistency theory, Gerhard Lenski (1954, 1956, 1966), explains the process this way:

When people rank higher on one status dimension than on another, they will emphasize their highest claim to rank and deemphasize their lowest. Thus, in presenting themselves to others, they will expect to be judged according to their highest status. Others, however, in seeking to maximize their own position, will tend to respond to them according to their lowest status.

Consider the case of university professors. They seek to be treated by others on the basis of their advanced educations. But many people outside universities ignore this claim and treat professors as having low status on the basis of their lack of power and wealth. Similarly, the Jewish industrialists in Germany in Weber's day sought to be ranked on the basis of wealth but often were treated with disdain on the basis of their ethnicity. A somewhat parallel example exists in the United States and Canada in the denial of status to "hick" millionaires—people lacking education and social graces who hit it rich and then unsuccessfully attempt to move in fashionable circles. The phrase "the vulgar rich" expresses the status inconsistency of many wealthy people.

Lenski argued that persons who are denied the social rank that they believe they deserve become antagonistic toward the rules governing status in their society. Consequently, persons suffering from status inconsistency will favor political actions aimed against upper-status groups; that is, they will support liberal and radical parties and proposals. This very important theoretical conclusion helped explain why persons of considerable social standing often seem to turn their backs on their own class interests and support the claims of the less privileged.

To test status inconsistency theory, Gary Marx reasoned that in the 1960s all wealthy, famous, or highly educated African Americans suffered from status inconsistency. Therefore, according to the theory, upper-status African Americans ought to be *more* militant about changing racial conditions than were African Americans with consistently low statuses. So, Marx predicted that African American bankers and physicians, for example, would be more radical than African American janitors and housekeepers. Research based on large national samples confirmed his hypothesis (G. Marx, 1967).

Similarly, as members of an ethnic minority, Jewish bankers and industrialists in North America and Europe ought to have a record of support for liberal and radical parties, and they

do (Cohn, 1958). The theory also explains the strong preference of wealthy and powerful American Catholics for the Democratic party (Baltzell, 1964). And few groups have demonstrated a greater propensity for left-of-center politics than college and university faculties, not only in North America but worldwide.

In opting for political responses, people suffering from status inconsistency place the blame on others, on the "system." This is especially likely for those whose inconsistent status is the result of a group characteristic such as race, religion, or ethnicity, for these are status characteristics over which the individual has no control.

The notion of status inconsistency is possible *only* if we accept Weber's view that there are multiple bases for rank in societies. If, like Marx, we accept only one basis for rank, then of course it is impossible to conceive of inconsistency. However, no matter how we conceptualize stratification systems, questions arise about how people gain their positions. And this brings us to the topic of social mobility.

SOCIAL MOBILITY

Societies differ greatly in the amount of upward and downward movement that goes on within their stratification systems. In some societies few people rise above or fall below their position at birth; in others there is a great deal of mobility. The amount of mobility in societies depends on two factors. First, the *rules* governing how people gain or keep their position may make mobility difficult or easy. Second, whatever the rules, *structural* changes in society can influence mobility.

RULES OF STATUS: ASCRIPTION AND ACHIEVEMENT

Chapter 2 introduced the two primary rules by which societies determine status. *Achieved status* is a position gained on the basis of merit, or achievement. *Ascribed status* is a position based on who you are, not what you can do. When a society uses ascriptive status rules, people are placed in status positions because of certain traits beyond their control, such as family background, race, sex, or place of birth.

In all known societies, both achievement and ascription operate, but societies differ in which rule dominates. In medieval Europe, for example, one's status was predominantly based on ascription. Persons born of the nobility

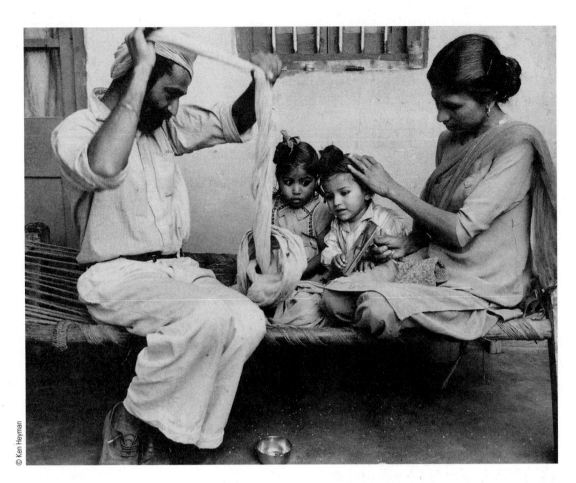

Fifty years ago the adult status of these Indian children would almost certainly have been exactly the same as that of their parents. Today, India's caste system is not as strictly observed as it used to be. However, it is still very likely that these children will inherit their parents' caste position.

were likely to remain in high positions; persons born of peasants were likely to remain in low positions.

When ascription is the overwhelming basis for status, we often speak of *caste* systems. Traditional society in India, for example, was composed of dozens of castes. Each person's caste group was defined by the caste he or she was born into, and each group was restricted to certain occupations. All the filthy and demeaning jobs, such as garbage collecting, were reserved for one caste, whose members were permitted to hold no other occupations. Similarly, highly skilled occupations, such as goldsmithing, were the exclusive right of another caste. However, even in caste systems, some people managed through luck and talent to rise above their origins; great prowess as a soldier in particular was often a ticket to higher status. And some highborn persons managed to fall to low positions because of misbehavior or incompetence.

Achievement is the primary basis of status in the United States, Canada, and other advanced industrialized nations. The majority of North Americans are socially mobile; they rise above

or fall below the positions of their parents. Nevertheless, ascription plays a significant role in these societies, too. Although African Americans and women are not wholly excluded from upper-status positions, they are underrepresented in such positions and overrepresented in lower ones. One reason for this is discrimination based on race and sex.

Social mobility is much more frequent when achievement rather than ascription is the primary basis for status. But societies also differ in the amount of social mobility that occurs because of the direction of structural changes in their overall status systems.

STRUCTURAL AND EXCHANGE MOBILITY

When the proportion of upper-status positions in a society increases, some upward mobility is inevitable. Because more openings exist at the top for the present generation than existed for the previous generation, some children of lower-status parents must be placed in high positions to fill them.

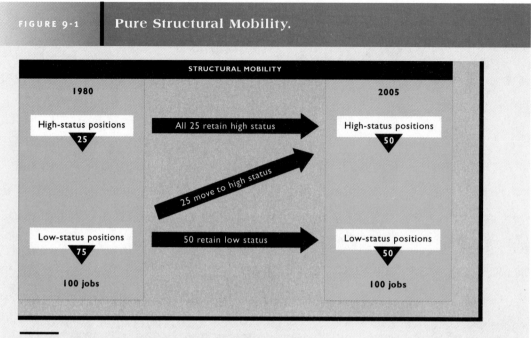

FIGURE 9-1   Pure Structural Mobility.

Structural mobility occurs because of changes in the ratio of upper-status to lower-status positions. When there is an increase in positions at the top, some people will have to be upwardly mobile to fill them. There has been a great deal of structural mobility in industrialized nations during the past century.

A simple example will make this clear (Figure 9-1). Suppose that in 1980 a society had 100 jobs, 25 of which were high-prestige, high-paying jobs and 75 low-prestige, low-paying jobs. However, the economy changed. As machines replaced humans on the job, there was less need for people to do unskilled work and a greater need for people to do highly skilled work (such as designing and maintaining the machines that replaced unskilled laborers). Thus, in 2005 this society still contained 100 jobs, but 50 were high-status positions.

Suppose further that in 1980 each of the 100 people holding the 100 jobs had one child who replaced him or her in the economy. Then, by 2005, 25 children of persons who held low positions would have entered high positions because 25 low positions had been eliminated from the status system and 25 high positions had been created. Thus, a third of the offspring of low-status people in 1980 become upwardly mobile simply as a result of changes in the status system. The reverse would have occurred if the proportion of high positions had decreased rather than increased.

Social mobility that results from changes in the distribution of statuses in society is called **structural mobility.** Note that structural mo-

bility occurs *regardless of the rules governing status.* Had our hypothetical society been based on ascriptive status rules, the outcome would have been the same. For example, if the demands of a particular occupation exceeded the capacity of the appropriate caste group, members of other castes would have filled the openings. Therefore, we can learn little about the rules governing status in a society merely by knowing its rate of mobility. We must also determine whether that mobility is structural or exchange mobility.

Mobility that is not structural is called **exchange mobility.** The word *exchange* indicates a trade-off. In exchange mobility some people rise to fill positions made available because other people have fallen in the status system. Let us reconsider a simple illustration in Figure 9-2. Suppose that the proportion of upper-status positions did not increase between 1980 and 2005. Thus, in 2005 there are still 25 upper-status positions and 75 lower-status positions. The only way that the offspring of persons in lower positions can gain an upper position is if someone born of upper-status persons moves to a lower status, thus making room at the top.

There is little exchange mobility when the ascriptive status rule operates, but if status depends on achievement, there is a fair amount

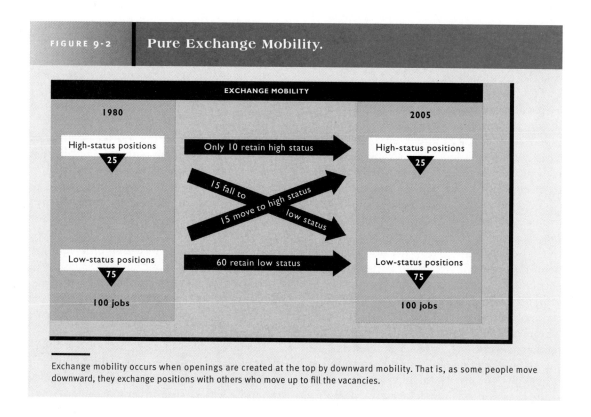

FIGURE 9-2   Pure Exchange Mobility.

EXCHANGE MOBILITY

1980 — High-status positions 25 — Only 10 retain high status → 2005 High-status positions 25

15 fall to low status

15 move to high status

Low-status positions 75 — 60 retain low status → Low-status positions 75

100 jobs   100 jobs

Exchange mobility occurs when openings are created at the top by downward mobility. That is, as some people move downward, they exchange positions with others who move up to fill the vacancies.

of such movement. Many children of talented and ambitious parents do not inherit the talent or ambition that earned their parents a high rank. Conversely, many children of low-status parents are more talented and ambitious than their parents are. Hence, if the status system is truly based on achievement, then we can expect a good deal of reshuffling in each generation. In Chapter 10 we shall examine mobility in the United States and other industrial nations.

CLASS CULTURES AND NETWORKS

Even in quite open and fluid stratification systems, class membership consists of considerably more than material assets, or wealth. In addition classes tend to develop and maintain distinctive cultures. These typically consist of styles of speech, etiquette, dress, body language, interests, information, and tastes (Coleman, 1990). French sociologist Pierre Bourdieu (1984) coined the term **cultural capital** to distinguish between the economic aspects of class and these powerful cultural assets. Bourdieu pursued at length how cultural capital adds to the advantages or disadvantages of one class vis-à-vis another, impedes upward mobility, and is transmitted from parents to children as if it were inheritable wealth. Some instances of

cultural capital are obvious, as in the case of academic credentials and titles, and some are so subtle that few other than members of the class can detect them (such as the peculiar use of "ain't" by the British upper class).

Bourdieu's research found that the culture of the upper social classes is more oriented toward abstract thought and formal reasoning and is more concerned with art, literature, and "intellectual" leisure activities. Lower-class cultures are more narrowly focused on the concrete, on matters of fact, and on the necessities of life. Notice the resemblances to Piaget's cognitive stages (Chapter 6). In similar fashion upper-class speech more often employs abstract ideas, uses more tenses, and more often refers to other times and places, while lower-class speech is more direct, more oriented to the present, and more frequently consists of incomplete sentences. Not surprisingly, these differences appear at an early age. Not only do upper-class children come to school better prepared in terms of knowing such things as numbers and the alphabet, but they come knowing more of the "educated" culture that the schools attempt to impart. Their homes are filled with books and magazines, there are real paintings and prints on the walls, they have been to concerts, they are familiar with computers. Within the limits of their age, they already speak correctly according to the grammatical rules of "proper" usage. They not only know about distant places, but they may already have visited them.

Thus, classes not only reproduce themselves by passing on wealth to the next generation; they also reproduce themselves culturally. Indeed, upper-class culture often functions as a sort of code that lets members immediately recognize one another in ways entirely invisible to members of lower classes. Moreover, class conflict often consists of more than disputes over wealth—disputes rage over culture, too. Indeed, *classes often amount to subcultures.*

Classes also differ in terms of their social networks. As noted in Chapter 2, cosmopolitan networks are inherently more powerful than are local networks. Having far greater reach, they can mobilize allies and support far and wide. This means, of course, that members of cosmopolitan networks will have advantages vis-à-vis the placement of members in the stratification system. But equally critical in terms of privilege is the access of cosmopolitan networks to far more information, which allows them to sustain rather different and more elaborate cultures than local networks can. Put another way, class differences in cultural capital are rooted in network differences.

Working-class people tend to have quite local networks, consisting mainly of relatives and people whom they have known for many years— often since grade school. This is especially true in older cities such as those in Europe and the eastern part of the United States, where many people live in the same house or on the same block their whole lives. Working-class women tend to remain within their neighborhoods, even for shopping. Working-class men also prefer to socialize in the neighborhood, often spending much of their leisure time with their old buddies at a local tavern. Because everyone knows the same people, everyone *knows* the same things. Because everyone likes everyone else, everyone tends to *believe* the same things (Homans' law of agreement, Chapter 3). And whenever the group confronts problems for which past experience has not prepared them, they lack links to people who may have a solution. This lack is worsened by the tendency of local networks to depend primarily on direct contact with *people* for all, or nearly all, of their information and to attend only to very local media. Thus, someone out of work might ask friends about opportunities in other cities, but if their friends don't know and if the local media do not happen to run a report on this topic, the person will seldom go to the library to read out-of-town newspapers or pursue other *impersonal* avenues of information.

However, if the cost of local networks is the relative isolation of the group, the cost of cosmopolitan networks is the relative isolation of the individual. Those in more privileged strata have many acquaintances but very few friends; their relationships tend to be brief, superficial, polite, and often entertaining, but rarely do they involve deep emotions. People in the higher social classes also have local networks. But usually, these are very small, consisting of little more than immediate family and an old friend or two. Consequently, unless they are famous, members of cosmopolitan networks are far less likely to have funerals, mostly because it would be hard to find enough mourners.

To sum up, many of the rather obvious differences in class cultures are the result of nothing more profound than differences in their networks, a point we will pursue further in Chapter 10.

Thus far, we have examined concepts of stratification, but now we must confront the basic theoretical issues: Why are societies stratified, and can stratification be eliminated? As noted at the beginning of this chapter, these issues have been debated for centuries. In the remainder of this chapter, we shall see that the debate continues among contemporary sociologists. However, before examining the three principal modern sociological theories of stratification, it will be useful to relate the contemporary debate to its roots in the nineteenth century.

MARX AND THE CLASSLESS SOCIETY

Karl Marx lived and wrote during a period of extremely rapid social change, which produced a great deal of social displacement and individual suffering. Across Europe the Industrial Revolution was in full swing. Great factories dotted the countryside, belching forth noxious smoke as well as an unprecedented volume of production. Until that time, most human beings had lived on farms or in tiny villages. But soon most people were living in cities, and it seemed that most would end up sweating over machines in factories.

To many, trading a life in a cow barn for a life in a factory was not real progress. The success-

During the nineteenth century, the word *industry* mainly meant huge factories and plants, powered by coal-burning steam engines that belched immense clouds of smoke and pollution. Marx believed that life in these "Satanic mills" was so intolerable and degrading that it would soon produce a revolutionary proletariat. Here, in Pittsburgh, Pennsylvania, in 1903 the huge Homestead steel plant (soon to become United States Steel) pours out grime—while inside the huge mill sheds, the fathers and older brothers of these boys suffer from intense heat. Recently, Pittsburgh was rated as one of the most livable cities in North America; a lot has changed since this picture was taken.

ful rise of industrial unions was yet to come. Many people, including Marx, found existing conditions intolerable, and out of their anger and frustration came an immense burst of utopian social thought.

A **utopian** constructs plans for an ideal society. Thousands of utopian schemes have been proposed over the past two centuries alone, and many have even been tried. During the nineteenth century, for example, scores of groups in the United States set up their own separate communes, where they attempted to enact a utopian living plan (Noyes, 1870; Nordhoff, 1875). Some of these groups, such as Oneida and Amana, were based on a common set of religious beliefs. Others, such as New Harmony, embraced socialism. As we saw in Chapter 4, the religious communal groups tended to last much longer, but they all broke up in the end. Nevertheless, whether religious or socialist or both, these groups reflected a deep discontent with the quality of life in nineteenth-century industrial societies (Kanter, 1972).

Within this context of rapid social change, widespread social unrest, and rampant utopian thought, Marx set out to create a scientific theory of history. His aim was to show how the inevitable forces of history would produce a revolution of the proletariat, which in turn would produce socialism and a communist society. After years of writing, Marx concluded that the coming revolutionary societies would be "classless"; that is, everyone would belong to a single social class.

Marx's conclusions, which were joyfully received by many, sharply contrasted with the predictions of **anarchists,** who were a major force in radical thought and politics at the time. The anarchists had reached a conclusion similar to that held by modern sociology: Stratification is an inescapable feature of human societies. But unlike most modern sociologists, the anarchists reacted to this conclusion by proposing to dispense with society. Indeed, they earned their name by proposing to "smash the state" and to live without social organization. However, because anarchists failed to explain convincingly how humans could live apart from society without also living in constant fear, danger, and disorder, most people found their solution worse than the problem. Then, in a few sentences, Marx claimed to prove the anarchists incorrect by showing how both society and equality were possible.

To see how he arrived at this conclusion, we must recall how Marx defined social class. As stated earlier in this chapter, Marx conceived of only two classes in modern industrial societies, and these classes differed on the basis of the ownership (or nonownership) of the means of production. By Marxist definition, classes exist only because privately owned property exists. This alone separates people into bourgeoisie and proletariat. Thus, to achieve a classless society, the *private ownership of the means of production*

must be abolished. Then, the bourgeoisie ceases to exist, and everyone becomes a proletarian.

But how can this be done? If the state seizes ownership of the means of production, then no person will own these means. Moreover, because everybody will belong to the same class, there will be no class conflict.

In accordance with Marxist doctrine, communist regimes came to power as they claimed their societies to be classless because they had placed ownership of the means of production in the hands of the state. These claims have long caused heated debate among people in noncommunist nations because communist societies remained very noticeably *stratified*. For example, party members in the former Soviet Union had luxuries unavailable to other citizens. How could this have been a classless society?

DAHRENDORF'S CRITIQUE

The most penetrating analysis of Marx's claims about the creation of classless societies was written by Ralf Dahrendorf (1959), a leading conflict theorist of stratification. Later in this chapter, we shall examine elements of his work. Here we concentrate on his argument that Marx's claim about classless societies was a "magic trick" that didn't mean what it appeared to mean.

Dahrendorf pointed out that everything Marx said about communist societies being classless was true, *but true only by definition.* Marx defined *class* according to the ownership of the means of production. If the state owns all means of production, then everyone falls into one class *as Marx defines class.*

But, Dahrendorf continued, notice that Marx never said this would produce *unstratified* societies. Indeed, communist societies have one class only in the limited sense expressed by Marx, but Marx did not say that everyone would have equal power or equal prestige. Moreover, it does not even follow that people in communist societies would be equal economically. For, Dahrendorf asked, *who is the state?* Is it really all of the people? Or is the state in fact controlled by political specialists?

Here Dahrendorf turned to Weber. What would happen to Marx's prediction of a classless society if he had based his definition not simply on the ownership of the means of production but also on its *control?* People do not have equal control over the means of production, even if private ownership is outlawed. Human affairs still require organization and direction. Someone still must manage each factory, for example. Because Weber argued that variations in the control of the means of production are often more significant than technical ownership, the socialization of industry clearly does not create classless societies as Weber used the term. Instead, people still differ greatly in their *control* over the means of production. Those who run the government in effect "own" the means of production.

Indeed, Dahrendorf argued, Marx did not escape the anarchists' conclusion at all. He only resorted to a solution by definition that did not alter the brute reality of stratification. As Weber himself noted, Marx was only proposing to replace the capitalist boss with a communist boss, and there is no reason to suppose that this substitution would reduce stratification.

Interestingly enough, Dahrendorf's critique of Marx had been anticipated in 1896 by a young Italian sociologist, who also anticipated major elements of both modern functionalist and conflict theories of stratification.

MOSCA: STRATIFICATION IS INEVITABLE

In his book *The Ruling Class* (1896), Gaetano Mosca (1858–1941) laid out a three-step "proof" that societies must be stratified. His first proposition was that *human societies cannot exist without political organization.* Mosca used the term *political organization* in the broadest sense to mean all forms of coordination and decision making in human activities. This proposition simply recognizes that human society is impossible if everyone runs around helter-skelter. Instead, group life requires mutual undertakings, and the actions of individuals must be directed and coordinated. Put another way, to have societies, humans must take collective actions.

Mosca's second proposition resembles the conclusion the anarchists reached: *Whenever there is political organization (or society), there must be inequalities in power.* Here Mosca simply recognized that coordination requires leaders, and leaders by definition have greater power than their followers; to lead, a person must be able to give orders and have them obeyed. Thus, according to Mosca, differences in power are built into the basic social roles by which societies are created and maintained—an insight that anticipated modern functionalist theories.

By themselves these two propositions establish that societies will always be stratified in terms of power. But Mosca carried his reasoning one step further and anticipated modern conflict theories of stratification. He argued in his third proposition that *because human nature is inherently self-serving, people with greater power will use it to exploit others and therefore to gain material advantages.* Thus, given the existence of power inequalities, material inequalities will always exist, too. Mosca concluded that stratification is not something human societies can avoid but instead is an inescapable feature of collective life. For advancing this view, many of Mosca's contemporaries dismissed him as a cynic and a reactionary. But today, more than a century later, his ideas have become part of the sociological mainstream. Indeed, as we now examine in detail modern theories of stratification, it will be obvious how fully Mosca anticipated them.

THE FUNCTIONALIST THEORY OF STRATIFICATION

The modern functionalist view of stratification is most closely identified with the work of Kingsley Davis and Wilbert E. Moore (1945, 1953). The key to their **functionalist theory of stratification** is in seeing society as a system of roles or positions. Inequality or stratification exists in societies because it is built into these roles and into the problem of filling them adequately.

Davis and Moore began by arguing that positions in society differ in the degree to which they are functionally important. That is, poor performance in some roles is more damaging to the society than is poor performance in some other positions. For example, while it is true that a society engaged in a war requires both soldiers and generals, a general is in a position to make more devastating errors than is any given soldier. Remember that for every famous general who won a battle that he should have lost, there was a general on the other side who lost a battle he should have won. History tends to be unkind to societies that appoint generals who snatch defeat out of the jaws of victory (Fair, 1971).

Some positions are inherently more important to the system, and Davis and Moore argued that some are also inherently more difficult to fill adequately. These positions require qualities that are naturally rare or that require a considerable preliminary investment in time, training, and effort. For example, some positions in a society require occupants with very high intelligence or great tact or other characteristics that are always in short supply in any population. Others—surgeons, for example—also require many years of training. Extensive training is always potentially in short supply, for new people constantly must begin training to fill future needs.

Thus, all societies face a general problem of motivation—"to instill in the proper individual the desire to fill certain positions, and, once in these positions, the desire to perform the roles attached to them" (Davis and Moore, 1945). How can this be accomplished? Davis and Moore argued that the only way to produce this kind of motivation is to adjust the reward system. Theirs is a supply-and-demand argument. To ensure an adequate supply of the right people, it is necessary to attach higher rewards to the positions that are most important and hardest to fill. Why would anyone want to become a general if the rewards were the same as those of a private?

Furthermore, it isn't enough to find some people who want to be generals; it is also important to attract the right kind of people to the position of general. Stratification, therefore, exists because the positions in society differ in their importance to the system and because it is necessary to ensure that competent people fill the most important positions. Indeed, people also vary in their ability to perform important roles. Hence, as Davis and Moore (1945) put it, stratification or social inequality

is an unconsciously evolved device by which societies insure that the most important positions are conscientiously filled by the most qualified persons. . . . Those positions convey the best reward, and hence have the highest rank, which a) have the greatest importance for the society and b) require the greatest training or talent.

Differential rewards, according to Davis and Moore, prevent less essential or less important positions in society from competing with the more important for scarce talents.

The data in Table 9-2 suggest that the essential principle of the functionalist theory is widely accepted. In surveys conducted in seventeen nations, respondents could agree or disagree that "no one could be expected to study for years to become a lawyer or a doctor unless they expected to earn a lot more than ordinary workers." Nowhere did less than a majority agree, and in most nations the level of agreement was overwhelming. Of course, this does not demonstrate that Davis and Moore are correct. But it does show that the principle that jobs requiring special talent and training necessitate higher wages is generally taken for granted.

REPLACEABILITY

Many have attacked Davis and Moore for attempting to justify social inequalities—for arguing that people get pretty much what they deserve in life. Davis and Moore rightfully responded that they were not trying to justify stratification but merely to explain it. They no more chose to make stratification an inevitable part of society than physicists chose to make

TABLE 9-2	"No One Could Be Expected to Study for Years to Become a Lawyer or Doctor Unless They Expected to Earn a Lot More Than Ordinary Workers."		
NATION	**AGREE (%)**	**NATION**	**AGREE (%)**
Germany	88	**United States**	**70**
Bulgaria	87	Slovenia	68
Austria	86	Norway	65
Australia	80	Russia	65
Poland	78	Philippines	63
Italy	73	**Canada**	**62**
New Zealand	73	Czech Republic	62
Sweden	72	Hungary	60
Great Britain	71		

Source: Prepared by the author from the International Social Survey Program, 1992.

gravity a part of the physical universe. Nevertheless, Davis and Moore's explanation of stratification has been criticized justifiably because it comes dangerously close to being circular (Stinchcombe, 1968). The trouble stems from their inability to define the notion of functional importance adequately.

Davis and Moore based their analysis on the proposition that positions or roles in society differ in functional importance—that is, in their consequences for the continued operation of society. This is vital to their argument that stratification results from the need to motivate the most qualified people to take these positions. Unfortunately, they found it difficult to establish that one position is more important than another except on the basis of how hard it is to fill that position. That is an inadequate standard.

Today, for example, it is very hard to fill the position of housekeeper; to ensure an ample supply would require wages to be set at a level most potential employers are unwilling to pay. Clearly, it does not follow that this position is extremely important. But should we then count it as of low functional importance because we are not willing to pay much to fill the position? To rate the importance of a position on the basis of how much people are paid for filling it leads into a trap. Then one must argue that rewards differ because of functional importance, but that is to say that rewards differ because rewards differ, since functional importance has been defined as a difference in rewards. This is a tautology, or circular argument.

Fortunately, positions can be ranked according to functional importance without leading to contradiction or tautology. A position is of high functional importance to a society depending on its **replaceability**—that is, to the degree that either the position itself or its occupants are hard to replace. Let us return to the example of housekeepers. The salary people are paid to clean other people's houses is low, even when those willing to take such jobs are in short supply, because the position of housekeeper is very replaceable. That is, their employers are fully able to take over the functions of the housekeeper, and when they are sufficiently motivated (by the financial savings involved), they will tend to do so.

A position is highly replaceable when its functions can be performed by people in many other positions. Thus, hospital janitors are highly replaceable because all other hospital workers could perform their job. Orderlies are next most replaceable because doctors and nurses could perform their functions. Doctors are least replaceable because, presumably, not even nurses could fully take on their hospital duties.

People are highly replaceable when little skill is needed to perform their particular roles. Little training or skill is required to mop floors, for example, and thus, people who hold such positions are always potentially in competition with all other workers for their own positions. On the other hand, very few people have the talent and the training needed to be surgeons, and so, few people can compete for these positions. Thus, positional replaceability is the dominant basis of functional importance. People in highly replaceable positions also tend to be individually highly replaceable. Given the replaceability notion, the functionalist theory can easily be viewed as a supply-and-demand argument about the existence of stratification.

A TOY SOCIETY

People who study the operation of systems often employ models to try out different arrangements to see what happens. Sometimes these models are small replicas of the actual system, but often, they are very simplified versions of the system. The latter models omit many components of the original system to study a few features of the system more closely. In a sense such a simplified model is an educational toy, for as George Homans (1974) has aptly put it, such a toy "should be taken lightly like the serious thing it is. . . . At the very least the toy may show how inadequate our assumptions and formulations are."

To help explain stratification systems, we can learn much by closely inspecting a very simplified toy society. Imagine a spaceship on a long voyage that is suddenly forced to crash-land on a tiny asteroid. Four people live through the crash. Let's call them Ay, Bee, Cee, and Dee. To survive, they will need four things: food, air, water, and heat. The survivors differ greatly in their technical skills. As shown in Table 9-3, Ay is able to produce all four critical products; Bee can produce everything but air; Cee can produce food and heat; and Dee can produce only heat.

Clearly, Ay is irreplaceable. If Ay dies or stops making air, then everyone will die. Moreover, if Ay stops *sharing* air with the other three, then they will die. Yet Ay can live without any of the others. To an exaggerated degree, Ay is like a

TABLE 9-3	Abilities of Space Survivors to Provide Needs			
	ABILITY TO PRODUCE			
SURVIVOR	AIR	WATER	FOOD	HEAT
Ay	X	X	X	X
Bee		X	X	X
Cee			X	X
Dee				X

doctor in relation to the rest of a hospital staff—he or she can do all the jobs.

Conversely, consider the situation of Dee. Everyone can get along without Dee, while he or she can't live without aid from the others. Clearly, Ay can make the most favorable exchanges with the other survivors. Ay has a monopoly on air and needs nothing from anyone else. He or she may decide not to bother making heat and let Dee provide heat in exchange for air. But Ay will not trade air for heat if it takes more effort to make air for Dee than for Ay to make his or her own heat. Similarly, Ay may bargain with Bee and Cee, perhaps to get water from Bee and food from Cee, in return for air. As a result of this bargaining process, Ay will always end up with more food, water, and heat than the others have. This is because Ay will supply his or her own food, water, and heat, and refuse to trade air, whenever the terms of exchange do not favor Ay.

Now let's introduce time into this model. Like all humans, these marooned travelers will eventually die. Let's assume, however, that they have children so that our toy society has a future. If the next generation is to survive, then these children must be prepared to take over the vital productive functions. Finding a replacement for Ay is clearly imperative—without an air maker, all will die. But suppose that air making requires a rare talent or a long period of arduous training. Someone with talent must be motivated to become an air maker or to invest the time and effort needed to learn air making. How can this best be ensured? The most efficient way is to make every child desire to be an air maker. Thus, the only person with the potential for air making will not be wasted by becoming a heat maker. The way to do this is to highly reward air making so that people will aspire to this occupation. Indeed, we have already seen that air making will be highly

rewarded as long as it is the least replaceable skill.

This toy society reveals the essential points of the functionalist theory of stratification. We can see how specialization and exchanges between specialists result in stratification. Of course, many aspects of real societies have been omitted from this toy society. Among these are forces that limit the potential exploitation of Cee and Dee by Ay and Bee. These will be discussed in the section on the conflict theory of stratification.

THE SOCIAL EVOLUTION THEORY OF STRATIFICATION

Chapter 4 pointed out that all functionalist theories depend on an evolutionary premise. Indeed, recall that Davis and Moore referred to stratification as "an unconsciously evolved device." No functionalist theory completely explains any phenomenon unless it is connected to an evolutionary theory. Thus, we need to know how social stratification evolves.

The fundamental premise of all social evolutionary theories is that humans will retain that culture which they believe is rewarding. Humans persist in efforts to find ways to gain rewards—to find procedures and implements that will achieve the desired results. Those that don't seem to work will be discarded; those that appear to work best will be preserved. If humans preserve culture, then as time passes culture will accumulate—that is, over time humans will possess a more complex culture (Lenski, 1966, 1976).

As culture becomes more elaborate, something important happens: Any given individual can master less of it. Why? The human mind can learn only a limited amount of information. As culture becomes more complex, it soon exceeds the capacity of any individual to master all of it.

According to the **evolutionary theory of stratification,** when no single individual can master all aspects of a culture, specialization must occur. That is, individuals will master parts of their culture and enter into exchanges with others who have mastered other parts. In this way the accumulation of culture inevitably leads to a division of labor and therefore to stratification.

Some aspects of culture are more valued than others. That is, people desire or need certain things more than they do other things. As people and groups within a society specialize, some will be able to command higher prices for their goods and services than others can. Stratification means the existence of such inequalities.

Therefore, the accumulation of culture, because it results in cultural specialization, also results in stratification. The notion of replaceability is pertinent in determining the relative advantage of one specialty over another. Indeed, at this point a social evolutionary theory must draw on functionalism to assess how specialties will be valued in the stratification system. Furthermore, as we shall now see, a full understanding of stratification must also incorporate the important insights of conflict theory.

THE CONFLICT THEORY OF STRATIFICATION

Modern conflict theory provides a needed corrective to functionalist theory. Like the functionalists, most conflict theorists accept that stratification is unavoidable (Ossowski, 1963; Lenski, 1966; Harris, 1979). But the **conflict theory of stratification** adds several key insights about how stratification systems are subject to distortion. The first of these concerns how persons high in the stratification system will take advantage of their position to exploit others, a phenomenon that makes societies more stratified than can be accounted for by functionalist theory alone. The second of these insights concerns how the political process can be used to influence the stratification system, particularly by limiting replaceability.

EXPLOITATION

Modern conflict theories build on Mosca's third proposition: Humans pursue their own self-interests (Dahrendorf, 1968). It follows that people in a position to exploit others will tend to do so. Thus, if societies must be stratified, those on top will use their position to increase their rewards. As a result societies are always more stratified than they need to be (Lenski, 1966).

Let's reexamine our toy society. Ay is functionally irreplaceable, being the only air maker. While Ay alone can provide others with air, he or she can do without the others because Ay can also make water, food, and heat. According to the functionalist theory, inequality occurs in this toy society because it is vital to reward Ay highly

to ensure that someone will always be able and willing to make air. But conflict theorists point out that Ay is also a monopolist and therefore has immense power. Ay can set a very high price for air, limited only by the fact that, if the price is too high, the others will die for lack of food, water, or heat (having given too much to Ay) and therefore will be unable to pay Ay anything at all for air. It follows that Ay can set a higher price on air than would be needed simply to motivate air making. In fact, Ay can get more for air than the minimum price at which he or she would be willing to make air. The difference between the price that Ay actually gets and the minimum price at which Ay would sell air is **exploitation.**

When we look around at real societies, we frequently see examples of such exploitation. The OPEC nations, for example, banded together to monopolize oil supplies to inflate the price of oil far beyond the level needed to motivate people to drill, refine, and sell oil. Similarly, the earnings of physicians would probably have to decline substantially before the number and qualifications of students seeking entry to medical school would also fall. The functionalist theory can explain why doctors earn more than orderlies do, but it requires conflict theory to explain why the actual income gap between these two occupations is as large as it is.

THE POLITICS OF REPLACEABILITY

Looking once more at our toy society, we can discover another major omission. A monopoly like Ay's is unlikely to occur naturally. Such monopolies usually cannot exist unless power is being exerted to prevent others from competing. Members of the medieval nobility, for example, were able to monopolize the ownership of all land and therefore set the rents that peasants had to pay only because the nobility also monopolized military force. Their monopoly of military force did not occur because they were the only able-bodied men in medieval societies or because they were so much stronger and braver than all other men. It arose because they were able to control the most advanced military technology—armor, war horses, and weapons—against which other men were helpless. The nobility prevented other men from receiving military training. Indeed, when the crossbow was invented, it was outlawed because it enabled people with little training and no expensive equipment to attack the knights. Thus, coercion was the basis of the knighthood's monopoly. And coercion underlies most (if not all) forms of monopoly.

The implications of this fact are important because they suggest that the principle of replaceability, so central to the functionalist theory, is also subject to distortion—that power may be used to control and manipulate replaceability artificially. This means that political power may be used to exaggerate or minimize the degree of stratification in societies. To examine this possibility, consider two examples of how groups can use power to decrease a position's replaceability artificially: professions and unions.

Professions first attempt to establish their positions as irreplaceable and then make their members irreplaceable in the position (Freidson, 1973). The claim to irreplaceability is based on some form of expertise that no other position can perform. This tendency to monopoly is a natural outgrowth of the proficiency a position develops (or is thought to have developed).

Consider the medical profession. In the nineteenth century, medical doctors were not very proficient. Not until about the turn of the century, by some estimates, were people better off going to a doctor than not going. Most treatments and medicines used in the nineteenth century were useless, and some were quite harmful. During this period of limited proficiency, the position of doctor faced serious competition from other positions in performing the function of healing. Other healing practitioners such as osteopaths, homeopaths, food faddists, faith healers, patent medicine sellers, and the like claimed to be able to perform the needed medical function as well as or better than medical doctors could. Eventually, physicians could demonstrate that their ability to heal was greater than that of people in these other positions. Demands for consumer protection against health frauds intensified, and in the end medical doctors were given a monopoly on the right to prescribe drugs and perform surgery (later the osteopaths joined with the medical doctors).

Once they have a monopoly on expertise and are thus highly irreplaceable, professions tend to seek control over who can fill their positions. In part this is because only they possess the essential knowledge and skills of the position, so they alone can pass them on to others. This, of course, makes it possible for professions to control who and how many enter the professions—and thus to control supply and increase their rewards beyond what successful recruitment would require. That is, doctors can limit the supply of doctors. They can also limit the freedom of nondoctors to perform medical tasks. To the extent that they are successful in limiting supply and preventing competition, doctors increase their irreplaceability, and like Ay, they can force others to pay higher prices for their services.

Unlike the medieval knighthood, however, professions such as medicine do not possess the means to force other people to do as they are told. Therefore, the monopolies created by professions are always subject to external checks. The government, which has a monopoly on the means of force, may refuse to protect a medical monopoly on prescriptions, for example. Or political power may be used to standardize or reduce fees charged by a profession or to force an increase in the supply of trained members admitted to the profession. If the rewards for being a doctor get too low, then the quality and supply of physicians will decline. But if the rewards rise too high, then they may be reduced by political decisions that the profession cannot control.

Unions are also efforts to create monopolies—to use power to decrease replaceability. To see how unions do this, let us return to our toy society. Suppose our simple four-person society were greatly enlarged and that there were many more Cees and Dees than Ays and Bees. Let us also assume a democratic political process so that the wishes of the Cees and Dees count. It is easy to see that these "lower-class" persons might use their collective political power (or even coercive force) to strike better bargains with the Ays and Bees. Again, the key to an improved bargaining position is making positions less replaceable.

The standard union tactic is to use contracts and laws to prevent other positions from performing the main function of their own position. That is, if Dee can prevent the other three members of our four-person society from producing heat, despite the fact that they could produce it, then his or her position is equal to Ay's.

Craft union rules have precisely this effect. In construction work, for example, union contracts clearly specify what tasks belong to which position, and other positions are barred from performing such tasks. In extreme instances a crew of carpenters and a crew of plasterers may wait all day for an electrician to come and turn on a switch that any normal adult could have managed. In this way the actual high replaceability of a position is converted into an artificial state of very low replaceability.

As with professions, the second goal of unions is to control the conditions for entering positions. This is accomplished by mandatory union membership—the closed shop—and by entrance and apprenticeship procedures governing membership.

As we shall see in Chapter 10, unionization has been a major force in reducing the degree of stratification in industrialized democracies. For example, the difference in average pay between blue-collar and white-collar workers has been drastically reduced. Thus, the politics of replaceability work not only for business monopolies and the professions but also for factory workers. Indeed, the business monopoly has been subjected to the strictest regulation.

CONCLUSION

In Chapter 4 I suggested that although there are a number of distinctive theoretical schools in macrosociology, to me they seem quite compatible. Here we have seen that each of these theoretical approaches explains something important about stratification that the others leave unexplained. Functionalist theory explains why various occupational specialties are differentially rewarded—because some workers are much less replaceable than others. Evolutionary theory explains how specialization arises in societies and thus supplements the functionalist theory. Conflict theory adds a vital point—that power will be exploited to increase or decrease stratification, and thus, stratification will reflect the outcome of conflict among groups in a society.

Because the aim of this chapter is to present the conceptual and theoretical tools sociologists use to analyze stratification, it has necessarily been quite abstract. To apply this material to concrete issues, Chapter 10 examines how stratification systems operate in a variety of human societies.

Review Glossary

Terms are listed in the order in which they appear in the chapter.

means of production Everything, except human labor, that is used to produce wealth.

bourgeoisie Marx's name for the class made up of those who own the means of production; the employer or owner class.

proletariat Marx's name for the class made up of those who do not own the means of production; the employee or working class.

lumpenproletariat Literally, the "ragamuffin proletariat"; the people on the very bottom of society, whom Marx labeled "social scum."

class consciousness The concept Marx used to identify the awareness of members of a class of their class interests and enemies.

false consciousness A term that Marx applied to members of one class who think they have common interests with members of another class.

subjective class The class to which a person thinks he or she belongs.

property The term many sociologists use to identify what Weber called *class*. Property includes all economic resources and opportunities owned or controlled by an individual or a group.

prestige Social honor or respect; synonymous with Weber's term *status*.

power The ability to get one's way despite the opposition of others; synonymous with Weber's term *power*.

status inconsistency A condition in which a person holds a higher position (or status) on one dimension of stratification than on another. For example, an uneducated millionaire displays status inconsistency.

status inconsistency theories Theories built on the proposition that persons who experience status inconsistency will be frustrated and will therefore support political movements aimed at changing the stratification system.

structural mobility Mobility that occurs because of changes in the relative distribution of upper and lower statuses in a society.

exchange mobility Mobility that occurs because some people fall in the stratification system, thereby making room for others to rise.

cultural capital Assets based on knowledge, style, speech, tastes, and the like, which can be used to "purchase" privileges and power.

utopian One who tries to design a perfect society.

anarchists Followers of a political philosophy that regards the state as inevitably repressive and unjust and who therefore propose to destroy the state and live without laws or government.

functionalist theory of stratification A theory that holds that inequality is built into the roles of any society because some roles are more important and harder to fill, and to ensure that the most qualified people will seek to fill the most important positions, it is necessary to reward these positions more highly than others.

replaceability A measure of the functional importance of a role based on the extent to which other roles can substitute for or take on the duties of that particular role. For example, a doctor can easily substitute for an orderly, but the reverse is not so.

evolutionary theory of stratification A theory that holds that because culture accumulates in human societies, eventually it happens that no one can master the whole of a group's culture. At that point cultural specialization, or a division of labor, occurs. Since some specialties will be more valued than others, inequality, or stratification, will exist.

conflict theory of stratification A theory that holds that individuals and groups will always exploit their positions in an effort to gain a larger share of the rewards in a society, and therefore, societies will often be much more stratified than functionalism can explain. Put another way, this theory holds that the stratification system of any society is the result of conflicts and compromises between contending groups.

exploitation All profit in an exchange in excess of the minimum amount needed to cause an exchange to occur.

professions Occupational organizations that can prevent their functions from being performed by those not certified as adequately trained and qualified in an extensive body of knowledge and technique.

unions Occupational organizations that can prevent their functions from being performed by others on the basis of contractual rights.

Suggested Readings

Bendix, Reinhard, and Seymour Martin Lipset, eds. 1966. *Class, Status, and Power.* 2nd ed. New York: Free Press.

Dahrendorf, Ralf. 1959. *Class and Class Conflict in Industrial Society.* Palo Alto, Calif.: Stanford University Press.

Freidson, Eliot, ed. 1973. *The Professions and Their Prospects.* Beverly Hills, Calif.: Sage.

Lenski, Gerhard. 1966. *Power and Privilege.* New York: McGraw-Hill.

Sociology Online

www.socstark10.com

GO TO THE INTERNET AND TYPE: www.socstark10.com.

↗**CLICK ON:** 2000 General Social Survey

✓ **SELECT:** BOOKS@16 About how many books were there around your family's house when you were 16 years old?

↗**CLICK ON:** Analyze Now
Recall Bourdieu's discussion of cultural capital.

TRUE OR FALSE: Being raised in a home with many books is an example of inherited cultural capital.

TRUE OR FALSE: People who earn high incomes are more likely than those who earn less to have been raised in a home having many books.

↗**CLICK ON:** Select New Data Set

↗**CLICK ON:** The 50 States

✓ **SELECT:** NO PHONE: Percentage of occupied housing units with no telephone.

↗**CLICK ON:** Analyze Now
Which state has the most phone-less homes? Which has the least? Why do you suppose this difference exists?

↗**CLICK ON:** Select New Data Set

↗**CLICK ON:** Nations of the Globe

✓ **SELECT:** WORKER OWN: Percentage who believe business and industries should be owned by the employees who elect their managers.

↗**CLICK ON: ANALYZE NOW**
Which famous social theorist would have been most likely to agree that business and industry ought to be owned by the employees? In what two nations do more than half of the public currently agree? Can you notice anything these two nations have in common that might help explain this?

USING INFOTRAC COLLEGE EDITION

GO TO THE INTERNET AND TYPE:
www.InfoTrac-college.com

↗**CLICK ON:** Register New Account
You will be asked to enter your Access Code and to create and enter a User Name and Password. Having done so,

↗**CLICK ON:** InfoTrac College Edition or Log On.

Every year, much is written about class and status, so you can find articles that expand on topics covered only briefly in the chapter. You may wish to identify the theory of stratification used by various authors.

SEARCH TERMS:
Social Class
Social Status

In addition, I have selected a specific article that will usefully supplement the chapter: *Five and a Half Utopias*. You can find it by searching for this title. You may read the article on the screen or print it using the usual print commands. If you also go to www.socstark10.com and click on the InfoTrac College Edition icon, you can read an explanation of why I selected this article and find several questions that will help you connect the article to material in this chapter.

CHAPTER

10

Comparing Systems of Stratification

I N AN OLD COMEDY ROUTINE, one man asks another, "How's life?" "Compared to what?" is the reply. This piece of foolery reveals the basis of all scientific investigations. Virtually every important question sociologists ask requires comparisons.

Suppose we ask, "How stratified is Mexico?" The first reply must be, "Compared to what?" To answer that question requires us either to compare Mexico as it was at several points in its history (thus permitting an answer such as "less stratified than it used to be") or to compare several societies (thus permitting an answer such as "less stratified than India is").

This chapter examines stratification systems by comparing different kinds of societies. In this way the concepts and theories of stratification covered in the previous chapter can be applied and illustrated.

We shall examine stratification and social mobility in three major types of societies. First, we shall examine the simplest form of human society: the small band of hunter-gatherers. Next, we shall examine the more complex agrarian societies and see why they are so much more stratified than simple societies. Finally, we shall examine stratification in modern industrial societies, focusing on the United States and Canada.

The way societies are ordered in this chapter—from the least to the most modern—implies social evolution. It is true that agrarian societies developed out of simple hunting and gathering societies and thus appeared later. It is also true that industrial societies grew out of earlier agrarian societies. However, such an evolutionary process is neither inevitable nor uniform. Many, probably most, simple societies did not evolve into agrarian societies; indeed, hunting and gathering societies still exist. Most

CHAPTER OUTLINE

SIMPLE SOCIETIES

AGRARIAN SOCIETIES
Productivity
Warfare
Surplus and Stratification
Military Domination
Culture and Ascriptive Status
Box 10-1 Stirrups, Saddles, and Feudal
Domination

INDUSTRIAL SOCIETIES
Industrialization and Stratification
Social Mobility in Industrialized Nations

A CLOSER VIEW
Lipset and Bendix: Comparative
Social Mobility

A CLOSER VIEW
Blau and Duncan: The Status
Attainment Model

A CLOSER VIEW
Biblartz, Raftery, and Bucur:
Mothers and Mobility

A CLOSER VIEW
John Porter and Colleagues: Status
Attainment in Canada
Female Status Attainment

A CLOSER VIEW
Michael Hout: Trends in Status
Attainment

A CLOSER VIEW
Bonnie H. Erickson: Networks
and Power in Canada

A CLOSER VIEW
Yanjie Bian: Networks and Success
in China

CONCLUSION

REVIEW GLOSSARY

SUGGESTED READINGS

SOCIOLOGY ONLINE

agrarian societies did not become industrialized; indeed, some began to industrialize and then lapsed back into agrarianism (Chirot, 1976; Tainter, 1988). In Part V we shall examine social change and consider why and how societies evolve or fail to do so. For now simply keep in mind that the concept of social evolution is surrounded by uncertainties.

SIMPLE SOCIETIES

Societies covering large areas and containing large populations are relatively recent. For most of human history, societies were very small, usually having only about fifty members (Murdock, 1949). Indeed, the largest simple societies rarely exceeded several hundred members and then only in extremely favorable environments (Kroeber, 1925). These simple societies often wandered over large areas, but the territory they inhabited at any given moment was small.

There were several kinds of simple societies if we classify them by how they made their living. (I shall use the past tense in describing these societies, but keep in mind that some still exist.) The majority were **hunting and gathering societies.** Rather than living in a fixed spot, they moved in search of game and edible plants. Slightly more advanced simple societies lived by herding animals. They, too, moved about as their animals required new grazing areas. Other simple societies mastered elementary gardening and thus tended to be less nomadic than the herders or the hunter-gatherers were. Nevertheless, they tended to stay in one spot just long enough to grow one crop and then moved on. Only about 10 percent of simple societies managed to live in a fixed location (Lenski, 1966).

Both the herding and the gardening societies were wealthier than the hunter-gatherers and had a slightly more complex division of labor. Therefore, they also tended to be more stratified. However, because hunter-gatherer societies represent the most primitive level of human existence, they reveal the most elementary forms of stratification. For this reason we shall look at them closely.

The major fact of life in hunting and gathering societies was the threat of death. These societies had few members because they could not support more. Traveling on foot, they could not cover much distance in a day. Hence, they could feed only the number of members who

Neg #17023, photo: R.M. Anderston, American Museum of Natural History

could survive on the food found within so restricted a range.

They were not very deadly hunters. Most of them probably lacked weapons even as primitive as the bow and arrow and relied instead on stones, clubs, spears, and traps. In part this was because they didn't produce enough to support specialists who could develop better weapons (Lenski, 1966). Furthermore, edible vegetation is usually seasonal and in most places not abundant. Because they could not preserve food very well, if at all, they experienced chronic famines and only occasional feasts. Even when meat was abundant, it spoiled quickly, leaving the group again without food. Colin Turnbull (1965) has reported that Pygmies in the dense jungle of the Congo deplete an area of fruit and game (scaring away more than they catch) within about a month and then move on.

These simple societies frequently lost the battle to survive. This is evident when we consider how slowly the population of humans on earth grew prior to the development of agrarian societies. Recent archeological discoveries in Africa indicate that humans existed more than 3 million years ago. Yet only 10,000 years ago, after at least 3 million years of reproduction, it is estimated that there were fewer people on earth than live in New York City today (Davis,

1971; Harris, 1979). Ten thousand years ago, the human population was still growing so slowly that it would have taken another 60,000 years to double in size. Today, the world's population is doubling about every 37 years, and our problem is not too few people but too many. Population growth was the result of increased food production, which widened the margin of survival enough to ensure against constant crises and an appalling death rate. For most of our time on earth, however, the constant problem was simply to survive.

Under such circumstances human societies were not very stratified in terms of property (Harris, 1979). Possessions were limited to what could be carried from place to place. Because there was never much more than enough wealth to go around, the wealthiest could accumulate little more than could the poorest. Furthermore, with almost no role specialization—no full-time leaders, for example—power was also greatly equalized (Fried, 1967; Harris, 1979). Indeed, the ability to coerce others was quite limited and thus provided little basis for power. If, for example, some hunters began forcing others to give them their game, then there was little to prevent those being exploited from deserting the band. Nor could the strongest hope to live merely by taking from others because

Inuit on the move to a new campsite along Coronation Gulf in northern Canada during the winter of 1911. All hunting and gathering societies are very small, but environmental conditions caused Inuit societies to be tiny. Danish explorer Knud Rasmussen explained: "It took a large stretch of ground to provide the single individual with the necessities of life; the fewer the hunters the better were the chances, so they migrated along the coasts in little flocks."

© Lorne Resnick/Stone/Getty Images

Nomadic societies have very few possessions because of the constant need to move on. If you cannot own more than you can take with you, you will never have very much.

unless everyone worked hard at getting food, survival was jeopardized for all. If you cannot produce more than enough for yourself, then you cannot be exploited easily.

However, hunting and gathering societies were nevertheless stratified to a significant degree. Usually, the primary bases of stratification were age and sex. Adults held power over children, and men dominated women.

However, *within* age and sex groups, simple societies were not very stratified. Men were essentially equal in terms of possessions and political power, as were women. Stratification within each sex was based mainly on achievement. Some men dominated others because they were bigger or smarter, better hunters or braver

fighters, or more persuasive in group decision making. These distinctions showed up as differences in prestige and power rather than in material inequalities. Similarly, among women, prestige differences on the basis of skill at domestic tasks, wisdom, persuasiveness, fertility, and perhaps physical appearance undoubtedly existed.

The rule seems to be that the smaller, poorer, and less secure a human society is, the less it is stratified. Universal poverty was the basis of equality among hunters and gatherers. But as soon as humans became more productive and better organized, their societies became more unequal. Herding societies and simple horticultural (gardening) societies were more stratified than were hunter-gatherer societies. When humans finally settled in one place and began to grow crops, inequality increased enormously (Cohen, 1977; Harris, 1979). Suddenly, gaps in status appeared as great as those between slaves laboring in a field and emperors living opulently in palaces.

Table 10-1 shows the immense impact of settlement on stratification. The data are based on the 186 societies commonly referred to as the Standard Cross-Cultural Sample (see Chapter 4). No industrialized societies are included, and pains have been taken to make sure that all variables for any given society are from the same time period. As the data show, nomadic societies are not very stratified: Only 10 percent are

TABLE 10-1	Fixity of Residence and Stratification		
DEGREE OF STRATIFICATION	**FIXITY OF RESIDENCE**		
	NOMADIC	**SEDENTARY**	**PERMANENT**
Low	89.8	74.3	46.1
Medium	10.2	25.7	24.5
High	0.0	0.0	29.4
	100.0%	100.0%	100.0%
N =	(49)	(35)	(102)

Source: Prepared by the author from the Standard Cross-Cultural Sample.

classified as having a medium level of stratification. Unlike nomads, sedentary societies are not constantly on the move, but neither do they stay in one place. Some rotate seasonally through a regular set of fixed locations. Some create semipermanent settlements only to be forced frequently to leave them for ecological or ritual reasons (the death of a headman, for example). Sedentary societies also aren't apt to be highly stratified. But when societies have a permanent location, many of them develop medium and high levels of stratification.

AGRARIAN SOCIETIES

The development of agriculture changed the world (Pfeiffer, 1977). No longer did humans wander the earth eking out a hand-to-mouth existence. They were able to settle in permanent locations and therefore could construct better shelters and accumulate possessions. More important, life ceased to be a constant struggle to find food.

As time passed, the technology of agriculture improved. With each improvement human society became more complex. The basis for this growing complexity was *surplus food production*. When a family could produce enough food to feed others as well as themselves, two new social phenomena became possible: *specialization* and *urbanization*.

As long as everyone had to take part in the quest for food, no one could devote full time to other pursuits. Thus, during most of human existence, there were no full-time priests, political leaders, or other specialists. But when farmers could produce some surplus, others could be supported who made their shoes, forged their tools, conducted their religious ceremonies, and guided their political affairs. Furthermore, people not needed for food production could congregate in a central place and thus create cities.

PRODUCTIVITY

With the invention of plows and effective animal harnesses, agricultural productivity became so great that some people were freed from farming. At that point **agrarian societies** appeared, some of which became great empires and perhaps the first real human civilizations. However, the number of persons who didn't farm, even in the most advanced agrarian societies, was small. Gideon Sjoberg (1960) estimates that no agrarian society had fewer than 90 percent of its members engaged in farming, and usually 95 percent were needed. Still, even by freeing one in twenty from farming, people could pursue a great variety of specialized tasks and produce an elaborate social structure.

It is important to realize that the famous cities of historical agrarian societies were not large by modern standards (see Chapter 19). In fact, even by 1800, when industrialization was already well under way, fewer than fifty cities in the world had more than 100,000 residents. Nevertheless, they were cities, not farming villages, because their residents did not grow their own food. Until agriculture became relatively efficient, there were no cities.

WARFARE

For a long time, it was believed that hunter-gatherers seldom had wars and that when they did, the "fighting" was mostly for show— sometimes opponents placed more emphasis on "counting coup" (touching an enemy in battle) than on killing (Boas, 1897; Blick, 1988; Ferguson, 1984, 1992). I repeated these claims about hunter-gatherer conflicts in the first eight editions of this textbook, but it turns out not to be so. The truth is that hunter-gatherers frequently engaged in very bloody warfare and often tried to kill everyone on the other side, including women and children (Bamforth, 1994; Keeley, 1996).

Art Resource, NY

Even very primitive agriculture made human societies incredibly richer than before. Here we see people in Senegal grinding wheat in a way similar to that used thousands of years ago by the earliest agrarians. In addition to giving them a reliable food supply, agriculture makes it possible for people to have the large houses that can be seen in the background and a great many possessions beyond the means of hunting and gathering societies.

With the rise of agrarian societies, things got even worse. Warfare was (and is) chronic in agrarian societies, partly because they offer an abundance of loot. Thus, in a historical survey of eleven European nations during their centuries as agrarian societies, Pitirim Sorokin (1937) computed that they had been at war 46 percent of the time. Gerhard Lenski (1966) reports that Sorokin's figures are probably too low because he ignored many minor conflicts. Marc Bloch (1962) characterized agrarian European history as "the state of perpetual war."

Table 10-2 lets us take a more global view of agricultural development and external warfare. Marc Howard Ross (1983) coded levels of external warfare for eighty-four of the societies included in the Standard Cross-Cultural Sample. Societies have been rated as constantly at war if they are at war at least part of every year.

Warfare is scored as common if a society gets into a war at least every five years. Societies having wars less often were rated as having occasional warfare. More than 80 percent of the fully agrarian societies have a war at least once every five years, and two-thirds of them fight every year. One-third of the societies lacking agriculture are constantly at war, and nearly half of them fight occasionally. Constant warfare contributes greatly to the extreme stratification of agrarian societies.

SURPLUS AND STRATIFICATION

In an agrarian society, we can first apply Marx's notions about the ownership of the means of production. In simple societies the means of production were primarily physical and therefore personal. But when the means of production are material things, they can be conceived of as property. Fields are different from land that is simply out there to travel over in search of game; fields must be cleared, planted, cultivated, and harvested. Thus, the question arises, Whose field? The answer to that question determines who is rich and who is not. In addition to land, many other forms of potential property existed, such as farming tools, livestock, seed, permanent houses, and the food stored from the last harvest. Because wealth was substantial, some people became much wealthier than others (Harris, 1979).

The increased wealth of agrarian societies combined with (and permitted) a complex division of labor that produced even greater stratification. We have seen that it is hard for one hunter to be much wealthier than another. But as Chapter 9 demonstrated, persons in a particular occupation can have bargaining advantages over persons in other occupations. Because durable wealth existed in agrarian societies, bargaining advantages could be stored up to produce increasingly greater inequalities.

Indeed, because human labor in agrarian societies was so productive, other humans could be exploited. When people can barely produce enough to feed and clothe themselves, others cannot live off their labor. But in agrarian societies a few could live off the labor of many. This capacity for labor to produce surplus—that is, for peasants to grow more food than they required to live—was the basis for the marked inequalities found in such societies.

The ability to produce a surplus raises the possibility of *humans becoming property*. When labor produces little or no surplus, there is no profit in owning another person. But when labor produces surplus, then by owning another person, one can possess all of the surplus that the other person can produce. In fact, when people are sufficiently productive, they can support those whose task is to keep them enslaved. With the emergence of slavery, the human being became a means of production (Patterson, 1982).

The production of surplus also makes government possible. Full-time specialists who coordinate the activities of societies can be supported. Thus, at the apex of every agrarian society was a ruling elite whose members "neither spin nor reap" but who lived by extracting surplus production in the form of taxes and tributes from the peasants.

Table 10-3 presents the strong link between agricultural productivity and stratification. No societies lacking agriculture have a high degree of stratification, while nearly half of the fully agrarian societies are highly stratified. Conversely, nearly 90 percent of societies that have not yet developed agriculture are rated as having a low degree of stratification as compared with about a third of the most agrarian societies.

MILITARY DOMINATION

In agrarian societies the elite held power by dominating the military force. The production of surplus also made this possible by enabling some members of society to become soldiers and to specialize in developing and using military technology. As these specialists began to appear, a gap opened between the military capacity of the average man and that of the specialist soldier. Marc Bloch (1962) pointed out that as long as the average male citizen is a potential soldier who lacks nothing "essential in his equipment," he can resist being exploited. In simple societies the weapons of war were much like those used for hunting, and all able-bodied adult males were therefore armed and capable of battle.

But with surplus production the means of warfare became specialized and differed from hunting weapons. Soon a special group of men existed who had been trained from childhood in the tactics and techniques of fighting and who possessed superior weapons and equipment for war. The medieval peasant, armed with a hayhook, an axe, or a knife, was simply no match for the heavily armored knight. A great deal of wealth was needed to equip one knight with armor, lance, long sword, and an adequate horse.

TABLE 10-2	Agricultural Development and Warfare			
	LEVEL OF AGRICULTURAL DEVELOPMENT			
FREQUENCY OF EXTERNAL WARS	NONE	LOW	MEDIUM	HIGH
Constant	33.3	57.1	56.0	66.6
Common	23.8	7.1	12.0	16.7
Occasional	42.9	35.8	32.0	16.7
	100.0%	100.0%	100.0%	100.0%
N =	(21)	(14)	(25)	(24)

Source: Prepared by the author from the Standard Cross-Cultural Sample and from Ross (1983).

TABLE 10-3	Agricultural Productivity and Stratification			
	LEVEL OF AGRICULTURAL PRODUCTIVITY			
DEGREE OF STRATIFICATION	NONE	LOW	MEDIUM	HIGH
Low	86.8	78.5	63.5	38.6
Medium	13.2	17.9	31.7	15.8
High	0.0	3.6	4.8	45.6
	100.0%	100.0%	100.0%	100.0%
N =	(38)	(28)	(63)	(57)

Source: Prepared by the author from the Standard Cross-Cultural Sample.

Thus, surplus provides the conditions necessary for a ruling class to monopolize military capacity (Lenski, 1966).

Because agrarian societies were based on military rule, they also tended to be expansionist. Armies are expensive to maintain, even when they are not fighting. Rulers therefore often paid for their armies by having them seize or plunder neighboring people—hence the chronic warfare found in such societies. Moreover, chronic warfare also centralized power within the ruling elite, who surrendered power to a king or an emperor (Lenski, 1966).

CULTURE AND ASCRIPTIVE STATUS

Often, ruling elites did not evolve within agrarian societies but were imposed from without. As the Mongols did in China, frequently a group of nomadic raiders, who lived by repeatedly plundering a farming region, would decide to settle down and become permanent exploiters of the region by ruling over it and gathering taxes. For this reason ruling elites in agrarian societies

BOX 10-1   STIRRUPS, SADDLES, AND FEUDAL DOMINATION

STRATIFICATION IN AGRARIAN SOCIETIES arose from the ability of the elite to monopolize military capabilities. When the weapons of war are like those for hunting, and when the average man has both the training and the essential equipment for going to war, the ability of the elite to coerce him is very limited. The immense inequalities of agrarian societies arose as a group of specialists in waging war emerged. In medieval European societies, this military monopoly developed as the unexpected consequences of two seemingly small innovations—stirrups and the Norman saddle.

After the fall of the Roman Empire, western Europe was a collection of tribal kingdoms that were not very stratified. One such tribe was the Franks, whose territory included most of modern France. Like the other tribal kingdoms, Frankish society was composed largely of free and independent farmers. When war came, all able-bodied men took their shields, swords, and spears and formed a huge host of infantry. They wore no armor and depended on their number and high morale to gain victory. They were part-time farmer-soldiers. As such, they could resist not only invaders but also their king.

Then came stirrups and the Norman saddle. Prior to this time, mounted soldiers could only use the strength of their arms to deliver a blow because any attempt to charge head-on and drive home a long lance would merely vault the rider off of his horse. This was because they had no stirrups to brace against, and they rode with light, almost flat, pad saddles or even bareback. Therefore, cavalry was little used. The Romans, for example, used mounted soldiers only for scouting duties and to

pursue fleeing enemies. But with stirrups to brace against and the support of a saddle with a very high pommel and cantle (the latter being curved to partly encase the rider's hips), it became possible for cavalry to charge at a full gallop and to put the weight of horse and rider behind their lances.

As its name suggests, the Norman saddle was developed in Normandy and spread across western Europe during the seventh and eighth centuries (Hyland, 1994). The stirrup reached Europe from China at about this same time (White, 1962).

Soon a few mounted Franks began to show up for battle armed with long lances. As these mounted tactics evolved, defenses also evolved. For protection against the lances, riders began to wear heavier armor. This required them to ride very big, strong horses. Soon the knight in armor had arrived, but only in limited numbers.

During the seventh century, a few of the richest Franks began to report for military duty wearing armor and mounted on horses. The rest came on foot as always. Possibly the first charge by knights took place in a battle between the Franks and the Saxons in 626 (Montgomery, 1968). But another century passed before the real military superiority of the knights was recognized. In 733 Charles Martel led the Frankish host against the invading Saracens, Muslims who had conquered Spain and sought to push their rule north. At the Battle of Tours, the Franks routed the Saracens and pushed them back across the mountains into Spain. The turning point in the battle came when the Saracen ranks collapsed under the weight of a thundering charge by mounted knights. Lacking stirrups, the mounted Saracens could not withstand the lance attack. Victory convinced Charles Martel that knights held the key to military supremacy. He began to transform the Frankish host from an army of foot soldiers to an army of armored cavalry.

This ended the importance of farmer-soldiers in Frankish military operations. They could not afford to arm as knights. History records that in 761 a Frank named Isanhard "sold his ancestral lands and a slave for a horse and a sword" (White, 1962). To buy weapons and a suit of armor in the eighth century cost the equivalent of twenty oxen, enough for ten plow teams, at a time when only well-to-do farmers owned such a team. Moreover, a knight needed an unusually big, fast, well-trained horse—a horse, as Field Marshal Viscount Montgomery (1968) put it,

Here Frankish knights, braced in their new stirrups and Norman saddles and wearing protective chain-mail armor, sweep over Saracen horsemen, as shown in the stained-glass choir windows of St. Denis Abbey. The Saracen rider on the right has no stirrups; his sword and shield are useless against the long lance of the Frankish knight.

Giraudon/Art Resource, NY

strong enough to carry him when fully-armed, sufficiently trained not to bolt or panic in battles, and fast enough to take part in a charge at full gallop. Such a horse had to be specially bred and trained.

The famous breeds of huge horses we see pulling beer wagons in ads today were not developed to pull loads but to carry knights wearing armor so heavy that they had to be hoisted up by a derrick and lowered onto the saddle. Moreover, a knight could not go off to war with only one horse; he needed spare mounts. He also needed someone to hold his spare mounts during the battle, to help him get into his armor, to hoist him onto his horse, and to transport his supplies. Thus, each knight needed a retinue of aides and servants. These people also had to have mounts and some weapons. Furthermore, horses had to be fed grain, not just grass. To feed one war-horse for a year required the grain crop of several peasant farms. Finally, to fight as a mounted knight took years of training. Knights had to be free to dedicate their lives to military training from early childhood, freedom only a few could have (White, 1962).

To have an army of knights required the transformation of Frankish society. Few could afford to be knights, and even a king could not afford to equip and support many. But the peasants could. So each knight was given title to a tract of land and the authority to tax all who lived on it in return for his service as a knight when called upon. This is the political system known as *feudalism*.

In feudal societies land ownership is based on military obligations. The ruler grants title to large areas in return for the fulfillment of a military quota. A great lord, for example, might have been obligated to provide several hundred knights when called upon. To do so, he assigned portions of his estate to lesser nobles in return for promises to provide some portion of this force. They in turn could assign land to others for a pledge of service. The result was that the great mass of society was taxed to support a relatively small number of knights who ruled the people. The people could be greatly exploited because they did not possess the essential equipment and training necessary to resist the knights. With the rise of the knight, the average Frank became a heavily taxed peasant, subject to coercion, in a command economy.

Knights held their control of land as a hereditary right, but only so long as they fulfilled their military obligations, for knighthood was not a ceremonial title. War was chronic, and woe unto him who failed to fulfill his obligations to his lord when ordered to duty. Excerpts from a summons sent by Charlemagne, the great emperor of the Franks, to his nobles in 806 suggest how Frankish society had changed since the days of the infantry host and hint at the seriousness of the feudal system of obligations:

> You shall come to Stasfurt on the Boda, by May 20th, with your "men" prepared to go on warlike service to any part of our realm that we may point out; that is, you shall come with arms and gear and all warlike equipment of clothing and victuals. Every horseman shall have shield, lance, sword, dagger, a bow and a quiver. On your carts you shall have ready spades, axes, picks and iron-pointed stakes, and all other things needed for the host. The rations shall be for three months. . . . On your way you shall do no damage to our subjects, and touch nothing but water, wood and grass. . . . See that there be no neglect, as you prize our good grace. (quoted in Montgomery, 1968)

Of course, as the Franks used their knights to extend their rule over new territory, their neighbors soon followed their example. For centuries feudalism held all Europe in thrall.

© Bettmann/Corbis

As is correctly shown in this drawing of Roman cavalry, they had no stirrups or adequate saddles. Mounted soldiers like these could only strike with the strength of their arms. Had they tried to charge with a lance, they would have been vaulted off their horses.

Scala/Art Resource, NY

The productivity gains made possible by the invention of agriculture had the ironic consequence of making possible extreme levels of exploitation, including slavery. Egyptian slaves, like these depicted in a Pharaoh's tomb, could be held in bondage because they could produce an economic surplus—a surplus that made it possible and worthwhile to enslave them.

were often of a different ethnic background and occasionally spoke a different language from that of their subjects. Thus, cultural differences reinforced stratification: Those on the bottom had to cross a cultural as well as an economic barrier to rise in the system.

Moreover, even if the ruling elite was not a different ethnic group, cultural differences between members of the elite and their subjects soon appeared. A major difference between any elite and the masses is leisure. The rise of civilization—the development of an elaborate culture—is rooted in the existence of a leisure class (Veblen, 1899).

People who tilled the fields or dug in mines from dawn until dusk had little time to study the heavens, compose poems, pursue theology, or invent new tools. Such activities require spare time, which can be provided only by surplus production that frees people from labor. The elite alone had leisure and were therefore the primary creators of culture. Because they created culture, the ruling elite developed a culture much more elaborate than the culture of those whom they ruled.

The most visible cultural differences between the elite and the masses involved speech, etiquette, protocol, and even body language (Braudel, 1981). Thus, the leisure of the elite was translated into a huge array of interaction cues that strongly influenced how prestige was displayed, protected, and passed on. A peasant who donned fine clothes and attempted to impersonate a gentleman had little chance of succeeding; he would talk wrong, walk wrong, and act wrong. The cultural barrier between the elite and the masses made it virtually impossible for a person not born and raised in the elite to fit in, even if she or he somehow managed to achieve high rank. This cultural wall not only severely limited upward mobility but also restricted contact between the elite and the masses, thus widening the gap between them. Indeed, in agrarian societies the elite often came to believe that they belonged to a superior human species.

As a result of these factors, agrarian societies were extraordinarily stratified. The overwhelming majority toiled endlessly and had but the barest necessities of life. Braudel (1981) reported that in agrarian societies around the world, the peasants were (and are) physically stunted by their meager diets. These usually consisted of nothing but boiled grain (rice, wheat, oats, or corn) or bread and vegetables in season. Meat and dairy products were the rarest of luxuries. Indeed, the dramatic change in diet resulting from industrialization was the primary reason for the marked changes in the size of humans, discussed in Chapter 5. Peasants in agrarian societies had to work very hard and received very little, and their lives were restricted. Sometimes they were slaves. Usually, they were bound by law and custom to remain on the land where they were born and to accept their lot.

Above the huge mass of peasants in agrarian societies were artisans and merchants, who were well off by comparison but who had little power, property, or prestige compared with the nobility—the actual ruling elite. The elite was a tiny proportion of the overall population, rarely more than 2 percent and usually much less (Lenski, 1966). At the apex of the elite was a king or an emperor, surrounded by a splendid court, who possessed immense power and incredible wealth. The ruler took as much as one-fourth of the society's production, and the ruling elite as a whole may have had as much as half (Lenski, 1966). Thus, while poor people in

industrial societies are far richer than average people in agrarian societies, the most powerful people in agrarian societies are far wealthier than the richest members of industrial societies. No person in North America has the wealth that an average Egyptian Pharaoh took with him to his tomb, to say nothing of the many palaces staffed by thousands of servants he left behind. And Catherine the Great of Russia personally owned about 27 million serfs—peasants held in virtual slavery who farmed her immense estates. The most successful agrarian societies in history—ancient Egypt, Rome, Byzantium, Persia, China, the Inca and Aztec empires, and the nations of medieval Europe—were the most highly stratified societies in the history of the world.

INDUSTRIAL SOCIETIES

The historical succession of human societies clearly shows that whenever human beings are able to produce more, their societies become more unequal. The hunting and gathering nomads had almost nothing and were relatively egalitarian. Simple farming technology increased the degree of stratification. As fields grew larger and life became more secure, the

Soon agrarian societies produced sufficient wealth so that ruling elites could live in immense splendor. In medieval society feasting was a major social activity for the wealthy, and the amounts sometimes consumed were extraordinary. For example, when a new Prior of Canterbury was installed in 1309, his guests consumed 36 oxen, 100 hogs, 200 piglets, 200 sheep, 1,000 geese, 793 chickens, 24 swans, 600 rabbits, 9,600 eggs, and huge amounts of bread and wine. Needless to say, the prior's peasants had to tighten their belts that winter while their flocks and herds recovered.

Art Resource, NY

gap between those on the bottom and those on the top became immense. Undoubtedly, the conditions of life of the average person improved as societies became more complex; even though most of their crop was taken from them, medieval peasants ate and lived better than hunter-gatherers (Braudel, 1981). But the life of the average person became progressively inferior to the life of the powerful. Ironically, the more productive human labor became, the more laborers could "pay" someone to oppress and exploit them.

Thus, nineteenth-century scholars expected that the staggering increases in productivity made possible by industrialization would make societies even more stratified. Marx, for one, predicted that the ruling class would become ever smaller and ever richer while everyone else was crushed into the proletariat. He believed that wages would fall to the lowest possible level and lead to industrial slavery: masses of workers forced to toil at their machines in return for only the barest essentials. But it didn't turn out that way.

The trend toward greater stratification in response to greater productivity suddenly reversed itself. Instead of all but a few being crushed into the lowest working class, the middle classes expanded rapidly, and unskilled labor jobs began to disappear. The gap in the standard of living between the top and bottom levels of society narrowed. Welfare programs were instituted to place a "safety net" under society to prevent privation. Social mobility greatly increased. Thus, status was based much less on ascription than on achievement.

Table 10-4 lets us see this pattern in global perspective. The data are based on the fifty most populous nations of the world. Some of them are highly industrialized; some are still primarily agrarian societies. We can measure the degree of stratification in each society as the percentage of total national income that goes to the richest 10 percent of families. That is, the more stratified the society, the larger the portion of income the elite will receive. Notice the very strong *negative* correlations between this measure of stratification and five measures of industrial development and modernization. These measures show that in effect, where there is more wealth to be had, it is being shared more equitably. For example, in nations where more people own TVs, a smaller share of national income goes to the richest 10 percent.

Why did these patterns emerge? Why did stratification decline as societies industrialized? And what does the stratification system look like in modern industrial societies? To answer these questions, we shall examine the Canadian and American stratification systems, partly because they have been the most exhaustively studied and partly because the basic aspects of stratification in democratic nations are similar.

INDUSTRIALIZATION AND STRATIFICATION

To explain why industrialization led to less stratification, we must combine insights from functionalist and conflict theories of stratification, as summarized in Chapter 9. Industrialization caused two major social changes that reduced stratification.

The first change is that industrialization raised the level of skill and training required to perform the average job (Lipset and Bendix, 1959; Wallerstein, 1974). Or as Peter Drucker (1969) put it, **industrial societies** are not based on getting people to work harder but rather to "work smarter." Laborers with picks and shovels undoubtedly work much harder to build a road than bulldozer operators do, but by working smarter rather than harder, one bulldozer driver does the work of hundreds of manual laborers. This is what the term **industrialization** means: using technology to make work much more productive. (One indicator of how technology and industrialization go hand in hand is presented in Table 10-5.)

However, industrialization changed more than tools and work techniques. It also changed the skills needed to do the work. For example, illiterates can do an excellent job of shoveling dirt. And shovelers do not need to know any

TABLE 10-4	Correlations between Stratification and Industrialization	
		CORRELATIONS WITH PERCENT OF TOTAL NATIONAL INCOME GOING TO RICHEST 10 PERCENT OF FAMILIES
Level of economic development		−.61
TV sets per 1,000 population		−.65
Per capita gross domestic product		−.64
Telephones per 1,000 population		−.66
Average life expectancy		−.56

Source: Prepared by the author from *Nations of the Globe,* an electronic data base created by MicroCase Corporation, now distributed by Wadsworth Publishing Company.

TABLE 10-5	Personal Computers in Selected Nations
	PERSONAL COMPUTERS PER 100 PEOPLE
United States	**58.5**
Sweden	50.7
Switzerland	50.2
Norway	49.1
Singapore	48.3
Australia	46.5
Denmark	43.2
Canada	**39.0**
Great Britain	33.8
Germany	33.6
Japan	31.5
France	30.5
Israel	25.4
South Korea	19.0
Greece	7.0
Mexico	**5.1**
Russia	4.3
China	1.6
Cuba	1.0
Algeria	0.6
India	0.5

Source: *Statistical Abstract of the United States, 2001.*

arithmetic. Nor do they need any special training to use their tools or maintain them. This is not true for bulldozer operators. They need to be able to read and follow directions on how much dirt to remove and where to put it. They need training to operate their machines skillfully and to maintain and repair them.

To the extent that occupational positions require education and training they are *less replaceable.* Laborers employed to shovel dirt can easily be substituted for those employed to hoe fields, but a bulldozer driver cannot be replaced so easily by operators of modern farm machines. It follows from the functionalist theory of stratification that as positions become less replaceable, their relative rewards increase.

In the long run, industrialization has made the average worker less replaceable. Compared with workers in agrarian societies, a much greater investment of time and money must be expended to prepare the average industrial worker for his or her job. Not until industrialization changed the nature of work did any society attempt to teach everyone to read, write, and

do arithmetic. Indeed, technological societies can exist only with an educated labor force. As a result the average worker can produce much more than before. More important, the training necessary to do most jobs has enabled workers to demand a higher level of reward.

This leads to a second consequence of industrialization: The average worker is more powerful and thus more able to resist coercion. As a result wages and profits are determined by bargaining rather than by force. In agrarian societies force or the threat of force could be used to get people to work hard. For example, overseers with whips could keep slaves busy digging ditches. But it is much more difficult to force people to work smarter—slaves did not seek ways to be more productive. Moreover, as members of a society are educated and trained, it becomes increasingly costly to coerce them because education inevitably increases awareness and raises aspirations (Lipset and Bendix, 1959). For this reason it was illegal to educate blacks in many parts of the American South before the Civil War (Sowell, 1978).

Industrialization has gone hand in hand with the rise of democracy. As we shall see in Chapter 17, some social theorists argue that economic exploitation by the ruling elite had to be curtailed for industrialization to occur; thus, democratization caused the Industrial Revolution. Even if this argument is not valid, industrialization clearly caused the proliferation of democracy; for example, it greatly expanded the proportion of the population allowed to vote. As the gap in skill and training required of average jobs and the most demanding jobs has decreased, so have the differences in power between the holders of these occupations.

According to conflict theory, people exploit their power to influence their rewards. Thus, as large sectors of the population in industrialized societies became more powerful, they demanded a larger share of the economic pie. Let us recall the toy society discussed in Chapter 9. There we saw that under conditions of political freedom, the Cees and the Dees could use their numbers to obtain a better bargain with the Ays and Bees. Thus, the decrease in stratification in industrialized societies can be seen as the outcome of ongoing conflict over who will get what. For example, automobile factory workers receive a lower wage than they would like but higher than management would like to pay. Both sides compromise on the basis of the relative costs of a strike.

In modern industrial societies, there has been a rapid trend toward more complex and skilled jobs, while unskilled jobs have been disappearing. The result has been to make the average worker *less replaceable* and therefore to decrease the wage differentials across jobs.

© Jean Gaumy/Magnum Photos

To sum up, industrialization could yield incredible levels of production only by changing the nature of work. Thus, it not only led to an immense rise in the possible standard of living for the average person but also made such a rise necessary. For the average worker to become so much more productive, he or she had to receive far more education and training. This in turn made the average worker much less replaceable and hence more powerful. Therefore, it became necessary to satisfy workers' demands for a larger share.

Keep in mind that this does *not* mean that industrialized societies are unstratified. The rich and the poor, the powerful and the weak, the famous and the unknown exist in all societies. Two things distinguish industrialized from agrarian societies. First, industrialized societies are less stratified—there is a smaller gap between the top and the bottom. Second, even the poor in industrialized societies are better off than rather well-off people in agrarian societies. With the rise of industrial societies, for the first time since humans settled down and began to build complex civilizations, the trend toward increased social inequality was reversed.

Industrialization did more than simply reduce stratification; it also dramatically changed the *rules* governing stratification. Status was no longer primarily ascriptive but came to depend more on achievement. This change stimulated (and was stimulated by) social mobility. Indeed, to really understand stratification systems in industrial nations, the best thing to study is social mobility: How common is it for people to rise or fall in the stratification system, and why and how do they do so?

SOCIAL MOBILITY IN INDUSTRIALIZED NATIONS

Historically, North America has been regarded as the land of opportunity, where hard work could lead to success regardless of family background. Alexis de Tocqueville, a French aristocrat who visited the United States during the 1830s and wrote the classic study *Democracy in America* (1835), was surprised to find not only that Americans frequently achieved great wealth and power despite humble beginnings but also that they were proud to have done so. In Europe, Tocqueville noted, upwardly mobile people tried to hide their humble origins. Tocqueville, like nearly all other observers, assumed that upward mobility was unusually common in the United States compared with other industrializing nations.

Alexis de Tocqueville.

During the 1930s and 1940s, American sociologists began research on social mobility and found that many Americans did indeed rise above their social origins, while others were downwardly mobile. But did this reflect a high rate of mobility? Any answer required a *comparison*, and studies comparing social mobility in America with that in other countries were lacking. Let us see how sociologists pioneered comparative studies of social mobility, thereby opening issues to which an immense amount of research has been directed ever since.

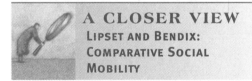

A CLOSER VIEW
LIPSET AND BENDIX: COMPARATIVE SOCIAL MOBILITY

In the 1950s Seymour Martin Lipset and Reinhard Bendix, two sociologists at the University of California, Berkeley, set out to see how much social mobility there was in the United States compared with other industrial nations. Such a study had only just become possible because of the spread of public opinion polling to many industrial nations after World War II. Often, these studies asked respondents their occupation and sometimes the question "What was your father's occupation?" (Until recently,

because so many women did not work or held only short-term employment, mobility studies covered only males.)

Lipset and Bendix (1959) collected opinion poll studies from a number of industrialized nations and calculated for each nation the proportion of sons whose occupations were of higher status than the occupations of their fathers, and vice versa. This allowed them to compare the rates of upward and downward mobility for these nations.

They found a great amount of social mobility in *all* the industrialized democracies studied. In fact, the total amount of social mobility was virtually identical (Figure 10-1). These findings seemed to refute the widespread belief that the United States has an unusually open stratification system. Thus, Lipset and Bendix concluded that the key question was not why there was so much social mobility in the United States but why there was so much mobility in all of these societies despite great differences in their cultures and social histories.

Lipset and Bendix argued that status based on individual achievement is inherent in the technological demands of industrial societies. Efficiency is the dominant concern in these societies, and therefore, less capable offspring of successful parents are necessarily displaced downward by more talented people moving up from below.

Also, a great deal of the upward mobility Bendix and Lipset observed in these societies is *structural mobility*. That is, industrialization has greatly increased the proportion of higher-status occupations and has correspondingly decreased lower-status occupations. As we saw in Chapter 9, when this occurs, many people must be upwardly mobile just to fill the demand. Figure 10-1 reveals that in the societies studied by Lipset and Bendix, a much larger proportion of people had risen than had fallen in occupational status.

When structural change forces a great deal of upward mobility, people's attitudes are likely to change. That is, when upward mobility is common and not at the expense of people born into high-status families, it is hard to oppose it or to discriminate against those who have risen.

As pioneers, Lipset and Bendix had to conduct their research with serious limitations. Lacking large samples, they were limited to very crude measures of social mobility. They were forced to define mobility as movement between only two general levels of occupations: persons holding white-collar (or nonmanual) jobs and

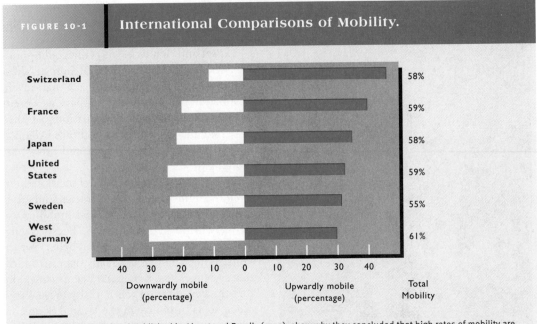

FIGURE 10-1   International Comparisons of Mobility.

	Total Mobility
Switzerland	58%
France	59%
Japan	58%
United States	59%
Sweden	55%
West Germany	61%

40 30 20 10 0 10 20 30 40

Downwardly mobile (percentage)   Upwardly mobile (percentage)   Total Mobility

These data, gathered and published by Lipset and Bendix (1959), show why they concluded that high rates of mobility are characteristic of industrialized nations in general, not just the United States. They found little difference in the total amount of mobility in these six nations, but significant differences in the amount of downward mobility, with the former West Germany being higher than others and Switzerland being lower. Later research showed, however, that the United States does differ from other industrial nations in its rate of long-range (rags-to-riches) upward mobility.

those holding blue-collar (or manual) jobs. A man was scored as upwardly mobile if his father had been a factory worker and he was an office clerk. Thus, small differences between the occupations of a father and his son could register as upward mobility. These crude measures also prevented Lipset and Bendix from assessing dramatic changes in status.

A decade later, much better data on social mobility became available. Although these confirmed Lipset and Bendix's conclusion that mobility is high in all modern, industrial societies, they also showed that, compared with Europe, there was something unusual about social mobility in the United States. **Long-distance mobility** involves huge upward or downward shifts in status—as when an unskilled laborer's child becomes a brain surgeon or president of a large corporation. Peter M. Blau and Otis Duncan (1967) reported that 1 of every 10 sons of manual workers in the United States ends up in an elite managerial or professional occupation. By comparison only 1 Italian in 300, 1 Dane in 100, and 1 Frenchman in 67, makes such a leap. Thus, what separates Americans of humble origins from their European counterparts is not

upward mobility per se but rather the much greater chance of going all the way. This may help explain the findings in Table 10-6, which show that Americans are less likely than people in some other industrialized nations to object to current income differences. Only 27 percent of Americans strongly agree that income differences in their country are too large. In contrast, more than half of Bulgarians, Italians, and Russians agree with this statement.

If the frequency of long-range mobility convinces you that there is a substantial opportunity to rise to the top of the income pyramid, perhaps you will not be motivated to lower the rewards for those who make it. Such an interpretation is also consistent with the results shown in Table 10-7. Few Americans think they need to come from a wealthy family to get ahead, nor do many think they need political connections. In contrast, substantial percentages in other industrialized nations *do* think a person needs a wealthy background and political connections. But Americans are more likely than people in any of these nations to rate "hard work" as very important for getting ahead. While the majority in each nation believes in hard work, the Italians are barely more likely to rate it as very important than they are to think one needs political connections.

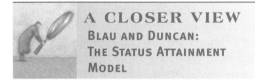

A CLOSER VIEW
BLAU AND DUNCAN: THE STATUS ATTAINMENT MODEL

Aside from the question of *how much* social mobility goes on in modern societies, for decades sociologists sought answers to a second important question: What determines an individual's chances of achieving upward mobility? Then two sociologists realized that this wasn't quite the right question to ask.

Blau and Duncan (1967) pointed out that the answer to this question is so simple that it isn't meaningful. They wrote:

The lower the level from which a person starts, the greater is the probability that he will be upwardly mobile, simply because many more occupational destinations entail upward mobility for men with low origins than for those with high ones.

That is, people's odds of rising are greater the lower their origins, and their odds of falling are greater the higher they start. The folks on the bottom can't go down; the folks on the top can't go up. But, Blau and Duncan pointed out, this trivial answer reveals that the wrong question is being asked: No one would really think that the best way to get a high-status position is to be sure to start on the bottom. So, Blau and

Duncan argued, the question is not, How are people mobile? but, How do people attain their statuses? Put that way, we can see how people acquire a status with or without being mobile.

Blau and Duncan conducted a landmark research study to provide answers to their revised question, and ever since then their results (and other work of this kind) have been known as the **status attainment model.** The study was based on interviews with a national sample of 20,700 selected American men age twenty to sixty-four. The U.S. Bureau of the Census conducted the interviews in 1962, using questions prepared by Blau and Duncan.

TABLE 10-6	"Differences in income in my country are too large."		
NATION	STRONGLY AGREE (%)	NATION	STRONGLY AGREE (%)
Bulgaria	80	Austria	34
Italy	53	New Zealand	29
Russia	52	**United States**	**27**
Slovenia	48	**Canada**	**25**
Hungary	44	Sweden	24
Germany	40	Norway	22
Poland	39	Australia	17
Great Britain	35	Philippines	9
Czech Republic	35		

Source: Prepared by the author from International Social Survey Program, 1992.

TABLE 10-7	How Do People Get Ahead in Life?		
	"FOR GETTING AHEAD IN LIFE, HOW IMPORTANT IS:"		
NATION	"COMING FROM A WEALTHY FAMILY" VERY IMPORTANT (%)	"HAVING POLITICAL CONNECTIONS" VERY IMPORTANT (%)	"HARD WORK" VERY IMPORTANT (%)
Italy	40	55	57
Hungary	34	30	61
Austria	30	43	65
West Germany	25	23	60
Great Britain	22	7	84
Australia	18	15	83
Switzerland	15	25	65
United States*	**14**	**9**	**89**
Netherlands	11	7	67

Source: Prepared by the author from International Social Survey Program, 1987.

Note: Surveys were conducted before the reunification of Germany.

*Oversampled cases omitted.

Not surprisingly, Blau and Duncan found that it's better to start at the top than at the bottom of the stratification system. Men whose fathers held high-status jobs had increased odds of holding a high-status job, too. However, the correlation was considerably weaker than many would have expected. Moreover, the primary mechanism linking the occupational status of fathers and sons turned out to be education.

The higher a man's status, the more years of education his son is likely to receive. And there is a very strong correlation between education and occupational status: Many high-status jobs such as doctor and lawyer have advanced-degree requirements. With education controlled, Blau and Duncan found that the influence of family background (father's occupation) on status attainment is rather modest.

Blau and Duncan's work prompted many other sociologists to study status attainment. An interesting twist was added by Christopher Jencks and his colleagues (1972), who sought to measure the effects of family background on occupational achievement by comparing sets of brothers. Jencks found that brothers were nearly as different in their levels of status attainment as were randomly selected pairs of men from the general population. For men drawn at random and paired, their average prestige scores (see Chapter 16) differed by twenty-eight points on a ninety-seven-point scale. For brothers the average difference was twenty-three points. Similarly, random pairs of men had an average difference in annual income of $6,200 a year, while the average for brothers was $5,600— nearly as great. Thus, brothers who grow up in the same home and are raised by the same parents end up nearly as different in terms of status attainment as do unrelated men who grow up in different homes.

A more recent study collected data on all members of a large sample of families over a ten-year period and traced the status attainment of family members as they left home and found their place in the occupational structure. The study included an extra-large sample of very poor families. Yinon Cohen and Andrea Tyree (1986) were therefore able to gain an unusually detailed basis for comparing factors governing upward mobility among the very poor with those operating among people from more favorable backgrounds. Once again the data confirmed a lot of mobility: Two-thirds of the children from the poorest families escaped from the bottom 20 percent income group, and over one-third made it to the upper 40 percent income group. Although we have long known that education is a key to status attainment, Cohen and Tyree found that education has greater importance for status attainment for people from poor homes than it does for others.

Cohen and Tyree's data forced sociologists to recognize something new: Marital status is the main determinant of family income, regardless of a person's background. Getting married and staying married have more to do with family income than does any other factor because affluence today is mainly the result of two-earner families. Or as Cohen and Tyree put it, "The probability of escape from poverty for single men and women, especially the latter, is considerably less than for married men and women."

Finally, these results have been fully confirmed by the latest results of the Wisconsin Longitudinal Study (Warren et al., 2002). Begun in 1957 with a random sample including 10,317 women and men who graduated from high school that year, with periodic follow-up studies, the data now trace this age cohort from high school to the point when many have retired. The results show that occupational achievement is due to two fundamental factors: education and cognitive ability. Over the course of a lifetime, the effects of education decline (being most important early in getting people placed in jobs), while the effects of cognitive ability persist, influencing promotions even late in life.

It seems appropriate to qualify the findings about status attainment at this point. Granted, family background has little or no impact on status attainment among persons having the same amount of education, but that finding can be interpreted as telling us *how* family background influences success, not that background doesn't matter. That is, family background seems not to result in much favoritism toward the offspring of upper-status parents nor in discrimination against those from lower-status families. In that sense, then, the finding that education negates the effects of family background on success suggests that the stratification system is open and "fair." But we also must recognize that children from upper-status families tend to enjoy a substantial educational advantage. That is, they are likely to attend better schools, to complete more years of schooling, and to get better grades. As we shall see in Chapter 16, such educational advantages translate into substantial occupational and economic returns. Thus, anything that influences the status of a child's family is apt to influence that child's adult status. This fact recently led researchers to an interesting new finding about status attainment.

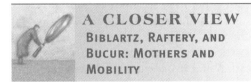

A CLOSER VIEW
BIBLARTZ, RAFTERY, AND BUCUR: MOTHERS AND MOBILITY

A standard research finding is that people raised in a two-parent family achieve higher status than do those raised in a female-headed household. One of the factors involved is income—female-headed families have less income than do two-parent (and often two-earner) families. However, a substantial difference persists when comparisons are made between the children of female-headed and two-parent families having the same income. Another factor is that there tends to be less supervision of children in female-headed families than when two parents can share the responsibility for keeping track of children. In addition, many psychologists have spun quite elaborate accounts of how children suffer when their home lacks a male role model.

Then one day Timothy Biblartz and Alexander Bucur of the University of Southern California and Adrian Raftery of the University of Washington wondered whether gender was a significant factor or whether the real effect was simply due to having only one as compared with two parents. That is, for a variety of reasons including how the courts award child custody, most children not living in two-parent families end up with their mothers. But what about kids growing up with only their fathers? If there are no income differences, are they really better off for having a male role model? Or might they in fact be worse off for lack of emotional warmth, assuming that women are more affectionate with children? Turning to a very large survey including enough children growing up in father-only families to permit comparisons, Biblartz, Raftery, and Bucur (1997) did the appropriate analysis. The results were remarkable.

Children raised in a mother-only household have higher status attainment than do children growing up in a father-only household! So much for the importance of a male role model.

Children reared in two-parent families do better than those in one-parent families of either gender, but at least in terms of status attainment, the divorce courts seem wise to award children to the mother. The reason seems to be that, other things being equal, women are more attentive and responsible parents.

A CLOSER VIEW
JOHN PORTER AND COLLEAGUES: STATUS ATTAINMENT IN CANADA

During his lifetime, John Porter (1921–1979) was regarded by many as Canada's leading sociologist. He was born in Vancouver and spent most of his career at Carleton University in Ottawa. During the early 1960s, Porter grew determined to establish nationwide survey research as a component of Canadian social research. He was especially interested in launching studies of mobility and status attainment in Canada, for he had concluded that his nation had a needlessly rigid stratification system.

In 1964 Porter finally succeeded in raising the funds needed to conduct interviews with a nationwide sample of Canadians. To help carry on the research, Porter recruited Peter C. Pineo, who then took the lead in writing up the results (Pineo and Porter, 1967). Their initial study focused on perceptions of occupational prestige (see Chapter 16). But it also set the stage for a full-scale Canadian replication of Blau and Duncan's American study of status attainment, for as soon as Blau and Duncan's first articles appeared in 1965, Porter set his sights on doing a Canadian version. He at once began a campaign to raise funds—an effort that took him nearly a decade. Porter's target was to collect data from 44,000 Canadians over age seventeen. As the project grew, Porter and Pineo recruited others to aid them: Frank E. Jones, Monica Boyd, John Goyder, and Hugh A. McRoberts. Finally, in 1973 the study was launched under the auspices of Statistics Canada (Pineo, 1981; Vallee, 1981).

The Canadian study was much more than simply a replication of Blau and Duncan's work. A most important step was the inclusion of women. As noted, mobility and status attainment studies had been limited to male respondents because so many women were not employed outside the home, while many women who were employed worked only from time to time and did not pursue long-range occupational goals. But as the baby boom began to ebb, patterns of female employment began to change, and by the 1970s it was both feasible and important to include women in such a study.

Porter and his colleagues expected that their results would be very different from Blau and Duncan's. Canadians, including many nonsociologists, had long believed that there was much less social mobility in Canada than in the United States—that opportunities for occupational advancement were much more restricted in Canada than in most other industrialized nations. As Porter wrote in his major work, *The Vertical Mosaic* (1965), "Canada has not been a mobility oriented society," a situation that he blamed partly on a lack of educational opportunities. In addition Porter—along with most Canadian observers—believed that ethnicity and immigrant status played unusually prominent roles in the Canadian stratification system. Indeed, he used the term *vertical mosaic* to suggest a stratification system constructed of distinctive ethnic strata.

The results of the massive 1973 survey produced some stunning surprises (Pineo, 1976, 1977; Boyd et al., 1981). Canada was *not* an exception to Lipset and Bendix's proposition that social mobility is high in all industrialized nations. Canada's rate of social mobility is almost identical to that of the United States. Moreover, the Blau and Duncan status attainment model for the United States is duplicated in Canada. Table 10-8 compares the correlations between father's and son's occupational prestige for Canada and the United States. It is exactly .40 in both national studies. The table also shows the correlations between an individual's education and his occupational prestige. The results are .60 for the United States and .61 for Canada.

Now notice something else in Table 10-8. For Canada the correlations are repeated for various subgroups, letting us examine the impact of ethnicity—long thought to be so potent in Canadian stratification. Thus, native- and foreign-born men can be compared, as can English- and French-speaking men. This comparison reveals that the father-to-son transmission of status is the same within all four groups, as is the correlation between a man's education and his occupational prestige. This does not mean that these groups are equal in terms of their occupational prestige (although the differences in average occupational prestige scores turned out to be small); it does mean that the process of status attainment is very similar for all four.

How did John Porter react to evidence that some portions of his most famous book were wrong or at least overstated? Like a scientist. Peter Pineo (1981) recalls:

I began writing up the material and sent advance copies of everything to Porter. He reacted quite cheerfully to it all and . . . volunteered that he would like to collaborate [in writing final drafts of the findings].

Subsequent status attainment studies in Canada have confirmed the findings of the 1973 research project. Thus, in a detailed study of men in Ontario, Michael D. Ornstein (1981) found that ethnicity as such plays a very slight role in status attainment, especially if one takes into account generational differences in education and language acquisition. He also demonstrated that the statistical model that most closely fits status attainment in Canada is extremely similar to the model for the United States.

TABLE 10-8	Status Attainment in the United States and Canada

| | FATHER'S OCCUPATIONAL PRESTIGE |
SON'S OCCUPATIONAL PRESTIGE	(CORRELATION)
United States	.40
Canada	.40
Native-born	.40
Foreign-born	.40
English-speaking	.40
French-speaking	.41

| | SON'S EDUCATION |
SON'S OCCUPATIONAL PRESTIGE	(CORRELATION)
United States	.60
Canada	.61
Native-born	.62
Foreign-born	.63
English-speaking	.62
French-speaking	.61

Source: Prepared by the author from Blau and Duncan (1967) and Boyd et al. (1981).

FEMALE STATUS ATTAINMENT

When the Canadian sociologists analyzed their data on female status attainment, they also found some surprising results (Boyd et al., 1981). First of all, native-born Canadian women with full-time jobs come from higher-status family backgrounds than do their male counterparts. On the average their fathers have nearly a year more of education and hold higher-status occupations. Second, the average native-born Canadian working woman has a higher-status occupation than do similar males. Finally, the correlations between women's occupational prestige and their fathers' education (.26) and occupational prestige (.22) are much *lower* than for men. Moreover, these same findings have turned up in American studies; it has now become standard practice to include women in status attainment research (Treiman and Terrell, 1975; Featherman and Hauser, 1976). How can these patterns be explained?

First of all, women are less likely than men to hold full-time jobs and are especially unlikely to work the lower their job qualifications. For many married women, especially those with young children, low-paying jobs offer no real economic benefits; the costs of working (including child care) are about equal to the wages paid. As a consequence low-paying, low-status jobs are disproportionately held by males. This fact accounts for women having jobs of higher average prestige. But women are also underrepresented in the highest-prestige jobs. As a result their occupational prestige is limited to a narrower range than that of men, which reduces correlations with background variables. That the average working woman's father has more education and a better job than does the father of the average employed male can be understood in the same terms. More qualified women come from more privileged homes; the daughters of the least-educated and lowest-status fathers aren't in full-time jobs.

In fact, the *husbands* of working women have occupations with higher than average prestige. This is because of a very high correspondence between the occupational prestige of husbands and wives when both are employed full time (Hout, 1982). People who marry tend to share very similar levels of education and similar family backgrounds. Indeed, as we shall see in Chapter 13, divorce and remarriage contribute to the similarity of husbands and wives in terms of occupational prestige (Jacobs and Furstenberg, 1986).

These findings must not cause us to overlook the fact that women long were excluded from many occupations and are still underrepresented in elite managerial and professional careers. What they do show, however, is that within the special conditions outlined here, female status attainment does not differ much from that of men.

A CLOSER VIEW
MICHAEL HOUT: TRENDS IN STATUS ATTAINMENT

For a long time, sociologists have recognized that a very substantial amount of the social mobility going on in industrial nations was structural (see Chapter 9). That is, many people end up in higher-status occupations than their parents simply because the number of lower-status occupations has been declining while the number of upper-status jobs has been expanding. A major portion of this shift has involved a great decline in farming jobs and an enormous increase in highly skilled technical jobs.

However, in recent years it has become evident that structural mobility is on the decline: There just aren't many unskilled jobs left in the American economy, so increasingly, people cannot be the children of parents holding such jobs. For example, there are very few people left in farming jobs, and most of these are farm owners. So, a generation ago, when a lot of structural mobility was being generated by the children of farm laborers as they moved to the city and found better jobs, now such people are scarce.

Other things being equal, a decline in structural mobility would necessarily reduce the amount of mobility taking place in a society and thus could increase the competition for higher-status positions. These possibilities caused some social scientists to express concern that exchange mobility might decline, too. They suggested that when there is a lot of mobility going on, employers will hire and promote on the basis of merit. However, when upper-status people become sufficiently worried that their children will be downwardly mobile, they might respond by favoring one another's children when making hiring decisions. As a consequence, status might become more closely linked to family background, thus reversing the trend toward an increasingly open and democratic stratification system.

But things that only might happen also might not happen. So Michael Hout, a sociologist at the University of California, Berkeley, decided to find out if any real trends were developing in the American stratification system. He based his research on the General Social Survey, a survey conducted annually with a national sample of Americans. Hout used the surveys conducted in 1972 through 1985 and limited his study to men and women who were in the labor force and who were age twenty-five to sixty-five.

His findings were very clear: While structural mobility has been declining, there has been a corresponding *increase in exchange mobility* so that the overall amount of mobility has remained the same. Thus, the correlation between family background and occupational attainment *declined by one-third* between 1972 and 1985. Moreover, this was true for both men and women.

Thus, the American stratification system has been becoming more, rather than less, open. Hout also discovered that these trends primarily are the result of the growing proportion of the labor force with college degrees. Hout stated flatly: "College graduation cancels the effects of background." Thus, as the proportion of Americans with college educations increases, the fairness of the stratification system also increases.

The primary results of status attainment research in the United States and Canada, and subsequently in many other nations, sustain the arguments first developed by Lipset and Bendix. High rates of mobility and a decline in ascribed status do appear to be inherent in the technological demands of industrialized societies, therefore making performance rather than connections primary in occupational attainment. Indeed, Lorne Tepperman (1976) has been able to create a formal model of shifts in social mobility as societies undergo industrialization, predicting the shape of stratification systems and of mobility rates over time.

But if status tends to depend more on achievement than on ascription in all industrialized societies, not all persons in North America (or elsewhere) have been freed of ascriptive status rules. Until recently, racial and ethnic discrimination has excluded some groups of North Americans from most of the better jobs. In the next chapter, we shall examine this matter in depth.

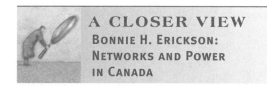

A CLOSER VIEW
BONNIE H. ERICKSON: NETWORKS AND POWER IN CANADA

In Chapter 9 we noted how classes differ in terms of the structure of their networks and that having cosmopolitan networks is a source of class privilege. According to Pierre Bourdieu (1984), culture is stratified, and the culture of the elite is regarded as the most distinguished culture not because it necessarily is superior but because those who rule get to say what is culturally superior or inferior. In keeping with these theoretical views, the early studies of class culture stressed the advantages people gain from familiarity with what often is called "highbrow culture," for this serves as a primary marker of class distinctions (Bourdieu, 1984). **Highbrow culture** primarily consists of acquired tastes—the appreciation of art, literature, music, furnishings, and food and wine that requires experience and instruction. Abstract expressionist paintings, opera, classical music, jazz, Shakespeare's plays, the novels of James Joyce, Persian rugs, and snails in garlic butter are examples of highbrow culture in Western societies. Consistent with the discussion of class networks in Chapter 9, access to highbrow culture is a function of cosmopolitan networks.

Recently, however, a sociologist at the University of Toronto began to doubt this view of class culture as a set of distinctive layers with highbrow culture characterizing the upper classes and "mass culture" characterizing the lower classes. Narrowing her focus to "forms of culture that can be used to advantage in seeking a better class position or in conducting class relationships," Bonnie H. Erickson (1996) proposed that "cultural inequality is not so much a hierarchy of tastes (from soap opera to classical opera), as it is a hierarchy of *knowledge* (from those who know little about soap opera or opera to those who can take part in a conversation about both)." Put another way, the cultural advantage of the upper classes is not merely possession of a *different* culture but of *more* culture. As Erickson put it, "The most widely useful form of cultural resource is *cultural variety*." And the upper classes possess far greater cultural variety because of an aspect of their social networks that Erickson calls **network variety,** defined as the number of classes, status groups, occupations, and cultures included in one's social network. That is, the extent of a person's cultural variety is the result of her or his network variety. Obviously, cosmopolitan networks will encompass far greater variety.

To test her theoretical conclusions, Erickson conducted a survey of persons involved in the private security industry in Toronto—a total of 150 firms. She chose this sample to compare people all involved in the same sector of the economy and to compare owners and managers with lower-level supervisors and with ordinary workers. Each respondent was asked a series of questions about whether they knew someone employed in specific occupations or positions:

Now I am going to ask you whether you know anyone in a certain line of work at all in the Toronto area, for example, whether you know any lawyers. Please count anyone you know well enough to talk to even if you are not close to them.

Each respondent was asked about being acquainted with people in nineteen categories including, in addition to lawyers, business owners outside your own company, doctors, engineers, professors, schoolteachers, bankers, insurance brokers, accountants, carpenters, electricians, locksmiths, plumbers, detectives, and so on. Table 10-9 shows the results. Respondents were separated into three "classes" based on their work positions: Upper includes owners and managers; middle includes supervisors; and lower consists of employees. The first line in the table shows the mean number of the nineteen occupational categories to which each class group was connected by acquaintance. The higher their class, the greater the diversity of their network.

In addition respondents were asked a number of questions to assess their knowledge of various cultural topics, including sports, art, books, and magazines. Mean scores on each topical test also are shown in Table 10-9. The results support Erickson's position that class advantages consist of cultural variety rather than of cultural differences. That is, upper-class respondents not only scored higher on knowledge about art, books, and business magazines, but higher also on knowledge of sports, and they were as informed about mass magazines. When it comes to culture, bosses can hold their own in conversations on topics of interest to their employees, but the reverse tends not to be true. It is not, as Bourdieu (1984) supposed, that bosses know of Picasso and Rembrandt while workers know

TABLE 10-9	Class, Networks, and Culture		
	UPPER	MIDDLE	LOWER
Mean number of cross-occupational ties	14.0	10.4	9.4
Mean score on name recognition test about:			
Sports	7.0	6.4	5.9
Art	4.9	3.4	3.2
Books	5.4	3.8	3.7
Business magazines	2.8	1.6	1.2
Mass magazines	3.2	3.1	2.9

Source: Recalculated by the author from Erickson (1996).

yesterday's sports scores and the latest celebrity scandals according to *The National Enquirer.* Instead, Erickson's study showed that bosses also know all the scores and are as well informed about the marital life of movie stars or Oprah's current weight.

Erickson also found that sports were the primary "common topic" of conversation at work, all classes being conversant on the subject. The topic of sports, she wrote, is

popular in all class levels and widely seen as something in common with others at work. Sports discussions help to build cooperative ties across class levels. But, at the same time, women and foreign-born people know much less about sports, so they are marginalized in the informal networks that both keep companies integrated and help further individual careers.

Thus, Erickson concluded that although culture is a very important source of power, it is not the direct result of class differences but of "social network diversity." Every contact we have with someone in a different occupation or other culturally distinct situation "provides a channel of access to [this] distinctive cultural repertoire. . . . Social networks are the continuing adult education of culture, and diversified networks are the liberal arts programs teaching a little of almost everything."

A CLOSER VIEW
YANJIE BIAN: NETWORKS AND SUCCESS IN CHINA

In his famous study of the "strength of weak ties," described in Chapter 2, Mark Granovetter discovered that Americans more often find jobs through their acquaintances than through their close friends and relatives. While members of one's local network to whom one has strong ties may care far more about helping you find a job, usually they will be less able to do so than will members of an extended cosmopolitan network, united by only weak ties. But what if hiring is not based primarily on qualifications but rather on connections? What if acquaintances do not exchange information on job openings, reserving it instead for their friends and relatives? Indeed, what if a major consideration in hiring decisions concerns the strength of ties between those hiring and those being hired?

Recently, Yanjie Bian (1997), a sociologist at the University of Minnesota, decided to find out by repeating Granovetter's study in China. He based his research on a survey of 949 adult residents selected randomly in the third largest city in China, Tianjin. Respondents were interviewed at length about their job history and how they had found their first job. When he analyzed the data, Bian found that most had obtained their jobs through a strong tie, not a weak one. He interpreted this finding by noting that the strength of weak ties primarily has to do with the transmission of *information*. That is, American workers *found out* about a job opening through their weak ties. Then they got the job by applying for it and being hired on the basis of their qualifications. In China, however, it is *influence*, not information, that is critical to getting a job. Most Chinese obtained their first job from someone who felt an obligation to the applicant's family and close friends. Obviously, influence is transmitted more effectively via local networks (consisting of strong ties) than via cosmopolitan networks. Hence, "weak ties" are only strong in societies where achievement outweighs ascription in determining status attainment. A more recent study based on a survey of 6,473 Chinese adults (Li and Walder, 2001) expanded Bian's results, identifying Communist party membership as the key network placing Chinese in desirable positions. Indeed,

young Chinese gain party membership through their family connections, and this membership in turn determines their access to advanced education and to placement in promising career paths.

CONCLUSION

Chapter 9 presented the basic concepts and theories sociologists use to understand stratification. In this chapter we have used these concepts and theories to examine how stratification systems work in different societies. We have seen that the relative lack of stratification among primitive hunter-gatherers was an equality of poverty. The rise of civilization and the rise of social inequalities went hand in hand. As societies became more productive and complex, the masses of humanity were increasingly exploited by a tiny elite. But this trend was reversed with the advent of modern industrial societies, despite expectations that these would become the most stratified societies in history.

We have also seen that industrialization increased the skills required for most occupations and therefore decreased the replaceability of the average worker. This improved the bargaining position of the average worker and led to a reduction in inequalities. Moreover, the development of democracy in industrial societies has severely limited the ability of the elites to use force to exploit others. In this way status based on ascription has given way to status based on achievement in industrial societies.

But these changes have not affected everyone in these societies, at least not at the same time. Thus, while some people can achieve high status, others continue to be ascribed a low status, especially those who belong to certain racial or ethnic groups. In the next chapter, we shall apply principles of stratification and an understanding of stratification systems in modern industrial nations to analyze ethnic and racial conflict.

Review Glossary

Terms are listed in the order in which they appear in the chapter.

hunting and gathering societies The most primitive human societies; their rather small numbers of members (often fewer than fifty) live by wandering in pursuit of food from animals and plants.

agrarian societies Societies that live by farming. Although these were the first societies able to support cities, they usually require that about 95 percent of the population be engaged in agriculture.

industrial societies Societies with economies based on manufacturing in which machines perform most of the heavy labor.

industrialization The process by which technology is substituted for manual labor as the basis of production.

long-distance mobility Mobility that occurs when an individual or group rises from the bottom to the top of the stratification system.

status attainment model The process by which individuals achieve their positions in the stratification system.

highbrow culture Culture that consists primarily of acquired tastes—the appreciation of art, literature, music, furnishings, and food and wine that requires experience and instruction. Abstract expressionist paintings, opera, classical music, jazz, Shakespeare's plays, the novels of James Joyce, Persian rugs, and snails in garlic butter are examples of highbrow culture in Western societies.

network variety The number of classes, status groups, occupations, and cultures included in one's social network.

Suggested Readings

Blau, Peter M., and Otis Dudley Duncan. 1967. *The American Occupational Structure*. New York: Wiley.

Bloch, Marc. *Feudal Society*. 1962. Chicago: University of Chicago Press.

Boyd, Monica, John Goyder, Frank E. Jones, Hugh A. McRoberts, Peter C. Pineo, and John Porter. 1981. "Status Attainment in Canada: Findings of the Canadian Mobility Study." *Canadian Review of Sociology and Anthropology* 18:657–673.

Giddens, Anthony. 1975. *The Class Structure of Advanced Societies*. New York: Harper & Row.

Patterson, Orlando. 1982. *Slavery and Social Death: A Comparative Study*. Cambridge, Mass.: Harvard University Press.

Sociology Online

www.socstark10.com

GO TO THE INTERNET AND TYPE: www.socstark10.com.

CLICK ON: The 50 States

✓ **SELECT:** % IN UNION: Percentage of workers who are union members.

CLICK ON: Analyze Now
Is there a substantial regional pattern? If so, can you suggest a reason?

CLICK ON: Select New Data Set

CLICK ON: Nations of the Globe

✓ **SELECT:** $ RICH 10%: The percentage of total national income received by the richest 10 percent.

CLICK ON: Analyze Now
Look at all the nations in which the richest 10 percent receive more than 40 percent of all income. How many of them are advanced industrialized nations?

CLICK ON: Select New Variable

✓ **SELECT:** INEQUALITY: A Gini Index of inequality (deviation from equal distribution of income or consumption): 0=perfect equality, 100=perfect inequality.

CLICK ON: Analyze Now
Examine nations with scores of 45 or higher. How many of them are advanced industrialized nations?

How does the chapter explain these findings?

USING INFOTRAC COLLEGE EDITION

GO TO THE INTERNET AND TYPE:
www.InfoTrac-college.com

CLICK ON: Register New Account
You will be asked to enter your Access Code and to create and enter a User Name and Password. Having done so,

CLICK ON: InfoTrac College Edition or Log On.

Because so many parents want their kids to rise in the stratification system, there is a lot of discussion of mobility in popular periodicals. How well informed are the authors?

SEARCH TERMS:
Occupational Mobility
Social Mobility

CHAPTER

11

the benny goodman jazz quartet

On Easter Sunday in Chicago in 1936, three of these men made history when the famous "King of Swing," clarinetist Benny Goodman, along with drummer Gene Krupa, gave a concert with Teddy Wilson on piano. Although musicians had long been playing in integrated bands in the recording studios, this was the first time African American and white musicians appeared together before a paying audience. The reaction to the trio was so favorable that soon Goodman expanded the group by adding Lionel Hampton[1] on vibes. Although Goodman continued to lead his big band, the quartet usually played part of the program and made many hit records.

Racial and Ethnic Inequality and Conflict

THE YEAR IS 1862. The scene is the outback of Australia. Two tiny bands of natives have crossed paths. No one sees familiar faces in the other group, and everyone is tense. Then spokesmen for each group begin to discuss their ancestry, examining the family trees of both groups minutely. This is not a ritual form of greeting; its outcome may decide the fate of all present. Everyone hopes that some mutual relatives will turn up in the examination, for only if the groups are related can they be friends. If they are not relatives, then they are strangers. Their language does not distinguish between the word for stranger and the word for enemy: Strangers *are* enemies. If the bands are strangers, then there is nothing to do but flee or fight. The only good stranger is a dead one.

Fear and loathing of strangers are not peculiar to the culture of Australian aborigines. They are almost universal human traits. The outsider, the foreigner, the stranger has always provoked anxiety, suspicion, hatred, and dread in human beings (Williams, 1947). Such responses are especially likely *when the strangers are noticeably different because of cultural or physical differences.* Inventing social and cultural means to neutralize conflict between groups with noticeable differences has been one of the major tasks of modern societies.

Let's return to 1862, to the western prairies of North America. Many Native American tribes still move across their ancestral hunting grounds in pursuit of the great buffalo herds. Each tribe is small, often having no more than 1,000 members. Each small tribe holds strongly negative beliefs about the other tribes, and conflicts among tribes are frequent. However, all of the Plains tribes reserve their greatest contempt and hatred for the Utes, who live in

---

1. On the very day that I began to revise this chapter for the ninth edition, news came that Lionel Hampton, the last
   living member of the quartet, had died at age ninety-four. He continued to perform in public until the end.

CHAPTER OUTLINE

INTERGROUP CONFLICT
  Race
  Ethnic Groups
  Cultural Pluralism

PREOCCUPATION WITH PREJUDICE
  Allport's Theory of Contact

  A CLOSER VIEW
  The Sherif Studies

SLAVERY AND THE AMERICAN DILEMMA

STATUS INEQUALITY AND PREJUDICE

  A CLOSER VIEW
  Edna Bonacich: Economic Conflict
  and Prejudice

  Exclusion
  Caste Systems
  Middleman Minorities

IDENTIFIABILITY

EQUALITY AND THE DECLINE OF PREJUDICE
  The Japanese Experience in North America
  Japanese Americans
  Japanese Canadians

**MECHANISMS OF ETHNIC
AND RACIAL MOBILITY**
  Geographical Concentration
  Internal Economic Development and
  Specialization
  Development of a Middle Class

HISPANIC AMERICANS

**GOING NORTH: AFRICAN AMERICAN
"IMMIGRATION" IN THE UNITED STATES**
  African American Progress
  Integration

  A CLOSER VIEW
  Firebaugh and Davis: The Decline
  in Prejudice

  Barriers to African American Progress
  Box 11-1 Do White Fans Reject
  "Too Black" Teams?

CONCLUSION

REVIEW GLOSSARY

SUGGESTED READINGS

SOCIOLOGY ONLINE

the foothills of the Rockies. The Utes have darker skins than the other tribes, and they are universally loathed and described as ugly. Everyone thinks it natural to kill Utes whenever possible. Meanwhile, white settlers are moving onto the plains. They cannot tell one tribe from another, and to them all Indians are savages, thieves, drunkards, and killers. In time they begin to say that the only good Indian is a dead Indian.

Also in 1862, the Chinese and Japanese are just beginning to have extensive contact with Europeans. They find this contact unpleasant. European customs seem barbaric, and to make matters worse, Europeans are very ugly. (Asians also complain that Europeans smell extremely bad. Asians bathed regularly, while Europeans seldom did so.[2]) Because of large, protruding noses and long, narrow faces, Europeans are called "dog faces." Europeans in turn find the faces of Asians oddly flat, and they refer to Asians as "slant-eyes."

Nor is 1862 a banner year for brotherhood elsewhere in the world. In the United States, the Civil War rages, in which more than 400,000 Yankees and Confederates will be killed. In part this war is fought to determine whether African Americans will continue to be held in slavery. Yet even if the majority of Northerners want to free the slaves, they do not intend to accept them into their society. Freed slaves are not permitted to vote or to testify in court in most northern states, and schemes to ship all African Americans back to Africa or to relocate them in Indian territory are popular.

Meanwhile, in Canada the American Civil War has caused a momentary halt to the flood of French-speaking Canadians leaving Quebec for the United States. The exodus, which eventually will total nearly a million people, is a direct result of political and economic repression imposed on the French by a government entirely controlled by English-speaking Canadians—a government installed by military force in the aftermath of a rebellion aimed at establishing Quebec as an independent republic. Many decades will pass before French-speakers regain political equality. As soon as the American Civil War ends, the French Canadian exodus will begin again—a barometer of the bitterness between the "two Canadas."

Over in Europe the Austrians hate the Slavs. The French hate the Germans. The Germans look down on the Poles. And the English look down on everyone, treating even the Scots and the Welsh as inferior foreigners. The Irish flee to America to escape the harsh conditions of life the English impose on them, but the Irish find they are not particularly welcome in America, either.

In 1862 Charles Darwin's new theory of evolution arouses the greatest intellectual excitement. It argues that complex biological species evolved from simple, primitive forms of life. The horse, for example, was once the size of the jackrabbit (and about as numerous), and apes and humans are biological cousins. The theory of evolution seems to answer a problem that has long vexed Europeans. Over the previous three centuries, Europeans explored the globe, and they now are poking into the last uncharted regions. Everywhere they go they find human societies with a lower level of technology than theirs. Indeed, they find society after society still living in the Stone Age. Why is ignorance so widespread? Why are Europeans so much more advanced? Evolutionary theory is used to support supremacist notions: The white race is at a higher stage of human evolution, and non-whites are more primitive species of humans. Europeans have long believed in their racial superiority; now they think they have a scientific basis for it. And if whites are more advanced biologically, should they not be careful to preserve their heritage and not dilute it by crossbreeding?

This chapter examines the causes of intergroup conflicts due to noticeable cultural or physical differences, particularly conflicts within rather than between societies. Specifically, it explores racial and ethnic differences as the basis for ascribed low status—that is, how such factors as race, religion, language, and customs generate inequalities among members of a society. Indeed, this chapter will argue that *inequalities in property, power, and prestige among different racial and ethnic groups in a society are the basis of intergroup conflict.* We shall see how such inequalities make particular groups into strangers, thereby provoking the irrational reactions typical of humans confronting strangers. We shall also see that as status inequalities between groups disappear, so do prejudice and hatred.

INTERGROUP CONFLICT

Chapter 2 introduced the basic concepts that sociologists use to examine intergroup conflicts, but some additional comments will prove useful here. This chapter discusses *intergroup* rather than *interracial conflict*, since much of the hatred, prejudice, and discrimination among groups is not based on race. Antagonisms between Protestants and Catholics in the United States during the nineteenth and twentieth centuries were not interracial, nor is the deadly and festering conflict in Northern Ireland today. Two centuries of conflict between English- and French-speaking Canadians have not involved racial differences. Nor are the continuing massacres in Africa based on race. Rather, in Africa it is people of the same race, but of different tribes, who are so eager to kill one another— conflicts based on culture.

The term **intergroup conflict** encompasses all such disputes, whether they are over culture or over skin color. However, this chapter is not concerned with intergroup conflict between two groups with an identical cultural and racial heritage, who may battle over politics or wealth. Our concern is only with intergroup conflicts based on *noticeable physical or cultural differences.*

RACE

A **race** is a human group with some observable, common biological features. The most prominent of these is skin color, but racial groups also differ in other observable ways such as eyelid shape and the color and texture of hair. They also differ in subtle ways that are not visible, such as blood type. Although race is a biological concept, racial differences are important for intergroup relations solely to the extent that people attach cultural meaning to them. Only when people believe that racial identity is associated with other traits such as character, ability, and behavior do racial differences affect human affairs. Historically, racial differences have typically been associated with cultural variances as well because persons of different races were usually members of different societies. Race, then, has usually been an accurate indicator of who is and who is not a stranger.

2. Queen Isabella of Spain, who funded Columbus, boasted that she had taken only two baths in her life—at birth and the day before her wedding. For a time during the eighteenth century, it was against the law in Philadelphia to bathe more often than once a month (Stuller, 1991).

Major trouble arises when different racial groups exist within the same society and people still assign cultural meanings to these physical differences. For one thing, members of one racial group usually cannot escape prejudice by "passing" as members of another racial group, although many light-skinned blacks can and do pass as whites. While an Italian could change his name to Robert Davis, join the Presbyterian Church, and deny his true ancestry, people cannot as easily renounce their biology.

This does not mean, however, that racial differences must always produce intergroup conflict. Biological differences may be unchangeable, but by themselves they are not important. It is what we *believe* about these differences that matters. And what we believe can change. The notion of a society that is color-blind simply refers to a society in which no cultural meanings are attached to human biological variations.

ETHNIC GROUPS

Ethnic groups are groups whose cultural heritages differ. We usually reserve the term for different cultural groups within the same society. By themselves cultural differences are not enough to make a group an ethnic group. The differences must be noticed, and they must both bind a group together and separate it from other groups. As Michael Hechter (1974) put it, an ethnic group exists on the basis of "sentiments which bind individuals into solidarity groups on some cultural basis."

The 2000 Census asked Americans their primary ancestry. Table 11-1 shows the distribution of persons who reported a European nation as their ancestry, omitting those making up less than 1 percent of the total population. Germans are the largest ancestry group in the United States, including more than one American in six. Next largest are the Irish and the English, followed by the Italians and the Poles. However, identifying oneself in terms of ancestry is not the same as maintaining an ethnic identity. Consider the more than 15 million Americans of Italian ancestry. Some of them can speak Italian; many cannot. Some like Italian food; some do not. Some are Roman Catholic; some are not. But the existence of an Italian American or Italian Canadian ethnic group does not simply depend on such cultural factors. What is important is that some of these people *think of themselves as sharing special bonds*—of history, culture, and kinship—with others of Italian ancestry. This makes them members of an ethnic group. But persons of Italian ancestry who do not identify themselves with their Italian heritage are not members.

Wsevolod Isajiw (1980) has pointed out that ethnic groups are "involuntary" groups in that people don't decide to join one as they might decide to join a fraternity or sorority. Rather, people are born into an ethnic group. However, unless they live within the confines of a relatively strict caste system, people often make a voluntary choice about *continuing* to belong to an ethnic group. In fact, as noted in Chapter 2, a substantial proportion of North Americans of Italian ancestry are not part of an Italian ethnic group (Alba, 1985). This is hardly limited to persons of Italian ancestry. After careful analysis comparing the 1990 Census data with those of 1980, Stanley Lieberson and Mary C. Waters (1993) found a substantial shift away from naming a European ancestry group and a very great increase in the proportions giving their ancestry as "American." They concluded that "If this holds in future decades, it would mean a growing ethnic population of 'unhyphenated whites.'"

CULTURAL PLURALISM

For a long time, people believed that intergroup conflicts in North America would be resolved through *assimilation:* As time passed, a given ethnic group would surrender its distinctive

TABLE 11-1	Primary* Ethnic Ancestry of Americans of European Descent	
	NUMBER (IN 1,000s)	PERCENT OF TOTAL POPULATION
German	42,885	15.2
Irish	30,528	10.8
English	24,515	8.7
Italian	15,724	5.6
Polish	8,877	3.2
French	8,310	2.9
Scottish	4,891	1.7
Dutch	4,542	1.6
Norwegian	4,478	1.6
Scotch-Irish	4,319	1.5
Swedish	3,998	1.4
Russian	2,657	1.0

Source: *Statistical Abstract of the United States, 2005.*

*Making up at least 1 percent of the total U.S. population.

On July 8, 1853, Commodore Matthew Perry led a squadron of four American warships into Tokyo Bay in defiance of Japan's laws excluding foreign contacts. After a period of negotiations, Japan agreed to treaties opening its ports to Western shipping. A Japanese artist drew these portraits of Perry (lower right) and his senior officers. The Asian reaction to the facial features of Westerners is clearly displayed in this set of drawings: The "dog face"—the protruding face with a large nose—is evident in each. Moreover, other than for minor details such as glasses and whiskers and variations in age, *all Westerners looked alike!* Notice, however, that the Japanese artist depicted all of these men with eyelids of typical Asian shape.

cultural features and disappear into the dominant American or Canadian culture. At that point people would no longer think of themselves as "ethnic," nor would others continue to do so.

Today, many once formidable intergroup conflicts have been resolved in North America. Yet the ethnic groups in question, mostly European, did not disappear. True, their ethnic identity differs from that of their forebears. Typically, they have lost their native language and their bonds with the old country. But their present culture retains some elements of the old—religious affiliation, for example—while integrating a new heritage based on the special experiences of the group in the United States or Canada (Glazer and Moynihan, 1970). The important point is that conflict vanished not because noticeable differences disappeared but because the differences became unimportant. Such conflict resolutions are called *accommodation*, not assimilation. The growth of mutual interests between conflicting groups enables them to emphasize similarities and deemphasize differences.

When intergroup conflict ends through accommodation, the result is ethnic or cultural *pluralism:* the existence of diverse cultures within the same society. That the United States is no longer a Protestant nation but rather a nation of Protestants, Catholics, and Jews, as well as followers of other faiths (plus nonbelievers), demonstrates cultural pluralism.

Obviously, accommodation and assimilation are not the inevitable outcomes of intergroup conflict. Conflict has sometimes been resolved by the *extermination* of the weaker group, as happened with the Jews in Nazi Germany, Catholics in Elizabethan England, Indians in the Caribbean and on the North American frontier, Armenians in Turkey, and various tribal minorities in black Africa today. Intergroup conflicts have also led to the *expulsion* of the weaker group. For example, Jews have often been expelled from nations, and Europeans were expelled from Japan in the sixteenth century. Following World War II, Pakistan expelled Hindus, India expelled Muslims, and Uganda expelled both Pakistanis and Indians, while Vietnam has driven out several hundred thousand Chinese. Finally, intergroup conflicts have been stabilized by the imposition of a **caste system,** whereby weaker groups are prevented from competing with the stronger, and through *segregation*, whereby a group is inhibited from having contact with others.

The history of the New World contains all of these methods of resolution: Groups have been assimilated, accommodated, exterminated, expelled, segregated, and placed in a low-status caste. This variety, plus the persistence of intense intergroup conflicts, makes the United States and Canada extremely important in the study of intergroup relations. North America is a huge natural laboratory for examining the dynamics of such conflicts and a useful means for overcoming them.

Such an examination inevitably arouses our emotions. When we examine the history of prejudice and discrimination in North America, we cannot—nor should we—avoid anger and frustration. However, it would be tragic if we let these feelings prevent us from appreciating the extent to which ethnic, religious, and racial hatred have been overcome. That, too, is a very important part of our history. In addition, as you read the chapter, you should keep in mind that prejudice and discrimination are not a peculiar American problem. Not only are these problems as old as history, but they are far more prevalent in many parts of the modern world than they are in North America.

Table 11-2 puts racial and religious prejudice in global perspective. National samples of adults in many nations were asked, "On this list of various groups of people, could you please indicate any that you would not like to have as neighbors." Column one shows the percentage in each nation who indicated they would not want people of another race as neighbors. In Bangladesh slightly more than seven people of ten reject persons of another race as do about two-thirds of Egyptians. Notice, too, that the upper end of the list is dominated by Asian, Islamic, and African nations; no Western European nation exceeds the 17 percent of Belgians who would not want neighbors of another race. Mexico is quite low with 15 percent, while only 8 percent of Americans and 4 percent of Canadians gave a racist response. Anti-Semitism remains quite common in many places, especially since the question about Jews as neighbors was not asked in Arab countries—86 percent in India, 41 percent in South Korea, 34 percent in Nigeria, and 28 percent in Japan said they didn't want Jews in their neighborhood. Unfortunately, too, anti-Semitism remains substantial in Eastern Europe: 28 percent in Bosnia, 25 percent in Poland, and 23 percent in Romania and Lithuania. However, in many nations opposition to Muslim neighbors is far higher than opposition to Jews.

PREOCCUPATION WITH PREJUDICE

Until very recently, social scientists regarded prejudice as the *cause* of intergroup conflict. Hostile actions against some racial or ethnic minority were believed to reflect the underlying hostile beliefs or prejudices that groups had about one another. Therefore, the urgent questions were: Why do people become prejudiced?

TABLE 11-2	"On this list of various groups of people, could you please indicate any that you would not like to have as neighbors."		
NATION	PEOPLE OF A DIFFERENT RACE (%)	JEWS (%)	MUSLIMS (%)
Bangladesh	72	21	—
Egypt	66	—	—
India	42	86*	31*
Saudi Arabia	38	—	—
Indonesia	35	—	—
South Korea	35	41	57
Vietnam	32	—	27
Turkey	32	16	—
Albania	31	17	30
Nigeria	31	34*	24*
Bulgaria	28	18	21
Algeria	28	—	—
Iran	24	—	—
Romania	24	23	31
South Africa	24	24	24
Philippines	21	—	26
Croatia	20	18	27
Zimbabwe	20	19	18
Jordan	20	—	—
Armenia	19	—	—
Dominican Republic	19	—	—
Malta	19	21	28
Macedonia	19	20	26
Uganda	18	22	14
Belgium	17	13	22
Taiwan	17	—	19
Slovakia	17	10	25
Belarus	17	15	27
Tanzania	17	—	13
Poland	17	25	24
Venezuela	16	—	—
Italy	16	13	17
Mexico	**15**	**19***	**17**
China	15	—	—
Estonia	15	11	22
Greece	14	19	21
Bosnia	13	28	13
Azerbaijan	12	—	—
Finland	12	9	19
Slovenia	12	17	23
Ireland	12	11	14
Japan	11*	28*	29*
Moldova	11	25	44
Spain	11	22	13
Ukraine	11	10	24
Czech Republic	10	5	15
Georgia	10	—	—
Lithuania	10	23	33
France	9	6	16

continued

TABLE 11-2	"On this list of various groups of people, could you please indicate any that you would not like to have as neighbors.", continued		
NATION	PEOPLE OF A DIFFERENT RACE (%)	JEWS (%)	MUSLIMS (%)
Switzerland	9	—	19
Chile	9	9	7
Great Britain	9	6	14
Norway	8	9*	19
Portugal	8	11	8
Russia	8	11	14
United States	**8**	**9**	**11**
Austria	7	8	15
Denmark	7	3	16
Uruguay	7	10	—
Luxembourg	6	8	14
Argentina	5	7	6
Australia	5	—	—
Latvia	5	5	15
Netherlands	5	2	12
Singapore	5	—	—
Canada	**4**	**4**	**7**
Germany	4	5	10
Sweden	3	2	9
Brazil	3	—	—
Iceland	3	4	12

Source: Prepared by the author from the World Values Surveys, 2001–2002.

* Data from the World Values Surveys, 1995–1996.

— Not asked.

and What can be done to cure them of their prejudice? Little attention was paid to social and economic relations among groups that might give rise to mutual hostility. Instead, prejudice was blamed on personality defects or ignorance. For a long time, social scientists searched the heads and hearts of people to discover what was wrong with them that caused them to be prejudiced. This search peaked in the mid-1960s.

An immense number of theories have been advanced to explain why people develop prejudices toward people of different racial or ethnic backgrounds. For several decades the most influential of these was the theory of the *authoritarian personality* (Adorno et al., 1950). Its proponents argued that some people are in effect oversocialized so that they accept the norms and values only of their own group and reject any variations. When such people are confronted with others whose norms and values differ from theirs, they become very anxious. To resolve this anxiety, they adopt the belief that all

who differ from them are inferior, sinful, inhuman, or otherwise objectionable. Prejudice, then, was seen as a defense mechanism against having to question one's own cultural heritage. Another theory blamed prejudice on feelings of personal inadequacy or low self-esteem; that is, people adopted prejudices to have someone else to look down on (Ackerman and Jahoda, 1950).

Many sociologists, myself included, reacted to the notion that prejudice was entirely in the head. Consequently, a huge number of studies were done to link prejudice to social influences on the individual, especially to channels by which prejudice might be learned or unlearned.

Again and again, researchers found that the more education and income a person has, the less likely that person is to be prejudiced against other racial and ethnic groups (Selznick and Steinberg, 1969; Quinley and Glock, 1979). Other studies found that the more religious a person is, the more likely that person is to be prejudiced against members of other faiths; however, religion had no effect on prejudice against persons of other races or ethnic groups of the same religion (Glock and Stark, 1966; Stark et al., 1971). It is significant that the most recent study has found that all forms of interreligious prejudice have effectively disappeared in the United States, except one: hatred of "Fundamentalists." However, anti-Fundamentalism is quite limited to well-educated Americans who lack an active religious affiliation (Bolce and De Maio, 1999).

Yet despite immense effort over several decades, prejudice researchers failed to explain compellingly *why* people are prejudiced. Moreover, it began to seem likely that prejudice was not the fundamental cause of intergroup conflict. Instead, it became increasingly evident that intergroup conflict causes prejudice. Thus, the proposition that curing prejudice will relieve conflicts reverses cause and effect. To the contrary, sociologists became convinced that only by resolving conflicts will prejudices subside.

In the remainder of this chapter, we shall examine why sociologists came to accept the proposition that racial and ethnic conflicts are rooted in status inequalities between groups and generate prejudice. As we shall see, this approach contradicts many popular beliefs about intergroup relations. For example, if prejudice were the result of ignorance and a lack of understanding, an obvious solution would be to break down the barriers that isolate one group from another. According to this view, strangers need to get to know one another. Thus, intergroup relations are improved by more frequent intergroup contact.

For decades that sounded like very good advice, and innumerable programs were instituted to "bring people together." Today, however, sociologists think that such efforts are doomed as long as real grievances between groups exist. In fact, increased contacts between groups can increase conflict, hostility, and prejudice. To see why, let's watch as several social scientists first began to develop this new theoretical approach.

ALLPORT'S THEORY OF CONTACT

Research in the 1940s and 1950s on prejudice seemed to support the popular belief that prejudice thrives in isolation and that getting people together thus will overcome prejudice. However, some social scientists began to doubt this conception of the problem. Foremost among them was Gordon W. Allport of Harvard, one of the world's most distinguished social psychologists at the time.

If contact is the answer to the problems of prejudice, Allport reasoned, then why didn't racial prejudice in the American South disappear long ago? One might argue that northern racists have simply never had a chance to get to know African Americans, but in the South, African Americans and whites have long been in close contact. Many white southerners, for example, grew up having much closer relations with an African American servant than with their own parents. In small southern towns, the two racial groups had been in close, regular, daily contact for generations. Yet prejudice was as strong in these towns as anywhere in North America. Why?

In 1958 Allport proposed his answers in a classic book, *The Nature of Prejudice*. Contact won't necessarily make relations between two groups better, he argued; often, it will make relations worse depending on the conditions under which that contact occurs.

According to **Allport's theory of contact**, prejudice will decrease if two groups with equal status have contact. But prejudice will increase or remain high if it occurs under conditions of *status inequality*, in which one group is dominant and the other subordinate. This accounts for the failure of race relations to improve in the South until recently. White merchants had contact only with African American customers, not with African American merchants. White children had contact primarily with African American servants, not with African American teachers or African American fellow students. Such contacts reinforced white people's views of themselves as superior. The inequality of the situation forced African Americans to submit and caused whites to perceive them as submissive.

This view of the effects of contact was supported by studies of the U.S. Merchant Marine. The prejudice of white seamen against African Americans did not decrease as a result of repeated voyages with African American cooks and mess stewards aboard. But it changed significantly after President Harry S Truman ordered the Merchant Marine to integrate, and white seamen voyaged with African American seamen of equal rank (Brophy, 1945).

However, even contact between groups of equal status does not always improve relations,

Contact between groups will decrease prejudice when it occurs under conditions of equal status and cooperation as illustrated by these American soldiers assigned to duty at Arlington National Cemetery.

A CLOSER VIEW
THE SHERIF STUDIES

In the 1950s Muzafer and Carolyn Sherif of the University of Oklahoma conducted a series of studies of young boys at summer camp. Their results vividly demonstrate how easily prejudice arises among groups. The Sherifs assigned young boys to living groups when they arrived for a two-week stay at summer camp and then manipulated their activities to test how prejudice arose and declined between groups. Posing as the camp janitor, Muzafer Sherif was able to wander about the camp at will in a nearly invisible role and eavesdrop on the boys.

In one experiment the Sherifs (1953) broke up existing friendships by assigning boys who were friends to different living groups. Living groups were then made the basic camp units for activities, and each living group was treated as a separate team in sports competition. During activities requiring frequent competition among living groups, the Sherifs found that hostile stereotypes characteristic of intergroup prejudices arose within several days. For example, one group soon labeled various other groups as "cry-babies," "cheats," and "sissies." These harsh feelings arose even when many boys in one group had close friends in another. When the Sherifs altered the contact so that the groups performed cooperative rather than competitive tasks, the negative stereotypes quickly subsided.

If such hostilities can be produced in a few days among young boys of similar background and with longstanding friendships, then is it any wonder that antagonism arises so easily in the real world between groups of strangers who are separated by truly noticeable differences and different experiences?

Allport pointed out. *Prejudice will intensify if the groups are engaged in competition* (poor whites competing with poor African Americans for unskilled jobs, for example), but *prejudice will decline if the groups cooperate to pursue common goals.* Thus, when white police officers work with African American police officers, their prejudice decreases (Kephart, 1957).

Since Allport's work was published, considerable research has supported his views (Sigelman and Welch, 1993; Schofield, 1995; Moody, 2001). Contact overcomes prejudice only when people meet on equal terms to cooperate in pursuing common goals. Contact accompanied by inequality and competition breeds contempt. It can even turn former friends into strangers.

Based on the important theoretical work by Allport and the amazing experimental results the Sherifs obtained, a new approach to intergroup conflict began to spread among social scientists. Sociologists were especially intrigued by the ways that Allport's theory could be applied to large social structures. For example, because prejudice among groups of unequal status is likely, most groups of new immigrants coming into contact with the established groups in a society are likely to encounter prejudice.

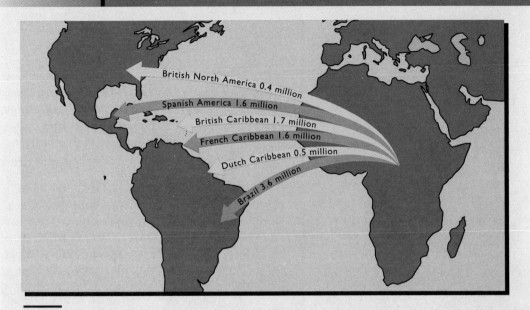

FIGURE 11-1   Patterns of the Slave Trade, 1500–1870.

British North America 0.4 million
Spanish America 1.6 million
British Caribbean 1.7 million
French Caribbean 1.6 million
Dutch Caribbean 0.5 million
Brazil 3.6 million

About 10 million slaves were transported to the New World between 1500 and 1870. Here we see the primary destinations of the slavers. Notice that only a small proportion, perhaps 500,000, were taken to North America. This is surprising because the first U.S. Census, taken in 1790, recorded 700,000 slaves plus some free African Americans, and by 1820 there were 1.5 million slaves in America. Indeed, by 1825 the United States had by far the largest slave population in the Western Hemisphere, because compared with slaves in the Caribbean and South America, slaves in the United States had low death rates and high birthrates. Historians attribute this to "better diet, shelter and general material conditions" (Patterson, 1982). Nevertheless, the United States was among the last Western nations to outlaw slavery.

Source: Adapted from Curtin, 1969.

As sociologists began to make these theoretical applications, many widely accepted beliefs were called into question. People had, for example, long assumed that racist attitudes led to slavery in the United States. Now it suddenly seemed more likely that slavery was the *cause* of the unusually virulent racist beliefs that had flourished in the American South. Before we discuss how economic inequalities or status differences cause prejudice, let us examine the institution of slavery.

SLAVERY AND THE AMERICAN DILEMMA

Until quite recently, slavery was nearly universal, found in nonnomadic societies all over Europe, Asia, Africa, and the Western Hemisphere. Ancient Rome and Greece were slave societies—slaves probably outnumbered the free population in Greek cities such as Athens (Finley, 1982) and constituted a majority of the Roman labor force (Grant, 1978). Slavery even existed among the

Northwest Indians long before Columbus sailed to the New World. Consequently, "few peoples in the world have not constituted a major source of slaves at one time or another" (Eltis, 1993).

The word *slave* derives from the word *Slav* because for centuries the Slavs were the primary source of slaves for European and Islamic nations. Constant raids on the Slavic tribes produced slaves "on a massive scale" and thus white, blue-eyed Slavs were "sold into bondage all across the continent of Europe and in the Ottoman Empire" (Sowell, 1994). In fact, Slavic slaves were so common that the Arabic word for slave also derives from the Arabic word for Slav (Lewis, 1990).

By the sixteenth century, Africa had become the major source of slaves, especially for Islamic nations, and following the development of plantation agriculture, Africa became the primary source of slaves for the Western Hemisphere (Figure 11-1). It must be noted that although millions more black slaves were imported by Islamic nations than came to the Western Hemisphere (Austen, 1979), this resulted in large black

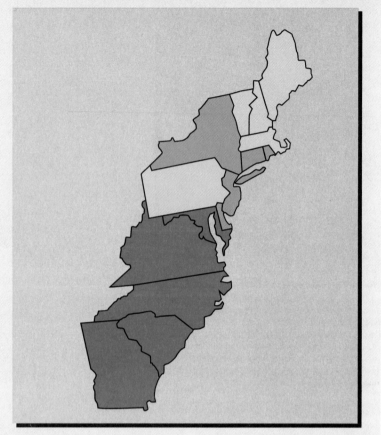

FIGURE 11-2   American Slaves, 1790.

Percent of Population Who Are Slaves, 1790

Massachusetts	0.00
Vermont	0.00
Maine	0.00
New Hampshire	0.11
Pennsylvania	0.85

North Carolina	25.55
Maryland	32.22
Georgia	35.52
Virginia	39.14
South Carolina	42.99

Connecticut	1.18
Rhode Island	1.45
New Jersey	6.19
New York	6.26
Delaware	15.06

Here we see the proportion of slaves in the population of the fifteen United States in 1790. There actually were seventeen slaves in Vermont that year (too few to raise the population percentage above 0.00), but there were none in Massachusetts or Maine—slavery was against the law there. Not surprisingly, the proportions are highest in the South; South Carolina, for example, was nearly half slave. But note that there also were substantial slave populations at this time in many northern states such as New York and New Jersey.

Source: U.S. Census, 1790.

populations only in the West. The difference was an almost complete lack of fertility among black slaves in Islamic societies (Lewis, 1990).

However, even though Africa had become the primary source of slaves, white slaves continued to be captured and sold until the end of the nineteenth century—in the 1820s approximately 6,000 Greeks were sold as slaves to buyers in Egypt, while thousands of Germans living along the lower Volga were taken away as slaves by Mongol raiders (Sowell, 1994). These white slaves in Asia were, of course, the exceptions, and most slaves in Asia were Asians, although typically of a different culture than that of their owners. Thus, untold millions of Asians were enslaved as a result of Manchu raids on China, Mongolia, and Korea, and slave raiders from the Philippines terrorized the South Pacific. The volume of slavery was substantially lower in Southeast Asia than in the Middle East—although, ironically, slavery still exists in parts of Southeast Asia.

In 1776, the year of the American Declaration of Independence, slavery still existed in Russia, Poland, Hungary, and in parts of Germany as well as in all of Asia, Africa, and North and South America (Smith [1776] 1937). In fact, "western Europe was the only region of the world where slavery [had] been abolished altogether" by the late eighteenth century (Sowell, 1994).

Ironically, in declaring their independence from British rule, the American Founding Fathers wrote: "We hold these Truths to be self-evident, that all Men are created equal." It is beyond dispute that these American patriots meant this statement. It is also beyond dispute that many of them, including Thomas Jefferson[3] who wrote these words, owned slaves (see Figures 11-1 and 11-2).

The doctrine of equal opportunity and the sanctity of individual freedom have long been fundamental American values. Yet these beliefs were maintained in the face of continuing slavery and later through decades of harsh discrimination against African Americans. How could Americans square their ideals with their practices?

In 1944 Gunnar Myrdal, a Swedish economist, published the results of a monumental study of American race relations. In it he focused on the stunning contradiction between

---

3. Jefferson made many efforts to outlaw slavery and once nearly succeeded in doing so in Virginia. However, he felt it irresponsible to free his slaves within a slave-owning society because he believed that they would only suffer worse fates.

New York Public Library

democratic ideals and racist practices. The title of his famous book termed this contradiction *An American Dilemma.*

Many Americans dealt with **the American dilemma** between democratic values and racism by rejecting racism. The Puritan descendants of New England were as opposed to slavery as they were to English colonialism, and they provided the grass-roots support for the Abolitionist movement that contributed to the outbreak of the Civil War and culminated in Lincoln's Emancipation Proclamation. But most Americans found another way out of this contradiction. They did not regard human slavery as proper, but they also did not consider African Americans fully human. If African American people are not fully human, then the ideal that "all men are created equal" does not apply. (The humanity of African American *men* was at issue. That *women* were omitted from this phrase was not then embarrassing.) As a result slavery in America was perhaps as much a *cause* of prejudice as a consequence of it.

But how could seemingly intelligent people become convinced that just because people have a dark skin they are not fully human? It wasn't just skin color, nor was it just that African Americans came from exotic, primitive cultures. Rather, it was the nature of the institution of slavery and the contact between slave and master that enabled whites to see slaves as inhuman: African Americans were required to behave as if they were in fact not fully human (Patterson, 1982).

Recall the conditions under which Allport believed contact would worsen prejudice and then consider slavery. There is no greater status inequality than that between master and slave. No group can be more dominant over another than slave owners over slaves. Slaves were not educated, and they rarely had been anywhere except on the plantation where they were born. They were traded and sold like prize livestock, and because they were African American, they could not easily run away and blend into the free population. They were required to be

Slaves plant sweet potatoes on a plantation on Edisto Island, South Carolina, in April 1862. The picture was taken during a period of occupation by the Union Army (which explains why some of the men wear parts of U.S. Army uniforms). Eight months later, the Emancipation Proclamation went into effect. But by then the Union troops had withdrawn, so these people remained slaves until the war's end.

subservient to whites, who dealt with their resistance severely. Because African Americans had no choice but to be wholly dependent, most African Americans appeared childlike to their masters. Whites denied them experience, knowledge, and literacy, thus keeping them ignorant.

If you regard people as livestock, raise them to act in such a fashion, and prevent them from acting in other ways, it is no surprise that they then reinforce your belief that they are livestock. Centuries of slavery had made African Americans and whites complete strangers. And the harsh racist prejudice employed to justify slavery has lived on to infect relations between African Americans and whites in our own time.

STATUS INEQUALITY AND PREJUDICE

Sociologists currently view status inequality as the cause, not the result, of prejudice and discrimination. Now we shall examine this line of reasoning in depth.

When two groups obviously differ in some cultural or physical characteristic (religion or race, for example) and encounter each other, these differences will dominate their initial perceptions: Each group will tend to magnify the differences and attribute unflattering traits to the other. However, whether these initial reactions subside or intensify depends on the conditions under which their contact continues.

When both groups live in the same society, contact is hard to avoid. According to Allport's theory, future relations between the two groups depend on whether the two groups are of equal status and whether the benefits of cooperation outweigh those of competition. When such groups are (or come to be) of equal status and when they benefit from cooperation, relations ought to improve rapidly, regardless of initial negative reactions. But *if their status is unequal (and so long as it remains unequal), then contact between these groups will worsen feelings.* Moreover, when two such groups are of unequal status, economic competition between them is virtually unavoidable.

From this line of reasoning, several conclusions follow. First, contact between culturally and racially distinctive groups of unequal status will increase prejudice and probably discrimination as well. Second, this prejudice will not subside until after the status inequality and economic competition have been eliminated. Finally, within any society, noticeable physical and cultural differences will produce prejudice only if they are associated with status inequality or with competition.

If these conclusions are correct, then efforts to overcome prejudice before overcoming status inequality are bound to fail. Instead, efforts should be directed at eliminating the inequality; then prejudice should subside.

To see how these processes work, let's consider the relations between Catholics and Protestants over the past century in the United States. For contrast we shall also examine relations between Protestants and Catholics in Northern Ireland today.

As discussed in Chapter 2, the arrival of large numbers of Catholic immigrants in the United States during the latter nineteenth century provoked bitter reactions among American Protestants. Like most immigrants, the Catholics arrived with little money or education and took unskilled labor jobs. For several generations Catholics were at or near the bottom of the American stratification system. Indeed, the average income of Catholic families lagged considerably behind that of Protestants until well into the twentieth century. This inequality fueled the bitter prejudices Catholics and Protestants held toward one another.

Economic inequality and discrimination placed Catholics in direct conflict with the Protestant majority. For Protestants, Catholic efforts to achieve economic parity represented a threat of loss, for Catholic gains would come at the expense of Protestant privilege. As we shall see later in this chapter, the exclusionary immigration laws the United States adopted in the 1920s were imposed by the Protestant majority in an effort to protect their relatively high wages from being undercut by Catholic immigrants willing to work for less.

Eventually, Catholics did gain economic parity with Protestants (Greeley, 1974). When that happened, the basis for conflict subsided. Catholics were no longer of lower status than Protestants; the competition over privilege had ended in a draw. It was then, not before, that bitter anti-Catholic and anti-Protestant prejudices began to subside. Indeed, they subsided so rapidly that few college-age Protestant and Catholic Americans today even know of them, although they were still widespread when their parents were young. As recently as the 1950s, books attacking Catholics as pagans and the enemies of democracy became bestsellers. Nonetheless, in 1960 a Catholic was elected president of the United States.

During the days of Catholic-Protestant conflicts in the United States, observers were convinced that these conflicts were caused by cultural differences—specifically, by differences in religion. Because these differences were unlikely to disappear, it was thought that prejudice would remain as well. However, because these religious differences do not provoke prejudice today, they must not have been the cause. Rather, the religious differences were used as indicators, or **markers,** of underlying conflicts over status. With the conflict resolved, these markers lost their ability to inflame antagonisms.

In his major work on intergroup relations, Stanley Lieberson (1980) explained how cultural and racial markers are made potent by status conflict:

I am suggesting a general process that occurs when racial and ethnic groups have an inherent conflict [competition over status between unequal groups]. Under the circumstances, there is a tendency for the competitors to focus on differences between themselves. The observers (in this case the sociologists) may then assume that these differences are the sources of the conflict. In point of fact, the rhetoric involving such differences may indeed inflame them, but we can be reasonably certain that the conflict would have occurred in their absence.

He then suggested a "thought experiment." Because Protestants in Northern Ireland refer to themselves as "Orangemen," while Irish Catholics identify with the color green, let's suppose that the Protestants in Northern Ireland had orange skins and the Catholics green skins. Clearly, these marked physical differences would become part of the rhetoric of hatred and prejudice in Northern Ireland, serving as additional markers of difference. Lieberson even suggested that such physical differences would play a secondary role in fueling the conflict. However, it would be silly to credit the orange and green "racial" differences as the real cause of the conflict because a bitter, murderous conflict over economic opportunity and status already exists without such differences. In similar fashion, Lieberson argues, physical differences between African Americans and whites, both real and imagined, enter into the rhetoric of interracial

conflict in the United States, but they are not the cause.

Lieberson (1980) concluded that fear of African Americans as economic competitors is the real cause of racial stereotypes. This means

that were the present-day [economic] conflict between blacks and dominant white groups to be resolved, then the race issue could rapidly disintegrate as a crucial barrier between the groups just as a very profound and deep distaste for Roman Catholics on the part of the dominant Protestants has diminished rather substantially.

Keeping in mind this overview of how economic inequality generates prejudice, let us now examine these matters in detail.

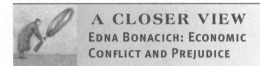

A CLOSER VIEW
EDNA BONACICH: ECONOMIC CONFLICT AND PREJUDICE

The roots of ethnic and racial antagonism usually lie in economic inequality and conflict. This is primarily because subordinate racial and ethnic minorities represent an economic threat to many members of the dominant majority. That is, the presence of a disadvantaged racial or ethnic group in a society makes available a supply of persons who can be hired for wages lower than those paid to majority group workers for the same job. Such persons will work for less—either because they lack the power to demand wages as high as those of the dominant group or because they have different economic motives for working.

Edna Bonacich, a sociologist at the University of California, Riverside, has examined the relationship between labor market conflict and racial and ethnic antagonism in considerable detail. She has identified four factors that often cause or require members of subordinate racial and ethnic groups to work for substandard wages (1972, 1975, 1976).

The first is a *very low standard of living.* Often enough, such groups migrate from a region with so low a standard of living that wages considered substandard by members of the dominant group are very attractive to them. Illegal Mexican workers in the American West and Southwest, for example, work for much less than the going wage paid to Americans, but they earn far more than they could in Mexico. Similarly, African Americans who migrated

from the rural South in the 1930s and 1940s improved their standard of living even though they were paid well below union wages (and often were employed as strikebreakers) in northern mines, mills, and factories. The same was true of Chinese and Japanese workers arriving on the West Coast at the turn of the twentieth century. Indeed, white, Protestant "Okies," who fled their farms for California in the 1930s, during the years of the Dust Bowl and the Great Depression, found substandard wages in California to be a great improvement over the abject poverty they faced when they lost their land.

This photo of a country store, taken in Gordonton, North Carolina, by Dorothea Lange in 1939, reveals the poverty and isolation of the rural South, which in comparison with other parts of the United States, made the South look like a foreign land. Here we can glimpse some of the complexities and contradictions of southern race relations at that time. The white store owner, standing in the doorway, seems to be on friendly terms with the African American men sitting on his front porch. Yet, while these men are free to enter the store and buy whatever they like, they may not sit down and relax inside: They can buy soda pop inside but must return to the porch to drink it because both whites and African Americans shop at this store. Had the store been for African Americans only, then they could have sat down inside. The poverty of the African American customers is evident, but so is that of the white owner of this dilapidated store. While whites enjoyed many social privileges denied African Americans, most whites as well as African Americans suffered from the economic and social deprivations that marked life in the rural South until recent times. Some of these young African Americans may have migrated north. If so, their journey was more like that of immigrants coming through Ellis Island than like that of people moving from one state to another.

Library of Congress

A second impediment to high wages pointed out by Bonacich is a *lack of information*. Immigrants from a very poor country may not know that they are being exploited. Or members of a subordinate group may not know about minimum wage laws, legal recourses for unpaid wages, and the like, often because they do not speak the language.

A third factor that limits the ability of ethnic and racial minorities to secure high wages is their *lack of political power*. They may lack citizenship and thus have no voting power or may be too few in number to force favorable reforms. In contrast, workers of the dominant group may have substantial political power and strong unions that enable them to demand high wages, which in turn make them vulnerable to wage competition from groups unable to unionize or to wield political power. This is especially true for unskilled jobs. As we saw in Chapter 10, unskilled positions have high replaceability, which always depresses wages unless political power can be used to create artificial conditions that lower replaceability. Because most immigrant workers in America were not skilled, they had little power, and the potential usefulness of any to strike was limited by the ease with which they could be replaced.

A contrasting case makes the point. Jewish workers founded some of the first successful unions in the United States in the late nineteenth century. They rapidly won contracts for better pay and shorter hours, conditions most industrial workers did not enjoy at that time (Howe, 1976). The key to the power of the Jewish trade union movement lay in the special

Library of Congress

Farm families from Oklahoma, bankrupted by the Dust Bowl and the Great Depression, loaded their remaining belongings onto old cars and pickups and set out for California. Despite being white Anglo-Saxon Protestants, they were greeted in the Golden West as undesirable foreigners— as "dirty Okies"—and many Californians tried to keep them out or drive them away.

skills of the Jewish workers, especially those in the garment industry. Factory owners lacked an alternative supply of skilled tailors and seamstresses. Interestingly enough, the employers from whom Jewish unions won their settlements also tended to be Jews. This success in one industry shows that Jews arrived in the United States already possessing the skills needed to make rapid economic progress.

The *economic motives* of subordinate racial and ethnic groups are the fourth factor in their low wages, according to Bonacich. Often, they intend only to be temporary workers. As a result they often accept low wages and poor working conditions, knowing they need not endure them forever. In addition temporary workers have little to gain and much to lose by striking or by refusing bad jobs in hopes of getting better ones. For workers with short-term economic goals, comparatively low wages are a greater advantage than are prospects of better wages.

For example, many Italian immigrants came as sojourners (temporary residents) to North America. They were young men who came to earn and save their wages and then return to rural Italy, where the much lower standard of living made their savings worth much more. For them to participate in a long strike over an unpleasant work environment or to go unemployed while searching for a higher-paying job conflicted with their goal: to save money as fast as possible and return home.

Similarly, H. A. Millis (1915) reported that Japanese workers in California displaced other workers not only because they accepted low

wages but also because they worked twelve to fourteen hours a day and on weekends. They did this so that they could save money faster and return to Japan sooner. And in Africa villagers will often work very cheaply because they plan to work just long enough to afford the price of a bride, a rifle, or a bicycle (Berg, 1966). Furthermore, as Bonacich pointed out, it is against the interests of management to pay high wages to sojourners, lest the workers achieve their short-term goals faster and leave that much sooner.

Clearly, then, the presence of racial and ethnic groups willing to work for substandard wages threatens the economic well-being of other workers and is likely to cause strong antagonism toward the subordinate group. Cheap labor threatens to make *all* labor cheap. Workers belonging to the dominant group must either work for less or lose their jobs to members of the subordinate racial or ethnic group, unless such economic competition can be prevented.

Two strategies have commonly been employed to prevent a group from competing with the dominant group by providing cheap labor (Bonacich, 1972). The first is *exclusion*. Members of the subordinate group are denied entry into the society or driven out if they have already entered. The second is to establish a *caste system* that limits the subordinate group to certain occupations, often the most menial and undesirable ones.

EXCLUSION

The influx of cultural and racial groups who are perceived as a threat to the dominant group's wages often causes a nativistic reaction: a demand that foreigners be kept out.

As already mentioned, massive immigration of Catholics during the late nineteenth century led to nativistic policies in American politics, often justified on vicious racial and ethnic grounds. Italians and eastern Europeans were seriously discussed as "inferior racial stocks" whose immigration threatened the "racial purity and superiority" of North Americans (Grant, 1916). Following World War I, Protestant opposition to the resumption of immigration led to immigration quotas designed to prevent Catholics and Asians from entering the country in significant numbers. American workers won these tight quotas over the opposition of big business, which wanted to keep its historic supply of cheap labor.

Similarly, a workingman's party spearheaded the drive to exclude Asians from the West Coast. As the socialist Cameron H. King Jr. wrote in 1908, this party swept California "with the campaign cry of 'The Chinese must go.' Then the two old parties woke up and have since realized that to hold the labor vote they must stand for Asiatic exclusion" (in Bonacich, 1972).

Laborers and farmers patrolled California and tried by illegal means to prevent Okies from entering the state during the 1930s (Mc-Williams, 1945). And it has almost always been white workers who have tried to drive out African Americans moving into the industrial regions of the North from the rural South.

CASTE SYSTEMS

Total exclusion has not always been possible. Under such conditions, dominant-group members have often tried to create a caste system that would restrict minorities to certain occupations, thus preventing them from competing for places in other occupations (Hechter, 1978).

As previously discussed, heredity is the basis of stratification in caste systems, and ascription determines status. Each caste group has exclusive access to certain occupations and positions in the society. Only those born to the highest caste are permitted to perform the most important occupations and hold the most powerful positions. Those born to the lowest caste must perform the most unattractive and servile work.

India is a classic instance of a very elaborate caste society. However, elements of a caste system can exist even in a society in which status is generally based on achievement rather than ascription. Within the dominant group of such societies, status may be based primarily on achievement, but one or more racial or ethnic groups may be ascribed a uniformly low status and thereby constitute a lower caste.

A low-caste status prevents a racial or ethnic group from competing with the dominant group for wages. A caste system may be created directly by strict rules about which positions a subordinate group may or may not hold or indirectly by excluding the subordinate group from entering unions or schools or from performing certain activities required to obtain better positions. For example, taboos against using drinking fountains or restrooms reserved for an upper caste may confine lower-caste members to a few limited physical locations and thereby to a narrow range of jobs. These subtle caste systems—long typical of race relations in the United States—constitute what Michael Hechter (1974) has called a **cultural division of labor,** whereby cultural or racial differences among members of a society are used as the basis for occupational placement.

MIDDLEMAN MINORITIES

It is important to realize that a cultural division of labor need not place an ethnic or racial minority on the bottom of a stratification system. Often, minorities have been used as "middlemen" in societies, serving as both links and buffers between the upper and lower classes. As Hubert M. Blalock (1967) and Bonacich (1973) have pointed out, **middleman minorities** often defuse potential class conflicts by becoming the focus of frustration and anger. In times of stress or unrest, Blalock noted, elites and the lower class often form a coalition against the middleman minority and vent their frustration on it. For example, many nations in feudal Europe permitted Jews to perform only certain middleman roles, such as moneylender, tax collector, and merchant. Thus, they were always potentially in conflict with both the nobility and the peasants. In hard times the nobility blamed Jewish tax collectors for decreased revenues but let them be the targets of peasant anger about high taxes. And in hard times both the nobility and the peasants were in debt to the Jewish moneylender or the Jewish merchant. Often enough, killing or expelling the Jews proved a most attractive way for both nobles and peasants to cancel their debts.

Chinese and Indian minorities frequently have formed middleman minorities in societies in Africa and Southeast Asia. They, too, have often been scapegoats for social conflicts within the dominant group (Blalock, 1967).

IDENTIFIABILITY

Status conflicts always cause bitterness. For example, it is common in high schools for groups to be clearly distinguished on the basis of school performance, popularity, and often, athletic skill and for there to be strong negative feelings between higher- and lower-status groups. This is true even when the groups are of precisely the *same* race, ethnic background, and religion—in French Catholic high schools in Quebec, for example. Indeed, we have seen how rapidly such barriers arise among little boys at summer camp.

When the conflicts truly affect one's life, and when conflicts occur between easily identifiable groups, they are even bitterer. For example, when people lose their jobs to strikebreakers or to those willing to work for lower wages, they are, of course, bitter. When the offending group is not set apart by cultural or racial differences, the anger must focus on the offending group as individuals, and their behavior is explained as the result of personal flaws. However, when the offending group has a clear group identification, bitterness can be directed toward *the group as a whole*. Then it is not a case of Harry, Marty, Mary, or Beth taking jobs but of those "dirty Swedes" or those "damn Irish" taking them. In this manner racial and cultural differences are infused with the passionate hatreds that status competition and conflict generate: What is merely different is made contemptible, and strangers are transformed into enemies.

Racial and culturally identifiable "enemies" frequently acquire derogatory names reflecting prejudice, hatred, and contempt directed toward them. In North America these names have included such terms as Wops, Hunkies, Micks, Krauts, Dagos, Gringos, Prots, Frogs, Breeds, Polacks, Rednecks, Kikes, Greasers, Wogs, Mackerel Snappers, Japs, Chinks, Chukes, and Niggers.[4] Indeed, as discussed earlier, one such group was known as Okies—people who moved west to California from the Oklahoma Dust Bowl during the 1930s. The prejudices against Okies were no different from those typically held against other racial and ethnic minorities (McWilliams, 1945). They were said to be dirty, to breed like rabbits, and to be superstitious, shiftless, sly, and ignorant. Yet the Okies were all white Anglo-Saxon Protestants, most of whom could trace their ancestry back to early colonial settlers. What made these people Okies? First, economic conflict did because they would take any job at any wage to feed their hungry families. Second, they had easily identifiable cultural traits: a rural southern dialect and country ways.

EQUALITY AND THE DECLINE OF PREJUDICE

Today, many of the richest farms in California are owned by people who arrived as Okies during the 1930s. Nobody calls these people Okies anymore, although they sometimes call themselves Okies as an expression of pride in how successful they have become.

If status competition and conflict fuel prejudice, then status equality causes the "tank to run dry." With status equality there is no longer anything to fight about. Once a subordinate group has achieved economic equality, for example, it no longer threatens as a source of cheap labor. And it no longer makes inroads into skilled occupations or into upper-status professional and managerial fields. The inroads have already been made.

Contact rapidly reduces prejudice under status equality because it occurs among equals with a mutual interest in improving society—in cooperating to make the economy more productive. Under these conditions, old hatreds and fears dissolve. This process is accelerated as younger generations with no experience of past conflicts become adults. Let's examine a recent example in detail.

THE JAPANESE EXPERIENCE IN NORTH AMERICA

The history of the Japanese immigrants and their descendants in North America is of special interest because both cultural and racial differences set them apart from the majority. Moreover, the Japanese immigrants were primarily farm laborers, literate but lacking education when they arrived. For decades they were the targets of intense prejudice and discrimination, especially in the Far West. How the Japanese overcame these barriers and achieved greater economic success than many immigrant groups of European origin provides an excellent foundation for the rest of the chapter. Although the

---

4. Many readers will be unable to connect some of these names with the ethnic or religious group to which they were applied. Good. You are living evidence of how rapidly prejudice can subside once status conflicts have been resolved. Most people over fifty could easily recognize these group epithets; they did not pass some of these words on to their children. Perhaps you will let your children be ignorant of the rest.

stories of the Japanese in the United States and in Canada differ only slightly, it will be useful to recount each separately.

JAPANESE AMERICANS

The Japanese never made up more than a tiny portion of the great tide of immigration to America. In 1907, their peak year of entry, only 30,000 arrived, making up just 3 percent of all immigrants for that year. However, the Japanese immigrants located mainly in Hawaii and California, causing great local uproar about a "yellow peril."

In California an alien land law was passed to prohibit Japanese from owning land. For a time the Japanese easily evaded this law by putting the land in the name of white neighbors or their American-born children. In 1920 a referendum was placed before California voters that attempted to close these loopholes and to exclude all aliens who were ineligible for citizenship from owning land. At that time only Asian aliens living in the United States were ineligible to become citizens. The law passed by a three-to-one margin, indicating the degree of public hostility to the Japanese. In time, however, court decisions and new evasion tactics defeated the intent of the law.

Meanwhile, in 1924 a federal law was passed that prohibited immigration from Japan and other Asian nations—an action that caused great offense in Japan and played a role in decisions leading to World War II. The immediate consequences of this Asian exclusion policy were a halt to immigration and a rapid transformation of the Japanese American population from being primarily foreign-born to being primarily American-born. This in turn led to rapid upward mobility by the Japanese Americans because no new, poor, unskilled immigrants arrived to depress the average status of the group.

Rapid Japanese American economic success was not limited to farming, although that was the usual occupation of the first generation. Early on, Japanese Americans discovered a great demand for their gardening skills, a form of self-employment requiring little more capital investment than a truck, mowers, and various hand tools (Jiobu, 1988). As early as 1928, there were 1,300 self-employed Japanese American gardeners in southern California (Kitano, 1969). Japanese Americans also set up

produce markets in the cities to sell their vegetables and fruits, and in 1929 there were more than 700 such markets in Los Angeles alone (Light, 1972). As late as 1940, the majority of Japanese American men in America were farmers. That same year about one-third of all commercial truck farm crops grown in California were produced by Japanese American farmers (Kitano, 1969; Petersen, 1971, 1978; Sowell, 1981). Japanese Americans also excelled in running other small businesses where unskilled labor, especially the labor of children, could be most productive. For example, by 1919 Japanese Americans owned almost half of the hotels and a fourth of the grocery stores in Seattle (Light, 1972).

The first generation of Japanese Americans worked hard to establish their own farms and small businesses. They were aided in this by the creation of their own credit associations, in which they pooled their savings and used them to finance one another in starting businesses or buying property (Miyamoto, 1939). But the first generation did something else equally important: They sent their children to school.

Table 11-3 compares the proportions of Japanese Americans enrolled in school at various ages in 1930 with native-born white Americans.

Japanese American children in California in about 1910 with their book bags suspended from a carrying pole. Although Japanese immigrants were not well educated and primarily worked in agriculture, from the very start they sent their children to school. Within a few decades, they far surpassed the average level of education of other Americans.

© Brown Brothers

TABLE 11·3	Enrollment in School at Various Ages, 1930	
AGE	NATIVE-BORN WHITES (%)	JAPANESE AMERICANS (%)
7–13	96.1	97.2
14–15	90.0	97.3
16–17	61.0	88.8
18–20	24.4	51.8

Source: Adapted from Hirschman and Wong (1986).

(Alaska and Hawaii were not states then and are not included in these statistics.) Among children age seven to thirteen, nearly everyone is in school. But by age fourteen a clear difference appears: 97.3 percent of the Japanese Americans are in school, while only 90 percent of native-born whites are attending. At age sixteen the gap has grown very large—88.8 percent versus 61 percent in favor of Japanese Americans. And when we examine people of college age, the Japanese American advantage is slightly more than two to one.

By 1940 these patterns had resulted in the average Japanese American male having gone to school about a year beyond high school, whereas the average white American male had attended school only slightly beyond the eighth grade! Later, this education would pay huge occupational dividends (Hirschman and Wong, 1984, 1986).

But first we must consider the war. On December 7, 1941, carrier aircraft of the Japanese Imperial Fleet made a massive surprise attack on the American fleet lying at anchor in Pearl Harbor. At the same time, Japan launched invasions of many parts of Southeast Asia and many Pacific islands. The war inflamed public emotions. Most Americans at that time agreed with President Franklin D. Roosevelt that December 7, 1941, was a "day that will live in infamy."

In the immediate aftermath of Pearl Harbor, concern mounted about the allegiance of the Japanese American population. Many American leaders expressed the belief that the Japanese Americans remained loyal to Japan and posed a threat as saboteurs and spies. Initially, 1,500 Japanese who had been outspoken supporters of Japan were arrested as potential security risks. Building on the already substantial public prejudice against the Japanese, war hysteria led to the rounding up of 100,000 Japanese Americans living on the West Coast—men, women, and children. They were relocated in a number of internment camps, to the applause of political leaders and writers across the political spectrum. All seemed united in the belief, expressed by General J. L. DeWitt, who directed the roundup, that "a Jap's a Jap, and it makes no difference whether he's a citizen or not." The blind prejudice of such views was later exposed by the thousands of Japanese Americans who fought in Europe during World War II, including the legendary 442nd Regimental Combat Unit, whose heroics made them the most decorated unit in the U.S. Army.

Internment brought economic disaster to the Japanese Americans living on the American mainland (oddly enough, only they were interned, while the 150,000 Japanese Americans in Hawaii, living close to Pearl Harbor, were not). Thousands of farms and small businesses were lost forever, while many others were sold for next to nothing by desperate people forced to leave for a camp. And bigotry flourished. A Gallup Poll conducted on the West Coast in December 1942, a year after Pearl Harbor, found that an overwhelming majority of white Americans in that region said they would never again hire Japanese or shop in stores owned by Japanese after the war was over.

At war's end the Japanese Americans resumed their efforts to find a place for themselves in American life. The immediate postwar American economy was booming, and there was great demand for highly educated people, especially those with technical or financial training. Now the huge educational advantages achieved by the Japanese Americans before the war translated into occupational achievement. By 1960, eighteen years after their internment and terrible financial losses, the wages of employed Japanese American men had nearly caught up with that of native-born whites, as can be seen in Table 11-4. In another ten years, they had exceeded that of whites, and the gap has increased ever since.

For the sake of comparison, the table also shows the relative income of Chinese American males. Like the Japanese Americans, the Chinese also have been unusually committed to education. Thus, by 1980 Asian Americans as a group were much more likely than other Americans to have college degrees. Equally important, their degrees are concentrated in fields such as engineering, science, medicine, and law— degrees that offer direct entry into the highest-paying occupations.

However, a continuing influx of Chinese immigrants masks the fact that the average American-born Chinese male also probably earns more than the average white (there has been little immigration from Japan).

As the Japanese Americans overcame economic inequality, attitudes toward them shifted rapidly. This can be demonstrated in several ways. The first is intermarriage. Perhaps nothing emphasizes the stress in relations between a majority and a disliked minority more than the question, Would you want your son or daughter to marry one? The answer is always a resounding no. In the 1920s only about 2 percent of all Japanese American marriages involved a non-Japanese spouse. By the 1950s this had climbed to about 20 percent (Levine and Montero, 1973) and soon after reached about 50 percent (Kihumura and Kitano, 1973; Heer, 1980; Rodriguez, 1999). Thus, by 1968, for every three babies born with two Japanese American parents, there were two born having only one Japanese American parent. By 1981 the numbers balanced, and since that time more infants have been born with only one rather than two Japanese American parents. In 1989, for example, there were three babies born having only one Japanese American parent for every two born with two Japanese

TABLE 11-4	The Post-World War II Earnings Recovery by Japanese Americans*			
	RATIO TO WHITE EARNINGS			
	1960	1970	1980	1990
Whites	100	100	100	100
Japanese Americans	91	101	105	120
Chinese Americans	84	92	89	90

Sources: Hirschman and Wong (1984) and Hirschman and Snipp (1999).

*Only for men age twenty-five to sixty-four.

American parents. These statistics reflect a dramatic change in attitudes.

JAPANESE CANADIANS

In Canada, too, the Japanese were only a small group—the 1901 census counted about 20,000, nearly all of them living in British Columbia. Also in Canada even this small number caused a panic about a "yellow peril." In 1895 Japanese Canadians were denied the right to vote. They also were excluded by law from most professions, from employment in the civil service, and from teaching positions. In 1907 white mobs

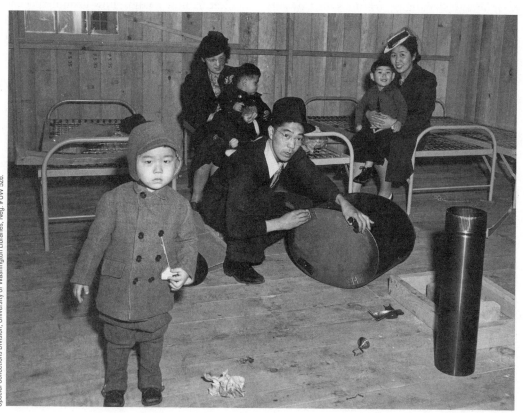

Irwin Yoshimura stares into the camera while his father attempts to set up an oil stove to heat the barracks to which the Yoshimura family and other Japanese Americans have been assigned at a temporary assembly point in Puyallup, Washington, in 1942. Notice that mattresses for the cots have not yet arrived. From Puyallup this group was sent to Moses Lake in eastern Washington where a more permanent detention camp was established for Japanese Americans. Persons of Japanese ancestry also were forced to leave the west coast of Canada. In both nations families like the Yoshimuras suffered severe financial losses as well as humiliation.

rampaged through the Japanese district in Vancouver intent on driving out "heathen Orientals." Soon after, the Canadian and Japanese governments reached an agreement to allow only 400 new immigrants to come to Canada each year. In 1928 a new agreement cut the quota to 150 a year.

As in the United States, many of the Japanese took up farming. Others settled in small fishing villages where they continued to sustain themselves—even though, during the 1920s, the government severely limited the number of Japanese granted licenses to fish. Also as in the United States, Japanese Canadian children went to school. By the 1930s the average Japanese born in Canada probably had more years of schooling than did most other Canadian ethnic groups (Ujimoto, 1976).

In the wake of the bombing of Pearl Harbor, Canadian authorities moved against Japanese residents, demanding that they immediately withdraw to more than 160 kilometers (100 miles) from the Pacific Ocean. About 20,000 were crowded into the livestock barns on the exhibition grounds in Vancouver, and from there they were taken to inland camps to spend the war years in ramshackle barracks built almost overnight.

In the United States, the Japanese suffered great financial damage from having to sell their property in haste or abandon it. In Canada the government confiscated Japanese-owned farms, homes, businesses, fishing boats, and the like and sold them. After the war the Canadian government attempted to deport about half of those interned but was prevented from doing so by public protests. Finally, in 1949 Japanese Canadians were granted the right to vote.

Despite all this the Japanese Canadians have achieved immense economic and social progress (Ujimoto, 1976). Making use of their educational advantages and newly granted freedom of occupational choice, they flocked into universities, the professions, the arts, and business administration. Today, Japanese Canadians have an average income higher than that of most white ethnic groups. And just as there has been a very rapid shift in attitudes toward the Japanese in the United States, so, too, have they gained social acceptance in Canada—as indicated by an intermarriage rate that is well above 50 percent (Adachi, 1978; Sunahara, 1981).

Thus, we see that despite deep prejudice, despite clear racial and cultural markers, and despite the added hysteria caused by war with Japan, the Japanese in the United States and Canada suddenly ceased to be strangers and enemies as they achieved status equality.

Let us now analyze how groups succeed in North America. How did Italians, Poles, Irish, Japanese, and the other immigrant groups rise from poverty to gain equality and acceptance?

MECHANISMS OF ETHNIC AND RACIAL MOBILITY

In the course of North American history, a great many different ethnic and racial groups have been on the bottom of the stratification system. Nearly without exception, while each was on the bottom, it was widely believed that their future was hopeless and that they would always be on the bottom. For example, at the time of the American Revolution, Benjamin Franklin wrote that the Germans who had recently settled in Pennsylvania would never fit in or contribute to American society.

But time and again these prophecies have been wrong, as group after group has escaped an ascribed low status to achieve equality and thereby eliminate prejudice. The proper Bostonians "knew" that the "drunken, lazy, superstitious

Officers of the Royal Canadian Navy confiscating the boat of a Japanese Canadian fisherman in Esquimalt, British Columbia, shortly after the bombing of Pearl Harbor. The area within 160 kilometers of the west coast of Canada was made off-limits to Japanese Canadians during World War II. Since this was where most of them lived, they were placed in detainment camps inland, and the government seized and sold their farms, businesses, homes, and property.

Public Archives, Canada/PA-37468

Irish" would never amount to anything. But a century later, their average income had surpassed that of Protestants of English origin in the United States (Greeley, 1974).

The waves of immigrants from southern and eastern Europe who began arriving after 1880 caused one of America's leading sociologists of the time, E. A. Ross, to write in 1914 that they were so racially inferior, they would drag the nation down, intellectually and morally:

It is fair to say that the blood now being injected into the veins of our people is "subcommon." [Many of these new kinds of immigrants] are hirsute, low-browed, big-faced persons of obviously low mentality. . . . Clearly they belong in skins, in wattled huts at the close of the Great Ice Age.

That the mediterranean peoples are morally below the races of northern Europe is as certain as any social fact. Even when they were dirty, ferocious barbarians, these blonds [northern Europeans] were truthtellers.

The Northerners seem to surpass southern Europeans in innate ethical endowment . . . but they will lose these traits in proportion as they absorb excitable mercurial blood from southern Europe.

The year Ross published this book, he served as president of the American Sociological Society and as professor of sociology at the University of Wisconsin. Ironically, in 1959 Ross's pioneering essays on social control were rescued from obscurity, edited, and republished by two leading American sociologists, Edgar F. Borgatta and Henry J. Meyer—an Italian and a Jew! Presumably, eastern and southern Europeans had biologically evolved an incredible amount in only fifty years, if one were to take Ross's 1914 judgments seriously.

If many groups have risen from the bottom of the stratification system in North America, how have they done so? Through what tactics or mechanisms have groups achieved their upward mobility? Three basic elements seem crucial. The first is geographical concentration. The second is internal economic development and occupational specialization. The third is development of a middle class (Wirth, 1928; Glazer and Moynihan, 1970; Sowell, 1981; Portes, 1987).

GEOGRAPHICAL CONCENTRATION

Discrimination has often forced subordinate ethnic, racial, and religious minorities to live in enclaves segregated from the surrounding dominant group. However, in addition to discrimination, the tendency for racial and ethnic minorities to concentrate in certain neighborhoods has been encouraged by the needs and desires of subordinate groups: to band together for self-help, to maintain familiar features of their native culture (such as churches, festivals, and traditional foods), and to use their own language in daily life (see Chapter 19).

Geographical, or neighborhood, concentration is often denounced as a barrier to better intergroup relations. Yet Allport's theory suggests that, at least in the early days of a subordinate group's American experience, concentration may help to minimize the negative consequences of contact. But geographical concentration may have additional benefits: When a group is concentrated in a few locations, its economic and political power is maximized.

Consider the Irish. Today, they make up about 8 percent of the U.S. population. Suppose that when they arrived in this country, they had scattered across the landscape. If they had spread out evenly, then everywhere they would have constituted 8 percent of the population, and nowhere would they have been more numerous. But the Irish congregated in a few major cities and in particular neighborhoods within those cities. As a result they quickly became a majority or a sizable minority of the local population. Hence, they soon exerted maximum pressure on local affairs affecting their interests, running the governments of many cities and using their power to further their economic and social interests.

In similar fashion the Japanese and Chinese congregated on the West Coast, where their numbers mattered. Indeed, from 1976 through 1980, three U.S. senators were Japanese Americans, although Japanese make up less than 0.4 percent of the U.S. population. In both Canada and the United States, Jews have achieved political influence beyond their numbers because they are geographically concentrated.

The importance of geographical concentration is perhaps nowhere so clear as in the case of French Canada. Far from being a state of mind or an ethnic subculture, the French in Canada are a virtual society. The great durability of French culture and the present political power of French-speaking Canadians are rooted in geography—in the existence of Quebec.

INTERNAL ECONOMIC DEVELOPMENT AND SPECIALIZATION

Geographical concentration also facilitates a group's ability to develop its own economic resources and institutions, which can then be used to finance further upward mobility. In the beginning most subordinate groups have difficulty getting credit to finance businesses or agricultural enterprises, and they desperately need financial resources not subject to outside control. To meet this need, immigrant groups typically pool their funds to provide start-up capital for purchasing or opening small stores and businesses within their own neighborhoods. Then, *by buying within their own community and by reinvesting the profits, immigrant groups gain economic freedom.*

Here geographical concentration can be vital. If a group is not concentrated in particular neighborhoods, then it will be difficult to shop only in stores run by group members. But when a group is concentrated, economic self-interest is reinforced by the convenience of shopping close to home.

Here, too, language may play a role. Unlike most groups, the Irish did not start their rise to equality by first acquiring neighborhood businesses. This may have been because the Irish spoke English. Polish or Japanese housewives were often willing to pay higher prices at small groceries run by Poles or Japanese because they wanted to shop where clerks understood their language. However, Irish housewives lacked this inducement to patronize Irish grocers.

This factor may also pose a major difficulty for African Americans attempting to run small shops in African American neighborhoods. Unlike Puerto Ricans and other Spanish-speaking groups in America, African Americans, like the Irish, have not tended to run small businesses. Although Spanish-speaking consumers may be reluctant to shop in larger stores where clerks do not speak Spanish, African American consumers have no language barrier hindering them from going where prices are lowest (Glazer, 1971).

A major step forward for most subordinate groups has been the founding of their own financial institutions. We have seen that the Japanese Americans created small credit associations and eventually founded their own banks. In Chapter 2 we traced the history of a small neighborhood bank that began as a source of loans for the Italians in San Francisco and eventually grew into the Bank of America. The small local savings bank was also a typical feature of Irish communities. In fact, Joseph Kennedy, father of President John Kennedy, began his remarkable finance empire by opening such a small savings bank.

In addition to seeking economic development and independence, many racial and ethnic minorities have taken advantage of the particular occupational opportunities existing when they arrived. As a result they have often been (and continue to be) highly overrepresented in certain specialized occupations. We have already seen how the Japanese Americans specialized in a few occupations: operating truck farms, gardening, and running small hotels and grocery stores. More recently, they are greatly overrepresented in engineering, optometry, medicine, and dentistry (Lieberson, 1980; Sowell, 1981). Such specialization maximizes the capacity for self-help: People already in an occupation can aid friends and relatives by showing them how to gain entry to the occupation (and perhaps even hiring them); indeed, such people can be the primary source of training for that occupation (Hechter, 1978).

Such occupational specialization has been typical of subordinate groups. For example, the Irish used their control of city governments to enter civil service occupations. As late as 1950, the Irish in the United States were three times more likely than other whites to be police officers and firefighters. In similar fashion in 1950 Italians were eight times overrepresented among barbers, Swedes were five times overrepresented among carpenters, Greeks were twenty-nine times overrepresented among restaurant operators, Jews were seventeen times overrepresented among tailors, and nearly all diamond cutters were Jewish (Lieberson, 1980).

This line of analysis has led to the formulation of an **enclave economy theory.** Most closely associated with Alejandro Portes and his colleagues (Portes, 1981, 1987; Portes and Bach, 1985; Portes and Manning, 1986), the theory stresses that ethnic and racial minorities can make more rapid initial economic progress when they create an enclave economy. Portes (1981) describes such economies as consisting of

immigrant groups which concentrate in a distinct spatial location and organize a variety of enterprises serving their own ethnic market and/or the general population. Their basic characteristic is that a significant proportion of the immigrant work force works in enterprises owned by other immigrants.

In part this scene from San Francisco's Chinatown at the turn of the twentieth century reflects the hostility and discrimination against Asians that was typical of the time and place. But Chinatown stood for more than segregation; it also helped the Chinese concentrate the economic and political resources of their community, which in time helped them escape from poverty and prejudice. The same pattern has governed the history of many racial and cultural minorities in Canada and the United States.

The advantages of this situation are that it creates an increasingly successful group of entrepreneurs within the ethnic community and that ties of ethnic solidarity force these business owners to give their workers a better deal than would employers from outside the enclave. Moreover, in an enclave economy, firms will promote members of the minority to supervisory positions from which they would tend to be excluded outside (Portes and Bach, 1985; Portes and Jensen, 1989).

DEVELOPMENT OF A MIDDLE CLASS

Progress by subordinate groups requires leadership and expertise. To create financial institutions requires group members who know how to run them. To use a concentrated population to influence local politics requires political leadership.

We saw in Chapter 2 that some immigrant groups arrived with a large middle-class contingent—Jews and, later, Cubans are examples. Both groups made extremely rapid economic progress because they already had the skills needed for effective community building. Moreover, both groups used higher education to provide their children with a rapid route for entering upper-income occupations not requiring much capital investment. Unlike buying a store or starting a bank, no initial investment is required for a person to earn a high salary as a scientist, physician, engineer, or accountant. The person need only obtain the appropriate education. That the Jews rapidly became the most highly educated group in North America accounts for the fact that they also have the highest average family income of any racial, religious, or ethnic group (Greeley, 1974; Heaton, 1986). As we have seen, the Japanese also used higher education to improve their status quickly.

We already have seen that initially the concentration of a racial or ethnic group helps to create internal economic resources, and one important use of these resources is to educate the next generation. However, there comes a time when a racial or ethnic minority must begin to achieve substantial assimilation if it is to continue its upward mobility. As Jimmy M. Sanders and Victor Nee (1987) have demonstrated in their study of Cuban and Chinese immigrants, ethnic solidarity and concentration are favorable to the growth of business and financial firms and thus to the self-employed, but group members earn higher salaries when they are employed outside the enclave, if they have achieved sufficient levels of cultural assimilation to compete. Here mastery of language is of primary importance. We may amplify this point by pausing to examine the most rapidly growing American minority: persons of Hispanic origins and culture. In so doing we can review portions of what has been covered thus far in the chapter.

HISPANIC AMERICANS

In the late 1980s, demographers estimated that Hispanics would outnumber African Americans by the year 2015 (Exter, 1987). But it happened much sooner than that—the 2000 Census found that 12.6 percent of Americans were Hispanic compared with 12.3 percent who were African Americans. The U.S. Bureau of the Census defines the Hispanic population on the basis of persons' self-identification of their origins or descent as "Mexican American," "Chicano," "Mexican," "Puerto Rican," "Cuban," "Central or Southern American," or "Other Spanish." In 2000 more than 35 million Americans selected one of those responses when asked (Table 11-5).

The geography of the Hispanic population reflects the fact that it is diverse. More than half of all Hispanics live in New Mexico, California, Texas, and Arizona (Figure 11-3). The overwhelming majority of these are people of Mexican origin or descent—they make up nearly 60 percent of all Hispanic Americans. New York also has a substantial Hispanic population, and these people are mostly Puerto Ricans, whereas Florida's Hispanic population is primarily of Cuban background.

As a result, it is quite misleading to regard "Hispanic" as the name of an ethnic group. Most persons of Hispanic descent interviewed in a re-

TABLE 11-5	American Minorities	
GROUP	NUMBER (IN 1,000s)	PERCENT OF POPULATION
Hispanic Americans	35,306	12.6
Mexican Americans	20,641	7.3
Puerto Rican Americans	3,406	1.2
Cuban Americans	1,243	0.5
Other Hispanic Americans	10,017	3.6
African Americans	34,658	12.3
Asian Americans	10,243	3.7
Chinese Americans	2,433	0.8
Filipino Americans	1,850	0.7
Asian Indians	1,679	0.6
Vietnamese Americans	1,123	0.4
Korean Americans	1,077	0.4
Japanese Americans	797	0.3
Other Asian Americans	1,285	0.5
American Indians and Alaska Natives	2,476	0.9
Pacific Islanders	399	0.1

Source: *Statistical Abstract of the United States, 2005.*

cent national survey indicated they preferred to be identified on the basis of national origin rather than referred to as Hispanic Americans. That is, most would rather be called Mexican Americans, Puerto Rican Americans, or Cuban Americans, while substantial numbers of those born in the United States preferred to identify themselves simply as Americans (Brodie et al., 2002).

These preferences are reflected in the marked differences in the situations of the three major Hispanic groups. For example, there are no "illegal" Puerto Ricans living in the United States, since all persons born in Puerto Rico are American citizens. Cuban Americans differ greatly from the other two groups in terms of economic circumstances: They have substantially higher median incomes and are far more apt to be in high-status occupations. One factor in the greater Cuban success is that it was predominately the more educated and successful people who fled from Cuba to escape the Castro regime. But a second very important aspect is that by now most Cuban Americans were born here. In contrast, more than two-thirds of Mexican Americans and Puerto Rican Americans are foreign-born (Brodie et al., 2002).

The impact of place of birth shows strongly in Table 11-6. Huge increases in household income and occupational status occur between the newly arrived foreign-born Hispanic immigrants and their children. The table also shows a major cause of these differences: *education*. The majority of foreign-born Hispanics have not graduated from high school; three-fourths of their children have. Few foreign-born Hispanics attended college; nearly half of their children have done so. Not surprisingly, research finds only trivial income differences between Hispanic and non-Hispanic males when comparisons are made between high school graduates, *if they are fluent in English* (Stolzenberg, 1990). And that raises an additional issue: assimilation.

As was clarified in Chapter 2, all immigrant groups are at an economic disadvantage when they first arrive because it takes time to acquire the skills needed to obtain higher-paying jobs. Of these, probably no single factor is as important as learning to speak English. Table 11-7 shows the belief that Hispanics don't learn English is nothing but a myth: 96 percent of those born here do. In fact, other data show that by the third generation, many Hispanics no longer can speak much Spanish (de la Garza et al., 1992). The fact that so many Hispanics have limited English skills is because the majority are foreign-born.

FIGURE 11-3   **Distribution of Hispanic Americans.**

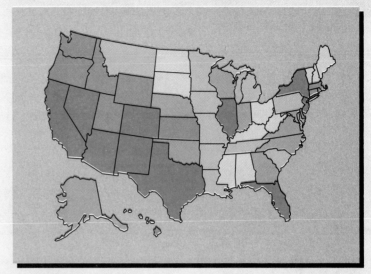

Percentage of Hispanic Americans

New Mexico	42.1
California	32.4
Texas	32.0
Arizona	25.3
Nevada	19.7
Colorado	17.1
Florida	16.8
New York	15.1
New Jersey	13.3
Illinois	12.3

Michigan	3.3
Arkansas	3.2
Pennsylvania	3.2
Minnesota	2.9
Iowa	2.8
Louisiana	2.4
South Carolina	2.4
Tennessee	2.2
Missouri	2.1
Montana	2.0

Connecticut	9.4
Utah	9.0
Rhode Island	8.7
Oregon	8.0
Idaho	7.9
Washington	7.5
Hawaii	7.2
Kansas	7.0
Massachusetts	6.8
Wyoming	6.4

Ohio	1.9
Alabama	1.7
New Hampshire	1.7
Kentucky	1.5
Mississippi	1.4
South Dakota	1.4
North Dakota	1.2
Vermont	0.9
West Virginia	0.7
Maine	0.7

Nebraska	5.5
Georgia	5.3
Oklahoma	5.2
Delaware	4.8
Virginia	4.7
North Carolina	4.7
Maryland	4.3
Alaska	4.1
Wisconsin	3.6
Indiana	3.5

Hispanics in New Mexico, California, Texas, Arizona, and Colorado are primarily Mexican Americans. In New York the majority are of Puerto Rican descent; in Florida Cuban Americans predominate.

TABLE 11-6	Income, Occupation, and Education by Place of Birth among Hispanic Americans	
	FOREIGN-BORN	**NATIVE-BORN**
Household Income above $30,000	35%	60%
In a White-Collar Occupation	32%	71%
Graduated from High School	45%	77%
Attended College	16%	42%

Source: Brodie et al., 2002.

TABLE 11-7	Assimilation of Hispanic Americans	
	FOREIGN-BORN	**NATIVE-BORN**
English-speaking	28%	96%
Describe Themselves as "Americans"	32%	90%
Roman Catholic	76%	59%
Protestant	16%	30%

Source: Brodie et al., 2002.

In similar fashion the table shows that among those born here, 90 percent prefer to describe themselves as Americans (as opposed to, say, Mexican Americans). This disposes another myth about Hispanics. Finally, the effects of living in the United States show up very clearly in terms of religious affiliation: Native-born Hispanics are about twice as likely as their parents to be Protestants.

GOING NORTH: AFRICAN AMERICAN "IMMIGRATION" IN THE UNITED STATES

Thus far, the focus of this chapter has been on racial and ethnic groups that are relatively recent immigrants. At first glance it might seem that theories and research which apply to them would not have much relevance for understanding the situation of African Americans. Members of this minority group are not recent immigrants: Most of them can trace their ancestry in the New World back more generations than can any white group. As early as 1680, most African Americans in North America were native-born, and by 1776 nearly all of them were (Fogel and Engerman, 1974; Patterson, 1982).

Hence, it may seem silly to apply our model of how new immigrants achieve success and acceptance in two or three generations to a group that has endured many generations of privation. However, a number of sociologists (Lieberson, 1973, 1980; Sowell, 1981) have rejected this conclusion. They argue instead that, in the ways that count, African Americans, American and Canadian Indians, and Inuit have, until recently, been nearly as isolated from the mainstreams of American and Canadian society as they would have been had they lived in another country. In analyzing their situations, then, it makes sense to regard these groups *as if* they were recent immigrants. Moreover, because of the legacies of slavery and their long history of suffering, African Americans offer the most significant and theoretically strategic case. So, let's take a sociological look at African American history.

Even today, about half of the African American population in the United States lives in the South. Until the 1940s the overwhelming majority of African Americans lived in the South. In addition the African American population was not merely southern but also very heavily rural.

World War II propelled a great wave of African American migration north. Recruiters for northern industries toured the South, sending trainloads of African Americans north to work in the defense industries. After the war, the movement north continued, eventually involving millions of people. As a result, whereas only a few decades ago most African Americans lived in rural areas, today most live in cities. More African Americans now live in Chicago than in the whole state of Mississippi, and more live in New York City than in any southern state.

Admittedly, African Americans crossed no national border when they journeyed from Tupelo to Chicago or from Macon to New York. But because of the conditions of life in the South until very recently, going north for African Americans was more like a journey from rural southern Italy to America than a journey from Chicago to Detroit.

Recent industrial development and population growth have transformed the American South so greatly that even many younger southerners are unaware that only a few years ago this region was mired in grinding poverty and backwardness. Indeed, while the legacy of slavery lived on in the discrimination against African Americans (they could not attend schools with whites or even drink from the same public water

Library of Congress

Roping the last of their baggage to the rear of their car, this African American family prepares to flee the poverty of rural Florida in 1940 to begin a new life in New Jersey. Although technically they were not immigrants, their journey crossed cultural gaps as great as those crossed by people who came from overseas.

fountains as whites), the most severe deficiencies of southern life affected *both* African Americans and whites.

Table 11-8 attempts to show the misery of the American South by contrasting conditions in Mississippi with those in Iowa—both highly agricultural states settled somewhat later in American history. Data for New York State offer an additional contrast.

Notice that in 1850 the majority of persons living in Mississippi were African American slaves, while Iowa and New York were by then free states. Few Americans anywhere went to college in 1850, but in the whole state of Mississippi that year, there were eleven college students. In Mississippi most animals used for traction (to pull wagons and farm equipment) were mules or oxen, not horses—a clear indication of backward agricultural technology.

By 1880 the majority of Mississippi citizens were free African Americans, but the state remained mired in ignorance and poverty. Nearly half of the population was illiterate—and the average grade school operated only seventy-seven days a year. Kids in Iowa went to school twice as long. Forty years later Mississippi's poverty was still acute. In 1920 the average farm in Iowa was worth more than *twelve times* as

much as the average farm in Mississippi. Nearly all the farms in Iowa had phones; most farms in Mississippi did not. The school year had lengthened greatly in Mississippi, but it still lagged behind Iowa and New York.

By 1940 African Americans made up a slightly smaller percentage of Mississippi's population and a larger proportion of New York's. Yet infant mortality was still almost twice as high in Mississippi as in Iowa and New York. With the Great Depression not yet fully abated, farm values and family incomes were low everywhere but were much, much lower in Mississippi. That year nearly all homes in northern states had radios. But in Mississippi more than a third could not listen to Jack Benny or "The Lone Ranger." By 1960 the poverty gap between Mississippi and the two other states had narrowed some. But still, while most of the nation watched football and old movies on TV, a third of those in Mississippi could only stare at the wall. By 1980 the gap had narrowed yet again, and nearly half of the state's residents were urban. Mississippi students were dropping out of high school at about the same rate as in New York—with the dropout rate being far lower in Iowa. By 2000 dropout rates had fallen dramatically in all three states, and Mississippi

TABLE 11-8	Southern Poverty, 1850–2000		
		MISSISSIPPI	IOWA	NEW YORK

	MISSISSIPPI	IOWA	NEW YORK
1850			
Population (in 1,000s)	607	192	3,097
Slaves	51%	0%	0%
Urban	2%	5%	28%
College students per 1,000	0.02	0.52	0.86
Mules or oxen among traction animals	54%	38%	28%
1880			
African American	52.2%	0.6%	1.3%
Illiterate	41.9%	2.4%	4.2%
Number of days in school year	77	148	179
1920			
Average farm value	$2,903	$35,616	$7,376
Farms with autos	5.5%	73.1%	35.2%
Farms with telephones	10.4%	86.1%	47.6%
Automobiles per 1,000	27.2	150.1	38.5
Number of days in school year	122	174	188
1940			
African American	49.2%	0.7%	4.2%
Infant mortality per 1,000 live births	61	37	37
School spending per pupil	$25	$78	$147
Median annual family income	$386	$746	$1,048
Average farm value	$1,632	$12,614	$6,180
City homes with a radio	61.5%	93.7%	96.7%
1960			
African American	42.0%	8.0%	9.3%
Income per capita	$1,208	$2,010	$2,718
Homes with a telephone	45.3%	89.2%	82.3%
Homes with a TV set	66.5%	89.2%	90.7%
1980			
African American	35.2%	1.4%	13.7%
Median family income	$14,922	$20,243	$20,385
High school dropouts	36.3%	12.0%	33.3%
Urban	47.2%	58.6%	84.6%
School spending per pupil	$1,610	$2,264	$3,180
2000			
African American	36.3%	2.1%	15.9%
Median family income	$26,677	$33,209	$35,410
Number of students per computer	14.4	8.1	10.4
High school dropouts	10.8%	6.6%	10.0%

students were not far behind in terms of access to computers.

The decline in the African American population of Mississippi and the increase in proportion of African Americans living in New York reflect the massive northward migration of southern African Americans in the past few decades. And these historical data help us under-

stand what African Americans were *leaving* and what they *brought with them* from the South—experience only as subsistence farmers and field hands, little schooling (which was of low quality), no technical training or modern job skills, and no experience of living in large cities. In these important respects, they were truly going off to a new country. Only their ability to speak English set them apart from most other immigrant groups.

AFRICAN AMERICAN PROGRESS

Stanley Lieberson (1973, 1980) has severely criticized sociologists assessing African American progress for failing to take into account the fact that African Americans are so newly arrived from the South. When sociologists analyze the progress of other racial and ethnic groups, they automatically control for the number of generations in this country. For example, when we compare the educational achievement or the income level of Italian Americans with that of German Americans, we first sort the data to eliminate differences resulting from the fact that one group has been in America longer. Thus, we are trying to compare first-generation Italians with first-generation Germans and so on.

Until recently, sociologists compared northern African Americans with northern whites without such generation controls. In fact, only in recent years have a substantial number of African Americans been born in the North, and relatively few African Americans are the children of parents born in the North. When such generational controls are taken into account, many African American–white differences are greatly reduced. The more generations that African Americans have lived in the North, the higher their education and the better their jobs (Lieberson, 1973, 1980; Sowell, 1981).

The fact that most African Americans lived in the South for so long has been a major impediment to their success. That many southern whites have recently followed the African Americans north emphasizes this point, for these white migrants have displayed many of the same disadvantages.

All of these matters aside, African Americans have been making rapid progress in education and income. A second important factor in assessing recent African American progress is age. Because African Americans have had a higher birthrate than whites in recent decades, the African American population is somewhat younger than the white population. For this reason comparing the average family income of African Americans and whites is also comparing the average income of younger and older families. Because people usually earn more as they get older, such comparisons can misleadingly suggest racial and ethnic differences in success that are merely due to age differences (Sowell, 1981).

TABLE 11-9	Relative Educational Gains by African Americans, 1960–2000*					
	COMPLETED HIGH SCHOOL (%)					
	1960	1970	1980	1990	2000	2003
African Americans	20.1	31.4	51.2	66.2	78.5	80.0
Whites	43.2	54.5	68.6	79.1	84.9	85.1
As a percentage of white rate:	46.5	57.6	74.6	83.7	92.5	94.0
	COMPLETED COLLEGE (%)					
African Americans	3.1	4.4	8.4	11.3	16.5	17.3
Whites	8.1	11.3	17.1	22.0	26.1	27.6
As a percentage of white rate:	38.3	38.9	49.1	51.4	63.2	63.0

Source: *Statistical Abstract of the United States, 1992, 2005.*

*Persons over age twenty-four.

In any event African Americans have made very substantial educational gains over the past two decades, as can be seen in Table 11-9. In 1960 most Americans over age twenty-four had not graduated from high school, but whites were more than twice as likely to have done so than were African Americans. Since then, while high school graduation has risen rapidly in both groups, the gap between them has also been closing fast—by 2003 African Americans were 94 percent as likely as whites to have a diploma. A similar, if less rapid, pattern holds for having a college degree.

In addition to differences in age, another factor distorting income comparisons between whites and African Americans has to do with the fact that there is a far higher percentage of female-headed families among the latter. In 1950 only about one of every ten African American families with children under the age of eighteen was headed by a woman. In 1991, 54 percent of these African American families lacked a father, as opposed to only 17 percent of white families. Female-headed families of any race or ethnicity have less than half the median income of married couples and far less income than families in which both the husband and wife are employed. In fact, substantial numbers of female-headed families must rely on welfare for their support. As a result, when the incomes of all white and African American households are compared, the latter have only 58 percent as much as the former. Much of the gap is due to the fact that African American families are more than three

To truly understand the American experience of African Americans, it is necessary also to understand the economic and social history of the American South—to know that until as recently as forty years ago, most southern whites as well as African Americans lived in rural poverty, received inferior educations, and suffered from poor nutrition. When white southerners like these little Georgia boys grew up, many of them also went north. And there, they too resembled immigrants and encountered many of the problems faced by African Americans from the South.

Margaret Bourke-White Estate/TimePix/Getty Images

TABLE 11-10	Income Gains by African American Working Couples	
	MEDIAN FAMILY INCOME IN 1990 DOLLARS	
	1967	2000
African Americans	$28,700	$50,758
Whites	$40,040	$57,242

Source: U.S. Bureau of the Census (2001).

times as likely to be headed by a female. When this factor is removed and the comparison is restricted to two-earner families, as in Table 11-10, the difference is far smaller—African American income is 89 percent that of whites. This is a significant shift since 1967, when African American working couples earned only 71 percent as much as white couples.

In Chapter 12 we shall attempt to understand why these sudden changes in family composition have so recently afflicted the African American family. We shall discover that a major cause is a shortage of African American men, especially young adults. African American women greatly outnumber African American men. And that means, of course, that for many African American women, the two-earner family is impossible.

But even with the poverty experienced by female-headed families, the economic situation of African Americans has improved greatly in recent decades. If we accept the notion of African Americans as immigrants, then they began to arrive in large numbers only forty years ago, and many came much more recently. Viewed this way, their record of achievement compares very favorably with that of most earlier ethnic and racial groups. This observation is in no way intended to minimize the centuries of African American suffering in North America or the continuing poverty of many African American citizens. But it does demonstrate the immense achievements of millions of African Americans—achievements that are overlooked by the news media, which tend to portray recent African American history as marked by unrelieved poverty, frustration, and helplessness.

Indeed, Thomas Sowell (1981) has bitterly attacked the common political rhetoric that racism prevents African American progress and that only a massive reformation of society can enable African Americans to gain equality as propaganda that dishonors and demeans African Americans. To accept such a position, he wrote, means that "some of the longest and hardest struggles for self-improvement must be denied—which is to say, history itself must be denied."

However, an accurate portrait of the situation of African Americans must also consider the many impoverished African American families and the special barriers that African Americans must overcome.

INTEGRATION

The image of race relations that the American national news media project is one of extremely segregated residential neighborhoods—that with the exception of a few token African American residents in some areas, American cities and towns consist of all-white and all-African American districts. As we shall see in Chapter 19, there is some truth to this media image, especially in the older industrial cities of the Midwest and Northeast. But there is also a great deal of variation among cities, with cities in the West being far less segregated. Moreover, the standard used to gauge integration is, as a practical matter, very hard to satisfy. As is discussed in Chapter 19, the U.S. Census Bureau does not consider a block or a neighborhood fully integrated unless its racial and ethnic composition is identical with that of the city as a whole. Thus, if a city is 80 percent white, 5 percent Asian, and 15 percent African American, then only blocks or neighborhoods having that exact 80-to-5-to-15 mix are classified as integrated. When this standard is used, few if any blocks or neighborhoods anywhere qualify as fully integrated. But if we use a less restrictive test of integration, a quite different picture emerges. The 1998 General Social Survey asked African Americans if there were "any whites living in this neighborhood" and asked whites if there were "any blacks living in this neighborhood." Of whites, 63 percent said theirs was an interracial neighborhood, while 82 percent of African Americans said whites lived in their neighborhood.

That African Americans are twice as likely as whites to live in an interracial neighborhood is due in large part to simple arithmetic. Since whites outnumber African Americans by almost nine to one, many whites could live in an all-white neighborhood even if every African American lived in an interracial neighborhood. The same principle applies to interracial friendships. Thus, a recent survey reported that 82 percent of African Americans said they knew a white person whom they considered "a fairly close personal friend," compared with 62 percent of whites who claimed a fairly close African American friend (Sigelman and Welch, 1993). That such large majorities of both groups claim to have an interracial friendship suggests that there has been a substantial decline in prejudice in recent years. Moreover, this seems to be reflected in a very rapid increase in interracial marriages (Kalmijn, 1993). Let's watch two sociologists find out if there actually has been a general decline in racial prejudice.

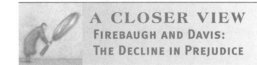

A CLOSER VIEW
FIREBAUGH AND DAVIS: THE DECLINE IN PREJUDICE

Glenn Firebaugh of Penn State and Kenneth E. Davis of Vanderbilt wanted answers to three questions: Has there been a steady decline in racial prejudice against African Americans in recent years? If so, is this because people who once were prejudiced have changed their attitudes, or is it simply that members of older, more prejudiced generations have been dying, to be replaced by members of younger, less prejudiced generations? Finally, is prejudice declining at the same rate in all parts of the country?

To find out, they utilized the General Social Survey, to which I often refer throughout this book. This survey is conducted annually, which made it possible for Firebaugh and Davis to compare answers to identical questions over the period 1972–1984. Their findings were clear (Firebaugh and Davis, 1988).

- Prejudice against African Americans declined substantially during this period. Moreover, the rate of decline accelerated throughout the period.

- Attitude change and generational replacement have *both* made major contributions to this decline—each accounting for about half of the observed shift.

- Prejudice has declined *more rapidly in the South* than in any other part of the nation! This is true even when the researchers disregarded whites who had moved to the South from other regions (recently, there has been very substantial migration to the South). Southern whites still are somewhat more likely to be prejudiced, but the gap between them and whites elsewhere is now small compared with even a few years ago.

Firebaugh and Davis's results are confirmed by a study of voting for the annual All-Star Baseball Team (Hanssen and Anderson, 1999). The starting members of the team are elected by the votes cast by millions of fans each year. Thirty years ago, black players were underrepresented when individual performance statistics are taken into account. By the early 1980s, this bias had disappeared entirely. Today, African American ball players are slightly overrepresented, despite the fact that only small numbers of black fans cast ballots. This is entirely consistent with the findings about basketball attendance (Box 11-1).

BARRIERS TO AFRICAN AMERICAN PROGRESS

Every racial and ethnic minority that began at the bottom of the American status structure had unique features that influenced how rapidly and easily it was able to rise. Although the general model of how groups become upwardly mobile in America applies to each, we have also paid attention to these special features. African Americans have been making considerable progress, but it is important to consider the special obstacles they are faced with.

THE LEGACIES OF SLAVERY Earlier in this chapter, we saw how the existence of slavery in the United States and the American ideal of individual freedom created a special American dilemma that gave rise to ardent racial prejudice. While all disadvantaged minorities in this country were hampered by prejudice, none has faced

---

BOX 11-1   DO WHITE FANS REJECT "TOO BLACK" TEAMS?

IF THE VISIBILITY of outstanding individual talent is the key to minority stardom (see Box 5-2), then fans should rapidly come to place much greater value on ability and performance than on race or ethnicity. But in re-

Michael Jordan walks on air. The immense popularity of the Washington Wizard's superstar was entirely consistent with the research results of Schollaert and Smith that white fans care about winning, not about the racial composition of a team.

AP/Wide World Photos

cent years there has been much media concern about the negative consequences for fan support of teams that have become "too black." The argument is that although sports fans readily embraced the first African American superstars, they have grown increasingly restive as teams began to have substantial numbers of African American players.

These worries have become quite vocal as professional basketball has turned into an African American–dominated sport. By the late 1960s, African Americans made up more than half of the players on the average National Basketball Association team, and today, the average team is about 75 percent African American (Schollaert and Smith, 1987). This has led both white (Halberstam, 1981) and African American (Edwards, 1982) observers to suggest that problems loom for the future of the NBA, and perhaps American basketball in general, because white fans can't identify with teams that are nearly all African American.

Through it all, however, as the NBA teams have grown increasingly African American, average attendance has risen. So what's the truth? Recently, two researchers who specialize in the sociology of sport decided to find out.

Paul T. Schollaert and Donald Hugh Smith of Old Dominion University, famous for its powerhouse women's basketball teams, decided to see if the racial composition of NBA teams influenced attendance. To do so, they gathered data on the percentage of African American players and total attendance for all NBA teams annually over fourteen years, from 1969 through 1982. Their aim such harsh prejudice and discrimination as have African Americans. That anti-African American prejudice has waned so substantially in the past twenty years reflects rapid African American economic and educational achievement. But prejudice and discrimination have also placed much greater burdens on African Americans than on other groups.

Moreover, centuries of slavery severed African Americans from all but traces of their traditional cultures and limited their sense of common identity. Thus, in recent decades African Americans have had to rediscover their cultural roots and construct a cohesive group sense to give them a common cause (Carmichael and Hamilton, 1967). The success of these attempts is reflected in the greatly increased identification with Africa and interest in African history and culture prevalent among African Americans in recent years.

NO HOMELAND African Americans also lack a homeland in the sense of having a specific nation or society of origin. This is because African Americans came from many different parts of Africa; moreover, until recently, these areas were not independent nations but colonies ruled by European powers. This situation has had major practical as well as symbolic consequences.

Most immigrants to the United States and Canada based their decisions about when and whether to immigrate on calculations about

was to see if shifts in racial composition correlated with shifts in attendance. But to do this, they also had to gather data on other factors as a guard against spurious findings. They argued that fan support is affected by winning. So they collected data on win-loss records. They also hypothesized that star players might boost attendance even when a team was not having an outstanding season. They measured this by the number of each team's players named to the All-Star Game each year. They also included data on annual ticket prices for each team.

In addition Schollaert and Smith proposed that attendance may be influenced by the location of the arena (suburban or central city) and by its seating capacity because this determines the number of tickets a team could possibly sell. They found they had to remove the Portland Trail Blazers from the analysis because the team plays in a small arena that has been sold out for every game for years—hence, there has been no variation in Portland's attendance.

Characteristics of the metropolitan area within which teams are located might also affect attendance. Schollaert and Smith thus included data on the population, reasoning that where there are more people there are more potential ticket buyers. Data on median income for the area also were included on the basis that fans with more money can buy more tickets. The percentage of African Americans in each area's population was noted to make sure that as a team increased its proportion of African American players, a rise in attendance of African American fans didn't offset and therefore hide a loss of white support. Finally, the number of other local major league sports franchises was included because these offer competition for fan support. (The Portland Trail Blazers attribute their annual sellouts in part to the fact that they are the only major league franchise in town.)

So what were the results of this study? *The racial composition of teams did not influence attendance.*

The major determinant of attendance in a given year is how many games the team is winning that year. The next most important factor is arena size. A close third is a winning record the year before. Next is the total population from which the team draws. Fifth is the number of All-Star players. Finally, median income of the area and ticket prices had weak effects on attendance. The percentage of the population that is African American, number of competing franchises, and location of the arena didn't matter.

Schollaert and Smith concluded that the widespread belief that white fans resent African American dominance of basketball is itself a myth rooted in "racism"—in this case a racial stereotype of whites as bigots.

Whatever the case, the fact remains that these results are consistent with the acceptance of African Americans in entertainment long before there even was an NBA. For example, whites flocked to hear the famous all African American big bands of the swing era, such as those led by Count Basie, by Duke Ellington, and by Lionel Hampton. In fact, these bands performed primarily in front of white audiences.

FIGURE 11-4	New Names and Fading Ethnicity.

Alan Alda	Alphonso D'Abruzzo	Ted Knight	Tadeus Władysław Konopka
Woody Allen	Allen Konigsberg	Cheryl Ladd	Cheryl Stoppelmoor
Pat Benatar	Patricia Andrejewski	Bruce Lee	Lee Yuen Kam
Robbie Benson	Robert Segal	Jerry Lewis	Joseph Levitch
Mel Brooks	Melvin Kaminsky	Hal Linden	Harold Lipshitz
Ellen Burstyn	Edna Gilhooley	Karl Malden	Malden Sekulovich
Cyd Charisse	Tula Finklea	Chuck Norris	Carlos Ray
Cher	Cherilyn Sarkisian	Bernadette Peters	Bernadine Lazzaro
Tom Cruise	Thomas Mapother	Stephanie Powers	Stefania Federkiewicz
Rodney Dangerfield	Jacob Cohen	Joan Rivers	Joan Sandra Molinsky
John Denver	Henry John Deutschendorf, Jr.	Jane Seymour	Joyce Frankenberg
Morgan Fairchild	Patsy McClenny	Martin Sheen	Ramon Estevez
Sally Field	Sally Mahoney	Talia Shire	Talia Coppola
W. C. Fields	William Claude Dukenfield	Suzanne Somers	Susan Mahoney
James Garner	James Bumgarner	Connie Stevens	Concetta Ingolia
Pee-Wee Herman	Paul Rubenfield	John Wayne	Marion Morrison
Larry King	Larry Zeigler	Raquel Welch	Raquel Tejada
Ben Kingsley	Khrishna Banji	Gene Wilder	Jerome Silberman

opportunities here versus opportunities in their homeland. Thus, immigration into North America fluctuated sharply in response to economic conditions, rising when the economy boomed and jobs were plentiful and falling during times of recession. African Americans never chose to come to America, nor did they choose when to become free and thus when and where to start to earn a living. In fact, African Americans had to scrape for their first jobs in the aftermath of the Civil War, in a South ravaged by defeat. While their migration north often reflected the desire for greater opportunity, here, too, African Americans were less able than were other immigrants to respond to fluctuations in opportunity.

As pointed out in Chapter 2, most ethnic and racial immigrants could go back to their homelands if things did not turn out well for them here. Indeed, during the Great Depression of the 1930s, about 500,000 immigrants went back to Europe (Handlin, 1957). African Americans lacked this option.

Finally, the homeland governments often aided many immigrants after they were here. For example, many governments provided clergy to staff churches in North America for immigrants from their nations. Other governments officially intervened with the U.S. or Canadian government to influence the treatment of immigrants from their countries. Nothing of the sort was available for African Americans in North America (Lieberson, 1980).

VISIBILITY Although the stereotypes of prejudice associated distinctive physical traits with many ethnic groups (that Jews could easily be recognized by their noses, for example), most members of these groups lacked such traits. For example, a study of Jewish men in New York revealed that only 14 percent had what was regarded as a "Jewish nose" (Fishberg, 1911). Moreover, while many immigrant groups had a distinctive culture and quite distinctive names, these traits could easily be changed, thus concealing one's origins and diminishing **visibility**. That Kirk Douglas was once known as Issur Danielovitch demonstrates the point (Figure 11-4). A Pole or an Italian could learn to speak unaccented English, change his or her name, join a Protestant church, and claim to be a WASP. Most African Americans are native speakers of English, have Anglo-Saxon names, and are Protestant, but they're still African American.

Racially different groups cannot pass as members of the majority, which may present a more difficult barrier. However, Japanese and Chinese have achieved equality despite being racially different, and most eastern and southern European immigrants did not discard their names, religions, or many of their distinctive cultural patterns en route to acceptance and equality.

NUMBERS Perhaps one of the greatest problems faced by African Americans is simply that they greatly outnumber all other previously disadvantaged racial and ethnic minorities. Thus, while several hundred thousand Japanese and Chinese sufficed to cause a "yellow peril" on the West Coast, African American migration north has involved millions of people.

Moreover, sheer numbers prevent African Americans from adopting some of the tactics other groups have used. Occupational specialization is one of these. As Lieberson (1980) put it, "Imagine more than 22 million Japanese Americans trying to carve out initial niches (in the U.S. economy) through truck farming!"

Thus, given that economic conflict is the primary factor in prejudice and discrimination, the large size of the African American population has made African Americans a greater threat to whites, especially in earlier times when African Americans constituted a huge source of cheap labor. When changes in the immigration laws stemmed the tide of cheap labor from abroad in the 1920s, African Americans became the only significant competitive threat to whites, especially whites holding unskilled labor jobs. Thus, the antagonism that once was spread across many groups—Italians, Poles, Japanese, as well as African Americans—focused primarily on African Americans; race, not religion or ethnicity, became the burning concern of organized hate groups.

Of course, the size of the African American population also presents opportunities. Once barriers to African American voting were removed, for example, African Americans became a potent political bloc, which in recent years has been reflected by a sharp increase in the number of African Americans holding elected office in the United States.

None of these barriers to African American progress seems so effective as to limit African Americans to economic and social inequality. Indeed, each has been substantially overcome, if with great pain and difficulty. Yet we need to see how each has made the road upward unusually difficult.

Of course, African Americans are not the only group still suffering inequality in North America; they are simply the second largest of these groups and the one whose North American experience constitutes the most shameful, painful chapter in American history. Thus, it is instructive to see the extent to which African Americans have fulfilled the dream expressed in the anthem of the Civil Rights Movement, "We Shall Overcome." If African Americans can retrace the steps taken by so many other disadvantaged groups in North American history, then there should be no doubt that other racial and ethnic minorities can do it, too.

CONCLUSION

The primary purpose of this chapter was to show that prejudice, discrimination, and conflict caused by racial and cultural differences are not recent phenomena or limited primarily to North America. These problems are best understood not on the basis of peculiar historical circumstances or even white bigotry but rather on the basis of sociological theories having universal application. Intergroup conflict fueled by racial and cultural markers is as old as human existence. So are the processes by which these conflicts are overcome.

That these conflicts cause such agony is reflected by the intensity with which social scientists

© 1990 Rick Reinhard

During World War II, the American armed forces were segregated, and African American soldiers usually were commanded by white officers—the few African American officers never commanded white troops. By the time of the Gulf War, African American officers (male and female) were common in the armed forces, and the nation's highest-ranking officer, Chairman, Joint Chiefs of Staff, was General Colin Powell. In 2001 he was chosen to be secretary of state. In 2004 he was replaced by Condoleezza Rice, an African American woman.

have sought to understand and prevent them. In my judgment the contents of this chapter demonstrate that this search has been fairly successful. Current sociological theories of intergroup conflict have dispelled much of the fog shrouding these hostilities. We can now say with some certainty why and how strangers become enemies and how and why these antagonisms will pass.

As we have seen, the key to intergroup conflict is status inequality. Contact between groups that are unequal in status and competing for status produces prejudice. Prejudice does not subside until status equality has been achieved. Keep in mind that two groups with equal status does not mean that all members of each group must be of the same status. Italians have achieved status equality with WASPs not because all Italians have become rich. Rather, groups are of equal status when the distribution of their members in the status structure is the same. Thus, when a person's race does not reliably indicate his or her status, then racial equality has been achieved. At that point members of different races are equally likely to be rich or poor.

Review Glossary

Terms are listed in the order in which they appear in the chapter.

intergroup conflict Conflict between groups that are racially or culturally different.

race A human group having some biological features that set it off from other human groups.

ethnic groups Groups that think of themselves as sharing special bonds of history and culture that set them apart from others.

caste system A stratification system wherein cultural or racial differences are used as the basis for ascribing status.

Allport's theory of contact Theory holding that contact between groups will improve relations only if the groups are of equal status and do not compete with one another.

the American dilemma Term used by Gunnar Myrdal to describe the contradiction of a society committed to democratic ideals but sustaining racial segregation.

markers (cultural or racial) Noticeable differences between two or more groups that become associated with status conflicts between the groups.

cultural division of labor A situation in which racial or ethnic groups tend to specialize in a limited number of occupations.

middleman minorities Racial or ethnic groups restricted to a limited range of occupations in the middle, rather than lower, level of the stratification system.

enclave economy theory Theory that proposes that the spatial concentration of an ethnic group permits it to create its own business enterprises, thus speeding the economic progress of the group.

visibility The degree to which a racial or an ethnic group can be recognized—how easily those in such a group can pass as members of the majority.

Suggested Readings

Alba, Richard D. 1990. *Ethnic Identity: The Transformation of White America.* New Haven, Conn.: Yale University Press.

Cobas, José A., ed. 1987. *The Ethnic Economy: Special Issue of Sociological Perspectives* 30:339–472.

de la Garza, Rodolfo O., Louis DeSipio, F. Chris Garcia, John Garcia, and Angelo Falcon. 1992. *Latino Voices: Mexican, Puerto Rican, and Cuban Perspectives on American Politics.* Boulder, Col.: Westview Press.

Driedger, Leo, ed. 1978. *The Canadian Ethnic Mosaic: A Quest for Identity.* Toronto: McClelland and Stewart.

Eltis, David. 1993. "Europeans and the Rise and Fall of African Slavery." *American Historical Review*, December.

Lewis, Bernard. 1990. *Race and Slavery in the Middle East.* New York: Oxford University Press.

Lieberson, Stanley. 1980. *A Piece of the Pie: Blacks and White Immigrants since 1880.* Berkeley: University of California Press.

Wilson, William Julius. 1987. *The Truly Disadvantaged: The Inner City, the Underclass and Public Policy.* Chicago: University of Chicago Press.

Winks, Robin W. 1971. *The Blacks in Canada: A History.* Montreal: McGill-Queen's University Press.

Sociology Online

www.socstark10.com

GO TO THE INTERNET AND TYPE: www.socstark10.com.

↗**CLICK ON:** 2000 General Social Survey

✓**SELECT:** %AFRI-AMER What is your best guess as to the percentage of the United States population that is made up of African Americans?

Referring to Table 11-5, in which of the three response categories to this question does the correct percentage fall?

Examine the responses by race and by region. Do these results suggest that people tended to base their estimates on their perceptions of their immediate environments?

↗**CLICK ON:** Select New Data Set

↗**CLICK ON:** The 50 States

✓**SELECT:** %NO ETHNIC: Percentage reporting no ethnic ancestry.

The 1990 Census form asked Americans about their ethnic origins. Some answered "German," some "Irish," and so on. But a substantial number said their ancestors came from many different places and they had no particular ethnic origin.

↗**CLICK ON:** Analyze Now
There is a substantial regional pattern. Why do you suppose that is the case?

↗**CLICK ON:** Select New Data Set

↗**CLICK ON:** Nations of the Globe

✓**SELECT:** MULTI-CULT: Multiculturalism? odds (chances out of 100) that any 2 persons will differ in their race, religion, ethnicity, tribe, or language group.

↗**CLICK ON:** Analyze Now

Does one continent stand out as high on multiculturalism? Has there been anything in the news in the past several years that might be related to this pattern? Is the United States one of the top fifty in terms of multiculturalism? Which two nations are the lowest?

USING INFOTRAC COLLEGE EDITION

GO TO THE INTERNET AND TYPE:
www.InfoTrac-college.com

↗**CLICK ON:** Register New Account
You will be asked to enter your Access Code and to create and enter a User Name and Password. Having done so,

↗**CLICK ON:** InfoTrac College Edition or Log On.

Each of these search terms will reveal an immense and thoughtful literature.

SEARCH TERMS:
Assimilation
Discrimination
Ethnic Relations

CHAPTER

12

into the wild blue yonder

In the earliest days of aviation, many daring young women became international celebrities for pioneering flights. Of all women fliers, E. Lillian Todd was perhaps the most unlikely. Here she is in 1909 preparing for takeoff in a Wright Brothers plane. This former stenographer, who had no engineering background, was the first woman to design an airplane. Later, she founded the Junior Aero Club of America and through it taught millions of hobbyists to build model planes.

Gender and Inequality

I**N NO KNOWN SOCIETY** have women equaled men in terms of power. If current trends continue, modern democratic societies may become an exception to that rule, but full equality of the sexes remains to be accomplished. To begin this chapter on gender inequality, it will be useful to survey the circumstances of women across the variety of human societies.

Table 12-1 is based on the Standard Cross-Cultural Sample of 186 premodern societies. In 88 percent of these societies, political leaders are *always* men—and in none are women equally likely to be political leaders. Kinship groups are very important in most of these societies, and in 84 percent kinship leaders also are *always* male. In only 10 percent of these societies do women have political rights equal to men. Keep in mind that these are only rights, not realities. For example, in 10 percent of these societies, women have the right to be political leaders, but they do not have the same probability of actually being a leader.

Now let's examine gender relations. Wife beating is common and accepted in 80 percent of these societies. In two-thirds there is an explicit view that men should and do dominate their wives. In fact, when men fail to dominate their wives, it is they, not the wives, who are the object of public scorn. As a result of these norms, in two-thirds of these societies, women have little or no control over their marital and sexual lives. Indeed, in most of these societies, wives are expected to show deference to their husbands, especially in public. Such deference may take the form of a wife walking several paces behind her husband or kneeling when she serves him food. Finally, the table shows that rape is rather common in these societies.

But what about modern societies? The answer depends on how modern. Generally speaking,

CHAPTER OUTLINE

AN OPERATIC INSIGHT

SEX RATIOS AND SEX ROLES: A THEORY
    Sex Ratios
    Causes of Unbalanced Sex Ratios

    Box 12-1 The "Gilded Cage"

    Ancient Athens: Too Few Women
    Ancient Sparta: Too Few Men
    Sex Ratios and Power Dependence

TESTING THE SEX-RATIOS
AND SEX-ROLES THEORY
    Trends in North American Sex Ratios
    Gender and Social Movements
    Women in the Labor Force
    The Sexual "Revolution"
    Sex Ratios and the African American Family
    Sex Ratios and the Hispanic American
    Family

CONCLUSION

REVIEW GLOSSARY

SUGGESTED READINGS

SOCIOLOGY ONLINE

the more modern the society in terms of industrialization and standard of living, the greater equality between the sexes and the fewer who support traditional definitions of sex roles. This is obvious in Tables 12-2 and 12-3.

Respondents in the sixty nations shown in Table 12-2 were asked whether a woman needed to have children in order to be fulfilled. Clearly, this item taps support for traditional sex roles. In some nations nearly everyone agreed—97 percent in Bangladesh. However, in many European nations as well as in North America, the majority rejected this statement—only 18 percent of Canadians and 14 percent of Americans agreed. But perhaps the most instructive feature of this table is that within countries men and women are in very close agreement. In a few nations, women were somewhat more likely to agree, and in a few others (Japan, for example), men were a bit more likely to hold this view. But in the overwhelming number of nations, differences were nonexistent or so small as not to be statistically significant.

Table 12-3 compares twenty-six nations on the basis of a measure of Female Empowerment, developed and computed by social scientists at the United Nations. It assesses the relative power of women in the nations, based on such things as the percentage of political offices held by women and their representation in high-prestige occupations. Looking at the table, we see that women have come close to being as powerful as men in Norway, with an index score of .908 (a score of 1.000 indicates complete equality between the sexes). One way to interpret the Empowerment Index is to convert it to a percentage, and thus, we can see that Norwegian women have about 91 percent as much power as men, Canadian women have about 79 percent, and American women have 77 percent. The UN began calculating these scores in 1994, and they have risen substantially in all of these nations since then. Norway's score in 1994 was .750, Canada's was .660, and the score for the United States was .620.

Many reasons have been suggested why the extent of gender inequality varies so much across time and spaces. Unfortunately, most of them have been quite unsatisfactory because they attempt to explain cross-cultural variations on the basis of individual psychology. Thus, for example, it often is claimed that society A treats women worse than society B because the men in society A are more sexist. But that's a circular argument. It says in effect that women are subjected to greater sexism where the culture is more sexist. This entirely misses the question that needs to be answered: Why are some cultures more sexist?

In pursuit of an answer to this question, this chapter examines in detail a very creative attempt to explain variations in gender relations. Then, using that theory as our model, we will examine the rise of modern feminism and examine changes in the modern status of women.

However, it is inefficient to isolate the topic of gender from major social processes and institutions. Consequently, you already have encountered the topic of differential gender socialization in Chapter 6, and in Chapter 7 you explored the remarkable gender differences in crime and delinquency. Gender will receive careful attention in many other substantive chapters as well. For easy reference all discussions of key gender issues are indexed in Figure 12-1.

TABLE 12-1	Dimensions of Gender Inequality in Premodern Societies	
POWER		**PERCENT OF SOCIETIES**
Political leaders are *always* male		88
Leaders of kinship groups are *always* male		84
Women have equal political rights		10
GENDER RELATIONS		
There is a significant amount of wife beating		80
There is an explicit view that men should and do dominate their wives		67
Women have little or no control over their marital and sex lives		64
Wives display institutionalized deference to their husbands		65
Rape is frequent		62

Source: Prepared by the author from the Standard Cross-Cultural Sample.

AN OPERATIC INSIGHT

One evening Marcia Guttentag and her sixteen-year-old daughter attended a performance of Mozart's *The Magic Flute*. It was the first time she had heard this opera sung in English rather than in the original German, and as she listened to the lyrics, she had one of those rare moments of insight that every scientist dreams of and few ever experience.

Guttentag noticed that the male leads sang again and again about the same thing: their urgent desire to find a wife and their determination to cherish her forever. The emphasis on lasting relationships immediately reminded her of the lyrics of songs popular in her youth, when Big Band singers crooned "Our love is here to stay."

After the opera, Guttentag asked her daughter if she had noticed anything odd about the music. "She said that the lyrics were strange because the men sang about wanting to make a lifelong commitment to one woman—a wife" (Guttentag and Secord, 1983). Both agreed that this was in sharp contrast to current popular music, which stresses "brief liaisons and casual relationships between men and women," a theme of "love 'em and leave 'em."

"Why?" Guttentag asked herself. Why such a shift in attitudes toward women? Then inspiration struck. Is it possible that suddenly there has been *a shift in the supply of women?* Could it be that in recent decades "a shortage of men" has developed? In the past did men typically face

© David Powers Photography, 1988

Francisco Araiza sings the part of Tamino in the San Francisco Opera Company's production of *The Magic Flute* in 1987. Hearing this opera sung in English led Marcia Guttentag to ask herself how societies would differ depending on whether they had a shortage or an abundance of women.

a shortage of eligible women? Are Papageno and Tamino in *The Magic Flute* facing an unfavorable **sex ratio** so that there are far too few women to provide each man with a wife? Could sex ratios

TABLE 12-2	Percentage Who Agreed That "a woman must have children in order to be fulfilled"						
NATION	TOTAL POPULATION	MEN	WOMEN	NATION	TOTAL POPULATION	MEN	WOMEN
Bangladesh	97	97	97	Chile	64	61	66
Indonesia	92	92	92	France	63	63	63
Nigeria	91	91	91	Portugal	63	59	67
Hungary	90	86	94	Uganda	63	56	69
Jordan	89	89	89	Dominican Republic	62	62	62
Egypt	88	88	88	Lithuania	57	51	62
Albania	87	84	89	Croatia	55	50	61
Philippines	86	81	91	Italy	53	51	55
Latvia	85	82	88	Taiwan	52	48	56
Pakistan	85	83	88	Venezuela	53	52	53
Bosnia	84	82	86	Argentina	51	49	53
Morocco	83	81	85	Germany	50	50	50
Vietnam	83	80	85	Iran	45	44	46
India	82	80	84	Japan	44	48	41
Ukraine	82	77	85	Spain	44	41	47
Armenia	80	80	81	**Mexico**	**43**	**42**	**44**
Georgia	79	76	82	Czech Republic	42	39	44
Algeria	79	80	77	South Africa	42	41	44
Romania	79	75	83	Puerto Rico	33	33	33
Russia	78	73	83	Belgium	32	34	30
Moldova	75	71	78	China	32	26	37
Zimbabwe	75	71	79	Austria	30	30	30
Turkey	75	73	77	Colombia	28	27	28
Belarus	73	73	73	Sweden	24	27	20
Bulgaria	70	66	74	**Canada**	**18**	**19**	**17**
Denmark	68	67	70	Great Britain	18	18	18
Estonia	67	61	72	New Zealand	16	15	16
Greece	67	63	70	**United States**	**14**	**16**	**12**
Saudi Arabia	65	63	67	Finland	12	12	12
Poland	64	65	64	Netherlands	7	7	7

Source: Prepared by the author from the World Values Surveys, 2001–2002.

TABLE 12-3	Female Empowerment Index for Selected Nations		
	1.000 – FULL EQUALITY WITH MEN		**1.000 – FULL EQUALITY WITH MEN**
Norway	.908	Japan	.531
Sweden	.854	Russia	.467
Netherlands	.817	Chile	.460
Australia	.806	Venezuela	.444
Canada	**.787**	Ukraine	.411
United States	**.769**	South Korea	.377
Spain	.716	Cambodia	.364
Ireland	.710	Iran	.313
Great Britain	.698	Turkey	.290
Israel	.614	Egypt	.266
Poland	.606	Bangladesh	.218
Italy	.583	Saudi Arabia	.207
Mexico	.563	Yemen	.123

Source: United Nations, *Human Development Report*, 2005.

FIGURE 12-1 Index to Discussions of Gender Issues in Other Chapters.

WOMEN AND SCIENCE
Chapter 6: pp. 169–170

CRIME AND DELINQUENCY
Chapter 7: pp. 184–185

EXTRAMARITAL SEX
Chapter 13: pp. 377–378

GENDERED FERTILITY
Chapter 18: pp. 526–527

HORMONES AND BEHAVIOR
Chapter 5: pp. 135–136

MOTHERS AND MOBILITY
Chapter 10: p. 273

HOUSEHOLD CHORES
Chapter 13: p. 383

LABOR FORCE PARTICIPATION
Chapter 16: pp. 451–453

PROFESSIONAL WOMEN
Chapter 16: p. 466

ONE-PARENT FAMILIES
Chapter 13: pp. 379–380

POLITICAL OFFICEHOLDING
Chapter 15: pp. 432–436

RELIGIOUS COMMITMENT
Chapter 14: pp. 393–395

REMARRIAGE
Chapter 13: pp. 380–383

SEX-ROLE SOCIALIZATION
Chapter 6: pp. 166–168

FEMALE STATUS ATTAINMENT
Chapter 10: p. 275

FIGURE 12-2 **Sex Ratios—Males per 100 Females.**

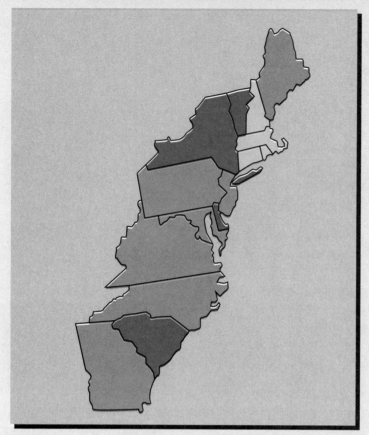

Massachusetts	96
Rhode Island	97
Connecticut	98
New Hampshire	101

Maine	104
Georgia	104
North Carolina	104
New Jersey	105
Virginia	106
Maryland	106
Pennsylvania	106

New York	107
South Carolina	109
Delaware	109
Vermont	113

1790: Males per 100 Females

Until the end of World War II, men outnumbered women in the United States. The first U.S. Census, conducted in 1790, found that there were 104 males per 100 females in the nation's population. But because men were more likely than women to migrate to frontier areas, women outnumbered men in three northeastern states: Massachusetts, Rhode Island, and Connecticut. In this chapter we shall see how the earliest women's movements occurred in states where females outnumbered males and why sex ratios played a role in these developments. As female mortality rates from childbirth-related illnesses began to decline rapidly, women began to outlive men and the sex ratio began to shift. In 1946, for the first time in history, women outnumbered men in the United States—there were 99.8 males per 100 females in the total population. The 2000 census counted 96.7 males per 100 females. Nevertheless, a distinctive geography of gender still exists. The ratio of males to females is highest in the West, men outnumbering women in Alaska, Nevada, Colorado, Wyoming, Hawaii, Utah, and Idaho. In contrast, women far outnumber men in the East, Rhode Island, New York, and Massachusetts having the least men relative to women.

lie behind the immense differences in relations between the sexes from one society to another and one era to the next?

Guttentag recognized this as the germ of a potentially immense theoretical breakthrough. So she devoted herself entirely to working out the theoretical implications of her insight and, at the same time, to collecting appropriate data from such diverse times, places, and peoples as ancient Greece, medieval Europe, Orthodox Jews in nineteenth-century Russia and Poland, eighteenth and nineteenth-century Americans, and finally, whites and African Americans in recent decades (Figures 12-2 and 12-3). And as she went forward, everything began to fall into place. Sex roles and relationships have long been immensely influenced by the ratio of males to females.

In this chapter we shall use Guttentag's work as the framework for a wide-ranging assessment of inequalities between men and women. Other important studies will, of course, be woven into the narrative. But this chapter will be shaped by one piece of scholarship far more than is true for any other chapter—for several reasons. First of all,

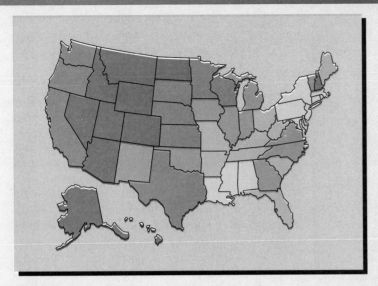

2000: Males per 100 Females

New Jersey	94.0
Connecticut	93.8
Louisiana	93.5
Mississippi	93.3
Pennsylvania	93.3
Maryland	93.2
Alabama	93.1
Massachusetts	92.9
New York	92.8
Rhode Island	92.3

Kentucky	95.5
Arkansas	95.1
Florida	95.0
Tennessee	94.8
Maine	94.7
West Virginia	94.6
Missouri	94.5
Ohio	94.4
South Carolina	94.3
Delaware	94.2

Alaska	107.1
Nevada	103.5
Colorado	101.2
Wyoming	101.1
Hawaii	100.8
Utah	100.6
Idaho	100.5
North Dakota	99.7
Montana	99.4
Arizona	99.3

California	98.9
Washington	98.9
South Dakota	98.5
Texas	98.3
Oregon	98.3
Minnesota	98.0
Kansas	97.6
Wisconsin	97.5
Nebraska	97.1
New Hampshire	96.9

Oklahoma	96.6
Georgia	96.6
New Mexico	96.1
Indiana	96.1
Virginia	96.1
Michigan	96.1
Vermont	96.1
Iowa	96.1
North Carolina	95.9
Illinois	95.7

I could not resist the opportunity to let you watch the birth of a really important idea, see it grow into a sophisticated theory, and then look on as it is confronted with stringent empirical tests. This opportunity was especially irresistible because the theory is so purely *sociological*. Sex ratios are not traits of individuals; they are *social structures*. When individuals move from a society with an excess of males to one with an excess of females, their lives change even though their gender does not.

My second reason is that Guttentag's work combines so much theoretical scope with such creative research that it presents the opportunity to use it not only to illuminate gender relations but also to provide a mid-book review of the sociological process as we have encountered it thus far. I thought you might find it helpful to see the entire sociological forest again before moving on to look at some more big trees in this forest. Finally, in this instance it seems more important to thoroughly understand a big idea about gender relations than to be exposed briefly to a flock of little ones, many of which also appear in the mass media frequently.

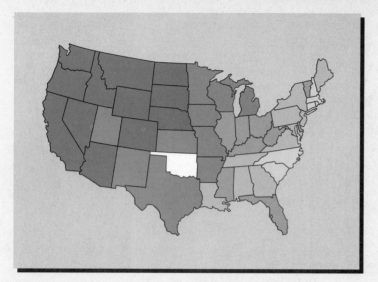

FIGURE 12-3 **Sex Ratios in 1880.**

1880: Males per 100 Females

Montana	258.7	Illinois	106.4
Arizona	231.1	Indiana	104.4
Wyoming	215.2	West Virginia	103.5
Nevada	207.9	Florida	102.5
Idaho	201.9	Delaware	102.2
Colorado	198.0	Kentucky	102.0
Washington	157.7	Ohio	101.9
North Dakota	155.6	Vermont	100.9
South Dakota	155.6	Mississippi	100.5
California	149.6		
Oregon	144.8		
Nebraska	122.6	Maine	99.8
		Pennsylvania	99.6
		Tennessee	99.5
New Mexico	117.1	Louisiana	99.5
Kansas	116.8	New Jersey	98.0
Minnesota	115.9	Georgia	97.9
Michigan	111.3	Maryland	97.8
Texas	111.1	Alabama	97.3
Iowa	109.2	Virginia	97.2
Missouri	108.3	New York	97.2
Arkansas	107.8		
Utah	107.2	South Carolina	97.1
Wisconsin	107.0	North Carolina	96.6
		New Hampshire	96.6
		Connecticut	96.5
		Massachusetts	92.8
		Rhode Island	92.7

In 1850 Horace Greeley, editor of the *New York Tribune*, wrote, "Go West, young man, and grow up with the country." Clearly, that's exactly what many young men did, whether or not they ever heard of Greeley. As the map shows, in 1880 there were really two Americas—at least in terms of gender. The West was male territory, the East was female. West of the Missouri River was a land of cowboys and miners where women were in short supply. Along the eastern seaboard, women outnumbered men. Moreover, these ratios were even much more extreme when only young single adults were counted. (In 1880 the area that was to become Oklahoma was officially designated as Indian Territory; its population was not enumerated by the census.)

SEX RATIOS AND SEX ROLES: A THEORY

Sad to say, three years after she began working out the connections between sex ratios and gender relations, Marcia Guttentag died suddenly. She left a book only half written. All might have been lost but for the fact that she had recently married Paul F. Secord, who is also a well-known sociologist. As a memorial to his wife, Secord set out to finish her work. In addition to his wife's drafts and notes, Secord was able to draw on countless discussions they had as she worked on the book. Eventually, however, Secord's contributions went far beyond simply writing up chapters Guttentag left unwritten. For he became as absorbed in the topic as she had been, and so he worked over the theory and added substantially to the supporting research. Finally, the book appeared: *Too Many Women? The Sex Ratio Question*, by Marcia Guttentag and Paul F. Secord (1983). The next year it received the American Sociological Association's Award for Distinguished Contribution to Scholarship.

SEX RATIOS

To examine the Guttentag and Secord theory, let's begin by imagining a world in which one gender greatly outnumbers the other in the age groups during which people usually marry and have children. Based on ordinary economic principles of supply and demand, we can assume that the gender in short supply has an advantage, that it can impose premium costs on members of the other gender in return for assenting to marriage.

Now assume that eligible females are the ones in short supply. How will they benefit from their advantage in the marriage market? Doesn't it make sense that when women have the advantage they can more easily resist male efforts to restrict their choices? For example, isn't it logical that a woman in this position could play one suitor off against another and thereby obtain great latitude in what she was allowed to say and do? Couldn't such a woman expect to have considerable sexual freedom where men greatly outnumber women—freedom, for example, to take a series of lovers, moving on whenever she got bored or disappointed?

Now, think about the reverse situation, in which women greatly outnumber men. Here men have the advantage, so women will have to pay the premiums. Isn't it likely that women in this situation will have to vie with one another to be pure, faithful, submissive, maternal—all the "virtues" associated with traditional female sex roles?

In fact, Guttentag and Secord's theory of sex ratios and sex roles predicts sex-role patterns opposite to those just outlined! And reality seems to agree with them: In societies around the world, women are clearly much more subject to the limits of traditional sex roles where there is an excess of men. The correlations shown in Table 12-4 indicate that in nations where men outnumber women, females lack power. They also are less likely to be employed outside the home, and the abortion rate is low. Reflective of a sexual "double standard," there is considerably greater approval of extramarital sex in societies having an excess of men, since it is taken for granted that this freedom is limited to males. How can this be? As we shall see, there are important constraints on the bargaining position of individuals other than the relative supply of persons of their sex.

But before we see how the Guttentag and Secord theory clarifies these issues, let's consider a question that you may already be wondering about: How does it happen that societies have too many men or too many women? Aren't sex ratios purely a function of biology?

CAUSES OF UNBALANCED SEX RATIOS

In most human populations, for every 100 female births there are from 105 to 106 male births (Matras, 1973). However, since males have higher rates of infant and child mortality, the sex ratio soon evens out if nothing else intrudes. But in most times and places, other things have intruded. Usually, these intrusions have produced an excess of men; throughout history most human societies have had a shortage of women. Sometimes, however, the intrusions produce a shortage of men.

GEOGRAPHICAL MOBILITY Extremely unbalanced sex ratios often are caused by periods of rapid and large-scale migration or immigration. In the typical instance, young men depart in pursuit of economic opportunity, leaving the young women behind. As a consequence there is an excess of women in the place from which the men are moving and a shortage of women in the place they are moving to. For example, the 1850 Census found that in California men outnumbered women by more than twelve to one

TABLE 12-4	Correlations between Sex Ratios and Sex Roles in 174 Nations
	MALES PER 100 FEMALES
Gender Power Ratio	−.309
Percent women in labor force	−.629
Abortion rate	−.506
Percent who approve of extramarital sex	.620

Source: Prepared by the author from *Nations of the Globe*, an electronic data base created by MicroCase Corporation, now distributed by Wadsworth/Thomson Learning.

(or 1,222 men per 100 women). The cause? The California Gold Rush that began in 1849. The Forty-Niners, as the prospectors were called, were nearly all younger men, most of them single. Meanwhile, back East the absence of these young men was clearly felt. In Massachusetts in 1850, the Census counted only 96.5 men per 100 women. That may not sound like much of an imbalance. But it was concentrated in the prime marriage age group; hence, for every ten women between age fifteen and thirty, there were fewer than nine men. Worse yet, as we shall see, shortages of men typically are greatly increased when only the unmarried population is examined. Hence, young women in Massachusetts in 1850 probably had only about three chances in five of finding a spouse. And as the American West continued to attract young men, the sex ratios in many states became even more unbalanced. Figure 12-3 shows that by 1880 there were fewer than 93 men per 100 women in Massachusetts and Rhode Island, whereas out West there were 258.7 men per 100 women in Montana, 231.1 in Arizona, and 215.2 in Wyoming. And again, the imbalance was concentrated among younger adults.

FEMALE INFANTICIDE The major cause of imbalanced sex ratios is female **infanticide.** *Most* human societies have systematically killed a substantial proportion of female infants—sometimes through selective neglect, but most often by smothering them or by abandoning them to die from exposure and dehydration (Petersen, 1975). In China, for example, census-takers found sex ratios as high as 430 boys for every 100 girls in some areas as late as 1870 (Ho, 1959). In India in 1846, a British enumeration in the Punjab found not one female child in 2,000 upper-status Bedi Sikh families with children (Dickeman, 1975).

BOX 12·1   THE "GILDED CAGE"

IN THE SUMMER OF 1897, word reached the West Coast cities of Canada and the United States that gold had been discovered on the Klondike River in Canada's far northern Yukon Territory. Within hours a gold rush began. Among the people crowding aboard the ships heading north (most of them from Seattle) were men of all ages and social background who had been afflicted with "gold fever," but very few women. And those women who did join in the rush to the Yukon faced an odd situation combining immense opportunity with very little freedom—as a song popular at the time put it, "I'm only a bird in a gilded cage."

Insights into sex roles in Dawson City can be seen in the photograph showing two very elegantly dressed women standing among more than two dozen men at the bar of the Monte Carlo Saloon. Here miners showered Margie Newman with gold nuggets as she danced on stage, and Roddy Conners once spent $50,000 to dance with Jacqueline and Rosalinde, two sisters who, like the women in the photograph, charged $1 for each circuit around the small dance floor. Here, too, a miner once paid $20 a bottle for enough wine to fill a bathtub so Cad Wilson, famous for her rendition of "Such a Nice Girl, Too," could bathe in it. Others, including Oregon Mare, Sweet Marie, and Flossie de Atley, expected to be treated to $60 pint bottles of champagne to sit at a man's table—and one miner spent $1,700 in a single night in this manner. But whether they were superstars or just saloon hostesses, the women in Dawson City were sex objects in the most literal sense. Lavished with gifts and attention, they were nevertheless treated like property. A popular saloon singer (it may have been the woman pictured on the right) was auctioned off to one of the city's instant millionaires for her weight in gold—more than $60,000 at the time, or over $500,000 in today's dollars.

National Museum of Canada, Ottawa

Pierre Berton Enterprises

Nor was female infanticide limited to non-Western societies. Female infanticide and abandonment were widespread in Europe even quite recently. William L. Langer (1972) reports, "In the eighteenth century it was not an uncommon spectacle to see the corpses of infants lying in the streets or on the dunghills of London and other large cities." In 1835 the French government officially recorded the abandonment of more than 150,000 infants. Although this total included only those found still alive, most died. Many male French infants were abandoned, too, but the majority were females (Langer, 1972).

HEALTH AND DIET Not only do males have higher infant and childhood mortality rates, but they also are subject to a higher rate of fetal deaths—female fetuses are more robust and have a higher rate of survival in the womb. This becomes especially significant among populations suffering from inadequate nutrition and lack of medical care. In such circumstances there will be little difference in the ratio of males to females at birth, and an excess of females will emerge in childhood—in the absence of female infanticide. Later in this chapter, we shall see that male fetal deaths have played a major role in producing the marked shortage of males among African Americans. This is also reflected in the disproportionate number of African American infants suffering from a low birth weight.

Another aspect of diet also must be noted: Societies that practice female infanticide often tend to provide a less adequate diet for women and girls, and this shows up in their mortality rates (El-Badry, 1969; Kennedy, 1972).

DIFFERENTIAL LIFE EXPECTANCY Because of the high mortality rates associated with childbirth, in many societies men have outlived women, adding to the excess male population. Today, however, in most societies women outlive men. Hence, even in populations that have a well-balanced sex ratio in early adulthood, a shortage of men begins to build up as age cohorts mature. In their later years, women greatly outnumber men in all advanced industrial nations. In the United States today, there are only two males for every three females over age sixty-five, and over age eighty-five women outnumber men by well over two to one. As a result most men over age sixty-five are living with a wife (78%), whereas only 40 percent of women over age sixty-five have a husband—51 percent of the women are widows compared with 14 percent of the men.

© Culver Pictures

Vincent de Paul, a Catholic priest, devoted much of his time and effort to rescuing infants abandoned in the streets of Paris in the seventeenth century. The religious society he founded, and which today bears his name, established and sustained the first shelters in Europe for foundlings, as abandoned babies came to be known. He was canonized as Saint Vincent de Paul in 1737.

WAR Many societies have experienced periodic shortages of young adult males because of heavy wartime casualty rates. In Chapter 18 we shall see that France suffered from a lack of young men following World War I. The Soviet Union had an acute lack of younger males following their heavy losses in World War II. During these periods sex-role norms often are relaxed until the usual sex ratio is restored.

SEXUAL PRACTICES It now appears likely that the extreme excess of male births among Orthodox Jews—which probably has persisted since biblical times—is the result of strict norms governing *when* intercourse occurs in relation to the woman's menstrual cycle. Because Talmudic law forbids intercourse during and for seven days following menstruation, intercourse is most likely to occur just prior to ovulation, and some evidence indicates that this is when the conditions in the woman's reproductive tract are most favorable to Y sperm and hence to a higher proportion of males being conceived (Guttentag and Secord, 1983). Similar patterns may exist, or have existed, in other cultures.

The Metropolitan Museum of Art

This fourth-century B.C. terra-cotta figurine of a Greek woman portrays the very concealing style of clothing worn by women in cities such as Athens where males greatly outnumbered them. In this same era, however, in the Greek city-state of Sparta, where women outnumbered men, women's dresses came only to the knee and often were sleeveless and low-necked.

In the long run, of course, our concern in this chapter is not with *why* unbalanced sex ratios occur but rather with how they shape sex roles and gender inequality. However, before we examine the way Guttentag and Secord's theory explains these connections, it will be helpful to deal with concrete examples. So, let's contrast two societies having similar cultural roots but very different sex ratios: the ancient Greek city-states of Athens and Sparta.

ANCIENT ATHENS: TOO FEW WOMEN

As with most premodern societies, Athens in its classic period (500 to about 300 B.C.) had a great excess of males, except for brief periods of heavy military casualties. The cause? Female infanticide—practiced by rich and poor families alike. It was rare even for large families to raise more than one daughter. A study of tomb inscriptions in nearby Delphi made it possible to reconstruct 600 wealthy Greek families. Of these, only six had raised more than one daughter (Lindsay, 1968).

Moreover, girls received little or no education and were raised to be quiet and submissive.

Typically, Athenian females were married at puberty, but their husbands were much older. Hence, quite aside from inequalities built into sex roles, marriages began with one partner having much greater maturity, experience, and education than the other had.

Although women were celebrated in poetry and drama as romantic love objects, under Athenian law women were classified as children, regardless of age, whereas males legally obtained adult standing at age eighteen. Hence, an Athenian woman was at all stages in her life the *legal property* of some man—first of her father, then of her husband, and should her husband die, of her son. If a man wanted a divorce, he needed only to order his wife from his household. If a woman wanted a divorce, she had to have her father or some other man bring her case before a judge. Women were denied all participation in politics: They could not hold office, vote, or serve on juries. Instead, the Athenian woman's role was that of mother and manager of the household (Pomeroy, 1975; Finley, 1982).

Two aspects of Athenian sexual patterns also are typical of societies with too few women. First, "respectable" women were isolated by extremely protective sexual norms. They were secluded from the sight of all men other than close relatives. They wore clothing designed to conceal. Virginity was expected of brides and absolute chastity of wives; unchaste wives could be sold into slavery, as could unmarried women who had lost their virginity. If a woman was seduced or raped, her husband was compelled to divorce her, while also having the legal right to kill the man who had violated his property rights (Lacey, 1968; Pomeroy, 1975).

The second aspect of sexual patterns in societies with a high ratio of males to females is the highly visible presence of "disreputable" women—running the gamut from common prostitutes to the courtesans and mistresses of the rich and powerful. The prostitutes of Athens wore provocative and revealing gowns, and some solicited openly on the streets. Males, of course, were not expected to be chaste and were thought to need additional sexual outlets (Pomeroy, 1975).

ANCIENT SPARTA: TOO FEW MEN

The ancient Greek city-state of Sparta offers a direct contrast to Athens. Far smaller than Athens or other prominent Greek city-states, Sparta was nonetheless the dominant power in

Greece for centuries. Why? Because of a standing army of unparalleled skill and tenacity, an army that for centuries observed "unfailing obedience to the rule never to retreat in battle" (Finley, 1982). Spartan mothers bidding goodbye to sons going off to battle always advised them to come home either with their shields or upon them—that is, win or die.

Upon reaching age eighteen, all able-bodied male Spartans went off for twelve years of full-time military service, after which they remained liable for recall until age sixty. The army had two primary missions. The first was to defeat any and all other armies that challenged it. But the second, and more important, mission was to maintain control over a subjugated peasantry known as helots. Because all male Spartans took up the same occupation—soldier—their society depended on outsiders to grow the food and perform the other needed domestic tasks. And with most of the men off in army camps, it often fell to Spartan women to manage the estates and oversee the helots.

Sparta practiced infanticide, but without sexual preference: Only healthy, well-formed babies were allowed to live. However, because males are more subject to birth defects and more apt to be sickly infants, the result was a slight excess of females from infancy, a trend that accelerated with age because of male mortality—including mortality from war.

If Sparta put similar value on male and female infants, it took a radically sex-differentiated approach to child rearing. At age seven all Spartan boys left home to be raised by the state in public, all-male dormitories—the toughest military boarding schools in history. This system made Sparta a highly sex-segregated society. Although men could marry at age twenty, they could not live with their wives until they left the army after age thirty. This meant that Spartan men had very limited contact with Spartan women from age seven through thirty and were, understandably, inclined to prefer the social company of men thereafter. However, Spartan men who did not marry were ridiculed in public and suffered legal penalties.

As is typical of societies where women outnumber men, Sparta offered girls as much education as it did boys. They spent much less time than Athenian women learning traditional women's work (cooking and sewing were consigned to slaves) and received a substantial amount of physical education and gymnastic training.

And a Spartan woman did not belong to anyone. If she divorced her husband, she took all of

Giraudon/Art Resource, NY

her own property and half of what the household had produced. If her husband was at fault, he was fined. Spartan women not only controlled their own property but also had the right to dispose of their husband's in his absence; and if a father failed in his duties concerning his children's property, control was transferred to his wife or mother. Indeed, it was thought that women had special abilities with financial matters, and it is estimated that women were the sole owners of at least 40 percent of all land and property in Sparta (Pomeroy, 1975).

Given their good educations and their economic power, Spartan women were not known for their reticence or submissiveness. In fact, they were famous throughout Greece for being outspoken and witty. Not one of the many women poets remembered in classical Greek literature came from Athens, but many were from Sparta.

Although Spartan culture emphasized childbearing (the army always needed more soldiers), sexual patterns in Sparta were the opposite of Athens. The women were not hidden away. In contrast to the heavy, concealing gowns worn in Athens, women in Sparta wore short dresses and went bare legged. When women are in oversupply, men tend to regard them in much less romanticized terms and place considerably less emphasis on virginity and chastity. Hence, while Athenian parents married off their daughters at puberty, in part for fear they might lose their virginity before marriage, Spartan women usually did not marry until they were twenty or older. Also illustrative of the more casual Spartan approach to sexuality, rape and adultery carried only monetary fines, and a husband would have been scorned for deserting a wife who had been raped. Moreover, there were too few men to provide a sufficient number of husbands and no males their own age around for Spartan women from age seven through thirty. No doubt many older men took advantage of this situation to demand casual sexual relations, as men have in modern times. Indeed, societies with an excess of women tend to be sexually permissive for both men and women.

To sum up: Where women are scarce, they are treated as precious property, but without rights of their own. Where women are in excess supply, there is much less gender inequality, but men are inclined to be less dependable as spouses and lovers. Now let's try to discover why.

SEX RATIOS AND POWER DEPENDENCE

Guttentag and Secord's theory has both a micro- and a macrolevel. At the microlevel it is an *exchange* theory and begins with the fundamental rational choice premise introduced in Chapter 1 and elaborated in Chapter 3: *People will seek to maximize their rewards and minimize their costs.* In other words, as people interact with one another, they keep score. Suppose a man and a woman are involved in a dyadic, or two-person, relationship in which they are exchanging affection, among other things. Although they may put it in more romantic terms, from time to time each must ask: Is this relationship paying off? Am I getting back as much as I am giving? How can I obtain more in return for less? Could I do better by finding a different exchange partner? From that last question, we see that no exchange between two humans is limited to them alone—social circumstances always intrude, limit, define, or otherwise affect their options. And sometimes social circumstances greatly influence the *relative power* of members of a dyad.

DYADIC POWER The capacity of each member of a dyad to impose his or her will on the other member is called **dyadic power**. A social circumstance that can greatly influence dyadic power is the sex ratio within which a dyad is located. As Guttentag and Secord (1983) explained:

When one sex is in short supply, all relationships between opposite-sexed persons are potentially affected in a similar way: The individual member whose sex is in short supply has a stronger position and is less dependent on the partner because of the larger number of *alternative relationships available to him or her* [emphasis added].

When women are in short supply, they can select from among several suitors who must vie for their favor. When men are in short supply, they can pick and choose from among several available women.

The concepts of power and of dependence allow us to isolate the key elements in dyadic relationships. Let us consider a dyad made up of A and B, or Alan and Betty. Both will have certain goals they wish to achieve within this relationship—perhaps each will be seeking to satisfy sexual and emotional desires. Recall that we have defined *power* as the ability to get one's way over the resistance of others (see Chapter 9). Thus, if one member of a dyad has superior power, he or she will be more successful in achieving personal goals within the relationship. And a member of a dyad with inferior power can be identified as the *dependent* member. Or to put this more formally: The dependence of one member of a dyad on the other is equal to the person's inability to achieve his or her goals *outside* the dyad (Emerson, 1962).

An unfavorable sex ratio causes **power dependencies** in dyadic relationships for members of the sex in excess supply. From this it follows that men will experience power dependence in their dyadic relations with women when women are in short supply; this unbalances their relative abilities to achieve their goals outside the dyad. The imbalance shows up in the romanticized and idealized images of women typical of societies with a shortage of women. But then why do women appear to be worse off when they are scarce and better off when they are in excess? To answer this question, we must look beyond dyadic relations.

STRUCTURAL POWER We have seen that a social structure—sex ratios—can shape power relations within dyads. Other social structures can play a similar role. Among these manifestations of **structural power** are organized activities by gender groups. Dependent members in dyadic relations may seek to improve their bargaining positions in a number of ways, as Guttentag and Secord pointed out. They may seek to develop techniques and cultural means to make themselves unusually attractive to the opposite sex. For example, where women are in short supply, men may pursue such romantic approaches to courting as serenading beneath their "fair damsel's" window or by sending poetry. Or men may fight duels or tattoo their bodies to gain female admiration. When men are in short supply, it will be women who will seek male admiration, and this often proves to be a very good thing for the cosmetics industry.

Another response by the gender in excess supply is to attempt to reduce their reliance on dyadic relations with the opposite sex, spending more of their time in same-sex dyads. Or they may withdraw from dyadic relationships, seeking their satisfaction in solitary activities as artists, writers, or religious ascetics. But if the imbalance between the sexes continues for a substantial period, "an appreciable number of individuals from the gender lacking dyadic power may well get together and organize various types of actions to correct the situation" (Guttentag and Secord, 1983).

Here Guttentag and Secord have used insights from *conflict* theories to explain macro-level responses to sex ratios. Recall from Chapter 4 that conflict theories emphasize the ways that various groups in societies attempt to utilize their power to *shape social structures to serve their own interests*. Historically, men have usually lacked dyadic power, so they have been motivated to shape the rules governing status to favor them. Moreover, in this circumstance men also have the power of greater numbers on their side. Thus, as men seek to offset the greater power of women at the dyadic level, their organized efforts lead to the elaborate culture of traditional sex roles. That is, men have combined to create social structures based on norms governing appropriate sex-role behavior—norms that serve to limit the dyadic bargaining power of women by limiting their ability to form additional dyadic relations. Chief among these norms are:

- Brides shall be virgins. This reduces the margins within which women can form alternative dyads.

- Wives shall be chaste. This reduces women's options to find alternatives outside the dyad, once it is formed.

- Women shall devote themselves to the roles of wife and mother. This tends to discourage both divorce and the pursuit of careers outside marriage. This typically leads to:

- Defining women as temperamentally unsuited to positions of power and authority, which facilitates male control of laws as well as customs.

Keep in mind that men do not need to send thought police around to enforce these norms. Instead, both males and females will be *socialized from birth into a culture in which these norms are taken for granted and seen as self-evident truths*. That is, as Chapter 7 clarified, most men and women will *internalize* these norms. Hence, women will be as quick as men to criticize people who deviate too far from standards of behavior appropriate to their gender, and parents will do their best to raise their daughters to be properly feminine and their sons to be properly masculine.

But what about societies with an excess of women? Typically, these societies have a long history of excess males, so when an excess of women develops, it does so within a culture sustaining traditional sex roles providing for male superiority. Initially, an excess of women will combine with a male-biased culture to cause a "sexual revolution." That is, when *both* structural and dyadic power favor men, they respond by exploiting their advantages to the fullest, discarding one dyadic partner for another, giving little to any one partner, and demanding the maximum benefits from each. Hence, norms valuing virginity disappear, and soon norms of chastity do, too. Moreover, as men seek more from women, it is to their immediate benefit if wives can provide economic as well as emotional and sexual rewards. Norms restricting women's roles begin to erode.

Meanwhile, women will be reacting to their new conditions and reassessing their options. As they gain less satisfaction and security from traditional sex roles—as wives and mothers find themselves cast aside to support themselves and to raise their families alone—and as they gain more exposure to the "man's world" of careers and decision making, women can be expected to exert organized efforts to restructure the sex-role culture.

Initially, this effort probably will take the form displayed by women's movements in modern societies—a demand for social, legal, and economic equality. But what will happen once this is fully achieved *if* an acute shortage of males continues? Will women eventually evolve and impose an elaborate new sex-role culture to limit the dyadic power of men? Could there be societies with a mirror image of traditional sex roles, with men enclosed by norms of virginity and chastity? Could househusband be the primary male role in societies of the future—especially since it no longer is necessary to breast-feed infants? No one knows.

TESTING THE SEX-RATIOS AND SEX-ROLES THEORY

Once Guttentag and Secord had worked out their theory of sex ratios and sex roles, they set out to test it against appropriate empirical evidence. Keep in mind that we can never prove a theory—we never run out of future opportunities to which it applies and hence of opportunities for it to fail. So scientists assess theories by doing their best to *disprove* them. Turning to history, Guttentag and Secord tested their theory against evidence from ancient Greece. As we have seen, they found that Athens displayed the patterns their theory predicts for societies with an excess of males, whereas Sparta was as predicted for societies with an excess of females.

Next, they asked whether the significant shifts in sex ratios that occurred in medieval Europe had produced the predicted shifts in sex roles. They found that the sex ratios had. When men greatly outnumbered women early in medieval times, women were highly prized and norms defining traditional sex-role relations were strong. In late medieval times, the excess of men gave way to an excess of women. Soon men were reluctant to marry, and sexual permissiveness flourished. A medieval women's movement also appeared on schedule, in parallel versions inside and outside religious institutions. As Guttentag and Secord reported: "Women struggled to gain economic, religious, and social places for themselves outside the traditional roles of wife and mother."

But the bulk of their efforts to test the theory rests on modern times, especially on modern developments in Europe and North America. These societies have recently produced an enormous amount of research and writing on contemporary sex roles, sex-role socialization, the feminist movement, and gender inequality. Does the evidence fit together as the theory would predict? To find out, let's take an in-depth look.

TRENDS IN NORTH AMERICAN SEX RATIOS

Table 12-5 shows trends in sex ratios in the United States over the past two centuries: a marked shift from an excess of men to an excess of women and a shift even more dramatic among African Americans than whites. Among whites, in 1910 there were almost 107 men for every 100 women. In 2000 there were slightly more than 96 men for every 100 women. For African Americans, an excess of females appeared by the middle of the nineteenth century, and by 2000 there were about 90 African American men for every 100 African American women.

A similar shift has occurred in Canada, but much more recently. In the 1961 census, men still outnumbered women in Canada (102.2 to 100). But the 1991 Canadian census reported only 95.7 men per 100 women. Meanwhile, Mexico remains a society with a slight excess of men, having 101 males per 100 females, but the ratio there has declined, too.

The reason for excess male populations in Mexico until recent times was a high female mortality rate associated with childbirth, while in Canada and the United States, the primary cause was immigration. Thus, while males congregated in western Canada and the United States, more males than females were arriving in the flood of immigration reaching both nations during the nineteenth and early twentieth centuries. Indeed, Canada's sex ratio remained male dominated later than did the U.S. sex ratio because of later patterns of heavy immigration.

Sex ratios based on total population do not reveal the full story. A primary aspect of the Guttentag and Secord theory has to do with sex ratios within the ages at which people desire to marry. Moreover, in most societies women tend to marry men somewhat older than themselves, which probably is a response to the fact that females pass through puberty sooner than males do.

In Canada and the United States, brides are on average nearly three years younger than their husbands. So, Guttentag and Secord computed the ratio of men age twenty-three to twenty-seven to women age twenty to twenty-four. They found that the sex ratios were much less favorable to women in their prime marrying years than the overall ratios would suggest. In 1980 there were 87 white males, age twenty-three to twenty-seven, for every 100 white females, twenty to twenty-four. Among African Americans, the ratio was only 72 males per 100 females in these age groups. The second thing Guttentag and Secord's data showed was the "marriage squeeze" American women faced during the 1960s and 1970s because of the baby boom. When a population grows rapidly, younger cohorts are always larger than the ones born before them. Since women from younger cohorts tend to marry men from older cohorts, women baby boomers faced an acute shortage of slightly older men to marry. In contrast, men their own age had an abundance of slightly younger women to choose among.

In recent years this pattern has been reversed. With a declining birthrate, cohorts entering adulthood are smaller than the cohorts preceding them, so the sex ratio in 1980 was much less unfavorable to young women than it had been a decade earlier.

TABLE 12-5	Males per 100 Females, United States, 1790–2004	
	WHITES	**AFRICAN AMERICANS**
1790	103.8	—
1800	104.0	—
1810	104.0	—
1820	103.2	103.4
1830	103.8	100.3
1840	104.5	99.4
1850	105.2	99.1
1860	105.3	99.6
1870	102.8	96.2
1880	104.4	97.7
1890	105.4	99.5
1900	104.9	98.6
1910	106.6	98.7
1920	104.4	99.2
1930	102.9	97.0
1940	101.2	95.0
1950	98.9	94.3
1960	97.3	93.3
1970	95.3	90.8
1980	94.8	89.6
1990	95.3	89.5
2000	96.6	90.2
2004	97.8	90.9

Sources: U.S. Bureau of the Census, 1975; *Historical Statistics of the United States*; *1980 Census of the Population, General Population Characteristics*, Part 1 (Washington, D.C.: U.S. Government Printing Office); and *Statistical Abstract of the United States, 2001* and *2005*.

© Culver Pictures

Geographical mobility was a major cause of the great oversupply of men in Canada and the United States until recently because it is young men who are most apt to move from one nation or region to another. This photo (circa 1890) of passengers crowding on deck to catch their first glimpse of Ellis Island captures a common sight—not a woman to be seen.

At the turn of the twentieth century, San Francisco had more than 130 males for every 100 females, and the imbalance was far greater among single persons age eighteen to thirty-five. This led to organized efforts to import brides. These women came all the way from England aboard the liner *California* by way of the new Panama Canal. Each had a fiancé waiting at the pier—an engagement arranged by mail. Since women were so scarce in San Francisco and the whole North American West, these women could expect to be idealized and pampered. But according to the theory of sex ratios and sex roles, they could also expect to be excluded from many occupations and spheres of power and decision making.

Library of Congress

Thus far, we have been dealing with national sex ratios. But humans experience life locally, not nationally. For example, although the United States still had more males than females in 1880, local sex ratios—those that determined the relative supply of husbands or wives individuals faced—usually differed greatly from the national average. Recall that in 1880 many states on the eastern seaboard had a substantial excess of women, whereas many western states had huge excesses of men. And these local conditions are the ones most pertinent to the Guttentag and Secord theory. Keeping that in mind, let us assess the shifting nature of sex-role relations in North America over the past century to see if it supports or contradicts predictions from the theory.

GENDER AND SOCIAL MOVEMENTS

Guttentag and Secord predict that when one gender group faces a substantial lack of dyadic power for an appreciable time, they will organize to seek ways to remedy their problem. According to the theory, the elaborate sex-role culture by which women are prevented from fully exercising their options was created by organized male efforts to remedy the lack of dyadic power inherent in the oversupply of men. Similarly, when men have the greater dyadic power, based on their inferior numbers, women can be expected to organize to change the balance of power. Guttentag and Secord found support for their prediction among Spartan women and in a women's movement in late medieval Europe. But what about modern times?

Recall the unfavorable sex ratios women faced in many eastern states throughout the nineteenth century. Shouldn't they have become activists? According to the theory, they should have. According to history, they did.

THE "WOMAN MOVEMENT" On July 19, 1848, a group of women gathered in Seneca Falls, a village in upstate New York, "to discuss the social, civil, and religious rights of women."

Library of Congress

On November 9, 1912, these women posed on a New York City rooftop before joining 20,000 other women in a march for suffrage. The *New York Times* described the event: "As for the parade itself, it was a line, miles long, of well-dressed, intelligent women, deeply concerned in the cause they are fighting for." Each of these women carries a lantern, as did the other thousands; when these were lighted at sundown, they created a spectacle that the *Times* described as "a rolling stream of fiery lava."

Led by Elizabeth Cady Stanton and Lucretia Mott, the group issued a "Declaration of Sentiments," modeled on the Declaration of Independence, proclaiming that "all men and women are created equal." Thus was born what was known throughout the latter half of the nineteenth century as the "woman movement," an effort to end the subordination of women. "Men, their rights and nothing more; women, their rights and nothing less!" was the motto of the weekly newspaper published by the National Woman Suffrage Association.

Although there were many factions and conflicting points of view within the woman movement, two somewhat contradictory positions were widely shared: (1) There was no difference between men and women, and (2) women possessed a superior moral nature from which the whole of society stood to gain. Hence, many of those active in the movement combined the drive for women's rights with a call for more general social reform and proposed that the one would result in the other.

As Jane Frohock expressed this: "It is woman's womanhood, her instinctive femininity, her highest morality that society needs now to counteract the excess of masculinity that is everywhere to be found in our unjust and unequal laws" (Cott, 1987). The tension between these two positions grew as time passed. In her superb historical study of American feminism, Nancy F. Cott (1987) summed up the situation as follows:

By the close of the century the spectrum of ideology in the woman movement had a see-saw quality: at one end, the intention to eliminate sex-specific limitations; at the other, the desire to recognize rather than quash the qualities and habits called female, to protect the interests women already had defined as theirs and give those much greater public scope.

However, underlying ideological ambiguities were submerged as the woman movement gave rise to the suffragist movement, dedicated to obtaining the right to vote for all American women. As Cott observes, this issue served as "a platform on which diverse people and organizations could comfortably, if temporarily, stand."

Back in 1913 these young suffragettes formed a pep band to publicize a rally in support of "VOTES for WOMEN," as their banners proclaim. As was typical of small-town life in those days, the rally was held outdoors in front of the courthouse. Preceded by a concert to help assemble a crowd, a series of speakers proclaimed the cause of female suffrage—and in these days before amplifying equipment, public speakers had to have strong voices. On the right is a poster advertising a feminist mass meeting in New York City, 1914.

© Culver Pictures

And this unity produced a rapid swelling of the ranks during the decade from 1910 to 1920, culminating in final ratification of the Nineteenth Amendment on August 26, 1920:

The right of citizens of the United States to vote shall not be denied or abridged by the United States or by any State on account of Sex.

Congress shall have power to enforce this article by appropriate legislation.

During the last few years of the drive for suffrage, women all over the nation took part. But in keeping with the Guttentag and Secord theory, the woman movement began and achieved greatest prominence in the northeastern states with their longstanding shortage of men. And it was also there that the next major development took place.

THE FEMINIST MOVEMENT It has become an article of faith among contemporary writers on the status of women that the suffrage movement derailed the woman movement. Once women received the vote, it is claimed, their efforts to achieve full equality and to reform basic sex-role relationships died out, not to reappear until 1963, when Betty Friedan's *The Feminine Mystique* aroused a revival of feminine militancy (Epstein, 1976; Rothman, 1978; Rapp and Ross,

1983). In fact, what had died out by the time women gained the right to vote was the "woman movement," and what replaced it was not apathy or quietude, but something called *feminism*.

The word seems to have first appeared in English usage around 1906. Before long it became a popular new buzzword. In 1913 a leading campaigner for women's rights wrote that the word *feminism* was "something so new that it isn't in the dictionaries yet" (Cott, 1987). By 1914 Henrietta Rodman, a militant New York City teacher, had organized the Feminist Alliance.

What did the word mean? Why did it suddenly become so popular? Cott (1987) points out that "woman movement" is ungrammatical and implies an agenda limited to matters of concern only to women, whereas the term *feminism* implies much broader intentions and the involvement of both women and men in fashioning new relationships.

Cott defines **feminism** as an ideology having three essential features: (1) opposition to all forms of stratification based on gender, (2) belief that biology does not consign females to inferior status, and (3) a sense of common experience and purpose among women to direct their efforts to bring about change.

Feminism sprang from the woman movement and the suffrage movement in the second decade of the twentieth century and has neither declined nor gone into suspension since then.

The feminism of the 1960s was a direct outgrowth of the movement that began to attract women in New York and Boston fifty years before (Cott, 1987). What misled so many observers was that in the beginning the feminist movement was very small, its early appeal being limited to women of two different and unusual backgrounds. The first were women active in leftist politics and union organization, often women employed in the garment industry. The second were young female intellectuals and artists ready to explore bohemian lifestyles—like the avant-garde women who were frequenting the Greenwich Village scene by the 1920s. These women typically were well educated, often at women's colleges, and frequently, their mothers had been active in the woman movement. Hence, while the suffragist movement rapidly attracted a massive, popular base, feminism began as a movement among small, often elite, groups and grew but slowly. When the suffragists ceased their huge marches, it seemed as if all organized interest in gender inequities had disappeared, too. Instead, the feminist movement continued to grow, fueled by two other major trends changing the circumstances of American women.

WOMEN IN THE LABOR FORCE

When men are in short supply, women increasingly find they need to become self-supporting. Early in the nineteenth century, as many young men headed West (in those days, the "Wild West" was in western New York State and Ohio), many young women in New England supported themselves in their homes by spinning wool into yarn. This practice became so widespread that the term *spinster* soon ceased to mean a female spinner and instead identified any unmarried woman over a certain age (Guttentag and Secord, 1983). But as the proportion of males declined and the number of women needing work increased, new inventions made it possible for huge textile mills to spring up in New England—drawn there partly because of the supply of young women workers. "New England textile factories from the start employed a vastly greater proportion of women than men" (Cott, 1977). Thus began a trend that in the course of a century has drawn the majority of North American women from their homes to hold regular jobs.

Table 12-6 shows this trend. In 1890 only 18.2 percent of American women were employed full-time outside the home. Over the next 100 years, female labor force participation (as the census calls it) increased regularly until now more than half of all women in the United States and Canada hold full-time jobs. In fact, the data in Table 12-6 tend to understate female labor force participation because the table is based on all women over age fifteen. When women over age sixty-four are removed from

TABLE 12-6	Percentage of American Females Employed- Full Time Outside the Home, 1890–2002*
1890	18.2
1900	20.0
1920	22.7
1930	23.6
1940	25.8
1950	29.9
1960	35.7
1970	41.4
1980	51.5
1990	57.5
2000	60.2
2002	59.6

Sources: U.S. Bureau of the Census, *Historical Statistics of the United States: Historical Times to 1970*, and *Statistical Abstract of the United States, 2001* and *2005*.

*Based on all women over age fifteen.

In periods of rapid social change, some things change faster than others. Here we see the Becker sisters branding cattle on their ranch in the San Luis Valley in 1894. They could demonstrate that "cowgirls" were able to do jobs long reserved for "cowboys," but they weren't ready to dress like cowboys. Instead, they had learned to rope and tie steers while dressed in the demure women's fashions of the time.

Colorado Historical Society

The scores of textile mills that sprang up in New England during the nineteenth century depended on a labor force of young single women like these. English novelist Charles Dickens reported in his 1841 *American Notes*, an account of his travels in America, that "these girls, as I have said, were all well dressed; and that phrase necessarily includes extreme cleanliness. . . . They were healthy in appearance, many of them remarkably so, and had the manners and deportment of young women; not of degraded brutes of burden. . . . I am now going to state three facts which will startle readers this side of the Atlantic very much. Firstly, there is a piano in a great many of the boarding houses. Secondly, nearly all of

Museum of American Textile History

these young ladies subscribe to circulating libraries. Thirdly, they have got up among themselves a periodical." What Dickens saw was the massive entry of young women from "respectable" backgrounds into the New England labor force in response to a marked shortage of eligible young men. Note how nicely they are dressed and groomed and how each is wearing an apron to protect her skirt.

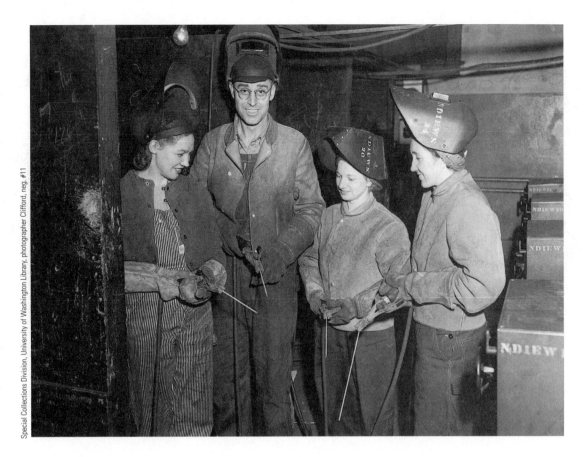

Many women entered the labor force during World War II, often taking jobs that previously had been done only by men. Here three women are being taught to weld at a ship-yard in Tacoma, Washington, in 1943. Some have suggested that this wartime experience caused a rapid shift in female employment patterns—that once out of the home earning good pay, women refused to go back to the role of house-wife. But most of them did go back, and their increased fertility rate created the baby boom. Consequently, Table 12-6 shows no unusual spurt in female employment between 1940 and 1950. The really rapid increase has taken place since 1960, as the baby boomers began to come of age and young single American women greatly outnumbered eligible men.

the base, then 71.7 percent of American women are in the labor force. Table 12-7 shows an even more dramatic and recent trend: the tendency for mothers of young children to go off to work every day. In 1950, with the baby boom in full flower, slightly fewer than 14 percent of women with children under age six held jobs. By 2000, more than 60 percent did.

Not only has the traditional division of labor between husband as earner and wife as house-keeper given way to the two-earner household, but occupations also have become much less gender segregated than they once were—although substantial differences persist. As recently as 1970, women made up only 33.9 percent of persons employed in the top jobs, those classified by the census as managerial and professional occupations, but held 46.3 percent of those jobs by 1991; and the trend continues upward. Within this category, males dominate the fields of engineering (8.2% female), architecture (17.1% female), and law (18.9% female). Among those fields most dominated by women are nursing (94.8%), elementary school teaching (85.9%), and librarianship (83.0%).

Women still dominate the occupation of secretary (99.0%), and men still dominate the skilled

blue-collar crafts such as auto mechanic (less than 1% are women) and carpenter (1.3% are women).

Although the majority of women now work outside the home, they bring home significantly less money than men do. This is a familiar newspaper feature story, but the "facts" usually are presented in a very crude and misleading way. The typical story is content to report that the average annual income for American women is only about half as much as that of men. But that figure includes the millions of women who are full-time homemakers and earn no income of

TABLE 12-7	Percentage of Married American Mothers with Children Under Age Six Employed Full Time Outside the Home, 1950–2002
1950	13.6
1960	18.6
1970	30.3
1980	45.1
1990	58.9
2000	62.8
2002	60.8

Source: *Statistical Abstract of the United States, 2001* and *2005.*

Library of Congress

In the summer of 1920, only weeks before final ratification of the Nineteenth Amendment gave American women the right to vote, something brand new began: beauty contests, known initially as "bathing beauty contests." Suddenly, the American woman went from floor-length skirts (when men ogled pretty ankles) and similarly concealing beachwear to bare legs and figure-hugging fabrics. Here we see the contestants in the first Annual Bathing Girl Parade held at Balboa Beach in 1920. A year later, the Miss America Beauty Pageant began at Atlantic City. If you look closely at the photograph, you will be able to detect many signs of the sudden revolution in women's wear and, of course, in the norms governing standards of female modesty. Clearly, many of these young women had never appeared in such revealing suits before, and they don't know how to pose in them to look their best. Instead, they look nervous, awkward, and very self-conscious.

their own. Sometimes the reporters are careful enough to base their figures only on men and women who worked year-round, full-time. In 1984 that would show women earning only 64 percent as much as men—better, but still a depressing picture of inequity. But this, too, is what Cynthia M. Taeuber and Victor Valdisera (1986) call a "simplistic" use of statistics. They point out that such a comparison neglects the fact that the women and men are not doing the same kinds of work. While women have entered many occupations once closed to them, they still are concentrated in the relatively lower-paying occupations, such as secretary, and remain very underrepresented in some high-paying jobs, especially the highly unionized, skilled, blue-collar occupations, such as mechanic, plumber, electrician, and machinist.

There can be no doubt that discrimination and false notions about women as the "frail" sex are involved in these patterns. However, many additional factors must be considered to really understand male/female income comparisons. One is that because female employment rates have risen rapidly, the average working woman is considerably younger than the average working man. That means the average male benefits from greater seniority and a longer opportunity to have accrued raises. When Taeuber and Valdisera analyzed 1984 income data, they found that among persons employed full-time, year-round, age eighteen to twenty-four, women's income was 86 percent that of men—and most of the remaining difference could be explained by differences in types of jobs.

Another aspect of the gender wage gap is turnover. Women change jobs more often than do men, and they take time out from the labor market more often. Again, this influences seniority and experience. Of employed persons, one of every nine men has worked at the same occupation for twenty-five years or more, whereas only one woman in twenty has done so. One reason for higher rates of female job turnover is maternity and child rearing. Another is the tendency of families to move when it benefits the husband's career, thus causing wives to quit jobs and seek new ones. Ironically, this move usually is the wise financial choice for individual families because the husband's salary is higher, but it tends also to cause husbands' salaries to remain higher.

A final consideration is sex-role socialization. One aspect of this is that women still tend to

The photo also reveals that no customs had yet been established about what "bathing beauties" should wear with their swimsuits. Many of these young women wear hats, but some don't. Many have on high heels; others are in flats or are barefoot. Some wear hose; others wear socks. One of the many ironies explained by Guttentag and Secord's theory is that women marching for equal rights and women parading in scanty costumes tend to go hand in hand. In the most sexist societies, women are romanticized, but male tendencies to regard them as sex objects are greatly muted by strictly enforced norms of modesty, virginity, and chastity. It is when women enjoy much greater freedom from male-imposed limits on their behavior that they also are, at least initially, most subject to being overtly paraded as sex objects.

shun some of the activities that lead to high incomes. For example, women have been far less likely to choose majors that lead to the highest-income professional and managerial jobs, spurning the sciences and engineering while greatly overenrolling in the humanities and education (Taeuber and Valdisera, 1986). In addition cultural definitions of the characteristics needed to be an effective leader are traits regarded as masculine—traits such as being ambitious, aggressive, tough, confident. In contrast, traits associated with femininity, such as being sensitive and understanding, are seen as inappropriate in leaders (Hollander, 1985). Women are less likely to be selected as leaders to the extent that people are socialized to perceive that women lack appropriate leadership traits. Note, too, that among men there has long been a significant bias in favor of selecting tall men for leadership slots (Deck, 1971). If short men are seen as lacking leadership skills, women also probably suffer from their smaller stature.

All of these factors help to account for the fact that when men and women have similar education and training and are employed in the *same* occupation, the men are more apt to hold managerial and supervisory positions and hence to earn higher salaries (Taeuber and Valdisera, 1986). Ironically, when people are asked to name traits they would prefer in a manager they would like to work *for*, they tend to favor more feminine traits, but reverse themselves and stress masculine traits when asked to describe a manager they would like to have working *for them* (Cann and Siegfried, 1987).

THE SEXUAL "REVOLUTION"

In 1900 in New England, about one in every four first births was to an unwed mother or to a woman who had married after becoming pregnant (Smith and Hindus, 1975). Fifty years earlier only about one birth in ten involved an unwed or premarital pregnancy. What was going on?

What was *not* going on is what the suffragette Christabel Pankhurst had proposed with her "notorious" slogan: "Votes for Women and Chastity for Men!" With an abundant supply of eligible women, men grew increasingly less concerned about sexually monopolizing one woman and became more interested in easy sexual access to several. Sexual norms began to change, and premarital sex became more common, as the data on first births demonstrate.

As always with changes in basic cultural patterns, at first the changes are slow and accompanied by a great deal of controversy as people who have been effectively socialized into the old

© Bettmann/Corbis

California Historical Society

The young women on the right model fashionable beachwear at the turn of the twentieth century. Bare arms were considered daring in those days, so it is likely that very few women actually bought suits like the one worn by the woman on the far right. Twenty years later, women suddenly appeared on the beach in suits that bared not only their arms but also their legs. A lot of people were very upset by this. When significant norms are challenged, bitter conflicts usually break out between those who want to preserve old norms and proponents of the new. Thus, in July 1922, two years after bathing beauty contests introduced "revealing" swimsuits, the women above were arrested for indecency and escorted from Chicago's Lake Michigan beach by police—including the policewoman who is between the two women in the foreground.

norms seek to enforce and defend them. Thus, in the early twentieth century, there was a lot of organized effort to stem the "rising tide of immorality." These efforts were directed not so much at people who broke sexual norms as at people who advocated doing so or who made a "public display" of sexuality through art, literature, or drama. "Banned in Boston" became a common phrase reflecting that city's strict censorship. And in the center of the controversy stood the feminists. For unlike the woman movement before them, they did not advocate ending the double standard of morality by restricting males but rather a complete sexual freedom for females. And they did so in writing and in public speeches. As Cott (1987) characterized the sexual views of the early feminists:

Seeing sexual desire as healthy and joyful, they assumed that free women could meet men as equals on the terrain of sexual desire just as on the terrain of political representation or professional expertise.

Today, these views seem commonplace, but in the America of the early 1900s, they were condemned as a dangerous, reckless advocacy of "free love" and an attack on basic institutions such as marriage and the family. And there is a certain amount of truth behind these charges. For there is a darker side to the moral stance of the early feminists: *the sexual exploitation of women.* As Cott (1987) put it:

Feminists were far from acknowledging publicly the potential for submergence of women's individuality and personality in heterosexual love relationships, or the potential for men's sexual exploitation of women who purposely broke the bounds of conventional sexual restraint. In private they saw, inevitably, travesties of their ideal.

And here was a tragic irony. As women battled to free themselves of dehumanizing limits—for example, to be allowed to wear comfortable and appropriate gym clothes and beachwear—they unleashed forces that tended to impose new forms of dehumanization. The new styles of the 1920s freed women from painful corsets and floor-length skirts, but they also put women on parade. In similar fashion, as women struggled to be free of the sexual double standard, they made themselves increasingly vulnerable to being divorced, deserted, and used by husbands and lovers.

© Bettmann/Corbis

Feminists today have made a major issue of the many ways in which women are treated as sex objects, especially in the mass media, and dehumanized and sexually exploited by such recent innovations as the singles bar. But this state of affairs is the very transformation in sex roles and sexual norms Guttentag and Secord predicted. Although it may be possible to change current patterns in gender roles and relations, these patterns are not simply accidents of history.

SEX RATIOS AND THE AFRICAN AMERICAN FAMILY

In the 1920s the great African American composer and performer Fats Waller popularized a song he wrote to tell women how to hold onto their men: "Find Out What They Like." The opening lines advise, "Find out what he likes and how he likes it, give it to him just that way." Meanwhile, the famous Bessie Smith replied for African American women with her bitter "Why don't you do right, like some other men do?" What these songs reflected was *not* a tradition of matriarchy and of female-headed families in African American culture, as too many have suggested. What Fats Waller and Bessie Smith were

Thomas "Fats" Waller (1904–1943) was one of the most prolific and influential composers of American popular music during the 1920s and 1930s. He turned out scores of hits, including "Honeysuckle Rose," "Ain't Misbehavin'," and "Squeeze Me." His tunes often were included in Broadway shows. Equally gifted as a pianist and singer, he was a pioneer of early radio—seen here during a 1934 broadcast on CBS. Although there is no reason to suppose Waller knew that African American women substantially outnumbered African American men in the 1920s, the lyrics of his songs often reflected the greater dyadic power of men—that it was up to women to hold onto men.

National Maritime Museum, San Francisco

In 1900 Captain William T. Shorey, his wife, and their two children posed for this family portrait in Oakland, California. Shorey was master of a whaling ship and the only African American captain on the Pacific Coast at this time. But if Shorey was exceptional in his occupational achievement, the Shorey family was typical in its composition: It included a father as well as a mother and children, as did *most* American families at this time, African American or white. Although today more than half of African American children are growing up in a female-headed family, this has been a recent development, not a cultural holdover from slavery days.

singing about was a trend that was influencing dyadic power relations between African American men and women—a trend that seventy years later has resulted in the majority of African American children living in a fatherless household. That trend was a growing shortage of African American men, especially of younger single African American men. To conclude this

chapter, we will examine how many of the current family-related problems that afflict African Americans today are the result of their extremely imbalanced sex ratio.

Recall from Table 12-5 that by the middle of the nineteenth century there already were more African American women than men. During the twentieth century, the trend continued so that each year the sex ratio among African Americans reflects more women and fewer men. Moreover, this shortage is even more acute among young African American adults: In 1980 there were 72 single African American males age twenty-three to twenty-seven for every 100 African American women age twenty to twenty-four. But even this fails to show how severe the shortage of African American males is. The availability of African American men is further reduced by the gender imbalance of interracial marriage patterns. African American men are at least twice as likely as African American women to marry someone of another race (Kalmijn, 1993).

WHY ARE AFRICAN AMERICAN MEN IN SHORT SUPPLY? The shortage of African American males begins at birth. Rather than having from 105 to 106 males born for every 100 females, the sex ratio is nearly even at birth for African Americans. The primary reason for this is a considerable gender imbalance in fetal deaths among African Americans, which probably reflects poor nutrition and health care, especially in previous generations. The next factor in African American sex ratios is infant mortality. Because African Americans have a substantially higher infant mortality rate than do whites or Hispanics, and because male infants are more likely to die, by the end of the first year of life the number of males has dipped below the number of females. As African American children grow up, the proportion of males continues to drop because of higher African American mortality rates and the especially high mortality of young African American men from accidents, drugs, and violence.

CONSEQUENCES FOR THE AFRICAN AMERICAN FAMILY One of the earliest consequences for African American women faced with an acute shortage of men was the need for employment. By 1900, when only 20 percent of all American women were employed outside the home, 43.2 percent of African American women were in the labor force compared with 14.6 percent of native-born white women.

But the primary consequence has been to make the father-mother-children family a receding memory among African Americans. Superficial scholarship to the contrary, the fact is that the African American family was not broken up by slavery, and sexual promiscuity was not the norm among African Americans on southern plantations. Indeed, the stable nuclear family and long-lasting marriages seem to have been typical, and African American slave families took a very protective attitude toward the chastity of their daughters (Fogel and Engerman, 1974; Guttentag and Secord, 1983). The signs of severe disruption displayed by the African American family are recent. In 1970 only 28 percent of African American families including minor children were headed by a woman, and two-thirds of African American children lived with both their parents. In 1997 more than half of African American families with children were headed by a woman and only 36 percent of African American children lived with both parents. In 1960 the **illegitimacy ratio** (the percentage of all births that occur out of wedlock) was 22 percent among African Americans; in 1996 it was 70 percent. Here is another way to see what has been happening to the African American family: The 1998 General Social Survey asked a large sample of Americans with whom they had lived when they were sixteen. Only 49 percent of African Americans (compared with 74 percent of whites) said they had lived with both their father and their mother at sixteen.

Recent research links the rise of the one-parent African American family directly to the shortage of potential husbands confronting African American women. The first link is obviously the simple lack of men relative to the number of women. Analyzing data for U.S. metropolitan areas, Mark Fossett and K. Jill Kiecolt (1993) found that variations in the African American sex ratio were highly correlated with the proportion of one-parent families among African Americans. Where relatively more potential husbands were available, more women married. Second, precisely as Guttentag and Secord predict, African American men are far less likely than other American men to desire to marry (South, 1993).

SEXUAL NORMS Guttentag and Secord's theory would predict not only that men would be less willing to marry when faced with a great excess of women but also that they would expect women to grant sexual privileges readily. The data on the African American family make it evident that sexual norms have shifted among African Americans

John Phillips/TimePix/Getty Images

In 1939, when John Phillips took this picture for *Life* magazine, this young bride had overcome a substantial shortage of potential grooms. In Georgia at that time, there were only 91.3 African American men per 100 African American women, and the sex ratio was even more unbalanced when only young single adults were counted. At this moment, as her mother adjusted her veil, this young woman was not thinking about statistics. But she may well have noted that many of her friends, waiting in the church, were having trouble getting the local men to marry and settle down.

in precisely the direction the theory anticipates. Clearly, when most infants are born to women without husbands, there is reason to expect a lot of sexual activity among people who are not married. Moreover, a growing literature indicates the pressures African American women feel from African American men to provide sex without commitment (Washington, 1975; Guttentag and Secord, 1983; Wilson, 2002).

To conclude this analysis, let us briefly examine the impact of sex ratios on family structures among Hispanic Americans.

SEX RATIOS AND THE HISPANIC AMERICAN FAMILY

Overall, Hispanic American women do not face a shortage of men. In the age group fourteen to twenty-four, there are 107.2 Hispanic American men per 100 women. In the age group twenty-five to forty-four, the ratio is 106.1 to 100. However, as we saw in Chapter 11, the overall figures hide some striking variations. Among Mexican Americans, the ratio of males to females is even slightly higher than the overall total. But among Puerto Rican Americans, the opposite is true: Women outnumber men to about the same extent as among African

American families are far less likely to be female headed (19% versus 43%), and the average Mexican American woman has more children and is much less likely to be divorced than is the average Puerto Rican American woman. Furthermore, Puerto Rican American children are about twice as likely to be born out of wedlock as are Mexican Americans.

CONCLUSION

Our understanding of the problems created by unbalanced sex ratios is only in its infancy. After all, not until 1983 had more than a handful of social scientists even heard of Guttentag and Secord's theory. But it has caused a flurry of research on a wide range of topics, including the involvement of women in crime. Where men greatly outnumber women, the latter are much less apt to commit crimes, and the police are more successful in solving cases of rape (South and Messner, 1987).

In my own work on the rise of Christianity, I discovered that in the early Christian communities females far outnumbered males, whereas the opposite was true of the surrounding pagan world. In full agreement with the theory, Christian women enjoyed far greater gender equality than did pagan women (Stark, 1995).

But do variations in sex ratios account for *all* of the variations in the status of women across societies and history? Certainly not! As with all scientific theories, Guttentag and Secord's theory rests on the underlying assumption that if "other things are equal," then variation in the status of women will be negatively correlated with the sex ratio. But other things are never equal (except in experiments), and clearly, some of these other things matter. Religion can be one of these other things. For example, the absolute prohibition against infanticide played a major role in producing the more favorable sex ratio among the early Christians. Another factor is the fit between the dominant modes of production and those of reproduction. As Randall Collins, Janet Salzman Chafetz, Rae Lesser Blumberg, Scott Coltrane, and Jonathan H. Turner (1993) noted in their effort to synthesize various theories of gender inequality:

Women who are pregnant, breast-feeding, or caring for young children adjust their productive and reproductive activities to make them mutually compatible; hence [in premodern societies] their economic contributions tend

© Corbis/Sygma

Muslim nations today exemplify how cultural rules traditionally have enclosed and oppressed women in response to extremely unbalanced sex ratios. This Arab husband and his son wear Western clothes; the wife still must wear a veil and hide her figure under a billowing robe while in public. Where women are in short supply, men tend to define them as too valuable to be seen, let alone to be free.

Americans. As for Cuban Americans, women also outnumber men, but this is entirely due to age: The Cuban population is relatively older, and it is only among older Cubans that women outnumber men.

Among Mexican Americans, men outnumber women partly because of differential immigration: More men than women come here from Mexico. In addition Mexican Americans have very low infant mortality rates (compared with African Americans and Puerto Rican Americans). When infant mortality is low, males will outnumber females. The causes of a shortage of males among Puerto Rican Americans seem similar to those among African Americans.

In any event, according to Guttentag and Secord's theory, the role of women among Mexican Americans should be more traditional than it is among Puerto Rican Americans. And the available evidence suggests this is true. Mexican to be greatest in food-gathering or domestic gardening modes of production, lowest in dangerous or long-distance activities of hunting, herding, fishing, or military plundering.

Even in the most advanced industrial nations, the link between reproduction and production remains, albeit in much reduced form.

A modern dual-career couple, taking advantage of reproductive technology [such as bottle-feeding] and commercial or collective childcare, still tends to find the mother giving more commitment to childcare and domestic responsibilities. (Collins et al., 1993)

Democratic politics is an additional factor. Where women enjoy political freedom and where public sentiments have an impact on law and custom, feminist movements are bound to appear and to influence social structures.

We have examined this process in the United States, and it will receive additional attention in Chapter 15.

My aim in this chapter was not to examine in detail all of the factors influencing the status of women. Rather, it was to explore a major recent contribution in this area and to let you see in detail how theories are formulated and to trace out the many and diverse concrete implications of a small set of theoretical statements.

Keep in mind, of course, that theories never can be proved. So, as researchers continue to test this theory, evidence may turn up that will cause us to greatly revise it or even to reject it. But there can be no doubt that Marcia Guttentag's trip to the opera will have lasting consequences. By calling our attention to the fact that societies often have a great excess of one gender, she has identified a matter of basic interest. All future attempts to understand gender relations must take sex ratios into account.

Review Glossary

Terms are listed in the order in which they appear in the chapter.

sex ratio The number of persons of one gender relative to the number of persons of the other gender, usually expressed as the number of males per 100 females.

infanticide The practice of killing infants soon after birth, often done by simply abandoning them out of doors.

dyadic power The capacity of each member of a dyad to impose his or her will on the other member.

power dependence The dependence of one member of a dyad on the other is equal to their *in*ability to achieve their goals *outside* the dyad.

structural power Power based on statuses within social structures.

feminism An ideology having three essential features: (1) opposition to all forms of stratification based on gender, (2) belief that biology does not consign females to inferior status, and (3) a sense of common experience and purpose among women to direct their efforts to bring about change.

illegitimacy ratio The proportion of all births that occur out of wedlock.

Suggested Readings

Collins, Randall, Janet Salztman Chafetz, Rae Lesser Blumberg, Scott Coltrane, and Jonathan H. Turner. 1993. "Toward an Integrated Theory of Gender Stratification." *Sociological Perspectives* 36:185–216.

Cott, Nancy F. 1987. *The Grounding of Modern Feminism.* New Haven, Conn.: Yale University Press.

Evans, Sara M. 1989. *Born for Liberty: A History of Women in America.* New York: Free Press.

Guttentag, Marcia, and Paul F. Secord. 1983. *Too Many Women? The Sex Ratio Question.* Beverly Hills, Calif.: Sage.

Pomeroy, Sarah B. 1975. *Goddesses, Whores, Wives, Slaves: Women in Classical Antiquity.* New York: Schocken Books.

Sociology Online

www.socstark10.com

GO TO THE INTERNET AND TYPE: www.socstark10.com.

↗**CLICK ON:** 2000 General Social Survey

✓ **SELECT: WIFE@HOME:** It is much better for everyone involved if the man is the achiever outside the home and the woman takes care of the home and family.

↗**CLICK ON:** Analyze Now

TRUE OR FALSE: Men are significantly more apt to agree with this statement.

How would you explain this finding?

TRUE OR FALSE: Traditional sex-role stereotypes seem to be dying out.

What table did you examine to find the answer?

↗**CLICK ON:** Select New Data Set

↗**CLICK ON:** The 50 States

✓ **SELECT: % FEM MD:** Percentage of physicians (M.D.s) who are female.

↗**CLICK ON:** Analyze Now
Is there a regional pattern to this variable? Is this consistent with the discussion in the chapter about where feminism began and why?

↗**CLICK ON:** Select New Data Set

↗**CLICK ON:** Nations of the Globe

✓ **SELECT: %FEM.LEGIS:** Percentage of parliamentary seats held by females.

↗**CLICK ON:** Analyze Now
Examine the nations where more than 30 percent are women. Do they have something in common? What about the nations having 0.0 percentage?

USING INFOTRAC COLLEGE EDITION

GO TO THE INTERNET AND TYPE:
www.InfoTrac-college.com

↗**CLICK ON:** Register New Account
You will be asked to enter your Access Code and to create and enter a User Name and Password. Having done so,

↗**CLICK ON:** InfoTrac College Edition or Log On.

Keep in mind that many discussions of gender issues are scattered throughout the book (see Figure 12-1).

SEARCH TERM:
Feminism

In addition, I have selected a specific article that will usefully supplement the chapter: Sex and the Marriage Market. You can find it by searching for this title. You may read the article on the screen or print it using the usual print commands. If you also go to www.socstark10.com and click on the InfoTrac College Edition icon, you can read an explanation of why I selected this article and find several questions that will help you connect the article to material in this chapter.

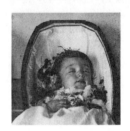

a prairie family

In 1903 this farm family in North Dakota's Red River Valley drove their buggy to Grafton to have a family portrait made. They wanted the picture not only for themselves but also to send to relatives in Europe. The young boy seated at the far left is my father. My grandfather, Andrew Stark, was born and raised in Sweden, and my grandmother was Norwegian. After coming to America, Andy made his living as a horse trader in South Dakota, but after he got married, he became a farmer. The Starks had a large family—two more sons would soon be born—but large families were typical in that time and place. I never knew my grandfather, but I still feel that I am part of his family.

The Family

SOCIOLOGICAL WRITING ON THE FAMILY has long been dominated by two themes: *universality* and *decline*. The theme of universality asserts that the family exists in all human societies. For a number of compelling reasons, people cannot live as solitary creatures, nor can human females raise their young by themselves as mother cats do. Hence, humans always live in groups containing adults of both sexes as well as children. Moreover, within any society, people form small clusters, called families, containing males and females, adults and children. Membership in these clusters usually is determined by common ancestry and sexual unions.

This definition of family is vague because sociologists and anthropologists have had much difficulty framing a more specific definition, given the amazing variety of social forms called families in different societies. Again and again,

more specific definitions of the family have been found not to apply in one society or another, thus destroying the claim that the family is universal. Yet all societies *do* seem to have families.

The second theme in modern sociological writing on the family is that, despite the universality of the family, in modern societies the family is in decline. Some claim that, thanks to modernization, the family has eroded dangerously: Families are now shrunken and unstable, and the modern family is increasingly unable to provide for the well-being of its members. Indeed, recent textbooks typically end discussions of the family with the question, Will the family survive? The answer rarely is anything more definite than maybe.

Despite the problems of definition, a great deal of historical, anthropological, and cross-cultural evidence supports the universality theme. The family *is* a fundamental social institution

CHAPTER OUTLINE

DEFINING THE FAMILY

FAMILY FUNCTIONS: AN INTERNATIONAL PERSPECTIVE
Sexual Gratification
Economic Support
Emotional Support

A CLOSER VIEW
Life in the Traditional European Family

Household Composition
Crowding
"Outsiders"
Child Care
Relations between Husbands and Wives
Bonds between Parents and Children
Bonds among Peer Group Members

MODERNIZATION AND ROMANCE

MODERNIZATION AND KINSHIP

MODERNIZATION AND DIVORCE
How Much Divorce?
When Romance Fades

GENDER AND EXTRAMARITAL SEX

A CLOSER VIEW
Trent and South: International Comparisons in Divorce

LIVING TOGETHER

THE ONE-PARENT FAMILY

REMARRIAGE

A CLOSER VIEW
Jacobs and Furstenberg: Second Husbands

A CLOSER VIEW
White and Booth: Stepchildren and Marital Happiness

HOUSEHOLD CHORES

CONCLUSION

REVIEW GLOSSARY

SUGGESTED READINGS

SOCIOLOGY ONLINE

occurring in all societies, although its particular forms differ substantially from place to place. Even the radical utopian communes of the nineteenth century did not succeed in eliminating the family as the basic unit of social relations (Nordhoff, 1875).

The theme of decline has seemed equally well supported by evidence. Statistics show that in all of the most modernized nations, the divorce rate has risen rapidly. This would seem to reflect the weakness of fundamental family bonds today. However, to know whether the modern family is really less able to fulfill its functions, we need to know whether the family in traditional societies fulfilled them better. For a long time, social scientists thought it self-evident that the traditional family did function better, and so they didn't bother to seek pertinent evidence.

Recently, however, much has been learned about families in the "good old days." This evidence seriously challenges the theme of decline. Be prepared to discover that family life in premodern times often was cruel and spiteful to an extent that will absolutely shock you. People often expressed happiness when their spouse died and were unmoved by the death of a child.

Thus, it can be argued that the family has become more important than ever during the past century and much better able to provide strong emotional attachments among its members than did families in traditional societies.

In this chapter we shall first wrestle with the problem of what the family is. Once we can define the family, we will see how it has changed in response to modernization and then be able to assess the theme of decline. To do this, we shall examine what the family was like in Europe several hundred years ago. How did families live? How did family members treat one another? How did they feel about one another? How distinct was the family from the larger community? Against this benchmark, we shall assess the modern family. Is family life better or worse than it used to be?

DEFINING THE FAMILY

Perhaps the most consistent efforts to define the family have drawn on the functionalist approach to sociological theory. A number of anthropologists and sociologists have started from the premise that if the family is a universal social institution, then it must do something vital for human beings, something not done as well or as easily by other institutions. In seeking to define the family on the basis of *what it does*, functionalists have attempted to specify a list of functions the family performs.

In 1949 George Peter Murdock formulated what may be the most influential of these definitions. He defined the family as "a social group characterized by common residence, economic cooperation, and reproduction." He added that the family "includes adults of both sexes, at least two of whom maintain a socially approved sexual relationship, and one or more children."

Murdock then spelled out four primary functions of the family: sexual relationships, economic cooperation among members, reproduction, and the educational function, by which he meant the socialization of infants and children. Murdock admitted that agencies and relationships outside the family may share in the fulfillment of these functions, but "they never supplant the family."

However, Murdock's definition soon came under attack. On the one hand, some critics pointed out examples of a society or two that seemed exceptions to one or more of Murdock's functions. For example, many claimed that economic cooperation is not an element of the family as it exists in a typical kibbutz (rural commune) in Israel. Instead, economic cooperation exists among all members of the kibbutz (Reiss, 1988). Among the Nuer of East Africa, some families contain no adult males—an older woman can adopt a younger woman and her children (Gough, 1974). On the other hand, some critics complained that Murdock's functions were too narrowly conceived. For example, they pointed out that sexual intercourse is but a small part of the emotional life husbands and wives share, and Murdock's sexual function completely ignores all the other strong, but nonsexual, relationships among family members.

Eventually, most sociologists of the family solved the problem by adopting a definition based on the notion of kinship and limited to the single function of child care. Other functions, such as those Murdock identified, are treated as necessary to *individual* well-being but not as necessary parts of the family. For example, while humans need emotional support and frequently rely on their families as their primary source, in some societies this function is fulfilled primarily from outside the family. This revised approach lets us see that the connections between the family and various functions are problematic. In fact, a major activity of family sociologists is to examine changes over time or variations from one society to another in which functions the family fulfills and how well it does so.

The standard definition of the **family** as a universal human institution is a small kinship-structured group with the key function of nurturant socialization of the newborn (Reiss, 1988). This definition attempts to specify both *who belongs* to a family and *what the family does*— its primary function. Let's examine each element of this definition in greater detail.

The phrase "kinship-structured group" is more satisfactory than one invoking "biological ties" because many societies define kinship on the basis of such things as common residence rather than biological relationship; indeed, some preliterate societies have rather mistaken notions about biology. But as Ira Reiss (1988) put it, "The tie of kinship is a special tie in every society of which we have any record." Yet if we define the family as a kinship cluster, then how can we justify that additional element of the definition which asserts that the key function of the family is "nurturant socialization of the newborn"? Is it always so? What about some future society that consigns all reproduction to government laboratories where eggs from female donors are fertilized with sperm from male donors and develop in a mechanical womb—a society where the infants are raised by trained nurses? Where's the family? And what happens to the universal applicability of this definition?

Here's the answer Reiss (1988), a leading family sociologist, gives:

Kinship, as we have defined it, is not dependent on biological connections, but rather is dependent on socially defined connections. In a society that assigned nurses to rear newborn infants, the ties between the nurse and the infants given to her/him would be defined as distinctive and special. Such ties of nurse and children, because they begin from infancy and last a period of years, would possess special emotional significance and would invest both the nurse and the children with special rights and duties relative to each other. What would such a small group of nurse and children be but a kinship group or a family? . . . The feeling of belonging to or "descending" from one another is the heart of the kinship notion. . . . The earliest memories of a child are of those who nurtured that child, and these memories are what comprise the feelings of descent.

Research might show that children raised in these "nurse and children" families did not do as well as children raised by a father and mother, but that finding would not change the designation of both forms as families.

The facts of human reproduction and maturation, and the extent of human culture, seem to require families to fulfill the nurturant socialization of infants. Human infants are born so helpless and take so long to mature that they require an immense amount of nurturance. A colt can stand within an hour, run before it is a day old, and race in the Kentucky Derby at the age of three. A human infant can learn to walk during its second year and takes many years to achieve physical and mental maturity. Moreover, even in the most primitive societies, there is a great deal to learn before one can adequately fill the role of adult. It is easier to supply such a large amount of nurturance if more than one adult is available to provide it—which is why it is so common to find families that include couples united in **marriage,** a formal commitment to maintain a long-term relationship involving specific rights and responsibilities.

Thus far, our efforts to define the family have been quite abstract. So, let's look at some specific aspects of family life in societies unlike our own.

Table 13-1 lets us examine variations in marriage and family patterns in premodern societies. The data are based on a set of 186 societies that make up the Standard Cross-Cultural Sample, which is a subset of the societies included in the *Atlas of World Cultures* described in Chapter 4. No modern industrialized societies are included

in the sample, and even the most advanced societies in this set are relatively undeveloped agrarian societies.

Indeed, the vast majority of these societies are small tribal groups—including several dozen Native American and Inuit societies. More than 75 percent have no written language, and many are seminomadic. Although many of these societies still exist, their value here is to offer us a view of cultures during earlier stages of development.

The first line in the table reveals that most premodern societies are not monogamous: Three-fourths permit men to have more than one wife. Most are patrilocal: Newlyweds take up residence with or near the groom's parents.

In three of four societies, it is rather easy for men to divorce their wives. In only 61 percent of these societies do husbands and wives share a "bed"—that being a mat, hammock, animal skins, or whatever it is people sleep on in a given society. Wherever they sleep, in the great majority of these societies, couples do not have privacy during sleep. Taboos against sex during menstruation are widespread, and in a third of cases, menstruating women actually are segregated from others (many groups maintain special menstrual huts to which women withdraw). Typically, men marry women who are considerably younger. Although couples eat together in most societies, in only a few societies do they spend much of their leisure time together. In half of these societies, men do no domestic chores. Finally, the **nuclear family** (a family containing only one adult couple) is the exception.

Because of the very high mortality rates and short life expectancy of persons in premodern societies, one or both parents frequently die before their children are mature. These factors encourage the formation of families containing more than one adult couple. That is, in societies such as these, nuclear families do not live apart from other relatives. Rather, the basic family unit includes several nuclear families; these are called **extended families.** Extended families can be composed in many different ways. For example, they can consist of an adult couple (the grandparents), their children, and the spouses and children of their children. Because people in some cultures often die before they become grandparents, the extended family often contains several brothers, their wives and children, and sometimes unmarried siblings.

Regardless of its composition, the extended family is larger than the nuclear family and

TABLE 13-1	Variations in Family Life in Premodern Societies	
		PERCENT OF SOCIETIES
Men may have multiple wives		77
Newlyweds settle closer to groom's family		69
Easy for a man to divorce his wife		75
Husbands and wives share the same "bed"		61
Couple has privacy when sleeping		28
Norm against sex during menstruation		82
Women are segregated during menstruation		34
Grooms are much older than their brides		70
Couples spend much leisure time together		27
Couples usually eat meals together		74
Men do no domestic chores		51
Nuclear family is typical		29

Source: Prepared by the author from the Standard Cross-Cultural Sample.

always contains more than one adult couple. Extended families would seem able to provide more effective and attentive child care and socialization than nuclear families can simply because more adults are available for these tasks. For example, one adult can watch the infants while the others go out to pick fruit or gather eggs.

The term **polygamy** is often used to describe "extended" families wherein one husband has many wives. This is not technically correct. The term *polygamy* properly is applied to all plural marriages, in which one person (of either sex) has multiple spouses. **Polygyny** is the term for marriages involving one man and multiple wives.

Table 13-1 showed that polygynous families were permitted in most premodern societies. But that does not mean they were common; in any of these societies, most men could not afford more than one wife, and in any event, rarely were there sufficient surplus women to provide multiple wives for many men. That is, if some men have many wives, other men can have none, unless surplus women are obtained by raids on other societies. Consequently, although even today most Islamic societies still permit men to take multiple wives, few do so. And as can be seen in Table 13-2, there is considerable disapproval of such marriages among modern Muslims—from 90 percent disapproving in Egypt down to 26 percent in Jordan. Notice, too, that only in Egypt and Turkey, where most people disapprove, are men somewhat more favorable toward multiple wives than are women. In six of the remaining eight, women are substantially less opposed than are men. Perhaps this reflects men's concern that many of them would be without wives were some men to have many. Perhaps, too, it reflects that women tend to support more traditional views of the family whatever they might be in a given culture.

FAMILY FUNCTIONS: AN INTERNATIONAL PERSPECTIVE

We have noted that many functions once included in the definition of the family now are seen as problematic. In many (perhaps even most) societies, these functions are fulfilled primarily by the family, but there are always many exceptions. Consequently, sociologists have found it useful to regard the links among various functions of the family, in specific times and places, as an appropriate research topic.

TABLE 13-2	Percentage in Ten Islamic Nations Who Believe a Man Should *Not* Be Permitted to Have More Than One Wife		
NATION	**TOTAL (%)**	**MEN (%)**	**WOMEN (%)**
Egypt	90	86	93
Turkey	84	77	92
Bangladesh	60	65	54
Indonesia	54	59	48
Pakistan	44	51	36
Iran	43	53	32
Algeria	42	59	27
Morocco	35	34	35
Nigeria	30	30	30
Jordan	26	38	16

Source: Prepared by the author from the World Values Surveys, 2001–2002.

For example, we can ask such questions as: Is the family in Canada and the United States providing as much emotional support as the family in Mexico? Or are Western families as close-knit and supportive as families in Japan? As families in medieval times? Because such questions and comparisons will form the major part of this chapter, let's examine some functions usually associated with the family and see how major nations compare on each.

SEXUAL GRATIFICATION

All societies have norms governing sexual behavior. Some impose very narrow limits on who may engage in sex, with whom, when, and how. Others have few restrictions. Within societies families tend also to play a role in the establishment of sexual norms. In general, the family provides for sexual intercourse between certain members and typically prohibits it between certain other members—a prohibition often referred to as the **incest taboo.** But beyond these broad parameters, the link between the family and sexual gratification varies greatly.

Such variation can be seen in Table 13-3. National samples in thirty-six nations were asked "Do you and your spouse share the same sexual attitudes?" More than two-thirds of Icelanders replied yes, as did 65 percent of Americans and 62 percent of Canadians. In many other nations, however, only a minority of married adults share sexual attitudes. And the Japanese are nearly off the scale on this question—only 21 percent said yes. Perhaps it is no surprise, then (data are not shown), that few

TABLE 13-3	"Do you and your spouse share the same sexual attitudes?"		
NATION	YES (%)	NATION	YES (%)
Iceland	71	France	55
Norway	70	Romania	55
Sweden	68	Nigeria	54
Switzerland	67	Chile	53
Turkey	67	Belgium	50
Denmark	66	Italy	46
Netherlands	65	Austria	45
United States	**65**	Estonia	45
Hungary	64	Finland	43
Argentina	63	Germany	42
Slovenia	62	Belarus	40
Canada	**62**	**Mexico**	**40**
Great Britain	61	Russia	40
India	60	Latvia	39
Ireland	59	Portugal	39
Bulgaria	59	Lithuania	38
Spain	58	China	31
Brazil	57	Japan	21

Source: Prepared by the author from the World Values Surveys, 1995–1996.

TABLE 13-4	"Is an adequate income very important for a successful marriage?"		
NATION	VERY IMPORTANT (%)	NATION	VERY IMPORTANT (%)
Nigeria	77	**Canada**	**40**
Chile	67	Iceland	39
Belarus	66	France	38
India	59	Lithuania	36
Turkey	58	Great Britain	35
Russia	57	Switzerland	35
Brazil	56	Latvia	34
Mexico	**55**	South Korea	34
Japan	55	Finland	32
Ireland	52	Estonia	31
Hungary	52	China	30
Portugal	49	Austria	30
Bulgaria	45	Italy	30
United States	**45**	Germany	26
Belgium	45	Netherlands	26
Romania	45	Sweden	23
Argentina	44	Norway	21
Spain	43	Denmark	11
Slovenia	40		

Source: Prepared by the author from the World Values Surveys, 1995–1996.

Japanese (28%) think that "a happy sexual relationship is very important for a successful marriage." In contrast, 93 percent of Nigerians, 73 percent of Canadians, and 69 percent of Americans think a happy sexual relationship is necessary. Clearly, the role of the family in providing sexual gratification differs a lot between Western nations and Japan. This variation turns up for many other family functions as well.

ECONOMIC SUPPORT

In most societies families serve as the primary economic units. Families contain **dependents**—such as infants, children, the elderly, and the disabled—who cannot support themselves. By sharing in a common economic unit, some family members provide care for their dependent kin. In primitive societies able-bodied adults will hunt and gather food and then feed the rest of the family. In modern societies some family members earn money to support the others. Moreover, in most societies decisions about the distribution of resources among family members are made within the family. Obviously, the family's role as a source of economic support can vary across societies and through time. As mentioned, the family is not the basic economic unit for residents of a kibbutz. And in modern society pension plans, insurance, welfare programs, and a variety of charitable and service agencies have assumed responsibility for supporting many dependents who once were the responsibility of their families.

Table 13-4 shows the percentages in each nation who believe that an adequate income is very important for a successful marriage. The less developed nations with relatively low per capita incomes (Nigeria, Chile, Belarus, India) dominate the top of this distribution. Japan is the obvious exception, tied with Mexico (55%) and just behind Turkey, Russia, and Brazil. In contrast, only relatively small minorities in Scandinavian nations think income is very important for a successful marriage. This probably is because these nations sustain unusually elaborate welfare systems that have replaced the family as the primary source of support for dependents.

EMOTIONAL SUPPORT

In many societies families serve as primary groups for their members, giving people a sense of emotional security, a sense of belonging and of personal worth. For North Americans the question might be posed: If our families won't love us, who will? When we are young, we expect our parents, brothers, sisters, and other relatives to appreciate us. Later, we expect this from spouses and even from our children. Of course, not all families fulfill these needs for their members—as divorce, child abuse, extramarital affairs, runaways, and other phenomena suggest. And in different times and places, there has been much less reliance on the family for emotional support.

Table 13-5 lets us examine such variations. One way to gauge the closeness of family relations is to ask people to report on their own families. Thus, people in twenty-one nations were asked about the relationship between their father and mother when they were growing up. Judged on this basis, family relationships are closest in South Africa, where 70 percent said their parents were "very close." Mexico (53%), Canada (51%), and the United States (49%) are quite similar. Japan (29%) is lowest, nearly matched by Germany (31%).

Thus far, we have seen that the Japanese differ from most other nations in terms of family functions. Only a tiny minority say they and their spouse share sexual attitudes, and most Japanese do not think "a happy sexual relationship" is important for a successful marriage. In contrast, the majority of Japanese think money is very important for marriages. Finally, the Japanese family is not bound by close emotional ties. Hence, the data in Table 13-6 should not be a surprise. When asked to express their overall satisfaction with their "home life," only 12 percent of the Japanese said they were "very satisfied." In comparison, 65 percent of Poles said they were very satisfied, as did 57 percent of Americans and 56 percent of Canadians.

That the Japanese are so dissatisfied with their home life can be understood on the basis of the extreme degree to which the family imposes upon the individual in Japan, as described at length in Chapter 8. But Table 13-6 shows that Japan is not unique—satisfaction with home life is very unusual in many societies. Moreover, when we compare nations at the bottom of the table with those at the top, we can recognize that it is the most traditional societies that are relatively lacking in satisfaction with family life. This is entirely at odds with the conventional wisdom that the family is strong in traditional cultures while it has been crumbling in the more modern nations. Perhaps, then, there is

TABLE 13-5	"During the time you were growing up, would you say your father and mother were very close to each other . . . ?"
NATION	VERY CLOSE (%)
South Africa	70
Belgium	63
Norway	59
Hungary	58
Ireland	58
Northern Ireland	58
Italy	55
Sweden	55
Mexico	**53**
Great Britain	53
Denmark	52
Canada	**51**
France	51
United States	**49**
Spain	48
Australia	48
Iceland	45
Netherlands	39
Finland	36
Germany	31
Japan	29

Source: Prepared by the author from the World Values Surveys, 1981–1983.

something faulty about the idea that the family is stronger in more traditional, less modern societies. To find out, let's look closely at how the family functioned in Europe a few centuries ago.

A CLOSER VIEW
LIFE IN THE TRADITIONAL EUROPEAN FAMILY

Not until quite recently did social historians and sociologists of the family begin to dig out reliable data on family life in times past, and not until the late 1960s and the 1970s did substantial reports on these efforts begin to appear (Laslett, 1965, 1977; Rosenberg, 1975). Up to that time, our notions about family life in, for example, seventeenth-century Europe came from novels, letters, diaries, and autobiographies written at the time. The trouble with these sources is that they reflect a narrow stratum of society: the wealthy and literate. Although they may shed light on how the privileged few lived and felt, they tell us very little about the lives of the vast majority.

TABLE 13-6	"Overall, how satisfied or dissatisfied are you with your home life?"		
NATION	VERY SATISFIED (%)	NATION	VERY SATISFIED (%)
Poland	65	Italy	40
Denmark	64	**Mexico**	**38**
Ireland	61	Nigeria	38
Switzerland	61	Slovak Republic	37
Chile	58	Czech Republic	34
United States	**57**	Slovenia	33
Canada	**56**	Spain	33
Brazil	55	France	31
Sweden	55	Germany	30
Finland	52	Russia	30
Great Britain	50	South Korea	28
Portugal	48	Belarus	26
Hungary	48	Romania	25
Iceland	47	Turkey	24
Belgium	46	India	22
Argentina	45	Lithuania	20
Netherlands	44	Bulgaria	18
Austria	44	Estonia	17
Norway	40	Latvia	15
China	40	Japan	12
South Africa	40		

Source: Prepared by the author from the World Values Surveys, 1995–1996.

Europe's peasants and urban laborers left no literary traces. To discover what the life of an average family was like in past times and places, scholars have had to laboriously reconstruct the period from tax records; lawsuits; parish records of baptisms, weddings, and funerals; and even information on gravestones. These labors proved to be worthwhile. The picture of traditional family life is far from the warm, intimate, loving, caring extended family that we have long celebrated.

For many sociologists the first real fruits of these historical searches came with the publication of Edward Shorter's *The Making of the Modern Family* in 1975. In it Shorter combined the research of many scholars to depict the traditional family and contrast it with the modern family. Shorter's book changed sociologists' views about the family. Let's see what he found out about the traditional European family.

HOUSEHOLD COMPOSITION

The first step in assessing family life is to know who is living with whom—that is, what the usual composition of a household is. From many studies of different parts of Europe, Shorter discovered that the extended family living in a single household was not typical except for the wealthy, both urban and rural. As we shall see, the typical household did include more than a nuclear family, but the additional members were often only temporary, such as lodgers and hired hands. Moreover, the traditional household was much smaller than had been assumed. While wealthy households often included ten or more people, most households had only five or six members (Table 13-7).

We know that in those days women gave birth to many children, often as many as eight or ten. How, then, could the normal household be so small? One reason was high infant and child mortality. One of every three infants died before the age of one, and another third died before reaching adulthood. Another reason is that children typically left the household to take full-time employment at ages that seem incredibly young to us.

In the eighteenth century, for example, children in western France left home to work as servants, shepherds, cowherds, or apprentices at

New historical research has found that preindustrial households contained many fewer children than had been supposed because children were sent off on their own at a young age. Three centuries ago in France, for example, children began to leave home to work as shepherds, servants, and apprentices at age seven or eight, and by age ten nearly all children had left home.

age seven or eight! By age ten virtually all children had gone off on their own. In England at this same time, children did not begin to leave home until age ten, but by fifteen nearly all of them had left. Keep in mind that people physically matured later in this period (see Chapter 5). These were little kids who were having to go it alone.

Of course, not all the children left. Eldest sons stayed home or returned home after a period of working elsewhere and one day took over the farm. In some places daughters remained home until they married. Nevertheless, the traditional household is remarkable for the small number of children living in it, especially given the large number who were born.

In addition the traditional household contained fewer adults than one might expect. High mortality meant that there were few elderly in the households, and many homes lacked either a father or a mother. In fact, female-headed households were as common in the past as they are today. The primary cause of such households today is divorce, and thus, the father often continues to see the children and to provide financial aid. Back then, the cause was death. The average married couple had only about ten years together before one died. As a result many people remarried; therefore, many children grew up with a stepparent and with half sisters and half brothers. Perhaps you've wondered why so many fairy tales involve wicked step-

mothers: because she was such a common part of life back when parents weren't especially nice even to their own children.

Thus, the image of the large extended families of preindustrial societies is based on wealthy households. These households were large because their rate of mortality was lower (the rich were much more likely to live to see their grandchildren, and more grandchildren survived), because their children were not pushed out to fend for themselves at young ages, and because many servants were considered part of the household.

TABLE 13-7	Average Household Size in Preindustrial Societies	
NATION	**YEAR**	**NUMBER OF PERSONS IN AVERAGE HOUSEHOLD**
British North America	1689	5.85
England	1599	4.75
France	1778	5.05
Germany	1687	5.77
Italy	1629	4.50
Japan	1746	5.50
Poland	1720	5.40
Scotland	1779	5.25
Serbia	1733–1734	4.95

Source: Laslett (1977).

There was no privacy in preindustrial households; most dwellings had but a single room in which all activities took place. Often, more than one family shared a one-room home. Imagine yourself growing up in this Flemish household painted by Pieter Breughel (1525–1569).

Giraudon/Art Resource, NY

CROWDING

Although the average traditional household was not large, even compared with modern households, it was crowded. The overwhelming majority of traditional European families lived in one room, where all indoor family activities took place. Rural families usually shared their one-room houses with livestock and poultry, while urban families frequently had a lodger or some other nonfamily member sharing their living space. Usually, the one room wasn't even very large. At night beds were arranged on the floor, and when people had mattresses, the beds were often crowded: Adults and children, males and females, family members and outsiders huddled together for warmth.

As late as the 1880s, when good census data were first recorded, half of the people in Berlin and Dresden still lived in one-room households. This situation seemed much more common throughout Europe earlier in the century. In Chapter 19 we shall see that far less crowding in households can still cause serious strains among family members. When American families have more than one person per room, husband-wife and parent-child relations become strained. With whole families crowded into one room, family relations in preindustrial times were simply terrible, as we shall see.

"OUTSIDERS"

Though much smaller than had been believed, many traditional households contained nonfamily members. Many rural households contained male and female teenagers who served as hired hands. Such outsiders were particularly common during the peak of the farming season. Urban households frequently included lodgers who paid to eat and live with a family. Often, several unrelated families shared one-room urban homes, forced into a common residence by poverty. Moreover, there was a considerable coming and going by these live-in outsiders. Families tended to have people they did not know well living temporarily in their midst.

Clearly, the traditional family lacked privacy and a well-defined boundary. Family members ate, slept, gave birth, engaged in sex, and argued not only in full view of one another but also in full view of a changing audience of outsiders. And the traditional family was under close observation by neighbors, too. Even rural families did not live far apart, each on their own farm as in Canada and the United States, but in cramped farming villages. These crowded living conditions undermined feelings of family unity.

CHILD CARE

We have seen that the traditional family was quick to send kids out on their own. This reflected more than mere economic necessity or the fact that unskilled children could perform productive labor in preindustrial economies. It also reflected an indifference toward children and neglectful child-care practices. Shorter put it bluntly: "Good mothering is an invention of modernization."

A good index of neglect and indifference is found in journals kept by local doctors. All of these doctors complained about parents leaving their infants and young children alone and untended for much of the day. Rashes and sores from unchanged swaddling clothes afflicted nearly all infants. Repeated accounts tell of children burning to death because they were left too close to an open hearth, and reports of unattended infants being eaten by barnyard pigs are frequent. In the part of France where silkworms were raised, a peasant proverb acknowledged that children were neglected during the busy season: "When the silkworms rise, the kids go to paradise." Indeed, throughout Europe rural infants were most likely to die during the harvest season, when they were most neglected.

Even when parents were around their infants, they ignored them. Mothers rarely sang or talked to their infants when they tended them, nor did they play games with them as the children grew older. In fact, mothers didn't even refer to children by name, calling a child "it" or, in France, "the creature."

Mothers frequently were unsure of their children's ages (Shorter reports a mother who said her son was either eleven or maybe fourteen), failed to recall how many children they had

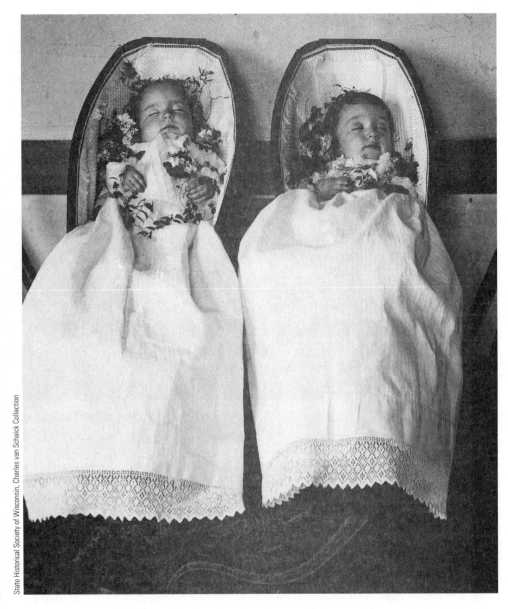

State Historical Society of Wisconsin, Charles van Schaick Collection

given birth to, and often gave the name of a child who died to the next one born.

Because of the high rates of infant mortality, it might be understandable that parents were somewhat reluctant to form intense emotional bonds with their babies. But in some parts of France, parents typically did not attend funerals for children younger than five, and there is widespread evidence that infant deaths often caused little if any regret or sorrow. Instead, parents often expressed relief at the deaths of children, and many proverbs reflected this attitude. Moreover, dead and even dying infants were often simply discarded like refuse and were frequently noticed "lying in the gutters or rotting on the dung-heaps."

Large numbers of legitimate infants whose parents were still living were abandoned outside

This poignant photograph of twins in their coffins, taken in Wisconsin in the 1890s, stands in dramatic contrast to the typical attitudes toward children in preindustrial families. Funeral portraits like this showed the determination of grieving parents to remember their lost children. In preindustrial times parents often were little interested in the death of a child—and sometimes simply reused its name for a later child.

Had photography existed 300 years ago, there would have been very few pictures like this one. Only in modern times has it become typical for parents to be deeply attached to their children.

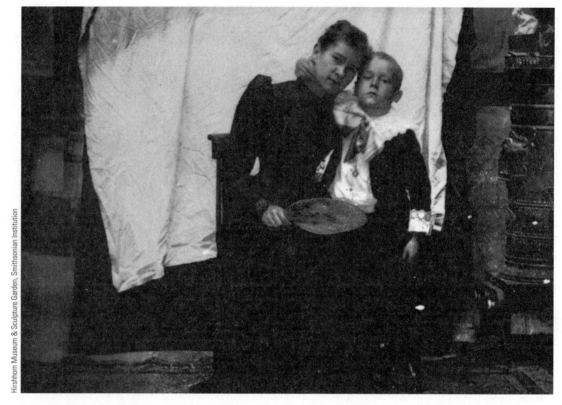

Hirshhorn Museum & Sculpture Garden, Smithsonian Institution

churches or foundling homes. Some scholars suggest that as many as half of the children abandoned in parts of France during the eighteenth century were abandoned by intact families. Additional indifference and neglect are evident in the large numbers of infants sent off to wet nurses, despite the well-known fact that such children faced much higher probabilities of death. Indeed, wet nursing became a prominent cottage industry outside of major cities in the eighteenth and nineteenth centuries, as families, especially the poor, sent away infants so that the mother would not be tied down by nursing and child-care responsibilities. An estimated one-sixth or more of all babies born in Paris in 1777 were shipped out to wet nurses.

Once at the wet nurse's, infants were often not nursed at all but fed a paste of grain and water and given little attention. In truth these homes were baby barns, crowded with infants. Parents seldom, if ever, visited. Deaths were often covered up so that payments could still be collected. But staggering numbers of these babies were never seen again. Nor, it seems, were they missed.

The extraordinary infant death rates in preindustrial Europe now become easier to understand. The general conditions of life and public health practices alone would have produced high mortality rates. But the actual rates were pushed even higher because of neglect and indifference. Indeed, as late as the 1920s, a government study in Austria attributed about 20 percent of infant deaths to "poor care."

Obviously, infants usually gained little emotional support from the preindustrial family. But what about other family members? What Shorter found seems as alien to us today as the idea of parents engaging in sex while sharing their bed with children and a lodger.

RELATIONS BETWEEN HUSBANDS AND WIVES

Only in modern times have most people married for love. In the "good old days," most married for money and labor—marriage was an economic arrangement between families. How much land or wealth did the man have? How large a dowry would the bride bring to her spouse? Emotional attachments were of no importance to parents in arranging marriages, and neither the bride nor the groom expected emotional fulfillment from marriage.

Shorter (1975) noted an absence of emotional expression between couples and doubted that more than a few actually felt affection. The most common sentiments seem to have been resentment and anger. Not only was wife beating commonplace but so was husband beating. And when wives beat their husbands, it was the husband, not the wife, who was likely to be punished by the community. In France a husband beaten by his wife was often made to ride through the village sitting backward on a donkey and holding the donkey's tail. He had shamed the village by not controlling his wife properly. The same practice of punishing the husband was frequently employed when wives were sexually unfaithful.

The most devastating evidence of poor husband-wife relations was the reaction to death and dying. Just as the deaths of children often caused no sorrow, the death of a spouse often prompted no regret. Some public expression of grief was expected, especially by widows, but popular culture abounded with contrary beliefs. Shorter reported the following proverbs:

The two sweetest days of a fellow in life,
Are the marriage and burial of his wife.

Rich is the man whose wife is dead
and horse alive.

Indeed, peasants who rushed for medical help whenever a horse or cow took sick often resisted suggestions by neighbors to get a doctor for a sick wife. The loss of a cow or a horse cost money, but a wife was easily replaced by remarriage to a younger woman who could bring a new dowry.

You may wonder how this view of wives squares with Chapter 12. Since women were in relatively short supply, would they not have been highly valued? Simply because men think of women as valuable "goods" does not mean they will love or even like their wives. Thus, in ancient Rome, where female infanticide resulted in far more extreme sex ratios than existed in preindustrial Europe, men found it difficult to relate to women, and many men preferred not to marry at all since prostitution flourished. As Beryl Rawson (1986) reported, "One theme that occurs in Latin literature is that wives are difficult and therefore men do not care much for marriage." In 131 B.C. the Roman censor Quintas Caecilius Metellus Macedonicus proposed that the senate make marriage compulsory because, although "we cannot have a really harmonious life with our wives," the society needed children and therefore men must marry (Rawson, 1986). As with the women in Athens, Roman women were hidden away as too valuable to associate freely. But they weren't often loved.

BONDS BETWEEN PARENTS AND CHILDREN

Besides the lack of emotional ties to infants and young children, emotional bonds between parents and older children were also weak. First, most children left the household at an early age. Second, when they did so, it was largely a case of "out of sight, out of mind." If a child ventured from the village, he or she was soon forgotten, not just by the neighbors but by the parents as well. All traces were lost of those who moved away. According to Shorter, a French village doctor wrote in his diary in 1710 that he had heard about one of his brothers being hanged but that he had completely lost track of the others.

Finally, even those children who stayed at home in the village did not come to love their parents. Instead, they fought constantly with their parents about inheritance rights and about when their parents would retire, and they openly awaited their parents' deaths. Shorter concluded that dislike and hatred were the typical feelings between family members.

BONDS AMONG PEER GROUP MEMBERS

Surely, people in traditional societies must have liked someone. Unfortunately for our image of traditional family life, the primary unit of society and attachment was not the family but the peer group. The family provided for reproduction, child rearing (such as it was), and economic support (often grudgingly), but emotional attachments were primarily to persons of the same age and sex *outside* the family.

Wives had close attachments to other wives, and husbands to other husbands. Social life was highly segregated by sex and was based on childhood friendships and associations. For example, a group of neighborhood boys would become close friends while still very young, and these friendships remained the primary ties of these people all their lives. The same occurred among women. While this no doubt provided people with a source of intimacy and self-esteem, it hindered the formation of close emotional bonds within the family.

A woman would enter marriage expecting to share her feelings not with her husband but with her peers. Men reserved intimate feelings for

The industrial age made romantic love possible. Of course, people in earlier centuries sometimes fell in love, but it was not typical, and certainly, love was rarely the basis for deciding whom to marry. Although modern people seem determined to fall in love, they also have discovered the pains of falling out of love.

families and lifelong friends. It was instead a nasty, spiteful, loveless life that no modern person would willingly endure. Indeed, as industrialization made other options possible, the family changed radically because no one was willing to endure the old ways any longer.

MODERNIZATION AND ROMANCE

In Part V we shall examine the why and how of the immense social changes that have resulted from modernization. Here we shall point out that life in modern societies is not simply better than in preindustrial societies but also *different*.

Nowhere is this clearer than in the transformations in family life. Where once a "happy couple" meant an absence of mutual antagonism, today that phrase is reserved for people who feel strong positive sentiments. We do not hope for a tolerable marriage; we seek love. Nor do we think it enough that parents do not hate, abuse, or neglect their children; we expect parents to love, nourish, and encourage them. We expect people to grieve when their parents die, not to be relieved that they are out of the way. In short we assume that families foster deeply felt emotional attachments. How did this transformation come about?

Quite simply, modernization radically changed the conditions of life, giving people the opportunity to seek individual happiness. Shorter has sketched a number of these changes. First, industrialization freed individuals from depending on inheritance for their livelihoods. Eldest sons no longer had to wait for their fathers' land; daughters no longer had to wait for a husband with land. Both sons and daughters could seek wage-paying work, especially in the rapidly expanding urban industries. Soon young people were heading off to the cities in droves.

This change allowed people to make their own marital choices, free from both parental approval and concern about keeping property in the family. People no longer had to delay marriage until their parents died or retired, and property concerns no longer dominated the choice of a spouse. As these matters became less important, other concerns emerged. And as men and women began to select their marriage partners, they began to seek people who appealed to them. Romantic attraction rapidly became

their peers, too. In this way the weak boundaries defining the household were perforated by primary relations beyond the family. Thus, outsiders determined much that went on within a household. Husbands and wives often acted to please their peers, not each other.

Of course, sometimes people loved their children, and some couples undoubtedly fell in love. But most evidence indicates that life in the preindustrial household was the opposite of the popular, nostalgic image of quiet, rural villages where people happily lived and died, secure and loved, amidst their large the basis for marriage. "I love you" became the precondition for asking, "Will you marry me?"

Of course, notions of romantic love were not discovered in modern times. Greek and Roman poets and dramatists wrote of love; indeed, the Trojan War was thought to have been fought because of Paris's all-consuming love for Helen. And love was a major theme of court poets and minstrels in the days of chivalry and knighthood. But until relatively modern times, few could afford to let love be the basis for marriage selection. The Puritans in England were among the first to stress the importance of romantic sentiments between husbands and wives, and this reflected their status as members of a newly affluent middle class who could afford to marry for love.

In fact, affluence explains much of modern family life. The average modern family is wealthy beyond the dreams of preindustrial families. One of the first fruits of this affluence was space and privacy. As rapidly as economic circumstances permitted, families sought sufficient household space to gain privacy from one another and to shield themselves from outsiders. While married couples today routinely and openly express affection in ways unthinkable in the past, they are able to keep their most intimate relations private.

Moreover, the rise of romantic love in marriage redirected the primary attachments of the individual to within the household. Husbands and wives now expect their relationships to take priority over attachments to peers. A popular song at the turn of the twentieth century proclaimed, "Those wedding bells are breaking up that old gang of mine." With modernization came the expectation that husbands would not remain "one of the boys."

This redirection of primary attachments to family members was facilitated by mobility. People now seldom remain in the same place throughout their lives. As people become adults, they move away and break ties to their peers. Even if an individual does not go away, most of the peer group does. A common observation today is that if you want a lifelong friend, you had better marry one. Husbands and wives have become the only consistent, permanent emotional attachments.

Romantic love between spouses has also affected parent-child relationships. We now commonly believe that children must be wanted or they should not be engendered. By the late nineteenth century, child care and good parenting became a major topic in book publishing

(Zuckerman, 1975). Attitudes toward children have changed so dramatically that it is now against the law to treat children in ways that were once customary.

MODERNIZATION AND KINSHIP

One of the widely noted "symptoms" of the decline of the family in modern times is the erosion of kinship bonds. We have seen that these perceptions are based in part on nostalgic illusion; the preindustrial family was not the warm, secure nest we once thought it was. Yet for many North Americans and Europeans, the extended family remains an ideal standard against which the nuclear family is seen as wanting. Reiss (1988) has noted the irony that this should be so even though most of these same people do not want to live with their parents or with their grown children.

Nevertheless, the question persists: Has kinship really taken on less importance among people in modern societies? To answer the question, we shall have to distinguish between the *quantity* and the *quality* of kinship bonds.

There can be no doubt that most Americans are much less likely to have as many close bonds to brothers and sisters, cousins, aunts and uncles, and nephews and nieces than was the case several generations ago. This is a simple result of the *decline in fertility*.

The preindustrial household contained relatively few people despite widespread impressions about family life in that period. However, in the nineteenth and early twentieth centuries, especially in the United States and Canada, the average woman had many more babies than she does today, and even with mortality considered, the average family was much larger. The result is that several generations ago the average person in the United States and Canada had many more relatives (Pullum, 1982). A simple example will make the point.

At the turn of the twentieth century, the average North American woman gave birth to more than three children, and when only those women who married are counted, their average was four children. So put yourself in an average family. If all members marry and each couple has the average number of children, then if you gave a party and invited your relatives, the gathering would include your two parents, twelve aunts and uncles (six by marriage), twenty-four cousins,

TABLE 13-8	Persons Who See Their Mother Every Day*		
NATION	PERCENTAGE	NATION	PERCENTAGE
Italy	63	Russia	38
Spain	59	Israel	31
Philippines	50	Germany	26
Slovenia	49	Norway	21
Hungary	45	Great Britain	19
Japan	44	**Canada**	**19**
Poland	41	Australia	18
Northern Ireland	39	Netherlands	18
Ireland	39	New Zealand	18
Austria	39	**United States**	**15**
Czech Republic	38	Sweden	14

Source: Prepared by the author from the International Social Survey Program, 1994.

*Includes only persons whose mother still lives.

three siblings, twelve nieces and nephews, and four children—a total of fifty-seven people.

In contrast, put yourself in an average family when the birthrate is two per female. Now your family reunion will draw two parents, four uncles and aunts (two by marriage), four cousins, one sibling, two nieces and nephews, and two children—a total of fifteen people. And that's part of what has happened to kinship bonds in modern life: There simply are far fewer of them. Other things being equal, a person born at the turn of the century had *six times* the probability of forming a close bond with a cousin, for example. So, modernization has greatly reduced the *quantity* of kinship bonds and in that way may have reduced the quality as well. But then again, maybe with so many kinship bonds, the chances were reduced of having any one of them become really close.

Table 13-8 offers a glimpse into family relations. National samples in twenty-two nations were asked how often they saw their mother (if she was still alive). In Italy nearly two-thirds replied that they saw her *every* day! The same is true for nearly half of the Japanese. In Russia and in Israel, about a third saw their mother every day. But only 19 percent of Canadians and 15 percent of Americans reported this level of contact. Of course, this item measures both the closeness of family ties and the proximity of children to their parents. One supposes that where most people see their mother daily she

must live very close by, or even with her children. That so few Americans and Canadians see their mothers daily reflects the high level of geographical mobility in both societies: Most people do not remain in the neighborhood in which they grew up. However, even in so highly mobile a society as the United States, most people do maintain ties with their parents, siblings, and other relatives. Thus, while only 15 percent saw their mother daily, an additional 47 percent reported seeing her a least once a month or more. In similar fashion the 1998 General Social Survey found that 52 percent of Americans spent a social evening with relatives at least once a week and only 4 percent said they never did so.

The picture that emerges here is not one of weak family ties. The American family has grown much smaller, but it seems to have stayed rather tightly knit. In societies such as the United States and Canada, where there is so much mobility and families are scattered, there is increased pressure on marriage to fulfill all of the functions of the family. That it appears able to do so for most Americans seems to be reflected in the fact that the most recent poll data show that 65 percent rate their marriage as very happy, and only 2 percent say it is not too happy.

But you may be asking, if people are so happy with their marriages, why are there so many divorces? Don't high divorce rates suggest a breakdown in ties between wives and husbands?

MODERNIZATION AND DIVORCE

Ironically, a high divorce rate probably indicates that the marital relationship has become *much more important than it used to be*. Back when most couples had weak emotional ties at best, they seldom divorced; now, although couples marry for love, they often divorce in anger and disappointment. Let's explore why divorce occurs and what it means for family life.

Divorce means the end of a marriage, but it does not necessarily mean the end of a family because two-thirds of divorces occur between people who have children. One parent (usually the father) leaves the household, but a family remains. Moreover, divorce does not mean that many people experiment with marriage and then opt for a single life. Fewer than 5 percent of North American adults at any given moment report their current marital status as divorced; about 75 percent who divorce remarry. Thus, millions of couples give up on their marriages but not on marriage itself. These statistics offer an important insight into *why* people get divorced.

Most people who get divorced report that their marriage ceased to provide adequate emotional satisfaction—that is, their relationship was no longer happy. That might mean that the current high divorce rate indicates a lot of unhappy marriages, but it could also mean that at any given moment the great majority of marriages are happy ones. How is this possible?

Over the past eighty years, divorce laws have become much less restrictive. The intention behind this legislation was to strengthen the family by permitting intolerable marriages to be dissolved. The rationale was that if the bad marriages are ended by divorce, most marriages will be good ones. Today, when many marriages end in divorce, it seems unlikely that people are enduring bad marriages to the same extent as when only a few got divorced. Indeed, it seems likely that people today become dissatisfied with marriages people would have deemed acceptable fifty years ago.

Marital satisfaction is partly a matter of comparison. In days when few people divorced, a couple comparing themselves with their friends might have rated their marriage as good. Today, the same couple might find theirs to be a poor marriage by comparison because the standard has risen: Marriages must be better to qualify as satisfactory when more unsatisfactory marriages are eliminated by divorce. Thus, as divorce rates rise, the average level of satisfaction in existing marriages should rise also. We have reason to suspect, however, that a substantial part of this perceived satisfaction is simply a "newness" or "variety" effect. That is, as people divorce and remarry, many may not find someone who suits them better but simply a replacement for a partner who had become too familiar.

HOW MUCH DIVORCE?

Everyone agrees that, like all industrialized nations, the United States has a high divorce rate. However, it is so difficult to calculate an accurate divorce rate that no one really knows what proportion of Americans end up in the divorce courts. The usual method of calculation, used in most countries, is to divide the number of divorces that occur in a given year by the total population to reflect the number of divorces per 1,000 persons. When calculated for recent times, that rate shows that divorces rose rapidly from 2.2 per 1,000 in 1960, peaked at 5.2 in 1980, and then dropped back down to 4.3 in 1997. Many specialists in the family think that in recent years the true divorce rate declined somewhat just as those data show. Unfortunately, the U.S. government stopped collecting data on the number of divorces as well as the number of marriages in 1998, so no further tracking of the rates is possible.

Admittedly, this way of calculating a divorce rate has many shortcomings and requires many corrections. For one thing, if the age composition of the population shifts substantially, this must be factored into comparisons. For example, in 1960 about 39 percent of the population was too young to marry (and therefore could not divorce), while in later years the population got older, a larger proportion married, and divorce became proportionately more possible. Thus, the upward shift from 1960 to 1980 may have been largely an increase in the proportion able to divorce and not a shift in the probability of a marriage ending in divorce. By the same token, in recent years the marriage rate has been declining which necessarily forces some decline in the divorce rate, since the unmarried cannot get divorced. Such measurement problems leave us somewhat in the dark as to trends in divorce.

A truly accurate divorce rate would be based on the final outcome of all marriages occurring in a given year. But to obtain it, we would have to wait until the last couple married in that year

TABLE 13-9	Percentage of American Women Whose First Marriage Ended in Divorce	
	DIVORCED AFTER:	
	5 YEARS OF MARRIAGE	15 YEARS OF MARRIAGE
All Women	20%	43%
Race/Ethnicity		
Asian	10%	23%
Hispanic	17%	42%
White	20%	42%
African American	28%	55%
Age at Marriage		
Less than 18	29%	59%
18–19	24%	49%
20–24	17%	36%
25 and over	8%	35%
Family Income		
Low	31%	65%
Medium	19%	40%
High	13%	31%
Religion		
Catholic	17%	37%
Conservative Protestant	18%	40%
Other (non-Christian)	17%	40%
Liberal Protestant	21%	44%
None	27%	56%
Parents		
Not divorced	17%	38%
Divorced	26%	52%

Source: Centers for Disease Control, 2002.

had either divorced or died. Thus, this year we would be about ready to calculate a divorce rate for those couples who married in the early 1930s. That would be an accurate rate, but it would leave us without knowledge of what had been going on during the past seventy-five years or so.

However, there is a way to estimate these outcomes without waiting to see how each marriage turns out (Preston, 1975). But the computation is highly complex, and thus far, the only available rates based on this method apply to the United States alone. What do they show? A dramatic increase in divorce. Of all weddings held in 1923, 19 percent ended in divorce. For weddings held in 1975 the projection is that 50 percent will result in divorce. Projections for more recent years (Weed, 1980) suggest a slight decrease—which also shows up in a recent decline in the rate of divorces per 1,000 population.

As sociologists examined the more accurate divorce rate, everyone's worst suspicions were confirmed. Marriage clearly has become a very unstable institution if half of all marriages end in divorce. For this fact was interpreted to mean that half of all Americans who get married will eventually get divorced. *But it isn't true!* Marriages are not people: Far more marriages break up than there are people who get divorced.

Because some people get married and divorced again and again, they contribute again and again to the statistics on broken marriages. Imagine four college roommates. Upon graduation each marries. Three never get divorced. But one marries and divorces three times. Hence, this group produced six marriages and three divorces—so half of all marriages of this group ended in divorce. Although this is correct, it also is true that 75 percent of these people never divorced.

Recently, social scientists have turned to surveys to attempt a more accurate estimate at which Americans get divorced. The most important of these surveys was conducted by the Centers for Disease Control in 1995. Interviews concerning their marriage and family life were conducted with a national sample made up of 10,847 American women age fifteen to forty-four. Some of the key findings are shown in Table 13-9. Of these women, 20 percent of their first marriages ended in divorce within five years, and by the end of fifteen years, 43 percent had been divorced. Most of these women had remarried, as will be discussed later in this chapter.

The data also show that not all women were equally prone to become divorced. Asian women were far less likely than others to have become divorced (23%), while African American women had the highest rate (55%). The younger they married, the more likely women were to divorce, and high income couples were those least apt to divorce. Of religious affiliations, Catholics were least likely to divorce, and people without a religion were the most likely. Finally, women were substantially less likely to divorce (38%) if their parents had remained married than if their parents had been divorced (52%).

While these findings cannot be generalized directly to people over forty-four or to men, they do suggest that at least half of current American marriages end in divorce. Why is divorce so common? The two reasons given the most emphasis by experts on the family are: (1) Romance is a highly perishable commodity, and (2) the opportunities to get divorced have increased.

WHEN ROMANCE FADES

We have seen that romance has become the basis for marriage in modern times. People now expect deep romantic sentiments to lead them into marriage and to sustain their marriages. Unfortunately, these feelings can fade and be difficult to revive. Because so many adults rely on their spouses for their deepest emotional ties, immense weight is placed on these romantic feelings. Even small tensions are easily magnified, for any discontent or threat to this primary attachment provokes anxiety. Indeed, romantic sentiments may suffer from too frequent assessment, and the slightest doubts can easily shatter that "special feeling." So, too, can the simple passage of time, especially when sexual attraction is the focal point of romance.

Studies suggest that sexual attraction, in and of itself, is based partly on novelty and tends to decline with time (Pineo, 1961). If this is an intrinsic feature of sexual attraction and not just a temporary aspect of current sexual patterns, then a decline in sexual attraction and satisfaction will permanently threaten marriages based on sexual attraction. That is, if sexuality is a primary basis for emotional attachment between husbands and wives, then marriages will tend to weaken as familiarity causes a loss of fervor. In this sense a good deal of divorce may reflect a form of swapping sexual partners.

In any event clearly most people do not equate love with sexual thrills. While a decline in sexual novelty may be at the root of many divorces, many other couples find that their relationships improve the longer they live together. Research shows that marital satisfaction is higher the longer a couple has been married (Campbell, 1975). Once again, this finding could partly reflect a bias of selection. As time passes, more of the less-satisfied couples get divorced. However, many couples report that their marriages have become more satisfying over time and that they are happier now than when they were just married.

GENDER AND EXTRAMARITAL SEX

One aspect of sexuality that clearly does play a significant role in divorce is the extramarital "affair." Despite lurid reports based on SLOPS studies, such as Shere Hite's (1987) claim that more than 80 percent of women who have been married ten years or more have engaged in ex-

TABLE 13-10	"Have you ever had sex with someone other than your husband or wife while you were married?"		
	MEN (%)	WOMEN (%)	TOTAL (%)
Yes	25	13	19
No	75	87	81
	100	100	100

Source: Prepared by the author from the General Social Survey, 2000.

tramarital sex, the great majority of Americans do not cheat on their spouses. In the 2000 General Social Survey, 19 percent admitted to an affair (the question was asked using the secret ballot technique explained in Chapter 4). However, men are substantially more likely than women to have an affair, as can be seen in Table 13-10. These gender differences in behavior reflect differences in attitudes as well. Table 13-11 shows that there is a great deal of variation across cultures in the extent to which people condemn extramarital sex—from more than 90 percent in Malta and India to only 26 percent in Spain. Notice, however, that in nearly every nation women are more inclined to oppose extramarital affairs than are men. For example, in Canada only 48 percent of men think affairs are never justified, compared with 60 percent of Canadian women.

Similar gender differences (not shown) exist on attitudes toward prostitution. Thus, while 36 percent of Canadian men say it is never justified, 48 percent of Canadian women say so. Gender differences about prostitution are especially large in Eastern Europe—in Slovenia 49 percent of the men and 71 percent of the women condemn prostitution.

Extramarital sex is a frequent cause of divorce: first, because people often abandon a spouse to marry their lovers and, second, because people often walk out when they discover their spouse is being unfaithful.

A CLOSER VIEW
TRENT AND SOUTH: INTERNATIONAL COMPARISONS IN DIVORCE

According to data published by the United Nations, among nations that publish reliable statistics, the United States has the highest per

TABLE 13-11	"Please tell me whether you think it can always be justified for a married person to have an affair, never justified, or something in between?"		
	PERCENT "NEVER JUSTIFIED"		
NATION	TOTAL POPULATION (%)	MEN (%)	WOMEN (%)
Malta	94	91	97
India	93	92	93
Iceland	78	73	84
Ireland	71	65	76
United States	**70**	**67**	**72**
China	70	67	73
Argentina	70	58	81
Romania	69	63	74
South Africa	68	63	72
Poland	65	61	69
Denmark	65	63	67
Brazil	63	49	77
Hungary	63	52	72
Norway	62	58	66
Chile	60	50	70
Coratia	58	31	65
Canada	**54**	**48**	**60**
Nigeria	54	55	53
Switzerland	54	48	60
Ukraine	54	47	61
Great Britain	54	50	58
Belgium	53	47	59
Finland	51	49	54
Sweden	50	53	47
Austria	49	44	53
Netherlands	49	44	53
Italy	49	42	56
Russia	48	45	50
Japan	46	43	49
Latvia	44	39	49
Slovenia	43	38	48
Mexico	**40**	**34**	**47**
Czech Republic	39	35	43
France	36	31	40
Slovak Republic	35	28	43
Estonia	33	27	39
Spain	26	21	30

Source: Prepared by the author from the World Values Surveys, 1995–1996, 2001–2002.

capita divorce rate in the world while Malaysia has the lowest rate: 0.02 per 1,000. (This excludes nations where divorce is prohibited.) Such great variation prompted Katherine Trent and Scott J. South, sociologists at SUNY Albany, to attempt to explain why the divorce rate is higher in some nations than in others (South and Trent, 1988; Trent and South, 1989).

They proposed the following hypotheses:

■ Modernization (urbanization and economic development) will reduce the importance of the family and increase the rate of divorce.

■ A rise in female labor force participation will free women from economic dependence on men, and this will make it easier for people to get divorced.

■ Divorce will fluctuate according to sex ratios: Where men greatly outnumber women, divorce rates will be very low, but where women outnumber men, divorce rates will be high.

■ Because the Catholic Church bans divorce, the divorce rate will be lower the larger the proportion of Catholics in a nation's population.

To test these hypotheses, Trent and South based their analysis on sixty-six nations for which adequate data were available.

Their results were quite interesting. First of all, modernization seems to cause a modest decrease in divorce rates during its early stages, but as nations continue to develop, the divorce rate soon begins to climb. Second, the divorce rate is higher where a larger proportion of women are employed outside the home. Third, the data strongly support the hypothesis based on Guttentag and Secord's theory (see Chapter 12) about the effects of sex ratios. To the extent that men outnumber women in the fifteen to forty-nine age group, divorce is lower. Where women outnumber men, divorce is high despite the fact that marriage rates are lower. That is, under these circumstances men are less willing to marry or to stay married.

Finally, Trent and South found no evidence in support of the hypothesis that the more Catholics, the less divorce. They acknowledged that several Catholic nations prohibit divorce and thus report no divorce rate. They noted, however, "that among countries permitting a legal divorce, the relative sizes of Catholic . . . populations do not seem to have a great influence on the divorce rate."

LIVING TOGETHER

In 1970 half a million American couples were living together without being married. Today, well over 4 million are doing so. Many observers of this trend have suggested that it ought to have positive benefits—that by living together people gain greater insights into their compatibility and therefore unsatisfactory unions can be ended without resort to divorce. This is thought to be especially beneficial if it causes unstable unions to be broken off before any children are involved.

But it also can be argued that people who live together without marrying are overselected from among those who tend to reject traditional values concerning marriage, who are less opposed to divorce, and who are reluctant to make lasting commitments. If this argument is true, then people who live together before they marry ought to be more prone to divorce, even if the most unsuitable matches have been weeded out short of the marriage ceremony.

A study of women in Sweden, where living together is common among young adults, lends strong support to the second view (Bennett, Blanc, and Bloom, 1988). Overall, women who lived with their eventual husbands prior to marriage were 80 percent more likely to get divorced than were women who did not. Even women who had lived with their spouse for over three years prior to marriage were 50 percent more likely to get divorced.

Data based on American couples show very similar patterns. Of women who lived with their future spouse before marriage, 51 percent had divorced by the end of fifteen years of marriage, while only 39 percent of women who had not cohabited with their spouse divorced within fifteen years (Centers for Disease Control, 2002).

Thus, a rising rate of living together would seem unlikely to result in a reduced divorce rate. But to the extent that living together primarily involves those with the highest probability of divorcing, it may not raise the divorce rate either. That is, these couples might have divorced anyway, even had they married before living together.

THE ONE-PARENT FAMILY

The primary concern raised by high divorce rates is not broken marriages but broken families. How does divorce affect children? In recent years researchers have devoted a good deal of study to this question. The results are quite consistent—children whose parents divorce have lower levels of well-being, the latter including such diverse things as academic

TABLE 13-12	Unmarried Motherhood		
NATION	BIRTHS TO UNMARRIED MOTHERS (%)	NATION	BIRTHS TO UNMARRIED MOTHERS (%)
Iceland	65	**Canada**	**26**
Sweden	54	Portugal	20
Norway	49	Netherlands	19
Denmark	46	Germany	18
France	39	Belgium	15
Finland	37	Spain	11
Great Britain	37	Italy	8
United States	**33**	Switzerland	8
Austria	29	Greece	3
Ireland	27	Japan	1

Sources: *Eurostat Yearbook*, 1999; *Statistical Abstract of the United States, 2002.*

achievement, behavior problems, delinquency, self-esteem, and psychological adjustment (Amato, 1993). And many studies have traced these problems to the fact that the children of divorced parents so frequently are raised in a one-parent home (Amato, 1993). It is important to see, however, that many one-parent homes are not the result of a divorce. Some are caused by the death of one parent, and a rapidly growing number occur when single women become mothers.

In 1960 only 5 percent of all births in the United States were to unmarried women. Since then, this has risen to 33 percent, which means that of every 100 babies born, 33 are to women without husbands. This same trend has gone on in many other nations, as can be seen in Table 13-12. In Iceland and Sweden, most births are to unmarried women. In Greece and Japan, unwed births are rare.

Some unwed births in the United States involve couples who are living together but who are not married. However, most are to women who live alone or with their parents. In the past many illegitimate children were put up for adoption. Today, most unwed mothers keep their children. Furthermore, in 90 percent of divorces, the children remain with the mother, and the combined result is a substantial number of female-headed households—households having children under age eighteen and headed by a woman.

The primary consequence of female-headed families is poverty—the lowest-income families are very disproportionately made up of mothers on welfare raising children. Moreover, even when female-headed families are not poor, they are much less likely than are other families to be affluent, for the major source of high family incomes is *two earners*.

Lack of income can have many negative effects on family life. But sociologists are increasingly concerned about another shortage that besets the one-parent family: *time*. When there is only one parent rather than two, supervision of children may be reduced greatly. In the case of divorce, when children must split their loyalties between parents, the result may be weaker attachments to each in comparison to attachments in two-parent families.

From Chapter 7 it should be clear that if one-parent families cannot sustain the same level of supervision and the same strength of attachments as can two-parent families, then research ought to find that the one-parent family is in greater risk of having delinquent children. Research on this hypothesis has produced mixed results over the years. Studies suggest that children in one-parent families are more prone to various forms of delinquency but that the differences are not great (Wilkinson, 1980; Gottfredson and Hirschi, 1990).

Research does find that *poor parenting*, regardless of the structure of the family, is a primary cause of deviant behavior among children. Put another way, it isn't how many parents a child has at home that matters; it's how effective they are at being parents that is of primary importance in how a child turns out (Wilson and Herrnstein, 1985; Gottfredson and Hirschi, 1990).

REMARRIAGE

Most people who divorce get remarried. Table 13-13 shows that 75 percent of American women remarried within ten years of their first divorce. That African American women (49%) are much less likely to remarry is consistent with the shortage of African American men, discussed at length in Chapter 12. Women also are more likely to remarry if they are less than twenty-five at the time of their divorce. Surprisingly, remarriage is very little affected by having children—77 percent without children remarried, while 70 percent with two or more children remarried, too. Conservative Protestants (82%) are more likely to remarry than are members of other religious groups, and those without a religion (66%) are least likely to remarry. Southerners (77%) are the most likely to remarry (despite large African American communities), while those living in the Northeast (68%) are least likely to do so.

But whom do people choose for a second marriage? Do they select differently the second time around? How well do second marriages work out? What is the impact of children from a first marriage on adjustments in a new marriage?

To answer questions like these, sociologists have begun to collect data on what Frank F. Furstenberg Jr. and Graham B. Spanier (1984) have called **conjugal careers**. Sociologists have long studied occupations in terms of careers—the pattern of employment individuals follow over the course of their working lives. Why not treat marriage the same way? Let's watch two sociological studies of conjugal careers as they shed light on the three questions we've posed.

TABLE 13-13	Percentage of American Women Who Remarried within Ten Years of Their Divorce
	PERCENT REMARRIED WITHIN TEN YEARS
All Women	75%
Race/Ethnicity	
White	79%
Hispanic	68%
African American	49%
Age at Divorce	
Less than 25	81%
25 and over	68%
Number of Children at Divorce	
None	77%
One	77%
Two or more	70%
Religion	
Conservative Protestant	82%
Other (non-Christian)	79%
Liberal Protestant	77%
Catholic	71%
None	66%
Region	
South	77%
Midwest	76%
West	75%
Northeast	68%

Source: Centers for Disease Control, 2002.

A CLOSER VIEW
JACOBS AND FURSTENBERG:
SECOND HUSBANDS

It is well known that first marriages involve women and men of quite similar social status. For example, Michael Hout (1982) found that couples who both work full time are very likely to hold jobs similar in prestige. That is, women who work in factories tend to be married to other factory workers; women professionals tend to have husbands who also have a profession. This is not surprising because the two most common places for married couples to have met is at school or at work. People who met in school tend to have the same levels of education and hence to qualify for the same levels of jobs. And people who meet at work tend to be doing the same sort of work. But what happens the second time around?

To find out, Jerry A. Jacobs and Frank F. Furstenberg Jr. (1986) of the University of Pennsylvania collected data from a national sample of American women over ten years. In 1967 two groups of women were interviewed. One sample was limited to women age thirty to forty-five. The second included only women age fourteen to twenty-five. Ten years later, these same groups were interviewed again. Then the data were assembled to form a marital or conjugal career for each woman up to 1978. By then 743 of the older women had been divorced or widowed and had remarried, as had 413 of the younger women. Jacobs and Furstenberg focused on these two groups in their study.

At first glance it appeared that women bettered their economic situation by remarriage: Second husbands had a higher average occupational prestige and income than did first husbands. But this was an illusion. Women are older when they remarry, and therefore, second husbands are older at remarriage than are first husbands. When the *current* economic position of first and second husbands was compared, the average turned out to be the same. On average, then, women marry second husbands who are no more, and no less, successful than their first husbands.

Within this overall finding for the average remarriage, however, there were some distinctive patterns applying to certain subsets of women. The sooner women remarried, the greater the similarity in the social status of their first and second husbands. When older women remar-

ried, they more often married men noticeably more or less successful than their first husbands; the older a woman is when she remarries, the smaller is the pool of available husbands because of the much lower life expectancy of men.

Children also play an important role in the economic success of second husbands. That is, a woman with children under age ten is usually unable to marry a second husband who is as successful as her first husband. As Jacobs and Furstenberg put it, women "who bring children with them into the second marriage are at a distinct disadvantage. The negative socioeconomic consequences of bringing children into the second marriage are substantial." Put another way, men will prefer a family of their own to assuming another man's family, ready-made. Other things being equal, the more financially successful the man, the more options he will have in choosing a wife, and he will tend to use those options to marry a woman without younger children.

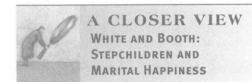

A CLOSER VIEW
WHITE AND BOOTH:
STEPCHILDREN AND
MARITAL HAPPINESS

If women pay a price for bringing children into a marriage, what happens after that? How do stepchildren influence relations between a couple following remarriage? To find out, Lynn K. White and Alan Booth (1985) of the University of Nebraska interviewed a national sample of married adults under age fifty-five in 1980. Then in 1983 they reinterviewed them—a total of 1,673 people.

White and Booth were aware of research showing that the divorce rate is higher in remarriages than in first marriages. So, they were especially interested in measuring aspects of the marital relationships of their respondents—how happy they were, how much fighting they did, and the like. And because they had interviewed people twice over a three-year period, they could contrast those who separated or divorced during this period with those who did not—and 7 percent had done so.

When they examined their results, their first and most striking discovery was that couples who remarry do not have a higher rate of divorce than people who are marrying for the first time if there are no stepchildren in the home or if it is the first marriage for the other partner. When it is a remarriage for both partners and when one or both bring children from a prior marriage to live in the home, the odds of divorce increase by 50 percent. And not surprisingly, couples with stepchildren in the home report a lot more tensions and problems with children than do couples with only their own biological children in the home. Apparently, the children feel it, too. White and Booth found that stepchildren leave home at a younger age than do biological children. Summing up their findings, White and Booth wrote:

A comparison of parents with stepchildren and those without shows a strong and consistent pattern: parents with stepchildren more often would enjoy living away from their children, perceive their children as causing them problems, are dissatisfied with their spouse's relationship to their children, think their marriage has a

---

negative effect on their relationship with their own children, and wish they had never remarried.

Several years later, this same sample of married adults—not just those who had divorced since the start of the study—was interviewed one more time. By then many of these couples no longer had children living at home, and what White wanted to know was how this had affected their marriage. Many observers have noted the serious emotional readjustments required of parents, especially mothers, when their children leave. In fact, this often is referred to as the "empty nest syndrome" (Bart, 1972).

But once again, what everyone "knew" turned out not to be true. White and a new collaborator, John N. Edwards, discovered that the psychological effects of an "empty nest" tend to be quite positive. That is, most parents experienced significant increases in marital happiness

TABLE 13·14	Among Married Couples, the Percentage wherein the Woman "Always" or "Usually" Performs the Chore		
NATION	LAUNDRY (%)	SMALL REPAIRS (%)	GROCERY SHOPPING (%)
Italy	98	11	57
Bulgaria	96	8	43
Japan	95	19	81
Russia	94	1	61
Czech Republic	94	42	64
Slovenia	94	12	45
Austria	94	10	53
Germany	91	4	42
Northern Ireland	91	5	58
Spain	91	15	69
Hungary	90	4	51
Ireland	88	8	67
Netherlands	88	4	58
Poland	88	6	55
Norway	82	3	40
Sweden	81	2	42
Great Britain	81	5	43
Philippines	80	5	66
Australia	79	8	59
Israel	78	13	36
New Zealand	75	4	59
Canada	**70**	**5**	**46**
United States	**69**	**6**	**49**

Source: Prepared by the author from the International Social Survey Program, 1994.

and life satisfaction when the last child had moved out. These findings are not simply based on parents' perceptions of increased happiness but rest on comparing their attitudes after their children were launched on their own with those they expressed during earlier interviews when the children were still at home. White and Edwards (1990) referred to this as a "postlaunch honeymoon."

Some of you have, no doubt, caused such a honeymoon for your own parents by coming to college. However, you should know that White and Edwards also discovered that simply having the kids leave is not enough to make parents happier. There also must continue to be a lot of contact with the kids. So, don't forget to call and write!

HOUSEHOLD CHORES

Recall from the beginning of the chapter that in premodern societies men tend not to help with household chores—in the majority of these societies, they did no chores at all. With the immense increase in the proportion of wives in the labor force in modern societies, there has been increased emphasis on the need for husbands to help out at home. In fact, judging from the amount of attention given this topic in the media, this would seem to be a major element of contention between modern husbands and wives.

Nevertheless, the division of labor within the home remains quite specialized in terms of who does what, as can be seen in Table 13-14. In each of these twenty-three nations, women do the laundry in the great majority of households. In many nations husbands virtually never even help with the laundry—American and Canadian husbands are the most likely to do so. On the other hand, in all of these nations, women very rarely do the small repairs around the home—Czech and Japanese women are the most likely to do so. However, in most nations grocery shopping is quite evenly divided between husbands and wives.

CONCLUSION

The modern family has many imperfections. Many families abuse or neglect their children. Arrest statistics show that family fights are common and that homicides frequently occur within families. Many of today's marriages will end in divorce. Yet these problems do not demonstrate that family life is deteriorating or that the family is less able to fulfill its functions. We have seen what the good old days were like, and as far as family life is concerned, they should be called the "miserable old days."

I have emphasized a historical and cross-national view in this chapter not to minimize concern about current family problems but rather to put them in perspective. It's not helpful to run in search of answers to why the world is going to pot if that is not really what's happening. Here the decades of sociological concern about the decline of the family seem instructive. For example, for years sociologists asked why marriage no longer worked without asking if a high divorce rate might reflect unceasing efforts by individuals to find marriages that do work. Preindustrial marriages may have lasted, but did they work? If we rate them by current standards, then they seldom did.

In similar fashion people too often assume that their own society is plagued with unique problems or is unusually prone to particular social ills. But the many comparisons we have examined in this chapter suggest that American, Canadian, and Mexican families are about average in terms of the way they perform the functions typically associated with family life.

Review Glossary

Terms are listed in the order in which they appear in the chapter.

family A small kinship-structured group with the key function of nurturant socialization of the newborn.

marriage A formal commitment between a couple to maintain a long-term relationship involving specific rights and duties toward one another and toward their children.

nuclear families Families made up of only one adult couple and their children.

extended families Families made up of at least two adult couples.

polygamy All plural marriages, wherein one person (of either sex) has multiple spouses.

polygyny The proper term for marriages involving one man and multiple wives.

incest taboo Prohibition against sexual relations between certain members of the same family.

dependents Family members unable to support themselves.

conjugal careers The histories of individuals in terms of marriages.

Suggested Readings

Laslett, Peter. 1977. *Family Life and Illicit Love in Earlier Generations.* London: Cambridge University Press.

Patterson, G. R. 1980. "Children Who Steal." In *Understanding Crime: Current Theory and Research*, edited by Travis Hirschi and Michael Gottfredson. Beverly Hills, Calif.: Sage.

Reiss, Ira L. 1988. *Family Systems in America*. New York: Holt, Rinehart and Winston.

Rosenberg, Charles E., ed. 1975. *The Family in History.* Philadelphia: University of Pennsylvania Press.

Shorter, Edward. 1975. *The Making of the Modern Family*. New York: Basic Books.

Sociology Online

www.socstark10.com

GO TO THE INTERNET AND TYPE: www.socstark10.com.

↗**CLICK ON:** 2000 General Social Survey

✓**SELECT:** FAMILY @16: Were you living with both your own mother and father around the time you were 16?

↗**CLICK ON:** Analyze Now

TRUE OR FALSE: The most educated Americans are significantly more likely than the less educated to come from father-mother families.

TRUE OR FALSE: African Americans are far less likely than whites to have grown up in father-mother families.

TRUE OR FALSE: Upper-income Americans are significantly less likely than those who earn less to come from father-mother families.

↗**CLICK ON:** Select New Data Set

↗**CLICK ON:** The 50 States

✓**SELECT:** %DIVORCE: Percentage of those 15 and over who currently are divorced (not remarried).

↗**CLICK ON:** Analyze Now

Is there a regional pattern to this variable? Which state is highest?

↗**CLICK ON:** Select New Variable

✓**SELECT:** MARRIAGE: Marriages per 1,000 population.
Is there a regional pattern to this variable?

Which state is highest? How might it be possible for a state to have a marriage rate so much higher than any other state?

↗**CLICK ON:** Select New Data Set

↗**CLICK ON:** Nations of the Globe

✓**SELECT:** TRUST KIN? Percentage who expressed complete trust in their family.

↗**CLICK ON:** Analyze Now

Which two nations are highest in trusting kin? Which nation is lowest? Does this suggest anything to you?

USING INFOTRAC COLLEGE EDITION

GO TO THE INTERNET AND TYPE:
www.InfoTrac-college.com

↗**CLICK ON:** Register New Account
You will be asked to enter your Access Code and to create and enter a User Name and Password. Having done so,

↗**CLICK ON:** InfoTrac College Edition or Log On.

Given such an immense and complex topic, the chapter skipped many interesting topics about the family. These search terms will lead you to some of them.

SEARCH TERMS:
Family Life Surveys
Family Research
Television and Family

CHAPTER

14

Three country churches, photographed by Dorothea Lange in South Dakota during the summer of 1938, illustrate the vigorous competition among denominations that occurs whenever a society has an unregulated religious economy. Many critics have used this photo as proof that denominationalism is a silly waste of resources—that local farm families would have been better off with one united congregation. But the people who attended these churches knew that the three differed substantially in their doctrines and forms of worship, and they thought these differences were important. And if the buildings looked much alike, that was because each was purchased as a precut kit, and the kit manufacturers provided for only minor differences, such as the shape of the steeple.

Religion

NOBODY KNOWS WHEN HUMANS first acquired religion. Unlike tools chipped from stone, cultural ideas do not lie secure for millions of years, awaiting the archeologist's pick. So, while we know that humans living over a million years ago made tools, we can only guess about their religion. However, there can be no doubt that our Neanderthal ancestors had religion at least 100,000 years ago because evidence of their faith has been unearthed. The Neanderthal buried their dead with great care and provided them with gifts and food for use in the next world. And deep in their caves, the Neanderthal built small altars out of bear bones. These relics make it clear that the Neanderthal believed in life after death and conducted ceremonies to seek the aid of supernatural beings. Such beliefs and practices are properly called religion, and all human societies since the days of the Neanderthal have had religion.

In this chapter we shall try to understand why religion is a vital part of human societies. What does religion do for people? How does it influence social life? Then we will explore the concept of a **religious economy:** the marketplace of competing faiths within a society. Although societies often claim to have only one faith (and sometimes use military force to keep competing faiths out), this is never really true. We shall see why not, why "underground" faiths exist even in the most repressive nations, and why these tend to erupt into significant movements whenever repression eases. Viewing the religious sector of societies as economies of faith permits us to examine how religious organizations influence one another. We shall see that in time the most successful religious organizations become increasingly worldly, a process called **secularization.** As this occurs, conditions become favorable for new organizations to break away and restore

CHAPTER OUTLINE

THE NATURE OF RELIGION
  The Gods
  Gods and Morality
  Gender and Religious Commitment

RELIGIOUS ECONOMIES

CHURCH-SECT THEORY

SECULARIZATION AND REVIVAL

INNOVATION: CULT FORMATION

CHARISMA

THE AMERICAN RELIGIOUS ECONOMY
  Secularization and Revival
  Secularization and Innovation

THE CANADIAN RELIGIOUS ECONOMY

CULT MOVEMENTS IN EUROPE

THE PROTESTANT EXPLOSION
IN LATIN AMERICA

EASTERN REVIVALS

THE UNIVERSAL APPEAL OF FAITH

CONCLUSION

REVIEW GLOSSARY

SUGGESTED READINGS

SOCIOLOGY ONLINE

a less worldly form of the conventional faith, a process known as **revival.** We shall also see how wholly new religions can arise in societies, a process called **religious innovation** or cult formation.

We shall use the model of religious economies to examine current conditions and trends in the United States. Then we shall apply the model to Canada, the nations of Western and Eastern Europe, and the nations of Latin America.

THE NATURE OF RELIGION

A most difficult problem facing sociologists of religion has been to define their subject matter. An adequate definition must include the vast array of faiths found in the world without including too much. As Georg Simmel urged in 1905, a general definition of religion must apply "alike to the religion of Christians and South Sea Islanders." It must isolate the common elements in Buddhism, Islam, and other faiths of modern times, as well as the faiths of our primitive ancestors, such as the Neanderthal.

For a few decades, this problem was solved by recognizing that all religions have one feature in common: They always involve answers to **questions about ultimate meaning,** such as, Does life have a purpose? Why are we here? Is death the end? Why do we suffer? Does justice exist?

It is characteristic of humans to ask such questions. Indeed, these questions must have troubled the Neanderthal, for they had accepted answers to some of these questions. Hence, religion has been defined as socially organized beliefs and activities offering solutions to questions of ultimate meaning. But that definition is too broad. It applies to communism as well as Catholicism. And it applies to a philosophical system that denies that there can be answers to questions of ultimate meaning. It is inconvenient to have a sociological concept that ignores differences between what are widely regarded as religious and antireligious positions.

In the end most sociologists of religion agreed that the term *religion* ought to be applied to only particular kinds of answers to questions of ultimate meaning—those that posit the existence of the supernatural. Defined this way, religion can invoke the power, wisdom, authority, and aid of the Gods, a capacity that nonreligious philosophies lack (Spiro, 1966; Berger, 1967; Stark, 1981; Stark and Bainbridge, 1985).

Thousands of years ago, our ancestors created sacred chambers by painting superb animal figures like these on the walls of natural caverns. Even modern visitors are awed by these exotic images and immediately sense the sacred intentions of those who used to come here at times of special religious significance.

THE GODS

If we closely examine the ultimate questions that humans keep asking, many of them clearly require a very special kind of answer. People do not usually ask if life has meaning; they ask, What is the meaning of life? Why does the universe exist? Why do I exist? For life to have meaning in this sense, history must be guided by intention. For this to be true, a consciousness capable of imposing intention on history must exist. In other words, if the universe is to have purpose, then it must have been created and directed by a conscious agent—a being capable of making plans and having intentions. Such a being has to be of such power, duration, and scale as to be beyond the natural world. That is, such a being must be supernatural.

Many questions of ultimate meaning, therefore, can be answered only by referring to the **supernatural**—to beings or forces beyond nature who are able to suspend, alter, ignore, and create physical forces. To believe, for example, that there is life beyond death is to accept the supernatural. To believe that earthly suffering is compensated in the world to come also requires belief in the supernatural. Some things that humans greatly desire cannot possibly be attained in this world but can come only from the Gods.

By defining **religion** as socially organized patterns of belief and practices that concern ultimate meaning and assume the existence of the supernatural, sociologists can isolate the essential element that sets religion apart from other aspects of social life and accounts for its universal appeal.

Systems of thought that reject the supernatural cannot satisfy the concerns of most people. Atheists can search for explanations of how the universe functions, but they cannot say that these functions have an underlying purpose. Communists can promise to reduce poverty, but they cannot offer an escape from death. In any society some people can accept the beliefs that the universe has no purpose, that what we gain in this life is all we shall receive, and that death is final. But as we shall see throughout this chapter, for most people this is not enough: Only religion can fulfill their needs, their hopes, their dreams.

Table 14-1 shows that most people in most nations say they are religious. Keep in mind that most who did not say they were religious were *not* saying they were "irreligious." Rather, many people who do not claim to be "religious" understand this question to be asking if they are "especially religious." But they are sufficiently religious so that they do not regard themselves as atheists (people who deny the existence of the

TABLE 14-1	Religiousness around the World				
NATION	"I AM A RELIGIOUS PERSON" (%)	"I AM A CONVINCED ATHEIST" (%)	NATION	"I AM A RELIGIOUS PERSON" (%)	"I AM A CONVINCED ATHEIST" (%)
Egypt	98	0	Taiwan	72	2
Nigeria	96	0	Iceland	72	3
Uganda	93	1	Denmark	71	5
Poland	92	1	Ireland	71	2
Tanzania	90	1	Latvia	71	3
Bangladesh	89	0	Indonesia	70	0
Zimbabwe	87	2	Chile	69	1
Pakistan	86	0	El Salvador	69	1
Peru	86	1	Serbia	68	3
Georgia	86	1	Armenia	67	4
Portugal	85	3	Ukraine	67	3
Brazil	85	14	Albania	65	5
Jordan	85	0	Belgium	65	7
Moldova	85	1	Slovenia	65	8
Azerbaijan	84	1	Finland	62	3
Italy	83	3	Netherlands	61	6
Iran	82	1	Russia	60	4
Macedonia	82	1	Spain	59	6
Puerto Rico	82	1	Australia	58	5
United States	**81**	**1**	Hungary	58	5
Argentina	81	3	Switzerland	54	4
Romania	81	1	Germany	52	7
Croatia	80	3	Uruguay	52	8
Turkey	80	1	New Zealand	48	5
Philippines	79	0	Norway	47	4
Venezuela	78	1	Bulgaria	47	6
Slovakia	77	4	France	44	14
South Africa	77	3	Colombia	42	1
Mexico	**76**	**2**	Czech Republic	41	8
Austria	75	2	Sweden	37	6
Malta	75	0	Great Britain	37	5
Greece	75	4	Vietnam	36	12
India	75	2	Estonia	36	6
Dominican Republic	74	3	Belarus	26	9
Lithuania	74	1	Japan	23	12
Bosnia	73	6	China	14	24
Canada	**72**	**4**			

Source: Prepared by the author from the World Values Surveys, 2001–2002.

supernatural), for these are few everywhere except in China, where 24 percent claim to be "a convinced atheist." The Chinese data are suspect because the communist government of China stresses atheism. In the 1980s surveys found comparable percentages of atheists in other communist societies such as the Soviet Union. As will be discussed later in this chapter, with the fall of repressive, antireligious regimes,

many people no longer felt the need to say they were atheists, as evidenced by the data shown here for Russia and other Eastern European nations. Hence, it is significant that in the 1996 World Values Survey, 38 percent of Chinese claimed to be atheists and only 6 percent said they were religious. Thus, in just six years, the number of Chinese atheists fell by more than a third and the number who said they were

© Michael Weisbrot/Stock Boston

religious more than doubled. What happened? During this time the Chinese government began a somewhat more relaxed policy toward religion.

In any event religion remains a potent social factor in most of the world.

GODS AND MORALITY

Religion functions to sustain the moral order. This classic proposition, handed down from the founders, is regarded by many as the closest thing to a "law" that the social scientific study of religion possesses. The fundamental idea is that humans inevitably ask: *What does the God or the Gods want or expect from us?* It has been assumed that the answer to this question always involves expectations concerning human behavior—a set of norms. That is, religions explain why certain norms exist, why they are right, and why they must be obeyed.

More than a century ago, one of the celebrated founders of the sociology of religion,

W. Robertson Smith, explained that "even in its rudest form Religion was a moral force, the powers that men revered were on the side of social order and moral law; and the fear of the gods was a motive to enforce the laws of society, which were also the laws of morality" (1889). Émile Durkheim, of course, argued that religion exists *because* it unites humans into *moral communities,* and while law and custom also regulate conduct, religion alone "asserts itself not only over conduct but over the *conscience.* It not only dictates actions but ideas and sentiments" ([1886] 1994). And according to Bronislaw Malinowski, "every religion implies some reward of virtue and punishment of sin . . . " (1935).

In one form or another, this proposition appears in nearly every introductory sociology and anthropology text on the market. But it's wrong. Moreover, it wasn't even handed down from the founders, at least not unanimously! Indeed, the founder of British anthropology, Edward Tylor, and the founder of British sociology, Herbert Spencer, both took pains to point

These Roman Catholic girls are taking their first Holy Communion—a sacrament that illustrates the capacity of religions to answer questions of ultimate meaning. Communion, also known as the Lord's Supper among many Protestants, symbolizes the belief that all souls may be saved and gain everlasting life through symbolic and literal acceptance of Jesus Christ.

out that only *some kinds* of religions have moral implications.

Tylor ([1871] 1958) reported, "To some the statement may seem startling, yet the evidence seems to justify it, that the relation of morality to religion is one that only belongs in" some societies. ". . . the popular idea that the moral government of the universe is an essential tenet of natural religion simply falls to the ground" since many religions are "almost devoid of that ethical element." Tylor did not claim that morality is absent from these societies but that, rather than resting on religion, moral and "ethical laws stand on their own ground of tradition and public opinion."

Although little noticed, through the years several anthropologists have reported that the particular tribe each had studied seemed not to connect religion and morality. For example, J. P. Mills (1922) noted that the religion of the Lhotas includes no moral code: "Whatever it be which causes so many Lhotas to lead virtuous lives it is not their religion." Summing up many such field reports, the celebrated Mary Douglas (1975) flatly asserted that there is no "inherent relation between religion and morality: there are primitives who can be religious without being moral and moral without being religious."

Tylor's observation that not all religions support the moral order should always have been obvious to anyone familiar with Greek and Roman mythology. The Greco-Roman Gods were quite morally deficient. They were thought to do terrible things to one another and to humans as well—sometimes merely for amusement. And while they were quite apt to do wicked things to humans who failed to propitiate them, the Gods had no interest in anything (wicked or otherwise) humans might do to one another. Instead, the Greek and Roman Gods only concerned themselves with direct affronts. For example, no religious sanctions were incurred by young women who engaged in premarital sex *unless* they immersed themselves in sacred waters reserved for virgins (MacMullen, 1981). Because Aristotle taught that the Gods were incapable of caring about mere humans (MacMullen, 1981), he could not have concurred that religion serves the function of sustaining and legitimating the moral order. In their day all of the classical philosophers would have ridiculed such a proposition as peculiar to Jews and Christians—and they would have been correct (MacMullen, 1981; Meeks, 1993; Stark, 1996b).

As will be seen, the proposition about the moral functions of religion requires a particular conception of supernatural beings as deeply concerned about the behavior of humans toward one another. Such a conception of the Gods is found in many of the major world faiths, including Judaism, Christianity, Islam, and Hinduism. But it appears to be largely lacking in the supernatural conceptions prevalent in much of Asia and in animism and folk religions generally.

It would seem to follow, therefore, that the moral behavior of individuals would be influenced by their religious commitments *only* in societies where the dominant religious organizations give clear and consistent expression to divine moral imperatives. Thus, for example, were proper survey data available, they should show that those who frequented the temples in Greco-Roman times were *no more* observant of the prevailing moral codes than individuals who were lax in their religious practice. As Tylor pointed out, this is not to suggest that societies in antiquity lacked moral codes but only that these were not predicated on religious foundations. Hence, the moral effectiveness of religions varies according to the moral engagement of their Gods, if any, because there are "godless" religions.

Not all religions propose the existence of a God or Gods—defined as *conscious supernatural beings.* The elite forms of Taoism and Confucianism conceive of the supernatural as a *supernatural essence*—an underlying mystical force or principle governing life but which is *impersonal, remote, lacking consciousness, and definitely not a being.* As explained in the *Lao-tzu,* the Tao is a cosmic essence, the eternal Way of the universe that produces harmony and balance. Although the Tao is said to be wise beyond human understanding and "the mother of the universe," it also is said to be "always nonexistent," yet "always existent," "unnameable" and "the name that can be named." Both "soundless and formless," it is "always without desires."

Supernatural essences may be ideal objects for meditation and mystical contemplation, but they are entirely lacking in moral relevance. An unconscious divine essence is unable to issue commandments or make moral judgments. Thus, conceptions of the supernatural are irrelevant to the moral order unless they are *beings*—things having consciousness and desires. Put another way, only beings can desire moral conformity. But even that is not sufficient. Gods can only lend sanctions to the moral order if they are *concerned* about, informed about, and act on behalf of humans. Moreover, to promote virtue among humans, Gods must be virtuous—they must *favor good* over evil. Finally, Gods will be effective in sustaining moral precepts, the greater their *scope*—that is, the greater the diversity of their powers and the range of their influence. All powerful, all seeing Gods ruling the entire universe are the ultimate deterrent.

Two conclusions follow from this discussion. First, the effects of religiousness on individual morality *are contingent on images of Gods* as conscious, morally concerned beings; religiousness based on impersonal essences or on amoral Gods *will not influence* moral choices. Second, participation in religious rites and rituals will have little or no *independent* effect on morality, for their effects will be contingent on the nature of the image of the supernatural toward which they are directed.

Recently, I conducted an elaborate research study to test these conclusions, based on data for the United States and thirty-three other nations (Stark, 2001b). The results were consistent and overwhelmingly supportive.

In each of twenty-seven nations within Christendom, the greater importance people placed on God, the less likely they were to approve of buying goods they knew to be stolen, to approve of someone failing to report that they had accidentally damaged an auto in a parking lot, or to approve of smoking marijuana. The correlations were as high in Protestant as in Roman Catholic nations and where average levels of church attendance were high or low. Indeed, participation in Sunday services (a measure of ritual activity) was only weakly related to moral attitudes, and these correlations disappeared or became very small when the God "effects" were removed through regression analysis. That is, God matters; ritual doesn't.

The findings are similar for Muslim nations, where the importance placed on Allah is very strongly correlated with morality, but mosque attendance is of no significance. In India, too, concern for the Gods matters, but temple attendance has no detectable effect on morality. But in Japan and China, where the Gods are conceived of as many, small, and not particularly interested in human moral behavior, religion is irrelevant to moral outlooks—concern about the God(s), visits to temples, and prayer and mediation all are without any moral effects.

These results show that, in and of themselves, rites and rituals have little or no impact on the major effect universally attributed to religion—conformity to the moral order. Thus, it seems necessary to amend the "law" linking religion and morality as follows: *Images of Gods as conscious, powerful, morally concerned beings function to sustain the moral order.*

GENDER AND RELIGIOUS COMMITMENT

One of the most intriguing questions about gender differences is, Why are women always more religious than men? Popular sayings from across the continents and the centuries universally attribute greater religiousness to women. Women were far more likely than men to convert to the early Christian movement (Stark, 1996b). American religious movements throughout the nineteenth and twentieth centuries typically overrecruited women (Stark and Bainbridge, 1985). And all contemporary research shows that women are more likely to hold religious beliefs and to engage in religious behavior than are men (Argyle and Beit-Hallahmi, 1975; Batson, Schoenrade, and Ventis, 1993; Miller and Hoffmann, 1995; Stark, 2002). Table 14-2 fully supports this research literature. In each of these nations, women are more likely than men to pray daily and to identify themselves as a religious person.

Many efforts have been made to explain these differences. It was long believed that gender differences in religion derive from the traditional female role; that is, as part of being socialized to care for other family members, women are raised to take responsibility for their religious needs, too—as reflected in the traditional German saying that women's work consists of "church, children, and cooking." Another somewhat similar explanation proposed that because many women do not work outside the home or pursue careers, they simply have more time to allocate to religious activities (Azzi and Ehrenberg, 1975). However, both explanations were rejected by research showing that career women are as religious as housewives, and both are more religious than their husbands or male peers (de Vaus, 1984; Cornwall, 1988). Furthermore, the genders differ not only in terms of time-consuming religious participation but also in terms of belief and actions, such as prayer, that do not impose demands on time.

More recently, differences were successfully traced to general male/female personality differences on the basis of differences *within*

TABLE 14-2	Gender and Religiousness in Selected Nations	
NATION	**WOMEN**	**MEN**
Percent who pray every day		
Morocco	78	57
Turkey	75	58
United States	**66**	**45**
Poland	63	38
Mexico	**58**	**39**
Argentina	52	23
India	52	39
Italy	52	23
Canada	**46**	**26**
Finland	32	15
Spain	32	12
Greece	31	27
Russia	23	7
Percent who say they are "a religious person"		
Jordan	94	75
Italy	89	71
United States	**86**	**76**
Canada	**78**	**66**
Mexico	**78**	**71**
Chile	77	59
Finland	72	50
Russia	71	47
Australia	65	50
France	50	38

Source: Prepared by the author from the World Values Surveys, 2001–2002.

genders. Thus, Edward Thompson (1991) found that the higher their score on a standardized measure of femininity, the more apt men were to be religious, while women who scored high on masculinity were less religious than women who scored lower. This led Alan S. Miller and John P. Hoffman (1995) to suggest that gender-based religious differences are like gender differences long observed for crime and delinquency. Recall that Gottfredson and Hirschi (1990) argued that criminal actions are part of a more general set of behaviors that offer short-term gratifications in exchange for risk and that men are far more likely than women to have weak self-control and thus to engage in these risky actions. That is, burglary, rape, and robbery, like drinking, smoking, speeding, taking drugs, and engaging in unprotected sex, are behaviors that produce immediate gratifications

but that are obviously risky and thus are behaviors avoided by people who are able to defer gratification in favor of greater long-term benefits. To this list of risky behaviors, Miller and Hoffman added irreligiousness, noting that "one can conceive of religious belief as risk-averse behavior and the rejection of religious beliefs as risk-taking behavior."

Miller and Hoffman's logic is in accord with a classic argument in theology known as "Pascal's wager." Blaise Pascal (1623–1662), a French philosopher, wrote that anyone with good sense would believe in God because belief is a no-loss proposition. He noted that God either does or does not exist and people either do or do not believe in God. Considering each of the four combinations involved here, Pascal reasoned that if God exists, then after death those who believe will gain the rewards promised by religion and avoid the punishments of nonbelief. In contrast, nonbelievers will miss out on the rewards and receive the punishments. On the other hand, if God does not exist, those who believe will simply be dead and will not receive either rewards or punishments. But they will be no worse off for having believed, for the same fate will await nonbelievers. Given these alternatives, the smart bet, or wager, is to believe, for a person has everything to gain and nothing to lose by believing.

But Pascal overlooked something—believers give up some gratifications here and now. Religious belief implies willingness to exercise self-control in this life, forgoing some immediate gratifications. Thus, if one is willing to risk that God does not exist, one can enjoy many immediate gratifications prohibited by religion. Since most sins also are illegal, and vice versa, the interests of sociologists of religion and of criminologists converge on the same set of behaviors that overwhelmingly are committed by males. People who are prone to risk the secular costs of immediate gratification also are prone to risk the religious costs of their misbehavior. If this analysis is correct, then the gender differences in religiousness have the same root as the gender differences in other forms of risky behavior. That is, if men are more likely to commit crimes because they "are differentially socialized to be risk takers, then they are also being differentially socialized to be less religious." When they analyzed appropriate data, Miller and Hoffmann (1995) found that within each gender, those scoring high on risk aversion were more religious. Moreover, when they compared men and women with an equal orientation toward risk, their religious behavior and belief did not differ.

Recently, it was realized that religions differ considerably in the level of risk they impose on being irreligious. Just as "godless" religions were found not to influence conformity to the norms, supernatural essences pose no risks to the irreligious. If differences in risk taking are the cause of gender difference in religiousness, then these differences ought to be small to nonexistent in Asia. And that's what the latest research demonstrated (Miller and Stark, 2002).

Of course, we still don't really know why women are more averse to risk taking. But this gender difference seems to hold in all cultures for which adequate data exist.

RELIGIOUS ECONOMIES

Early religions were local affairs. A tribe or very small society and its religion were one. A person was born into a religion as part of being born a member of his or her group. Although religion constantly changed even in small primitive societies, the idea of choosing a religion was as alien as the idea of choosing one's tribe or family.

As societies became more complex, they began to include several cultures and religions. Larger cities of ancient Egypt, Greece, and the Roman Empire contained a variety of different religions (Johnson, 1976; Meeks, 1983). In such cities people could compare religions, worry about which one was best, and regard religion as a matter of choice. Such a religious situation is best described as a *religious economy*. Just as commercial economies consist of a market in which different firms compete, religious economies consist of a market (the aggregate demand for religion) and firms (different religious organizations) seeking to attract and hold a clientele.

The notion of religious economies underscores the dynamic interplay of different religious groups within a society. This interplay accounts for the religious makeup of societies at any given time and explains why and how religions change.

As with commercial economies, a key issue is *the degree to which a religious economy is regulated by the state*. To what extent do free-market conditions prevail, and to what extent is the religious economy distorted toward monopoly by coercion? For reasons to be explained shortly, the natural state of a religious economy is **religious pluralism,** wherein many religious "firms" exist because of their special appeal to certain segments of the market (or population). However, as for a commercial organization, it is always in the interest of any particular religious organization to secure a monopoly. This can be achieved, and even then to just a limited extent, only if the state forcibly excludes competing faiths.

In medieval Europe states used coercion to create a monopoly for Catholicism. Anyone who deviated from orthodoxy was subject to punishment, including execution. However, even at the height of its power, the medieval Catholic Church was beset by dissent and heresy from all sides and never achieved full monopoly. Whenever and wherever state coercion wavered, competing faiths burst forth and prospered (Johnson, 1976). Nevertheless, regulation often made it difficult and dangerous for competing faiths, thus greatly reducing religious choice. Yet even medieval society is best understood in terms of its religious economy.

To understand why religious economies incline toward pluralism, we need to understand the major processes at work within them. We shall explore these processes by first seeing why a virtually endless supply of new and competing organizations exists in any religious economy.

CHURCH-SECT THEORY

In 1929 H. Richard Niebuhr published *The Social Sources of Denominationalism*. In this book he tried to explain why Christianity was fractured into so many competing denominations. Why weren't Christians content with one church? Why did they constantly form new ones?

The answer he proposed combined two concepts developed by Max Weber with elements of conflict theory. Weber had distinguished two kinds of religious organizations: churches and sects. **Churches** intellectualize religious teachings and restrain emotionalism in their services. They offer an image of the Gods as somewhat remote from daily life and the individual. **Sects** stress emotionalism and individual mystical experiences and tend toward fundamentalism, rather than intellectualism, in their teachings. They present their gods as close at hand, taking an active interest and role in the lives of individuals.

Churches and sects also differ greatly in terms of their network structures. Churches tend to be based on cosmopolitan networks, while sects

When members of a congregation actively participate in the services as these members of the Pentecostal Holiness Church in Port Wentworth, Georgia, are doing—when members seem to be really enjoying themselves—chances are the group is one that sociologists would classify as a sect. Congregations with more formal and sedate services usually are classified as churches.

© Polly Brown/Actuality, Inc.

tend to consist of intense local networks. This easily is seen in Table 14-3. Nearly half (47%) of members of Protestant sects say that four or five of their five best friends belong to their church congregation, and fewer than one in ten has no close personal friends in her or his congregation. In contrast, nearly half (43%) of members of Protestant "churches" have no close friends in their congregation and only about one in ten has four or five. This is not a function of proximity since the average sect member travels

much farther to attend services than does the average church member. These network differences greatly influence the experience of participants. Attending a church is more like being a member of an audience at a movie or lecture than being a member of a group. But for members of sects, religious participation offers a strong sense of community and solidarity.

Just as classes differ in terms of their networks, the network structures of churches and sects appeal to persons of different classes. Niebuhr argued that sects provide for the religious needs of people low in the stratification system—the masses. Churches provide for the religious needs of the middle and upper classes (McKinney and Roof, 1982). Class conflict, according to Niebuhr, underlies the religious conflicts that split Christianity into many different denominations.

Niebuhr stressed the unique ability of religion to make life bearable, even for those in misery. This is achieved by turning one's thoughts away from this world and stressing the primacy of the spiritual world. The more we believe that this life is only a brief prelude to the afterlife and that we shall find relief from our pains in the more perfect world to come, the more easily we can bear life's burdens. Indeed, religions commonly teach that if you

TABLE 14-3	Network Structures of Churches and Sects (Protestants Only)	
"Of your five best friends, how many are members of your church congregation?"		
	CHURCHES* (%)	SECTS** (%)
Four or five	12	47
Two or three	27	37
One	18	7
None	43	9
	100	100

Source: Prepared by the author from the *Study of Religion in American Life.*

*United Church of Christ, United Methodist, Episcopalian, Presbyterian, Lutheran (ELC).

**Assemblies of God, Church of God, Church of Christ, Nazarene, Seventh-day Adventist, Gospel Lighthouse, Foursquare Gospel, Pentecostal.

spurn material pleasures in this life, you will increase your rewards in the everlasting life to come—that the social order will be turned upside down in the next life, where "the first shall be last, and the last, first."

To make these views convincing, however, religions must resist the pleasures of the material world. A religious organization filled with members, especially its leaders, enjoying material pleasures is hampered in its efforts to serve the religious needs of the deprived.

The key to Niebuhr's theory is the proposition that *successful religious organizations always shift their emphasis toward this world and away from the next.* He argued that as religious organizations grow and become more popular, the proportion of middle- and upper-class members will increase. These members have much less need than the deprived to reject this world in favor of the next. Indeed, they will want to harmonize their religious beliefs with their own worldly success. In time these members will prevail, and the religious organization will cease to preach that material success in this life will be punished in the next. These faiths will cease to emphasize the spiritual and will portray the supernatural in ever more remote and less vivid terms. That is, the religion will become progressively worldly.

However, such a shift will erode the ability of the religious organization to satisfy the religious needs of the lower classes. This will lead to growing discontent. Eventually, *the masses will defect to form a new religious organization,* a sect, which emphasizes the original otherworldliness of the former organization.

Thus, Niebuhr proposed a dynamic cycle. Religions originate as sects designed to serve the needs of the deprived. If they grow and flourish, then these sects increasingly serve the interests of the middle and upper classes and are transformed into churches, thereby making them less effective in satisfying the needs of the poor. Then the conditions that prompted the original **sect formation** are re-created, a split occurs, and a new sect is formed. In time this sect, too, is transformed into a church, whereupon a new sect is born; thus, there is an endless cycle of the birth, transformation, and rebirth of new religious organizations. Niebuhr explained the existence of the huge array of Christian denominations as the result of countless cycles of this church-sect process.

In the almost seventy-five years since Niebuhr first sketched his **church-sect theory,**

© Paolo Koch/Photo Researchers

it has been much refined and elaborated (Wilson, 1959, 1961, 1970; Wallis, 1975). Niebuhr's definitions of church and sect were not clear or efficient, and in 1963 Benton Johnson proposed better ones. He suggested that church and sect are opposite poles on an axis representing the degree of tension between religious organizations and their sociocultural environment (Figure 14-1). Tension, as Johnson defined it, is a manifestation of deviance. To the degree that a religious organization sustains norms and values different from those of the surrounding culture, it is deviant, and tension will exist between its members and the outside world. In the extreme case, tension is so high that the group is hunted down by outsiders. Religious groups whose norms and values resemble those of the larger society have no tension. *Churches* are religious bodies with relatively low tension; *sects* are religious bodies with relatively high tension.

Niebuhr tended to limit this application of church-sect theory to religious organizations. But the theory is even more useful when it is applied to whole societies. Let's see how church-sect theory has recently been linked with the concept of religious economies.

Members of this Protestant congregation, located in a tiny Inuit village far above the Arctic Circle, belong to a Protestant sect. As is typical of sects, members are expected to devote a lot of time to their religion; most attend three times on Sunday and on Wednesday and Friday evenings, too. What the Glad Tidings Mission lacks in terms of a building is made up for in the excitement and emotional intensity of the services, which stress salvation and personal religious experience. According to church-sect theory, the poor and dispossessed seek the comforts of an intensely otherworldly religion, whereas the more affluent prefer a religion more accommodated to this world. Out of this tension comes a succession of sect movements.

FIGURE 14-1   Degrees of Tension between Religions and Society.

Degree of Tension with Sociocultural Environment

Low   Medium   High

Clergy who "fit right in" illustrate that some religious groups are wholly accommodated to their social environment. Clergy who demonstrate against pornography or abortion reflect that some degree of tension exists between their religious group and the social environment. Clergy who are jailed for refusing to submit to state regulation of church-sponsored grade schools show that religious bodies can exist at a level of high tension with their environment. The higher the tension between a group and its environment, the more sectlike it is.

SECULARIZATION AND REVIVAL

Social scientists and modern intellectuals have generally paid close attention to only one aspect of church-sect theory: They have noted the movement away from traditional Christian teachings by many of the largest and most respected denominations. Projecting these trends into the future, they have predicted that soon religion will disappear. Noting that these denominations continue to retreat from a vivid, active conception of the supernatural, many social scientists have concluded that this is because supernatural beliefs cannot be maintained in an increasingly scientific age, and therefore, human societies will soon, to paraphrase Sigmund Freud (1927), be cured of the infantile "illusion" of religion.

The process by which religion will disappear has been called *secularization* to indicate a turning away from religious and toward secular explanations of life. This process is regarded as far along and irreversible. Indeed, the distinguished anthropologist Anthony F. C. Wallace (1966) spoke for the majority of modern social scientists when he wrote:

The evolutionary future of religion is extinction. Belief in supernatural beings . . . will become only an interesting historical memory. To be sure, this event is not likely to occur in the next generation; the process will very likely take several hundred years. . . . But as a cultural trait, belief in supernatural powers is doomed to die out, all over the world, as a result of the increasing adequacy and diffusion of scientific knowledge.

Other scholars have theorized that science might provide a substitute for religion—people might perform solemn rituals patterned on those of religion but with an explicitly antisupernatural thrust (see the photo on page 399). In fact, the French philosopher Auguste Comte, who coined the term *sociology*, intended that sociology serve as the scientific substitute for religion. Thus, the assumption that religion is both false and doomed has been widespread among social scientists from the start. This assumption has led them to discover terminal symptoms in every sign of decline in religion but to ignore or be perplexed by every sign of vigor. For 300 years social scientists have predicted the triumph of secularization (Stark, 1999). Whenever they have confronted broad-based religious revivals, they have dismissed them as death spasms.

I must confess that as a young sociologist I largely shared these views. But as I did research on religious groups, I found it very difficult to square these claims with what I saw. For millions of people, faith was alive and well. Many sophisticated scholars appeared to have no problem in reconciling science with a belief in the supernatural. Could the secularization thesis be flawed? By

1980 I had concluded that it was—that secularization was but one aspect of religious change (Stark and Bainbridge, 1980, 1985; Stark, 1981). Other sociologists began to express similar views (Bell, 1980; Martin, 1981). Together, we began to apply church-sect theory to religious economies and to argue that the secularization thesis rested on a misperception—that social scientists had mistaken the obvious decline of once powerful religious organizations for a general decline of religion. In this sense they had seen only the transformation of some religious organizations into states of ever lower tension but had failed to note the reactions to this trend elsewhere in religious economies. A more comprehensive view of religious economies suggests that *secularization is a self-limiting process that leads not to irreligion but to a shift in the sources of religion.* One of the ways this occurs was already implicit in church-sect theory.

Church-sect theory suggests that many religious bodies are always in the process of becoming very worldly. That is, secularization should occur in all religious economies. But we must also expect the trend toward secularization to produce religious reactions: the formation of sects. We can call this process *revival.* As secularization weakens some organizations, new ones split off to revive less worldly versions of the faith. This helps to explain not only why some religious bodies have declined as the secularization thesis predicts they should but also why religion refuses to fade away: why, for example, a religious body like the Church of God in Christ can grow at an extraordinary rate in contradiction of the secularization thesis. The Church of God in Christ has moved into the market vacuum created by the secularization of once-dominant Protestant bodies. The result is a change in the source of religion—a shift in what religious group people turn to—not the demise of religion.

But there is a second response to secularization besides revival. Sometimes people do not revive the conventional faith by embodying it in new organizations. Sometimes they turn to new faiths altogether.

INNOVATION: CULT FORMATION

Sects are not new religions; they are new organizations reviving an old religion. They claim to have returned to a more authentic version of the traditional faith from which its parent organization has strayed. Thus, a set of churches and

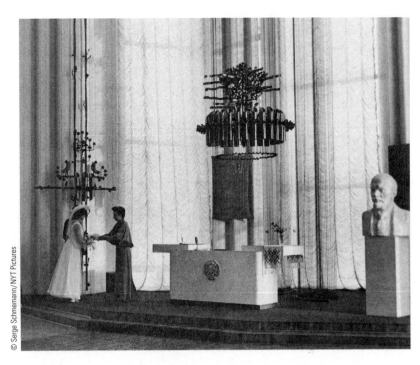

© Serge Schmemann/NYT Pictures

sects will form a single religious tradition. For example, most churches and sects in the United States and Europe are part of the conventional Christian religious tradition.

Sometimes, however, organizations appear that are based on religions outside the conventional religious tradition. This may occur by *importing* a faith from another society; Hinduism in the United States and Christianity in India are examples. New faiths also appear through *cultural innovation.* Someone may have new religious insights and then succeed in attracting followers.

New religions, whether imported or the result of innovation, are deviant and thus elicit unfavorable reactions from others. Like sects, they are in a high state of tension with surrounding society. But unlike sects, new religions cannot claim cultural continuity with conventional religious beliefs and practices.

The hostility usually directed at new religions is reflected in the name applied to them: cults. Sociologists use this term without prejudice to distinguish new religions from sects arising out of old religions. **Cults** are religious movements that represent a new or different religious tradition, whereas churches and sects represent the prevailing tradition in a society. The negative connotations of the word *cult* reflect the unusually high tension between these movements and their social environment.

All religions begin as cult movements. All of today's great world faiths once were regarded as

A Russian couple was married some years ago in a "socialist wedding palace" complete with an "altar" and a bust of Lenin. This reflected a seventy-year government campaign to eradicate religion in the USSR. However, as recent changes have swept through the country, it has become clear that religion remains a powerful force in Russian society. Sociologists of religion argue that religion without the supernatural is no religion at all and therefore that efforts like this can always be expected to fail.

weird, crazy, foolish, and sinful. How Roman intellectuals in the first century would have laughed at the notion that a messiah and his tiny flock in Palestine, an obscure corner of the empire, posed a threat to the mighty pagan temples. But from obscure cult movements have risen not only Christianity but also Islam, Buddhism, and other faiths that today inspire hundreds of millions of faithful adherents.

Given their current rate of growth, the Church of Jesus Christ of Latter-day Saints—the Mormons—may be repeating this pattern of a meteoric rise from obscurity to world significance. In 1830 this faith began with six members: the three Smith brothers, the two Whitmer brothers, and Oliver Cowdry. Today, there are more than 10 million Mormons, and even if they continue to grow at a somewhat slower rate, there will be at least 265 million Mormons worldwide a century from now (Stark, 1984, 1993).

New religions appear constantly in all societies (Stark and Bainbridge, 1985). Nearly all of them fail. To succeed, many things are required, but the primary necessity involves *opportunity*. That is, for new firms to make their way against large, long-established firms in a religious economy, the older firms must be failing to serve the needs of a significant number of people. People do not abandon a faith that satisfies them to embrace a new faith: *New faiths prosper only from the weaknesses of old faiths.*

Sometimes new faiths find opportunity because of great social crises that overwhelm conventional faiths. For example, plagues or natural disasters may cause a sudden loss of confidence in conventional faiths (Wallace, 1956; Stark, 1992a). Wars can have the same effect, especially on the losing side: New faiths repeatedly swept through the native tribes of North America as their efforts to resist white encroachments failed (Mooney, 1896), and many new religions have flourished in Japan since World War II (McFarland, 1967; Morioka, 1975).

But a major opportunity for new faiths results from *the excessive secularization of the old.* That is, after many cycles of the church-sect process, an entire religious tradition may lose its ability to provide a plausible faith for a substantial portion of the population. Such moments are rare, but when they occur, new faiths quickly rise in influence, just as Christianity overwhelmed a highly secularized and complacent paganism. Thus, secularization prompts two reactions that restore religion: revival and innovation (**cult formation**). Rather than being a symptom of the death of religion, *secularization provides the impetus for religious change.*

CHARISMA

We saw in Chapter 3 that people join new religious movements primarily because of their attachments to members of those movements; when a sufficient proportion of a person's attachments are to members of some religious group, the person is likely to accept that religion. That tells us something not only about how religious movements recruit and grow but also about *how they begin.*

As long as only one person accepts a new religious message, no religious movement exists. To launch a social movement, this person must convince other people to share his or her beliefs and join. Many people each year believe they have discovered a new faith, but only a few of them can convince others to join. What characterizes those who can attract followers?

Pondering the special gifts of religious founders, Max Weber credited them with charisma. **Charisma** is a Greek word meaning "divine gift." Weber used it to indicate the ability of some people to inspire faith in others, to get others to believe their message. For contemporary sociologists of religion, *the basis of this gift is an unusual ability to form attachments with others.* Just as people join religions out of attachments to members, in the beginning people accept the claim of a founder of a new faith because they develop very strong attachments to the founder. All studies of new religions report that the founders possessed remarkable gifts for interpersonal relations (Wallis, 1982).

Moreover, as we saw in Chapter 3, founders of new faiths typically turn first to those *with whom they already have strong attachments in their quest for converts.*

Having outlined the dynamic character of religious economies and the processes by which sects and cults form, we can now analyze religion in the United States. Then we shall see how the same dynamic patterns occur in a variety of other nations.

THE AMERICAN RELIGIOUS ECONOMY

Europeans have long marveled at both the diversity of religions and the high levels of participation in religious activities in the United States. Max Weber, for example, noted that Americans gladly contributed sums of money to their churches that would shock people in Europe. Others wondered how so many faiths could exist side by side.

It is true that the American religious economy is very diverse, as more than 1,500 separate denominations exist in the United States (Melton, 2002). Nor are they all small; twenty-two American denominations enroll 1 million members or more each (Table 14-4). Church attendance is high: In any given week, about 40 percent of Americans attend services. Moreover, almost two-thirds (about 63%) are official members of a local congregation or parish.

However, the American religious economy is unusual only in that it is an exceptionally free market with little regulation. Within this economy the three major processes of secularization, revival, and innovation are well developed and related.

SECULARIZATION AND REVIVAL

Many major religious bodies in the United States have become highly secularized in the sense that they no longer present traditional versions of their faith or emphasize the supernatural. Table 14-5, based on the General Social Surveys for 1996 through 1998, shows one indicator of this shift. I was able to combine these national samples because the very same question was asked about the Bible in each year. Combining the samples provided enough cases to characterize the beliefs of members in the larger denominations.

Looking at the table, we see that only small minorities of members of the Unitarian-Universalist Church, the United Church of Christ, the newly merged Evangelical Lutheran Church in America, the Episcopal Church, the United Presbyterian Church, and the United Methodist Church expressed their belief that "the Bible is the actual word of God and is to be taken literally, word for word." A century ago most members of these denominations would have affirmed this statement. But to find majorities holding this traditional tenet of Christian doctrine today, we must look to the lower half of the table, to groups that are best described as sects.

If secularization weakens the holding power of religious organizations, then those denominations at the top of Table 14-5 ought to be showing signs of decline. On the other hand, if

TABLE 14-4	American Denominations with 1 Million Members or More	
		MEMBERS IN 2004
Roman Catholic		67,259,768
Southern Baptist Convention		16,439,603
United Methodist Church		8,251,175
Church of Jesus Christ of Latter-day Saints (Mormons)		5,503,192
Churches of God in Christ		5,499,875
National Baptist Convention, USA, Inc.		5,000,000
Evangelical Lutheran Church in America		4,984,925
National Baptist Convention of America, Inc.		3,500,000
Presbyterian Church (USA)		3,241,309
Assemblies of God		2,729,562
National Missionary Baptist Convention of America		2,500,000
Progressive National Baptist Convention		2,500,000
Lutheran Church–Missouri Synod		2,488,936
Episcopal Church		2,320,221
Jehovah's Witnesses		2,303,015*
Churches of Christ		1,500,000
Pentecostal Assemblies of the World		1,500,000
Greek Orthodox Archdiocese of America		1,500,000
United Church of Christ		1,296,652
Baptist Bible Fellowship International		1,200,000
Christian Churches and Churches of Christ		1,071,616
The Orthodox Church in America		1,000,000

Source: *Yearbook of American and Canadian Churches, 2005.*

*Memorial Attendance (see Stark and Iannaccone, 1997).

TABLE 14-5	American Denominations and Literal Faith in the Bible	
Percent of members who agree that "the Bible is the actual word of God and is to be taken literally, word for word"		
Unitarian-Universalist	7	
United Church of Christ	12	
Evangelical Lutheran Church in America	22	
Episcopal Church	22	
United Presbyterian Church	25	
United Methodist Church	31	
Jehovah's Witnesses	51	
Church of Christ	56	
Southern Baptist Convention	57	
Church of the Nazarene	58	
Assemblies of God	68	
United Pentecostal Church	69	
Church of God	81	
All Protestants	**43**	
Roman Catholics	**20**	

Sources: Prepared by the author from National Opinion Research Center, General Social Surveys, 1996–1998.

TABLE 14·6	Some Growing and Some Declining American Denominations*		
	MEMBERS PER 1,000 U.S. POPULATION		
DENOMINATION	1960	2000	PERCENT CHANGE
Christian Church (Disciples)	10.0	2.7	−71
United Church of Christ	12.4	5.0	−60
Episcopal Church	18.1	8.2	−55
United Methodist Church	58.9	29.8	−49
Presbyterian Church (USA)	23.0	12.7	−45
Evangelical Lutheran Church in America	29.3	18.2	−39
Unitarian-Universalist	1.0	0.8	−20
Roman Catholic	233.0	221.7	−5
Southern Baptist Convention	53.8	56.3	+5
Church of the Nazarene	1.7	2.2	+35
Seventh-day Adventist	1.8	3.1	+72
Foursquare Gospel	0.5	0.9	+80
Church of Jesus Christ of Latter-day Saints (Mormons)	8.2	18.2	+122
Jehovah's Witnesses**	1.4	3.5	+150
Assemblies of God	2.8	9.1	+225
Church of God (Cleveland, Tenn.)	0.9	3.1	+244
Church of God in Christ	2.2	19.5	+786

Sources: *Yearbook of American Churches, 1962,* and *Yearbook of American and Canadian Churches, 2001.*

*American members only.

**"Publishers" only.

secularization is inevitable as science triumphs over faith, then the denominations holding to Bible literalism ought to be the ones in decline.

Table 14-6 offers compelling evidence for the notion that as religious bodies deemphasize the supernatural, they seem less able to satisfy religious needs. The United Church of Christ, Presbyterians, Episcopalians, Methodists, and Lutherans all have experienced substantial losses in their shares of church membership. But in the middle of the table, the sign turns from negative to positive, reflecting growth rather than decline. Beginning with the Southern Baptists, denominations are significantly on the rise. Indeed, the higher the proportion of members who take the Bible as the literal word of God, the higher the group's rate of growth. What we see happening here is the interplay between secularization and revival. As some denominations are eroded by secularization, new sects erupt and seize the opportunity to attract members to a less secularized faith, thus reviving and revitalizing the religious tradition.

Sect formation is very common in the United States. At least 417 American-born sect movements currently exist, and probably hundreds of others have existed in times past (Stark and Bainbridge, 1981). Most sects are very small: 28 percent have fewer than 500 members today. However, some are very large: The Southern Baptists, with more than 16 million members, are the largest Protestant body in the nation.

Sect formation revives the conventional religious *tradition* and reflects efforts by church members to remain church members: Those committed to the religious tradition but who find their church has become too worldly form sects to restore the otherworldliness of that tradition.

This analysis suggests that membership in conventional religious groups will be highest where sects are most active. This is precisely what contemporary data show. Sect movements are clustered in those states where membership in Christian churches is highest. Sects are very underrepresented in parts of the country where overall church membership and attendance are low (Stark and Bainbridge, 1980). That is, sects move not into market openings where there is general religious inactivity but only into those where people are active but dissatisfied. I shall expand on this point as we examine the contrasting patterns of cult success.

SECULARIZATION AND INNOVATION

If sects represent efforts by the churched to *stay* churched, then cult movements represent efforts by the unchurched to *become* churched. That is, cult movements arise where both sects and churches fail to satisfy the religious market. To see this more fully, let's examine the geography of religion in contemporary America.

THE "UNCHURCHED" BELT The American South is frequently called the "Bible Belt," where religion, especially evangelical Protestantism, is unusually strong. In many of our studies, however, William S. Bainbridge and I have failed to find evidence of such a belt (Stark and Bainbridge, 1985, 1997). The South does not differ from most other regions of the nation in terms of church membership rates, church attendance rates, belief in God, or belief in life after death.

Although the South may not be a Bible Belt, the Far West is certainly an unchurched belt. As

FIGURE 14-2   The Social Geography of Church Membership.

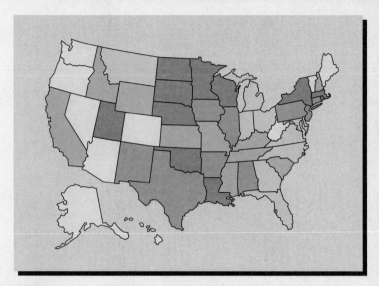

Utah	74.7
North Dakota	73.2
South Dakota	67.8
Massachusetts	64.1
Rhode Island	63.5
Minnesota	61.7
Oklahoma	60.8
Wisconsin	60.4
New York	60.4
Louisiana	58.8
Nebraska	58.8

Iowa	58.5
New Mexico	58.2
Pennsylvania	57.9
Connecticut	57.9
New Jersey	57.7
Arkansas	57.1
Texas	55.5
Illinois	55.3
Alabama	54.8

Mississippi	54.6
Kentucky	53.4
Missouri	51.7
Tennessee	51.1
Kansas	49.4
Idaho	48.5
New Hampshire	47.7
South Carolina	47.6
Wyoming	46.7
California	46.1

North Carolina	45.4
Ohio	44.9
Georgia	44.8
Montana	44.7
Maryland	43.3
Indiana	42.9
Michigan	41.8
Virginia	41.6
Florida	41.1
Delaware	40.6

Arizona	39.9
Colorado	39.5
Vermont	39.1
Maine	36.4
Hawaii	36.2
West Virginia	35.9
Alaska	34.3
Nevada	34.3
Washington	33.0
Oregon	31.3

Church membership is different from religious preference. If asked their religion, more than 90 percent of Americans state a preference for some particular religious body. However, many who say they are Baptists, Catholics, or Jews, for example, do not actually maintain a current membership in any local religious group—about 63 percent of Americans maintain local memberships. This map shows the percentages in each state who actually belong to a local congregation. The variation is huge—from a low of 31.3 in Oregon up to 74.7 in Utah. There also is a very distinct regional pattern to church membership rates. The western states are very low in comparison with most of the rest of the nation—seven of the ten lowest states are in the West. Moreover, through the early 1980s, California was in the lowest ten, too. But its rate was raised by the large number of immigrants; foreign immigrants, including those from Asia, are overwhelmingly active in a church. It also should be noted that contrary to popular beliefs about a "Bible Belt," the South is not outstanding as to church membership—Louisiana is the only southern state in the top ten, and having a large Catholic population, it is not typical of the South. On the other hand, Massachusetts, Rhode Island, and New York are in the top ten.

is clear in Figure 14-2, church membership is far lower in the Pacific region than in any other region.[1] The Far West also has low rates of church attendance. Clearly, the conventional churches have failed to take root along the shores of the Pacific. However, it is church membership, and not religious belief, that is missing in the Far West. Westerners are nearly as likely to have faith in God and believe in life after death as people elsewhere in the country. Thus, the average westerner believes in the supernatural but lacks a church affiliation that gives form and expression to his or her beliefs. This

should provide an ideal market opportunity for cult movements able to form attachments with unchurched West Coast "believers."

THE GEOGRAPHY OF CULT MOVEMENTS

It should be no surprise that religious innovation is far more common and successful in the Far West than elsewhere in the nation. Figure 14-3 shows the distribution of the headquarters of cult movements active in the United States today. The states of the Far West tower above the rest. Nor is the West's affinity for novel religion and mysticism a recent development. Data

1. The rates in Figure 14-2 underestimate church membership. They have been corrected so that they accurately reflect regional differences in church membership, but the figures are somewhat depressed because of omissions in reporting. If these rates were summed, they would produce a national church membership rate of 58 percent, whereas the correct rate for the United States is about 62 percent.

FIGURE 14-3   Cult Movement Headquarters per Million Population.

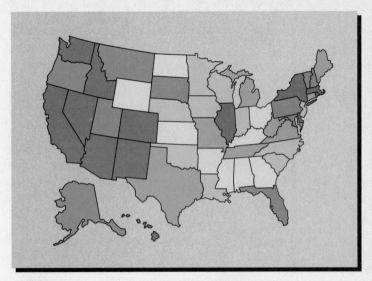

New Mexico	7.3
California	5.8
Nevada	4.2
New York	3.4
Colorado	3.3
Arizona	3.3
Massachusetts	2.7
Hawaii	2.7
Washington	2.5
Idaho	2.0
Illinois	2.0
Rhode Island	2.0

Minnesota	0.7
Iowa	0.7
Louisiana	0.7
Wisconsin	0.6
Nebraska	0.6
South Carolina	0.6
Oklahoma	0.6
New Jersey	0.5
Indiana	0.5
North Carolina	0.5

New Hampshire	1.8
Oregon	1.8
Vermont	1.8
Alaska	1.8
Utah	1.7
Delaware	1.5
South Dakota	1.4
Montana	1.3
Maryland	1.3
Florida	1.3
Pennsylvania	1.3

Kansas	0.4
Arkansas	0.4
Ohio	0.3
Georgia	0.3
Alabama	0.2
North Dakota	0.0
Kentucky	0.0
Mississippi	0.0
Wyoming	0.0

Connecticut	1.2
West Virginia	1.1
Virginia	1.1
Missouri	1.0
Michigan	0.9
Texas	0.8
Tennessee	0.8
Maine	0.8

These rates are based on the location of the headquarters of each of hundreds of cult movements listed in J. Gordon Melton's (2002) *Encyclopedia of American Religions*. This measure understates the strength of new religious movements in America because several of these movements are very large and have centers in many states, but only the state where they are headquartered gets credit. No state has no cult centers, but four do lack a headquarters.

collected in a special 1926 U.S. Census tally of religious groups showed that even then church membership was comparatively low in the Far West and cult membership very high. This was true even as long ago as 1890 (Stark and Bainbridge, 1997).

A major cause of low church membership rates on the West Coast is constant and rapid population movement (Welch, 1983). When people move frequently, as has always been common in the West, they abandon attachments to all social organizations—not just churches but also fraternal clubs, hobby groups, veterans organizations, PTAs, political groups, and the like. Arriving in a new place, they also find it difficult to reestablish such connections in communities where many others are also newcomers and transients. Instead, they are likely to form attachments to people who are not attached to such organizations. Recall from Chapter 3 that after moving to San Francisco, the Unificationists began to grow again only when they discovered how to locate and build attachments with other newcomers to the city. In places where unattached newcomers abound, new movements have a much greater opportunity to grow than in places where the population is settled and attached. And unlike most of the conventional churches in the Far West, cults actively search out new members.

In a sense the disorganization of the Far West caused by population instability is somewhat like that created in times of crisis, when religious changes are likely to occur. Wars and natural disasters result in dramatic social disorganization. Constant population movement provides a less obvious but persistent form of the same thing.

However, the fact that cults attract members in all parts of the nation shows that secularization itself, even when not assisted by disorganization, produces a market for new faiths. Indeed, data on who joins cult movements demonstrate this factor.

WHO JOINS CULTS? The belief that secularization is leading to the demise of religion assumes that people who discard conventional faiths have embraced rationalism and no longer find supernatural beliefs plausible. Thus, sociologists have interpreted increases in the proportion of people who say "none" when asked their religious affiliation as very significant evidence of the trend to irreligion.

I also long assumed that people who claimed no religious affiliation were primarily nonbelievers. I was extremely surprised, therefore, when one of my studies showed that far from being secular humanists or rationalists, people who say they have no religion are those most likely to express faith in unconventional supernatural beliefs (Bainbridge and Stark, 1980, 1981). These people were many times more likely to accept astrology, reincarnation, and various psychic phenomena and to value Eastern mysticism.

Subsequently, my colleagues and I obtained the results of surveys of members of various contemporary cult movements: Unificationists, Hare Krishnas, Scientologists, witches, and several groups studying yoga. In each case we found an extraordinary overrepresentation of persons who had grown up with parents claiming no religious affiliation. Most of the other members had parents who were not active members of any faith, although they had a nominal affiliation (Stark and Bainbridge, 1985).

Thus, *to the extent that large numbers of people grow up in irreligious homes* (that is, in times and places where large numbers have drifted away from the conventional faiths), *large numbers of potential converts to cult movements will exist.*

THE CANADIAN RELIGIOUS ECONOMY

A comparison of the Canadian and American religious economies reveals striking similarities and differences. In both nations church membership is relatively high: about 63 percent in the United States and about 61 percent in Canada. But in Canada this membership is concentrated in only a few denominations. In the

TABLE 14-7	Leading Canadian Denominations
DENOMINATION	CANADIAN POPULATION (%)
Roman Catholic	43.6
United Church of Canada*	9.6
Anglican Church of Canada	6.8
Muslim	2.0
Jewish	1.1
More than 200 other denominations	19.9
Total	83.0

Source: Calculated by the author from the 2001 Canadian Census.

*Created in 1925 by a merger of the Presbyterians, Methodists, and Congregationalists, plus some independent congregations.

United States, the largest twenty-three denominations enroll about half the population; in Canada two denominations account for about half (Table 14-7). Clearly, the Canadian religious economy is far less diverse than that of the United States.

Harry W. Hiller (1978) has suggested that one reason for less diversity is that the close ties between church and state in Canada produced policies and institutional arrangements designed to "discourage religious experimentation." But another reason for less apparent diversity in Canadian religion is the merger of Canadian Presbyterians, Methodists, and Congregationalists into the United Church of Canada back in 1925; these remain separate denominations in the United States. In part it is also because the large African American population of the United States tends to belong to separate denominations. And partly it is because the Roman Catholic Church enrolls nearly twice the proportion of Canadians as Americans. Moreover, Canada's Catholics are more highly concentrated. As Kenneth Westhues (1976) has pointed out, only 9 percent of American Catholics live in a neighborhood where they make up a majority of the population, while almost 60 percent of Catholic Canadians live in an overwhelmingly Catholic neighborhood.

Faced with such limited choices, Canadians have grown increasingly restless about their religious affiliations—many have dropped out of the United Church of Canada and the Anglican Church of Canada, while expressing high levels of religious interest and commitment (Bibby, 2002). There has been some growth by the more conservative denominations, but nothing like the growth these groups have sustained in the United States.

TABLE 14-8	Cult Members in Canada		
	NUMBER OF MEMBERS		
	TOTAL	MALES	FEMALES
Paganism	21,080	7,825	13,255
New Thought	3,955	1,620	2,375
Spiritualist	3,295	1,060	2,235
New Age	1,530	545	985
Swedenbourgian	1,015	430	585
Satanist	855	715	140
Atheism	18,600	12,500	6,100

Source: Prepared by the author from the 2001 Canadian Census.

Unlike the American Census, the Canadian Census asks religious preference. Canadian membership figures for various cult movements are shown in Table 14-8. Several significant things can be read in the table. First, except for Satanism, these groups overrecruit women, which is entirely consistent with the earlier discussion of religion and gender differences. In fact, that men far outnumber women among Satanists seems in keeping with research relating lack of religiousness among males to the general tendency toward risky behavior, as does the high ratio of males to females among self-identified atheists.

A second important finding has to do with the rather small numbers involved in these groups, given that they receive a great deal of media attention. The membership figures attributed to new religious groups in the media usually are wildly exaggerated, and most new religious movements are, and remain, very small. For example, a Toronto magazine estimated that there were about 10,000 Hare Krishna members in that city when in fact there were but 80 full-time members (Hexham, Currie, and Townsend, 1985). In similar fashion, based on the amount of media coverage they are given, who would have guessed that only 1,530 Canadians would give their religious preference as New Age?

Despite being few in number, Canadian cultists are located where secularization theory would predict—in those places where the conventional faiths are weakest. In Canada, just as in the United States, "where" is the West. Overall, 6.2 percent of Canadians reported on their 2001 Census forms that they had no religious affiliation. But in western cities, such as Vancouver and Victoria, about a third said they had none, in comparison with many eastern cities having fewer than 3 percent unchurched. And the western cities have many times the cult membership rates of cities in eastern Canada (Stark and Bainbridge, 1997).

CULT MOVEMENTS IN EUROPE

If cults flourish where conventional churches are relatively weak, then parts of Europe ought to be extremely fertile ground for new religious movements. For example, in Iceland and Denmark, weekly church attendance is less than 4 percent, and in Sweden only 6 percent are in church on an average Sunday—rates far below those found even in the Canadian and American West. If secularization leads to innovation, then these nations ought to be awash with new religions. Yet for a long time, European sociologists of religion claimed that North Americans flocked to such groups but that Europeans found them of no interest. As it turned out, however, it was only European sociologists, not Europeans, who took no interest in cult movements. In fact, such groups are far more numerous in many European nations than they ever have been in North America.

To test this hypothesis, I assembled data from various published directories (Stark, 1993). The results are shown in Table 14-9. Overall, these European nations have *twice* as high a rate of cult movements as does the United States! Moreover, many European nations have rates many times that of the United States—Switzerland has a rate about ten times higher. Of even greater significance is that while the American rates are based on very accurate counts (Melton, 1993), most of the European rates reflect substantial undercounting—the sources available to me were not very complete. Consider Italy. The directory I used included 66 Italian cults for a rate of 1.2 per million. A subsequent census of cults conducted by the Center for Studies of New Religions in Turin, Italy, placed the actual number at 353, giving Italy a rate of 6.0 per million, more than three times higher than the United States (Introvigne et al., 2001).

So much, then, for claims by European scholars that, compared with the United States, Europe has few cult movements. Rather, as the theory would predict, much of Europe is awash in nonstandard religious movements.

THE PROTESTANT EXPLOSION IN LATIN AMERICA

Several years ago David Martin, a sociologist at the London School of Economics, told one of his English colleagues that he was planning to write a book about Protestantism in Latin America. The response he received was, "A very small book, surely" (Martin, 1989). When Martin's *Tongues of Fire: The Explosion of Protestantism in Latin America* appeared in 1990, there was nothing small about it. It not only ran to 352 pages but attracted a great deal of attention in the media as well. For the fact is that the steady and rapid growth of pluralism in Latin America, and the successful entry of highly competitive sects, had gone unnoticed in both scholarly and media circles. Indeed, most of the scholarly world assumed that such changes were impossible. Some agreed that Catholic Liberation Theology had a bright future in Latin nations, but a successful outbreak of evangelical Protestantism was dismissed as absurd—hence the haughty reactions Martin experienced when he began his study.

Ironically, while Martin deserves great credit, his was only the second book published on this subject in 1990 (Stoll, 1990), and neither book was all that timely. A very good study could have been done thirty years earlier, for evangelical Protestants had already achieved "liftoff" by 1960. For example, they had converted about 12 percent of the population of Chile by 1960. Given the very low rates of participation by nominal Catholics in Latin America (Barrett, 1982), rather small evangelical "minorities" will make up a very significant proportion of those active in religion. Hence, the growth during the 1950s was far more significant than it might have appeared to outsiders. Moreover, much was being written about the rapid gains of Protestantism in Latin America as early as the 1950s. However, because it appeared only in sectarian publications, the millions of Americans who regularly contributed funds to support missions to Latin America were the only ones who knew all about the "explosion of Protestantism" south of the border. It took the scholarly world thirty or forty years longer to catch on.

In any event Protestant groups—most of the Pentecostal variety—are sweeping over most of the continent. In Chile an estimated 22 percent are now *active* in evangelical Protestant

congregations, as are at least 20 percent in Guatemala and 16 percent in Brazil (Martin, 1990; Stoll, 1990; Chesnut, 2003). Moreover, if the present rates of conversion hold for the next twenty years, Protestants will be the majority in many Latin American nations.

Why is this happening? Primarily because over the past few decades government coercion against non-Catholics declined and a pluralistic religious economy became a possibility. Into this vacuum came a flood of Protestant missionaries representing a number of vigorous North American sects. Lacking a protected and subsidized status, these groups had to aggressively seek members to survive. In the beginning the missionaries provided the leadership for these groups. But in recent decades local leaders have taken charge. Masses of people in Latin America are finding the attractions of higher-tension sects as appealing as do people elsewhere and have taken the opportunity to join. And the process by which they join is the network

TABLE 14-9	Cult Movements in Europe and America	
NATION	**CULT MOVEMENTS PER MILLION POPULATION**	**NUMBER OF CULT MOVEMENTS**
Switzerland	16.7	108
Iceland	12.0	3
United Kingdom	10.7	604
Austria	7.9	60
Sweden	6.8	57
Denmark	4.5	23
Netherlands	4.4	64
Ireland	3.9	14
West Germany	2.5	155
Belgium	2.4	24
Norway	1.9	8
Greece	1.5	15
Italy	1.2	66
Portugal	1.0	10
France	0.9	52
Finland	0.8	4
Spain	0.7	29
Poland	0.5	17
Europe*	3.4	1,317
United States	1.7	425

Source: Stark, 1993.

*Total based only on the nations listed in the table.

pattern discussed in Chapter 3; that is, one person joins and soon brings in his or her family members, friends, or co-workers.

Many sociologists of religion now recognize that the greater the number of competing firms in a religious economy, the greater the proportion of the population who will be active in religious groups. That is, many specialized firms can, together, satisfy a far greater range of religious needs and tastes than can one or very few religious groups. Indeed, to the extent that one religious group dominates a religious economy, it will tend to be lazy and satisfied with low levels of participation. Writing in 1776 about established religions in general and the Church of England in particular, Adam Smith noted their lack of "exertion" and "zeal":

The clergy, reposing themselves upon their benefices, had neglected to keep up the fervor of faith and devotion of the great body of the people; and having given themselves up to indolence, were incapable of making vigorous exertion in defence even of their own establishment.

For centuries it was claimed that everyone or almost everyone in Latin America was Catholic. But if one looked very closely, one could notice widespread religious apathy—few attended mass on Sunday. Government repression prevented the church-sect process from functioning to satisfy those dissatisfied with the monopoly faith, and this tended to hide the actual weakness of the Catholic Church in Latin America. Ironically, the sweeping Protestant successes will probably do much to reinvigorate Catholicism. In several new studies, I have found that Catholics are far more active where they are in the minority than in so-called Catholic nations—the great vigor of American Catholicism being a case in point (Stark, 1992b, 1997; Stark and McCann, 1993). In having to respond to Protestant challenges, the Catholic Church in Latin America may well find that it is far stronger (in terms of member commitment) when it is far smaller (in terms of claimed membership).

EASTERN REVIVALS

The collapse of the Soviet Union had many remarkable consequences, not the least of which was to reveal the abject failure of several generations of dedicated efforts to indoctrinate atheism in Eastern Europe and the former Soviet Union. Almost immediately after they seized power in Russia in 1917, the communists embarked on a massive campaign to stamp out all traces of all religions. "Scientific atheism" was made a required part of the educational curriculum, beginning on the first day of school. Nearly all churches, synagogues, and mosques were closed, many were destroyed, and many others were converted to other uses, such as museums of scientific atheism. A few places of worship (most of them Russian Orthodox churches) were placed under very strict state control and allowed to remain open, but all others were closed. It is estimated that only about 5,000 religious congregations of all faiths (fewer than the current number in the state of Kentucky) still existed in the Soviet Union by the 1960s, an extraordinary decline. For example, in 1917 there were more than 20,000 Muslim mosques in the Soviet Union. By 1965, only 351 remained open (Ro'i, 1996).

In addition there was a great deal of official as well as informal discrimination against persons suspected of being religious. As Andrew Greeley (1994) explained:

If one wanted to get ahead . . . one either professed atheism and stayed away from churches or kept one's religious propensities a secret. . . . Never before in human history has there been such a concerted effort to stamp out not merely a religion but all trace of religion. . . . Atheistic Communism thought of itself as pushing forward the inevitable process of secularization in which religion would disappear from the face of the earth.

The communist regimes imposed by the Russian army on most of Eastern Europe at the end of World War II also instituted efforts to eliminate religion. Although enforcement efforts generally were less vigorous and brutal than those initiated in Russia following the revolution, they were as fully intended to wipe out religion. For example, in Hungary one of the first acts by the government in 1945 was to nationalize all church lands and confiscate all religious school buildings. This stripped the churches of their historical financial basis. To prevent the replacement of these funds by member contributions, it was ruled illegal for the churches to accept such support. Shortly thereafter, the government ordered the dissolution of fifty-three Catholic religious orders, excluding only several orders of teaching nuns, who were limited to recruiting two novices per year. The authorities also frequently arrested priests and members of various religious groups on the grounds that they corrupted the morals of young people by talking to them about religion. It was believed that such measures would only need to be temporary. Thus, when sentencing seven Jehovah's Witnesses, the Hungarian Court explained: "We shall lock you up for ten years, and when those ten years are up, our People's Republic will be stronger than it is now, and the people will be ideologically trained and immune to your trying to influence them with the Bible. Then we shall be able to release you" (*Yearbook*, 1996).

With the breakup of the Soviet Union and the collapse of communist regimes in Eastern Europe, religious repression subsided. So, after many decades of vigorous efforts to stamp out religion, what were the results? Is the average Russian and Eastern European the "scientific atheist" the schools were expected to create? Hardly, as a glance back at Table 14-1 will show. Atheists are few in Eastern Europe and Russia, and large numbers identify themselves as religious. This is not what surveys would have found when the communists were still in power, for then many people were (rightfully) too suspicious of survey interviewers to tell them the truth. Consequently, Sergei Borisovich Filiatov and Dmitri Efimovich Furman (1993), who have access to old government-conducted surveys of the Soviet Union, describe the results of recent polls as showing "an abrupt growth in religiosity and the disappearance, equally abrupt, of atheism." For example, a survey conducted in the major cities of Russia in 1990, after the Russian religious revival was well under way, found that 24 percent still said they were atheists. The same researchers found that only 8 percent gave this response two years later (Vorontsova and Filiatov, 1994). It seems entirely plausible that a similar rapid drop in the percentage of atheists would take place in China if the government were to permit religious freedom.

However, Soviet educational efforts to root out religion were not without some interesting side effects. Recall that whenever conventional supernatural beliefs are weakened or lack organized expression, unconventional beliefs will prosper, as we have already noted for the western parts of the United States and Canada. Applied to Russia, this principle would lead us to expect nontraditional beliefs to be very popular as a result of the repression of conventional

TABLE 14-10	Nontraditional Beliefs in Fifteen Major Russian Cities	
		YES (%)
I believe in the "evil eye"		67
I believe in astrology		56
I believe in UFOs		46
I believe in the Abominable Snowman		37

Source: Adapted from Vorontsova and Filiatov, 1994.

churches. Table 14-10 powerfully supports this expectation. Keep in mind that this survey was *not* based on those who attended the most backward schools in rural areas but on residents of the fifteen largest cities in Russia, who attended the best schools available (Vorontsova and Filiatov, 1994). Of these best-educated Russians, two-thirds believe in the "evil eye." More than half believe in astrology. Nearly half believe in UFOs, and more than a third believe in the Abominable Snowman. So much for a generation of scientific atheists!

In any event massive religious revivals currently are under way in the nations of the former Soviet Union and in the former Soviet-controlled nations of Eastern Europe as well. Some sense of the size of this revival comes from Dagestan, one of the republics in the new Russian Federation. In 1917 there were 1,702 mosques in Dagestan. In 1988, as the Soviet era drew to a close, there were but 27. By 1992 there were 800 mosques in Dagestan; in 1994 nearly 5,000 (Bobrovnikov, 1996).

THE UNIVERSAL APPEAL OF FAITH

At the start of this chapter, we examined the unique capability of religion to satisfy basic human needs. As long as people want to know what existence means, as long as they are prone to disappointment, suffering, and death, the religious impulse will not be stilled. Only religions, only systems of thought that include belief in the supernatural, can address problems of this magnitude.

From this line of analysis, we can see that Niebuhr left a vital element out of his church-sect theory. In stressing the needs of the deprived and the lower classes for an otherworldly faith, he failed to note that in the face of some of life's greatest questions, all human beings are deprived. No one, neither the rich or the poor, can achieve immortality in the natural world. And both rich and poor seek to find meaning in existence. The rich as well as the poor join religions. Granted, the rich tend to prefer more worldly churches, but there comes a point at which a religion can become too worldly, too emptied of supernaturalism, to serve either rich or poor. Thus, rising, vigorous, otherworldly religions attract the rich as well as the poor. Although well-educated and successful people tend not to join sects that are in a very high state of tension with the environment, they are often overrepresented among cult converts. In fact, the average cult convert these days is not a social outcast lacking education and good job prospects. Rather, the average convert is unusually well educated, with excellent career potential (Stark and Bainbridge, 1985).

CONCLUSION

For three centuries social scientists have confidently predicted the end of religion. Each new generation of social scientists has expected that their children, or surely their grandchildren, would live in an irreligious society.

Yet religion has not gone away. Granted, many of the great religious organizations of today may be fated to slide into oblivion. But to notice only their decline and to ignore the vigor of new religious organizations and of new religions in general is to look only at sunsets and never at the dawn. In the long course of human experience, many religions have come and gone, but religion has remained.

Oddly enough, while social scientists have awaited the end of religion, they have been content to teach that religion has been a universal social institution, found in all societies. They attribute this universality to the ability of religion to serve worldwide human needs. Thus, to expect religion to vanish meant that such needs would vanish or at least that a new institution such as science would replace religion.

However, this implication ignores the unique aspect of religion and the fundamental differences between religion and science. As discussed early in this chapter, some things that humans seem to desire can come only from the Gods. As long as such desires exist, religion will exist to satisfy them. Moreover, the supernatural claims of religion are, in their purest form, immune to scientific disproof. Scientists can send cameras and detection equipment through space to inspect the planets for signs of life, but they cannot send probes to test for life after death.

Review Glossary

Terms are listed in the order in which they appear in the chapter.

religious economy The set of competing faiths, and their adherents, within a given society or geographical area of a society.

secularization The process by which particular religious organizations become more worldly and offer a less vivid and less active conception of the supernatural.

revival Movements within religious organizations, or the breaking away of new organizations, to reaffirm less secularized versions of a faith (*see* sect formation).

religious innovation The appearance of new religions in a society either by founding a new faith (*see* cult formation) or by importing a new faith from another society.

ultimate meaning, questions about Questions about the very meaning of life, the universe, reality—for example, Does life have purpose? Is death the end? Why do we suffer?

supernatural That which is beyond natural laws and limits.

religion Any socially organized pattern of beliefs and practices concerning ultimate meaning that assumes the existence of the supernatural.

religious pluralism The existence of several religions in the same society.

churches Religious bodies in a relatively low state of tension with their environment.

sects Religious bodies in a relatively high state of tension with their environment but which remain within the conventional religious tradition(s) of their society.

sect formation The breaking off of a group from a conventional religion to move into a higher degree of tension with the environment.

church-sect theory The proposition that, in time, successful sects will be transformed into churches, thereby creating the conditions for the eruption of new sects.

cults Religious movements that represent faiths that are new and unconventional in a society.

cult formation The process by which a person or persons with new revelations succeed in gathering a group of followers.

charisma The unusual ability of some religious leaders to influence others.

Suggested Readings

Barker, Eileen. 1984. *The Making of a Moonie: Brainwashing or Choice.* Oxford: Basil Blackwell.

Martin, David. 1990. *Tongues of Fire: The Explosion of Protestantism in Latin America.* Oxford: Basil Blackwell.

Melton, J. Gordon. 2002. *The Encyclopedia of American Religions.* 7th ed. Detroit: Thomson/Gale.

Stark, Rodney. 1996. *The Rise of Christianity: A Sociologist Reconsiders History.* Princeton, N.J.: Princeton University Press.

Stark, Rodney, and Roger Finke. 2000. *Acts of Faith: Explaining the Human Side of Religion.* Berkeley: University of California Press.

Warner, R. Stephen. 1993. "Work in Progress toward a New Paradigm for the Sociology of Religion in the United States." *American Journal of Sociology* 98: 1044–1093.

Sociology Online

www.socstark10.com

GO TO THE INTERNET AND TYPE: www.socstark10.com.

⌐ **CLICK ON: 2000 General Social Survey**

✓ **SELECT: AFTERLIFE?** Do you believe there is a life after death?

⌐ **CLICK ON: Analyze Now**

Is there a significant education effect?

Is there a significant age effect?

How do these findings fit with the discussion of secularization in the chapter?

⌐ **CLICK ON: Select New Data Set**

⌐ **CLICK ON: The 50 States**

✓ **SELECT: % CATHOLIC:** Percentage of the population who give their religious preference as Catholic.

⌐ **CLICK ON: Analyze Now**

Is there a regional pattern to this variable? Which states are more than 50 percentage Catholic? Which states are below 5 percent?

⌐ **CLICK ON: Select New Variable**

✓ **SELECT: % JEWISH:** Percentage of the population who give their religious preference as Jewish.

Is there a regional pattern to this variable? Which state is highest? Which states have rates of 0.0 percent?

⌐ **CLICK ON: Select New Data Set**

⌐ **CLICK ON: Nations of the Globe**

✓ **SELECT: %CHRISTIAN:** Percentage Christian.

⌐ **CLICK ON: Analyze Now**

Is there a region that is very lacking in Christians?

⌐ **CLICK ON: Select New Variable**

✓ **SELECT: % MUSLIM:** Percentage Muslim

⌐ **CLICK ON: Analyze Now**

Does this help explain the regional pattern of Islam?

⌐ **CLICK ON: Select New Variable**

✓ **SELECT: JEHOV.WIT:** Number of Jehovah's Witnesses (publishers) per 1,000,000.

⌐ **CLICK ON: Analyze Now**

In which nations have Jehovah's Witnesses been more successful than in the United States?

USING INFOTRAC COLLEGE EDITION

GO TO THE INTERNET AND TYPE:
www.InfoTrac-college.com

⌐ **CLICK ON: Register New Account**
You will be asked to enter your Access Code and to create and enter a User Name and Password. Having done so,

⌐ **CLICK ON: InfoTrac College Edition or Log On.**

If you are interested in religious groups, these search topics will be helpful.

SEARCH TERMS:

Falun Dafa

Religions

Sects

Freer Gallery of Art, Smithsonian Institution, Washington, D.C.

CHAPTER

15

Through most of recorded history, societies have been ruled by a hereditary elite. The woman seated on the portable throne in this 1902 photograph is Tz'u-hsi, the dowager empress of China. She is surrounded by the imperial eunuchs, who served as her bearers as well as her advisors.

The empress was not merely a symbolic ruler like the current queen of England. A mere nod from her could cause anyone or everyone else in this photograph to have his head chopped off, and it required little more effort on her part to send armies into battle or to outlaw "foreign" religions. Moreover, even as great famines raged through China, killing millions, the imperial court dined on rare delicacies.

Tz'u-hsi opposed all proposals to modernize China or to democratize the government and gave her approval to the extremists whose uprising against foreigners is known as the Boxer Rebellion. In 1908 Tz'u-hsi died under mysterious circumstances and was succeeded by Pu Yi, a very young boy, whose life was the subject of the Academy Award-winning film *The Last Emperor*.

Politics and the State

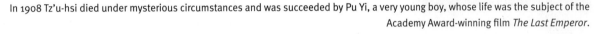

IN MEDIEVAL ENGLAND tenants could freely use all uncultivated land of a lord's estate as pasture for their livestock. This pasture land came to be known as the "commons" because it was used by the common people, and their right to use the land was known as the "freedom of the commons."

For centuries this arrangement worked well, and its fatal design flaw remained unnoticed. Freedom of the commons worked only as long as wars, disease, and natural disasters greatly limited the number of common people on an estate, thereby keeping down the size of the herds put to graze on the commons. However, when the rural population of England began to grow rapidly in the eighteenth century (see Chapter 18), a crisis suddenly arose. To see why, let's examine the situation from two points of view: that of each individual herd owner and that of the common people as a whole.

With free access to the commons, each individual herd owner always faces a decision of whether to add more animals to the herd. Operating as a rational human being, each herd owner asks, "Will I be better or worse off by adding more animals?" Suppose it is a question of adding another dairy cow to a herd of ten. Adding an eleventh cow will mean more milk, cream, and butter for the herd owner, who gets all of the benefits from the herd. Of course, adding a cow increases the demand placed on the commons to provide adequate feed. However, because all herd owners use the commons, this additional cost will be *shared equally*. Because a herd owner receives all of the profits from an expanded herd while paying only a portion of the costs of additional grazing, an individual herd owner will always benefit by adding animals even though this causes overgrazing. Thus, we can predict that when herders share a commons, they

CHAPTER OUTLINE

A CLOSER VIEW
Messick and Wilke: "The Tragedy of the Commons" in the Laboratory

PUBLIC GOODS AND THE STATE

FUNCTIONS OF THE STATE
Box 15-1 The Rewards for Free Riding
Rise of the Repressive State

TAMING THE STATE
The Evolution of Pluralism
Elitist and Pluralist States
Pluralism or Power Elites?

DEMOCRACY AND THE PEOPLE
A CLOSER VIEW
George Gallup: The Rise of Opinion Polling

THE AMERICAN "VOTER"

FEMALE CANDIDATES
A CLOSER VIEW
Hunter and Denton: Do Canadian Voters Reject Women for Parliament?

A CLOSER VIEW
Jody Newman: American Women as Candidates

IDEOLOGY AND PUBLIC OPINION

ELITE AND MASS OPINION
A CLOSER VIEW
Herbert McClosky: Elites and Ideology

A CLOSER VIEW
Philip E. Converse: Issue Publics

Organizations and Ideologies

CONCLUSION

REVIEW GLOSSARY

SUGGESTED READINGS

SOCIOLOGY ONLINE

inevitably will follow a strategy in which each adds animals as often as possible, making each individual herd as large as possible. As long as external constraints limit population and grazing, the system will not proceed to its tragic outcome. But when limits on expansion are removed, the result will be destruction of the commons and the ultimate economic ruin of all herd owners.

This analysis was first published in 1833 by William Forster Lloyd, an English mathematician who was trying to explain the crisis that had arisen in England over freedom of the commons (Hardin, 1964). For by then "the tragedy of the commons" had been recognized. The individual herd owners, unleashed by population growth, had overgrazed the commons to the point of massive erosion and destruction of the land's productive capacities. No matter how reduced the pasturage, any individual herder still was better off with more animals, no matter how skinny, than with a smaller number of equally skinny livestock.

What happened then? The lord of the estate, often joined by his most affluent tenants, enclosed the commons and withdrew free access. Thenceforth, grazing rights were strictly controlled, and in fact, most common people lost access to the commons. Because the same people now owned both herds and lands, the basis for computing self-interest was changed. When they opted to increase the size of their herds, they also bore the full cost. Hence, decisions had to reflect the need to preserve the pastures. Under these conditions the pastures recovered. Once again there was an abundance of dairy products and meat. Those with grazing rights grew rich and powerful—at the cost, however, of much individual freedom of choice. Freedom of the commons was replaced by their enclosure and control by an elite.

In this chapter we shall examine the great contradiction in human affairs revealed by what has come to be called "the tragedy of the commons." Many things vital to humans as *social* beings conflict with things vital to humans as *individuals.* To provide for the common or collective good, people often are forced to surrender considerable control over their lives to leaders and governments.

Unfortunately, as we shall see, this surrender often results in much misery, when leaders use their power and authority to repress, exploit, and even enslave their people. Indeed, through most of history, the state has been an institution of repression. Thus, we confront one of the

British Crown Copyright/RAF Photograph

A century before this aerial photo was taken, there were no tree-lined fence rows crisscrossing this English landscape. Instead, this was one huge pasture known as the commons because the common people had the right to graze their herds here. When rapid population growth began, the problem of administering "public goods" was revealed by the overgrazing of the commons. In response the landowner put up fences, planted trees, and henceforth denied his tenants the right to graze their herds here.

oldest dilemmas of political philosophy: how to have a state and keep it tame. As we examine theories for limiting the state's power and increasing individual freedom, we discover two fundamental kinds of state: the elitist state, which tends toward tyranny, and the pluralist state, which tends to permit considerable individual freedom.

We approach these matters by watching a famous equity experiment that assessed aspects of the tragedy of the commons problem. Next we move to the more macrolevel and assess an application of the same logic to the problem of *public* or *collective* goods. We then see how the state provides for public goods and explore means for preventing the state from being captured by a repressive and exploitative "power elite." The latter part of the chapter is both less macro and less theoretical. It concentrates on political behavior in democratic nations, especially the United States and Canada. We examine the rise of opinion polling and its implications for political decision making. Then we watch studies of voter reactions to female candidates for elective office. Finally, we assess

the role of ideology in forming individual political opinions.

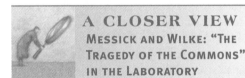

A CLOSER VIEW
MESSICK AND WILKE: "THE TRAGEDY OF THE COMMONS" IN THE LABORATORY

Several years ago a group of researchers at the University of California, Santa Barbara, figured out how to re-create the situation of the freedom of the commons in a social experiment. Led by David M. Messick and Henk Wilke (Messick et al., 1983), the team recruited undergraduate students as subjects. When the subjects arrived at the laboratory in groups of six, each was placed in a semiprivate booth containing a computer terminal. Then Messick and his colleagues explained that the group would participate in a "harvesting" game in which each person would remove points from a common pool. Each would receive a dime for every point he or she harvested. After each turn, the pool would be replenished at a variable rate but would never rise above 300 points—the level of the pool at the start of the game. Players could harvest from zero to ten points whenever it was their turn. They could continue to take points as long as any remained in the pool.

Subjects were told that the game had two goals. First, players should try to accumulate as many points as possible; the group of six with the highest total would receive a cash bonus at the end of the experiment. Second, players should make the resource last as long as possible to maximize the number of turns during which points still could be harvested. The parallel with the real world was spelled out as the experimenters told subjects that the game involved a choice between "taking all that one can or wants at the moment—at the risk of using up the resource—or taking a little on many occasions, with the possibility of getting more in the end because the pool is able to replenish itself." Finally, subjects were told they would be able to watch each player's choices as they occurred; these would appear on their computer screen.

The game began with each player in the same situation as herd owners in the days of freedom of the commons. Just as each herd owner had to choose whether to take a larger share of the commons' grass, these subjects had to choose between sustaining the pool by taking fewer points or maximizing their sure winnings by taking a dollar's worth of points every time they got a chance, since the cost of depleting the pool would be equally shared by all. The computer recorded what each subject did on each turn.

However, the players weren't really interacting. Instead, the computer was in control, so the feedback each player got about what others were doing and the condition of the pool was manipulated experimentally. Two *independent* variables were manipulated in this experiment. The first was *level of harvest*. One-third of the subjects were led to believe that so much was being taken in each turn that the size of the pool was dwindling rapidly; this was the *overuse* condition. The second third were informed that so little was being taken in each turn that the pool was growing; this was the *underuse* condition. The final third were shown results in which the pool remained of constant size; this was the *optimal* condition.

The second independent variable was *equity*. Half of those in each of the three harvest level groups were shown that members of the group were taking relatively similar numbers of points each turn; this was the condition of high equity. Those in the inequitable condition were shown

that members were taking very different numbers of points, that some were taking a lot more than others on each turn.

After ten turns the subjects were informed that the first part of the experiment was over. Each was told that in the past some groups had suggested that they could have done better if they had a leader who took points for the whole group and who then allocated points to each group member. Subjects were given the choice of continuing to withdraw points individually, as they had been doing, or electing a leader to control point withdrawals and distribution. No matter how the vote actually went, the subjects next were informed that the majority had voted to elect a group leader.

So each subject was asked to rank all six group members in terms of their preferences for who became leader. Each player had been identified as a color during the first part of the experiment, so colors were the identifications used in the voting procedure. Of course, the election was rigged. *Every* subject was told she or he had been elected leader. So for another ten trials, each subject took points from the pool and awarded a share to each player—unaware that all the others were doing the same.

What were the results? First, subjects in all groups tended to increase their harvests over time. The longer they played, the more they moved to maximize individual gain. Second, subjects took larger harvests and increased their harvests more rapidly when they believed there was great inequity across players—when they thought some players were taking a lot more than others. Third, players increased their harvests more in the underuse than in the overuse condition.

The independent variables also greatly influenced voting behavior. The overwhelming majority of those in the overuse condition voted to put harvesting in the hands of a leader—to opt in effect for government regulation. Only minorities in the other use conditions voted to have a leader. Equity also influenced vote: People in the highly inequitable groups were much more anxious to have a leader.

Finally, how did subjects behave once they became leaders? First, they decreased the size of the harvests to the level of optimal use, taking only the rate at which the pool was being replaced. So, just as the landowners in England restored the productive capacity of the commons, so did the leaders in this experiment. Second, they were equitable in assigning shares to the other five players. Third, like the English lords, the leaders in the experiment gave themselves larger shares than they gave others. That is, given leadership power, they moved to translate their advantage into cash—even at a dime a point.

In this remarkable experiment, we see why humans accept leadership and create governments, and we also see the risks entailed in doing so. For just as conflict theories of stratification anticipate (see Chapter 9), *human beings will tend to use power to exploit others.*

PUBLIC GOODS AND THE STATE

Mancur Olsen (1965) has argued that governments are unavoidable features of human societies. His conclusion rests on the simple point illustrated in the tragedy of the commons: To provide for the best interests of the group, coercion (the use of force) is necessary. To create public or collective goods, the interests of the individual and the interests of the group collide.

It wasn't sufficient for the English landlords to inform their tenants that they were overusing the commons. They knew it. But how could the overuse be stopped? Not by persuasion, for it will always be in the best interests of each individual to let the others be persuaded while he or she continues as before. So the lords had to put up fences, drive out tenant herds, and keep them out. In the end the English government had to enforce the property rights of the lords against the demands of the evicted herd owners.

There are many **public** or **collective goods** vital to the survival and welfare of human societies. First, we must be secure from harm by other members; we can't risk living together if we constantly must fear for our lives and possessions. Second, we must be secure against harm from external dangers—from being attacked or looted by members of another society. Third, certain resources and services must be provided that cannot be supplied by voluntary individual actions. For example, a dam to prevent floods requires that many people contribute wealth, time, and energy.

The crucial aspect of public goods is, of course, that the best deal for any individual is to reap the benefits without sharing the costs—that is, to be a "free rider." This problem remains chronic because so often it is impossible to withhold selectively the benefits of a specific public

Courtesy of Lynne Roberts

Any military cemetery demonstrates the inherent problem of creating public goods. If a society is to remain safe from external attack, some members must be induced to serve in the armed forces. Here what's best for the group collides with what's best for the individual. For any individual the best choice is to benefit from the protection provided by others while he or she remains out of danger. Just as long as their side won the war, men like Bill Cooper, John P. Okun, and Charles M. Magee would have been better off enjoying civilian life and keeping their names off monuments like these. But if they had all stayed home, North America might be under Nazi rule today.

In the spring of 1989, thousands of students from Beijing University gathered in Tiananmen Square to demonstrate for democracy. As the world watched on TV, the students chanted "Long Live Freedom, Long Live Democracy" and erected a copy of the Statue of Liberty. For a few weeks, it looked as if China's leaders might make democratic reforms along the lines demanded by the students. But repressive political elites seldom yield power as long as they have the means to resist.

© Abbas / Magnum Photos

good. When an enemy attack is beaten off, all members of a society benefit, but a warrior profits more if, while the others go off to battle, he stays home and avoids the risk of death or injury. Were such a strategy possible, public goods would not be created because (to pursue the example) *all* the warriors would stay home.

Olsen concluded that because it is always against an individual's self-interest to contribute to the public good, there will be no public goods unless means exist to *force* individuals to do their share. Often, this force need not be used, but its *potential* use must always be a credible threat.

The **state** or government is the organized embodiment of political processes within a society—the means by which decisions are made and social life is directed and regulated (Goldstone, 2000). The state arises in human societies because of the need to provide a credible threat to force the creation of public goods. As we shall see, such threats require that the state organize and *monopolize* the use of coercion. Indeed, Max Weber defined the state as consisting of such a monopoly: "One can define the state sociologically only in terms of the *means* peculiar to it . . . namely the use of physical force" (Weber, [1921] 1946).

Only through organized coercion can people assure themselves of public goods. And therein lies the greatest of all social dilemmas. We must have certain kinds of public goods to exist. To get

them, we must create organizations capable of coercing us. By doing that, however, we set up the possibility, even the probability, that those who control the means of coercion will act in their own interest rather than for the public good. As a result political leaders may use the government monopoly on coercive force for their personal benefit. Thus, peasants driven to create a government capable of defending them from bandits have often found themselves victims of a ruling elite. Members of this elite turn the coercive powers of the state against the peasants and tax them severely so that the elite can live in luxury.

FUNCTIONS OF THE STATE

As we have seen, Max Weber argued that the essence of a state or government is that it "claims the monopoly of the legitimate use of physical force" within its boundaries. Robert Nozick (1974) has developed this definition more fully:

A state claims a monopoly on deciding who may use force when it says that only it may decide who may use force and under what conditions; it reserves to itself the sole right to pass on the legitimacy and permissibility of any use of force within its boundaries; furthermore it claims the right to punish all those who violate its claimed monopoly.

BOX 15-1   THE REWARDS FOR FREE RIDING

PEOPLE OR GROUPS WHO BENEFIT from a collective activity without contributing to its creation are known as *free riders*. Their existence is inherent in the contradiction between the interests of individuals and the interests of the group. James S. Coleman illustrated this with a simple example involving multiple-person exchanges, as shown in Figure B-1.

Here are four exchange partners, each having the choice of contributing $8 to a group enterprise or contributing 0. For every $8 contributed, the return will be $12. The *total return* will be divided *equally* among the four. If everyone puts in her $8, then the total return will be $48 and each person will get back $12 for her $8, for a profit of $4, as is shown in the second line of the table.

Now, suppose Ann realizes that if she contributes nothing, and spends the $8 on pizza, she will still get back $9 if the others each put in their $8. Of course, this $9 return for no investment means that the other three players only make $1, getting back only $9 for each $8 invested. Soon, Betty catches on to Ann's game, so she also fails to contribute. As a result she and Ann each have a profit of $6, while Cindy

and Dana each lose $2. At this point Cindy decides to cheat, too. When only Dana puts in her $8, Ann, Betty, and Cindy each make $3, while Dana loses $5. As long as even one player invests, the others profit without cost.

The question arises, How can groups prevent members from taking unfair advantage of the contributions of others? And that is precisely where norms come in. In this four-person situation, each member knows that the others will be offended if she cheats. To the extent that there is a high level of group solidarity, each player will be honest. However, when large groups are involved, informal group pressure loses its effect. In fact, in large groups it often is difficult to keep track of who contributes what. Thus, an important proposition of exchange theory is: The larger the number of individuals required to participate if a public good is to be provided, the less likely any individual is to participate (Heath, 1976). From this it follows that the larger the group is, the greater will be the need for formal methods to monitor contributions and to enforce conformity.

FIGURE B-1   The Profits of Free Riding.

	ANN	BETTY	CINDY	DANA	TOTAL RETURN
Invests	$8	$8	$8	$8	$48
Profit/loss	+$4	+$4	+$4	+$4	
Invests	0	$8	$8	$8	$36
Profit/loss	+$9	+$1	+$1	+$1	
Invests	0	0	$8	$8	$24
Profit/loss	+$6	+$6	−$2	−$2	
Invests	0	0	0	$8	$12
Profit/loss	+$3	+$3	+$3	−$5	

Source: Adapted from Coleman (1990).

After weeks of hesitation, the Chinese communist government decided there would be no democratic reforms, and the army was ordered to clear student demonstrators from Tiananmen Square. When the students resisted, the army opened fire and thousands died. In the most dramatic confrontation of all, this lone student blocked the passage of tanks down the Avenue of Eternal Peace, shouting "Why are you here? You have done nothing but create misery." But the tanks rolled on.

© Stuart Franklin/Magnum Photos

To understand why the state's monopoly on force arises, it helps to see how the state uses this monopoly to secure certain public goods. In some cases the government must use coercion to prevent coercion by individuals. For example, one primary collective good is security against becoming the victim of coercion by other individuals, such as robbers and kidnappers.

Indeed, precisely here is where anarchist proposals to dispense with the state break down. Without a collective good such as security against harmful actions by other group members, people would not be able to remain within the group. That is, without organized coercion to prevent private coercion, humans would live in a condition that Thomas Hobbes described as "the war of all against all." In his book *Leviathan*, published in 1651, Hobbes tried to describe what life would be like in a condition of anarchy:

Hereby it is manifest, that during the time men live without a common power to keep them all in awe, they are in that condition which is called war . . . where every man is enemy to every man. . . . In such condition, there is no place for industry; because the fruit thereof is uncertain: and consequently no [agri]culture . . . no society; and which is worst of all, continual fear, and danger of violent death; and the life of man, solitary, poor, nasty, brutish, and short.

Hobbes argued that there could be no freedom where there was no security of person or property. An individual's freedom to live and to benefit from his or her efforts requires that limits be placed on the freedom of others. I am not free to live unless you are not free to kill me. I am not free to create unless you are not free to take my creations from me. When the attempts of individuals to coerce each other are not held in check, the possibility of preserving group life is destroyed.

Thus, a primary function of the state is to preserve internal order, to make life predictable and secure. However, the state also exists to provide the collective good of external security. In fact, the state serves to provide all collective goods—things its members could not individually provide for themselves, from irrigation ditches to armies to legal codes.

These functions of the state were clearly understood by those who wrote the preamble to the U.S. Constitution:

We the People of the United States, in Order to form a more perfect Union, establish Justice, insure domestic Tranquility, provide for the common defense, promote the general Welfare, and secure the Blessings of Liberty to ourselves and our Posterity, do ordain and establish this Constitution of the United States of America.

TABLE 15-1	Agricultural Development and the Scope of Political Structures			
	LEVEL OF AGRICULTURAL DEVELOPMENT			
SCOPE OF POLITICAL STRUCTURE	**NONE**	**LOW**	**MEDIUM**	**HIGH**
Local only	81.6	57.1	34.9	22.8
Semistate	18.4	39.3	57.2	36.8
State	0.0	3.6	7.9	40.4
	100.0%	100.0%	100.0%	100.0%
Number of Societies	(38)	(28)	(63)	(57)

Source: Prepared by the author from the Standard Cross-Cultural Sample.

RISE OF THE REPRESSIVE STATE

According to our definition, all societies, even the tiniest and most primitive hunting and gathering societies, have a state. However, in small simple societies, the state is no more than a loosely organized authority structure based on kinship and age: The "ruler" is more like the head of a large family than like a king or a president. That is, in small preliterate societies, some persons hold the authority to settle disputes and to oversee the creation of public goods. But there are no full-time leaders. Relations between the individual and those having authority are direct and personal. When someone is seriously deviant, the person or persons having authority administer punishment, usually by physical force. Because of these features, many anthropologists and sociologists refer to these as "stateless societies," applying the name *state* only when there are full-time specialists who exercise the functions of the state.

In reality, however, the dividing line between stateless societies and those having states is somewhat hazy. So, societies included in the Standard Cross-Cultural Sample have been classified in terms of the *span* or the *scope* of their internal political authority. Eighty-two of the 186 societies in the sample are classified as having only local political control. That is, political authority is limited to one place or social unit, such as a specific village, rather than being exercised over several such units. These societies best fit the definition of a stateless society, although a few of them do support full-time leaders. An additional seventy-five of these societies can be called semistates, for they sustain political leaders with a span of control that extends over multiple units and that may consist of two levels of administrators—that is, one set of leaders who in turn have superiors. Finally, twenty-nine of these societies qualify as fully developed states.

The existence of states, as with the existence of all specialization within societies, rests on the development of agriculture. For it is only the capacity to produce surplus food that allows people to specialize in work unrelated to obtaining food; in stateless societies the leaders must provide for their own subsistence. Table 15-1 shows this relationship. In societies lacking agriculture, 81.6 percent have only local political structures. This decreases sharply as the level of agricultural development rises; and where there is relatively well-developed agriculture, just 22.8 percent of the societies have only local political structures.

The emergence of political specialists is accompanied by the emergence of military specialists, who provide the monopoly on force that is the essence of the state. And when military specialists exist, they tend to be busy. Table 15-2 shows that even societies with local political structures generally are warlike; nearly half fight a war every year. But the typical agrarian state, with its cohorts of military specialists, is hardly ever at peace—and in none is war a rarity.

Not only does the agrarian state use its military power against its neighbors, but the state's monopoly on force is a constant threat to its citizens as well. A king, for example, will use his soldiers to provide for the common defense and to ensure public order, but he will also use them to exploit his subjects. Table 15-3 shows that the amount of stratification is low in 93.9 percent of societies with only local political authority, and none of these societies has a high degree of stratification. No agrarian state has a low degree of stratification, and 69 percent are highly stratified. This is the feature of the state that has long

TABLE 15-2	The State and Warfare		
	SCOPE OF POLITICAL STRUCTURE		
FREQUENCY OF EXTERNAL WARFARE	**LOCAL ONLY**	**SEMISTATE**	**STATE**
Constant	46.2	57.9	71.4
Common	15.3	15.8	14.3
Occasional	10.3	2.6	14.3
Rare	28.2	23.7	0.0
	100.0%	100.0%	100.0%
Number of Societies	(39)	(38)	(7)

Source: Prepared by the author from the Standard Cross-Cultural Sample.

TABLE 15-3	The State and Stratification		
	SCOPE OF POLITICAL STRUCTURE		
DEGREE OF STRATIFICATION	**LOCAL ONLY**	**SEMISTATE**	**STATE**
Low	93.9	53.4	0.0
Medium	6.1	33.3	31.0
High	0.0	13.3	69.0
	100.0%	100.0%	100.0%
Number of Societies	(82)	(75)	(29)

Source: Prepared by the author from the Standard Cross-Cultural Sample.

driven people to despair. How can the state be made bearable?

TAMING THE STATE

For more than 2,000 years, a dominant question in political thought has been how to limit the powers of the state. The nature of the problem seems clear enough. The state must exist. This means that the state will always have the potential to use its coercive powers to exploit and repress its citizens. How can the abuse of these coercive powers be limited without weakening the ability of the state to fulfill its necessary functions?

The Greek philosopher Plato thought the solution was creating a special class of philosopher-kings: persons who were trained to be fair and restrained in their use of state power. However, it seems too much to expect people who have been placed in positions of immense power to practice self-restraint. Indeed, as Lord Acton claimed in a widely quoted maxim, "Power corrupts, and absolute power corrupts absolutely."

Rejecting hopes that the state could be tamed by putting people of high character in power, political theorists searched for social arrangements that would limit the power of leadership positions, regardless of the character of the leaders. By the eighteenth century, political thinkers had concluded that two things were necessary to tame the state. First, *a clear set of rules* (procedures and laws) had to be established that defined the limits of state power and the manner in which that power could and could not be exercised. But rules by themselves mean nothing. Some of the most repressive regimes on earth have model constitutions that guarantee all sorts of rights and freedoms to citizens. Such guarantees do not mean anything unless they are embodied in a structure designed to ensure that the rules are observed. What kind of structure can have this effect?

This leads to the second requirement for taming the state: *a structure in which power is widely dispersed among many powerful groups.* In this way no one group can pursue its own interests without regulation; every group is checked by the other powerful groups acting to preserve their own interests. This aspect of taming the state was not so much discovered as it was something that just slowly evolved by trial and error. In fact, only after a considerable dispersal of power had occurred in England did political theorists recognize the principle involved. So, let's see what happened and how it works.

THE EVOLUTION OF PLURALISM

English democracy evolved from a single principle: the right to private property. The English believed that the essential feature of state repression was the use of coercion to deprive people of their property. If this could be prevented, then the state could be brought under control. In particular, if the king could be prohibited from taking any person's property without that person's permission, then the king could not squeeze all the economic surplus out of the people to provide for his luxuries. Nor could the king afford to go to war unless he had the support of the people, for he would lack the needed funds. Thus, the English believed that the state could be tamed if taxes could not be imposed or collected without the approval of those being taxed.

This view is in direct conflict with the Marxist theory of the state, which holds that private property is the root of all repression and exploitation by the ruling class. Marxists argue that the right to private property and the state's use of coercion to protect private property are the means by which the ruling class becomes and remains rich and exploits the masses. Marx claimed that only if property rights are abolished can the masses be liberated (see Chapter 9). Yet the fact remains that in Marxist states there is very little individual freedom, whereas in all nations with free elections and substantial individual liberties, the state guarantees the property rights of individuals.

This contrast between capitalist and Marxist states would not surprise those who helped to create democracy in England or those who wrote the U.S. Constitution. Long observation of European feudalism had convinced them that when property rights are not secure, the masses, not the rich, suffer most. They recognized that the repressive state mainly confiscates the property of the powerless—the peasants, small merchants, and artisans—because the wealthy rule the state and receive the wealth seized by the state.

There was in fact a second workable solution to the tragedy of the commons: to place ownership of herd and land in the same hands *by dividing* the commons into pastures, each owned by one of the herd owners. In this way the ecology of the commons would also have been restored because now the full costs of herd expansion would fall on the herd owners. Of course, this division would have violated the property rights of the landowners. But enclosure also violated a property right in the sense that, for centuries, the decision to labor in the lord's fields had been calculated on the basis of the value of grazing rights on the commons. Enclosure represented unilateral withdrawal of a valuable portion of this exchange relationship and, furthermore, destroyed the value of the evicted herds.

With whom did the government side? With the wealthy and powerful. Had *all* property rights been accorded protection, the common people would have gotten a much better bargain. Guaranteeing the property rights of everyone offers greater protection to those with the least property.

But the pressing question was, How could property rights be protected from the state? The solution is not simply a law or a constitution but a particular kind of social structure in which a number of powerful factions, or elites, restrict one another's ability to use the state's coercive power. The state can be tamed only when political power is dispersed among groups with diverse interests. Such a situation is described as **pluralism.**

Pluralism developed in England partly by accident. In 1215 King John found himself unable to control the nobility. To remain on the throne, he was forced to sign the Magna Carta, a contract in which he agreed to impose no taxes on the nobility except when they freely agreed to be taxed. This led to the creation of the House of Lords, wherein the nobility gathered periodically to vote on tax requests from the king.

In time the right to have one's property secure against seizure by the king was extended to property owners who were not members of the nobility. They began to send elected representatives to the House of Commons, where they also gave or withheld approval of the king's tax requests. The power to control the king's revenues proved to be the power to control the government. If the House of Lords or the House of Commons did not like a policy, it could withhold funds until it was changed.

Moreover, neither house of Parliament was dominated by a single group with identical interests. Policies favorable to some nobles often affected others adversely; policies good for merchants were often bad for shop owners or farmers. Thus, besides English kings having to depend on the two houses for their revenues, decisions within each house required a *coalition of groups* and therefore *a compromise of competing interests.* Governmental decision-making processes involved increasingly diverse groups and interests.

Of course, English rulers occasionally attempted to destroy these limits on their power and restore the absolute power of the throne. However, these efforts were always thwarted because too many people had too much to lose should the king regain control. Therefore, if one faction of nobles wanted to restore an unlimited monarchy, others combined to block them.

It has been noted that the existence of the English Channel also played an important role in weakening the powers of English monarchs (Hintze, 1975). The channel prevented European wars from extending into England. Thus, the king could not use external military threats as grounds to create and maintain a large professional army that then could be used to repress anyone who opposed him. Indeed, England's defense was based on maintaining a powerful navy to control the English Channel—and a navy cannot be used for internal repression.

The American Revolution occurred primarily because the English failed to extend their principle of no taxation without representation to their North American colonies. The English tried to levy taxes on the colonies, but they did not grant the colonies representation in the House of Commons, where they could influence tax policies.

Once free of English rule, the Americans had to create their own system of government. By this time, however, political philosophers, especially the Scottish rationalists, had analyzed why and how the English system worked. Thus, the men who wrote the U.S. Constitution did so with considerable understanding of the essential issues.

James Madison, the principal designer of the U.S. Constitution, believed that democracy always faces two threats. The first is a *tyranny of the minority:* the historical danger that a privileged few would capture the state and use its coercive powers to repress and exploit the many. This classic problem of taming the state had given rise to the English form of democracy. But Madison was also concerned about a *tyranny of the majority:* the danger that a majority of citizens would use the machinery of representative government to exploit and abuse minorities. Here Madison was mindful that even the English democracy persecuted religious dissenters and that coalitions of interest groups sometimes exploited weaker groups. Indeed, Madison was concerned that people who achieved great wealth be as secure in their property rights as anyone else, and he feared that the mass of citizens might use their superior numbers to impose discriminatory taxes on the rich and thereby escape paying taxes themselves. Put another way, Madison believed the poor could be as selfish as the rich.

To block both kinds of tyranny, Madison developed a system of government in which powers were widely distributed and procedures made somewhat cumbersome. Each of the three branches of our federal government—executive, legislative, and judicial—has the power to nullify actions taken by the other two. This is called the system of checks and balances in government. Madison hoped to make it possible for minorities to block actions against them, at least for a long period, and for substantial majorities to be required to take any action, thereby blocking minorities from controlling the government.

The system of checks and balances has dominated the American political process for more than 200 years. It has not always produced ideal results, and sometimes the United States has been less democratic than Madison had hoped. Despite these defects, the state has remained relatively tame. Moreover, many of the worst violations of individual liberty have been corrected within the system.

The important point, however, is that the U.S. Constitution, like all constitutions, is meaningless without support from political institutions and, indeed, without a general willingness to play by the rules. Some of the most repressive regimes on earth have magnificent written constitutions that solemnly guarantee all sorts of freedoms, while the British have provided a model of democracy without ever having written down most of their constitution. And Canada has sustained democracy both with and without an extensive written constitution. For a long time, aside from various acts of the British Parliament defining some aspects of Canadian governing principles, Canada primarily relied on the "unwritten" British constitution. In 1981 the Canadian government secured the right to frame and amend its own constitution without the approval of the British Parliament—a process referred to as "patriation." That same year the Canadian Supreme Court noted a number of documents and acts as parts of the constitution, and then in April 1982 Queen Elizabeth II came to Ottawa to sign the new Constitution Act, which begins with a very detailed list of rights and freedoms guaranteed to all Canadians.

Thus far, we have seen how the dispersal of power plays a fundamental role in taming the state through the process wherein one powerful faction is played off against another. Now let's more clearly develop the underlying theory.

ELITIST AND PLURALIST STATES

States are of two essential types: elitist and pluralist. In an **elitist state,** a single elite group rules. Sometimes power struggles go on within this elite—for example, when there is a dispute over who is to be the new head of state or the new party leader. But power resides almost totally with this single elite, which controls the state and can therefore use its coercive powers as it sees fit.

The elitist state is the most common type. The agrarian societies examined in Chapter 10 were elitist states, and most societies today are elitist states, regardless of what their constitutions say. Whenever only a single political party is permitted and when rule is passed on by power struggles within the one-party elite, that

In 1990, as they cast their ballots in a village in northern Hungary, these peasant women were among the few Hungarians who were not taking part in a free election for the first time. It had been forty-three years since it had been legal for anyone other than the handpicked candidates of the Communist party to appear on the ballot. These women seem very pleased to have a choice once again.

© AP/ Wide World Photos

state is elitist, even if it calls itself a democracy or a people's republic.

It is almost impossible for an elitist state to avoid being tyrannical. Plato's hypothetical unselfish philosopher-kings have not appeared. In short the elitist state is an untamed state wherein the state's coercive powers are used to repress and exploit people.

In a **pluralist state,** rules governing state power are maintained by the existence of many competing elites. The pluralist state has been tamed. That does not mean that all persons living in such a state have an equal amount of power in decision making; no such society exists. Indeed, the word *pluralist* refers to several (plural) elites. The state is tamed because power is dispersed among many contending minorities, or elites, each of which can secure some, but not all, of its desires.

The elitist state is ruled by a single minority; the pluralist state is ruled by shifting coalitions of many minorities. Robert Dahl (1956) explains the difference this way:

If there is anything to be said for the processes that actually distinguish democracy from dictatorship, it is not discoverable in the clear-cut distinction between government by a majority and government by a minority. The distinction comes much closer to being one between government by a minority and government by *minorities.*

Thus, Dahl conceded that in no states do majorities rule directly. Rather, in democratic states numerous minorities or interest groups represent the interests of most citizens: business interests, labor interests, religious interests, racial and ethnic interests, regional interests, and so on. The constant struggles and shifting coalitions among these groups prevent any one from imposing its will on the others. Should business become too powerful, for example, labor unions and consumer groups may unite to hold it in check. Or should labor become too powerful, consumer groups may side with business. Should the eastern states seek to exploit other regions, these regions may unite to protect themselves. And so it goes.

However, democracy is not the cause of pluralism. Rather, *pluralism is the mechanism that sustains democracy* as each of many political blocs act to preserve its right to influence decision making. Indeed, in some nations that do not

Direct democracy is possible only for very small groups, such as this group of New Englanders gathered for a town meeting. For larger groups representative democracy is required.

claim to be democracies, the existence of pluralism limits the power of the state—a pattern often found in Latin American countries ruled by military governments. In some nations that do claim to be democracies, the absence of pluralism results in little freedom for the individual; China is an example.

PLURALISM OR POWER ELITES?

The pluralist thesis has been hotly contested by some sociologists, especially by Marxists. They claim that the democracy of Western nations is mostly an illusion, that there really aren't a variety of independent elites contending with one another; instead, pluralist appearances notwithstanding, these nations are being run by a tiny ruling elite.

The most forceful and famous proponent of the ruling elite thesis was C. Wright Mills (1916–1962), who taught at Columbia University. In his book *The Power Elite* (1956), Mills charged that the United States is effectively ruled by a small set of influential people who, together, hold the preponderance of power. Mills identified these people as a **power elite** rather than as a "ruling class" because their power often is not based primarily on their economic positions. Instead, their power derives from the positions they hold in three major kinds of organizations that, according to Mills, dominate American life: the military, the

government bureaucracy, and the large corporations. These leaders don't actually get together and plan and conspire to run the nation, but Mills argued that they serve the same interests. Moreover, Mills claimed that these people tend to have such similar social backgrounds that they almost intuitively come to the same conclusions about what ought to happen. The power elite, said Mills, is male, Protestant, and from old East Coast families; its members tend to be educated in the Ivy League and to belong to the same set of clubs and organizations. They also serve together on many corporate boards and government commissions.

The Power Elite was a very controversial book and has been the focus of much conflict among sociologists. From the beginning many, such as David Riesman (1961), argued that Mills missed the variety of significant conflicts of interest among those said to form the power elite. For example, what's good for the banks often is bad for manufacturing firms. Price controls on oil may be good for the auto industry but bad for oil companies. Or high oil prices may be great for Texas and Alberta and bad for Michigan and Ontario.

Despite such criticisms, much empirical research also demonstrates that those in the highest reaches of government and business do tend to be people with similar backgrounds and many social ties to one another. Does that mean they constitute a single elite? The debate continues. The more widely accepted view seems to

© George Bellerose/Stock Boston

be that no single elite dominates in Western democracies, but neither do "the people" make most of the decisions.

DEMOCRACY AND THE PEOPLE

The essence of popular democratic political theory is that government should be "of the people, by the people, and for the people." But how can this be accomplished, and would it really be in the public interest? Clearly, it is impossible to run even a modest-sized city along the lines of the New England town meeting, where citizens gather to speak their piece and, by majority vote, make all town government decisions. Such a direct democracy requires tiny populations. Moreover, a substantial number of citizens do not take part in town meetings, and thus, government is not by *all* the people. Even if those who fail to attend were rounded up and forced to take part, it is not clear that government would be conducted more wisely or even with greater concern for the interests of all. Many people seem too little interested in politics or too uninformed to make any responsible contributions to decision making.

Table 15-4 shows the percentage of the adult population in each of forty-one nations who said that politics is important in their lives. Most people in Vietnam (78%), Tanzania, China, and Japan regard politics as important, but only slightly more than half of Americans (57%) think so, and very few people in Pakistan (14%) think so.

Thus, because of indifference as well as limits of practicality, democracies rest on the principle of representative government. Free elections are held to select persons to govern on behalf of the rest. Should these elected officials stray too far from the public will, they can be turned out of office at the next election. But even this solution is not perfect if the concern is to represent everyone because many citizens do not vote or otherwise participate in the political process.

Table 15-5 examines political participation by way of petitions. In some nations most people say they have signed petitions—89 percent in New Zealand, 81 percent in the United States, and 73 percent in Canada. But in many other nations, very few have ever signed a petition—only 4 percent in Pakistan and Zimbabwe. Low rates of literacy may explain some of the differences.

In the remainder of this chapter, we shall explore political participation and political opinion in Canada and the United States to see how democratic political processes shape the state. However, before turning to these matters, it will be useful to see how the recent development of public opinion polls made it possible to determine how the people feel about major issues and to discover who takes part in the political process.

A CLOSER VIEW
GEORGE GALLUP: THE RISE OF OPINION POLLING

Modern democracies are founded on the belief that those elected to represent the people will actually do so. This does not mean that they ought to be rubber stamps for public opinion. In fact, political leaders often earn great respect when they risk popularity to abide by their principles. Nevertheless, in a representative government, elected officials should at least know the people's feelings on various issues. Obtaining this knowledge does not seem difficult today—it is virtually impossible to open a newspaper or watch the nightly news on TV without learning the results of the latest opinion poll.

Nevertheless, until recently, most elected officials could only guess about public opinion, even on major issues, and often their guesses were wrong. Those representing a minority viewpoint can often instigate massive letter-writing campaigns, public demonstrations, and editorial support so that they appear to be representing the majority. But before public opinion polling, it was very difficult to see the contrivance of such campaigns. In fact, some of the earliest polls were as misleading as publicity campaigns in support of particular points of view.

Chapter 4 mentioned that the most famous SLOPS study was a "poll" conducted by the *Literary Digest* before the 1936 presidential election. Although it was based on millions of postcards returned by persons selected from various commercial mailing lists, it predicted a landslide victory for Alf Landon. Instead, Franklin Delano Roosevelt was reelected by a landslide. Meanwhile, in 1936 another poll on the election appeared in a number of newspapers. Unlike the *Literary Digest*'s poll, it was based not on millions of respondents but on fewer than 2,000; yet it correctly predicted an

TABLE 15-4	"How important is politics in your life?"		
NATION	"IMPORTANT" (%)	NATION	"IMPORTANT" (%)
Vietnam	78	Ukraine	38
Tanzania	71	Malta	38
China	68	Russia	38
Japan	68	Switzerland	38
Netherlands	57	El Salvador	38
United States	**57**	Azerbaijan	37
Philippines	55	Greece	37
Sweden	55	Bosnia	36
Algeria	52	Iceland	36
Nigeria	52	France	35
Bangladesh	51	Italy	34
South Korea	51	Great Britain	34
Brazil	51	Venezuela	34
Armenia	50	Colombia	32
South Africa	50	Belgium	32
Egypt	50	Albania	32
Australia	50	Ireland	32
Jordan	47	Chile	31
Singapore	47	Czech Republic	31
Mexico	**46**	Bulgaria	31
Georgia	46	Belarus	30
Norway	45	Poland	30
Taiwan	44	Croatia	29
Iran	44	Montenegro	28
New Zealand	44	Morocco	25
Peru	44	Romania	25
Denmark	42	Serbia	25
Zimbabwe	42	Argentina	24
Austria	41	Latvia	24
Canada	**41**	Spain	22
Lithuania	41	Estonia	21
Turkey	41	Finland	20
India	40	Hungary	18
Indonesia	39	Slovenia	15
Dominican Republic	39	Pakistan	14
Germany	39		

Source: Prepared by the author from the World Values Surveys, 2001–2002.

easy Roosevelt victory. This poll was conducted by the American Institute of Public Opinion (AIPO). AIPO had been founded in 1935, and on October 20 of that year, its first weekly report on public opinion about current issues appeared in several newspapers. This proved to be a very popular feature, and soon it was carried by scores of leading papers across the country. A year later, AIPO's correct prediction of the election made it an authoritative source of informa-

tion on public opinion. The name of the president and founder of the American Institute of Public Opinion soon became well known: George Gallup.

George Gallup received a doctorate from Northwestern University in 1928 and was head of the department of journalism at Drake University from 1929 to 1931. He returned to Northwestern for a year as professor of journalism and advertising but then went to New York

TABLE 15-5	"Have you ever taken political action by signing a petition?"		
NATION	**HAVE DONE SO (%)**	**NATION**	**HAVE DONE SO (%)**
New Zealand	89	Egypt	20
Sweden	87	Puerto Rico	19
United States	**81**	Chile	19
Australia	78	Estonia	19
Great Britain	78	Colombia	18
Canada	**73**	El Salvador	18
Belgium	68	Latvia	18
France	67	Armenia	17
Norway	65	Uganda	16
Switzerland	62	Moldova	15
Netherlands	59	**Mexico**	**15**
Ireland	58	Albania	15
Japan	57	Dominican Republic	14
Slovakia	57	Algeria	14
Austria	56	Georgia	14
Czech Republic	56	Turkey	14
Denmark	55	Hungary	14
Iceland	53	Venezuela	14
Italy	52	Bangladesh	13
Greece	49	Morocco	13
Germany	48	Ukraine	13
Brazil	47	Russia	11
South Korea	47	Taiwan	10
Finland	47	Tanzania	10
Israel	38	Azerbaijan	9
Croatia	33	Bulgaria	9
Malta	33	Romania	9
India	24	Philippines	9
Spain	24	Belarus	8
Argentina	22	Jordan	8
Lithuania	22	Singapore	7
Poland	22	Nigeria	6
Portugal	22	Indonesia	5
Bosnia	21	Vietnam	5
Peru	21	Pakistan	4
South Africa	21	Zimbabwe	4

Source: Prepared by the author from the World Values Surveys, 2001–2002.

to become director of research for Young and Rubicam, a leading advertising agency. In 1935 he founded the polling organization that has come to be known as the Gallup Poll.

Gallup's aim was to provide frequent reports of public opinion on major political, social, and moral issues. As he put it in his first newspaper report in 1935, his was a "nonpartisan fact-finding organization which will report the trend of public opinion on one major issue each

week. . . . The results of these polls are being published for the first time today in leading newspapers—representing every shade of political preference."

Gallup's first report demonstrated the difficulties of gauging public opinion without conducting a poll. In 1935 most of the world was in the midst of the Great Depression. Millions were out of work, many banks had failed, factories were closed or running at very reduced levels, and thousands of homeless people had taken to the highways in search of a livelihood. In the United States, after three years in office, Franklin Delano Roosevelt's "New Deal" had contributed little toward economic recovery. As a result a widespread campaign was begun to greatly increase government spending to feed and clothe the needy and to stimulate the economy. Countless public speakers claiming to represent the public demanded that Roosevelt increase federal spending. Many members of Congress joined in these demands, and press accounts frequently echoed the cry that "the people demand action now."

Along came the Gallup Poll. The first Gallup Poll ever published reported national responses to the question "Do you think the expenditures by the Government for relief and recovery are too little, too great, or just about right?" Only 9 percent of Americans thought the government was spending too little, 60 percent thought it was spending too much, and 31 percent thought the current level of spending was about right. Whether or not increased public spending would have helped recovery, clearly those who supported it as representing the will of the people were incorrect.

This was only the first of many instances in which Gallup findings revealed widespread misperception and misrepresentation of public opinion. Two months later, Gallup reported that despite the fact that most members of Congress, encouraged again by many organized political groups, wanted to decrease military appropriations, the public overwhelmingly wanted them increased. For example, 7 percent of the American public wanted a smaller budget for the Army Air Corps, while 74 percent wanted a larger budget. Apparently, the public was more concerned about the massive armaments programs then under way in Nazi Germany than were members of Congress.

In time political leaders and the mass media learned how hard it was to judge public opinion without taking a poll, and polling became a major industry. But the Gallup Poll, which now has affiliates in more than fifty nations, remains the most influential source of information on political and social issues.

THE AMERICAN "VOTER"

As a result of more than sixty years of election surveys, we now know a great deal about the American voter (Kuechler, 2000). For example, the wealthier and more educated people are, the more likely they are to support Republican candidates. Men are a bit more likely to vote for Republicans, while women slightly favor Democrats. African Americans vote for Democrats by a margin of more than four to one. For generations Catholics voted for Democrats, and Protestants (outside the South) favored Republicans, but today, Catholics and Protestants have similar party preferences and southern voters have swung from the Democrats to the Republicans. But perhaps the most significant facts have to do with the general lack of interest in politics. Recall from Table 15-4 that most Americans said politics wasn't an important part of their lives. In Table 15-6 we see that they were telling the truth. Only 49 percent of Americans bothered to vote in the 1996 presidential election. Only about one in ten wore a political button or put a bumper sticker on their vehicles during the 1996 campaign or made a campaign contribution. Fewer than one in four could name their member of Congress. When presented with the name William Rehnquist, very few (8%) knew he was the current Chief Justice of the Supreme Court. Lack of interest seems to account for this lack of information: Only one American in three reads a newspaper every day or watches either local or national TV news.

FEMALE CANDIDATES

The nomination of Geraldine Ferraro for vice president of the United States in 1984 brought out many conflicting points of view. There

TABLE 15-6	The American "Voter"	
PERCENTAGE WHO:		(%)
Voted in 1996 presidential election		49
Displayed a campaign button or bumper sticker		10
Made a campaign contribution		9
Could name her or his member of Congress		23
Recognized name of the Chief Justice of the Supreme Court		8
Read a newspaper daily		31
Watch TV news daily:		
National		29
Local		33

Source: Prepared by the author from the *American National Election Study, 1996.*

© Corbis

Geraldine Ferraro, the first woman to be nominated for vice president of the United States, poses with Democratic party presidential nominee Walter Mondale at the podium of the House Chamber of the Minnesota Legislature. After the Mondale-Ferraro ticket was resoundingly rejected by the voters, some commentators blamed the defeat on prejudice against women candidates. Research shows, however, that voters do not discriminate against women running for office.

seemed to be widespread agreement among Americans that it was a welcome step toward the full participation of women in positions of power and influence. In fact, 78 percent of Americans told Gallup interviewers they approved of the nomination. Still, there was a lot of disagreement over the net effects on voters of Representative Ferraro's candidacy. Many suggested that it would make little or no difference. Others claimed it might well get Walter Mondale elected. And others saw negative effects on voters. Then, in the aftermath of President Reagan's landslide victory, many grumbled that a lot of people hadn't really voted for him but simply had rejected a woman for so high an office.

Clearly, however, voters in some other democracies will vote women into high office, as the careers of Margaret Thatcher, Indira Gandhi, Golda Meir, and others attest. Are American voters unusually sexist? If so, how do some American women get elected to high office? And what about voters in other Western democracies, such as Canada and Australia? Do they vote against women? As it happens, an excellent research literature already exists on these mat-

ters. So let's watch while researchers in Canada and the United States investigate how voters in those nations respond to female candidates.

A CLOSER VIEW
HUNTER AND DENTON: DO CANADIAN VOTERS REJECT WOMEN FOR PARLIAMENT?

Canadian women received the right to vote in 1918 and the right to stand for elected office at the federal level in the following year. In 1921 the first woman was elected to the House of Commons. For decades very few women served as M.P.s (Members of Parliament). In 1980 only 5 percent of the 282 M.P.s were female. In 1991 there were 43 women in the House of Commons, or 15 percent. (In comparison, women make up 10 percent of the U.S. House of Representatives.)

Why so few female M.P.s? Do voters reject them? Examinations of election results seem to show that they do. In the election of 1979, the average male candidate for the House received 8,494 votes, while the average female candidate

Until recently, only a few Canadian women had walked the corridors of power here in the Parliament Buildings in Ottawa. Studies of election returns show, however, that it is not voters who discriminate against female candidates. Their difficulty is getting nominated by their party.

Capital Press Services/Miller Services, Ltd.

received only 4,493. In the election held the next year, the average male received 7,967, and the average female received 3,500. Overall, women seeking election to the House received only about half as many votes as did their male opponents. Moreover, male candidates were three times as likely as female candidates to win elections when women ran against men. But these figures hide most of what really keeps women out of the House.

Alfred A. Hunter and Margaret A. Denton (1984) of McMaster University noted that many uncontrolled variables raise the possibility that these gender effects are *spurious*, that something other than gender is producing these results.

Because of the existing underrepresentation of women in the House, a much higher proportion of the male candidates are *incumbents trying for reelection*. In any election incumbents have many advantages, including name familiarity and the fact that voters already have responded favorably to them. A second important factor is

that in recent years Canadian political parties have shown a tendency to be more likely to nominate women during times when the party has been less successful at the polls; in other words parties that have been losing nominate women for office. The time order is important here. The data do not show that parties start losing because they nominate women but that *after* they have been losing they increase their rate of female nominations. And candidates representing losing parties will tend to lose—another factor in the low number of votes women receive. To make matters worse, all parties in Canada disproportionately nominate women to run for "lost cause" seats where voters have a record of overwhelming support for another party.

To see if these factors could be the real causes of the low voter support for women, Hunter and Denton undertook an analysis of all House races in the 1979 and 1980 elections. For each race they noted the gender of each candidate, the party, and whether one candidate was an incumbent. They also examined the voting record for each seat in the previous three elections to construct a measure of "competitiveness": the odds in favor of any candidate based on votes his or her party gained in the past.

Hunter and Denton found that differences in the vote-getting abilities of male and female candidates "disappeared entirely" when the candidates were equated in terms of party, incumbency, and competitiveness. The apparent link between gender and votes is in fact spurious, as Hunter and Denton had suspected. It's not the voters who are keeping the proportion of women in the House so low. But if not them, who? Party political elites. As Hunter and Denton (1984) put it:

The problems which women aspiring to elected federal office experience largely occur before election day, most notably in their difficulties in securing nominations in the first place, and beyond this, in gaining nominations which carry a reasonable prospect of victory.

Their findings parallel those found in the United States. Having controlled for incumbency and party in recent elections to the U.S. House of Representatives, R. Darcy and Sarah Schramm (1977) reported:

The evidence indicates that the electorate is indifferent to the sex of congressional candidates. . . . Why then are so few women serving? The answer lies in the recruitment and nomination processes.

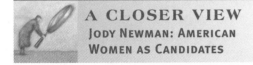

A CLOSER VIEW
JODY NEWMAN: AMERICAN WOMEN AS CANDIDATES

The Nineteenth Amendment, which gave American women the right to vote, was passed by Congress on June 4, 1919, and was ratified by the states on August 18, 1920. But at that time many Americans, women as well as men, still opposed giving women the vote, and many who approved still had reservations about electing women to political office. Thus, in 1937, when

New Paramount /Public Archives of Canada/PA-127295

the Gallup Poll first asked the question, only a third of Americans said they would be willing to vote for a woman for president (Table 15-7). However, support for a female presidential candidate has increased consistently through the years, and today, 94 percent say they would be willing to vote for a woman.

Nevertheless, many observers, especially media political "experts," claim that Americans discriminate against female candidates. For proof they usually point to the lack of women in elected offices and assume that this is caused by voter preferences for male candidates. But there are other plausible reasons women may not be fully represented among officeholders. One of these is simply that journalists, party leaders, and potential candidates *believe* voters prefer male candidates. Thus, party leaders may be reluctant to support female candidates when their party has a chance to win. Political action committees may be reluctant to fund female candidates. And the most qualified women may be reluctant to run for office, believing that female candidates face a disadvantage with voters. Therefore, the lack of women officeholders could merely reflect that voters don't often have the opportunity to vote for female candidates— that too few women run for office.

Agnes Campbell Macphail became the first woman elected to the Canadian House of Commons when she won a seat representing a rural Ontario district in the election of 1921— the first federal election in which Canadian women had the vote. Macphail first sat as a member of the Progressive party but later became an independent. She also was the first woman appointed a member of the Canadian delegation to the League of Nations. Macphail served as an M.P. until 1940, when she was defeated, partly because of her reluctance to support Canada's entry into World War II.

TABLE 15-7	A Woman for President, 1937–2000

If your party nominated a woman for president, would you vote for her if she were qualified for the job?*

YEAR	YES (%)
1937	34
1949	48
1955	52
1967	57
1972	70
1977	77
1986	84
1990	86
1993	91
1996	93
1998	94
2000	94

Sources: Prepared by the author from the Gallup Poll and National Opinion Research Center, General Social Surveys.

*Several slightly different wordings of the question were used prior to 1972.

To resolve these issues, Jody Newman (1994), a researcher for the National Women's Political Caucus, conducted a very well-designed study. First, she obtained the results of all elections for state legislatures in 1986, 1988, 1990, and 1992. Next, she obtained data for all elections of U.S. representatives and senators from 1972 through 1992. Finally, she included all elections for governor since 1972. In all she had data on 50,563 candidates for office.

Her next step was to compare the success of male and female candidates overall and against one another. The results were both clear and compelling: "a candidate's sex does not affect his or her chances of winning an election."

Newman's data showed that incumbents have a huge advantage in elections. Given that current officeholders are disproportionately men, incumbents also are disproportionately male. Consequently, as Newman pointed out:

The reason people may think that women are less likely to win is that most incumbents are men, and incumbents enjoy a huge advantage over challengers and open seat candidates. But when men running as incumbents were compared to women running as incumbents, men running for open seats to women running for open seats, and men running as challengers to women running as challengers, men had no advantage over women.

For example, in races for open seats in state legislatures, when a man ran against a woman, the woman won 52 percent of the time. And when a man ran against a woman for an open seat in the U.S. House of Representatives, the man won 51 percent of the time. The problem of female underrepresentation in office does not lie with the voters. Newman concluded:

As this study makes clear, the major reason that there are so few women in public office is simply that there haven't been many women running for office—particularly for open seats, where the chances of winning are good.

As a result of Newman's research, this soon may change.

IDEOLOGY AND PUBLIC OPINION

As I mentioned at the start of this book, all social scientists have the advantage of intimate familiarity with their objects of study; no astronomer can be a comet, but every sociologist is a person. Ironically, this familiarity is also the source of much frustration to social scientists because too often we expect everyone else to function as we do. Nowhere is this more evident than in the area of research on public attitudes and opinions. Time and again, a researcher has framed a hypothesis linking some set of beliefs or opinions—a linking that seemed obvious—only to have the data offer little or no support. Some examples from a recent survey based on my own students in introductory sociology will illustrate the point. Among the many opinion questions I asked the students were these two developed by Robert Wuthnow (1976):

If one works hard enough, one can do anything.
___ Agree strongly
___ Agree somewhat
___ Disagree somewhat
___ Disagree strongly

The poor simply aren't willing to work hard.
___ Agree strongly
___ Agree somewhat
___ Disagree somewhat
___ Disagree strongly

Wuthnow not only thought people who agreed with one would agree with the other (and vice versa) but also thought the statements formed part of a very general meaning system, or political ideology, on which people would base a whole spectrum of their beliefs and opinions. However, among my students these attitudes were hardly correlated at all. It turned out that they weren't correlated in Wuthnow's sample of the population of San Francisco either (Stark and Bainbridge, 1985).

In the same questionnaire, I asked students to agree or disagree with the following statement, also developed by Wuthnow:

It is good to live in a fantasy world now and then.

___ Agree strongly
___ Agree somewhat
___ Disagree somewhat
___ Disagree strongly

It seemed likely to me that people who agreed with this statement would be more apt to also report they liked to read novels—using reading as a way to enter a fantasy world. Not so. Nor was this item correlated with liking to read spy and detective novels. It wasn't even correlated with liking to read science fiction.

These examples help us to see a very basic problem for social scientists. We prefer to believe that every human is, underneath it all, a philosopher. As the celebrated anthropologist Clyde Kluckhohn (1962) put it, there is a "philosophy behind the way of life of every individual and of every relatively homogeneous group at any given point in their histories." Kluckhohn identified these philosophies as "world views." Wuthnow (1976) explained it this way:

People adopt relatively comprehensive or transcendent, but nonetheless identifiable, understandings of life which inform their attitudes and actions under a wide variety of conditions.

Wuthnow called these understandings "meaning systems." When such meaning systems are very prominent in a person's thought or in the discussions within a group, they sometimes are called ideologies. An **ideology** is a connected set of beliefs based on a few very general and abstract ideas. An ideology is used to evaluate and respond to proposals, conditions,

and events in the world around us. An ideology, then, is essentially a theory about life. If its content is primarily political, an ideology will consist of a few abstract assertions to explain why and how societies ought to be run. Hence, when faced with a specific issue, a person seeks to derive an answer from his or her ideological premises, not simply on a pragmatic issue-by-issue basis.

For years social scientists had not the slightest doubt that only a few people of deficient intellect lacked worldviews or that most people based their political reactions on an ideology. In fact, in one of my earliest papers, I flatly asserted that "all men and all human groups have . . . a worldview . . . furnishing them a more or less orderly and comprehensible picture of the world." Today, I am equally embarrassed by my use of "men" instead of "humans" and by the claim that we all are amateur philosophers.

I was wrong because I was misled in a way that many social scientists are misled: Most of us and most of those we meet are amateur philosophers, and many have a very noticeable political ideology. However, as social scientists have begun to do research on attitude and opinion surveys based on general populations, they have been forced to learn that although some people base their beliefs and actions on a worldview, and some people base their politics on an abstract ideology, a lot of normal and competent people don't. In fact, it appears that most people don't. Moreover, a substantial part of the population in any society ignores most of the "issues," reserving attention for occasional matters of great urgency or with special personal implications.

We can see, then, why Laurie Ekstrand and William Eckert found that religious affiliation or church attendance, among other factors, did not influence willingness to support a female political candidate (1981). It also helps us understand why they had expected to find such an influence. In framing their hypotheses, they wrote: "The tenets of religious fundamentalism seem inconsistent with support for female candidates. Fundamentalism urges a traditional role for women." Clearly, they were assuming that religious students were guided by an ideology—that each was a young Jerry Falwell. If this is so, then people who support prayer in public schools, for example, can be expected to oppose the Equal Rights Amendment, and they probably also will oppose both abortion and pornography. This set of expectations would clearly be consistent with the public position of conservative Protestant groups such as the Moral Majority.

The reasoning is cogent, but the conclusion is wrong. It is not only contrary to how student subjects acted in an experiment but also inconsistent with the political behavior of the general public. According to the 1990 General Social Survey, most Americans (61%) oppose busing students to achieve racial balance, yet most (57%) say they wouldn't mind living in a neighborhood that was 50 percent African American and only 10 percent say they would not vote for an African American candidate for president. The majority (56%) favor school prayer, but 58 percent also favor making birth control devices available to teenagers fourteen to sixteen even if their parents object; meanwhile, 93 percent want new laws against pornography. Perhaps even more important, most Americans who hold this pattern of beliefs do not see their views as inconsistent or muddled because they are not trying to remain in tune with an ideology.

A final example may clarify how willing people are to express seemingly contradictory opinions. In the 1990 General Social Survey, which asked about voting for a woman for president, respondents were also asked to agree or disagree with the following statement: "Most men are better suited emotionally for politics than are most women." One might suppose that people who agreed with that statement would be unlikely to endorse a woman for president. In fact, 67 percent of those who agreed said that they would, nevertheless, vote for a woman for president. The survey also asked people to agree or disagree with an even stronger statement about the proper political role for women: "Women should take care of running their homes and leave running the country up to men." But of those who agreed with this statement, 54 percent still insisted they would support a woman for president. Sociologists might think these attitudes form a tight little logical package, but clearly many Americans do not.

In a classic study of American voters, Philip E. Converse (1964) could classify only about 3 percent as basing their decisions and opinions on an ideology. Another 12 percent he classified as making some use of an underlying political ideology. About half of the voters took a mildly issue-oriented approach to politics, and the rest seemed to ignore all policy, issue, and ideological matters. These results imply that *some* people take a very different approach to politics than do others and that only *some* people participate in the democratic political process at all.

ELITE AND MASS OPINION

Converse (1964) recounts that a young scholar became interested in the rise of the abolitionist movement in the northern United States: how antislavery, abolitionist beliefs spread and shaped political opinion, and how this in turn fostered the new Republican party and culminated in the election of Lincoln and—soon after the Civil War broke out—in the Emancipation Proclamation. He was aware that the American Anti-Slavery Society never attracted more than 200,000 members, or about 3 percent of the adult population outside the South. So, to see how support for the abolitionist movement was translated into the nearly 2 million votes needed to elect Lincoln, he had to trace the informal channels through which abolitionist sentiments had spread beyond the confines of the Anti-Slavery Society. In other words he wanted to show how opposition to slavery had become an increasingly significant part of informal political discussion and of public opinion during the decade leading to the war. To do this, the young scholar analyzed the contents of many large collections of personal letters saved by various families in Ohio—letters written during the 1850s and 1860s. But his study never was published or even completed because the young sociologist found no references at all to abolition in any of the letters. This forces the conclusion that mass support for Lincoln and, eventually, for the war was based on many factors, but concern about the plight of the slaves in the South was not one of them (although that may have changed once the war got going). However, a small elite committed to abolition had sufficient influence to see that antislavery policies won. Moreover, the abolitionist ideology seems to have infused this elite with a sense of single-mindedness and dedication that got results.

This example does not reflect an isolated case. Leaders of social movements and political organizations usually display patterns of opinion consistent with a basic outlook or ideology.

A CLOSER VIEW
HERBERT MCCLOSKY: ELITES AND IDEOLOGY

Studies of opinions and attitudes not only show that ideologies and worldviews typically are limited to small elites within a society but also offer some clues about why this is so. Primary among these is the pressure toward intellectual consistency placed on some people because of the special positions they occupy and the roles they play. We can best see this by watching Herbert McClosky (1964) conduct a classic study of roles and ideologies.

McClosky arranged to distribute very lengthy questionnaires to 3,020 delegates and alternates to the Republican and Democratic presidential nominating conventions in 1956 (the Republicans renominated President Dwight Eisenhower while the Democrats renominated Adlai Stevenson). Then McClosky commissioned the Gallup Poll to distribute this questionnaire to a national sample of 1,500 American adults. The questionnaire included scores of items on political philosophy, positions on current national issues—the whole range of political concerns of the time. McClosky expected the answers to many of the questions to combine into distinct ideological clusters—to be very highly intercorrelated.

When McClosky began to examine his results, striking differences appeared. Among the delegates, the expected correlations held strongly. For the general public, the correlations were weak. When he separated the general public according to their levels of education, McClosky found much stronger correlations among those who had attended college. But even they displayed much lower correlations than did either group of delegates.

In a now-classic paper, McClosky explained that delegates to political conventions are selected for active participation in political affairs in their home states; many of them hold state and local political office. Every day they talk about politics, often seeking to convince others to agree with them and to support them. That's part of the politician's role. But it also places their political views under constant inspection by others, many of whom disagree with them or want to change the politicians' minds. A continual part of this interactive process involves pointing out inconsistencies: "If you say you favor this, how in the world can you also favor that?" The result of this pressure is that members of political party elites will tend to have political views that are highly internally consistent. Indeed, the easiest way to accomplish consistency is to adopt a political ideology and let it guide your specific issue positions.

Even most college-educated persons do not experience the pressures toward ideological consistency felt by professional politicians—hence, their lower correlations. However, because they are well educated, they will be more sensitive than the less-educated to more obvious inconsistencies. Moreover, many other occupational roles associated with higher education also place their occupants under pressures to form consistent worldviews. The clergy are an obvious example. So, too, are college professors, journalists, and intellectuals generally—explaining why social scientists have had so much trouble recognizing that not everyone shares their taste for detailed worldviews. Better-educated persons of all occupations are particularly likely to take an active part in politics, just as the Anti-Slavery Society was highly overrecruited from among the educated. As suggested, anyone who gets active in politics will feel constraints toward being more ideological, even when being active means no more than being interested and paying attention.

A CLOSER VIEW
PHILIP E. CONVERSE: ISSUE PUBLICS

The most severe shock to social scientists' conceptions of public opinion occurred when **panel studies** began. In these surveys the same sample of respondents is interviewed several times. One purpose of panel studies has been to chart the ebbs and flows in the fortunes of candidates as a campaign progresses. In a series of panel studies conducted during presidential campaigns by the Survey Research Center at the University of Michigan, some questions about political attitudes and opinions were repeated several times. As a result the same person's answers could be compared over time. The shock came when these comparisons exposed the fact that a substantial part of the population has no political opinions. On major issues some voters display a crystallized opinion that is stable over time. For example, if asked whether the post office should be sold to private business, they have a firm view that is the same on April 15 as on May 1. However, a very substantial proportion of respondents will give a series of utterly inconsistent responses over time. They are not constantly changing their minds; instead, many people simply offer impulsive answers that have no lasting significance to them. So over time, their answers form a meaningless, random pattern of responses. Converse (1964) summed it up this way:

Large portions of an electorate do not have meaningful political beliefs, even on issues that have formed the basis for intense political controversy among elites for substantial periods of time.

Thus, Converse found that on any given issue only a subset of the public will have opinions or interest. He coined the term *issue publics* to identify those who take interest in and who participate at least as observers in discussions of an issue. He identified issue publics in the plural, rather than define a single public who has political knowledge and interest, because people are inconsistent even in their lack of

beliefs and interests. People will belong to one issue public and not another depending on how much that issue directly affects them personally. Some take part only in issue publics devoted to retirement and old-age issues. Others attend only to agricultural issues. Still others are interested only in the space program (Taylor, 1983).

ORGANIZATIONS AND IDEOLOGIES

One reason many people lack political ideologies is that people rarely invent their own. Ideologies are intellectual creations, often involving many different authors and interpreters. Moreover, ideologies must be sustained, protected, and promulgated. In effect only elites can create and preserve ideologies. Thus, for an ideology to have general impact on a culture, it must be successfully popularized by an elite. Organizations based on a specific political ideology often take the form of a political party. However, many political parties are not committed to a single or sharply defined ideology. Where parties are less ideological, the public will be less ideological, but where many highly ideological parties compete for support, a larger proportion of the public will be ideological, too.

If political parties are a common source of political ideologies, then people whose politics are based on an ideology also will be apt to identify closely with a political party that advocates their ideology. Table 15-8 shows how people in eighteen nations responded when asked if they considered themselves to be close to a particular political party. In only two nations, Sweden and Italy, did the majority say they were close to a party. Presumably, most people who say they are not close to any party do not base their political views on an ideology. In that case most people in most of these nations are not ideological—only 30 percent of Mexicans, 27 percent of Americans, and 22 percent of Canadians regard themselves as close to a party. Keep in mind, too, that not all parties sustain an ideology. Few observers think the two major American parties are especially ideological; hence, even those who closely identify with the Democrats or the Republicans might lack a clear ideology. In contrast, parties in Europe tend to be quite ideological.

A major barrier to the highly fragmented and ideological political spectrum found in much of Western Europe is the geographical basis of representation used in both Canada and the

TABLE 15-8	"Do you consider yourself to be close or not to any particular political party?"	
	NATION	CLOSE (%)
	Sweden	91
	Italy	59
	West Germany	49
	Iceland	43
	Denmark	43
	Norway	39
	Japan	37
	Mexico	**30**
	Spain	29
	United States	**27**
	France	26
	Australia	25
	Canada	**22**
	Great Britain	21
	Belgium	20
	South Africa	20
	Ireland	19
	Northern Ireland	19

Source: Prepared by the author from the *World Values Survey, 1981–1983.*

Note: Surveys conducted before the reunification of Germany.

United States. In France, for example, if a party gets 5 percent of the vote nationwide, it automatically is assigned 5 percent of the seats in the parliament. In the United States and Canada, a party gets seats only by coming in first in a given congressional district or parliamentary riding. In principle a party could get 49 percent of the votes across Canada or the United States and get no representation because in every district some other party got 51 percent of the vote. Thus, there is no political future in building a highly ideological party aimed at appealing to a narrow interest group within an electorate—which has sealed the fate of highly ideological parties. Instead, a party must seek to appeal broadly. As a result successful parties are coalitions of many internal issue publics. This feature encourages parties to make a lot of internal compromises to satisfy their disparate membership. Such parties often will try to appeal to many of the same interest publics. The result is parties that tend toward the middle of the political spectrum and, therefore, tend to resemble one another. Voters always have a choice, but their choices are always limited.

CONCLUSION

This chapter began with an examination of the problem of public goods: how individuals must surrender some of their freedom to sustain organized social life. The experiment that simulated the tragedy of the commons further illustrated this necessity. In the end social life requires that the individual be subject to coercion by the group. The state arises as specialists appear in societies whose task is to monopolize and use the means of coercion. Unfortunately, those who control the state are apt to abuse their power and exploit the rest of the society. Thus, *the* political question always is, How to have a state and keep it tame—how to keep it from being repressive and exploitative? We then saw how the slow diffusion of power in England offered an answer to this question. If power can be dispersed among a number of elites with conflicting self-interests, they will keep the state tame by constantly shifting coalitions to limit one another's power. This is the principle of *pluralism*. Some sociologists dispute the belief that democracy exists in Europe and North America, claiming instead that a small interlocking power elite decides what is going on. But other sociologists regard this claim as oversimplified.

We then looked at the role of the public in democratic societies. First, we saw how the invention of public opinion polling has made it possible to discover what the public thinks about issues. Next, we examined how voters in Canada and the United States react to female candidates for high office, finding that they seem entirely willing to accept them. Then, we considered more closely the role of ideology in public opinion, finding that it primarily influences political discourse among members of political and occupational elites. Finally, we saw that successful political parties in North America tend to seek support from many elites and issue publics and, therefore, tend toward the middle of the political spectrum.

Review Glossary

Terms are listed in the order in which they appear in the chapter.

public or **collective goods** Things necessary for group life that individual members of a society cannot provide for themselves and that require cooperative actions by many members.

state The organized monopoly on the use of force (or coercion) within a society; synonymous with government.

pluralism, pluralist state A system or a society in which power is dispersed among many competing elites who act to limit one another's power and therefore minimize the repression and exploitation of members.

elitist state A society ruled by a single elite group; such states repress and exploit nonelite members.

power elite A term C. Wright Mills used to identify an inner circle of military, government, and business leaders he believed controls the United States.

ideology A connected set of strongly held beliefs based on a few very abstract ideas, used to guide one's reactions to external events; for example, a political ideology is used to decide how societies ought to be run.

panel studies Public opinion surveys that interview the same respondents several times.

Suggested Readings

Dahl, Robert. 1956. *A Preface to Democratic Theory.* Chicago: University of Chicago Press.

Gallup, George H. 1972. *The Gallup Poll: Public Opinion 1935–1971.* 3 vols. New York: Random House.

Gallup, George H. 1985. *The Gallup Poll: Public Opinion 1972–1984.* New York: Random House.

Newman, Jody. 1994. *Perception and Reality: A Study Comparing the Success of Men and Women Candidates.*

Washington, D.C.: National Women's Political Caucus.

Nozick, Robert. 1974. *Anarchy, State and Utopia.* New York: Basic Books.

Olsen, Mancur. 1965. *The Logic of Collective Action.* Cambridge, Mass.: Harvard University Press.

Sociology Online

www.socstark10.com

GO TO THE INTERNET AND TYPE: www.socstark10.com.

↗**CLICK ON:** The 50 States

✓ **SELECT:** % FED LAND: Percentage of all land owned by the federal government.

↗**CLICK ON:** Analyze Now

Is there a regional pattern to the variable? What impact might these differences have on regional political attitudes?

↗**CLICK ON:** Select New Variable

✓ **SELECT:** % VOTED 96: Percentage of voting age population who voted in the 1996 presidential election.

↗**CLICK ON:** Analyze Now

Is there a regional pattern to this variable? Which state is highest? Which state is below 40 percent? Does that seem to fit in with earlier findings concerning this state?

↗**CLICK ON:** Select New Data Set

↗**CLICK ON:** Nations of the Globe

✓ **SELECT:** %TURNOUT: Percentage of eligible voters who voted in the most recent parliamentary election.

↗**CLICK ON:** Analyze Now

How many nations have higher turnouts than the United States? Can you suggest reasons America has rather low voter turnouts?

USING INFOTRAC COLLEGE EDITION

GO TO THE INTERNET AND TYPE:
www.InfoTrac-college.com

↗**CLICK ON:** Register New Account
You will be asked to enter your Access Code and to create and enter a User Name and Password. Having done so,

↗**CLICK ON:** InfoTrac College Edition or Log On.

As noted in the chapter, voting has been one of the most intensely studied aspects of modern life. The relationship between voting and the power of political elites has been endlessly disputed by critics and supporters of democracy.

SEARCH TERMS:
Elite (Social Science)
Voting Research

CHAPTER

16

The Interplay between Education and Occupation

I
N SIMPLE SOCIETIES CHILDREN don't go off to school in the morning and their parents don't go off to jobs. Yet, children still get educated and work gets done. As with the family, religion, and politics, all human societies have educational and economic institutions. The forms often differ, but the basic functions are always performed in an organized way.

In hunting and gathering societies, children are educated by their parents and older siblings. Also, by tagging after their fathers and uncles, the boys slowly learn to perform the work expected of adult males, such as fishing. The girls learn their adult tasks from helping their mothers. Yet, even in such primitive circumstances, the link between education and occupation is direct and powerful. Until people are sufficiently educated, they cannot fulfill their economic responsibilities.

In modern societies the link between education and occupation is so obvious and important that we can often guess a person's education from knowing his or her occupation, and vice versa. We can be sure that lawyers and doctors spent many years in school and that unskilled laborers probably had minimal education.

Generally, the more education people have, the more they earn and the higher their occupational status. Indeed, if we know people's education and occupation, then we can often deduce many other things about them: how they vote; what kind of TV shows they watch; what kind of neighborhood they live in; what their tastes in food, clothing, art, automobiles, magazines, and music are; and even what kind of sporting events they attend.

Because of the interdependence of education and occupation, this chapter examines the interplay between these two social institutions.

CHAPTER OUTLINE

OCCUPATIONAL PRESTIGE

THE TRANSFORMATION OF WORK

THE TRANSFORMATION OF THE LABOR FORCE
Women in the Labor Force
Professional Women
Unemployment
Job Satisfaction

THE TRANSFORMATION OF EDUCATION

A CLOSER VIEW
"Educated" Americans

DO SCHOOLS REALLY MATTER?

A CLOSER VIEW
Barbara Heyns: The Effects of Summer

HIGH SCHOOL TODAY

HOMESCHOOLING

A CLOSER VIEW
Heyneman and Loxley:
School Effects Worldwide

DOES EDUCATION PAY?

A CLOSER VIEW
John W. Meyer: Theory
of Educational Functions

CONCLUSION

REVIEW GLOSSARY

SUGGESTED READINGS

SOCIOLOGY ONLINE

We shall begin by examining the dramatic shifts in the number and kinds of occupations produced by modernization. In Chapters 9 and 10, we saw that these changes resulted in a great deal of structural mobility. They also prompted major changes in the educational system, and these in turn led to more changes in the occupational structure. While the general connection between education and occupation is obvious, many of the links are subtle.

OCCUPATIONAL PRESTIGE

Because our occupations play a central role in our lives, we have very clear and sensitive notions about which jobs are "better" and which are "worse." A long series of studies of **occupational prestige** have demonstrated these distinctions.

Back in 1947 Paul Hatt and Cecil North presented a national sample of American adults with a list of eighty-seven occupational titles (Reiss, 1961). Each respondent was asked to rate the "general standing" of each job as excellent (a rating of 5), good (4), average (3), somewhat below average (2), or poor (1). From these numerical weights, an average score was computed for each occupation. Because of the computational method used, each occupation received a score between 20 and 100.

Table 16-1 shows the list of occupations and their scores. U.S. Supreme Court justice heads the list, followed by physician, nuclear physicist, scientist, government scientist, state governor, U.S. cabinet member, and college professor. At the bottom are garbage collector, street sweeper, and shoe shiner. Americans and people in other nations have rated this list of occupations many times, and the results are very stable over time and place. Notice how many of the higher-prestige positions require a college education or even postgraduate study.

In 1962 the Hatt and North study was repeated. The results were identical: The correlation was .99 (Hodge, Siegal, and Rossi, 1964). In a follow-up study, these same researchers measured the occupational prestige of an expanded list of occupations, 200 in all. This study prompted John Porter to seek funding for a replication in Canada, as mentioned in Chapter 10.

Having secured the needed funding and recruited Peter C. Pineo as his collaborator, Porter adapted the list of occupations to suit Canadian respondents. Most of the occupational

TABLE 16-1	Occupational Prestige Scores		
SCORE	OCCUPATION	SCORE	OCCUPATION
94	U.S. Supreme Court justice	72	Policeman
93	Physician	71	AVERAGE
92	Nuclear physicist	71	Reporter on a daily newspaper
92	Scientist	70	Bookkeeper
91	Government scientist	70	Radio announcer
91	State governor	69	Insurance agent
90	Cabinet member	69	Tenant farmer who owns livestock and machinery and manages the farm
90	College professor		
90	Member, U.S. Congress	67	Local labor union official
89	Chemist	67	Manager of a small store in a city
89	U.S. Foreign Service diplomat	66	Mail carrier
89	Lawyer	66	Railroad conductor
88	Architect	66	Traveling salesman for a wholesale concern
88	County judge		
88	Dentist	65	Plumber
87	Mayor of a large city	63	Barber
87	Board member of a large corporation	63	Machine operator in a factory
87	Minister	63	Owner-operator of a lunch stand
87	Psychologist	63	Playground director
86	Airline pilot	62	U.S. Army corporal
86	Civil engineer	62	Garage mechanic
86	State government department head	59	Truck driver
86	Priest	58	Fisherman who owns his own boat
85	Banker	56	Clerk in a store
85	Biologist	56	Milk route man
83	Sociologist	56	Streetcar motorman
82	U.S. Army captain	55	Lumberjack
81	Accountant for a large business	55	Restaurant cook
81	Public school teacher	54	Nightclub singer
80	Building contractor	51	Filling station attendant
80	Owner of a factory that employs about 100 people	50	Coal miner
		50	Dock worker
78	Artist whose paintings are exhibited in galleries	50	Night watchman
		50	Railroad section head
78	Novelist	49	Restaurant waiter
78	Economist	49	Taxi driver
78	Symphony orchestra musician	48	Bartender
77	International labor union official	48	Farmhand
76	County agricultural agent	48	Janitor
76	Electrician	45	Clothes presser in a laundry
76	Railroad engineer	44	Soda fountain clerk
75	Owner-operator of a printing shop	42	Sharecropper who owns no livestock or equipment and does not manage farm
75	Trained machinist		
74	Farm owner and operator	39	Garbage collector
74	Undertaker	36	Street sweeper
74	City welfare worker	34	Shoe shiner
73	Newspaper columnist		

Source: Hodge, Siegal, and Rossi (1964).

titles needed no adjustment (biologist, airline pilot, or musician). A few others needed minor editing: "state governor" was changed to "provincial premier" and "member of the United States House of Representatives" to "member of the Canadian House of Commons." In the end Pineo and Porter (1967) used 174 occupational titles common to the two nations. They added several more to assess regional variations within Canada ("whistle punk" and "cod fisherman" were two of these). Finally, they added two nonexistent occupations to see if people would respond "don't know" or just go ahead and rank them: "biologer" and "archaeopotrist." At this point Pineo and Porter had the list translated into French, using elaborate procedures to attempt to make the lists equivalent.

What were the results? First of all, the correlations between the American and Canadian results were a resounding .98; people on both sides of the border have the same ideas about what constitutes a good or a bad job. Second, expected differences between French- and English-speaking Canadians did not show up. "In spite of the stereotype of French Canadians as placing greater value on artistic pursuits than on others," they did not rank artistic occupations higher than did English-speaking respondents. Finally, substantial numbers did respond "don't know" rather than rank "biologer" or "archaeopotrist." The majority, however, assigned them above-average ranks, probably because both made-up names clearly suggest scientific professions and are easily associated with biologist and archeologist. One wonders if made-up names sounding like manual-labor jobs would have done so well ("deltahumper" or "snoodkelper," for example).

Not only do Americans and Canadians agree about occupational prestige, but people all over the world also appear to do so. Similar results have been reported for Germany, Great Britain, Japan, New Zealand, the former Soviet Union, Ghana, Guam, India, Indonesia, the Ivory Coast, and the Philippines (Hodge, Siegal, and Rossi, 1964; Hodge, Treiman, and Rossi, 1966). Because of these similarities, sociologists suspect that people of all nations are familiar with the occupations found in industrial societies and the relative importance of these occupations.

Further research has determined why people rate various occupations high or low. Peter Blau and Otis Dudley Duncan (1967) found that if they took the average education of persons in a particular occupation and combined that with the average income of those persons, they could accurately predict the occupational prestige score that people would give that occupation. In other words the more training an occupation requires and the more pay it offers, the greater its public prestige. This suggests that people rate a job by its importance. They seem to assume that no one will put in many years to prepare for a job that is unimportant and that society will not pay high salaries to get people to do unimportant work.

That prestige ratings reflect education and training indicates that the process of obtaining a particular occupational status begins when we are young. How much and what kind of education we receive are the primary factors determining our occupational opportunities. This touches on the matter of differential socialization discussed in Chapter 6. There we saw that much of socialization is geared to the specific roles a person is expected to play. One of the most important roles is the child's future occupation. Thus, from an early age, children tend to receive socialization appropriate to certain occupations. Children who display little academic aptitude tend to be placed in educational tracks that end with high school and lead to manual occupations. More academically talented children are tracked into college preparatory courses and groomed for technical and professional occupations. Education and occupation are thus intimately associated, and the interplay between them begins early in life.

THE TRANSFORMATION OF WORK

Peter F. Drucker (1969) examined how changes in work dramatically increased productivity to make modern standards of living possible. He pointed out that such achievements are not the result of harder work. Surely, modern workers do not work harder than their grandparents—they probably do not even work as hard. However, they get better results from their work because, as Drucker put it, "they work smarter."

Technological innovations have made it possible to work smarter. In times past ten laborers may have worked a week to dig a hole for the foundation for a new house. Today, one operator with an earthmover digs the same hole in a few hours. This operator is obviously not working harder than those who dug with shovels.

Kansas State Historical Society

Having a machine and the technical knowledge to use it, one modern worker possesses the strength of many manual laborers.

However, it is more than just machines that enable us to work smarter. Applying any knowledge to work produces smarter work. Consider the accomplishments of Frederick W. Taylor, who originated time and motion studies of work. Taylor applied scientific principles to increase the efficiency of even very unskilled tasks.

One of Taylor's most famous experiments, conducted in 1899, involved teaching a group of unskilled laborers to shovel sand efficiently. He selected a man named Schmidt and began to work on increasing the rate at which Schmidt could shovel sand. The first thing Taylor did was to experiment with shovels of different sizes. He argued that the shovel should not be too big, or else workers would tire rapidly; then the amount of rest time they would need would offset the extra amounts they shoveled while they worked. On the other hand, if the shovel was too small, the work would be inefficient. Through trial and error, Taylor selected the right-sized shovel for Schmidt. Taylor then experimented with Schmidt to find the best technique for using it; the goal was to find the technique maximizing the sand shoveled and minimizing the energy expended. Finally, Taylor determined the most efficient intervals for rest periods. He found that it was better for men to rest frequently for short periods than to rest occasionally for long periods.

Soon Schmidt was able to shovel more sand per shift than the combined amount shoveled by the two strongest men not prepared by Taylor. As a result of his increased production, Schmidt earned more than twice as much as the other workers. Then, the whole shoveling team was trained in the new method, and each laborer earned a much higher salary than before. Not only did they shovel much more sand, but they were also much less tired at the end of the day. All of this resulted not from using machinery but from what came to be known as **scientific management:** the application of scientific techniques to improve work efficiency.

These farmers were already working smarter back in the 1890s. Using horse-drawn wagons and a newly invented shucking machine, this crew picked and piled 10,000 bushels of corn. Nevertheless, today, one farmer driving a modern harvester could outwork many such crews.

Spring housecleaning a century ago included hanging all of the bedding out to air. This housewife in Seattle is pumping a pail of water for scrubbing. If she wants it to be hot, she will have to heat it on her kitchen range. On laundry day she will have to carry many pails of water into the house and heat them. After scrubbing the laundry by hand, she will carry it out to hang on the lines to dry. It will take her most of the next day to do the ironing.

Historical Photography Collection, University of Washington Libraries

Although today a host of experts is still carrying on Taylor's approach to the study of work, his attempts to revolutionize manual labor were short-lived. Unskilled manual work has been taken over almost entirely by machines. As machines have replaced people in the most repetitive, dirty, and sweaty jobs, people have turned more and more to "knowledge" work, including using knowledge to design, maintain, and operate these machines.

Consequently, the *kind* of work people are most likely to perform in industrial nations has changed remarkably. In 1900 fewer than 20 percent of North Americans had white-collar jobs; most people did manual labor on farms and in factories. Today, there are more white-collar than blue-collar workers, and fewer jobs involving manual labor are available each year.

Whereas manual workers manipulate "things" for a living, white-collar workers manipulate information. In Drucker's judgment we are changing from a primarily industrial economy to a "knowledge" economy. In fact, the most rapidly expanding job categories over the past century have been at the top—professional, technical, and managerial occupations. Today, one of four working Americans holds a job of this type. We are working smarter all the time, and this has changed *who* is working and *why* they work.

THE TRANSFORMATION OF THE LABOR FORCE

In highly industrialized economies, not only has what people do for a living changed, but the proportion of people who are employed also has dramatically altered. In 1870 about 40 percent of North Americans over age sixteen were in the **labor force.** In 2000 nearly two-thirds were in the labor force. This expansion of the proportion of Americans and Canadians in the labor force occurred despite the fact that much smaller proportions of both young people and old people are working today. Now most North Americans finish high school or college before entering the labor force, and most people must retire at age 70 (most retire before then). In 1870 few went to high school or college, and therefore, most started work young; people rarely retired as long as they could continue to work. How did an enormous expansion of the work force occur despite these changes? Women joined the labor force.

WOMEN IN THE LABOR FORCE

In 1900 few women in North America or anywhere else were employed outside the home. Today, about seven of ten American and Canadian women are full-time members of the labor force (Table 16-2).

Why have so many women gone to work? There are probably many reasons. As we saw in Chapter 12, the rapid expansion of women's participation in the labor force first occurred on the East Coast, where a highly unfavorable sex ratio forced many women to support themselves rather than relying on husbands to support them. This factor in turn attracted industries that were able to use this newly available work force.

A second reason women went to work is reduced fertility. Freedom from long years of pregnancy and child rearing has given women more opportunity to pursue a career. A third reason is increased freedom from housework. Women in 1900 made their own soap, spent Monday washing clothes, and needed most of Tuesday to iron them. The modern home has a washer and a dryer, commercial detergents, and clothes need little or no ironing. Reduced demands in the home have enabled women to take outside jobs.

However, another key reason for the massive entry of women into the labor force has been a change in the kinds of work available. The shift to knowledge work has lured many women out of the home. Modernization has eliminated muscle power as a major requirement for jobs and a major source of energy; many jobs today are entirely "mental." In Chapter 19 we shall examine how technology revolutionized farming. Here we need only note that farming today requires a great deal of sophistication but only a modest amount of strength. In fact, few jobs today require more strength than the average woman possesses. And it is clear in Table 16-2 that it is culture, not physical capacity, that accounts for most international differences in female participation in the labor force. Here we see the percentage of women age fifteen to sixty-four who are in the labor force. More than eight Icelandic women of ten are employed full time outside the home, as are seven of ten American and Canadian women. In southern Europe fewer than half of

TABLE 16-2	Female Labor Force Participation (Selected Nations)		
NATION	PERCENT WOMEN 15–64 IN THE LABOR FORCE	NATION	PERCENT WOMEN 15–64 IN THE LABOR FORCE
Iceland	82.6%	France	63.8%
Switzerland	77.4%	South Korea	55.6%
Norway	76.4%	Spain	50.9%
Denmark	76.0%	Greece	49.2%
Sweden	75.5%	Italy	47.8%
United States	**70.7%**	**Mexico**	**41.6%**
Canada	**70.5%**	Turkey	26.5%
Japan	64.4%	Saudi Arabia	5.1%

Source: *Statistical Abstract of the United States, 2005.*

the women work, and in Iraq and Saudi Arabia, hardly any women are in the labor force.

A final reason women work is obvious: for money. The old cliché "Two can live as cheaply as one" was probably never true. But it surely is true that two can live better when they both have jobs. Today, in most North American upper-income families, both adults are employed. In fact, the average two-wage family has an income about 40 percent higher than the average family in which only the husband works. Of families with incomes in the top 20 percent, only one in four depends on just one salary.

PROFESSIONAL WOMEN

According to the 1890 Census, there were 3,202 female physicians practicing in the United States. At that time very few doctors had graduated from medical school, and most qualified through apprenticeship programs. But shortly after the turn of the twentieth century, new rules went into effect, and only graduates of accredited medical schools were allowed to practice medicine. The medical schools did not admit women, and the woman doctor soon disappeared. Slowly, some medical schools did begin to admit an occasional female applicant, but the number of women physicians stayed very small. It was the same story in the legal profession. Law school admission policies kept the number of female lawyers very low.

In 1960, when the feminist movement began to stir, only about one person of twenty graduating from medical school was a woman, and women made up only 2.5 percent of law school graduates. As can be seen in Table 16-3, these

TABLE 16-3	Percentage of Professional Degrees Awarded to Women	
	MEDICINE	LAW
1960	5.5	2.5
1970	8.4	5.4
1980	23.4	30.2
1990	34.2	42.2
2000	42.7	45.9
2002	44.4	48.0

Source: National Center for Education Statistics, 2003.

percentages rose only slightly during the next decade. Then during the 1970s enormous changes took place. By the end of that decade, in 1980 women were obtaining 23.4 percent of medical degrees and slightly more than 30 percent of law degrees. These percentages have continued to rise rapidly. In 2002 nearly half of those graduating from medical or law schools were women. Moreover, women now receive well over half of the degrees in several other professions: veterinary medicine (72%), optometry (56%), and pharmacy (66%). Dentistry (38%) and theology (33%) have the lowest percentages of women earning professional degrees.

UNEMPLOYMENT

The term **unemployed** is applied not to everyone who is not employed but only to those of legal working age who are without jobs and seeking work. There are several ironies about unemployment that news reports often overlook. Unemployment sometimes rises when jobs are more plentiful and declines when jobs are scarcer. This is because people frequently decide to look for work when they believe they are more likely to find a job.

When many jobs are available, people are drawn into the labor force. Rapid increases in the number seeking jobs but who have not yet found them cause the unemployment rate to rise. Conversely, when jobs are thought to be scarce, many people cease looking and thus are not counted in the unemployment rate. Some of this volatility in the supply of persons seeking work is due to married women and young people who are still living at home; both groups tend not to look for work when jobs are hard to get.

An important component of unemployment bears little connection to the health of the economy: Several percent of the population are always seeking their first job or in the process of switching jobs. If people spend several weeks or more trying to get a first job or find a new one, then they will contribute to the unemployment rate. In addition some of the unemployed are people who routinely cease working so that they can collect unemployment benefits; that is, some workers take periods of unemployment as planned vacations.

These kinds of unemployment do not cause much social concern because the time without a job is brief and often voluntary. It is the long-term, chronic unemployment of many North Americans that causes concern because it results in poverty. Moreover, this unemployment is concentrated in certain areas, such as the Appalachia region of the United States or the Atlantic provinces of Canada, and in certain segments of the population, especially minority groups. For example, unemployment afflicts a higher proportion of African Americans than whites, and urban African American teenagers in the United States often have shockingly high rates of unemployment—sometimes as high as 40 percent.

Why are rates of unemployment higher among African Americans? Undoubtedly, discrimination plays a role, especially in skilled manual occupations (Lieberson, 1980). But a major cause is the dwindling supply of unskilled labor jobs. This is particularly evident in teenage unemployment, for only teenagers who are high school dropouts can be counted as unemployed; people enrolled in school are not counted as unemployed no matter how hard they seek work. Unfortunately, school dropouts are not qualified for most available jobs.

In Chapter 11 we examined the remarkable increases in African American education over the past several decades. African Americans today are about as likely as whites to enter college. But they remain more likely than whites to drop out of high school.

Today, most young Canadians and Americans graduate from high school, and only a minority in each nation has fewer than nine years of schooling. Even forty years ago, this would have been regarded as a spectacular achievement. But changes in the nature of work make it likely that the million or so Americans and Canadians who drop out each year before completing high school will have trouble finding and keeping jobs in the years ahead. There simply are fewer jobs for people lacking education. Hence, although in the past new immigrants could get started in North America by doing unskilled labor, there is little opportunity to do so today; in fact, today's immigrants tend to be well educated. Any group that has difficulty keeping its teenagers in school carries a special burden.

JOB SATISFACTION

Most employed Americans like their jobs. The 1998 General Social Survey asked, "On the whole how satisfied are you with the work you do—would you say you are very satisfied, moderately satisfied, a little dissatisfied, or very dissatisfied?" About five of ten (48%) said they were very satisfied, and another 38 percent said they were moderately satisfied, while only 3 percent said they were very dissatisfied. Of course, people with the highest-prestige jobs and the highest salaries tended to be the most satisfied, but even most people in low-status, low-paying jobs were at least moderately satisfied.

Table 16-4 reports another way the 1991 General Social Survey measured job satisfaction. All employed persons in the sample were asked, "If you were to get enough money to live as comfortably as you would like for the rest of your life, would you continue to work or would you stop working?" Seven of ten said they would keep working. There is no statistically significant gender effect, while racial and ethnic differences are small—Hispanic Americans being the most inclined to keep working even if they had no financial need for doing so. It turns out to be the very youngest Americans who are the most inclined to keep working, but here, too, the differences are small. It is not surprising that the more satisfied they were with their current job, the more likely people were to say they would not quit. What is surprising is that more than half of those who said they were very dissatisfied with their jobs still opted to work, as did two-thirds of those in the jobs having the least occupational prestige.

THE TRANSFORMATION OF EDUCATION

Clearly, no society can shift from an economy based on manual labor to one based on knowledge unless its people are educated—illiterates cannot process information. The vast transformation of work in industrial societies was based in part on vast changes in educational systems and practices.

TABLE 16-4	"If you were to get enough money to live as comfortably as you would like for the rest of your life, would you continue to work or would you stop working?"
	PERCENTAGE WHO WOULD CONTINUE TO WORK
Everyone	**67**
Gender	
Men	69
Women	66
Race/Ethnicity	
White	67
African American	67
Hispanic	69
Age	
18–29	76
30–39	69
40–49	67
50–65	55
Over 65	63
Education	
Less than high school	66
High school graduate	66
Some college	71
College graduate	69
Graduate educated	66

Source: Prepared by the author from the 2000 General Social Survey.

In 1647, only twenty-seven years after they had landed at Plymouth Rock, the Puritans of the Massachusetts Colony enacted a law embodying the very radical idea that all children should attend school—at the time almost no children went to school anywhere in the world. The Massachusetts School Law required that in any township having fifty households, one person must be appointed to teach the children to read and write, and the teacher's wages were to be paid either by the parents or the inhabitants in general. Furthermore, in any township having a hundred or more households, a school must be established, "the master thereof being able to instruct youth so far as they may be fitted for the university." Any community that failed to provide these educational services was to be fined "till they shall perform this order." As word spread that Massachusetts had passed a compulsory school law, it often was taken as further evidence that the Puritans were crazy.

From these rustic beginnings, the ideal of public schools for all children became part of

American culture—as settlers moved west, they took the "one-room schoolhouse" with them. Nevertheless, even 150 years ago, in most of the world, including Europe, most children were not schooled. Education was reserved for an elite few. That America—still largely a frontier—was able to contribute so many important inventions to the Industrial Revolution during the nineteenth century is now seen as a result of its educational efforts. Moreover, as the Industrial Revolution spread, policies of mass education spread with it. Still, it was not until well into the twentieth century that even most Americans attended high school. In 1920, 42.9 percent of all Americans age sixteen or seventeen were enrolled in school; that is, fewer than half of those of high school age were still students. Females (45.5%) were more likely than males (40.3%) to stay in school, and regional variations also were marked. More than half of teenagers in the Mountain (57%) and Pacific regions (55.4%) were still in school. The Middle Atlantic states were lowest (32.3%), closely followed by New England (39.0%).

It was not until after World War II that the majority of Americans began to finish high school: The class of 1948 included 52.9 percent of its age group. The class of 1999 included more than 70 percent of its age group.

Table 16-5 shows the average number of years of schooling for a number of nations. The connection between education and economic development is very clear. In the United States and Canada, the average person has gone slightly beyond high school. Most nations of Western Europe are close behind. But in the rest of the world, people receive substantially less education. For example, the average Russian has only nine years of schooling. In Italy, Romania, Spain, and Portugal, the average is below eight years. China (4.8) and Mexico (4.7) are similar, and the average person in Afghanistan and Niger attended school for less than a year.

If industrialization has led to mass education, it has not yet led to the great emphasis on higher education that has developed in the United States and Canada. In both nations public universities were founded in the nineteenth century with the aim of making college educations available to more citizens. And in both nations private and denominational institutions also play a substantial role in higher education. But in both nations the real explosion in higher education has occurred over the past fifty-five years. In 1950, 1,863 American colleges and universities enrolled 2.2 million students. By 1993 more than 14 million students were attending 3,638 American colleges and universities. Today, the majority of young Americans and Canadians enroll in college, far beyond the percentage of college-attenders in any other nation—in no nation in Europe do more than a third begin college, and 20 percent is typical.

But if the United States and Canada excel in terms of the quantity of higher education, what about quality? Here it may be useful to distinguish between education and schooling. The two words often are equated, but **education** refers to what a person has learned, while **schooling** refers to time spent in an organization (a school) dedicated to educating people and which often confers formal degrees and certificates on those who complete a period of enrollment. Although we often make assumptions about how well educated people are on the basis of how much schooling they have had, we realize that a lot of education is not the result of schooling and that schooling may not be very successful in educating.

As can be seen in Table 16-6, although American students spend more years in school than people in other nations, their time seems less well spent than in many nations. In 2003 a very

TABLE 16-5	Average Number of Years of Schooling (Assorted Nations)		
NATION	**YEARS OF SCHOOLING**	**NATION**	**YEARS OF SCHOOLING**
United States	**12.3**	Portugal	6.0
Canada	**12.1**	Ecuador	5.6
France	11.6	Kuwait	5.4
Norway	11.6	China	4.8
Great Britain	11.5	Iraq	4.8
Germany	11.1	**Mexico**	**4.7**
Japan	10.7	Bolivia	4.0
Denmark	10.4	South Africa	3.9
Israel	10.3	Saudi Arabia	3.7
Russia	9.0	Turkey	3.5
South Korea	8.8	Zimbabwe	2.9
Ireland	8.7	Egypt	2.8
Poland	8.0	India	2.4
Cuba	7.6	Pakistan	1.9
Italy	7.3	Haiti	1.7
Romania	7.0	Nigeria	1.2
Greece	6.9	Uganda	1.1
Spain	6.8	Afghanistan	0.8
Peru	6.4	Niger	0.1

Source: Prepared by the author from *Nations of the Globe*, an electronic database distributed by Wadsworth, 2002.

large educational assessment program was conducted in forty nations: Fifteen-year-olds were tested in logic, mathematics, and science. Table 16-6 shows the combined mean score achieved by students in each nation. The American students finished in twenty-eighth place, about the same as students in Russia and Italy, and far behind students in most of the developed nations, surpassing only students in less developed nations such as Thailand, Turkey, Mexico, and Tunisia. In contrast, students in Canada did quite well, finishing tied for seventh place.

To better understand why American students do not measure up in international comparisons, let's take a closer look at the "education" of Americans in general and of college graduates in particular.

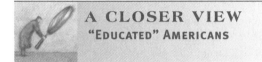

A CLOSER VIEW
"EDUCATED" AMERICANS

Several years ago, Congress commissioned the Educational Testing Service[1] (ETS) to conduct the first comprehensive study of American literacy. The term *literate* was very broadly defined as a synonym for "educated, cultured." ETS applied a more precise definition to *literate*, defining it as

---

using printed and written information to function in society, to achieve one's goals, and to develop one's knowledge and potential.

Under the direction of Irwin Kirsch, the ETS staff developed test items to be administered to a national sample including more than 26,000 randomly selected Americans over age fifteen. The test items were grouped into three primary forms of literacy: quantitative, documentary, and prose. Each of these forms of literacy was graded on the basis of five levels of achievement. Finally, more than 400 trained interviewers conducted the survey during an eight-month period in 1992.

When the initial results were released in 1993, they aroused considerable concern about the "failure of the schools," for approximately half of all Americans scored in the lower two levels of each form of literacy. These findings were especially disappointing because the questions were so extremely easy. The first item

TABLE 16-6	Academic Achievement by Fifteen-Year-Olds in Forty Nations
NATION	**MEAN PROFICIENCY TEST SCORES: COMBINED LOGIC, MATHEMATICS, AND SCIENCE**
Finland	547
Hong Kong (China)	546
Korea	543
Japan	543
Liechtenstein	530
Macao (China)	528
Netherlands	527
Canada	**527**
Australia	526
New Zealand	526
Belgium	521
Switzerland	520
Czech Rupublic	518
France	514
Sweden	508
Germany	506
Iceland	505
Denmark	502
Ireland	502
Austria	501
Hungary	498
Slovakia	495
Luxembourg	493
Poland	492
Norway	490
Spain	485
Latvia	485
United States	**484**
Russia	479
Italy	474
Portugal	468
Greece	458
Serbia	431
Uruguay	424
Thailand	424
Turkey	422
Mexico	**391**
Indonesia	372
Brazil	372
Tunisia	363

Source: Calculated by the author from a report by the Organization for Economic Cooperation and Development, 2004.

that follows is an example of the questions used to measure the second level of prose literacy, while the second example measures the second level of quantitative literacy.

---

1. These are the same people who administer the Scholastic Achievement Test (SAT), formerly the Scholastic Aptitude Test.

EXAMPLE 1

A manufacturing company provides its customers with the following instructions for returning appliances for service: *When returning appliance for servicing, include a note telling as clearly and specifically as possible what is wrong with the appliance.*

A repair person for the company receives four appliances with the following notes attached. Circle the letter next to the note which best follows the instructions supplied by the company.

A The clock does not run correctly on this clock radio. I tried fixing it, but I couldn't.

B My clock radio is not working. It stopped working right after I used it for five days.

C The alarm on my clock doesn't go off at the time I set. It rings 15–30 minutes later.

D This radio is broken. Please repair and return by United Parcel Service to the address on my slip.

It certainly does not require a lot of sophistication to select C. But one American in five can't.

EXAMPLE 2

Theater Trip

A charter bus will leave from the bus stop (near the Conference Center) at 4 p.m., giving you plenty of time for dinner in New York. Return trip will start from West 45th Street directly following the plays. Both theaters are on West 45th Street. Allow about 1½ hours for the return trip.

Time:	4 P.M., Saturday, November 20		
Price:	"On the Town"	Ticket and bus	$11.00
	"Sleuth"	Ticket and bus	$8.50
Limit:	Two tickets per person		

The price of one ticket and bus for "Sleuth" costs how much less than the price of one ticket and bus for "On the Town"?

Yes! All you do is subtract $8.50 from $11.00. But 22 percent of American adults can't answer the question—many seem to become confused because of the presence of information not relevant to the question.

To qualify at the *highest* level of quantitative literacy, people had only to answer questions such as the one shown in Figure 16-1. The ETS report characterized tasks at level five as requiring people "to handle two or more arithmetic operations in sequence." In this instance the correct answer is obtained by multiplying $156.77 (the monthly payment for a $10,000 loan) by 120 (the duration of the loan in months) to obtain the total amount repaid ($18,812.40). Then, the amount borrowed ($10,000) is subtracted from the total amount paid, giving the total amount of the interest charges ($8,812.40). Only *4 percent* of the adult U.S. population can perform at this level!

Once the initial findings had been reported, Paul E. Barton and Archie Lapointe (1995) of ETS analyzed the data to assess the connections between education and literacy. If anything, the results were even more disappointing. Many commentators had blamed the poor showing by the general population not on the schools, but on people who drop out of school and on new immigrants having a limited ability to speak English. But when it was discovered that the *majority* of persons born in the United States and with a four-year college degree score in the three lower levels of quantitative and prose literacy and in the lower two levels of documentary literacy, questions about the quality of instruction can't be dodged.

DO SCHOOLS REALLY MATTER?

Concern over the quality of education reflects the assumption that schools play an important role in what students learn. It might seem self-evident that schools are critical to the educational process. However, beginning in the 1960s, many critics, including many prominent social scientists, started to argue that schools have little effect on what people learn or on how well people are prepared for jobs.

Some critics pointed out that a great majority of people in colonial times could read, write, and do arithmetic even though few attended school, even grade school. Others noted that schools seem unable to overcome differences in background: Students from privileged homes do well in school, whereas those from disadvantaged backgrounds do poorly. They concluded that schools simply certify the educational advantages or disadvantages that students bring to school (Jencks et al., 1972).

These views of the ineffectiveness of schools were lent some support by a huge study conducted by James Coleman and his associates (1966). The U.S. Congress commissioned Coleman to assess the nation's schools and determine which aspects of schooling were the most valuable. As he began his research,

Coleman expected to find that African Americans suffered from attending poor-quality schools, and he hoped to prompt massive federal aid to correct the inequity (in Silberman, 1971).

What he found was startling. First, there was little difference in the quality of schools African Americans and whites attended in terms of expenditures per student, age and quality of the buildings, libraries, class size, and teacher training. Second, these aspects of school quality had no detectable impact on student achievement scores. Thus, lavish expenditures during the 1950s and early 1960s to upgrade schools had accomplished nothing in terms of actual education. Whether students went to school in ramshackle buildings or modern ones, attended large classes or small ones, or had fancy labs or makeshift equipment didn't matter. Nor did it matter if their teachers had advanced degrees or just two-year teachers college certificates.

Coleman could only conclude that school was simply a place where students learned in proportion to the educational qualities of their homes, neighborhoods, and peer environment. Still, Coleman's report contained one finding often overlooked in subsequent discussions: How well students from any background did in school was correlated with how well their teachers did on a vocabulary test. This suggests a link between declines in student achievement scores and a corresponding decline in the quality of teachers as measured by test scores.

So the questions persist: Do the schools actually accomplish anything? Do kids actually learn in school? For most social scientists, it seemed evident that people do learn in school, for even children from the most privileged homes are usually not taught at home to read, write, or do arithmetic. Most kids must be learning these things in school, if they learn them at all. Yet good evidence of the effectiveness of schools was lacking.

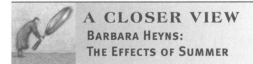

A CLOSER VIEW
BARBARA HEYNS:
THE EFFECTS OF SUMMER

How can we see if schools have a real impact on learning? The most obvious way would be to randomly assign some children to attend school and others to stay home and then to compare the results. But that would be both illegal and immoral. Because schooling at the elementary

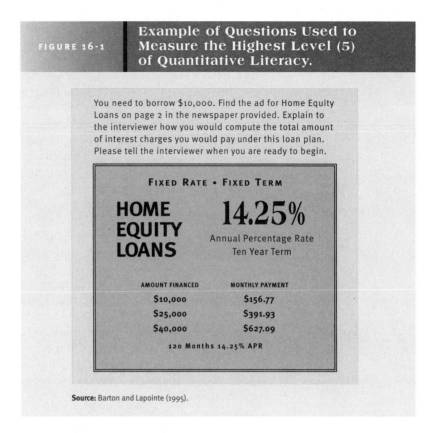

FIGURE 16-1

Example of Questions Used to Measure the Highest Level (5) of Quantitative Literacy.

You need to borrow $10,000. Find the ad for Home Equity Loans on page 2 in the newspaper provided. Explain to the interviewer how you would compute the total amount of interest charges you would pay under this loan plan. Please tell the interviewer when you are ready to begin.

FIXED RATE • FIXED TERM

HOME EQUITY LOANS

14.25%
Annual Percentage Rate
Ten Year Term

AMOUNT FINANCED	MONTHLY PAYMENT
$10,000	$156.77
$25,000	$391.93
$40,000	$627.09

120 Months 14.25% APR

Source: Barton and Lapointe (1995).

level is universal in the United States and Canada, we can't seek out students who do not go to school and compare their achievements with those who do. Faced with this problem, Barbara Heyns (1978) came up with a brilliant solution.

Kids don't go to school all year. Why not compare the learning that occurs during the school year with that occurring during summer vacation? In this way summer learning can serve as a basis for estimating what children might learn if they did not go to school.

Heyns gave verbal achievement tests to 2,978 students enrolled in Atlanta schools. They were tested at the start of the fifth grade, at the end of the fifth grade, at the start of the sixth grade, and again at the end of the sixth grade.

Heyns's results gave strong evidence that school matters—but much more to some kinds of children than to others. On the average, children in Atlanta learned much less during their summer vacation than they did in an equivalent time period during the school year. Their verbal achievement scores rose much more rapidly on a monthly basis over the school year than over the vacation. However, children from higher-income families learned about as much during vacation as during the school year. Children from the most deprived backgrounds actually lost ground during the summer: Their scores were not as high in the fall as they had been the previous spring.

What Heyns found means that, rather than merely maintaining differences children bring to school, schools greatly improve the situations of poor children. Differences in the rates of learning between African Americans and whites and between higher- and lower-income children were very small while school was in session, but when school was out, the kids from privileged backgrounds sprinted ahead. Schools therefore minimize initial background advantages by enabling the disadvantaged to keep up. However, schools can accomplish this only during the school year. The long summer vacations characteristic of most American schools undo much that is accomplished with underprivileged children during the school year. In the summer the academic effects of students' backgrounds reassert themselves. Moreover, these summer effects accumulate, so that as children advance through the grades, the children of more advantaged families get further and further ahead of the others.

Surprisingly, Heyns found that attending summer school did not prevent summer learning losses. Atlanta has a massive summer school program (a fourth of the students enroll), and children from disadvantaged backgrounds are especially likely to enroll. However, attendance at summer school had no influence on summer learning. Heyns concluded that this was because the summer school programs were oriented toward recreation rather than the regular curriculum. Students overwhelmingly said they went to summer school because it was fun.

What did the kids from the more advantaged homes do during the summer that caused them to continue to learn? Heyns examined many possibilities, including vacation trips and participation in organized summer activities such as sports or camps. But only one activity had real impact: reading. As Heyns (1978) put it:

The single summer activity that is most strongly and consistently related to summer learning is reading. Whether measured by the number of books read, by the time spent reading, or by the regularity of library usage, reading during the summer systematically increases the vocabulary test scores of children.

She estimated that every four books read over the summer produced an additional right answer on verbal achievement tests.

Heyns also discovered that a major factor affecting reading, independent of a student's background, was the distance from the student's home to the nearest public library. Eighty percent of the students who lived within seven blocks of a library used it regularly. Among children living more than seven blocks from a library, visits fell rapidly. Thus, in showing that schools matter, Heyns showed that libraries do, too.

Heyns's findings also suggest that schools might be much more effective if the school year were extended. The long summer vacation was instituted back when most kids were needed to help on the farm, and it has persisted long after this need has vanished. Indeed, children in Japan and many other industrial nations attend school throughout the year with short breaks.

Heyns's study was a major breakthrough. First, it showed how to study school effects. Second, it demonstrated conclusively that school does affect how well educated students become. Moreover, her study prompted other researchers to try other ways to assess the impact of schooling. Karl Alexander, Gary Natriello, and Aaron Pallas (1985) set out to extend Heyns's analysis to high school students by comparing the cognitive development of those who remained in school with those who dropped out. They tested a sample of 30,000 sophomores from 1,000 high schools and retested them two years later, when those who had remained in school were in the last months of their senior year. Then the sociologists computed the increase (or decline) in cognitive test performance for all respondents over the two-year period. They found that dropouts had a much smaller increase than did those who remained in school. Moreover, consistent with Heyns's findings, dropping out had the most severe negative effects on students from the most disadvantaged backgrounds.

HIGH SCHOOL TODAY

Several years ago, the U.S. Department of Education launched a huge study of high school by collecting data on 30,030 sophomores and 28,240 seniors. Two years later, 14,825 of the sophomores were restudied, including more than 2,000 dropouts. Follow-up questionnaires also were obtained from 11,995 of those first studied as seniors, who by then were two years beyond graduation. Additional follow-ups came throughout the decade.

The massive database made up of all the questionnaires these students filled out, their scores on dozens of standardized tests, and a huge array of other data including the courses they took and the jobs they held after school is known as the "High School and Beyond Study." I have done some analyses of these data to offer you a window into life in the contemporary high school. Since most of you were so recently in high school, the data may not surprise you as often as they did me.

Even I wasn't surprised by Table 16-7: Homework matters. Students who didn't do their homework mostly got bad grades. Moreover, those who didn't study as sophomores were more than four times as likely to drop out as were students who studied at least five hours a week.

I admit to being surprised by Table 16-8. First of all, I was surprised because so few sophomores spend even an hour a day during the week on homework. I was considered a very lazy student in high school, even though I always put in at least an hour a day on my studies. Second, given all of their advantages, I had expected white students to be more likely to do homework than are minority students. But that's not true. All the differences are essentially trivial except for one: Asian American students are twice as likely to put in at least an hour a day. Moreover, despite family economic advantages, whites are not so different from African Americans, Hispanic Americans, and Native Americans when it comes to dropping out, but again Asian Americans are very different. These findings are very consistent with what we discovered about educational achievements of Asian Americans in Chapter 11.

Public dissatisfaction with the public schools has become increasingly widespread and vigorous. There are constant complaints reported in the media that many people graduate without ever having learned to read even at an elementary level; indeed, some well-known professional athletes have admitted spending four years in college without anyone realizing they could not read. The schools also are blamed for high dropout rates that have in fact been rising over the past several years. Finally, even teachers have expressed great alarm about the disorder and lack of discipline in the schools.

In response to these complaints, educational leaders typically blame inadequate funding for current school problems. But if this is the cause,

© Susie Fitzhugh/Stock Boston

many have asked why the Catholic and other church-related schools achieve so much better results, since they spend far less money per student. Let's explore this debate.

Table 16-9 shows that students in public schools are far less likely to study and far more likely to drop out than are students in Catholic and private schools. As for the matter of disorder and lack of discipline, students in public schools are far more likely to complain about a lack of discipline and order in their schools. Finally, students in public schools are far less likely to say they expect to go to college.

Of course, when these contrasts are discussed, it often is suggested that these are spurious relationships. That is, Catholic and private schools are not better; rather, they simply have superior students to work with, as the public schools are stuck with the kids that Catholic and private schools expel or refuse to admit in the first place.

Research suggests that these girls will make as much educational progress during the summer vacation as they would during a similar period of attending school—because they are reading books while school is out. Children who don't read fall behind during the summer.

TABLE 16-7	The Effects of Homework					
	HOURS SPENT ON HOMEWORK EACH WEEK					
	NONE	LESS THAN 1	1–3	3–5	5–10	OVER 10
Percentage with an A average	1	4	6	10	16	27
Percentage with an average of D or lower	46	24	16	9	5	5
Percentage who dropped out before senior year	28	17	12	9	6	6

Source: Prepared by the author from the "High School and Beyond" database.

TABLE 16-8	Studying and Dropping Out by Race and Ethnicity				
	WHITES*	AFRICAN AMERICANS	HISPANIC AMERICANS†	NATIVE AMERICANS	ASIAN AMERICANS‡
Percentage of sophomores who do an hour or more of homework every day	27	24	20	24	52
Percentage who dropped out before senior year	11	14	19	18	5

Source: Prepared by the author from the "High School and Beyond" database.

*Non-Hispanics only.

†Hispanic Americans may be of any race.

‡Chinese, Japanese, Koreans, and Vietnamese.

TABLE 16-9	Contrasting the Effects of Different Kinds of Schools on Sophomores			
	PUBLIC	CATHOLIC	ELITE PRIVATE	OTHER PRIVATE
Percentage who do an hour or more of homework daily	26	47	94	50
Percentage who dropped out before senior year	18	2	0	9
"TO WHAT EXTENT ARE THE FOLLOWING DISCIPLINARY MATTERS PROBLEMS IN YOUR SCHOOL?"				
Students cut classes Percentage "often"	61	15	3	12
Students refuse to obey instructions Percentage "often"	30	14	8	13
Students get into fights with each other Percentage "often"	27	9	3	5
Students attack or threaten to attack teachers Percentage "often" and "sometimes"	22	6	7	5
Percentage who expect to attend college	59	78	95	76

Source: Prepared by the author from the "High School and Beyond" database.

However, in another major study, James S. Coleman, Thomas Hoffer, and Sally Kilgore (1982) disputed that claim. When they compared students from similar backgrounds, the Catholic schools still far surpassed the public schools in terms of the educational achievements of their students.

The data in Table 16-10 support Coleman and his colleagues and suggest that these school effects are not spurious. Reading down the figures we can see that, in both public and Catholic schools, students from higher-status families are less likely to drop out. But reading across the table we can see that, at all status levels, dropout rates are far higher in public schools.

The same finding holds for race and ethnicity. In each group the public school dropout rate towers over that for Catholic schools. It may be noteworthy, too, that the majority of African American students in Catholic schools are Protestants.

One can find the public schools in America falling far short of fulfilling their responsibilities without concluding that schooling is a waste of time. However, large numbers of American parents have decided that *going* to school is not the right choice for their children's education.

HOMESCHOOLING

How many young Americans are being schooled at home? No one really knows. The U.S. Department of Education estimated that 850,000 were being homeschooled in 1999. The president of the National Home Education Research Institute estimated the number at 1.3 million for that same year. Other sources suggest the number may be as high as 2 million. Whatever the number, all agree that it is increasing very rapidly.

The educational establishment is very opposed to parents keeping their children out of "regular" schools and raises three primary objections. The first is that homeschooling simply adds to the advantages already enjoyed by children of affluent families. But it does further harm to public education as the more affluent abandon the schools to students who have more difficult educational disadvantages. The second objection, somewhat in contradiction to the first, is that homeschooled children receive inferior educations. The third is that young people need to go to school in a setting that allows them an adequate opportunity to socialize and have new friends.

TABLE 16-10	Catholic/Public School Dropout Differences Withstand Controls	
	PERCENT WHO DROPPED OUT BEFORE SENIOR YEAR	
	REGULAR PUBLIC SCHOOL	CATHOLIC SCHOOLS
Family status*		
Lowest	28	7
Lower	16	2
Higher	13	0
Highest	8	1
Whites	17	1
African Americans	22	2
Hispanic Americans	23	6

Source: Prepared by the author from the "High School and Beyond" database.

*In quartiles.

A recent national survey of 57,278 households engaged in homeschooling, conducted by the National Center for Educational Statistics (2001), challenged the first objection. Parents who homeschool their children do not have higher family incomes than families whose children are in schools. What they do have is a parent who does not work or one who works at home and is able to devote the time and supervision needed to educate children at home. Parents who homeschool also tend to have large families, and therefore, they often involve older children in helping the younger ones with their lessons. Homeschoolers are more likely to be white and non-Hispanic (75%) than are students enrolled in schools (65%), but the difference is far smaller than any expert had assumed.

As to the second objection, homeschooled students substantially outscore schooled students on the SAT (McCusker, 2002). Moreover, their better performance in college has drawn so much attention that admissions offices have begun to seek homeschoolers (Cloud and Morse, 2001). As to the third objection, parents of homeschooled children usually arrange for their kids to spend time with other homeschoolers. For example, field trips for the homeschooled occur regularly in larger communities.

At present no more than three or four students of every hundred are being homeschooled. But ten years ago it was fewer than one in a hundred. Will the trend continue? Perhaps. But it seems more likely that something finally will be done about the inadequacy of the public schools.

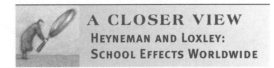

A CLOSER VIEW
HEYNEMAN AND LOXLEY:
SCHOOL EFFECTS WORLDWIDE

Because most sociological research takes place in a few of the most industrialized nations, many findings are based on a limited range of variation. Studies of school effects conducted in the United States, Canada, Great Britain, or France are restricted to a narrow range of variation in school facilities and quality. There probably isn't a public school in any of these nations without a TV, VCR, and computer. In addition, although none of these nations is without poverty, even the poor are affluent compared with most people in many other nations. Hence, Canadian, American, or British children may not be as dependent on schools for their learning as are children in less developed nations, which is consistent with the finding within the United States that schools do the most for kids from the more disadvantaged homes.

This line of reasoning led Stephen P. Heyneman and William A. Loxley (1983), sociologists on the staff of the World Bank, to analyze data on school effects by using a series of studies of school effects in twenty-nine nations. In all, over 300 different researchers were involved, working in more than a dozen languages, with a total sample of more than 10,000 schools, 50,000 teachers, and 260,000 students.

What did they find? First, kids in the wealthier, more industrialized nations learn more during the same number of school years. In part this difference in learning may be because of better schools. It probably also reflects the whole "package" of benefits that comes with a much higher standard of living.

Second, the poorer the nation, the less student backgrounds influence school performance, probably because education is a scarce resource in these nations. As Heyneman and Loxley (1983) pointed out: "Scarcity creates competition for school places from the onset of grade 1, and at a level of intensity unknown in wealthy countries until college or, in the case of the United States, until graduate school. This scarcity is well understood within both rich and poor families." Thus, family background tends to be canceled out in the process of gaining a place in school; only the better students get in or remain.

Finally, the poorer the nation, the greater the economic returns for getting an education. While there is a positive correlation between education and income in all nations, the correlations are higher in the poorer countries, where it remains relatively unusual to obtain many years of schooling.

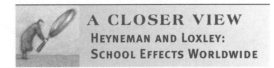

DOES EDUCATION PAY?

We have just noted the economic benefits of education, especially to people in less developed nations. Moreover, in Chapter 10 we saw the immense role education plays in the occupational achievement of Canadians and Americans. And a look at Table 16-1 will reveal the obvious connection between education and occupation: It is impossible to enter most of the highest-prestige occupations without attending college and often graduate school as well. In this sense education may not guarantee success, but it tends to be necessary for it. Without enough education many opportunities simply are unavailable.

Nevertheless, a number of people question whether education really pays off. Caroline Bird (1975) has argued that most people would come out ahead if they skipped college, invested the money saved thereby, and went right to work after high school. Why is there doubt about the economic value of education?

One reason is the rapid rise in the relative earnings of skilled blue-collar workers, such as plumbers, electricians, long-distance truckers, and tool and die makers. Many college graduates end up in lower-paying occupations than these, and college is of no advantage for entry into these skilled trades. In fact, going to college would be a waste of earning years for people planning to enter these occupations. But perhaps the primary reason people question the economic importance of education is simply that in advanced industrial nations a college degree is not worth as much as it used to be.

When relatively few people earned college degrees, they possessed a scarce occupational qualification. Now many people can earn degrees, and therefore, a degree is not a certain ticket to success. When 5 percent graduate from college, then only 5 of 100 people can compete for a job requiring a college degree. But when a third earn degrees, 33 of 100 people compete for those jobs requiring degrees. Hence, the decline in the value of a college education is the result not of colleges ceasing to prepare people for careers but of colleges preparing so many people for careers.

Like most children, this boy probably wonders from time to time whether homework is worth it. Getting a lot of education won't guarantee him a good job, but lack of education would exclude him from most of the better-paid occupations.

As the level of education has risen in the industrial nations, the relative advantage of completing a given level of education has declined. If people today want to have the same educational advantage that their parents had, then they must stay in school longer than their parents did.

French sociologist Raymond Boudon (1974) has created elegant mathematical models of this process of educational "deflation." As he pointed out, however, we must realize that such deflation applies to all educational levels, not just the top. That is, not only is a college degree of less value than it used to be but so is an eighth-grade education. The child of a school dropout who also drops out of school will have a harder time finding and holding a job than the dropout parent did.

Table 16-11 reveals the immense impact of education on income in the United States today. Here, data from the *Statistical Abstract of the United States* supply sufficient evidence for a more focused look at the link between education and income. Usually, such a table includes all adults. The trouble with that inclusion is that the group without high school degrees includes a disproportionate share of people over age sixty-five whose incomes have declined substantially because of their retirement. Similarly, the categories of those who have completed college

TABLE 16-11	Education and Mean Annual Income, Persons Age 45–55	
EDUCATION		**MEAN ANNUAL INCOME**
Not a high school graduate		$ 21,231
High school graduate		$ 31,251
Some college		$ 40,225
College graduate		$ 60,680
Master's degree		$ 67,096
Doctorate		$112,538
Professional		$126,230

Source: *Statistical Abstract of the United States, 2005.*

and graduate school include a disproportionate share of young people who have just entered the labor force and whose incomes have yet to catch up with those of skilled blue-collar workers who have been employed and gaining seniority since leaving high school. For this reason Table 16-11 has been restricted to persons in the prime earning years, age forty-five to fifty-five. Moreover, family income is greatly increased when there are two earners. So the table has been limited to individuals.

The data are rather dramatic. Simply by having enrolled in college, you probably have ensured that you will earn substantially more per year than your high school classmates who did not enter college. If you complete your degree, you may increase your annual wages by a third. A master's degree raises incomes even higher. And people with professional degrees earn more than four times as much as a person with a high school diploma.

The very tight connection between income and education, on the one hand, and the very rapid rise in the average level of education, on the other, has had unfortunate side effects for several disadvantaged groups, especially African Americans and Hispanic Americans. As Randall Collins (1979) has noted, the "inflation" of academic credentials has partly offset the economic value of gains in educational attainment these groups have made. Let's see how that happened.

In 1970, 11.3 percent of whites, 4.4 percent of African Americans, and 4.5 percent of Hispanic Americans were college graduates. Over the next eighteen years, African Americans and Hispanic Americans increased their levels of education dramatically. So, in 1988, 11.3 percent of African Americans and 10 percent of Hispanic Americans were college graduates. But they didn't catch up much because, by 1988, 20.9 percent of white Americans were college grads. By the time African Americans and Hispanic Americans were about as likely as whites were in 1970 to have a college degree, the market value of that degree had declined.

Ironically, Collins pointed out, it is really doubtful that the immense inflation in higher education means that colleges actually impart training vital for performing many of the jobs that now require degrees of applicants. Obviously, colleges impart needed skills to engineers, physicians, accountants, and scientists. But it is not at all clear that anything learned in college bears directly on performance in many other jobs for which college graduates are hired but then must be given substantial additional training—most sales positions, for example. Many of the specific job skills that colleges do provide probably could be acquired more rapidly and effectively through on-the-job training.

Collins therefore describes the expansion of higher education in America as the creation of a *"credential society"*—because of what he regards as a misplaced stress on limiting many occupations to those possessing specific occupational credentials or licenses. For example, it is impossible to obtain a college teaching position in the United States nowadays without a master's degree and very difficult to do so without a doc-

torate. Either requirement would have excluded Albert Einstein. Conversely, even the most famous faculty members of the best universities are prohibited from teaching in high schools unless they go back to college to earn a teaching certificate.

Why so much stress on credentials? Collins noted the way in which credentials serve as a sort of currency through which occupational positions are allocated. As the systems of educational credentials expanded, Collins (1979) argued:

The value of any particular kind and level of education came to depend less on any specific content that might have been learned in it, and more and more upon the sheer fact of having attained a given level and acquired the formal credential that allowed one to enter the next level [or ultimately to pass the requirements for entering a monopolized position].

Put another way, the point of a teaching credential is not so much to provide classroom skills as to control entry to the teaching profession. By getting state legislatures to make it illegal to teach without an authorized teaching credential, education departments of universities seized control of the replaceability of teachers. Recall from Chapter 9 that power can be used to make positions much less replaceable than they really are, thus inflating their status and rewards. In similar fashion credentials such as barber, beautician, and real estate licenses can be used to limit the numbers of people in a given occupation.

A considerable debate has raged among sociologists about the extent to which schools are used to educate as opposed to being used to *allocate status*. Many Marxist sociologists claim that the primary function of schools in American society is to re-create the class structure in each generation by designating people as successes or failures on the basis of racial, cultural, or class characteristics and hence assigning each individual to a status level (Illich, 1970; Bowles and Gintis, 1976). Proponents of these **allocation theories** of education claim that educational requirements and credentials are meant to screen out those who lack the opportunity to attend college or who rebel against the prevailing rules governing status allocation.

Because there is some justification for suspecting that educational credentials do not simply certify that one has received vital training, allocation theories speak to a significant question: What is the function of education in our society? However, the extremist rhetoric in which allocation theories typically are stated had tended to deafen sociologists to this underlying question. Then in the late 1970s, a sociologist at Stanford University drew upon allocation ideas to formulate a more general theory of the functions of education as a social institution.

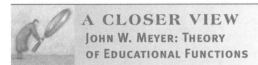

A CLOSER VIEW
JOHN W. MEYER: THEORY OF EDUCATIONAL FUNCTIONS

John W. Meyer (1977) began his theory by accepting the traditional view of educational socialization: that, through schooling, individuals increase their knowledge and competence, which in turn increases their abilities to perform adult roles. But he then added the insight that *levels of education, in and of themselves, are social statuses.* That is, aside from any other status individuals hold, they have a distinct status based on their amount of formal education: high school graduate, college graduate, Ph.D., and the like. Schools, then, can be seen as institutions empowered, or chartered, by society to grant statuses to individuals.

Moreover, a major aspect of schools as socializing agents is to encourage individuals "to adopt personal and social qualities appropriate to the position [or statuses] to which their schools are chartered to assign them." As with all positions in society, educational statuses come equipped with roles. A major effect of education is that *people learn to play the role appropriate to the status that their school confers on them.*

This proposition allowed Meyer to explain why variations in school quality seem of little or no importance in the attitudes, values, opinions, and behavior of graduates. Research shows that the amount of formal schooling a person completes has a great effect on a wide variety of personal qualities and characteristics: from the way people vote to their religious commitment. If this is a consequence of the content of actual instruction, then people who attended very high-quality schools ought to differ from those who attended low-quality schools. However, research has failed to turn up such differences (or they are extremely small). Instead, graduates of elite colleges resemble graduates of obscure schools much more closely than they resemble people who did not graduate from college.

According to Meyer's theory, this finding is to be expected if the real impact of schools is to admit people to a particular educational status. For then, all schools chartered to convey that status ought to have similar socializing effects. Indeed, Meyer pointed out that school quality is seldom of much importance in assessing a person's claim to a given status. Graduates of all North American high schools, for example, have the occupational rights reserved for high school graduates—no one asks if their high school was a good one. By the same token, a college degree satisfies the requirements to claim the status of college graduate, whether the degree was from Harvard or North Dakota State College in Valley City.

Thus, Meyer argued that the most powerful socializing property of schools is the ability to confer statuses that are recognized in society at large and that people who acquire a given status tend to perform the role attached to it in a similar fashion. Indeed, Meyer cited research showing that people adopt personal qualities appropriate to a given educational role on admission to a school chartered to grant that status: They often begin to do so on acceptance to such a school, even before they have attended (W. L. Wallace, 1966; Benitez, 1973).

Meyer also argued that socialization into these roles does not stop when people leave school. Instead, *people continue to act out the role attached to their educational statuses throughout their adult lives, regardless of their occupation.* Occupational success often varies over time, as do family relations and even geographical location, but a "college graduate" or "high school dropout" is an unchanging status once school is finished. People continue to respond to an individual's educational status, and the individual maintains the role appropriate to that status. Indeed, John Irwin (1970) found that educational statuses even count among inmates in prison, where people with college degrees or postgraduate training are frequently sought out for advice and information.

Meyer was not content to view educational institutions as allocating status only to the degree the occupational system allows. Allocation theorists have argued that educational institutions simply allocate people into positions determined by the occupational system; for example, medical schools are chartered to produce doctors only insofar as the occupational structure has recognized this occupation and the occupational group (doctors) has granted this power to schools. Meyer argued instead that *the educational system has the power to create new occupations, even elite occupations, and to control the placement of these occupations in the occupational structure.* This is possible because educational institutions, especially universities, play a leading role in defining new knowledge, developing new techniques and technologies, and giving these techniques legitimate occupational standing.

In other words many of the most highly paid, highest-status occupations in contemporary society exist because universities invented them, defined their worth, and determined the conditions under which people could enter them. There were no economists until universities established the science of economics and legitimized its claim to special competence. Nor were there sociologists, geneticists, or even football coaches until universities created a special body of knowledge and began training people to use it.

In this way Meyer (1977) undercut the narrow view of allocation theories: that education is a passive servant of the stratification system. Instead, "education helps *create* new classes of knowledge and personnel which then come to be incorporated in society." That is, the expansion of the education system increases the "number of specialized and elite positions in society."

As educational achievement has risen, the occupational structure has rapidly expanded at the top (more professional, managerial, and technical positions) and contracted at the bottom (fewer unskilled labor jobs). Meyer's theory helps explain how industrial nations have increasingly become "knowledge" economies.

Finally, Meyer argued that the rising level of mass education has expanded the proportion of the population regarded as having citizenship responsibilities, capacities, and rights. The larger the proportion of the population who are educated, the harder it is for elites to exclude them from decision making or to ignore their economic demands. Here Meyer parted company with the allocation theorists, who argue that education serves only elite interests. Instead, Meyer noted, the primary emphasis in modern education is on mass education—on providing the maximum number of people with the opportunity to be educated and to gain entry to elite occupations.

Meyer's theory offers a more comprehensive and intelligible view of the way in which educational institutions fit into society. Because schooling is largely confined to childhood and early adulthood, sociologists have tended to view educational institutions as having only early socialization effects, in much the same way that the cultural determinists (see Chapter 6) attempted to attribute adult personality wholly to child-rearing patterns. By recognizing how educational statuses continue to have socializing effects throughout a lifetime, Meyer was able to explain why these effects endure. It is not that what we learn in school lasts forever. Rather, we take with us from school an educational status that continues to influence our chances and experiences in life.

CONCLUSION

Education remains crucial to occupational achievement, but as more people get more education, a given level of education becomes less valuable. This is because educational and occupational institutions remain somewhat independent. Although a shift to a knowledge economy can occur only with an increase in the supply of educated people, such an increase does not automatically create more knowledge jobs for them to fill. Nevertheless, as John Meyer pointed out, schools can create new occupations, including new elite occupations. But the number of people aspiring to these occupations can still exceed the supply of positions available. A person may have the training and desire for a given occupation, but that alone does not create a position for that person to fill. On the other hand, as Raymond Boudon has pointed out, aptitude and motivation often suffice to gain an education. Thus, the supply of college-educated people, for example, may increase beyond the positions available in the economy, or at least beyond the level of upper positions.

In the nineteenth century, high school graduates qualified for most teaching jobs. Until the 1930s a person who attended a teachers college for two years was well qualified to teach. Today, most elementary and high school teaching jobs are reserved for people with master's degrees. The primary reason for this change has

© Michael Grecco/Stock Boston

A couple poses with their son after taking part in graduation exercises at Harvard University. Clearly, this man and woman will benefit from holding degrees from such a high-prestige school, but as John Meyer has pointed out, the more important thing is not what school they graduated from but that they have a college degree from somewhere.

been an increase in the proportion of educated people.

We also have seen that, despite research casting doubt on the contribution of schools to the learning process, schooling does matter. The effects of schooling are greatest for students from the most disadvantaged homes and those living in the poorer nations.

Finally, we have seen that the connections between educational institutions and the rest of society are neither simple nor restricted to in-school instructional effects. Education has strong and lasting socialization effects because it creates permanent statuses. Consequently, people tend to perform roles appropriate to their educational status, even when they follow an occupation that is above or below the occupational status usually associated with their level of education. The school-dropout millionaire never fully sheds that dropout status and is likely to continue to enact that role in some ways. Similarly, the person with a doctorate who drives a cab will often be "Doc" to the other drivers.

That people with the same amount of formal schooling hold the same educational status helps to explain why variations in school quality have little or no effect on behavior: These people are equal in the way that continues to matter most in their lives.

The educational structure does not simply allocate people among occupational categories. It often creates new occupations and alters the occupational structure. The power of universities to determine the social significance of a body of knowledge is so great that Meyer speculated about what might happen if universities began to offer accredited courses and degrees in astrology: Companies would soon begin to hire staff astrologers (perhaps placing them in the economic forecasting department), and government grants for research in astrology would soon be forthcoming. Professional astrologers have failed to achieve high prestige despite their ability to attract large numbers of clients because they have been unable to convince universities that they possess a body of valid knowledge.

Clearly, then, educational institutions are not passive servants of elites and the occupational structure. Perhaps this is most easily illustrated by a norm among faculty at major research-oriented universities. They never list their occupation as "teacher," even though they teach. Instead, they call themselves chemists, economists, or sociologists—people who pursue a creative, technical profession. In this way they claim an occupational status in the world beyond the classroom—in the world of adults, not students. Because the modern university does not simply prepare people to perform various professions, but discovers and analyzes the knowledge on which these professions are based, faculty members feel justified in treating their teaching functions as secondary.

Review Glossary

Terms are listed in the order in which they appear in the chapter.

occupational prestige The respect people receive on the basis of their job; often refers to a score on a standard system for rating occupations.

scientific management The application of scientific techniques to increase efficiency.

labor force Those persons who are employed or seeking employment.

unemployed Persons of legal working age who are not enrolled in school, who do not have a job, and who are actively looking for one.

education What a person has learned.

schooling Amount of time spent in an institution dedicated to educating and which often confers degrees and diplomas on those who complete a period of enrollment.

allocation theories Theories that argue that the primary function of schools is to allocate status, to place students in the stratification system, rather than to train them.

Suggested Readings

Boudon, Raymond. 1974. *Education, Opportunity, and Social Inequality: Changing Prospects in Western Society.* New York: Wiley.

Collins, Randall. 1979. *The Credential Society: An Historical Sociology of Education.* New York: Academic Press.

Heyns, Barbara. 1978. *Summer Learning and the Effects of Schooling.* New York: Academic Press.

Hodge, Robert W., Donald J. Treiman, and Peter H. Rossi. 1966. "A Comparative Study of Occupational Prestige." In *Class, Status, and Power,* 2nd ed., edited by Reinhard Bendix and Seymour Martin Lipset. New York: Free Press.

Meyer, John W. 1977. "The Effects of Education as an Institution." *American Journal of Sociology* 83:55–77.

Sociology Online

www.socstark10.com

GO TO THE INTERNET AND TYPE: www.socstark10.com.

↗**CLICK ON:** 2000 General Social Survey

✓**SELECT: LIKE JOB?:** On the whole, how satisfied are you with the work you do?

↗**CLICK ON: Analyze Now**

TRUE OR FALSE: Men are significantly more likely than women to have a job they find satisfying.

How does this result square with the chapter?

TRUE OR FALSE: Educated people are significantly more likely than the less educated to find their jobs satisfying.

Does this result surprise you? How could it be explained?

↗**CLICK ON: Select New Data Set**

↗**CLICK ON: The 50 States**

✓**SELECT: STUDN/COMP:** Number of students per computer in school.

↗**CLICK ON: Analyze Now**

Keep in mind that the higher a state's rate, the *fewer* computers it has relative to student enrollments.

TRUE OR FALSE: The large urban states are far ahead of the more rural states in terms of student access to computers.

ⓘ USING INFOTRAC COLLEGE EDITION

GO TO THE INTERNET AND TYPE:
www.InfoTrac-college.com

↗**CLICK ON: Register New Account**
You will be asked to enter your Access Code and to create and enter a User Name and Password. Having done so,

↗**CLICK ON: InfoTrac College Edition or Log On.**

Perhaps the most interesting recent development in American education is the rise of homeschooling, while telecommuting has been of equal significance in the world of work.

SEARCH TERMS:
Homeschooling
Telecommuting

In addition, I have selected a specific article that will usefully supplement the chapter: Tales of Suburban High. You can find it by searching for this title. You may read the article on the screen or print it using the usual print commands. If you also go to www.socstark10.com and click on the InfoTrac College Edition icon, you can read an explanation of why I selected this article and find several questions that will help you connect the article to material in this chapter.

CHAPTER

17

Social Change: Development and Global Inequality

ONCE IT WAS DIFFICULT to convince people that anything in life changed, for the pace of change was so slow that change was not apparent in one lifetime. Today, at least in modern societies, it is difficult to convince people that some things haven't changed and that some kinds of social change are not likely.

Yet many previous chapters have stressed continuity as well as change in social life, suggesting that some forms of societies are probably impossible. For example, Chapters 7 and 8 suggested that we shall never create societies without crime and other forms of deviance. Chapter 9 concluded that unstratified societies are impossible. Chapter 13 proposed that the family will persist and that the nuclear family has been made more important by the rise of urban, industrial societies. Chapter 14 examined the dynamics of religious change, which make it

seem unlikely that religion will disappear from societies despite the periodic decline of some religious organizations.

Basic elements of social life limit the scope of social change—not all changes are equally likely or even possible. And social conditions influence the pace of change. For most of human history, change took place very slowly. In the past few centuries, change has been very rapid, but only in some parts of the world. While some societies launch space rockets, others have not yet learned to make tools from metal. These observations lead us to the central sociological questions about social change: Why does it occur? What factors stimulate or retard change?

In the first part of this chapter, we shall examine general principles of social change. First, we shall assess internal sources of change—things that occur inside societies that cause them

CHAPTER OUTLINE

INTERNAL SOURCES OF SOCIAL CHANGE
Box 17-1 Modest Milestones
of Modernization

Innovations
Conflicts
Growth

CHANGE AND CULTURAL LAG

EXTERNAL SOURCES OF CHANGE
Diffusion
Conflict
Ecological Sources of Change

THE RISE OF THE WEST

MARX ON CAPITALISM
Capitalism
Precapitalist Command Economies

THE PROTESTANT ETHIC

THE STATE THEORY OF MODERNIZATION

DEPENDENCY AND WORLD SYSTEM THEORY
Elements of a World System
Dominance and Dependency
Mechanisms of Dependency

A CLOSER VIEW
Jacques Delacroix: Testing the
Dependency Theory

DIMENSIONS OF GLOBAL INEQUALITY

GLOBALIZATION

CONCLUSION

REVIEW GLOSSARY

SUGGESTED READINGS

SOCIOLOGY ONLINE

to change. Next, we shall see how external forces can produce change within societies.

Then we will apply these general principles in examining the truly dramatic changes that have taken place in the world during the past several centuries—social and economic changes that are summed up by the term *modernization*. **Modernization** is the process by which agrarian societies are transformed into industrial societies (see Chapter 10). Each of the remaining chapters in this book deals with particular aspects of modernization: the relationship between modernization and population trends, the rise of urban societies, new forms of organizations required to cope with and direct modernization, and the role of collective behavior and social movements in resisting and speeding modernization.

In the latter half of this chapter, we shall attempt to understand the causes of modernization, paying particular attention to the rapid changes that transformed Europe and North America. We will also try to clarify why so many societies have thus far failed to modernize. Is it possible to have a world in which all societies are modern, or has the modernization of some been at the expense of others?

Implicit in the differences across nations in their degree of modernization is the fact of global inequality. People in some nations not only are less modern than those in others, but they *have* less. Theories of modernization, then, are theories of global inequality. Consequently, toward the end of the chapter, we will explore the dimensions of global inequality. Finally, we assess the prospects for globalization. Will the "global village" replace the mosaic of cultures and conflicts that always have characterized life on this planet?

INTERNAL SOURCES OF SOCIAL CHANGE

In Chapter 4 we examined societies as social systems. In a system connections exist among the parts so that changes in one part cause reactions in other parts. Because social systems consist of self-conscious, active human beings, internal changes are always taking place, and often a particular change will have far-reaching consequences, which people frequently fail to anticipate. Let us examine certain kinds of common activities within social systems to see how they produce changes.

BOX 17-1	MODEST MILESTONES OF MODERNIZATION

1747	Raincoat: François Fresnau, Cayenne, French Guiana	1901	Safety razor: King Camp Gillette, Boston, Massachusetts
1829	Baked beans: Mrs. L. M. Child, Boston, Massachusetts	1904	Air conditioning: Willis H. Carrier, Buffalo, New York
1830	Lawn mower: Edwin Budding, Stroud, England	1908	Athletic supporter: O. G. Overby, Brooklyn, New York
1839	Bicycle: Kirkpatrick Macmillan, Courthill, Scotland	1913	Crossword puzzle: Arthur Wynne, New York City
1841	Detective fiction: Edgar Allan Poe, Baltimore, Maryland	1925	Motel: Arthur Heinman, San Luis Obispo, California
1848	Chewing gum: John Curtis, Bangor, Maine	1929	Tape recorder: Kurt Stille, Berlin, Germany
1849	Dry cleaning: M. Jolly-Bellin, Paris, France	1930	Scotch tape: Richard Drew (3M Corp.), St. Paul, Minnesota
1850	Blue jeans: Levi Strauss, San Francisco, California	1930	Frozen food: Clarence Birdseye, Springfield, Massachusetts
1853	Potato chips: George Crum, Saratoga Springs, New York	1933	Chocolate chip cookie: Ruth Wakefield, Whitman, Massachusetts
1861	Flush toilet: Thomas Crapper, London, England	1933	Comic book: Eastern Color Company, Waterbury, Connecticut
1867	Barbed wire: Lucien B. Smith, Kent, Ohio	1933	Parking meter: Carlton Magee, Oklahoma City, Oklahoma
1882	Electric fan: Schuyler Skaats Wheeler, New York City	1938	Ballpoint pen: Laslo Biro, Budapest, Hungary
1883	Hot dog: Anton Ludwig Feuchtwanger, St. Louis, Missouri	1945	Microwave oven: Percy LeBaron Spencer, Waltham, Massachusetts
1886	Coca-Cola: Dr. John Pemberton, Atlanta, Georgia	1948	Frisbee: Fred Morrison, Los Angeles, California
1896	Ice-cream cone: Italo Marcioni, Atlantic City, New Jersey	1964	Frozen pizza: Rose Totino, Minneapolis, Minnesota
1899	Aspirin: Felix Hoffman, Leverkusen, Germany	1980	Post-its: Arthur Fry, Minneapolis, Minnesota
1901	Hair permanent: Marcel Grateau, London, England		

INNOVATIONS

The most obvious thing that happens inside societies is that people have new ideas and change how they do something. When the Quakers implemented the idea of using prisons as a substitute for physical punishment (see Chapter 8), many other aspects of society were affected, including the criminal justice system, which expanded greatly. Three basic kinds of new ideas, or innovations, frequently cause social change.

NEW TECHNOLOGY New technology is a major source of social change. Chapter 10 traced the immense social changes produced by the invention of agriculture. Chapter 19 will explain how the automobile revolutionized the structure of cities and the character of urban life. Some historians suggest that the major revolutionary event in 1776 was not the rebellion of the American colonies against Great Britain but rather the perfection of the steam engine.

Note that new technology does not change societies by itself. It is the *response* to the technology that causes change. Often, new technology appears and goes unused for a very long time. For example, the Romans fully understood how to use windmills and waterwheels to replace muscle power for various kinds of work, but they made no use of this technology. Similarly, the Chinese had gunpowder centuries before Europeans did but did not exploit its military potential. And the Aztecs put wheels on many children's toys but did not use the wheel for transportation; they carried goods on their backs rather than pulling them in wagons.

NEW CULTURE Not only machines change the world. Beliefs and values can also produce dramatic social change. Indeed, many sociologists and historians have argued that rapid technological changes in Western societies were stimulated by acceptance of the **idea of progress** (Nisbet, 1980). That is, unlike most societies that did not

Neg #326744, Photo Botlin/American Museum of Natural History

NEW **S**OCIAL **S**TRUCTURES New forms of social structure can also be the result of invention. Chapter 20 is devoted to understanding the invention, application, and evolution of formal organizations as a new mode of social structure designed to cope with and direct modernization. Indeed, in the twentieth century, the search for technological innovation was substantially directed by formal organizations such as the corporation, government, and the university.

Because roles are a basic social structure, changes in roles and the creation of new roles often cause other social changes. Changing sex roles have stimulated many changes. For example, the fact that the majority of women are now in the labor force has changed the family (see Chapter 13), and the fact that women can now choose from a wide range of occupations has greatly increased the proportion of men teaching the lower grades of elementary school. In a similar fashion, the development of full-time specialists in combat was a major factor in the repressive character of government in agrarian societies (see Chapter 10).

CONFLICTS

Innovation is not the only major internal source of change. Much change is produced by conflicts among groups within societies. Chapters 9, 10, and 11 discussed many such conflicts among classes, racial and ethnic groups, and different regions. Thus, the Civil Rights Movement of the 1950s and 1960s not only removed many barriers to African American participation in the mainstream of American society but also changed many other aspects of society. Indeed, the end to official southern racism was essential to the rapid economic growth now taking place in the South. Or in an earlier time, conflict among the many Protestant denominations seems to have been the basis for making separation of church and state part of the U.S. Constitution, with the consequence of creating a competitive religious economy that generates high rates of church membership.

GROWTH

As we shall see in the remaining chapters in this book, population growth has been a major engine driving modern social change. Large populations present new problems that demand new modes of social organization. For example, as pointed out in Chapter 15, small populations

For a long time, anthropologists wondered why civilizations as sophisticated as those of the Incas and the Aztecs had failed to discover the wheel. Then toys like this one were discovered in Aztec ruins in Mexico— little dogs on wheels so children could pull them. Suddenly, the mystery deepened: Given that these civilizations did know about wheels, why did they use them only on toys and not to move heavy loads? Especially since they constructed good roads.

expect technological progress and often turned away from it, in Europe during the seventeenth and eighteenth centuries, people widely accepted that such progress was not only possible but also certain. Such confidence inspired ever more determined efforts to achieve technological and scientific progress, and as these efforts bore fruit, they inspired even greater confidence and effort. To a considerable extent, Europe made progress because it believed in progress and underwent rapid social change because Europeans wanted change.

In recent times belief in progress has waned in some parts of the industrialized world while gaining greater credibility in some of the less developed nations. This shows up very clearly in Table 17-1, which gives the percentage in each nation who believe that scientific advances will help rather than harm "mankind." Nations where the largest majorities express faith in scientific progress overwhelmingly are less developed nations that are relatively lacking in terms of local scientific establishments: Nigeria, Egypt, China, Vietnam, Morocco, and the like.

TABLE 17-1	"In the long run, do you think the scientific advances we are making will help or harm mankind?"		
NATION	**WILL HELP (%)**	**NATION**	**WILL HELP (%)**
Nigeria	85	Belarus	49
Egypt	79	South Korea	49
China	77	Albania	49
Vietnam	75	India	48
Morocco	72	Indonesia	48
Armenia	70	Argentina	47
Georgia	68	**Mexico**	**45**
Iran	67	Dominican Republic	45
Bangladesh	67	El Salvador	45
Philippines	67	Sweden	42
Jordan	66	Chile	41
Iceland	64	Switzerland	40
Turkey	62	Norway	39
Zimbabwe	62	Ireland	38
Brazil	62	Croatia	37
Venezuela	60	Great Britain	36
Azerbaijan	56	Austria	34
Australia	55	Taiwan	34
Lithuania	54	Colombia	33
United States	**54**	Spain	32
Bosnia	53	Italy	30
Algeria	52	New Zealand	26
Canada	**51**	Japan	23
Germany	50		

Source: Prepared by the author from the World Values Surveys, 2001–2002.

may permit direct democracy in which all citizens have the opportunity to participate in decision making. However, such procedures are impossible for large populations, which require new models of democracy such as representative government.

Similarly, large cities must be constructed very differently from small cities (see Chapter 19). The simple growth of cities has produced great social change. As we shall see in Chapters 18 and 19, the earliest stage of modernization—industrialization—caused rapid population growth, which in turn spurred even more rapid industrialization.

CHANGE AND CULTURAL LAG

Social change involves complex patterns of response because a change in one part of society forces changes in other parts. For example,

a sharp decline in the birthrate during the 1960s soon caused a crisis in American schools. Suddenly, there were too many teachers and classrooms for the number of students. Many teachers had to be let go, and many schools closed. Ex-teachers had to find new occupations, and some use had to be found for unneeded school buildings. In addition colleges reacted to the lower demand for teachers by cutting back their number of education majors. Thus, some students who had planned to become teachers had to rethink their career plans.

Beyond changing the schools, the reduced birthrate forced many readjustments elsewhere in society. Industries and stores that specialized in products for infants and young children had to respond to rapidly declining sales. A leading baby-food company launched new products designed for the elderly, for example; TV stations reduced the amount of children's programming. More recently, the lower birthrate has been reflected in smaller numbers of teenagers,

Plans to import modern technology often go astray. A rail line was laid near the village in which these women live. But as long as the rails serve only as a path and basket-carriers continue to serve as the primary means of transportation, these people have made no progress toward modernization.

© Werner Bischof / Magnum Photos

with a corresponding decline in sales of acne medications.

This is but a sketch of the most direct effects of the reduced birthrate (which resulted from changes in the family and in female employment). But none of these reactions was immediate. The schools did not readjust as soon as the number of births declined. In fact, they took no action until the number of elementary school students had seriously dropped. And the colleges did not reduce the number of new teachers they trained until massive unemployment confronted their graduates.

There can be considerable delay before a change in one part of a society produces a realignment of other parts. During such a period of delay, parts of a society can be badly out of harmony, such as when education departments continue to pump out waves of new graduates after there are no employment opportunities for them. William F. Ogburn (1932) described such periods as **cultural lags.** According to Ogburn, cultural lags are times of danger for societies because severe internal conflicts can result.

An excellent example of cultural lag will be discussed at length in Chapter 18. This involves the sudden introduction of pesticides to tropical

nations following World War II. Sprayed from airplanes, the pesticides killed disease-carrying insects, especially the mosquitoes that spread malaria. Within weeks infant and child mortality rates dropped rapidly—in some places by as much as half. But although greatly improved survival rates removed the need for parents to have large numbers of children, centuries of high mortality had firmly molded their cultures to prefer large numbers of children. So, as the children kept coming, and surviving, a huge population explosion occurred—far sooner than it would have been reasonable to expect major cultural changes in response to the new situation.

EXTERNAL SOURCES OF CHANGE

Unlike the solar system, social systems exist not in a vacuum but within a social and physical environment. Interaction with that environment is a major cause of change within societies as well as a major factor limiting the kinds of changes that occur.

Assume that a number of small societies exist in close proximity, such as the many small societies of Native Americans that once existed side by side. Assume that change is going on within each. Each change causes other parts within the societies to adjust in response to cultural lag.

But here internal changes can have external implications. That is, some cultural or social arrangements may weaken the ability of a society to withstand external threats, either from other societies or from the physical environment. Thus, some changes that may have been effective adaptations to internal social needs can be maladaptive to external demands. Societies that change in such directions are not likely to survive as, indeed, many societies have not.

What has just been outlined is the process of *social evolution* (see Chapter 4). All theories of social change imply evolutionary mechanisms such as this. All social change is subject to external constraints, and external factors frequently produce internal changes.

DIFFUSION

We have seen that innovation is a major source of change. But innovations, whether in the form of new weapons, new customs, or new religions, are more often imported from other societies than developed independently within a society. As the famous anthropologist Ralph Linton (1936) put it:

The number of successful inventions originating within . . . any one . . . society . . . is always small. If every human group had been left to climb upward by its own unaided efforts, progress would have been so slow that it is doubtful whether any society by now would have advanced beyond the level of the Old Stone Age.

More rapid progress has been possible because societies borrow innovations from one another. This transfer of innovations is called **diffusion.** Anthropologists and many other scientists have specialized in tracing the routes by which innovations have spread, or diffused, from their point of origin to other societies. Box 10-1 traces the diffusion of the stirrup from Asia to Europe and the role this played in the rise of feudalism. When Marco Polo returned to Italy from his journey to China, he brought back (among other things) the noodle, which was the basis for the development of many forms of pasta popular in Italy and elsewhere in the West

today. The horse was brought to the Western Hemisphere by Spanish explorers and spurred immense changes in Native American societies. Corn, tomatoes, turkeys, and peppers diffused from the Americas back to Europe and Asia. Gunpowder was invented in Asia and then spread around the world. And central to this chapter is the rapid diffusion of modern European culture and technology throughout the world and the impact of this diffusion.

Diffusion is not something that only "used to" happen—it still happens all the time. But with modern communications, it happens very rapidly. Television is an example. TV is an American invention, based on a series of patents by Allen B. DuMont, Philo T. Farnsworth, and Vladimir K. Zworykin. In 1931 RCA conducted the first experimental TV broadcasts from a transmitter at the top of the Empire State Building in New York City—then the tallest building in the world. The experiments were a success, but it wasn't until 1939 that RCA began regular broadcasts to New York City from this same transmitter tower. The delay was caused by the Great Depression—very few people were able to afford TV sets (which were much more expensive then than now). In fact, even in 1939 the broadcasts might as well have been regarded as experimental since there were fewer than 2,000 private sets in the city.

The growth of television was further delayed by World War II, and so it was not until the late 1940s that the medium really began to spread—and not until well into the 1950s that it was available in most major American cities. As can be seen in Table 17-2, by 1955 there were 22 TV sets for every 1,000 Americans (compared with 839 radios per 1,000). Canada was second, with 15.2 sets per 1,000, and Great Britain third. In the rest of the world, however, TV sets were still extremely scarce. Even in the highly industrialized nations of Europe, very few people had sets—fewer than 1 per 1,000 people in most places. And outside Europe, virtually no one had a set—even in Japan only 3 people of every 10,000 had one, while in many nations there were neither TV sets nor TV stations.

But as of nearly fifty years later, that's all changed—the diffusion of television is an accomplished fact. While the United States and Canada still have the most sets per 1,000 population, many other nations have nearly as many, and virtually all nations have TV stations. Keep in mind that these nations did not launch crash research programs to learn how to make TV transmitters, cameras, or receivers. Had that

TABLE 17-2	The Diffusion of Television and the Prevalence of TV Viewing		
NATION	TV SETS PER 1,000 POPULATION		WATCH THREE OR MORE HOURS PER DAY (%)
	1955	2003	
United States	22.0	938	42
Canada	15.2	691	35
Great Britain	13.0	950	42
Belgium	1.7	541	19
France	1.0	632	16
Denmark	1.0	859	11
Netherlands	0.9	648	14
Mexico	0.8	282	18
Italy	0.8	494	18
Russia	0.7*	538	14
Australia	0.3	722	45
Japan	0.3	785	31
Ireland	0.1	694	40
Iceland	None	509	8
Indonesia	None	153	33
Pakistan	None	79	17
Egypt	None	229	35
Nigeria	None	68	34

Sources: Data on TV sets per 1,000 from the *Statistical Abstract of the United States, 1959*; International Telecommunications Union, *World Communications Indicators 2004*. Data on TV viewing were prepared by the author from the World Values Surveys, 1981, 1995–1996, 2001–2002.

*Then the Soviet Union.

been necessary, there still would be many nations without TV. Instead, most nations simply bought their equipment from abroad or purchased the rights to use or manufacture the existing technology (some nations do not honor foreign patents and simply copied the technology without payment or permission).

The last column in Table 17-2 shows the percentage of people in each of eighteen industrialized nations who report that they watch three or more hours of TV on the average day. What the nations where people watch the most TV have in common is the English language. This may reflect the fact that the most popular shows, even in non-English-speaking nations, are made in the United States, Great Britain, and Canada, and that English-speaking audiences watch more hours of TV because of the availability of more hours of popular shows. Indeed, many nations have only one channel, controlled by the government. The shows broadcast on these channels seem sufficiently popular to get people to buy TV sets, but not so popular as to get them to spend a lot of hours watching.

CONFLICT

Threats from other societies frequently are sources of social change. The rapid evolution of firearms and artillery in Europe, once gunpowder reached it from Asia, occurred because Europe was divided into scores of feuding societies, each needing to match or exceed the military capacity of its neighbors. In contrast, the relative lack of conflict within the vast Chinese empire made the development of similar innovations unnecessary.

As we saw in Chapter 14, grave external threats often prompt religious innovation. For example, as repeated efforts by Native American tribes to fend off westward development by European settlers failed, often they concluded that their difficulties stemmed from a faulty religion, from worshiping the wrong gods, or from worshiping them in the wrong way, and new religious movements flourished.

ECOLOGICAL SOURCES OF CHANGE

Changes in the physical environment often produce social change. Concerns about depleting

© David Phillips

Bear Chief, as he posed for a photographer in Montana in 1910, illustrates the process of cultural diffusion. In Bear Chief's time, it was almost impossible to imagine Plains Indians without their fast ponies or firearms—they were regarded as among the finest light cavalry in the world. But Native Americans only had horses for a few generations (horses were brought to the Western Hemisphere by the Spaniards), and they never did make their own firearms or ammunition. The invention of firearms, of course, rested on the diffusion of gunpowder from China to Europe. By the same token, it is equally impossible to imagine Italian food without the tomato, but Italians had no tomatoes until they were obtained from Native Americans.

natural resources and polluting the environment have caused many recent changes in the United States. Droughts and natural disasters have often prompted massive social changes. Similarly, more favorable ecological changes have also prompted change.

The interplay between environment and society is well illustrated by the great Viking expansions that began about A.D. 900. The Vikings conquered or colonized Russia, large sections of northern Europe, Ireland, parts of England and Scotland, Iceland, and Greenland. This Viking "explosion" was probably caused primarily by several centuries of unusually warm weather (Mowat, 1965; Sawyer, 1982).

Warm weather meant much more abundant crops in Norway and Sweden, and more food meant population growth. Soon this larger population lacked land, and younger sons who could expect no inheritance set out to seek their fortunes by raid and conquest. As it happened, they had the sailing and fighting skills to succeed. And the fair weather made it possible to travel across the North Atlantic to such places as Iceland, Greenland, and even North America.

In time the weather turned cold once more, and the Viking population declined. The North Atlantic was once again shrouded in fog, battered by storms, and filled with icebergs. The Viking ships stopped going to Iceland and

These Eskimo boys in northern Greenland (about 1900) demonstrate the survival advantages of hunter-gatherers in the frozen North. When a shift in the climate made it impossible for the Vikings in Greenland to farm, they perished. Meanwhile, the Eskimos continued to get their clothing from animal hides and their food from hunting and fishing.

Neg. #232240, photo: D.B. MacMillan/American Museum of Natural History

Greenland. The Vikings in Iceland adjusted to the new conditions and survived centuries of isolation, but the Viking settlements in Greenland slowly died out. By the time new explorers from Europe visited Greenland again, only ruins and graves remained. Yet the Greenland Eskimos, who had been there long before the Vikings came, survived the shifts in the climate and still live in Greenland today. Their culture was able to readjust to the frigid climate that made farming impossible for the Vikings (McGovern, 1981).

In summary both internal and external forces produce social change, and both can cause societies to break down. Europeans in the eighteenth century were probably correct in believing that change was inherent in all societies. They were probably wrong, however, in their faith that change always means progress. Surely, the Native American peoples, despite gaining the horse, firearms, and other new technology from European settlers, did not find that change meant progress. In the end change destroyed their societies. Nor did northern and western Europeans find that improved weather brought only progress, for it also brought fleets of Viking raiders down upon them.

Keeping in mind these principles of social change, we may now assess the causes of the dramatic set of social changes known as modernization.

THE RISE OF THE WEST

A thousand years ago, Europe was a collection of small agrarian societies. Changes came so slowly that most people encountered nothing during their lifetimes that had not been familiar to their parents and grandparents. But beneath this tranquil surface, an era of unprecedented change was building up (Diamond, 1998; Landes, 1998). Within a few centuries, these small societies began to combine into larger states, while European technology suddenly raced far ahead of that of the rest of the world.

In the midst of this period of great progress, Europeans set out to explore, colonize, and trade with the rest of the world. Everywhere they went they found themselves possessed of superior technology. Indeed, when Western fleets began to voyage to China, long known in the West for its advanced civilization, they found a backward nation unable to defend itself against a few ships armed with cannons (Mendelssohn, 1976; McNeill, 1982).

Thus, the questions arose: *Why had China not kept pace? Why had social change been so rapid in Europe?* Indeed, why was so much of the world so little advanced—why were some societies still huddled in the Stone Age, while in Europe machines were replacing human labor?

These were and continue to be dominating questions in the study of social change. Moreover, new questions have arisen: *Why has continued exposure to Western technology had so little effect in some parts of the world? Can all societies become modernized? If not, why not? If so, how?* You will recognize that these questions not only preoccupy sociologists interested in modernization and social change but also are among the leading international political questions of our time.

There are four major bodies of theory about why the West suddenly produced the Industrial Revolution and sprinted ahead of the rest of the world. Because these theories range from Marxism to free-market economic theories, they stand in vigorous, basic disagreement. Yet, perhaps surprisingly, they agree on their initial assumption. Each attributes modernization, the rise of industrialized Europe, to the same basic source: the development of a particular pattern of economic relations called *capitalism*. Indeed, no conservative economist has ever heaped more praise on capitalism as the source of modernization than did Karl Marx, even though he devoted his life to planning for the overthrow of capitalism. So, it is fitting to begin our assessment of modernization by seeing why Marx thought it was the result of capitalism. Then we shall closely examine the nature of capitalism itself.

MARX ON CAPITALISM

Marx envisioned a new world in which all people could enjoy a good life, where no one would be hungry or homeless, and where everyone would have freedom, dignity, and security. Throughout the ages many people have longed for such a world. But in the mid-nineteenth century, Marx believed such a thing had become possible for the first time. In earlier times equality could exist only as the equality of poverty. Humans simply were not sufficiently productive to provide themselves with comfort. But as Marx surveyed the immense flow of products from Europe's new and booming industries, he became convinced that societies were finally capable of providing everyone with a good life. The potential wealth was there; it had only to be shared.

Marx did not believe that the Industrial Revolution itself caused Europe's new explosion of productivity. Instead, he thought that this new technology was the result of something more basic: a new mode of economic arrangements that unleashed the full productive potential of human beings. He called these economic arrangements *capitalism*.

Before the rise of capitalism, Marx wrote, humans had been victims of their own "slothful indolence." That is, people tried to avoid work and did not try to find ways to be more productive. In fact, many historians have been struck by the short workdays of medieval society and the careless farming methods of medieval peasants (see Thomas, 1979; Braudel, 1981). Similarly, anthropologists have long noted the casual attitudes toward work among primitives, as compared with modern work norms. Marx believed that Europe's great leap forward occurred because people suddenly began to work harder and smarter as a result of the inducements of capitalism. Capitalist society, he wrote in the *Communist Manifesto* (1848), was

the first to show what man's activity can bring about. It has accomplished wonders far surpassing Egyptian pyramids, Roman aqueducts, and Gothic cathedrals; it has conducted expeditions that put in the shade all former Exoduses of nations and crusades.

Moreover, capitalism cannot help but produce endless technological innovation: It must be "constantly revolutionizing the instruments of production." This is because capitalism has stripped away the traditional bases of relationships and left only one "nexus between man and man," that of "naked self-interest."

Here we encounter a great irony. Marx believed that the reason capitalism could be so productive was precisely the reason it ought to be destroyed. He believed that capitalist economies produced their economic miracles by degrading and alienating humans, both from one another and from themselves. By pursuing self-interest alone, humans ruthlessly exploit one another, according to Marx. Thus, he argued, capitalist societies not only were the first with the productive capacity to overcome poverty and exploitation but also were incapable of doing so.

Indeed, Marx believed that capitalist societies could only become increasingly unequal, eventually consisting of a tiny ruling elite (the bourgeoisie), possessed of incredible wealth and power, and a huge mass of "wage slaves" (the proletariat), sweating out their lives in dismal factories. Thus, Marx proposed communist revolutions, in which the masses would seize collective ownership of all means of production and turn the immense capacities of modern industrial societies to the benefit of all (see Chapters 9 and 10).

Although Marx believed capitalism had been the cause of modernization, he believed *the benefits of modernization could be separated from this initial cause*—that communism could replace capitalism once modernization was sufficiently developed. But what is capitalism? And what is the secret of its economic power?

CAPITALISM

Current dictionaries define **capitalism** as an economic system based on private ownership of the means of production and a system by which people compete to gain profits. Such a definition fails to reveal the feature of capitalism that differentiates it from other kinds of economies. For example, there was much private ownership of the means of production (farms, tools, and ships) in ancient Rome and medieval Europe, and there was competition for wealth. But these were not capitalist economies. What is unique about capitalism is its reliance on a *free market*.

In a free market, freely made choices of individuals set prices and wages. That is, people decide for themselves what price they will charge or pay. They cannot be forced to buy or to sell, to hire or to become employed. Each is free to make the best possible bargain.

Prices and wages are set by supply and demand. Competition among many people selling some commodity forces prices down. Competition among people wanting to buy something in limited supply forces prices up. Competition among people wanting to be employed forces wages down. Competition among persons wanting to hire forces wages up. *The essence of this system is that everyone seeks to maximize personal gain and that such gains can be accumulated in the form of private property, secure from arbitrary seizure by the government.*

In a free market, individuals benefit by being more productive. If one farmer works longer hours in his fields, then at the end of the year, he will have more wealth than his neighbors who worked less. Moreover, the free market rewards innovation and the reinvestment of wealth. A farmer who finds a way to plow better or faster will become richer. A farmer who saves some of his profits and uses them to buy more land, more cattle, or better machinery will become richer. In this way capitalist economies motivate everyone to try to become wealthier, and these collective efforts increase production; when there is more to be had, standards of living rise. Marx believed that capitalism had unleashed such productivity in Europe that it would soon be possible to eliminate all poverty.

PRECAPITALIST COMMAND ECONOMIES

Precapitalist societies do not rely on free-market principles and individual economic self-interest. Instead, they are **command economies.** That is, some people decide what work is to be done and command others to do it. The lord of a medieval estate decided which fields to plow, when, and what to plant and then ordered his peasants to do it. An emperor decided to build a road and ordered workers to assemble and set to their tasks. The weakness of command economies is that *those doing the work have nothing to gain by doing it well.* A slave may avoid the overseer's whip by doing just enough, but a slave will not eat better or gain possessions by working harder or discovering ways to become more productive. Lack of worker motivation caused Marx to scorn the slothfulness of such economies.

The goal of command economies is consumption. Everyone tries to consume what they can before someone else takes it away from them. Thus, there is no motive to produce surplus. Indeed, as we saw in Chapter 10, when individuals lack secure property rights, their surplus production is simply taken from them and used to support others who suppress and exploit them. *The secret of capitalism is to reward surplus production:* Permitting people to keep their wealth encourages them to seek wealth and to curb their consumption so that they can reinvest their wealth to create more wealth.

An episode from ancient China clearly reveals the productive superiority of capitalist over command economies. In the late tenth century, an iron-smelting industry rapidly developed in the central Chinese province of Hunan. By 1018 these iron smelters were producing more than 35,000 tons a year, an incredible achievement for the time. This iron industry was not the result of royal command. Instead, private individuals had recognized the great demand for iron and the vast supplies of ore and coal in Hunan, and they realized that the smelted iron could easily be transported to distant markets over an existing network of canals and navigable rivers.

Having invested in foundries, these Chinese industrialists were soon reaping huge profits from their enterprises, much of which they reinvested to build more foundries. Production rose rapidly. The availability of large supplies of iron soon led to the introduction of iron agricultural tools, which in turn rapidly increased food production in China. Thus, China began to industrialize many centuries before Europe's great leap. Then, as suddenly as it had begun, it all stopped. By the end of the century, only tiny amounts of iron came down the rivers from Hunan, and soon the foundries were forgotten ruins.

What happened? The imperial court had noticed that some commoners were getting rich by manufacturing iron and considered this undesirable. So, the government taxed away their earnings, declared a monopoly on the sale of iron, and took over the smelters. Workers had flocked to work in the smelters, where they earned more than they could as peasants, but now work was commanded of them. The motive for working died out, as did the fires in the smelters (McNeill, 1982).

If capitalism caused the rise of the West, what caused capitalism? Here Marx was relatively silent. He argued that it was invented by the bourgeoisie, who used it to overthrow the old medieval nobility. But he said very little about why and how the bourgeoisie developed capitalism. Thus, it was left to Max Weber to attempt the first general explanation of the rise of capitalism.

THE PROTESTANT ETHIC

The goal Weber set for himself was to explain why capitalism developed where and when it did

and why it failed to appear (except for the brief instance cited in ancient China) in other societies that had achieved a stage of economic development similar to that of precapitalist Europe. In Weber's judgment the essential question was, How had Europeans gained the self-discipline to cease unrestricted consumption while increasing production? Many societies have learned to curb consumption. Ascetic religions, for example, have led many people to spurn material things. But this has been accomplished by destroying their interest in *creating* material things. In Europe, however, people curbed their consumption while working all the harder to produce. How was this possible?

Weber believed the answer lay in the Protestant Reformation. In his famous book *The Protestant Ethic and the Spirit of Capitalism*, he argued that the religious ideas produced by Protestantism had motivated people both to limit their consumption and to pursue maximum wealth. People soon discovered that reinvestment was the fastest road to wealth.

The **Protestant Reformation** began in Germany when Martin Luther (1483–1546) asserted that the church was not needed to mediate between a person and God. Rather, each person should seek his or her own salvation through direct relations with God. In Switzerland John Calvin (1509–1564) carried this notion much further. According to Weber, Calvin argued that God was entirely unknowable. Therefore, no person could achieve salvation by appealing to God or by obeying the Ten Commandments. Who would be saved and who would be damned had been decided by God at the beginning of time and could not be altered. This doctrine was called predestination: Our futures were predestined by God.

But then, how could one be sure of being saved? Indeed, what motive was there not to sin? Weber claimed that Calvin provided an answer that put extreme pressure on the individual to lead an exemplary life: You can never be certain you are saved, but there are clues that indicate who is elected by God for salvation—persons whose lives are above reproach and who succeed in life.

No longer was work merely a calling to be endured; now it was seen as a glorification of God and more of an end in itself. The successful worker was the successful servant [of God], and money became a metric for the measurement of grace. (Demerath and Hammond, 1969)

John Calvin.

Art Resource, NY

In addition Calvinist Protestants condemned the most conspicuous forms of consuming wealth as sinful. They believed that you could not show the world that you were rich by a great display of your wealth, and therefore, you could not display that you were one of God's chosen. You could show your success only by visible productive activities. And because you could not consume much wealth, why not use it to generate even greater wealth and that much greater certainty that you would go to heaven?

Strict predestinarian views of salvation did not last long as a dominant theme in Protestantism, but the actions they set in motion did. It soon became popular Protestant doctrine that one could actually earn one's way into heaven. Economic zeal became the road to heaven.

From these cultural developments, Weber argued, capitalism blossomed. Soon the religious roots of capitalism were no longer needed, for capitalism became a secure ideology in its own right—the **spirit of capitalism.** The spirit then spread through both Protestants and Catholics and stimulated the Industrial Revolution.

Weber quoted at length from Benjamin Franklin to show how deeply belief in the importance of saving and reinvestment had become embedded in Western cultures.

Weber did not argue that the **Protestant ethic** was the sole cause of the rise of capitalism. Nor did he ignore interaction between developing commercial activities and developing religious doctrines. He merely argued that the development of these religious and economic ideas gave rise to the Industrial Revolution.

Although Weber's work on the Protestant ethic is widely admired as a classic work of sociology, it never was highly regarded by historical economists and has been severely criticized by religious scholars as well. Historical economists have found that the Protestant areas of Europe did not develop capitalism, or industrialize, sooner than did the Catholic parts of Europe (Samuelsson, 1957; Delacroix, 1992). This suggests that Protestantism had little or nothing to do with the rise of the modern world. Religious scholars reject Weber's representation of Calvin's theology, thereby refuting that portion of his theory. According to religious scholars, Weber seems not to have actually read Calvin or even any of his contemporary interpreters, but based his characterization of Calvin's theology on later writers who were opposed to Calvin's views and misrepresented him.[1] Because Calvin did not propose the theological ideas that Weber identified as the source of the Protestant ethic, the link to Calvinism identified by Weber can't be correct (Samuelsson, 1957). The ideas Weber attributed to Calvin eventually did gain support in some Protestant European circles, and some of them appeared in the Westminster Confession on which Presbyterianism was founded. But these theological ideas appeared too late to have played an important role in the rise of capitalism or in the spread of industrialization. Some recent writers on the origins of capitalism find the basis of the spirit of capitalism not in Protestantism, but in Christian theology more generally (Chirot, 1994).

Whatever the case, today even Weber's greatest admirers think that he emphasized religious values too much and economic and political factors too little. However, there is widespread agreement that Weber was addressing the right question. If, as it appears, a shift in how people regarded wealth was an important part of the

---

1. I am grateful to Irving Hexham, the distinguished religious scholar, for calling this to my attention.

development of capitalism, then we need to know the origins of such a shift.

THE STATE THEORY OF MODERNIZATION

A third line of social theory suggests that Marx and Weber were both correct but that both views are limited. Proponents of this perspective agree with Marx that capitalism led to the rapid modernization of Europe, but they agree with Weber that the rise of capitalism itself must be explained if we are to account for modernization. They disagree with Weber that Protestant ideology led to the development of capitalism, arguing instead that both Protestantism and capitalism were produced by something more basic in European history: the taming of the state.

Building on classical economic theories, these social scientists propose that capitalism could develop *only as the state became tamed*; moreover, capitalism *will always develop* when the state is tame (North and Thomas, 1973; Nozick, 1974; Chirot, 1985, 1994; Landes, 1998). Because of the centrality of the state in this explanation, it is called the **state theory of modernization.**

The state theorists argue that the critical event in European history was the limitation of government power in several European nations, especially England and Holland. This resulted in capitalist economies, which in turn gave rise to the Industrial Revolution, or modernization.

The argument is very simple. When a repressive state exists, so will a command economy. Few persons in such societies benefit by being more productive, for the state supports itself by confiscating all surplus production, which is then consumed by the ruling elite. Under these conditions, as Marx recognized, it would be foolish to curb consumption or try to produce more because of the state's insatiable appetite. The most powerful religious ideas could not cause people to act like capitalists in such societies. Even if some people do, they will soon find their work is in vain, as did the Chinese iron makers.

Daniel Chirot (1985) has pointed out that the untamed state always stifles economic development because *it cannot resist.* Ruling elites always overreproduce themselves and therefore need "to extract increasing amounts from their subjects." Comparing agrarian ruling elites with parasites, Chirot noted that they could not long resist the temptation to expropriate the wealth of merchants and any other visible source of

funds to support their always-expanding numbers. Periodically, of course, these practices led to bankruptcy; they never led to industrial development.

However, if the powers of the state are limited so that private property is secure from seizure, if profits are not subject to overtaxation, and if people are free to pursue their economic self-interest, capitalism becomes an attractive and viable option: The more people work and save, the better off they will be. Thus, whenever the state is prevented from seizing property, a free market, or capitalism, will develop. Rapid economic and technological progress then becomes very likely. When people are more productive, more wealth exists. In seeking to be more productive, huge numbers of people will seek more effective technology.

Marx was inclined to take technological progress for granted, seeing it as the natural result of human curiosity. The state theorists do not. They point to the historical fact that for long periods little, if any, technological progress occurred, whereas at other times new technology rose rapidly. Why? The answer they offer is that in some times and places, it is not worthwhile for people to develop new technology because they will not benefit from it. Why build a windmill so you can grind much more grain than you have been grinding by hand if your increased production will simply provide more flour for the nobility and no more for you?

Chirot (1985) explained it this way:

Economic rationality means that economic actors are willing and able to make reasonable predictions about their return on investments. Social and political systems which are arbitrary, or which do not guarantee property rights, or which do not protect key economic actors, are not conducive to economically rational behavior. This hardly means that in nonrational circumstances most people behave foolishly or illogically, only that predictability becomes so difficult that, to protect themselves, economic actors take measures which have little bearing on maximizing the productive power of their investments.

Since patent records have been kept, inventions have not been developed at a steady rate; many more patents are applied for during economic booms than during recessions (Schmookler, 1966). This suggests that people tend to invent things when it is profitable to do so. Historians of technology argue that it was the creation of patent laws that truly spurred invention during the Industrial Revolution (Jewkes, Sawers, and Stillerman, 1969). Patent laws protect an inventor's right to profit from his or her invention. Neither individuals nor firms would risk years of effort and the large investments needed to perfect many inventions if others could then simply steal their results.

Once the state had been tamed in parts of Europe, free markets sprang up, and the state theorists argue, from then on the development of modern industrial societies was virtually certain (North and Thomas, 1973). Moreover, some state theorists claim that the taming of the state was the primary cause of the Protestant Reformation (Walzer, 1963).

In Chapter 14 we saw that religious pluralism is the natural state of religious economies—that different kinds of people have different religious needs, so that the market is served best by a variety of faiths. When the state does not try to create a monopoly for one faith, many faiths will exist. Thus, religious variety sprang forth in those European nations where the power of the state was restricted. However, state theorists go beyond attributing religious pluralism to the taming of the state. They suggest that the rise of capitalism prompted specific Protestant doctrines, thus reversing Weber's argument. When people pursue their own economic self-interests, their sense of individualism is heightened. As a result people want to deal with God directly rather than through a religious hierarchy. The Protestant emphasis on each person seeking his or her own understanding of God and his or her own salvation was thus compatible with daily economic activities.

DEPENDENCY AND WORLD SYSTEM THEORY

Each of the three explanations of modernization just examined seeks the causes of the Industrial Revolution *within* societies. Marx believed that the development of capitalism within nations led to the Industrial Revolution. Indeed, Marx denied that nations could achieve communism without first passing through a capitalistic phase. Similarly, Weber tried to show that capitalism arose only where the Protestant Reformation had first planted an ethic promoting hard work, saving, and reinvestment. State theorists argue that capitalism arose only in nations that had first tamed the state. But a fourth body of modernization theory looks not to changes within a nation but rather to changes in relationships among nations as the causative force. This explanation is called **world system theory** or **dependency theory.**

Most advocates of world system theory claim to be Marxists, although the first extended statement of this view was written not by Marx but by J. A. Hobson, an English economist. In his *Imperialism,* published in 1902, he charged that industrial European nations looted their colonies by forcing them to sell their raw materials too cheaply and to buy manufactured goods at too high a price. In 1915 V. I. Lenin, who soon was to lead the Russian Revolution, borrowed Hobson's argument (as well as many of his statistics) and published them in a book called *Imperialism, the Highest State of Capitalism.* Ever since then, communist writers have claimed that Western capitalist nations not only exploit the less developed nations of the world but actually prevent them from modernizing. Some years ago, this approach gained serious advocates among American sociologists. The man most responsible for recruiting sociologists to world system theory is Immanuel Wallerstein.

ELEMENTS OF A WORLD SYSTEM

In *The Modern World System,* Wallerstein (1974) elaborated on the view that the modernization of the West was paid for by its less fortunate neighbors. That is, the causes of the Industrial Revolution are to be found not within individual nations but rather in the *relations among nations that unite them into a single social system.* He therefore set out to examine in detail the world system existing in the sixteenth century, during which Europe began to develop capitalism and to industrialize.

In Wallerstein's judgment the crucial development in the sixteenth century was the growth of an international economy that was not politically united. Through this economy some nations extracted wealth from other nations without having to resort to military force. Wallerstein argued that this interaction differed from all previous forms of international relations. In the past nations had extracted wealth from other nations by plunder or by forcing them to submit to political control as part of an empire. Thus, for a long period, Rome and before it Egypt dominated huge empires in which the threat (and often the use) of coercion extracted taxes and tributes.

Such empires were inefficient. Their command economies incurred such great military and administrative costs that they probably lowered the standard of living of all but the ruling elite. Any border troubles or internal rebellions raised the costs of maintaining the empire and strained available resources. When these costs could no longer be met, the empire became unstable and eventually collapsed.

What was unique about developments in Europe, according to Wallerstein, was that economic relations developed among nations whereby some could exploit others without paying the huge costs of running an empire. Thus, a few nations in western Europe were able to finance their rapid industrial development by extracting wealth from their neighbors without the need to plunder or dominate them through military force. Wallerstein called this international economy in sixteenth-century Europe the "modern world system," even though it was far from worldwide in scope. His term emphasized that this was an international social system.

Within this world system, Wallerstein argued, *stratification exists among nations.* A few nations form an upper class, some a lower class, and a few a middle class. A nation's class position is determined by its place in a geographical division of labor.

Wallerstein called the dominant, or upper class, nations in a world system **core nations.**

They have highly diversified economies and are the most modern and industrialized. Core nations also have the strongest internal political structures marked by stable governments and little internal class conflict. As did Lenin, Wallerstein argued that core nations are stable because they can provide a very high standard of living for their workers and thus in effect buy their cooperation. Core nations also have a large middle class and permit considerable political freedom and individual liberty.

At the bottom of world systems are nations that Wallerstein identified as **peripheral nations.** Many are located far from core nations. Typically, they have weak internal political structures and a low standard of living for workers. Because of their high potential for political instability and class conflict, they are ruled by repressive governments. Peripheral nations have highly specialized economies, typically relying on the sale of a narrow range of raw materials (such as food, ore, fiber, petroleum, or timber) to core nations.

A few nations in a world system may display features of both core and peripheral nations. Their economies are more diversified than those of peripheral nations but are more specialized than those of core nations. Wallerstein called these **semiperipheral nations.**

Proponents of world system theory argue that these African peasants must toil with old-fashioned methods because the developed nations prevent industrialization in the less developed nations.

DOMINANCE AND DEPENDENCY

To explain the relations among nations in a world system, Wallerstein applied a Marxist analysis of class relations within nations. Thus, he argued, the core nations act as an upper class exploiting the peripheral nations, which are the lower class. Core nations, like upper classes, are wealthy because they extract all surplus production from the periphery. They do this by dominating trade relations with the peripheral nations, making them dependent and distorting their economies so that peripheral economies cannot develop into modern societies.

Wallerstein used this model to explain the unevenness of modernization and industrialization in Europe. He admitted that he could not explain why some nations in western Europe initially got the jump on others and became core nations. But once some had done so, he claimed, they forced the nations of eastern Europe to accept a peripheral position. In this fashion he explained why nations such as Poland and Hungary began to industrialize but then stopped and regressed to agricultural nations that lived by exporting food to western Europe in return for manufactured goods. Because of their late start, Poland and Hungary could not successfully modernize against western European competition and thus became dependent on the West.

Although Wallerstein's book dealt with Europe during the sixteenth century, his real interest (and the main interest of many other sociologists now using world system theory) was current international relations. The two primary questions are these: (1) Why is so much of the world so little modernized despite centuries of trade and contact with more advanced nations? (2) Can modernization ever become worldwide? Proponents of world system theory answer that trade with advanced nations has prevented the less developed nations from developing and that unless these nations escape the control of advanced capitalist societies, they can never develop properly.

MECHANISMS OF DEPENDENCY

The world system theory specifies a number of mechanisms by which less developed nations are made dependent on and are dominated by the advanced nations (Frank, 1969; Galtung, 1971; Wallerstein, 1974; Chase-Dunn, 1975; Chirot, 1977).

The fundamental mechanism is that the less developed nations are *dominated by foreign firms and investors that control their economies.* This in turn has several consequences. First, profits flow back to the developed nations rather than being reinvested in the local economies: Foreign firms and investors spend or invest their profits back home. This denies the underdeveloped economies the capacity to grow.

Second, foreign firms and investors control what economic activities take place in underdeveloped nations. Generating manufacturing capacity is not in their interest; rather, they profit more by selling manufactured goods to underdeveloped nations and extracting raw materials from them. Therefore, foreign domination means that a peripheral nation remains a supplier of raw materials unable to become industrially self-sufficient.

Third, nations that specialize in exporting raw materials must remain poor. World demand for raw materials is relatively inelastic because of the small or negligible population growth in the advanced nations (so their need for raw materials does not increase). Thus, a nation cannot increase its wealth by increasing the amount of raw materials it exports. Also, world prices for raw materials are subject to manipulation by speculators in the advanced nations and often fall so low as to create economic depressions in supplier nations.

According to the world system theory, nations that specialize in the export of raw materials also develop very distorted economies. Their development tends to be limited to small enclaves of workers employed in the export industries, while the rest of the nation remains undeveloped. The result is a dual economy that confines all modernization to the export sector, while providing neither incentives nor resources for modernizing the rest of the country (Frank, 1969; Paige, 1974; Chase-Dunn, 1975).

A CLOSER VIEW
Jacques Delacroix: Testing the Dependency Theory

Intellectuals and many political leaders in the less developed nations find the world system theory most appealing. It tells them that the lack of progress in their nations is not their fault but is imposed on them by the developed nations. However, many social scientists, especially those in the advanced nations, have rejected these views.

Harbor scenes like this in Cotonou, Benin, in West Africa, encourage belief in the dependency hypothesis: When a nation is a primary exporter of unprocessed food and raw materials, its modernization is retarded. That nations lacking industry do not export manufactured goods is hardly surprising.

Although the world system theory offers an elegant model of international economic relations, many critics have argued that the real world is much more complex. The world system theory neatly divides the world into poor, underdeveloped nations that export raw materials and rich, modernized nations that export processed manufactured goods. However, critics point out that exceptions to this scheme abound. The United States and Canada are rich, modernized nations, yet they export more raw, unprocessed foodstuffs than the rest of the world combined. Taiwan's economy is dominated by foreign investors, yet rather than specializing in the export of raw materials, nearly 90 percent of its exports are manufactured goods (Barrett and Whyte, 1982). Moreover, research designed to support the dependency claims of world system theory was either unsuccessful or poorly designed and executed. Thus, a young graduate student at Stanford University in the mid-1970s saw an opportunity to do some important sociology.

Jacques Delacroix (1977) wanted to test the dependency portion of world system theory: *that specialization in the export of raw materials prevents modernization.* His first task was to select valid measures of the major concept: modernization. He selected two measures that reflect slightly different aspects of modernization. The first of these is **gross domestic product (GDP),** which is the final market value of all goods and services produced during the year. GDP per capita is calculated by dividing GDP by the total population. This is not to suggest that everyone shares equally in a nation's GDP, but basing the rate on population eliminates variation across nations based only on population size. India's GDP is far larger than Switzerland's, but when the effects of population size are eliminated, Switzerland's per capita GDP reveals that it is a far richer nation than India.

For his second measure of modernization, Delacroix selected high school enrollment rates. A nation's level of school enrollment, especially above the grade school level, represents both an investment in modernization and modernization itself. Only educated populations can deal with technology and modern culture.

Next, Delacroix formulated two hypotheses derived from the dependency portion of world system theory: (1) If dependency theory is correct, then we ought to discover that per capita GDP increases more slowly in nations to the extent that they specialize in exporting raw materials, and (2) by the same token, we ought to find that dependent nations—those specializing in exporting raw materials—lack the resources to expand their educational systems, especially above the primary grades. In effect the rich

Courtesy of Lynne Roberts

Scenes like this in the port of Seattle, where huge grain ships line up to take on loads of wheat, call the dependency thesis into question. Similar scenes occur in all major American and Canadian ports because these two nations export more unprocessed foodstuffs than the rest of the world put together. Yet, contrary to dependency theory, the United States and Canada are not "underdeveloped" nations.

nations should be increasing per capita GDP and high school enrollments while the poor nations remain stagnant or advance much less rapidly.

Delacroix knew that the dependency hypothesis could not be tested properly at only one point in time. No one disputes that some nations are less developed than others. The issue is whether these nations are catching up in terms of modernization. Clearly, changes over time must be examined. So Delacroix obtained data for 1955 and 1970 for fifty-nine nations. He then examined changes in his dependent variables: changes in per capita GDP and changes in secondary school enrollment as these were related to his independent variable, the proportion of a nation's exports that are unprocessed raw materials.

What did he find? No relationships. Nations specializing in raw material exports showed as much increase in per capita GDP as did nations specializing in the export of manufactured goods. Secondary school enrollment grew as

much in the nations exporting raw materials as in those exporting manufactured goods. Delacroix found no support for the dependency hypothesis.

But he did find something else. Secondary school enrollments and increases in per capita GDP were strongly correlated. This led Delacroix to suggest that modernization is influenced primarily by internal processes rather than external processes of the world system. That is, nations that devoted substantial effort to educating their populations improved their standard of living, no matter what role they played in the world import-export system. Educational policy is decided within nations, not imposed on them by their trading partners. Indeed, Delacroix concluded that even the least developed nations retain considerable "freedom to maneuver" in determining domestic policies and thus retain substantial opportunity, as well as responsibility, to establish their own patterns of modernization.

Delacroix's research showed that a major source of modernization is secondary school enrollment. To the degree that a nation invests in its young people by keeping them in school, its modernization is speeded up. These Ivory Coast schoolchildren are their nation's primary asset.

© Marc & Evelyn Bernheim/Woodfin Camp

Since Delacroix's landmark study, a great deal of additional research has been done, and some modifications of his findings have been suggested. Using data on the expansion of education and industrial development in France, covering the period 1825–1975, Jerald Hage, Maurice A. Garnier, and Bruce Fuller (1988) found a clear linkage between the two but also proposed that the impact of education on industrialization was subject to three conditions, or thresholds. First, for the expansion of education (an increase in the proportion of the population being educated and the length of their schooling) to spur industrial development, there must be some standardization in the character and quality of education. Second, for an effect to occur, "a sufficiently large number of people must have received some education." That is, a rise from 2 to 4 percent in the proportion of all children entering first grade will not matter. Only when a substantial number of people are being educated can "liftoff" occur in the industrial sphere. Third, there must be substantial correspondence between what the schools teach and what skills industry requires. Turning out large numbers of students versed primarily in classical literature and music will not create a labor force well suited to fulfill technological occupations.

Of the many other studies of dependency theory conducted since Delacroix's work was published, some have claimed that even when dependency does not retard development, it sustains a form of development that fails to raise the standard of living for the masses—that in dependent nations the fruits of development benefit only the rich (Wimberly and Bello, 1992).

However, these claims were dealt a devastating blow by new research. Glenn Firebaugh and Frank D. Beck (1994) tidied up the many methodological and conceptual shortcomings of previous studies and substituted carefully specified, formal models of dependency theory in place of the ad hoc "collection of variables" assembled in previous studies. Then they tested each model on data for sixty-two nations over a thirty-three-year period, 1956–1988. The results were powerful and clear. As Firebaugh and Beck put it: "The effects of economic growth on national welfare are large and robust, whereas the effects of dependence are hard to find." That is, rather than being limited to the rich, economic development in the Third World improves the quality of life for everyone. Finally, Firebaugh and Beck reaffirmed Delacroix's finding that investments in secondary education greatly speed economic development.

DIMENSIONS OF GLOBAL INEQUALITY

Although much of this chapter has been devoted to explaining why there is global inequality, these discussions do not adequately reveal the extraordinary extent of these differences. So, let's pause here to examine some international statistics and contemplate the range of life situations they represent.

Table 17-3 compares nations in terms of their per capita gross national income (GNI), which can serve as a gauge of the standard of living enjoyed by the average person. Switzerland has a slightly higher GNI than does the United States, and Japan's is a bit lower. Then comes a very substantial drop to Sweden, Great Britain, and Hong Kong, each having a GNI per capita only about two-thirds that of the three leaders. Overall, however, there are few surprises about which nations are high or low. What you may find surprising is the range between the top and the bottom. Everyone knows that the average American is much wealthier than the average Ethiopian. But 354 times wealthier? What that means is that for every $354 worth of goods and services that the average American can afford, the average Ethiopian has $1 to spend.

Money isn't the only measure of standard of living. The best single measure of how well people live is how *long* they live. Hence, Table 17-4 ranks selected nations on the basis of their average life expectancy—the number of years the average newborn can expect to live. In the more industrialized nations, life expectancy exceeds seventy-five years. But as the level of economic development declines, so does life expectancy. In the poorest nations, it is less than fifty years—less than forty in some. In human terms this means that the average baby born in Japan can expect to live 2.5 times as long as the average baby born in Mozambique.

Poor nations have low life expectancy partly because they lack medical facilities, but even more important causes are widespread malnutrition and poor sanitation. For example, in terms of daily calorie consumption, the average American (3,699) consumes twice as many as does the average person in Ethiopia (1,856).

GDI tells us the size of each nation's economic pie, but it does not tell us how that pie is actually shared among citizens. And it is how the pie is shared that is another major factor in quality of life of the average person in a society. Put another way, it isn't just the general lack of wealth that causes so much misery in the least developed nations; it's that little as there is, it is very unequally distributed. Keep in mind that if the richest 10 percent of the population received only 10 percent of the national wealth, there would be no rich people—something that Chapter 9 suggests is unlikely or impossible. Thus, the issue is not inequality per se but the *degree* of inequality. And such inequality is

TABLE 17-3	Gross National Domestic Income per Capita for Selected Nations		
NATION	**GNI PER CAPITA IN U.S. DOLLARS**	**NATION**	**GNI PER CAPITA IN U.S. DOLLARS**
Switzerland	36,170	Saudi Arabia	8,530
United States	**35,400**	**Mexico**	**5,920**
Japan	34,010	Argentina	4,220
Sweden	25,970	Russia	2,130
Great Britain	25,510	Iran	1,720
Hong Kong (China)	24,690	China	960
Netherlands	23,390	India	470
Germany	22,740	Pakistan	420
Canada	**22,390**	Sudan	370
France	22,240	Cambodia	300
Singapore	20,690	Nigeria	300
Australia	19,530	Uganda	240
South Korea	9,930	Ethiopia	100

Source: *Statistical Abstract of the United States, 2005.*

TABLE 17-4	Life Expectancy in Selected Nations		
NATION	LIFE EXPECTANCY (IN YEARS)	NATION	LIFE EXPECTANCY (IN YEARS)
Japan	81.6	Poland	75.6
Australia	81.0	Russia	68.3
Canada	**80.7**	India	66.1
France	80.3	Pakistan	64.9
Netherlands	79.6	Ghana	55.6
Great Britain	79.2	Afghanistan	45.1
United States	**78.4**	Kenya	43.7
Taiwan	78.1	Ethiopia	40.1
Argentina	76.9	Zimbabwe	34.6
South Korea	76.8	Mozambique	32.4
Mexico	**76.3**		

Source: *Statistical Abstract of the United Sates, 2005.*

greatest in the poorest nations. For example, in Mozambique 56 percent of all national income goes to the richest 10 percent. However, there is far greater equality in the wealthier nations: In Canada the richest 10 percent receive 30 percent of the income, and in the United States they receive 28 percent. A detailed portrait of income inequality within nations was part of the online exercise for Chapter 10.

Keep in mind that the richest people in the poorest nations are not foreigners employed by multinational banks and industries, as world systems theory would imply. Rather, these inequalities are of domestic origins—the primary exploitation of poor people in Mozambique is by rich people in Mozambique. These local facts of life are what York W. Bradshaw and Michael Wallace had in mind when they wrote in their recent book, *Global Inequalities* (1996), that "Students and scholars typically become much less supportive of world-system theory after they conduct research and live in poor countries for a period of time."

The good news is that poor nations probably have the internal resources needed to develop far more productive economies and that as they do so many local problems will be modified. For example, as noted by Firebaugh and Beck, as the pie grows relatively larger, it is shared more generously. The bad news is that local problems often sustain themselves at the expense of development. As Bradshaw and Wallace remarked, some nations remain poor because they "have squandered their resources, suffered from abysmal leadership, and been drained by corrupt governments and businesses."

Finally, the best news is that the standard of living of the average person increased remarkably in all nations. And contrary to constant reports in the press and even from "experts," when properly defined and analyzed, income inequality among nations has been *declining* during the past generation (Firebaugh, 2003). This absolutely contradicts dependency theory.

GLOBALIZATION

The newest intellectual buzzword is *globalization*. Everyone seems certain that soon we all will be living in one big "global village"—a concept originated by Marshall McLuhan (1964), a famous and eccentric communication theorist. There are three main elements to globalization, as defined in Chapter 2. The first has to do with the development of rapid global communications, the capacity to move not only messages but people and things from anywhere to anywhere. The second aspect of globalization involves the creation of a global economy wherein economic activities anywhere in the world are felt everywhere. Finally, and often regarded as a result of the first two aspects, a relatively uniform world culture will emerge so that we all speak the same language, share the same values and norms, and sustain a common fund of knowledge, as if we were residents of the same community—hence the imagery of the global village.

It is obvious that the age of global communications has arrived. It is possible to place a long-distance telephone call to or from nearly anywhere on earth, swiftly and cheaply. The same is true for TV signals—CNN is seen everywhere that local governments allow privately owned satellite dishes. By the same token, tourists are everywhere, as are Coca-Cola and Big Macs. Economic globalization is less developed than global communications, but it is extensive, as we have seen in previous sections.

But will global communications and economic ties lead to the development of a global culture? Popular opinion to the contrary, many social scientists say no. Thus, they dismiss the fact that many Japanese like Big Macs and many Americans like sushi as superficial in light of the very deep cultural differences between the two nations. Indeed, critics of the global village assumption note that extensive communications and trade have not yet produced a common culture even in relatively small areas with extensive common histories. For example, ethnic and regional separatism has, if anything, intensified in many parts of the world in recent times as evidenced by the breakup of the Soviet Union, brutal warfare in the Balkans, linguistic separatism in Canada, and tribal massacres in Africa. Samuel P. Huntington (1996) has pointed out that modernization in terms of industrial development is compatible with many strong cultural patterns that are unlikely to blend. For example, contrary to the expectations of many early social scientists, each of the world's major religions has proven to be quite compatible with industrialization, but these religions remain relatively incompatible with one another. Thus, even if there is a lot of communication and commerce between, say, India and Pakistan, India will continue to have a quite distinctive Hindu culture while Pakistan will remain Muslim.

Finally, there are fundamental theoretical reasons to expect that local networks are inevitable, simply as a function of proximity, and that local networks always will generate and sustain a distinctive culture as well as a strong sense of solidarity. If the "self" is the essence of our individuality, and if social scientists are correct that the self is socially constructed on the basis of face-to-face relationships, then the self requires local networks. John H. Simpson (1996) concludes, therefore, that the self is incompatible with globalization since the idea of global primary and secondary groups is preposterous—there can be "no such thing as *the* global village." Indeed, Simpson points out that the term *global village* is a contradiction. The intimacy and group solidarity implied by the term *village* are inseparable from "small" and "local," which also are inherent in the term. To speak of a global village is to speak of local globalism, or of a worldwide local network. Simpson suggests that the anticipation of cultural globalism is based on the ethnocentric assumption that others will want to become like us, overlooking the fact that we don't want to become like them—and that everyone's "we" is someone else's "them." And so long as there continue to be primary groups, that will be true.

CONCLUSION

However capitalism arose and modernization occurred, the fundamental fact remains that some nations are much more modernized than others and that the less developed nations desire the quality of life that modernization provides. To achieve this quality of life, they may have to develop "modern" values of thrift and reinvestment, as Weber proposed; develop capitalist economies, as Marx believed; and tame their states, as state theorists argue. And they may have to pay close attention to their opportunities in the world trade economy, as world system theorists suggest.

But perhaps there is another factor to consider—one that may threaten not only the ability of less developed nations to modernize, but the capacity of even the most fortunate nations to maintain their quality of life: *rapid population growth*. At the end of the twentieth century, the news media gave intensive coverage to calculations that the sixth billion human being had arrived on earth, and all these reports were filled with dire warnings that the earth could not long sustain such growing numbers. Amidst all the sensationalism, it went almost unnoticed that leading "demographers" dismissed the alarms about population growth as hopelessly out-of-date. What are demographers? How can they be complacent about population growth? Read on.

Review Glossary

Terms are listed in the order in which they appear in the chapter.

modernization The process of industrialization, economic development, and technological innovation by which a culture achieves high standards of living and maximizes control over the physical environment.

idea of progress The philosophical doctrine that technological and social progress is inevitable.

cultural lags Periods of delay following a change in one part of a society when other parts of the society have not yet readjusted.

diffusion The process by which innovations spread from one society to another.

capitalism An economic system based on free-market exchanges and individual property rights with the result that any individual can benefit from becoming more productive.

command economies Economic systems wherein property rights are not secure and much productive activity is based on coerced labor.

Protestant Reformation The separation of Protestant Christians from the Roman Catholic Church during the sixteenth century; usually associated with the founding of the Lutheran Church in Germany.

spirit of capitalism According to Weber, a nonreligious version of the Protestant ethic; values favoring hard work, thrift, and the importance of economic success.

Protestant ethic According to Weber, doctrines holding that economic success reflects God's grace.

state theory of modernization Theory that wherever the power of the state to seize private property is curtailed, free markets will appear, capitalism will develop, and modernization will occur as a result of mass efforts to become more productive.

world system theory (or **dependency theory**) Theory stating that some nations become modernized by exploiting other nations and that their continuing exploitation prevents less developed nations from becoming fully modernized.

core nations According to Wallerstein, those most modernized nations, having diversified economies and stable internal politics, that dominate the world system.

peripheral nations According to Wallerstein, those nations in the world system that are forced to specialize in the export of unprocessed raw materials and food to the core nations and that must import manufactured goods. This makes them dependent on the core nations, which in turn force them to adopt economic and social policies that prevent them from modernizing.

semiperipheral nations Wallerstein's term for nations that fall in between core and peripheral nations, being more industrialized than the latter and less industrialized than the former.

gross domestic product (GDP) The final value of all economic activities within a society. When GDP is divided by the total population of a society, the result is per capita GDP.

Suggested Readings

Chirot, Daniel. 1985. "The Rise of the West." *American Sociological Review* 50:181–195.

Firebaugh, Glenn. 2003. *The New Geography of Global Income Inequality.* Cambridge, Mass.: Harvard University Press.

Hage, Jerald, Maurice A. Garnier, and Bruce Fuller. 1988. "The Active State, Investment in Human Capital, and Economic Growth: France 1825–1975." *American Sociological Review* 53:824–837.

Landes, David S. 1998. *The Wealth and Poverty of Nations.* New York. Norton.

McNeill, William H. 1976. *Plagues and Peoples.* New York: Basic Books.

North, Douglass C., and Robert Paul Thomas. 1973. *The Rise of the Western World: A New Economic History.* Cambridge: Cambridge University Press.

Stark, Rodney. 2005. *The Victory of Reason: How Christianity Led to Freedom, Capitalism, and Western Success.* New York: Random House.

Wallerstein, Immanuel. 1974. *The Modern World System.* New York: Academic Press.

Weber, Max. 1958. *The Protestant Ethic and the Spirit of Capitalism.* New York: Scribner's.

White, Lynn, Jr. 1962. *Medieval Technology and Social Change.* London: Oxford University Press.

Sociology Online

www.socstark10.com

GO TO THE INTERNET AND TYPE: www.socstark10.com.

↗**CLICK ON:** 2000 General Social Survey

✓**SELECT:** MA WRK GRW? Did your mother work for as long as a year, while you were growing up?

↗**CLICK ON:** Analyze Now

TRUE OR FALSE: It is becoming significantly more common for people to have had a working mother.

In what ways do you think this will cause social change?

↗**CLICK ON:** Select New Data Set

↗**CLICK ON:** The 50 States

✓**SELECT:** POP GROWTH: Population growth (decline) 1980–1996.

↗**CLICK ON:** Analyze Now

Which state has the highest growth rate? How might this help explain other things you have discovered about this state? (Hint: See the discussion of anomie in Chapter 7.)

↗**CLICK ON:** Select New Data Set

↗**CLICK ON:** Nations of the Globe

✓**SELECT:** IND GROWTH: Industrial production growth rate.

↗**CLICK ON:** Analyze Now

TRUE OR FALSE: It would seem unlikely that the less developed nations ever could catch up in terms of industrial production.

USING INFOTRAC COLLEGE EDITION

GO TO THE INTERNET AND TYPE:
www.InfoTrac-college.com

↗**CLICK ON:** Register New Account
You will be asked to enter your Access Code and to create and enter a User Name and Password. Having done so,

↗**CLICK ON:** InfoTrac College Edition or Log On.

The recent "Battle of Seattle" involving demonstrations against the World Trade Organization raised many claims concerning theories of modernization. Many recent articles in the world system do the same.

SEARCH TERMS:
World System
World Trade Organization

In addition, I have selected a specific article that will usefully supplement the chapter: Computers and the Pursuit of Happiness. You can find it by searching for this title. You may read the article on the screen or print it using the usual print commands. If you also go to www.socstark10.com and click on the InfoTrac College Edition icon, you can read an explanation of why I selected this article and find several questions that will help you connect the article to material in this chapter.

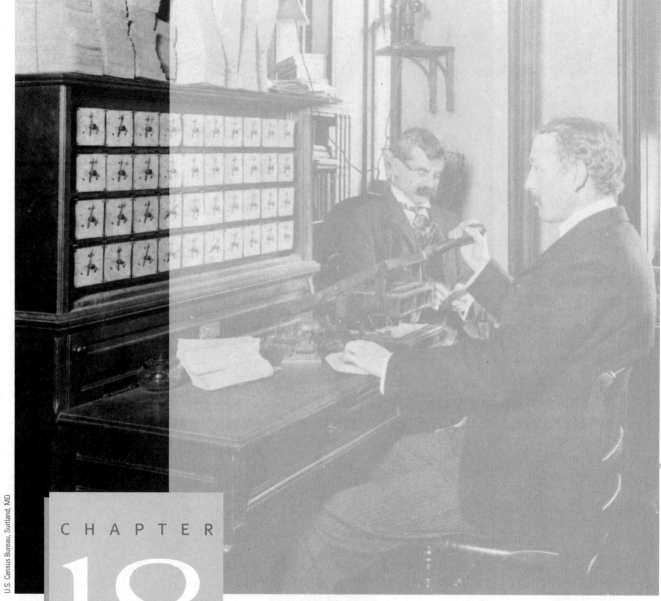

CHAPTER

18

Population Changes

IN 1066 WILLIAM, DUKE OF NORMANDY, landed in England with a small army and, having defeated King Harold in the Battle of Hastings, seized the English throne. After a few years, he began to suspect that his new subjects were not paying their proper amount of taxes. So, William the Conqueror, as he now was called, decided to find out what his conquest really consisted of. As described by a resentful Anglo-Saxon chronicler:

He sent his men all over England into every shire and had them find out how many hundred hides there were in the shire, or what land and cattle the king himself had in the country, or what dues he ought to have in twelve months from the shire. Also he had a record made of how much land his archbishops had, and his bishops and abbots and his earls, and . . . what or how much everybody had who was occupying land in England, in land or cattle, and how much money it was worth. So very narrowly did he

have it investigated, that there was no single hide nor a yard of land, nor indeed (it is a shame to relate but it seemed no shame to him to do) one ox nor one cow nor one pig was there left out, and not put down in his record: and all those records were brought to him afterwards. (quoted in Hallam, 1986)

This massive document came to be known as the *Domesday Book* (pronounced "DOOMS-day") because it made William's English subjects think of the accounting at the Last Judgment, described in the Bible. Today, historians consider the ***Domesday Book*** unsurpassed in medieval history for its thoroughness and for the speed at which it was assembled.

William was not the first ruler to count his population and his property. The Pharaohs of Egypt began doing so as early as 2500 B.C. The Old Testament records that once the Israelites were safely out of Egypt, "The Lord spoke to

CHAPTER OUTLINE

DEMOGRAPHIC TECHNIQUES
Rates
Cohorts
Age and Sex Structures

PREINDUSTRIAL POPULATION TRENDS
Famine
Disease
War

MALTHUSIAN THEORY

MODERNIZATION AND POPULATION

THE DEMOGRAPHIC TRANSITION

A CLOSER VIEW
Kingsley Davis: Demographic
Transition Theory

THE SECOND POPULATION EXPLOSION
Box 18-1 The Life Cycle of the Baby Boom
The Sudden Decline in Mortality
High Fertility and Cultural Lag

THE POPULATION EXPLOSION WANES
Economic Development
Numeracy about Children
Contraception and Wanted Fertility

THE CRISIS OF DEPOPULATION
Gray Nations
Culture and Immigration
Low Fertility and Gender Bias

CONCLUSION

REVIEW GLOSSARY

SUGGESTED READINGS

SOCIOLOGY ONLINE

Moses, saying 'Take a census of all the congregation of the people of Israel, by families.'" The emperors of Rome also counted and assessed their subjects frequently—Augustus ordered a count of the entire empire in 28 B.C., in 8 B.C., and again in A.D. 14. Why all this interest in counting people? To estimate tax revenues and military power.

To know how much revenue a tax can produce, a government needs to know how many people will be paying it. Moreover, only by knowing how many people live in a particular district can the central government be sure that local officials are not embezzling. For example, a local tax official might report to the imperial government that there are 5,000 tax-paying families in his district when there are in fact 8,000, thus enabling him to pocket the taxes paid by 3,000 families.

By the same token, only by knowing the number of able-bodied males of military age can a ruler estimate how large an army he can raise. In the days of tiny hunting and gathering societies, it was easy to count noses and know that fourteen warriors were available for battle. But when societies grew to include tens of thousands of people, counting noses became a major task.

Thus, from ancient times governments instituted a **census:** a population count, often broken down into useful categories such as sex, age, occupation, marital status, and the like. The *Domesday Book* was a census. But even though William's agents conducted what was probably the most efficient census taken in medieval times, it took them two years to assemble the *Domesday Book*, and gathering information was costly. Moreover, the *Domesday Book*, like any census, soon became outdated.

As we shall see, populations often change rapidly. Therefore, it may be necessary to redo a census frequently to have accurate information. However, because censuses are very expensive, governments are reluctant to conduct them. Generally, that has meant muddling along with badly outdated information. But over time, ways have been found to gauge population changes between censuses. And because populations so often fluctuate in size, governments encouraged inquiry into why this occurs. Thus was born the science of *demography.*

The word *demography* is formed from the Greek word *demos*, meaning "people," and *graphy*, meaning "description." Doing **demography** means describing the people. Whenever you read in the newspaper about such things as the marriage rate, the divorce rate, or population decline, you are reading about the work done by demographers.

In this chapter we shall examine fundamentals of modern demography, a major area of specialization within sociology. We shall also see the important role of population shifts in prompting or impeding social change. The chapter's primary objective is to assess theories of population change.

Historically, human population trends have undergone *six* dramatic shifts. These shifts are the main subjects of the major theories of demography. Thus, we shall examine historical population patterns while assessing the major theoretical achievements of demographers. Before we trace patterns of population change, however, we should understand a few basic technical tools used to monitor and describe human populations.

DEMOGRAPHIC TECHNIQUES

As a population changes, a census becomes out of date. For a long time, governments tended to lack accurate information on the size and composition of their populations. Then one day some unremembered bureaucrat hit upon an ingenious way to keep track of what was going on. It became the law to record all births and all deaths with the government. This procedure became common among societies, and governments created bureaus of vital statistics to keep an accurate count of these registrations. These records made it possible to update the census each year. (In many countries birth and death registration was facilitated by the clergy, who had kept local records for religious purposes.)

Let us consider a hypothetical nation that has just conducted a census showing that it has 10 million citizens. The next year 500,000 babies are born and 400,000 people die. By subtracting deaths from births, we determine that the population has undergone a net increase of 100,000. Adding this to 10 million, we know that this nation now has 10,100,000 citizens. Furthermore, by dividing the year's growth (100,000) by the total population of a year before (10 million), we can see that the population grew by 1 percent in a year. By making these computations each year, we can find out the size of the population and its rate and direction of change without taking a new census.

Suppose that each year 100,000 more people are added to this nation's population. Is this rate of growth the same year after year? No. As the population grows, each year 100,000 people constitute a smaller percentage of the existing population and thus represent a smaller percentage increase. We therefore know that population growth is slowing down. Because it is often vital for governments to know how fast their populations are growing, they pay great attention to the percentage of annual growth, or the **growth rate,** of the population.

Even what appear to be small growth rates can cause populations to grow at a breathless pace. For example, a population that is growing by 3 percent per year will double in size in twenty-three years and increase tenfold in only seventy-seven years. The growth rate is computed this way:

$$\frac{\text{Net population gain (or loss)}}{\text{Size of population}} \times 100 = \text{Growth rate}$$

For any given year, the net population gain (or loss) takes three variables into account: (1) the increase (or decrease) in births, (2) the increase (or decrease) in deaths, and (3) the increase (or decrease) due to migration.

A record of all births and deaths permits some crude insight into why a population changes in size. A population can grow because births are increasing, because deaths are decreasing, because people are migrating into a region, or for all three reasons. Similarly, a population can shrink because of a decline in births, an increase in deaths, a loss of people who move away, or for all these reasons. Assuming that migration is constant, suppose a government wants a larger population. It would need to know whether encouraging more births or combating disease is the more appropriate course.

Obviously, in answering this question, simply knowing the numbers of births and deaths over several years would not help much. To compare, say, births for two years, we need to take into account the fact that the total population for these years is different, making a direct comparison between the number of births meaningless. Again the solution is to compute a percentage— the ratio of births to the population total for each year. Such percentages are called *rates*.

RATES

We can compute a **crude death rate** by dividing the total number of deaths for a year by the total population for that year. A **crude birthrate** can be computed by dividing the total number of births for a year by the total population for that year. By using rates we can make meaningful comparisons between years in which the total population differed in size. For example, in 1910 there were 2,777,000 live births in the United States; in 1996 there were 3,915,000 live births. Although there were a million more births in 1996, fertility was much higher in 1910: There were fewer than half as many people in 1910 as there were in 1996 to produce those births. Thus, newborns added only 1.5 percent to the total population in 1996 in contrast to 3 percent in 1910.

Crude birth and death rates give only limited information. They do not take into account some important factors. For example, most members of a society cannot bear children—males cannot, nor can prepubescent or postmenopausal females. What will happen if the proportion of fertile females in a population changes? A change in the crude birthrate will occur even if fertile women are reproducing at exactly the same rate as before. Therefore, demographers try to avoid using crude birthrates. Instead, they prefer a **fertility rate,** which is the total number of births, divided by the total number of women within a certain age span. Most nations, including the United States and Canada, base their fertility rates on women age fifteen to forty-four.

The importance of using a fertility rate rather than a crude birthrate is demonstrated by the immense contrasts between the two rates in Canada. Table 18-1 compares the correlations between the crude birth and fertility rates and a number of important independent variables.

Use of the crude birthrate would lead to the conclusion that people have more children where incomes are higher, where religious affiliation is lower, and where men outnumber women. One also would conclude that neither telephones per 1,000 people nor the abortion rate influences births. But this would be absolutely incorrect. When we actually measure *fertility*—how many infants the average woman is having in various provinces—we discover that affluence, as measured by income and by phones, has a huge negative impact, as does the abortion rate, whereas religion has an immense positive effect. That the proportion of males doesn't really correlate with fertility underscores that only women bear children.

Just as not all members of a society can reproduce, not all members of a society are equally likely to contribute to the death rate in a given year. It is crucial to know who is and who is not contributing to the death rate. The *crude death rate* does not reflect these subtleties; it is simply the number of deaths per 1,000 in the general population. Demographers are mainly interested in **age-specific death rates.** These are computed by separating the population by age categories and computing the number of deaths per 1,000 of each age group.

Obviously, everyone dies. However, *when* people die greatly affects future population growth—and virtually every aspect of society. A comparison of crude death rates tells us that in 1900 nearly twice as many Americans per 1,000 (17.2) died as in 1996 (8.8). But an examination of age-specific death rates reveals that a massive shift occurred in the age at which people died. In 1900, 162.4 infants (age one year and younger) per 1,000 died. In 1996 only about a twentieth as many infants (7.6 per 1,000) died. In 1900 the birthrate was much higher than in 1996; however, as we have seen, far fewer of those born grew up to reproduce. The birth of 1,000 infants had less of a long-range impact on the population in 1900 than it does today.

Demographers concern themselves with a great many other rates. You have already seen a number of these in previous chapters. For example, we have discussed rates of crime, marriage, divorce, disease, and illegitimacy. All of these are constructed in the same way as fertility and death rates. Crude rates are based on units of 1,000 (or sometimes 10,000) persons in the total population. Other rates are specific to certain relevant groups within the population.

TABLE 18-1	Contrasting Canada's Crude Birthrate and Fertility Rate (correlations based on provincial rates)	
	CRUDE BIRTHRATE	FERTILITY RATE
Median income	.61	−.52
Percent of population with religious affiliation	−.28	.71
Telephones per 1,000	−.09	−.90
Percent of population that is male	.91	.10
Abortions per 1,000 live births	.12	−.80

Source: Prepared by the author from *Canada: An Electronic Data Base in MicroCase Format, 2002,* distributed by Wadsworth.

COHORTS

One of the main uses of demographic data is to provide a basis for long-term planning. A government may project future conditions by saying, "If the present birthrate holds steady, then we will have a population twice as large thirty years from now. We must make provisions to house, employ, and feed these additional people." A critical unit in such planning is the birth cohort, or age cohort. A **birth cohort** consists of all the persons born in a given time period, usually one year. The interesting feature of cohorts is that although they may get smaller as time passes, they never get any larger. At the end of 2000, there were no more people born in that year. Therefore, if we know the number of persons included in the 2000 birth cohort, and if we have accurate age-specific death rates for our population, then we can predict the size of that cohort as it passes through all the stages of life, from infancy to old age. For example, by subtracting the figure based on mortality expected by age six, we know the total number of children who will be entering first grade in 2006. Our society thus has some time to adjust the number of classrooms and teachers accordingly. Likewise, by subtracting the figure based on the probable mortality between birth and age eighteen, we know how many eighteen-year-olds there will be in 2018. If we can predict what percentage of these people will choose to attend college, then we can predict the size of the freshman class of 2018 at the end of 2000.

When fertility and mortality abruptly shift, governments often become obsessed with the future implications of a birth cohort. For example, the fact that the size of the cohort cannot be increased caused grave concern among French military planners. In 1870 France was totally defeated by Germany. For the next forty years, the French were determined to gain revenge. But while France and Germany had approximately the same-sized populations in 1870, Germany had 65 million people to France's 40 million by 1914. In part this came about because Germany annexed new territory during this period. But in part it was because France had a lower—and declining—birthrate.

Thus, where it mattered, in the age cohorts of males suitable for military service, the Germans had nearly twice the number as France had in 1914 (7.7 million to 4.5 million). According to William Shirer (1969), "Gloomy prophets in Paris did not see how France could escape

© Bruce Davidson / Magnum Photos

In the eyes of his proud father, this baby boy is the newest member of the family. In the eyes of the neighbors, he is the newest resident of this mining village. In the eyes of the British census department, he is another member of the birth cohort made up of all infants born in Wales that year.

another military debacle if the Germans chose to attack again." In 1914 the Germans did attack and just failed to win the war in the first six weeks. Finally, in 1918 the Germans were defeated, but only because millions of Russian, British, Canadian, Italian, and American troops came to the support of France.

From a demographer's point of view, France's revenge on Germany for having been defeated in 1870 was gained at dire future cost. Of French men age eighteen to twenty-eight, three of every ten died in the war and did not make their expected contribution to future fertility. More than a million more came home from war badly maimed and disabled and also did not reproduce to the extent they might have. In addition there were at least 2.4 million fewer births between 1915 and 1919 in France than there would have been had the young men not been off in the trenches. Thus, in 1933, when the birth cohort of 1915 was old enough for military service, nearly 500,000 potential new soldiers came of age in Germany, while fewer than 190,000 did so in France. Worse yet, the depressed level of the number born during World War I led to a depressed number of women of childbearing age in the 1930s. It was clear that Frenchmen of military age would be in even shorter relative supply by the late 1940s. Furthermore,

French troops rushing to take part in the great battle at the Marne River, where the German invasion was halted in 1914. Although the French eventually were on the winning side, their losses in World War I resulted in tiny birth cohorts for the next generation.

Imperial War Museum

the French fertility rate continued to drop throughout the 1920s and 1930s, while the German rate stayed high. In terms of potential manpower, France faced an ever-greater military disadvantage with Germany.

In 1940 the Germans attacked again, and this time they did defeat France in the first six weeks. The small French birth cohorts of 1915–1922 were overwhelmed as French military planners had long feared and as German military planners had anticipated for years.

AGE AND SEX STRUCTURES

Small birth cohorts occurring in France had two major effects on population size. First, *they were too small to replace their parents' cohorts*, and the size of the population declined as older cohorts died. Second, when these smaller cohorts reached reproductive age, *they produced fewer children than a larger cohort would have.*

This limit on population size directs our attention to another important demographic factor: the distribution of people of various ages and sexes within a population. Obviously, a population in which the majority are elderly or are males has a far lower potential for growth than does a population in which the majority are

females of reproductive age. Therefore, to predict future trends in a population, we need to know not only fertility and mortality rates but also the **age structure** and **sex structure** (or distribution) of that population.

Normally, populations fall into one of three age and sex structures, depicted in Figure 18-1. The first is an **expansive population structure** and is characteristic of present populations in the underdeveloped nations. The expansive structure is shaped like a pyramid—each younger cohort is progressively larger. Such a population grows very rapidly. At any given moment, there are many more people who have not yet begun to reproduce than there are people who have. This situation has serious implications for growth, even if fertility suddenly falls. For even if couples suddenly limit their families to only enough children to replace themselves, the population will continue to grow until each of the increasingly larger, younger cohorts has gone through the reproductive period.

Aside from immigration, that is what has been causing the population to grow in the United States during the past twenty years. Fertility has fallen to the level of replacement, but the population continued to grow while the massive birth cohorts of the post-World War II

FIGURE 18-1 | **Population Structures.**

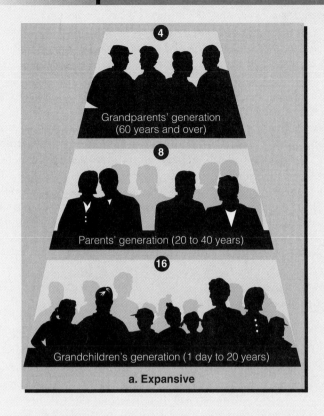

a. Expansive

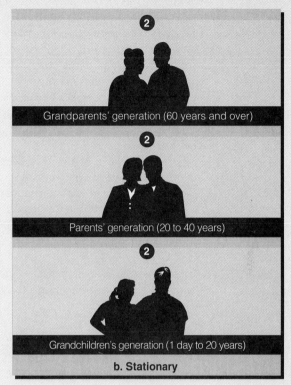

b. Stationary

(a) To depict an expansive population structure, we will assume that the average female gives birth to four children, which means that the grandparents' generation is only half as large as that of their sons and daughters. That is, two couples produce eight offspring. Now, let us assume that the four children in one family marry the four children in the other and that each couple also produces four children. Once again the population is doubled, and the grandparents end up with sixteen grandchildren. (b) In the stationary population structure, couples have only two children, and thus, each generation simply replaces itself, and the grandparents end up with only two grandchildren. (c) In a constrictive population structure, couples have fewer than two children—in this case only one. That means it takes two couples in the grandparents' generation to produce two offspring who marry and produce one grandchild.

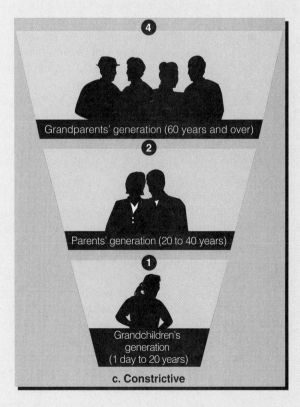

c. Constrictive

"baby boom" passed through their reproductive years. Growth is slowing down as the relatively smaller birth cohorts of the 1970s take over reproduction. As that happens, the age and sex structures of the United States may resemble the **stationary population structure** shown in panel b in Figure 18-1. The base of this structure is in proportion to the other cohorts. Thus, there are only enough infants and children to make up for early mortality, and each cohort entering the reproductive period is the same size as previous cohorts. Therefore, the population does not grow.

Finally, panel c in Figure 18-1 depicts a **constrictive population structure**. The bottom of this structure is smaller than the middle, indicating that in the future fewer people will enter reproductive ages. Such a structure reflects a declining population, now typical of many European nations.

Armed with these elementary demographic concepts, we can now examine the history of human population trends.

PREINDUSTRIAL POPULATION TRENDS

Primitive societies often had difficulty maintaining their populations. Women had to bear many children to ensure that several would survive to have children of their own. Historians believe that in primitive groups about 50 percent of all children died before the age of five (Petersen, 1975). For over several million years, the human population grew so slowly that only a few more people were added every 100,000 years. Kingsley Davis (1976) estimates that only 10,000 years ago, after at least 3 million years of human reproduction, there were only about 5 million human beings on the face of the earth, a figure about equal to the population of Chicago. For most of human existence, we were an endangered species.

But then came the *first great shift* in human population trends: Beginning about 10,000 years ago, we began to increase our numbers rapidly. This was caused by the development of agriculture. As humans ceased being nomadic hunters and gatherers and settled in one place to grow crops, life became more secure. There was a lot more food, and it was regularly available. With a better diet, we became healthier, and our reproductive rates finally began to outstrip death rates. More babies lived to maturity,

and the population began to grow more rapidly. Davis (1976) estimates that 8,000 years later, by A.D. 1, the worldwide human population was about 300 million. That means 295 million more people (or sixty times as many) had been added in 8,000 years than had been amassed over the previous several million years. (In the past 2,000 years, the population has increased by more than 1,300-fold to a world total of 6 billion.)

Nevertheless, although the human population began to grow rapidly, growth was not continuous. The population trends of agrarian societies fluctuated, as rapid growth was followed by rapid decline. By 1700 the world population was only about twice as large as it had been in A.D. 1. What caused these periodic declines in the population? A sudden increase in the death rate. Every so often, human populations were decimated by famine, disease, or war.

FAMINE

Though the invention of agriculture permitted a rapid increase in the human population, agrarian societies were extremely vulnerable to crop failures caused by drought, storms, or blight (plant disease). Famine could cause an immense number of deaths directly through starvation and indirectly through undernourishment, which made people more vulnerable to disease. Europe suffered from severe periodic famines until the twentieth century. Often, these famines were confined to one country or to one region of a country, but they sometimes affected most of the continent, as they did from 1315 to 1317, in the 1690s, and again from 1708 to 1709 (Petersen, 1975). Mortality rates soared and the population was seriously reduced. The last great famine in Europe took place when blight destroyed the potato crops in Ireland in 1845. As Chapter 2 pointed out, millions of Irish immigrants fled to the United States to escape starvation, but perhaps a million others, unable to flee, died as a result of the blight.

Nevertheless, Europe was much less vulnerable than Asia was to this cause of widespread death. In part this may have been because European agriculture was less productive. In good times European agriculture could not support great population growth; therefore, in bad times there were fewer people in danger of starvation. Asian farming, on the other hand, was very efficient but was vulnerable to the highly variable and unpredictable monsoon rains. Asia suffered more than Europe from severe droughts but had more plentiful crops between droughts. Famine was chronic in Asia (Petersen, 1975), and Walter Mallory (1926) found that a serious famine has been recorded in some part of China almost yearly for the past 2,000 years.

William Petersen (1975) reported on one of China's more recent and severe famines:

One of the worst famines of modern China struck four northern provinces in 1877–78. Communications were so poor that almost a year passed before news of it reached the capital. Cannibalism was common, and local magistrates were ordered "to connive at the evasion of laws prohibiting the sale of children, so as to enable parents to buy a few days' food." The dead were buried in what are still today called "ten-thousand-men holes." From 9 to 13 million, according to the estimate of the Foreign Relief Committee, perished.

Similarly, Davis (1951) calculated an immense loss of life due to famine in India during the 1890s by contrasting the rates of population increase for several decades:

In the previous decade [India's population] grew 9.4 percent, and in the following decade 6.1 percent. If the 1891–1901 decade had experienced the average rate of growth shown by these two decades, it would have grown by 7.8 percent instead of 1 percent. The difference is a matter of some 19 million persons, which may be taken as a rough estimate of loss due to famines.

Davis also pointed out that a great deal of food was shipped to India to help overcome the famine. Had this famine occurred in premodern times, millions more would have died.

Nevertheless, a much more recent famine in China overshadowed both of these nineteenth-century disasters. This famine occurred from 1958 through 1961 during a period of immense internal turmoil known as the Great Leap Forward. Angry peasants and disastrous government agricultural plans led to a famine so widespread that demographers now estimate that some 30 million premature deaths occurred, "approximately 17.3 million over age 10 and 12.2 million under age 10" (Ashton et al., 1984). Far fewer would have died had the Chinese government not chosen to keep the famine a secret from the outside world; the government made no attempt to buy grain abroad until 1961, after the millions had perished. A similar situation has

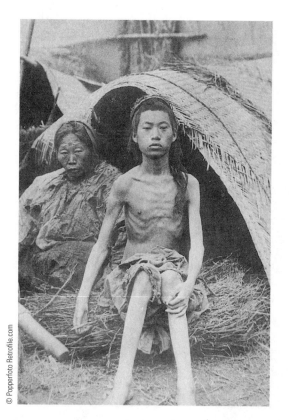

A famine victim in China's Hunan Province at the turn of the twentieth century. Famine was a chronic cause of high mortality rates in China for more than 2,000 years and struck again in the 1960s when the Cultural Revolution disrupted farming and food distribution.

been occurring in North Korea during the past several years. People are starving because the government is so concerned to keep the people ignorant of the outside world that it has been unwilling to allow relief supplies from abroad to be distributed.

DISEASE

A second major cause of a sudden rise in mortality is the outbreak of deadly, contagious diseases. One of the worst was bubonic plague, known as the Black Death, which often thinned the populations of agrarian societies in Europe and Asia. The worst outbreak began in Constantinople (today known as Istanbul, Turkey) in A.D. 1334. In fewer than twenty years, the Black Death mowed down millions in Europe and Asia; some estimates of plague deaths run as high as 40 percent of the total population of Europe and Asia (Figure 18-2). After the plague thousands of villages in Europe and Asia stood completely uninhabited and were never resettled (McNeill, 1976).

Smallpox was another epidemic killer. In 1707, 31 percent of the population of Iceland died of smallpox. Shortly after the first Spanish expedition reached the New World, smallpox and measles epidemics wiped out as many as

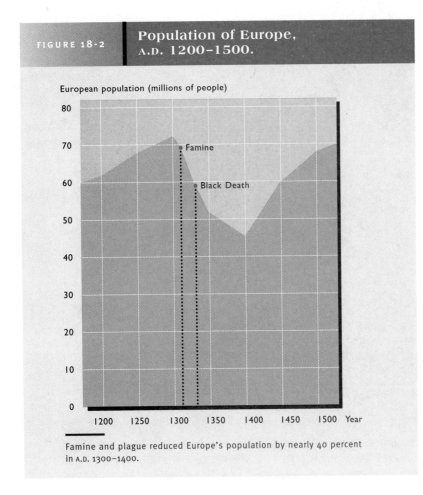

FIGURE 18-2

Population of Europe, A.D. 1200–1500.

European population (millions of people)

Famine

Black Death

Famine and plague reduced Europe's population by nearly 40 percent in A.D. 1300–1400.

Bibliotheque Nationale, Paris

This drawing is from a manuscript published by Henri de Mondeville in 1314, the first book on human anatomy that was based on actual dissections. Lacking such research, physicians could make little progress in understanding disease because they had virtually no knowledge of how the body functioned.

three-fourths of the inhabitants of Mexico and the West Indies (McNeill, 1976; Diamond, 1998).

WAR

Throughout recorded history innumerable societies have been ravaged and even destroyed by war. A case in point is the Thirty Years' War, which embroiled the nations of northern Europe from 1618 to 1648. The war was fought partly as a result of the Protestant Reformation, and religious antagonisms made it especially savage. By the end of the struggle, only 6,000 of 35,000 peasant villages survived in Germany, and an estimated 8 million Germans had perished (Montgomery, 1968). An even more devastating war was the Taiping Rebellion in China (1851–1864), in which prisoners were slaughtered and farms were burned (Ho, 1959). In one area of 6,000 square miles, no trace of human life remained. Earlier estimates that 20 to 30 million people were killed are now considered too low! A century later, the populations in the four provinces that fought the war were estimated to be still 14 percent below the number living there when the fighting began (Petersen, 1975).

MALTHUSIAN THEORY

Modern social science was born in the eighteenth century with the publication in 1776 of Adam Smith's economic treatise, *An Inquiry into the Nature and Causes of the Wealth of Nations*. This book not only was the start of economics but also led to the first demographic theory. In attempting to account for economic changes, Smith found it necessary to consider population patterns. This led him to a famous proposition: "Men, like all other animals, naturally multiply in proportion to the means of their subsistence." In other words human populations grow or decline according to *the availability of the necessities of life, especially food.* Eighty years later, Charles Darwin adopted this proposition of Smith's to help formulate the theory of evolution. Long before then, however, Smith's proposition prompted Thomas Robert Malthus to construct the first theory of population change.

Malthus was born in 1766 and, like Smith, was a Scot educated in England. At twenty-two he became a clergyman to realize "the utmost of my wishes, a retired living in the country." However, he soon became fascinated with Smith's economic theories, especially in using them to explain the growth and decline of human populations. In 1798 he published a short book, *Essay on the Principle of Population*. For the rest of his life, he continued to revise and expand this book through seven editions. The last edition was published after Malthus died in 1834.

Although the book eventually became very long, its central arguments are brief and easy to follow. Surveying the population patterns in Europe over many centuries, Malthus detected a repetitive cycle of rapid growth followed by rapid decline. He set out to explain this cycle.

His first clue came when he noticed that human (and animal) populations have the capacity for rapid exponential growth. That is, they need not show an **arithmetic increase** (1-2-3-4-5) but easily can show an **exponential increase** (1-2-4-8-16). For a human population to double every generation, all that is necessary is for every couple to raise four children. This is not difficult for people to achieve. Indeed, in Malthus's day the average North American woman gave birth to more than seven children during her lifetime.

If a population doubles each generation, it grows at an astonishing rate. For example, suppose we start with a single human couple. If they doubled their number (produced four children), and if each successive generation did so, then in only thirty-two generations a population of 8.4 billion would be achieved (well above the present world total). If we assume that a generation takes thirty years to grow up and reproduce, then in only about a thousand years, the population could have gone from an Adam and Eve to far more than today's world population.

Considering these figures, Malthus realized that something prevented the population from doubling every generation. For it was obvious that "in no state that we have yet known, has the power of population been left to exert itself with perfect freedom." Nowhere did humans exist in the numbers that would have resulted from unchecked doubling.

But what checks population growth? Smith had suggested available subsistence, and Malthus concluded that Smith was right. He reasoned that for the food supply to double in one generation would be "a greater increase than could with reason be expected." For the food supply to double again over the following thirty years "would be contrary to all our knowledge of the properties of land." Thus, unlike population, the means of subsistence cannot be increased exponentially. Here Malthus found an

© Bettmann/Corbis

Thomas Malthus.

inevitable tension. He made calculations to show what would happen to the gap between population and subsistence if both grew at the optimum rates:

The human species would increase as the numbers 1, 2, 4, 8, 16, 32, 64, 128, 256; and the subsistence as 1, 2, 3, 4, 5, 6, 7, 8, 9. In two centuries, the population would be to the means of subsistence as 256 to 9; in three centuries as 4,096 to 13, and in two thousand years the difference would be almost incalculable.

A shortage of subsistence, or famine, is always the ultimate check on population. But famine was not the only thing that kept the population down. As populations became denser, they were more vulnerable to epidemics: first, because greater contact among people made it easier for disease to spread and, second, because people weakened by hunger are less resistant to disease. Furthermore, increased populations taxed resources, which often led to clashes over land and resources. Malthus identified the limits to growth we discussed—famine, disease, and war—as **positive checks** that kept populations proportionate to the food supply.

From these insights Malthus created a theory that accounted for the population cycles

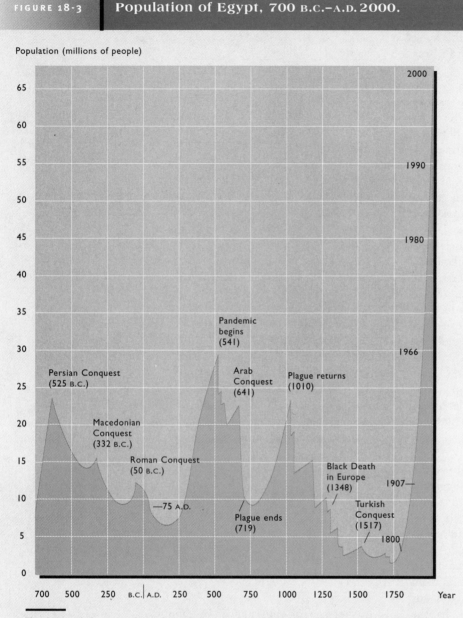

FIGURE 18-3   Population of Egypt, 700 B.C.–A.D. 2000.

The characteristic saw-tooth population pattern on which Malthus based his theory of population is revealed by centuries of Egyptian demographic history. A rapidly growing population repeatedly and rapidly shrank as external conquests caused agricultural shortages or as plagues broke out. By 1800, as Malthus worked on later editions of his theory, the Egyptian population began a long, unbroken rise.

the positive checks "until the population is sunk below the level of food; and then the return to comparative plenty will again produce an increase" in the population, and the pattern will be repeated (Figure 18-3).

Malthusian theory regards mortality as the fundamental variable determining population size. When mortality declines, populations grow. When mortality increases, growth slows or is reversed. You may wonder why Malthus was not concerned with fluctuations in fertility, since populations obviously grow or decline if the birthrate rises or drops. More to the point is the question of why rational and self-conscious people don't see the cause of their periodic misery—runaway population growth—and thus limit their fertility. Malthus believed that fertility could not be controlled. He granted that sometimes "vice" in the form of abortion or birth control can temporarily reduce fertility. He also granted the possibility of "moral restraint," or chastity, as a check to fertility. But he doubted that humans could control their sexual natures sufficiently to prevent rapid population growth whenever the food supply permitted it. He therefore did not consider changes in fertility significant in determining changes in population patterns. From his reading of the past, he believed that human fertility *was always high* and that only mortality rose and fell.

Although the **Malthusian theory of population** predicts that

occurring in agrarian nations. Essentially, his theory states that populations will always rise to the level of subsistence. Thus, when a new farming method is introduced, when there is a period of good crops, or when new land is cleared and put into production, population will grow. However, *the population growth will always tend to rise slightly above the supply of food*. This happens because initial food shortages causing poorer diets will not be sufficient to slow growth. Eventually, however, population pressure will activate one or more of

populations will follow cycles of growth and decline, these cycles do not prohibit a long-range upward, or even downward, trend. If the supply of food is increasing, then at the end of each cycle the population will be slightly larger, or just above what the new, larger food supply can support. Conversely, if the long-term trend is a reduced food supply, then the population will be smaller after each cycle.

Malthusian theory seemed to explain the facts of early agrarian societies quite well. The population had grown over the long run, as had the food supply. Yet the population remained subject to the cycle of ups and downs, and humans suffered from the activation of the positive checks of famine, disease, and war, just as Malthus had postulated. But even as Malthus was writing, the *second great shift* in population trends was under way, which seemed to defy Malthusian theory. The Industrial Revolution was taking place, accompanied by extraordinary population growth in Western nations without activating the positive checks Malthus predicted. A new optimism grew. Most people began to believe that by mastering technology humans could overcome the forces of fertility and mortality. Soon many social scientists considered Malthus's views outdated and pessimistic.

MODERNIZATION AND POPULATION

To most people, the phrase "Industrial Revolution" suggests machines and factories. But the Industrial Revolution first affected agriculture. Indeed, the age of the factory and the growth of large industrial cities were possible only because of the modernization of agriculture. Recall from Chapter 10 that in agrarian societies about 95 percent of the labor force is needed on the farms to grow enough food. Today, although fewer than 4 percent of Americans and Canadians farm, they grow enough food not only to feed all the rest of us but also to make North America by far the largest exporter of food in the world. As we shall examine in Chapter 19, the industrialization of agriculture caused this dramatic change. With machines replacing draft animals and hand labor, and with better plant and animal varieties, new techniques of crop rotation and field design, and the use of chemical fertilizers, food production soared. As Malthus would have predicted, so did the population.

In England, where the Industrial Revolution began, the population was three times larger by 1841 than it had been in 1700. As modernization spread across northwestern Europe, so did rapid population growth. In 1650 Europeans (including those living overseas) made up 18 percent of the world's population. In 1920 they made up 35 percent (Davis, 1971).

This growth was possible only because there was enough food. Indeed, the specter of famine suddenly disappeared from Europe: People began to eat far better despite there being millions more mouths to feed. This is demonstrated by the virtual disappearance of nutritional-deficiency diseases as causes of death in western Europe. Scurvy, a dreaded disease produced by a lack of vitamin C, was once a major cause of death. In less than a century, it became so rare that in 1830 a leading English physician failed to recognize its symptoms (Drummond and Wilbraham, 1957). Indeed, as we saw in Chapter 5, the improved diets resulting from modernized agriculture caused a revolution in the patterns of human growth in modern societies.

In addition to providing much more food, modernization resulted in more effective protection against disease. Public health measures caused the mortality rate to drop rapidly. Vaccination and inoculation campaigns prevented huge numbers of deaths, especially among children. For centuries smallpox was a dreaded killer. In 1980 World Health Organization officials announced that it no longer existed on earth. Of perhaps even greater importance were modern sanitation measures. Sewers, sewage treatment, and the availability of safe drinking water saved huge numbers of lives (as we shall see in Chapter 19).

Thus, the increase in food and sanitation measures greatly reduced mortality. As a result, during the eighteenth and nineteenth centuries, the population grew so large and so fast that we now speak of this as the first population explosion.

In the wake of these changes, many began to regard Malthusian theory as outdated. Modernization seemed to have given societies the capacity for unlimited growth because food supplies could expand as quickly as population grew. However, others suspected that modernization had simply postponed the day of reckoning, when Malthus's positive checks would again strike. Before this proposition could be tested, however, a *third great shift* in human population trends occurred. And this shift seemed to discredit the Malthusian theory once and for all.

THE DEMOGRAPHIC TRANSITION

According to Malthus, population size is determined by fluctuations in mortality because human fertility always remains high. This seemed to fit the patterns of population growth and decline observed up to Malthus's time. In the beginning the Industrial Revolution affected mortality almost exclusively. Increased food supplies and the conquest of many diseases caused

My grandfather may have been a member of this crew threshing wheat in North Dakota's Red River Valley in the 1880s. Although the crew is very large and the machinery antiquated compared with modern harvesting methods, these men were able to produce food at a level their own grandfathers would have found unbelievable. The result of the early industrialization of agriculture was a population explosion in Europe and North America.

Minnesota Historical Society

mortality to fall and therefore the population to grow. But then what Malthus said would not happen began to happen: Fertility began to decline in the more modernized nations.

We have seen that the population of England tripled between 1700 and 1841. Growth continued for a few more decades but at an increasingly slower rate, until by 1930 the population stabilized. However, growth was not halted by increased mortality. Instead, as mortality continued to decline, fertility also began to decline in the 1860s. Growth ceased by the 1930s because fertility no longer exceeded mortality.

Similar patterns were occurring on this side of the Atlantic as well. From the middle of the nineteenth century, the fertility rates of Canadian and American women also began a steady decline. By 1940 fertility in the United States was nearly at replacement level, while the average woman in Canada was having only 2.8 children. This was above the replacement level but still relatively low. **Replacement-level fertility** occurs when the number of births each year equals the number of deaths. In the most modern societies, replacement fertility occurs when the average woman has only slightly more than two children—one to replace herself, one to replace her husband, and a slight excess to make up for infant mortality. Replacement-level fertility produces **zero population growth** as soon as the age structure has adjusted.

Most industrialized nations achieved replacement-level fertility rates by the 1930s. Then,

after World War II, population stability was upset by a brief population explosion (see Box 18-1). However, fertility soon began to drop once more, and by the late 1960s, it was down to replacement level in most industrialized nations. Their populations continued to grow for a few years more, however, as the baby boomers passed through the reproductive age.

Modern nations have undergone a radical change in population patterns—described as the **demographic transition.** This transition involves a change from the age-old pattern of high fertility and high but variable mortality to a new pattern of low mortality and fertility. The demographic transition seemed to prove that Malthus's theory of population was no longer valid: In the modern world, people did control their fertility and thus averted the suffering that occurs when population size is determined wholly by mortality.

A CLOSER VIEW
KINGSLEY DAVIS:
DEMOGRAPHIC TRANSITION
THEORY

Why had this happened? How had modernization led to a decline in fertility? In 1945 Kingsley Davis (1909–1997), one of the most important contemporary sociologists (and my most formidable and stimulating teacher), proposed a theory of the demographic transition.

Library of Congress

Children were an economic benefit on pre-industrial farms: This New England farmer would have had a difficult time plowing with his ox team without his son to walk in front and goad the oxen. Although this picture was taken in 1899, long after the earlier picture of harvesting in North Dakota, the farming methods in use are pre-industrial. New technology was adopted much sooner in the frontier regions of the Great Plains than in the long-settled Northeast.

Acknowledging that Malthus's theory still seemed applicable to less modern parts of the world, Davis attempted to isolate those aspects of modernization that Malthus had not anticipated.

Davis argued that modernization naturally leads to conditions encouraging low fertility. People in modern societies had fewer children because they no longer wanted large families. One reason was that with mortality so greatly reduced, especially infant and child mortality, families no longer needed to have many children to ensure that some lived to adulthood. A major factor in high fertility is the tendency of families to "stockpile" children in anticipation that many will die young. Thus, in Rwanda 20.1 percent of all children born alive die before the age of five, and parents have many children to ensure that some will survive into adulthood.

On preindustrial farms and even in preindustrial crafts and manufacture, child labor is valuable. For example, small children on preindustrial farms can more than earn their keep by feeding chickens, gathering eggs, herding animals, milking, pulling weeds, and helping with household chores. Under such circumstances, larger families tend to be wealthier than smaller families. But as we shall examine in detail in the next chapter, a major aspect of modernization was a shift of population from farms to cities. In cities the labor of children is of much less value than on a farm. In fact, city kids are a financial drain.

Drawing on the choice premise that is basic to all social science, Davis argued that as large families became a cost rather than a benefit, people changed their conceptions about how many children they wanted. His stress on choice is important because the great reduction in fertility in modern times occurred *before* most modern birth control devices were invented.

Indeed, anthropologists have found that techniques for limiting fertility are known and often practiced in even very primitive societies (Ford, 1952; Harris, 1979). Moreover, historical demographers have learned that Malthus was simply wrong in his belief that humans never restricted their fertility. Throughout European history, long before the Industrial Revolution, fertility was tightly controlled whenever conditions made reduced fertility a reasonable course of action. For example, fertility often fell during economic depressions (Simon, 1981). The demographic transition reflected not new contraceptive technology but new conditions that influenced what people chose to do.

As outlined by Davis, the initial consequence of modernization is a sudden drop in mortality, which causes rapid population growth. This is because there is a delay before fertility begins to drop. The shifts from rural to urban living and from unskilled to skilled labor must also occur before the conditions that depress fertility prevail.

Davis's **demographic transition theory** has been subjected to a great deal of refinement and testing in the half century since it was first published. For example, demographers now include a number of "thresholds" of modernization that must be crossed before fertility is substantially reduced.

A listing of these thresholds by Bernard Berelson (1978) helps clarify the theory:

1. More than half of the labor force is not employed in agriculture.

2. More than half of the persons age five to nineteen are enrolled in school.

3. The average life expectancy reaches sixty years.

4. Infant mortality falls to 65 deaths per 1,000 infants.

5. Eighty percent of females age fifteen to nineteen are not married.

6. Per capita GNP reaches $450.

7. At least 70 percent of adults can read.

Table 18-2 tests the demographic transition theory by analyzing data from each of the 174 nations with populations of 200,000 or more. As can be seen, there are huge and significant correlations between measures of each threshold and fertility (no data were available on the marriage age of females).

When he published these thresholds, Berelson (1978) pointed out that they were not meant as precise rules, nor would a nation have to fulfill all seven to undergo the demographic transition. He suspected that meeting any three or four might suffice to cause a decline in a nation's fertility.

Several years later, Phillips Cutright and Lowell Hargens (1984) showed that Latin American nations which meet the two thresholds of 70 percent adult literacy and sixty-year life expectancy have declining fertility. Unfortunately, many of the nations of the world fall short of all of these thresholds.

This fact has troubled demographers and produced doomsday predictions about a "population bomb." Many dire predictions have been highly publicized during the past thirty years, each claiming that a huge new population explosion may soon reactivate Malthus's positive checks. Although many nations had failed to become sufficiently modernized to reduce their fertility, they had become modernized enough to reduce their mortality. Indeed, just about the same time that Davis published his theory, the *fourth great shift* in population trends occurred: massive, unprecedented population growth in the less developed nations.

THE SECOND POPULATION EXPLOSION

As we have just seen, the first population explosion occurred in Europe and North America as a result of modernization. The industrialization of agriculture and advances in public health reduced mortality. Eventually, however, rapid population growth was halted by a decline in fertility. While the demographic transition changed basic population patterns in the modern nations, preindustrial population patterns persisted in the rest of the world: high fertility checked by high mortality. Then, suddenly, mortality quickly fell in less developed nations, initiating the greatest population explosion in history.

Shortly after the end of World War II, demographers noticed rapid population growth nearly everywhere on earth. Growth soon halted in the most modern nations, but elsewhere growth rates continued to rocket. Throughout the 1950s and 1960s, the population in most of the less developed nations grew by rates of from 2 to 3.5 percent a year. Such rates mean that populations double in size every twenty to thirty-five years. In other words populations can increase enormously in a short time. For example, in the early 1970s, the world's population was growing at a rate that would double its size every thirty-seven years. If that rate were to hold for only 200 years, then instead of 6 billion people on earth, as there are now, there would be 157 billion!

TABLE 18-2	Demographic Transition Thresholds and Fertility	
THRESHOLD		**CORRELATION WITH FERTILITY RATE**
1. Percentage not employed in agriculture		−.32
2. High school enrollment		−.57
3. Average life expectancy		−.84
4. Infant mortality rate		−.85
5. Economic development		−.81
6. Literacy		−.80
7. Calories per person		−.72

Source: Prepared by the author from *Nations of the Globe: An Electronic Data Base in MicroCase Format,* 2002, distributed by Wadsworth.

BOX 18-1   THE LIFE CYCLE OF THE BABY BOOM

YOU UNDOUBTEDLY HAVE HEARD A LOT about baby boomers—possibly a lot more than you care to hear. But for all the talk about boomers, seldom does anyone explain what a boomer is and what all the fuss is about. Here's the story.

In May 1946, nine months after the surrender of Japan brought World War II to an end, demographers noticed that the number of births in the United States was up that month by 10 percent. The next month fertility continued to rise. By October births were up 50 percent. And by the end of the year, an all-time record of 3.4 million births had been established—one baby had been born every nine seconds.

These statistics were widely publicized and prompted many jokes about returning war veterans. But no one took this surge in fertility very seriously. During 1946 the director of the U.S. Bureau of the Census explained that the U.S. population might climb as high as 163 million by the year 2000 but that the current spurt in the birthrate was a brief and freakish postwar event. Demographers agreed that the American population could never come close to the 200 million mark—for the demographic transition had already taken place, and it was final. Indeed, most of what was written about fertility in 1946 and 1947 was concerned with a fertility deficit that would cause the population to decline rapidly.

How little the experts knew. For the "baby boom" following World War II was not a brief event. The birthrate remained high for almost twenty years; not until 1965 did it drop back to the level of 1940. The unthinkable 200 million mark was passed in 1968.

Moreover, the baby boom was not limited to the United States. Canadian fertility followed the same path, shooting up in 1946 and coming back to prewar levels in the early 1960s. In some parts of Europe, a baby boom took place, too, but usually on a much smaller scale.

The baby boom was far more than a set of birth statistics. The reality was that nations with relatively few infants and children suddenly were filled with them. As high fertility persisted, these huge birth cohorts of infants and young children began to cause major social changes. In fact, the baby-boom-age cohorts, those born between the mid-1940s and the early 1960s (and who are now in their forties and fifties), continue to have immense impact simply because there are so many of them. Let's retrace the impact of the baby boomers on North American societies and then anticipate their influence on the future.

Early Days

If academic demographers took a while to grasp the meaning of the rapidly rising fertility rates in the aftermath of the war, American business was not slow to see that radical changes were being wrought in basic consumer market patterns. Consider the following statistics for the United States (Jones, 1980).

* Sales of baby food rose from 270 million cans in 1940 to 1.5 billion in 1953.
* Sales of toys grew from $84 million a year in 1940 to $1.25 billion by the early 1950s.
* Business boomed for companies that bronze-plated baby shoes.
* Diaper sales doubled and redoubled, as did sales of washing machines.
* In June 1946 Pocket Books published *The Common Sense Book of Baby and Child Care* by Benjamin Spock, M.D. Unadvertised, unpromoted, and unreviewed, the 35-cent book sold 4 million copies by 1952 and has now sold well over 30 million copies.

New parents not only were rushing to buy food, clothing, and toys for their infants but also were seeking a comfortable environment in which to raise them: The baby boom accompanied a massive expansion of the suburbs. Between 1950 and 1970, the suburban populations of Canada and the United States doubled. More than 80 percent of the population growth during this period was located in the suburbs, as were 85 percent of all the homes

(continued)

© Bettmann/Corbis

BOX 18-1   THE LIFE CYCLE OF THE BABY BOOM (CONTINUED)

built then. Suburban life influenced other consumption patterns. It spurred automobile sales by making the two-car family necessary. Sales of lawn furniture and backyard barbecues zoomed, and hot dog sales tripled in the decade of the 1950s.

The average American and Canadian family, now living in the suburbs with young children, changed its entertainment patterns. Mom and dad no longer went dancing, and the era of Big Bands and popular ballrooms came to a crashing end. Moreover, mom and dad stopped going to movie theaters (and during the 1950s large numbers of huge, luxurious, downtown movie palaces closed their doors). Instead, they loaded the whole family in the station wagon and went to the drive-in. In 1948 there were fewer than 500 drive-in theaters in North America. A decade later, there were more than 4,000, with everything from playgrounds to Laundromats. When they weren't at the drive-in, the baby boom family was beginning to sit in front of a flickering little box in their living rooms—to watch television.

The baby boom kids were the first to grow up with television, the first to sing "M-I-C-K-E-Y M-O-U-S-E," the first to in effect live in a movie theater. From the very start, television has been shaped to an extraordinary extent by and for the baby boomers. Early

television was inundated with programming for little kids, and the grown-up shows mirrored the lives of the suburban baby boom family: *Ozzie and Harriet, Father Knows Best,* and *Leave It to Beaver.*

Then one day the baby boom kids headed off for school. When they got there, they found they were unexpected. In 1952–1953 the kids born in 1946–1947 turned six and began first grade; their group was much larger than the group a year ahead of them! As they passed through the school system, followed by even larger waves of students, they burst the seams of the system. By 1964 one of every four North Americans was enrolled in the lower grades or in high school.

Because demographers had claimed, through the first years of this population explosion, that it would be brief and unimportant, steps had not been taken to provide space and teachers for these throngs who suddenly showed up in the classroom. Many schools went to double-shift schedules, and everywhere class sizes were huge. Crash school construction programs began everywhere; during the 1950s California opened a new school every week. Because they were in such demand, teachers found their salaries rising rapidly.

© Jeff Smith

Looking back, we can see that the schools were not able to cope with the baby boom and that the quality of education that members of these cohorts received probably was seriously inferior to that of people slightly older. In 1964 the first crop of baby boomers graduated from high school and took the Scholastic Aptitude Test (SAT). Their average score was lower than that of the year before. And every year after that the average SAT score in the United States fell, until 1982 when the last boomers had graduated.

The growing baby boom kids had a major impact after school, too. Their tastes registered in a series of fads. For example, in 1955 kids reacted to an episode on the Walt Disney TV show, starring Fess Parker as Davy Crockett, and went out to spend more than $100 million in seven months, buying imitation coonskin caps (real coonskin prices jumped up by 3,200 percent), toy muzzle-loading rifles, T-shirts, and thousands of other items with a Crockett motif. In 1958 they rushed to buy more than 20 million Hula-Hoops in a few months.

The Teen Boom

The baby boom eventually produced crops of teenagers who continued to cause dislocations and rearrangements in the world around them. The sale of skin medications skyrocketed. So did record sales. With these sales came a dramatic shift in what music was popular and in the range of music performed. Although the baby boomers did not invent rock 'n 'roll, they turned it into what was, for several decades, the *only* popular music—the maturing baby boomers had ears only for "their" songs. An immense repertoire of popular songs suddenly disappeared. (This music began to reappear in the 1990s.)

Of course, the baby boomers soon did to colleges and universities what they had done to the public schools. Never before had so many people attended college. More important, never before had so high a proportion of people attended. The consequence was a massive increase in the number and the size of colleges and universities and in the number of faculty members.

Moving On

Thus far, we have looked only at the impact of the front end of the baby boom, but the baby boomers left important consequences behind them as well. For example, the boom in elementary education turned to a bust in the 1970s as the small birth cohorts behind the baby boom left the nation with far too many schools and teachers. A similar, but less sharp, depression also hit higher education as large cohorts have been followed by smaller ones. Moreover, since the baby boom, fertility rates have fallen to all-time lows, and baby product manufacturers have struggled to survive in shrunken markets. The amount of children's television programming has declined as well.

In June 1984 perhaps the most symbolic event in the baby boom's passage through the life cycle came in the announcement by Levi Strauss & Co. that it was shutting eleven manufacturing plants and laying off 3,200 workers. This brought to twenty-seven the number of plants the company had closed in less than two years. What did these plants make? The famous Levi's blue jeans. Why did the plants close? Because the baby boomers outgrew the age for wearing blue jeans, according to company officials. There was a sudden drop in the size of the population age eighteen to twenty-four, the primary jeans consumers. In the 1960s and early 1970s, the baby boomers had made Levi Strauss one of America's most successful companies. Then they left Levi Strauss with far too much production capacity.

Now other companies enjoy a huge upsurge in sales. The hot new products suddenly are those associated with anxieties about getting old—many of the baby boomers are over fifty. The runaway sales figures for exercise videotapes and equipment reflect the sudden new market for fitness. Creams to keep the skin soft and youthful have taken over the markets once dominated by acne remedies.

The Future

Of course, the wrinkle creams and the fitness programs will not work in the long run: The baby boomers will continue to age. As they do they will continue to shape society. Businesses and products directed at the elderly ought to thrive. But perhaps the major impact will be an overload on the economy as we find that the retired population outnumbers those who are working. How will today's cohorts of college freshmen manage to provide adequate goods and services for such a huge number of dependents? It seems likely that many of the baby boomers won't retire or won't retire as young as most people do today. For as the baby boomers begin to reach these ages, a labor shortage will develop (unless increased immigration greatly expands the population younger than the baby boomers). Such a labor shortage will give many baby boomers a chance to stay in their jobs beyond the age of sixty-five or seventy. But will they want to?

In future years you will hear much discussion of these issues and of what policies and programs ought to be pursued. But the reason for all these problems remains a simple one: For a few years following World War II, young Canadian and American couples turned their backs on the demographic transition and produced relatively large families in a short period of time. This resulted in some immense age cohorts, and ever since, they have formed a disproportionate population bubble moving through the life cycle, distorting the system as it goes.

A generation ago, many of these children would already have been dead and others would have died before reaching adulthood. The introduction of modern sanitation and public health measures into the less developed world, during and just after World War II, caused a dramatic decline in death rates—and produced the second great population explosion.

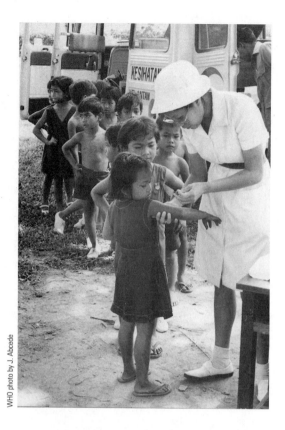

WHO photo by J. Abcede

Understandably, these population projections frightened demographers, as well as a lot of other people, and world population patterns received widespread publicity. For example, people were shocked when demographers pointed out that if Mexico continued to grow at the rate maintained during the thirty years following World War II, its population would increase from 63 million to 2 billion by the year 2080. Obviously, long before 2 billion people lived in Mexico, disaster would strike—if nothing else, massive starvation would set in.

In the wake of the second population explosion, predictions of world calamity received great publicity. Oddly enough, some of the most obviously faulty projections attracted the greatest attention and acceptance. A small book called *The Limits to Growth* (Meadows et al., 1972), which predicted that the world would run out of most raw materials and food in only a few years, sold over 4 million copies, despite its arguments having been dismissed as incompetent in virtually every scientific periodical (see Simon, 1981).

Still, very competent demographers and economists were influenced by the widespread anxiety about rapid population growth and warned of impending calamity. I confess that I joined in this dismal chorus. In a textbook on

social problems I published in 1975, I ended the chapter on population by telling students that they would be hearing of terrible famines in the underdeveloped nations for the rest of their lives. That will not be the message of this chapter. Apparently, Western social scientists forgot something important: Populations do not have babies—only people have babies. And people turn out to be a lot smarter than we sometimes assume. So let's examine the population explosion in the less developed nations to see why and how it occurred and how people have recently responded to it.

THE SUDDEN DECLINE IN MORTALITY

Whereas the brief **baby boom** in Western nations after World War II was caused by a rise in fertility, the population explosion in the less developed nations was not. There, fertility had always been high. The population explosion was caused by a sudden and dramatic plunge in the mortality rate.

The most important feature of this decline in mortality is that it was produced primarily not by internal changes in these nations but by external forces: *Low mortality resulted from a few elements of modern technology imported from developed nations.* Thus, while the decline in mortality experienced in the West had developed slowly, in the less developed nations it came suddenly and often without significant changes in other parts of these societies. The result was a period of extreme cultural lag, as we shall see.

To grasp just how the second population explosion took place, consider the case of a single nation. At the end of World War II, Sri Lanka (then known as Ceylon) had a very high mortality rate typical of agrarian societies in warm latitudes. The average life expectancy was only forty-three years, and the crude death rate in 1945 was about 22 deaths per 1,000.

Then, in 1946 the World Health Organization provided a small sum of money that Sri Lanka used to buy liquid pesticide from Switzerland and to hire some freelance pilots with surplus U.S. Army planes to spray the country. Thus, in the most literal sense, a modern, low mortality rate fell upon Sri Lanka from the sky. Disease-carrying insects, especially the dreaded malaria-carrying mosquitoes, were nearly wiped out in a few days.

As a result Sri Lanka's mortality rate fell by more than 40 percent within a few months.

Understandably, everyone in Sri Lanka thought it was wonderful to have greatly increased life expectancy. Because this program could be continued simply and cheaply, it was continued. By 1954 Sri Lanka's crude death rate had fallen to about 10 per 1,000 per year. By 1975 it was down to 7.7—a rate lower than that of most modernized nations (partly due to Sri Lanka's very young population). As a result of its reduced mortality, Sri Lanka's population began to grow very rapidly.

In other less developed nations, similar changes took place: Insects were sprayed, swamps and stagnant pools of water where insects bred were drained, populations were inoculated against dangerous communicable diseases such as smallpox, and safe drinking water was provided. Everywhere the mortality rate plunged. From 1940 to 1965, the mortality rate fell by 59 percent in Mexico, 63 percent in Puerto Rico, 44 percent in Egypt, and 72 percent in Taiwan.

But something besides imported public health technology was involved in the second population explosion: the modernization of agriculture. Modern chemistry provides not only insect killers but also weed killers and fertilizers. Agricultural biologists have made rapid progress in breeding faster-growing, disease-resistant, higher-yielding varieties of basic food plants. The less developed nations also imported these plants, along with farm machinery and modern irrigation techniques. Thus, the food supply increased in all parts of the world. Indeed, despite the rapid growth of population, food production increased even more rapidly; today, people in most nations eat much more and better food than they did thirty years ago. This change also reduced mortality.

HIGH FERTILITY AND CULTURAL LAG

The demographic transition theory suggests that the long-term effect of modernization is population stability: Eventually, fertility will balance out with mortality. The question posed about the population explosion in the less developed nations is whether they will have the time to pass through the demographic transition before their populations become so huge that mass starvation results.

The demographic transition in the West was gradual, based on broad internal social changes, and the first population explosion was slow compared with the second. In the West infant mortality fell slowly for a number of generations, and people had a long time to adjust to the changed conditions. So, a gradual downward trend in fertility was sufficient to prevent catastrophic rates of growth. Some people argue that mortality has fallen so fast in the less developed nations that people have had no time to adjust.

This resultant cultural lag has been the focus of the grave anxieties about the second population explosion. Indeed, billions of dollars have been spent to popularize family planning and to reduce fertility in the less developed nations. However, for a long time, it looked as if these programs would fail and that fertility would not be checked in time. This prompted not only projections of disaster but also the frequent depiction of people in the less developed nations as virtual animals, incapable of reasonable behavior. For example, William Vogt (1948), in the first best-selling book on the population explosion, blamed it on "untrammeled copulation" by people in the less developed nations, whom he characterized as the "backward billion." Many others asserted that fertility in these less developed nations was outside the realm of decision making (Simon, 1981).

Amidst all this hue and cry, people in many of the less developed countries began to respond reasonably to their new conditions. The cultural lag began to be reduced. As happened in the modernized nations, fertility began to decline in most of the less developed nations. Thus, the *fifth great shift* in population patterns began to develop.

THE POPULATION EXPLOSION WANES

By the early 1970s, some leading demographers began to detect fertility declines in many of the less developed nations—especially in the larger ones. Initially, many discounted these declines as too little and too late to avoid disaster. As time passed, however, the true dimensions of rapid fertility declines became evident. As can be seen in Table 18-3, by the mid-1960s, just as the population-explosion scare spread through the developed world, fertility rates began to plummet in those nations where population growth had been the most explosive.

The eight Asian nations shown in the table have almost half (48%) of the world's population. In each the fertility rate has fallen precipitously.

TABLE 18-3	Fertility in Selected Nations, 1965–2003			
		FERTILITY RATE		
	2001 POPULATION IN MILLIONS	1965*	1975*	2003
ASIA				
China	1,273	6.4	3.7	1.7
India	1,030	6.2	5.4	2.9
Indonesia	2 28	5.5	5.0	2.5
Pakistan	145	7.0	7.0	4.4
Bangladesh	131	6.8	6.6	3.2
Philippines	83	6.8	5.5	3.3
Thailand	62	6.3	4.5	1.7
South Korea	48	4.9	3.3	1.2
LATIN AMERICA				
Brazil	174	5.7	4.4	2.0
Mexico	102	6.7	5.5	2.5
Colombia	40	6.5	4.4	2.6
Argentina	37	3.7	3.3	2.3
Peru	27	6.7	5.3	2.7
NORTH AFRICA				
Egypt	69	6.8	5.4	3.0
Morocco	31	7.1	6.3	2.9
Algeria	32	7.4	7.3	2.2
SUB-SAHARAN AFRICA				
Nigeria	127	6.9	6.6	5.4
Ethiopia	66	5.8	5.9	5.6
South Africa	44	6.2	5.3	2.2
Sudan	36	6.7	6.7	5.1
Tanzania	36	6.6	6.4	5.2
Kenya	31	8.0	8.1	3.3

Sources: The World Bank, *Social Indicators of Development, 1990* (1991); *The World Bank Atlas* (1992); Population Reference Bureau, *World Population Data Sheet, 2004.*

*Available data closest to this year.

Thus, in Thailand in 1965, the average woman gave birth to 6.3 children; in 2003 to only 1.7 children—that is fewer than is needed to replace the current population. The current fertility rate is *below* replacement level in South Korea and China, too, which means that the population in these three nations soon will actually begin to decline. Keep in mind that populations that have grown rapidly will continue to grow for a time after the fertility rate has been reduced to or below replacement levels. That is because when populations are growing, each new birth cohort is larger than the one before it (see Figure 18-1). Thus, after fertility declines, the largest birth cohorts are yet to pass through their reproduc-

tive years, and even if they merely replace themselves, the population will increase further.

Fertility also has fallen rapidly in Latin America. Thus, back in 1965 the average woman in Brazil bore 5.7 children. Today, she has 2.0 children, which is a bit below replacement level, given Brazil's rate of infant mortality. Across North Africa, too, fertility has fallen fast and far. In the nations of sub-Saharan Africa shown in Table 18-3, fertility has declined, but in most of them, it remains quite high. While continued population growth here may have undesirable local consequences, the fact that these are relatively small nations limits their global impact.

These boats, which belong to a remote tribe of Indians in Mexico, symbolize the cultural gap responsible for the population explosion in the less developed world. When modern levels of mortality were combined with preindustrial levels of fertility, the future of these societies was placed in jeopardy. When powerful modern outboard motors were combined with primitive wooden boats, the future of the passengers was placed in jeopardy.

© John Running / Stock Boston

In any event there has been a massive decline in fertility in most of the less developed nations—a decline as sudden and dramatic as it was unexpected. Claims that it already was too late to avert a world overpopulation disaster proved premature. How did it happen? How was fertility slowed so rapidly?

ECONOMIC DEVELOPMENT

Demographers involved in population control projects have long asserted that "economic development is the best form of contraception." Obviously, this slogan derives from demographic transition theory—that as nations develop modern economies, their populations will respond with modern fertility patterns. And the rapid decline in fertility rates, especially in Asia and Latin America, does seem to reflect substantial economic progress in these regions. Since China had the most dramatic (and unexpected) fertility decline, let's look more closely at what happened there.

In recent years China's gross national income (GNI) grew by an astounding 9.4 percent a year, exceeded only by South Korea's rate of 10 percent. Moreover, during this same period, China's per capita GNI grew by an average of 7.8 percent a year. Given China's economic gains, its fertility would be expected to decline. Moreover, the demographic transition theory is confirmed *within* China. That is, some of China's thirty provinces experienced far more economic development than did some others. Table 18-4 shows that fertility varies across the provinces in accord with Berelson's (1978) thresholds—the correlations all are very high and negative.

TABLE 18-4	Demographic Transition Thresholds and Fertility in China's Provinces	
THRESHOLD		**CORRELATION WITH FERTILITY RATE**
1. Percentage not employed in agriculture		−.66
2. Percentage high school graduates		−.74
3. Life expectancy		−.67
4. Low rate of infant mortality		Not available
5. Low rate of teenage marriage		Not available
6. Affluence:		
GNI per capita		−.74
Percentage of farms with TVs		−.65
Percentage of farms with phones		−.74
7. Literacy		−.67

Source: Prepared by the author from *Modern China: An Electronic Data Base in MicroCase Format, 2002*, distributed by Wadsworth.

Moreover, China accomplished its amazing fertility decline without fully achieving Berelson's thresholds. The Chinese have surpassed the thresholds for literacy, GNI, infant mortality, and life expectancy, but more than half of the Chinese labor force is still employed in agriculture and only 44 percent of those age five to nineteen are in school. Thus, Berelson's observation that a nation that met three or four of the seven thresholds might undergo the demographic transition—as China is doing—holds true. It also is worth noting that way back in 1978 Berelson predicted that China would bring its fertility under control by the year 2000; he thought that India might succeed in doing so by then as well.

However, there is another important element to China's fertility decline: coercion. Chinese couples can be fined or demoted in their jobs if they have a third child, and even women who become pregnant a second time often are compelled to have abortions. This may explain why China's fertility rate fell faster than India's, for example. By the same token, given that coercion has not been used to reduce fertility in other nations, major shifts in family size clearly can be achieved without it. In fact, many demographers now believe that fertility rates can be controlled not only without coercion but perhaps without nearly as much economic development as had been thought.

NUMERACY ABOUT CHILDREN

The term *numeracy* has recently come into use to identify the capacity to use numbers, just as *literacy* refers to the capacity to read and write. "Numeracy about children" refers to people having a clear notion of what size a family ought to be or how many children they would like to have. At issue is not the capacity to count one's children but the willingness to think about future fertility in terms of numbers. Interest in this phenomenon was prompted when survey studies conducted in sub-Saharan societies in the 1960s found that many women simply could not be persuaded to answer questions such as, "If you could choose exactly the number of children to have in your whole life, how many would that be?" In a 1963 survey of women in Ghana, 45 percent living in rural areas and 36 percent living in urban areas would not, or could not, give a numerical answer (van de Walle, 1992). Here is an example, taken from a recorded interview conducted in Bamako, the capital of Mali, with a twenty-eight-year-old mother of seven (van de Walle, 1992):

Q. Maimouna, how many children would you like to have in your life?

A. Ah, what God gives me . . . I cannot tell the number I will have in my life [laughs].

Q. It is true that God is the one who gives the child, but if God asked the number of children you wanted, how many would you say?

A. Oh, me, I cannot tell the number of children to God. What he gives me is good, that's enough. To say that I can stop and say the number, to tell God what to give me, I could not do so.

Q. Even if God asks you?

A. Even if He asks me, I cannot say it.

It is quite pointless to stress birth control to people who regard the results of natural fertility as the "right size" of a family, so it is fortunate that this view of family size seems to be disappearing rapidly. Thus, a survey conducted in Ghana in 1988 found only 13 percent of women unwilling to give a numerical answer.

Etienne van de Walle (1992) believes that Europeans did not think of family size in terms of the number of desired children until quite recently, perhaps as late as the 1800s. He bases his judgment on literature, noting that family size is not even discussed—neither the desirable number of children nor even the actual number—in older works. Montaigne, writing in the sixteenth century, noted that all but one of his children had died in infancy but does not bother to say that the number who died totaled five. But according to van de Walle:

In the novels of the nineteenth century, the situation is clearly different. Precise statements about numbers of children appear as part of the plot. . . . Balzac reproduces the calculus of conscious choice of one married woman in the following words:

"It is possible to have a dozen children in a marriage, by getting married at our age [17]; and if we had them we would commit twelve crimes, make twelve unhappy beings. . . . On the contrary, two children are two gifts of happiness, two blessings . . ."

However it happened, numerical notions of proper family size came to dominate thinking about fertility in the developed world, and surveys of people in these nations do not encounter nonnumerical answers. Instead, they find people ready and willing to express firm, numerical preferences about what size family they would like to have.

Recently, the World Values Surveys asked people in forty nations, "What do you think is the ideal size of a family—how many children?" The answers make it possible to calculate the average, or ideal, family size in each nation. If every family in each nation had precisely the same number of children they thought to be ideal, then the actual fertility rate and the ideal rate would be the same. In fact, aside from Nigeria and India, in all these nations the actual fertility was below, sometimes substantially below, the ideal. In Ireland, for example, people believe the ideal family size is 3.52 children, but the fertility rate is only 2.02—Irish families are smaller than the national ideal. In similar fashion Americans say they would like 2.7 children, but fertility is 2.12, while Canadians say they would like 2.77 children but are having only 1.6. In Mexico actual fertility (2.62) also is below the desired ideal (2.79). That people in Nigeria and India are having more children than they want is probably the case in many other less developed nations with high birthrates. Ironically, people in the most developed nations are, for a variety of reasons, having fewer children than they would prefer, while people in the less developed nations are ending up with more children than they would like. These findings challenge the belief long held by population experts that the crucial tasks are to get people to become numerate about family size and then to *want* smaller families. These surely are necessary conditions for reducing fertility, but they seem not to be sufficient as long as people are having more children than they want.

CONTRACEPTION AND WANTED FERTILITY

Population experts have a new slogan these days: "Contraception is the best form of contraception." This saying is based on studies done in nations where little or no economic development has taken place but where millions of people still would like to have fewer children. To deal with this phenomenon, demographers have developed measures of **wanted fertility**: the number of children a couple wishes to have. When wanted fertility is subtracted from total fertility, it becomes evident that in many nations—Mexico and South Africa are but two—wanted fertility is substantially lower than actual fertility.

A study by John Bongaarts (1990) showed that this phenomenon was especially true for those nations with the highest current fertility rates, such as the sub-Saharan nations. Thus, Bongaarts found that people in Kenya would like 1.2 fewer children than they end up with, in Ghana 1 fewer, in Liberia 0.9, and in Mali 0.6. Keep in mind that people in many of these nations still want large families. In Kenya, for example, the wanted fertility rate is 6.7 children. But the fact remains that millions fewer babies would be born if people in these nations did not have unwanted births.

It turns out that the major reason for unwanted births in the less developed nations has been the difficulty in obtaining contraceptives. For example, John Caldwell and his colleagues (1992) report that until 1988 people in Nigeria could obtain contraceptives only from government family-planning clinics. Moreover, only married people were eligible to purchase contraceptives, and many Nigerians, particularly men, were embarrassed to be seen entering or leaving these clinics. In 1988 the government decontrolled the contraceptive market, and contraceptives have since become available in all pharmacies and in many other small stores, including service stations. The result was an enormous increase in consumption as previously unmet demand found an acceptable supply. Indeed, the truly dramatic rise in contraceptive use was among unmarried women, previously denied any legal supply. Similar findings are reported from many other nations, and demographers now agree that a meaningful population decline has begun in sub-Saharan Africa (Caldwell et al., 1992).

THE CRISIS OF DEPOPULATION

Back in the 1930s, sociology textbooks expressed gloomy forecasts that the advanced industrial nations faced "depopulation." Because their fertility rates were rapidly declining, and had fallen below replacement levels, it seemed inevitable that the populations of most western European nations as well as the United States would soon begin to decline. Thus, in 1931 Louis I. Dublin predicted that the population of the United States would peak at about 154 million by the mid-1980s and then decline to 140 million by 2000 and to 76 million by 2100. In her influential book *The Menace of*

TABLE 18-5	Selected Nations with Fertility Rates Below Replacement Level		
NATION	**FERTILITY RATE***	**NATION**	**FERTILITY RATE***
Hong Kong (China)	0.9	Portugal	1.4
Armenia	1.2	Georgia	1.4
South Korea	1.2	Austria	1.4
Taiwan	1.2	Estonia	1.4
Belarus	1.2	Switzerland	1.4
Poland	1.2	**Canada**	**1.5**
Romania	1.2	Cuba	1.6
Slovakia	1.2	Belgium	1.6
Ukraine	1.2	Cyprus	1.6
Bulgaria	1.2	Australia	1.7
Czech Republic	1.2	China	1.7
Bosnia	1.2	Thailand	1.7
Slovenia	1.2	Sweden	1.7
Japan	1.3	Great Britain	1.7
Singapore	1.3	Norway	1.8
Hungary	1.3	Denmark	1.8
Latvia	1.3	Netherlands	1.8
Greece	1.3	Azerbaijan	1.8
Germany	1.3	Finland	1.8
Lithuania	1.3	Mauritius	1.9
Spain	1.3	France	1.9
Russia	1.4		

Source: Population Reference Bureau, *World Population Data Sheet, 2004.*

*Number of births to the average woman during her lifetime.

Under-Population, published in 1936, Enid Charles claimed:

In parts of Europe and America the population has already ceased to be capable of maintaining its numbers. It cannot be too clearly emphasized that this statement is not a prediction of future events . . . but a description of what is actually happening at the moment.

No one disputed Charles's claims. Consequently, the sudden and rapid rise in fertility rates that produced the post-World War II baby boom brought many well-known demographers and their dire predictions about "depopulation" into disrepute. Nevertheless, Frank Notestein (1950), one of the more prominent predictors of depopulation, stood his ground, dismissing the baby boom as a momentary exception to the long downward trend in fertility.

Today, it appears that Notestein was right. Beginning in the 1960s, the fertility rates in the industrialized nations turned downward, and soon they were again below replacement levels.

Moreover, fertility rates in these nations have continued to fall, thus ushering in the *sixth* great transformation in population trends.

Table 18-5 shows the most recent fertility rates for nations currently below replacement level. Under modern conditions, it requires a fertility rate of somewhere from 2.02 to 2.07 to replace the population (depending on prevailing rates of infant and child mortality). Here we see that nearly the entire developed world is well below replacement-level fertility, and therefore, the population will decline in these nations unless the lack of births is offset by immigration. Moreover, where fertility rates such as these prevail, the population decline will be quite rapid. Of even greater importance is that fertility rates in these nations have not stopped falling and in some nations the decline even seems to be accelerating. In Russia, for example, the fertility rate fell from 2.1 in 1983 to 1.4 in 2003. If these trends hold, the population of Russia will decline from 147.9 million in 1990 to 126.7 million in 2025 (Haub, 1995).

TABLE 18-6	"Having Children Interferes Too Much with the Freedom of Parents"		
NATION	AGREE (%)	NATION	AGREE (%)
Russia	59	Czech Republic	22
Spain	54	Slovenia	18
Bulgaria	51	Netherlands	13
Japan	42	**Canada**	**11**
Austria	36	Great Britain	10
Poland	36	Australia	9
Italy	32	Ireland	9
Israel	28	Northern Ireland	8
Germany	24	New Zealand	8
Hungary	23	**United States**	**8**
Philippines	23		

Source: Prepared by the author from the International Social Survey Program, 1994.

Why has fertility fallen so far, so fast? An answer can be discovered in Kingsley Davis's original formulation of the demographic transition theory. Let us quickly review the essence of this theory.

Davis assumed that choice plays an important role in fertility. In premodern societies, however, these choices are merely implicit. Because high fertility is in the self-interest of couples as well as their societies, the culture does not provide for "selecting" family size, as indicated by a lack of numeracy about children. However, modernization soon reveals that fertility is subject to choice as people recognize that large families no longer are necessary. Moreover, as children become a financial burden rather than an asset, people soon choose to have fewer children. Hence, Davis predicted that in modern societies fertility and mortality would strike a balance so that population size is stabilized. Davis did not anticipate that there might be *no lower limit* to the number of children the average family would choose to have. And that's what worries many demographers today—that fertility rates could in fact fall to zero, or close to it.

As can be seen in Table 18-6, rapidly declining fertility rates in the industrial nations suggest that couples are increasingly placing a higher value on their leisure and on a more affluent lifestyle than on children. Nearly two-thirds of Russian adults agreed that "Having children interferes too much with the freedom of parents." In Spain 54 percent agreed as did 42 percent of Japanese—all nations having very

low and declining fertility rates. In contrast, very few in Ireland (both North and South) or in the United States expressed this view.

However, it seems unlikely that the United States will join in the spiral of depopulation, at least not if current trends hold. Notice in Table 18-6 that very few Americans agreed that children interfered too much with parental freedom. Moreover, after dipping below the replacement level during the 1970s, the American fertility rate bounded back and is expected to remain above the replacement level for the foreseeable future (United Nations, 2001).

GRAY NATIONS

When fertility rates fall so far and so rapidly as they have in China and most of Europe, the population not only dramatically shrinks in size, but it rapidly gets older (see the constrictive population structure in Figure 18-1). If present trends hold, by 2050 about half of all Europeans will be past the age of sixty (compared with only 20 percent of Americans). Hence, not only will Americans by then outnumber Europeans; they will far more greatly outnumber them in terms of people in their prime productive years (United Nations, 2001). China is rapidly progressing along the same path. With fertility now well below replacement level, the Chinese population also is rapidly aging: Between 2010 and 2040, the proportion of Chinese over age sixty-five will rise from 7 percent to 25 percent and will increase even more rapidly in the years that follow (England, 2005). Why does this matter? Because an increasingly smaller work force must support an increasingly larger population of retired people—many of whom place very heavy demands on medical and social services. If nothing else happens, it appears that the average European worker soon will be supporting not only her or his family but at least one elderly person and probably more. But of course, something else *is happening*: immigration.

CULTURE AND IMMIGRATION

As their populations have been aging, Europe's industries have faced labor shortages. To offset these, Western European governments have encouraged immigration by people of working age. Some of these have come from Eastern Europe—Sweden's industries have long been dependent on Hungarian immigrants, for example. But most Eastern European nations now have their own problems with rapidly declining and aging populations. So, the majority of immigrants into European nations have been coming from North Africa and the Middle East. These new immigrants are Muslims. No one truly knows how many Muslims now live in Western Europe, but it is believed that the population of illegal immigrants is very large. Even official estimates of legal Muslim immigrants equal 7 to 10 percent of the population of France and about 4 percent in Germany, Belgium, the Netherlands, Sweden, Switzerland, and Great Britain. These numbers are growing rapidly not only because many new immigrants enter each year but because the fertility rates of Muslim immigrants are well above the replacement level.

Rapidly growing Muslim populations have resulted in growing antagonism and conflict in most of Western Europe (see the statistics on not wanting Muslim neighbors in Table 11-2). In part this conflict stems from the fact that Muslims have made it clear that they have no intention of assimilating into European culture. Thus, for example, a battle has been raging in France because the government has prohibited Muslim girls from wearing their traditional head-scarves to school. These conflicts have become even bitterer since 9/11 and the actions of Islamist terrorists within European Muslim communities—the Dutch are considering restrictive measures following the murder of a film producer by Muslims who labeled his work blasphemous. Some recent European critics of continued Muslim immigration have used the term "Eurabia" to describe a future wherein European culture is replaced by Islamic culture. Yet the stark fact remains that at current levels of European fertility, even without any Muslim immigration, there soon won't be many people to sustain a "European" culture—there already are more people who claim Norwegian ancestry in the United States than there are in Norway.

LOW FERTILITY AND GENDER BIAS

Modern medical technology not only makes it relatively safe and easy for people to limit their fertility; but it recently has made it possible for parents to limit their children on the basis of gender preferences. There are several techniques available.

The first involves various methods that determine the gender of a fetus. And just as many societies in the past engaged in infanticide to rid themselves of female infants, many parents today, once they know the gender of the fetus, are using abortion to limit their families to males. This practice has become widespread in the urban areas of China, greatly exacerbated by repressive government measures against having more than one child. For the same reasons, in Chinese villages lacking sophisticated medicine, female infanticide is reported to be prevalent. Consequently, China is raising far more little boys than little girls—in 1995 (the most recent year available) the sex ratio of Chinese children under age five was 118 boys per 100 girls and rising (*The Economist*, Jan. 1, 1999). As these cohorts come of age, there is going to be a severe shortage of wives in China—a situation that may bode ill for the status of Chinese women (see Chapter 12). Morever, China is not alone. Obvious gender biases in births have appeared throughout Asia. The sex ratio of newborns in South Korea was 113 boys to 100 girls in 1992 (Kristof, 1993). In India the practice of aborting female fetuses recently led authorities in Mumbai (Bombay) to ban prenatal ultrasound scans of women under age thirty-five (older mothers having a substantially higher chance of birth defects in their offspring) (Kristof, 1993). The ability to select children on the basis of gender need not, of course, produce an excess of males. That it is doing so in Asia reflects the very strong preferences for male children that persists in these cultures (Croll, 2001).

Indeed, in China the current gender imbalance of live births is 152 males to 100 females among second children and 160 to 100 among third children, which reflects the "desperation by parents as they have additional children after the first one or two are females" (England, 2005).

A second technique involves separating male sperm into those carrying Y chromosomes, which produce males, and those carrying X chromosomes, which produce females (Busch, 1993). The female then is inseminated with sperm of only one gender type. This technique, called MicroSort, was developed originally for use with cattle, but clinical trials with humans began several years ago and have been 91 percent successful when attempting to select females and 73 percent effective when seeking male sperm cells (Ostrom, 2002). But it seems unlikely to result in unbalanced sex ratios because American and Canadian couples overwhelmingly prefer to have both a boy and a girl. However, they also have a strong preference for having the boy first, which could lead to a preponderance of families having older brothers and younger sisters (Busch, 1993). The impact of such a pattern is hard to anticipate.

Recently, Valerie Hudson and Andrea Den Boer (2004) projected that by 2020 China will have from 29 to 33 million more males than females in the age group of fifteen to thirty-four. In response they noted that China's strict moral code has already "broken down" to permit the rapid growth of prostitution and a sex industry. Based on the theory of sex ratios considered at length in Chapter 12, one also would anticipate that men able to secure brides will increasingly confine them in restrictive norms. In addition Hudson and Den Boer fear that such an excess of males may encourage China to launch military adventures.

CONCLUSION

There have been six major shifts in human population patterns during the past 10,000 years. The first of these occurred when the invention of agriculture permitted the population to begin to grow rapidly. This period of growth, however, was marked by cycles of growth and decline, as population was periodically cut back by sudden rises in mortality. This led Malthus to formulate a theory of population based entirely on fluctuations in mortality.

But no sooner had Malthus published his book than a second great shift occurred. Industrialization stimulated a long period of uninterrupted growth, sometimes referred to as the first population explosion. After several centuries of rapid population growth in the industrial countries, a third great shift occurred—one Malthus had believed impossible. Population growth was halted by a decline in fertility. This is called the demographic transition, which resulted in stable populations in which low mortality was balanced by low fertility.

In the aftermath of World War II, a fourth population change took place, a population explosion in the less developed nations caused by decreased mortality and increased food supplies. This growth was so rapid that many people predicted a return to Malthusian conditions, in which terrible famines would halt population growth. These predictions failed to anticipate the fifth major shift in population patterns: sharply falling fertility in most of the less developed nations as they, too, underwent the demographic transition.

This fifth shift in population patterns took most scholars, even expert demographers, by surprise. Perhaps the most influential book on population problems was *The Population Bomb*, published in 1968 by Paul Ehrlich, a biologist at Stanford University. In it he wrote, "The battle to feed all of humanity is over. In the 1970s the world will undergo famines—hundreds of millions are going to starve to death."[1] In 1974 the famous novelist and scientist C. P. Snow told the *New York Times*, "Perhaps in ten years millions of people in the poor countries are going to starve to death before our very eyes. . . . We shall see them doing so upon our television sets."

And in 1975 I wrote in a textbook for college students:

The population explosion is not just someone else's problem. It threatens every nation and every person. There will surely be global famines and mass starvation. . . . For the rest of your life, you will be hearing of terrible famines.

Despite such hysterical predictions, per capita food production continued to increase worldwide, as it had been doing for decades

---

1. Ehrlich never has admitted his errors and continues to predict coming overpopulation catastrophes that are always taken seriously and widely publicized by the news media.

(Simon, 1981). Now we're at the start of the twenty-first century, and millions have not dropped dead of hunger. Instead, the world has continued to eat even better than in the 1970s. In fact, the starving Africans shown on television are not victims of overpopulation but of corrupt or malevolent governments. Thus, the famine in Somalia required the intervention of U.S. Marines to prevent local warlords from hoarding all of the food and stealing all of the aid sent from outside—there would have been no famine had there been political stability. The recent famine in Ethiopia had nothing to do with food supplies and everything to do with a government policy of starving its ethnic political opponents. In similar fashion, as discussed earlier in this chapter, the massive famine that struck China in 1958–1961 was not caused by overpopulation but could be better described as a political famine. As Basil Ashton (1984) and his colleagues put it:

The tragedy of this famine is that internal food redistribution was obviously limited and major international relief was never attempted. It would not be inaccurate to say that 30 million people died prematurely as a result of errors of internal policies and flawed international relations.

Demographic and economic forecasting is extremely difficult and frequently very wrong. The possibility of error is maximized when current trends are projected into the future without an underlying theory about relationships among these trends. Here we see that much more attention should have been given to demographic transition theory and much less to simple projections of fertility. The less developed nations not only experienced a huge decline in mortality because of modernization but also increased their agricultural production, thus increasing food supplies. In time modernization began to have the predicted effects on their fertility.

Reactions to the population explosion might have been more subdued had social scientists remembered that at many times in history humans have limited their fertility when necessary. If even primitive tribes have achieved low fertility when they wanted to, then we should have suspected that people in nations already somewhat modernized might possess similar abilities. And, of course, they did.

In this chapter I have tried to explain population trends. However, after all is said and done, the key to understanding population lies in the most basic premise of microsociology, introduced in Chapter 1. Human behavior is based on choice, and humans choose to do what they believe to be in their own best interests. When mortality is high, the reasonable family will have many children. But when mortality is low and children are not an economic asset, the reasonable family will have fewer children. But how many fewer?

That brings us to the sixth and most recent shift in population trends. The demographic transition did not (or has not yet) resulted in stable populations. Instead, throughout the industrialized world, fertility rates have continued to fall and now are well below the level needed for replacement. Will this continue? Perhaps. But as we have seen, population trends have a way of reversing themselves. Perhaps a *seventh* major shift lies ahead. Consider this possibility: In a society where many couples remain childless, most children will be born to parents who want and value children. Perhaps that could result in a generation of children, most of whom will themselves be socialized to desire children of their own, while the cultural preference to remain childless dies out for lack of heirs.

So, what lies ahead? No one really knows. And having learned my lesson from earlier unfounded predictions about population trends, I shall not speculate further.

Review Glossary

Terms are listed in the order in which they appear in the chapter.

Domesday Book Pronounced "doomsday" book, this was an outstanding medieval census conducted by William the Conqueror following his takeover of England in 1066.

census A population count, often recorded in terms of such categories as age, sex, occupation, marital status, and the like. The U.S. Census is conducted during the first year of each decade.

demography Literally, written description of the people; the field of sociology devoted to the study of human populations with regard to how they grow, decline, or migrate.

growth rate Population gains or losses computed by dividing the net gain or loss for a particular period by the population total at the start of that period.

crude death rate The total number of deaths for a year (or similar period) divided by the total population that year.

crude birthrate The total number of births for a year (or similar period) divided by the total population that year.

fertility rate The total number of births for a year divided by the total number of women in their child-bearing years (the U.S. Census bases this rate on all women age fifteen to forty-four).

age-specific death rates The number of deaths per year of persons within a given age range divided by the total number of persons within that age range.

birth cohort All persons born within a given time period, usually one year.

age structure The proportions of persons of various age groups making up a total population.

sex structure The proportions of males and females in a population.

expansive population structure An age structure in which each younger cohort is larger than the one before it; such a population is growing.

stationary population structure An age structure in which younger birth cohorts are the same size as older ones were before mortality reduced them; such a population neither grows nor declines.

constrictive population structure An age structure in which younger cohorts are smaller than the ones before them; such a population is shrinking.

arithmetic increase A constant rate of growth (or decline); the same number of units are added (or subtracted) each cycle, as in 1-2-3-4-5.

exponential increase A rate of growth (or decline) that speeds up as an increasingly larger number of units is added (or subtracted) each cycle, as in 1-2-4-8-16.

positive checks According to Malthus, famine, disease, and war—the primary factors that check or stop population growth.

Malthusian theory of population Theory stating that populations will always rise to, and then go somewhat above, the limits of subsistence and then will be reduced by the positive checks, only to rise again and be checked again.

replacement-level fertility (sometimes called **zero population growth**) Point at which the number of births each year equals the number of deaths.

demographic transition A shift in population trends from high fertility, controlled by high mortality, to one of low mortality and low fertility.

demographic transition theory Theory stating that the demographic transition was caused by modernization, which reduced the need for and the value of large numbers of children.

baby boom A brief period of high fertility in many Western industrial nations immediately following World War II.

wanted fertility The number of children a couple wishes to have.

Suggested Readings

Anderson, Margo J. 1988. *The American Census: A Social History*. New Haven, Conn.: Yale University Press.

Ashton, Basil, Kenneth Hill, Alan Piazza, and Robin Zeitz. 1984. "Famine in China, 1958–61." *Population and Development Review* 10:24–37.

Berelson, Bernard. 1978. "Prospects and Programs for Fertility Reduction: What? Where?" *Population and Development Review* 4:579–616.

Croll, Elisabeth. 2001. *Endangered Daughters: Discrimination and Development in Asia*. London: Routledge.

Cutright, Phillips. 1983. "The Ingredients of Recent Fertility Decline in Developing Countries." *International Family Planning Perspectives* 9:101–118.

Davis, Kingsley. 1945. "The World Demographic Transition." *Annals of the American Academy of Political and Social Sciences* 237:1–11.

Eberstadt, Nicholas. 1994. "Demographic Shocks in Eastern Germany, 1989–93." *Europe-Asia Studies* 46:519–533.

England, Robert Stowe. 2005. *Aging China*. Westport, Conn.: Praeger.

Haub, Carl. 1994. *Population Change in the Former Soviet Republics*. New York: Population Reference Bureau.

Wrigley, E. A. 1969. *Population and History*. New York: McGraw-Hill.

Sociology Online

www.socstark10.com

GO TO THE INTERNET AND TYPE: www.socstark10.com.

↗ **CLICK ON:** 2000 General Social Survey

✓ **SELECT:** CHILDREN: How many children have you ever had? Please count all that were born alive at any time (including any you had from a previous marriage).

↗ **CLICK ON:** Analyze Now
Examine the table for age. How does this fit with the discussion in the chapter?

↗ **CLICK ON:** Select New Data Set

↗ **CLICK ON:** The 50 States

✓ **SELECT:** INF. MORTL: Infant deaths per 1000 live births.

↗ **CLICK ON:** Analyze Now

Which two states have the highest infant mortality rates? Which two are lowest? What does that suggest to you about the causes of infant mortality?

↗ **CLICK ON:** Select New Data Set

↗ **CLICK ON:** Nations of the Globe

✓ **SELECT:** CONTRACEPT: Percentage of sexually active women in their child-bearing years who are using contraception.

↗ **CLICK ON:** Analyze Now
Is there an area of the world where few women use contraception? What does the text have to say about why they don't in this area?

USING INFOTRAC COLLEGE EDITION

GO TO THE INTERNET AND TYPE:
www.InfoTrac-college.com

↗ **CLICK ON:** Register New Account
You will be asked to enter your Access Code and to create and enter a User Name and Password. Having done so,

↗ **CLICK ON:** InfoTrac College Edition or Log On.

These search topics will lead you to extended discussions of key topics covered in the chapter. You might be especially interested in treatments of world population growth.

SEARCH TERMS:
Census
Fertility
Fertility Rate

In addition, I have selected a specific article that will usefully supplement the chapter: Demographics: The Population Surprise. You can find it by searching for this title. You may read the article on the screen or print it using the usual print commands. If you also go to www.socstark10.com and click on the InfoTrac College Edition icon, you can read an explanation of why I selected this article and find several questions that will help you connect the article to material in this chapter.

CHAPTER

19

These proud folks in Leavenworth, Kansas, at the turn of the twentieth century had no idea how the automobile was going to change the shape, density, and character of cities. They only knew how much fun it was to ride around waving at friends and squeezing the horn to startle horses pulling wagons and buggies. Nevertheless, they were as much involved in changing the shape and density of American cities as they were in having a good time. In this chapter we shall see why and how the automobile changed cities.

Urbanization

UNTIL RELATIVELY RECENTLY, the vast majority of human beings lived and died without ever seeing a city. The first city was probably founded no more than 5,500 years ago. But even 200 years ago, only a few people could live in cities; nearly everyone lived on farms or in tiny rural villages. It was not until the twentieth century that Great Britain became the first **urban society** in history—a society in which the majority of people live in cities and do not farm for a living.

Britain was only the beginning. Soon many other industrial nations became urban societies. The process of *urbanization*—the migration of people from the countryside to the city—was the result of modernization, which has rapidly transformed *how* people live and *where* they live. In 1900 fewer than 40 percent of Americans and Canadians lived in urban areas. Today, 79 percent of Americans and 78 percent of Canadians are urban residents, and only about 1 percent live on farms (the remainder live in small towns and villages).

Large cities were impossible until agriculture became industrialized. Recall from Chapter 10 that even in advanced agrarian societies, it took about ninety-five people on farms to feed five people in cities. That kept cities very small. Until modern times, cities were inhabited mainly by the ruling elite and the servants, laborers, craftsmen, and professionals who served them. Cities survived by taxing farmers and were limited in size by the amount of surplus food the rural population produced and by the ability to move this surplus from farm to city.

Over the past two centuries, the Industrial Revolution has shattered this balance between the city and the country. Modernization drew people to the cities and freed them to come by

CHAPTER OUTLINE

PREINDUSTRIAL CITIES
Limits on City Style
Box 19-1: Causes of Death in the City
of London, 1632
Why Live in Such Cities?

INDUSTRIALIZATION AND URBANIZATION
The Agricultural Revolution
Specialization and Urban Growth

METROPOLIS
The Fixed-Rail Metropolis
The Freeway Metropolis
Preferring a Decentralized Metropolis
Commuting

SUBURBS

URBAN NEIGHBORHOODS

A CLOSER VIEW
Park and Burgess: Ethnic Succession

A CLOSER VIEW
Guest and Weed: Economics and
Integration

SEGREGATION IN WORLD PERSPECTIVE

THEORIES OF URBAN IMPACT
Anomie Theories
Effects of Crowding
Macrostudies of Crowding

A CLOSER VIEW
Gove, Hughes, and Galle: Microstudies
of Crowding

CONCLUSION

REVIEW GLOSSARY

SUGGESTED READINGS

SOCIOLOGY ONLINE

making farmers incredibly productive. Today, instead of our needing ninety-five farmers to feed five city people, one American farmer is able to feed more than a hundred nonfarmers.

This chapter examines the urbanization of industrial societies. It explores in detail why and how the shift to urban life occurred. We shall also examine the structure of cities: why and how cities have grown and changed. Finally, we shall consider the impact of urban living on people. Have big cities worsened the quality of life, destroyed the intimacy of social relations, and undermined the health and sanity of human beings?

Before we turn to these questions, we need a basis for comparison. Let us therefore go back into history and examine what life was like in the famous cities of preindustrial times. What was it like in ancient Athens and Rome? What was it like in London and Paris when they had only 40,000–50,000 residents and before they had factories or freeways, subways or suburbs? In the movies these are usually depicted as wonderful places—cities without air pollution, traffic jams, or crowded sidewalks, cities where everyone knew their neighbors and very little crime occurred.

PREINDUSTRIAL CITIES

Until fairly recently, cities were small, filthy, disease-ridden, crowded, and disorderly, and they were dark and very dangerous at night. If that description clashes with your image of Athens during the Golden Age of Greek civilization, it is because popular history so often leaves out the mud, manure, and misery.

TABLE 19-1	Population of Major Cities in Preindustrial Europe	
CITY	**POPULATION**	**YEAR**
Amsterdam	7,476	1470
Berlin	6,000	1450
Brussels	19,058	1496
Geneva	4,204	1404
London	34,971	1377
Paris	59,200	1292
Pisa	9,940	1551
Rome	55,035	1526
Vienna	3,836	1391

Source: Russell (1958).

Foto Marburg/Art Resource, NY

Typically, preindustrial cities contained no more than 5,000–10,000 inhabitants (Table 19-1). Large national capitals usually had no more than 40,000 inhabitants and rarely more than 60,000 (Russell, 1958). Few preindustrial cities, such as ancient Rome, grew to populations as large as 500,000 and then only under special circumstances. Moreover, these cities rapidly shrank back to a much smaller size as slight changes in circumstance made it impossible to support them.

LIMITS ON CITY STYLE

A major reason cities remained small was poor transportation; food had to be brought to feed a city. With only animal and human power to convey it, however, food could not be transported very far. Therefore, cities were limited to the population that could be fed by farmers nearby. The few large cities of preindustrial times appeared only where food could be brought long distances by water transport. Ancient Rome, for example, was able to reach the size of

present-day Denver (and only briefly) because it controlled the whole Mediterranean area. Surplus food from this vast region was shipped by sea to feed the city's masses.

However, as the power of the empire weakened, Rome's population declined as the sources of food dwindled. By the ninth century, the sea power of Islam had driven nearly all European shipping from the Mediterranean, and the cities of southern Europe, including Rome, were virtually abandoned. In fact, Europe had practically no cities during the ninth and tenth centuries (Pirenne, 1925).

Disease also checked the size of cities. Even early in the twentieth century, cities had such high mortality rates that they required a large and constant influx of newcomers from the countryside just to maintain their populations. As recently as 1900, the death rate in English cities was 33 percent higher than that in rural areas (Davis, 1965). A major reason for the high mortality in cities was the high incidence of infectious diseases, which are spread by physical

Dutch artist Pieter Breughel demonstrated in this painting his awareness that the rise of the city depended on increased agricultural productivity. The city in the background could exist only because farmers like this plowman could produce enough surplus food to feed city people. When European peasants began to plow with horses rather than oxen, they could farm twice as much land, so cities got larger.

Herds of municipal pigs such as these were used as street-cleaning crews in many nineteenth-century cities and even in small towns. These "road hogs" did remove some garbage, but they left more.

State Historical Society of Wisconsin, Charles van Schaick Collection

contact or by breathing in germs emitted by coughs and sneezes. Disease spreads much more slowly among less dense rural populations (McNeill, 1976).

Disease in cities was also caused by filth, especially by the contamination of water and food. Kingsley Davis (1965) pointed out that even as late as the 1850s, London's water "came mainly from wells and rivers that drained cesspools, graveyards, and tidal areas. The city was regularly ravaged by cholera."

Sewage treatment was unknown in preindustrial cities. Even sewers were uncommon, and what sewers there were consisted of open trenches running along the streets into which sewage, including human waste, was poured from buckets and chamber pots. Indeed, sewage was often poured out of second-story windows without any warning to pedestrians below.

Garbage was not collected and was strewn everywhere. It was hailed as a major step forward when cities began to keep a municipal herd of pigs, which were guided through the streets at night to eat the garbage dumped during the day. Of course, the pigs did considerable recycling as they went. Still, major cities in the eastern United States depended on pigs for

their sanitation services until the end of the nineteenth century.

Today, we are greatly concerned about pollution, especially that produced by automobile exhausts and factories. But the car and the factory cannot match the horse and the home fireplace when it comes to pollution. For example, in 1900 horses deposited some 26 million pounds of manure and 10 million gallons of urine on the streets of New York City every week.

London's famous and deadly "fogs" of previous centuries were actually smogs caused by thousands of smoking home chimneys during atmospheric inversions, which trapped the polluted air. Indeed, the first known air-quality law, decreed in 1273 by England's King Edward I, forbade the use of a particularly smoky coal. The poet Shelley wrote early in the nineteenth century that "Hell is a city much like London, a populous and smokey city." In 1911 coal smoke during an atmospheric inversion killed more than a thousand people in London, and this incident led to the coining of the word *smog*.

Pedestrians in preindustrial cities often held perfume-soaked handkerchiefs over their noses because the streets stank so. They kept alert for garbage and sewage droppings from above.

Jacob A. Riis Collection, Museum of the City of New York

Nobody worried much about littering in cities where horse-drawn wagons were the primary means of transportation. The preindustrial city suffered from much worse problems of pollution than does the modern industrial city.

They wore high boots because they had to wade through muck, manure, and garbage. And the people themselves were dirty because they seldom bathed. Not surprisingly, they died at a rapid rate (see Box 19-1).

Population density also contributed to the unhealthiness of preindustrial cities. People were packed close together. As we saw in Chapter 13, whole families lived in one small room. The houses stood wall to wall, and few streets were more than ten to twelve feet wide.

Why was there such density when the population was so small? First of all, for most of its history, the city was also a fortress surrounded by massive walls for defense. Once the walls were up, the area of the city was fixed (at least until the walls were rebuilt), and if the population grew, people had to crowd ever closer together. Even cities without walls were confined. Travel was by foot or by hoof. Cities did not spread beyond the radius that could be covered by these slow means of transportation, and thus, the city limit was usually no more than three miles from the center (Blumenfeld, 1971).

Second, preindustrial cities could not expand upward. Not until the nineteenth century, when

structural steel and reinforced concrete were developed, could very tall structures be erected. Moreover, until elevators were invented, it was impractical to build very high. By expanding upward, people could have much greater living and working space in a building taking up no greater area at ground level. This could, of course, have meant that cities would become even more crowded at street level. They did not, however, because even modern high-rise cities have much more open space than did preindustrial cities, and as we shall see, newer cities have expanded primarily outward rather than upward.

Preindustrial cities were not only dirty, disease-ridden, and dense but also dark and dangerous. Today, we sometimes say people move to the city because they are attracted by the bright lights, and we joke about small towns where they "roll up the sidewalks by 9 P.M." The preindustrial city had no sidewalks to roll up and no electricity to light up the night. If lighted at all, homes were badly and expensively illuminated by candles and oil lamps. Until the introduction of gas lamps in the nineteenth century, streets were not lighted at all. Out in the dark, dangerous people lurked, waiting for

BOX 19-1 CAUSES OF DEATH IN THE CITY OF LONDON, 1632

IN 1662 JOHN GRAUNT, a London store-keeper, published the first set of vital statistics by compiling the weekly reports of deaths issued by the clerks in each London parish of the Church of England. Here we see the break-down for the year 1632—a year in which there were no major plagues. To get some idea of health risks in an era ignorant of germs and lacking antibiotics, notice that 5 percent (470) of the deaths in London in 1632 were from infected teeth. Graunt based his volume on all death reports from 1603 to 1624. During this twenty-year period, 229,250 people died in London, a number larger than the total population of the city at that time. Such a huge death rate was sustained in part by the extraordinary mortality rate for infants and children—more than half of whom failed to reach the age of six. But a large influx of newcomers each year kept the city from becoming uninhabited. Two years after Graunt's book was published, an outbreak of plague killed about a third of London's population; the year after that, the Great Fire destroyed 80 percent of the city.

The Diseases, and Casualties this year being 1632.

Abortive, and Stillborn	445	Grief	11
Affrighted	1	Jaundies	43
Aged	628	Jawfaln	8
Ague	43	Impostume	74
Apoplex, and Meagrom	17	Kil'd by several accidents	46
Bit with a mad dog	1	King's Evil	38
Bleeding	3	Lethargie	2
Bloody flux, scowring, and flux	348	Livergrown	87
Brused, Issues, sores, and ulcers	28	Lunatique	5
Burnt, and Scalded	5	Made away themselves	15
Burst, and Rupture	9	Measles	80
Cancer, and Wolf	10	Murthered	7
Canker	1	Over-laid, and starved at nurse	7
Childbed	171	Palsie	25
Chrisomes, and Infants	2268	Piles	1
Cold, and Cough	55	Plague	8
Colick, Stone, and Strangury	56	Planet	13
Consumption	1797	Pleurisie, and Spleen	36
Convulsion	241	Purples, and spotted Feaver	38
Cut of the Stone	5	Quinsie	7
Dead in the street, and starved	6	Rising of the Lights	98
Dropsie, and Swelling	267	Sciatica	1
Drowned	34	Scurvey, and Itch	9
Executed, and prest to death	18	Suddenly	62
Falling Sickness	7	Surfet	86
Fever	1108	Swine Pox	6
Fistula	13	Teeth	470
Flocks, and small Pox	531	Thrush, and Sore mouth	40
French Pox	12	Tissick	34
Gangrene	5	Tympany	13
Gout	4	Vomiting	1
		Worms	27

Christened { Males .... 4994 / Females ...... 4590 / In all .... 9584 } Buried { Males .... 4932 / Females ...... 4603 / In all .... 9535 } Whereof, of the Plague. 8

Increased in the Burials in the 122 Parishes, and at the Pesthouse this year 993
Decreased of the Plague in the 122 Parishes, and at the Pesthouse this year 266

---

victims. To venture forth at night in many of these cities was so dangerous that people did so only in groups accompanied by armed men bearing torches. Many people today fear to walk in cities at night. Still, it is much safer to do so now than it used to be.

WHY LIVE IN SUCH CITIES?

Knowing what preindustrial cities were like, one must ask why anyone willingly lived there and why a large number of newcomers were attracted to cities each year from rural areas.

One reason was economic incentive. Cities offered many people a chance to increase their incomes. For example, the development of an extensive division of labor, of occupational specialization, virtually required cities. Specialists must depend on one another for the many goods and services they do not provide for themselves. Such exchanges are hard to manage when people live far apart. Thus, skilled craftsmen, merchants, physicians, and the like gathered in cities. Indeed, cities are vital to trade and commerce, and most early cities developed at intersections of major trade routes.

This engraving by William Hogarth (1697–1764) depicts the many perils of the London streets at night, including the chamber pot being emptied from a second-story window, splattering two drunken Freemasons on their way home from a lodge meeting. The "wickedness" of cities has been a theme in literature and art through the centuries.

In addition to economic attractions, cities drew people because they offered the prospect of a more interesting and stimulating life. As Gideon Sjoberg (1965) noted, "New ideas and innovations flowed into [cities] quite naturally," as travelers along the trade routes brought ideas as well as goods from afar. Moreover, simply by concentrating specialists in an area, cities stimulated innovation not just in technology but also in religion, philosophy, science, and the arts. The density of cities encouraged public performances, from plays and concerts to organized sporting events.

Cities undoubtedly also enticed some to migrate from rural areas in pursuit of "vice." The earliest writing we have about cities includes complaints about rampant wickedness and sin, and through the centuries cities have maintained the reputation for condoning behavior that would not be tolerated in rural communities (Fischer, 1975). In part this may be because from the beginning cities have been relatively anonymous places. And preindustrial cities may have been even more anonymous, their size notwithstanding, than modern cities.

Consider that cities relied on large numbers of newcomers each year just to replace the population lost through mortality. As a result cities tended to abound in people who were recent arrivals and who had not known one another previously. Before modern identification systems, many people in cities were not even who they claimed to be; runaway sons and daughters of peasants could claim exalted social origins. The possibility of escaping one's past and starting anew must have drawn many to the cities. But this also meant that cities then were even less integrated by longstanding interpersonal attachments than modern cities.

In any event it was primarily adventuresome, single, young adults who constantly replenished city populations. E. A. Wrigley (1969) has computed that from 1650 to 1750 London needed 8,000 newcomers each year to maintain its population. The newcomers averaged twenty years of age, were unmarried, and came from farms. Most of these newcomers came from more than fifty miles away—at least a two-day trip at that time.

For all our complaints about modern cities, industrialization did not ruin city life. Preindustrial cities were horrid. Yet for many young people on farms, the prospect of heading off to one of these miserable cities seemed far superior to a life of dull toil. Then, as the Industrial Revolution began, the idea of going off to the city suddenly appealed not just to restless young people but also to whole families. Soon the countryside virtually emptied, as people flocked to town (Table 19-2).

INDUSTRIALIZATION AND URBANIZATION

Industrialization and urbanization are inseparable processes; neither could have occurred without the other. Industrialization made it possible for most people to live in cities. It also made it necessary: Industrialization requires the concentration of highly specialized workers.

To understand how industrialization both caused and depended on urbanization, let us first examine how the effects of the Industrial Revolution on agriculture made urbanization possible.

THE AGRICULTURAL REVOLUTION

Preindustrial farmers could support only a very small urban population, and that only by accepting a very low standard of living. Cities could exist only by coercing peasants to surrender their crops and livestock. Industrial technology changed all that. Suddenly, farm productivity soared to undreamed-of heights, and farmers became eager to sell their crops to the cities. Let us chart this change as it took place in the United States, since good records exist and since North American agriculture has become the most industrialized and productive in the world. An example of this productivity can be seen in Table 19-3, which shows the annual milk production by the average dairy cow in a number of nations.

The American dairy industry is highly industrialized. The average dairy farm has at least several hundred cows, and all milking is done by machines. The larger farms have milking machines installed on a huge carousel, and each revolution allows sufficient time for a cow to be milked. Consequently, it takes only one operator to place and release cows as each milking station comes by the gateway. In contrast, in many nations milking still is done by hand and takes about ten minutes per cow. But it isn't just the immense saving of labor that distinguishes industrialized from old-fashioned dairy farming. Cows in the industrial nations produce far more milk. Thus, the average American cow gives 7.5 times more milk than does the average cow in Brazil. Part of the difference is superior breeding and part is superior feeding. Nevertheless, the productivity differences are so great that milk is far cheaper in the United States than in Brazil.

To see more clearly how industrialization revolutionized agriculture in the United States,

TABLE 19-2	The Urban Migration
YEAR	PERCENT OF U.S. POPULATION LIVING ON FARMS
1820	72
1890	42
1911	34
1921	30
1931	25
1961	8
1970	5
1990	1.9
2000	1.1

Source: U.S. Census reports.

let's examine changes in the number of persons who could be fed by one farm worker (Table 19-4). In 1820, when almost no modern technology had yet appeared on the farms, a full-time farm worker could produce only enough food to feed 4.1 persons (including the farm worker). That left little surplus to send to the cities after farm families had fed themselves. But by 1900 the average American farm worker could feed 7.0 people. At the turn of the century, the average American farmer was feeding 5.3 Americans and 1.7 persons abroad. This was just the beginning, for by the middle of the twentieth century, farm productivity began to accelerate at an incredible pace. In 1960 the average American farmer was feeding 22.3 Americans and 3.5 persons living in other nations. By 1970 one farmer fed 39.9 Americans and 7.2 foreigners. In addition the farmer in 1970 was feeding each person more food than had the nineteenth-century farmer, despite the fact that the government, through various subsidy programs, was preventing modern American farmers from growing nearly as much as they could. Perhaps even more surprising is the fact that the modern farmer accomplishes these wonders by working far fewer hours and with much less physical exertion than did preindustrial farmers.

Back in 1800 American farmers worked 56 hours for every acre of wheat they raised (Table 19-5). In return they harvested an average of 15 bushels for each hard-worked acre. Today, American farmers farm their wheat fields while riding in the air-conditioned cabs of huge diesel tractors and self-propelled combines. It takes them an average of 2.8 hours a year to farm an acre of wheat. And they average 31.4 bushels from each acre—twice as much wheat in 5 percent as many hours of work.

Similarly, corn yields have more than tripled, while only 4 percent as much labor is required. Modern farmers give less than one-third the hours of attention to each of their milk cows as their grandparents did in 1910, but the cows now give three times as much milk. In 1910 it took ranchers 4.6 hours of labor to raise 100 pounds of beef. Today, ranchers work 1.3 hours to raise that much.

Many things have gone into this agricultural miracle: new machines, animal breeds, varieties of plants, weed sprays, fertilizers, crop rotation, drainage and irrigation systems—in short the application of science and engineering to farming. But the major effect was the huge reduction in labor. Only because fewer farmers could feed

TABLE 19-3	Milk Production per Cow
NATION	**KILOGRAMS OF MILK PER COW, PER YEAR**
United States	**7,000**
Sweden	6,530
Denmark	6,170
Canada	**5,380**
France	5,160
Germany	5,150
Great Britain	4,580
Ireland	4,320
Australia	4,240
Austria	3,900
Czech Republic	3,610
Argentina	3,270
Greece	2,890
Mexico	**1,650**
China	1,540
Peru	1,110
India	950
Brazil	930

Source: Selected by the author from U.S. Department of Agriculture, *Agricultural Statistics, 1993.*

TABLE 19-4	The Agricultural Revolution
YEAR	**NUMBER OF PERSONS SUPPLIED WITH FARM PRODUCTS BY ONE U.S. FARM WORKER**
1820	4.1
1900	7.0
1940	10.7
1950	15.5
1960	25.8
1970	47.1

Source: U.S. Census, *Historical Statistics of the United States* (1975).

greater numbers of people did it become possible for people to move to the cities and staff the great urban industries.

Recall from the previous chapter that the percentage of the population engaged in agriculture is one of the critical thresholds of modernization demographers use. Only when people are released from fieldwork can they live in cities and pursue industrial occupations (and as a result begin to reduce their fertility).

As recently as 1934, it took 225 hours of work per year to care for a flock of 100 laying hens. For every egg laid in the United States or Canada, a farmer put in more than a minute of labor. By 1970 farmers worked less than 15 seconds for each egg. Today, one worker can do what it would have taken more than 100 to do in 1934: care for as many as 100,000 hens in a huge "egg factory" like this one. Automatic conveyor belts keep the feeding troughs full and the cages clean, and the eggs roll out where they can easily be reached. All this farmer needs to do is drive his electric cart down the aisles and put the eggs into flats, ready for shipment.

United States Department of Agriculture

Modern self-propelled combines harvesting wheat near Pullman, Washington. With such a machine, a single farmer can harvest more land in several hours than huge threshing crews at the turn of the twentieth century could do in a week. According to estimates, there are enough of these giant combines at work in North America that together they could harvest an area the size of the state of Kansas in a single day.

United States Department of Agriculture

If city people today were asked to name the machines vital for the existence of cities, they would probably mention automobiles, computers, telephones, and the great machines used in heavy industry. In fact, modern cities depend on machines few people see: tractors, plows, cultivators, and harvesters. Without these our cities would be small, and most of us would spend our lives following horses across grain fields or riding them to round up herds.

SPECIALIZATION AND URBAN GROWTH

Industrialization requires urbanization because it depends on the coordinated activities of large numbers of specialized workers who must perform their tasks in a few central locations.

Industrialization depends on specialization—an elaborate division of labor—to simplify production. As an example, let us consider the industrialization of shoemaking. The preindustrial shoemaker was a skilled craftsman who spent several years learning the various steps in the process of making a pair of shoes. In a modern shoe factory, a worker need only learn to perform one simple task in this process to be productive. By the use of machines to perform some tasks and by the concentration of labor in the most time-consuming aspects of shoemaking, many more shoes can be made for the same amount of labor expended by traditional shoemakers.

One consequence of the use of machines is that shoes became much cheaper to consumers. A second consequence is that while the traditional shoemaker could locate his shop anywhere he could find customers, industrialized shoemakers must gather in one place where each can make his or her contribution to the complex manufacturing process.

Industrialization also depends on bringing together many highly trained specialists to achieve goals beyond the ability of single individuals. It takes many people with a variety of sophisticated skills to make computers or jet planes or to construct oil refineries. This, too, requires people to gather.

But industrialization also produces an elaborate division of labor, not just within organizations but also among them. Plant A gets parts from plant B and supplies its production to plant C. It is often efficient if these plants are close together. Indeed, in the early stages of industrialization, limited transportation made proximity vital.

If people must gather in large numbers to work, then they will also concentrate in the same area to live. This was especially true in days when most people walked to work. When people congregate in one place in large numbers and do not farm for a living, we call that place a city. Thus, urbanization and industrialization are inseparable. This is obvious in Table 19-6. In highly industrialized Hong Kong, everyone lives in an urban area, whereas in much less industrialized China, only 32 percent of the population is urban.

TABLE 19-5	Changes in Agricultural Productivity, 1800–1980		
PRODUCT	1800	1900	1980
WHEAT			
Hours of labor per acre (yearly)	56.0	15.0	2.8
Hours of labor per 100 bushels	373.0	108.0	9.0
Yield per acre (in bushels)	15.0	13.9	31.4
CORN			
Hours of labor per acre (yearly)	86.0	38.0	3.6
Hours of labor per 100 bushels	344.0	147.0	4.0
Yield per acre (in bushels)	25.0	25.9	95.2

PRODUCT	1910	1980
MILK		
Hours of labor per cow (yearly)	146.0	45.0
Hours of labor per 100 pounds of milk	3.8	0.4
Milk per cow (in pounds) yearly	3,842.0	11,000.0
BEEF		
Hours of labor per 100 pounds of meat	4.6	1.3
PORK		
Hours of labor per 100 pounds of meat	3.6	0.5
CHICKEN		
Hours of labor per 100 pounds of meat	9.5	2.9

Source: U.S. Census, 1975, 1981.

TABLE 19-6	Percentage of Population Living in Urban Places	
NATION	**PERCENT URBAN**	
Hong Kong	100	
Belgium	97	
Israel	91	
Great Britain	90	
Germany	88	
Australia	85	
Sweden	83	
South Korea	82	
United States	**79**	
Japan	79	
Canada	**77**	
Iraq	77	
France	76	
Mexico	**74**	
Russia	73	
Italy	67	
Poland	66	
Hungary	64	
Ireland	59	
Egypt	45	
Pakistan	37	
China	32	
India	28	
Afghanistan	22	
Ethiopia	18	
Uganda	14	
Rwanda	6	

Source: World Bank, *World Development Indicators, 2005.*

METROPOLIS

Suppose a young man left his father's farm and moved to Silo, North Dakota, population forty-three, where he got a job in a café. Is this an example of urbanization? Most people wouldn't think so. Even though the folks in Silo are not farmers, their way of life would not seem very urban to people used to large cities (in small North Dakota towns, people still do not lock their doors, and many leave their ignition keys in their cars). It seems we have some notion that to be urban a place must be larger than Silo. But how much larger?

There can be no "true" answer to that question; it is simply a matter of judgment. American demographers classify a locale as an **urban place** if it has a population of more than 2,500 people. The statement that nearly three of four North

Americans live in urban places means they live in communities larger than 2,500.

Obviously, many urban places are not cities, at least not in the sense that the term is normally used. The U.S. Census does not classify a community as a **city** unless it has at least 50,000 residents, but that standard can be very confusing. Many large communities are crisscrossed by political boundaries that separate them into many independent units; often, a major city is surrounded by dozens of smaller independent communities. Suppose one of these has fewer than 2,500 people. Should we classify its residents as a part of the rural population? What if most of the adults in this community commute to jobs in the heart of the major city? The political boundaries that divide large communities often have no relation to the actual social and economic boundaries. In fact, the notion of a city as a legal entity is faulty.

The word *city* once had a rather clear meaning, and a person either lived in a particular city or did not. But today, two strangers meeting on a plane may say they live in Chicago and Toronto, when they are not legal residents of either city. Instead, each lives in a **suburb,** a smaller community in the immediate vicinity of a city. Yet these travelers were truthful in the real, if not legal, sense. Cities do not simply stop at their legal boundaries but extend socially and economically into many adjacent communities. That's why we often speak of Greater Chicago or Greater Toronto—to identify this larger aspect of cities.

Back in 1910 the U.S. Census tried to find a more suitable definition of city than is provided by legal boundaries. Recognizing that many surrounding areas are functionally part of a central city, they substituted the term *metropolitan area* for city and began to lump suburbs with their central city as a single unit: the **metropolis.** This term was taken from classical Greek and is translated literally as the "mother city."

The first metropolitan areas had 200,000 or more residents, including the settled areas around a city. Then, beginning in 1940, to be a metropolitan area, the central city had to have 50,000 people, no matter how many more lived in surrounding areas.

Still, problems persisted. As Roderick McKenzie (1933) pointed out, "Only a part of the area that is economically and socially tributary to each of these central cities was included." What was needed was a way to identify the **sphere of influence** of a city: the area whose

Chicago Historical Society/Negative #ICWi-04191

inhabitants depend on the central city for jobs, recreation, newspapers, television, and a sense of common community.

Therefore, in 1950 the **Standard Metropolitan Statistical Area (SMSA)** was created. Around central cities with 50,000 or more people, counties are included in the SMSA if at least 75 percent of those working in the county do not hold agricultural jobs and if the county serves either as the residence or place of employment for at least 10,000 nonagricultural workers. Furthermore, at least 15 percent of the workers living in a county must commute to the central city or at least 25 percent of those working in the county must commute from the city.

Statistics Canada uses similar principles to define **Census Metropolitan Area (CMA).**

CMAs must have an urban core "or continuously built-up area" with a population of at least 100,000. Surrounding municipalities are included in the CMA if at least 40 percent of their labor force is employed within the boundaries of the urban core or if at least 25 percent of those employed in the municipality commute from residences within the urban core.

As both of these names suggest, sociologists no longer use the term *city* in their technical vocabulary but speak instead of a metropolis or metropolitan area. Thus, we agree with people from Oak Park, Illinois, when they tell strangers they are from Chicago. The industrialized city is no longer a tight, tidy, compact entity, contained within defensive walls. It sprawls hither and yon across the landscape—a reality better expressed by the term *metropolis*.

Fixed-rail transportation systems created cities with densely packed central cores. They also could create monumental traffic jams, as can be seen here at the intersection of Dearborn and Randolph Streets in Chicago in 1909. At least these trolleys were powered by electricity. In many cities coal-burning engines pulled trains to depots in the heart of the city, belching their contribution to the polluted air.

Library of Congress

New technology often requires new skills: These girls in a Brooklyn high school in about 1900 used time in gym class to develop their skills in boarding trolley cars. Their practice apparatus allows them to learn to grab and swing aboard the high side step.

But even the modern metropolis isn't completely formless. Cities exist as places to work and to live. Given these fundamental purposes, the shape and organization of cities have been determined by transportation.

Two basic forms of modern metropolises exist depending on the dominant forms of transportation when the metropolises grew: the **fixed-rail metropolis** and the **freeway metropolis.**

Much concern about cities and urban policies has been generated recently because many people think cities must be like those built before cars and trucks took over transportation from fixed-rail systems. These critics dislike the new form of the metropolitan area, especially the way the older form is changing to be more like the new. This dispute helps reveal basic aspects of urban sociology, so let's examine and compare these two basic urban forms.

THE FIXED-RAIL METROPOLIS

The preindustrial city was small and dense because people relied primarily on walking for transportation. Although industrialization

caused cities to expand, the continued reliance on foot transportation meant cities were cramped and workers were housed close to their factories. The development of rail transportation made it possible for cities to expand greatly in area.

First came horse-drawn trolleys running on rails. Then came electric- and steam-powered trolleys running on the same rails. These new modes of transportation were much faster than walking (or even than riding a horse) and were cheap enough that people could afford to live farther from work. Thus, cities began to expand outward—not evenly but only along the rail lines. People could travel only where the rails led and could live or work only out from the center of the city along the rail lines.

Riding on these fixed-rail mass transit systems was unpleasant. The cars were usually overcrowded and dirty. Still, it beat walking. Moreover, it enabled large numbers of city dwellers to escape apartments close to huge, noisy, dirty factories and move out where real estate was cheaper and life less hectic. As soon as railroads appeared in the 1840s, many wealthy people fled the cities and commuted from country estates or distant, luxurious communities.

As industrializing cities began to sprawl, they began to resemble spiders. From a dense center in which business and industry were concentrated, the metropolitan area expanded along narrow corridors, where the fixed-rail lines extended (Ward, 1971). Fixed-rail transportation requires many riders going from and to a small number of stops. Thus, rail lines were extended from the city center only as the population grew at the end of the line. People were not as cramped as before, but they still had to crowd together.

Moreover, fixed-rail transportation made the center of the city the focal point. Offices and stores were concentrated here because everyone could most easily travel to the center. Industry was also concentrated in the heart of the city, usually adjacent to the business section. Factories relied on rail transportation to bring in supplies and to carry out the finished goods to market. So, factory locations were also close to rail routes.

The metropolis that was built to suit fixed-rail transportation fits our image of the old industrial cities of the eastern parts of Canada and the United States and in Europe: dense cities where, in the very center, the streets are jammed with people going to shops by day and to restaurants, theaters, nightclubs, concert halls, and sports arenas by night. Whether the citizens called it the Great White Way, the Loop, or the Hub, this was the heart and soul of what everyone found exciting and sophisticated about the city.

Today, the central core of most such cities is in decay, abandoned by shoppers, by offices, by industry, and by nightlife. Billions have been spent to renovate and renew these central cores in hopes of luring department stores, offices, and nightclubs to return. Yet they do not recapture their old glory. And many newer cities never had such centers at all. Why?

THE FREEWAY METROPOLIS

Compared with rail lines, streets and roads are inexpensive to build and maintain. With the mass production of automobiles and trucks, the metropolis no longer had to resemble a spider. The empty areas between the rail lines became easy to reach. Moreover, people could settle in sparsely populated areas where no rail line could afford to run but where anyone with a car could easily go. The metropolis began to spread outward evenly, shaped more by geographical barriers (such as rivers, harbors, hills, and ravines) than by rail routes. And just as the car freed people to live where they liked, the truck freed industry to decentralize.

The center of a city had always been a constrictive location for industry and business. There was always a shortage of space for expansion and considerable congestion from the dense concentration of plants. These shortcomings were offset by the urgent need to be located on a fixed-rail line and to be able to send material from one local plant to another by slow horse-drawn wagons. But the truck ended the dependence of industry on rail transport. Today, more than 80 percent of all commercial transportation in North America is by truck.

Meanwhile, new technology forced industry to seek low-density locations in the suburbs. Assembly-line methods of manufacturing require long, low buildings and therefore considerable space. Machines for handling and stockpiling materials, such as forklifts, work best in one-story buildings. Thus, plants have shifted to outlying areas where land is plentiful and cheap. Indeed, with workers now commuting mainly by car (and 87% of American workers do so), business and industry often require parking lots that cover considerably

Library of Congress

more area than do their plants and office buildings.

And so the metropolis has become decentralized. Business has moved to outlying locations, as have many residents. Not surprisingly, so have many stores and shops; the shopping mall has replaced the old central core as the dominant retail locale.

Although decentralization has caused substantial changes in cities that grew large during the era of fixed-rail transportation, it has not caused similar upheavals in many western cities. These cities grew up after the automobile and the truck had already displaced the trolley and the train. They were never high-density cities revolving around a cramped but thriving central core. They were never shaped like spiders.

For decades literary easterners have scoffed at the decentralized cities of the West for not being "real" cities at all. Los Angeles, for example, has been described as a dozen towns in search of a city. Why in the world, they complain, don't westerners put their cities together so that one need not travel twenty miles down a

Here a fashionable New York woman boards a real trolley. Because these vehicles had no center aisle, people just slid into the seat from either side. At rush hour the outside of the car was lined with riders unable to find a seat.

The fixed-rail city required that large numbers of residents live in crowded, expensive housing. Whenever people have had the opportunity, most have abandoned such areas for life in the suburbs. Despite complaints that the suburbs are dull, the people who live there usually seem to be having too good a time to notice.

© Bas van Beck/Leo de Wys

freeway to get from the office to the theater or from one department store to another?

This charge is accurate enough. Los Angeles does not have a "downtown"; instead, it has at least eight downtowns. Indeed, as many people in decentralized cities have discovered, the time it takes to get somewhere matters more than the actual distance traveled. By avoiding the congestion of shopping in dense, central cores, millions of North Americans have demonstrated that they would rather zip along a freeway and find easy parking at a shopping center than take much longer to go a few blocks in heavy traffic. That brings us to the fundamental point in the dispute over what a real city is. Most people want cities to be decentralized.

PREFERRING A DECENTRALIZED METROPOLIS

Several signs show that people find life better in sprawling, decentralized cities than in the older, centralized ones. First of all, when the chance came to move out of the central city and to the outskirts, millions did so as fast as they could. Although the dense central core offers an exciting urban life, it also causes vast numbers of

people to live in cramped, unattractive housing and forces their children to play in the streets.

Second, people running large industrial and business firms have joined in the move to less dense areas, taking their plants and offices with them. Surely, the rich can maintain luxury and privacy even in dense central cores, but they were among the first to leave. As early as 1848, approximately 20 percent of the leading businessmen of Boston were commuting to their downtown offices by train from the suburbs (Ward, 1971).

Third, large numbers of people have been migrating from the old, high-density cities to the new, decentralized metropolitan areas of the South and West.

Fourth, people avoid public transportation whenever possible. No public mass transit system in the nation can attract enough riders at a high enough fare to break even. People seem willing to commute long distances, but they also seem to want the freedom of driving their own vehicles. Hence, they prefer a metropolis constructed with auto transportation in mind — a freeway city.

But we need not rely on these indirect signs. The Gallup Poll (1972) asked Americans, "If you could live anywhere you wanted to, would you prefer a city, a suburban area, a small town, or a farm?" Only 13 percent chose the city. In contrast, a third picked the suburbs, another third said a small town, and one-fifth said they would like to live on a farm. Indeed, of people who were living in cities, only one person in five preferred living there.

This might suggest that Americans long to undo urbanization and return to the good old days of living in small towns and on farms. But that isn't true. Gallup asked those who said they would like to live on a farm or in a small town how far from a major urban area they would like to be. Three of four said no more than thirty miles (Fuguitt and Zuiches, 1973). Thirty miles is little more than a half-hour drive from the city—hardly a retreat to the sticks. Indeed, as

Claude Fischer (1976) remarked about these findings, anyplace "within thirty miles of a large city is essentially a suburb. That seems to be what most Americans want."

COMMUTING

Of course, if people live in suburbs, many of them will need to travel some distance to their jobs. Hence, according to the 2000 Census, it takes the average American 25.5 minutes to go from home to work: 76 percent commute alone in their private vehicle, and another 12 percent carpool; 5 percent take public transportation to work, only a few more than walk (3%).

The amazing thing about commute times is that they have remained about the same from one census to the next (see Table 19-7), and they

TABLE 19-7	Average Travel Time from Home to Work in Minutes		
AVERAGE COMMUTE TIME IN MINUTES		AVERAGE COMMUTE TIME IN MINUTES	
United States	25.5	Michigan	24.1
Selected Cities		Minnesota	21.9
Atlanta	31.1	Mississippi	24.6
Chicago	31.2	Missouri	23.8
Los Angeles	28.7	Montana	17.7
New York City	35.1	Nebraska	18.0
Washington, D.C.	29.7	Nevada	23.3
States		New Hampshire	25.3
Alabama	24.8	New Jersey	30.0
Alaska	19.6	New Mexico	21.9
Arizona	24.9	New York	31.7
Arkansas	21.9	North Carolina	24.0
California	27.7	North Dakota	15.8
Colorado	24.3	Ohio	22.9
Connecticut	24.4	Oklahoma	21.7
Delaware	24.0	Oregon	22.2
Florida	26.2	Pennsylvania	25.2
Georgia	27.7	Rhode Island	22.5
Hawaii	23.8	South Carolina	24.3
Idaho	20.0	South Dakota	16.6
Illinois	28.0	Tennessee	24.5
Indiana	22.6	Texas	25.4
Iowa	18.5	Utah	21.3
Kansas	19.0	Vermont	21.6
Kentucky	23.5	Virginia	27.0
Louisiana	25.7	Washington	25.5
Maine	22.7	West Virginia	26.2
Maryland	31.2	Wisconsin	20.8
Massachusetts	27.0	Wyoming	17.8

Source: U.S. Census, 2000.

are much the same across states and major cities. Only in the most rural states does the average person take less than twenty minutes to get to work, and only in New York, New Jersey, and Maryland do people take even slightly more than thirty minutes. This similarity is not nearly as unusual as it might seem. It holds across modern nations, and moreover, it has been the typical limit on commute time over the many centuries of civilization, regardless of the method of transportation! Indeed, Yacov Zahavi has formulated a *law of constant travel time*, which proposes that people will accept only about a thirty-minute commute, or an hour a day, beyond which they will move (Zahavi and Ryan, 1980; Zahavi and Talvitie, 1980; Ausubel and Marchetti, 2001). Careful analysis of historical records, some going back 5,000 years, seems to support this "law." Back when people either walked, rode horses, or traveled in horse-drawn conveyances, very few commuted for longer than about thirty minutes each way. Zahavi explained this constant by pointing out that the day always consists of twenty-four hours; thus, time is equally limited for people everywhere and in all times.

SUBURBS

As noted earlier, a suburb is a smaller community in the immediate vicinity of a city, being part of a metropolitan area. As the name indicates, while these are not rural places or villages, they are somewhat less than urban, hence "sub" urban. Although suburbs have existed for a long time (even in Greek and Roman times, many rich people lived in separate communities several miles beyond the city gates), the "suburban era" truly began following World War II. Spurred by the baby boom and by very low-interest, government guaranteed mortgages, millions of Americans abandoned the cities and moved to new suburban communities. By the late in 1990s, the Unites States had become a "suburban nation" in that more than half of all Americans now live in a suburb.

From early on, social scientists and other social commentators condemned suburbs and ridiculed the lifestyle they believed existed there (Jackson, 1985; Fishman, 1987; Gains, 1991; Kunstler, 1993). They scorned them as merely "bedroom communities," in that residents worked elsewhere, and as colorless, boring, and uncultured. In many books, scores of movies, and a succession of "studies," the "burbs" were (and are) denounced as dreadful, inhuman places—as a "crabgrass frontier," a "teenage wasteland," and a "conspiracy" against women. Most recently, suburbs are under attack as urban sprawl.

Eventually, however, even urban intellectuals had to ask themselves why people remain so eager to move to the suburbs (Peterson, 1999). This led to the discovery that attacks on suburbia rested on a faulty comparison (Kelly, 1993; Baxandall and Ewen, 2000). The critics had contrasted life in suburbia with the life of wealthy sophisticates living in the city, finding that suburbia was woefully lacking in museums, theaters, nightclubs, art galleries, concerts, and the like. A proper assessment of suburban life does not compare it to that of rich urbanites but examines the lives suburbanites would live if they remained within the city—as renters of cramped apartments in rundown, crime-ridden neighborhoods, having no place for children to play. The simple fact is that most people prefer suburbs because they provide them with a more satisfying lifestyle, just as most people prefer to shop in malls that are easy to reach and which provide ample parking, rather than shop in the commercial core of the central city. Of course, not everyone who would like to move to the suburbs is able to do so.

URBAN NEIGHBORHOODS

The ethnic or racial neighborhood has always been a feature of the city. In the cities of the Roman Empire, various sections of cities were named according to the ethnicity of persons living there—because members of a minority group lived in the same neighborhood. In addition to racial and ethnic divisions, class and status have always differentiated city neighborhoods, with neighborhoods ranging from very wealthy to very poor.

Of course, cities are constantly changing. Thus, even in ancient times, what was a Jewish or a Greek neighborhood in one generation might be something else a generation or two later. And rich neighborhoods sometimes turn into slums, and vice versa.

In North America the rapid growth of cities and the huge waves of immigration produced complex neighborhood patterns. Our cities have abounded with neighborhoods occupied by a single racial or ethnic group: Many cities have (or have had) a "little Italy," a "Germantown," or a "Chinatown." And all cities have their wealthy areas and their slums.

Observers of the nineteenth-century American city noticed the remarkable turnover in the ethnic identity of neighborhoods. A slum area occupied by the Irish or Germans would suddenly change as these groups moved out and newer immigrants such as the Italians or the Jews moved in. In time these groups also were replaced, often by African Americans, Asians, or Hispanics.

Minority ethnic and racial groups concentrate in particular neighborhoods for several reasons. Members of a group are lured to the same neighborhoods in which their relatives and friends live—a place where their native language, their customs, their food, and their religion predominate. They are also pulled toward these neighborhoods because they can afford to live there. Higher housing costs and discrimination keep members of ethnic groups out of these other neighborhoods.

Recently, there has been a substantial decline in discrimination, in part because it is illegal in Canada and the United States to refuse to sell property or to rent to people on the basis of race or ethnicity. However, long before these legal measures took force, many groups that once were the targets of discrimination in housing escaped their segregated neighborhoods and became integrated. In the 1920s two of America's most famous early sociologists proposed an explanation of the process.

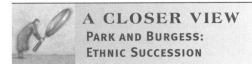

A CLOSER VIEW
PARK AND BURGESS: ETHNIC SUCCESSION

Robert E. Park (1864–1944) and Ernest Burgess (1886–1966) proposed that ethnic and racial segregation in cities was based primarily on economic and status differences. In their famous book *The City* (Park, Burgess, and McKenzie, 1925), they proposed a **theory of ethnic succession** that closely resembles the economic explanation of prejudice and discrimination outlined in Chapter 11. Park and Burgess argued that new immigrant groups huddle together in segregated neighborhoods upon arriving in America. However, as these groups begin to rise in the stratification system, these changes "tend to be registered in changes of location." That is, as new groups succeed in America, they move out of ethnic neighborhoods. This occurs, first, because they can afford to live in better neighborhoods; second, because they no longer are so tied to their traditional culture; and third,

because they have shed the stigma of low status: They are no longer regarded as undesirable neighbors.

Park and Burgess also accounted for the process of *succession*, whereby slum neighborhoods are successively occupied by the lowest-status groups of the time. For a long time, the Park and Burgess viewpoint dominated sociology. And a number of later empirical studies seemed to confirm it. Stanley Lieberson (1961, 1963) found that from 1910 through 1950, older European ethnic groups (such as the Germans) and newer ones (such as the Italians) increasingly lived together in the same neighborhoods. Karl and Alma Taeuber (1964) found similar trends.

However, in the wake of the racial confrontations of the 1960s, other sociologists suggested that the Park and Burgess model was inadequate, that it glossed over continuing ethnic inequalities, and that it did not apply to racial, as opposed to ethnic, neighborhood integration. Working with data for Toronto, Gordon Darroch and Wilfred Marston (1971) reported that individuals belonging to different ethnic groups still tended not to live in the same neighborhoods in Toronto, even when they were of equal status. Nathan Kantrowitz (1973) drew similar conclusions from New York City data, suggesting that even when ethnic groups escape the city, they tend to form segregated suburbs.

Once again, prejudice was judged the primary barrier to integration. Yet this did not square with the deemphasis on prejudice as being a major factor in race and ethnic relations, as we saw in Chapter 11. This was a serious discrepancy, and something was clearly wrong with one of these positions.

A CLOSER VIEW
GUEST AND WEED: ECONOMICS AND INTEGRATION

In 1976 Avery M. Guest and his student James A. Weed attempted to resolve this contradiction. First, they carefully reexamined what Park and Burgess had actually argued. They discovered that Park and Burgess were discussing group, not individual, upward mobility. That is, Park and Burgess did not suggest that as soon as a few members of a low-status economic group manage to become wealthy, they are welcomed in the best neighborhoods. Rather, they argued that when a particular group, such as the Italians, achieves economic parity with the majority, at that point their ethnicity will not be a barrier in choosing where to live.

This is a critical distinction. Darroch and Marston had shown only that higher-income members of a low-status group do not live in integrated neighborhoods, not what would happen when groups achieve status equality. But Guest and Weed, following Park and Burgess, argued that as long as a group's overall status is low, it will reflect on all members, including the more successful ones. As an example, they suggested that, in evaluating a neighborhood in which many Poles are living, a person of German descent will not ask, "Do *these* Poles earn as much as I do?" but "Are Poles as a *group* similar in status to Germans as a group?" If the answer is yes, then the German will move into the neighborhood. If the answer is no, the German will choose to live elsewhere. Group inequality, not individual comparisons, lies at the heart of prejudice and discrimination, Guest and Weed argued.

To test this view, Guest and Weed assembled data for Cleveland, Boston, and Seattle SMSAs. They chose these three cities because of their different histories and ethnic makeup, a choice that proved to be wise. Their first step was to determine the extent to which various racial and ethnic groups live in integrated neighborhoods.

The degree of segregation or integration of a neighborhood is measured by an **index of dissimilarity** (Taeuber and Taeuber, 1969). This index contrasts the racial and ethnic makeup of a neighborhood with the racial and ethnic makeup of the whole metropolitan area. If the racial and ethnic composition of a neighborhood is the same as that of the metropolitan area, then the neighborhood scores zero on the index—it is fully integrated. On the other hand, if a single racial or ethnic group lives in a neighborhood, while other racial and ethnic groups live in the metropolitan area, the neighborhood scores 100—it is wholly segregated. To measure the degree to which a particular racial or ethnic group is integrated or segregated in a city, sociologists compute the average dissimilarity scores for the neighborhoods in which this group lives. Guest and Weed's findings substantially supported the Park and Burgess theory and the status inequality approach to prejudice and discrimination.

Group neighborhoods differed as would be predicted from current status differences among

them. Persons of British origin, the people often known as WASPs, live in the most integrated neighborhoods. Others of northern and western European descent (such as Swedes, Germans, and Irish) are virtually as integrated. Groups arriving later from eastern and southern Europe (Czechs, Poles, Hungarians, and Italians) live in only slightly less integrated neighborhoods. Persons of Mexican descent live in quite unintegrated neighborhoods in Cleveland and Boston but in neighborhoods as integrated as those of eastern and southern Europeans in Seattle. African Americans and Puerto Ricans live in the least integrated neighborhoods, but again the differences are smaller in Seattle. Finally, and importantly, Asians live in neighborhoods as integrated as those of most people of European origin.

By comparing 1960 and 1970 data, Guest and Weed found that all neighborhoods had generally become more integrated. Asians had made the greatest gains, but African American neighborhoods had also become less solidly black.

These data appear to support the group mobility interpretation of Park and Burgess. Asians have recently made striking status gains, and they have become quite well integrated. Earlier status gains of eastern and southern European ethnic groups also appear in the breakup of the once solidly Italian, Polish, Hungarian, and Czech neighborhoods. Between 1960 and 1970, African Americans made substantial status gains, and their neighborhoods began to reflect these gains.

However, Guest and Weed were able to test the Park and Burgess model more rigorously. Using sophisticated statistical regression techniques, they found that when the effects of income differences among racial and ethnic groups are removed, relatively little neighborhood segregation based on race or ethnicity remains. That is, status inequality between groups seems to be the primary neighborhood barrier. As status inequalities disappear, so do racial and ethnic neighborhoods. At that point neighborhoods are identified only on the basis of class.

Guest and Weed also found that neighborhood patterns tend to persist. Hence, western cities such as Seattle are more integrated no matter which ethnic or racial group is examined. They suggest that one reason is because western cities are newer: Ethnic enclaves never got established as they had in cities that were settled long ago. Consider the following example. All cities are divided into census tracts. The Census Bureau tries to keep tracts about the same size

TABLE 19-8	The Ten Most and Least Segregated U.S. Metropolitan Areas				
METROPOLITAN AREA 10 *Most Segregated*	**INDEX OF DISSIMILARITY**		**METROPOLITAN AREA** 10 *Least Segregated*	**INDEX OF DISSIMILARITY**	
	2000	1990		2000	1990
Detroit	85.2	87.2	Albuquerque	28.0	36.3
Gary, Ind.	82.8	88.5	Orange County, Calif.	31.5	37.4
Milwaukee	82.4	82.7	Tucson, Ariz.	35.1	39.7
Chicago	79.2	82.9	San Jose, Calif.	35.4	39.7
Cleveland	77.6	82.4	Salt Lake City	37.0	50.4
Buffalo	77.4	80.0	McAllen, Tex.	39.5	49.5
Newark	77.1	80.4	Phoenix	39.5	48.0
New York City	74.9	76.0	Las Vegas	39.6	51.1
Cincinnati	74.5	76.7	Honolulu	40.4	44.3
St. Louis	73.9	77.3	Riverside, Calif.	42.0	41.6

Sources: U.S. Census, 2000; Trowbridge, 2002.

everywhere—around 5,000 residents. The 1980 Census found that there was not a tract in Seattle without African American residents. Although Seattle does have an African American neighborhood, only four tracts in 1980 were more than 50 percent African American and none was more than 85 percent African American.

Now consider Chicago. Chicago's community areas are about eight to ten times larger than census tracts. Nevertheless, the 1980 Census found that twenty-one of them had no African American residents and eight more had only 1 percent. Twenty community areas in Chicago were more than 90 percent African American, and eight were 99 percent African American. In Chicago most neighborhoods were either all African American or all white. Indeed, Chicago in 1980 was the most racially segregated city in America.

Guest and Weed suggested that cities like Chicago face the need to break down patterns of ethnic and racial neighborhoods established long ago, whereas cities like Seattle simply never had these patterns to contend with. In any event Guest and Weed showed that racial and ethnic succession, from ghetto to integration, continues in accordance with current theories of intergroup conflict.

Since the early 1980s, many demographers have used Taeuber's dissimilarity index to study segregation in American cities based on the U.S. Census of 1990 and 2000. Two very significant aspects turned up each time. First, in full agreement with the study by Guest and Weed, the newly built cities of the West and South are the least segregated—these cities have had no longstanding "black neighborhoods" to dismantle. This is easily seen in Table 19-8. All ten of the most segregated American cities are "old" industrial cities. All ten of the least segregated cities are in the West and Southwest. The second important aspect is that things are getting better, even if more slowly than might be hoped. In nineteen of these twenty cities, the dissimilarity score was lower in 2000 than in 1990. Salt Lake City led in this regard; its index score dropped form 50.4 to 37.0, while only Riverside, California, saw a slight increase.

SEGREGATION IN WORLD PERSPECTIVE

It is easy to suppose that segregation is a peculiarly American problem, but the truth is that segregation exists everywhere in the world where there is significant racial and ethnic diversity. For example, English cities have long had Irish and Welsh "ghettos" and now are developing very substantial segregated neighborhoods of non-White and Muslim immigrants. Similarly, segregated neighborhoods have appeared in most nations of Western Europe. Nor is segregation limited to societies with white majorities. In both India and Japan, for example, strict segregation of certain minorities enjoys official approval. Prejudice knows no racial, religious, or ethnic boundaries.

Recall from Table 11-2 that compared with people in many other nations, especially those in less developed nations, Americans and Canadians are far more willing to have people of different races and religions for neighbors. But nearly three-fourths of the people in Bangladesh and two-thirds of Egyptians would not want people of another race for neighbors.

To conclude this chapter, we must return to a basic question about city life raised in Chapter 1: Has urbanization harmed social relations and dulled our sensibilities?

THEORIES OF URBAN IMPACT

By the middle of the nineteenth century, educated people in Europe and North America recognized that rapid urbanization was under way—and they didn't like it. Cities were still unhealthy, squalid places. In 1841 the average life expectancy of men in London was five years less than in the rest of England. In the United States, the life expectancy of urban dwellers did not equal that of rural folk until after 1940 (Simon, 1981).

As cities grew, it seemed that their problems could only get worse. Many asked at what price people were being uprooted from their intimate, healthy, traditional lives in rural areas and crowded into impersonal, unhealthy, chaotic cities. Most people who raised this question were sure that the costs of urbanization would be devastating, even though they could see no way to stop the great migration to the cities.

One of the first social scientists to write in detail about the dangers of urbanization was Ferdinand Tönnies. In the mid-1880s he introduced the concepts of *Gemeinschaft* (community) and *Gesellschaft* (society or association) to capture the different qualities of life in preindustrial and industrial societies.

Gemeinschaft identifies the qualities of life Tönnies thought were being lost because of urbanization. It describes small, cohesive communities such as the farming village. People know one another well and are connected by bonds of friendship, kinship, and daily interaction. In such places people agree on the norms, and few people

fail to conform. In fact, such communities serve as primary groups for most of their members.

Gesellschaft is the exact opposite. People tend to be strangers, and they are united only by self-interest, not by any sense of common purpose or identity. There is little agreement about norms and much deviance. Human relationships are fleeting and manipulative rather than warm and intimate.

If we think of people as marbles, then in the *Gemeinschaft* the marbles are glued together into a solid piece, whereas in the *Gesellschaft* the marbles are constantly being tossed about in a revolving drum.

Following Tönnies, a long line of social scientists characterized urban life in such terms. Émile Durkheim wrote in 1897 that a primary consequence of urbanization is the breakdown of order: Urbanites live in a situation in which norms lack definition and force, a state he called **anomie.** In losing their attachments to others, people lose their primary source of moral judgment. As an early control theorist (see Chapter 7), Durkheim believed that conformity to the norms is caused by attachments, and thus urbanization, by destroying attachments, destroys the normative order. He therefore described the modern urbanite as adrift in a sea of normlessness (or anomie). Durkheim attempted to show that cities have much higher deviance rates than rural areas have.

ANOMIE THEORIES

Early American sociologists found Durkheim's theory of urban anomie very compelling. Perhaps this was partly because the great majority of them were raised on farms and in small towns, and Durkheim's position agreed with their own personal reactions to city life. In any event sociologists have long believed that cities are inimical to human relations and thus to the very basis of social life.

Louis Wirth (1938) made a major contribution to this position in a paper called "Urbanism as a Way of Life." In it he argued that city life forces the individual to become withdrawn from others. This occurs, first of all, because city people so often interact with complete strangers. Such interactions are necessarily impersonal, and this impersonality becomes a habit. Second, cities threaten to overload the people's senses, forcing them to shut out and ignore most of what is going on around them. We walk down streets filled with strangers, traffic flows past, store windows beckon, signs seek our attention, sirens and car horns blare, cell phones ring: The sights and sounds of the city would overwhelm us if we did not set up sensory buffers to filter out most of these stimuli. But in so doing, we become insensitive and unresponsive.

In 1939 two of Wirth's students published a study that convinced most American sociologists that the stress of city life caused mental illness. Robert E. L. Faris and Warren Dunham's famous book, *Mental Disorders in Urban Areas*, was based on Chicago's community areas. They stated their thesis in the book's first paragraph:

A relationship between urbanism and social disorganization has long been recognized and demonstrated. Crude rural-urban comparisons of rates of dependency, crime, divorce and desertion, suicide, and vice have shown these problems to be more severe in cities, especially the large rapidly expanding industrial cities. But as the study of urban sociology advanced, even more striking comparisons between different sections of a city were discovered. Some parts were found to be as stable and as peaceful as any well-organized rural neighborhood while other parts were found to be in extreme stages of social disorganization.

And what kinds of neighborhoods are disorganized? According to Faris and Dunham, these are the poor areas, which are afflicted with high rates of population turnover and hence low rates of attachments. As they phrased it, "Any factor which interferes with social contacts with other persons produces isolation." After a long and careful examination of maps and data, Faris and Dunham concluded that when humans are exposed to high rates of social disorganization, and especially as disorganization causes them to be isolated, they become mentally ill.

Table 19-9 is based on the data Faris and Dunham provided in their book and on census data published by Ernest Burgess and Charles Newcomb (1933). Among several measures of neighborhood affluence, Faris and Dunham used radio ownership—in 1930 just over 60 percent of Chicago households had a radio. Faris and Dunham did not use correlations to analyze their data—few sociologists did in those days—but the correlations shown in the table strongly support their conclusion that community areas lacking in radios were high in their rates of mental hospital

admissions. Life in the poor community areas of Chicago seemed too much for many residents to bear. Here, too, were the high rates of advanced syphilis (then an incurable disease that often destroyed mental capacities) and the high rates of alcohol abuse. During the next several decades, Faris and Dunham's findings were confirmed in other cities (Hare, 1956; Srole et al., 1962). Moreover, studies that found poor people to have higher rates of mental illness (Hollingshead and Redlich, 1958) were thought to support these ecological findings.

Nevertheless, from the very start, some social scientists rejected the claim that urban ecology causes mental illness. They argued instead that the mentally ill drift into the poorest areas of the cities after they become ill and that the mentally ill also drift downward in the stratification system because of their inability to get and hold jobs. This is called the "social drift" explanation of why mental illness is concentrated in slum neighborhoods and among persons of low social class (Dohrenwend, 1966; Turner and Wagenfeld, 1967).

If we look once more at Table 19-9, we can see evidence in support of social drift. Because advanced syphilis takes many years to develop, and because these are neighborhoods marked by population instability, it seems very unlikely that these victims contracted syphilis in these neighborhoods. A more likely scenario: As the disease began to take its toll on the mental abilities of these victims, they drifted into the slums. Similarly, people usually have been problem drinkers for some time before they are committed to hospitals. It seems reasonable to suggest that they, too, drifted in rather than to argue that

TABLE 19-9	Poverty and Social Pathology, Chicago, 1922–1931	
		CORRELATIONS WITH PERCENT OF HOUSEHOLDS OWNING RADIOS, 1930
1930–1931: Mental hospital admissions per 100,000		−.52
1922–1931: Average annual hospital admissions for advanced syphilis per 100,000		−.59
1922–1931: Average annual hospital admissions for alcoholism per 100,000		−.78

Sources: Calculated by the author from Faris and Dunham (1939) and Burgess and Newcomb (1933).

they became heavy drinkers only after they were exposed to the pressures of life in these neighborhoods. And if the alcoholics and syphilitics drifted into these neighborhoods, why not the mentally ill?

Indeed, the most striking characteristic of the most "socially disorganized" urban neighborhoods is that they are overrun by unattached male drifters. For example, there were 395 men per 100 female residents of the Chicago area with by far the highest rate of mental hospital admissions. Conversely, the community area with the lowest mental hospital admission rate had only 83 men per 100 female residents. Table 19-10 shows that the ratio of males to females is as good as radio ownership in identifying the neighborhoods with high rates of mental illness, advanced syphilis, and alcoholism. As the studies piled up, the social drift explanation dealt a fatal blow to the Faris and Dunham thesis.

Meanwhile, in the aftermath of World War II, Wirth's assertions about city life were incorporated into mass society theories to help explain why people responded to mass movements such as Nazism and communism. Mass society theorists argued that as isolated, unattached individuals, city people were easily attracted to mass movements, especially those that promised to restore order and provide followers with a clear sense of belonging.

Mass society theories were very popular for a few years, especially among psychologists. But they failed to survive sociological inspection. Mass society theorists were correct in arguing that a lack of attachments results in deviance and anomie (see Chapter 7). Where they went wrong was in the claim that anomie was characteristic of urbanites. Research found that people

typically maintain close attachments even in the largest cities (see Table 1-6). Human relations turn out to be much more durable than the early sociologists had supposed. Indeed, as we saw in Chapter 3 and will examine again in Chapter 21, people do not join social movements because they are loose marbles bouncing randomly in normless cities, but because they are attached to persons who already belong to the movement.

This is not to say there are no lonely, isolated people in the cities. There are, and many display the symptoms predicted by Durkheim and others: alcoholism, suicide, and criminal behavior. But most urbanites do not lack attachments, and the city does not have the destructive effects earlier sociologists believed it did.

EFFECTS OF CROWDING

What about the problem of "psychic overload"? This supposed effect of urban living was first identified by Louis Wirth, but it gained widespread attention in the 1960s as a potential hazard caused by population growth (Calhoun, 1962; Hall, 1966). Many critics of modern urban life have proposed that the population density of cities causes serious physical and mental pathologies. Noting that when rats are crowded into cages they behave extremely abnormally, many sociologists have issued doomsday predictions. For example, Peter Hall (1966) warned that increased urban density is an impending disaster "more lethal than the hydrogen bomb." Once again, sociologists turned to empirical research.

MACROSTUDIES OF CROWDING

The initial studies attempted to see whether neighborhoods with greater population density had higher rates of pathology than less dense neighborhoods had. The results did not support the psychic overload theory. People in dense neighborhoods were not more prone to alcoholism, mental illness, suicide, and other such problems than were people in less dense neighborhoods; in fact, city people were no more prone to these problems than rural people were (Fischer, 1975; Galle and Gove, 1978).

Sometimes the studies did find differences indicating crowding effects, but these proved to be spurious. That is, the kinds of people most likely to live in the most crowded places tend to have higher rates of pathologies wherever they live. Those who live in the most crowded neigh-

TABLE 19-10	Sex Ratios and Social Pathology, Chicago, 1922–1931	
		CORRELATIONS WITH MALES PER 100 FEMALES
1930–1931: Mental hospital admissions per 100,000		.61
1922–1931: Average annual hospital admissions for advanced syphilis per 100,000		.41
1922–1931: Average annual hospital admissions for alcoholism per 100,000		.62

Sources: Calculated by the author from Faris and Dunham (1939) and Burgess and Newcomb (1933).

© Evelyn Hofer

borhoods also tend to be poor, without families, or elderly or to have been mentally ill before they arrived in the neighborhood. When these characteristics of residents were taken into account, no crowding effects could be detected. Thus, density viewed at the macrolevel has no effect on people.

A CLOSER VIEW
GOVE, HUGHES, AND GALLE: MICROSTUDIES OF CROWDING

Although these studies disproved the dire assertions of impending doom from urban density, several sociologists thought that a more modest proposition—that *excessive crowding of a person's immediate environment* has negative effects—might still be valid. That is, the density of a neighborhood might not matter, but the degree to which people have privacy or "personal space" might matter a good deal.

Walter Gove, Michael Hughes, and Omer Galle (1979) designed a study to see if it mat-

ters that some people live in very crowded homes. They set out to discover whether crowding at the microsociological level matters. They reasoned that when a family lives in a home where there are several people to each room, it will be difficult for them to have privacy and to limit interaction with others. Therefore, people in crowded homes ought to experience a lack of privacy and an overload of demands from others. This might cause them to withdraw, both by staying away from home and by being unresponsive to other members of the household. Such withdrawal ought to have negative consequences for attachments and for mental health.

Gove and his colleagues selected more than 2,000 homes that varied in the number of persons per room. Analysis of the data, which were obtained by interviews with members of these households, supported their expectations. They found the following:

■ The more persons per room, the more people complained of a lack of privacy and of too great demands on them by others.

New York City office buildings tower over St. Patrick's Cathedral. Many people would think this symbolic of the impact of cities on religious life: that people who live on farms and in small towns are much more apt to take part in religion than are people in big cities. Many social scientists have described the ways in which the "sophistication" of cities erodes faith. But in fact, people in cities are the ones most apt to participate in religion. One reason is that they don't have to travel long distances to reach a church of their choice. Another is that there are many more choices available (Finke and Stark, 1988).

- People responded to crowding by with-drawing, both physically and emotionally.

- People in crowded homes had poorer mental health.

- Members of crowded homes had poor social relations with one another. There were more family fights, and husbands and wives were less satisfied with their marriages.

- Child care in crowded homes was poor. Parents expressed relief at getting the kids out of the home and were much less aware of where their children were and what they were doing when they were out.

- The effects of crowding began to show up when there was more than one person per room in a household.

Thus, we see that crowding can have nega-tive effects. Although it doesn't seem to matter how many people live in a neighborhood or even in a single home, it does matter how much room people have to find peace and quiet. When ten family members live in a ten-room home, they will be happier than when they live in a four-room home.

Although Gove and his colleagues found sup-port for microeffects of crowding, these effects are of little import for the more general fears about urban crowding. Few urban families live in crowded conditions—that is, with more than one person per room. In 1970 only 8.2 percent of households in America had more than 1.01 persons per room; the median household had almost two rooms per member. Moreover, crowding is declining, not increasing. In 1950 the median household had only about 1.5 rooms per member. In 2000 the median household had 2.2 rooms per occupant.

These changes partly reflect a decline in the average family size. But they primarily reflect the decentralization of cities and the resulting decreased density. The preindustrial city was ex-tremely crowded. The early fixed-rail industrial city was less crowded but much more crowded than the freeway city. Transportation has always been the key to density. As transportation changes enabled people to escape crowded, cen-tralized cities to get some elbowroom, they rushed to do so. They did not wait for sociolog-ical research to tell them that it was desirable to have a large enough home so that family members could find privacy.

CONCLUSION

We have seen that until very recently cities were small, unhealthy, filthy, dangerous, and crowded. Little wonder, then, that when rapid urbanization began in the nineteenth century, so many regarded it as tragic. However, the very processes of industrialization that prompted migration to the cities also transformed city life itself. Granted, many mistakes have been made in the design and administration of modern cities—such errors are inevitable when experi-ence is lacking. Yet if gloom-and-doom prophets such as Tönnies and Durkheim could visit a modern city, they would be dumbfounded not by the errors in planning or the problems that persist, but by the comfort, cleanliness, beauty, and tranquillity of our cities.

Of course, our cities have ugly, dirty, and dangerous neighborhoods. But even the worst parts of modern cities are an improvement on large sections of cities during the early days of the Industrial Revolution, to say nothing of the squalor and misery of preindustrial cities. On a walk through the most horrid urban neighbor-hood today, no one will see dead or dying infants lying on dung heaps—or even any dung heaps. This is hardly to suggest that there are no urgent urban problems; it is merely to give historical perspective to current concerns.

Perhaps the most important sociological lesson in this chapter is that the future is not always a simple extension of the past. Modern cities are not just big versions of older cities. As cities grew, they were greatly transformed. Indeed, in the next chapter, we shall see that growth alone is enough to revolutionize social structures and organizations.

Review Glossary

Terms are listed in the order in which they appear in the chapter.

urban society A society in which the majority of people do not live in rural areas.

urban place According to the U.S. Census, a community having at least 2,500 inhabitants, the majority of whom do not farm.

city According to the U.S. Census, a community having at least 50,000 inhabitants.

suburb An urban place in the immediate vicinity of a city.

metropolis A city and its sphere of influence.

sphere of influence (of a city) The area whose inhabitants depend on a city for jobs, recreation, newspapers, television, and a sense of community.

Standard Metropolitan Statistical Area (SMSA) According to the U.S. Census, a central city (with at least 50,000 residents) and all surrounding counties where 75 percent of the labor force is not in agriculture and where either 15 percent of the workers commute to the city or 25 percent of the workers commute from the city. Statistics Canada's **Census Metropolitan Area (CMA)** includes an urban core of 100,000 and surrounding communities if 40 percent of the labor force work in the urban core or 25 percent of those employed in the outlying area commute from the urban core.

fixed-rail metropolis A city whose form and size are determined by the routes of rail systems (trains and trolleys).

freeway metropolis A city that developed after the widespread use of autos and trucks freed city structures from rail dependence.

theory of ethnic succession Theory stating that ethnic and racial groups will be the targets of neighborhood segregation only until they achieve economic parity and that slum neighborhoods will therefore house a succession of ethnic and racial groups.

index of dissimilarity A measure of the degree to which a given ethnic or racial group lives in integrated or segregated neighborhoods; it compares the ethnic makeup of city blocks with the ethnic makeup of the city as a whole.

Gemeinschaft A German word meaning "community" and used to describe the intimacy of life in small villages.

Gesellschaft A German word meaning "society" or "association" and used to describe the impersonality of life in cities.

anomie A state in which norms lack definition and force, that is, in which people aren't sure what the norms are and don't greatly care.

Suggested Readings

Bourne, L. S., ed. 1971. *Internal Structure of the City.* New York: Oxford University Press.

Fischer, Claude S. 1976. *The Urban Experience.* New York: Harcourt Brace Jovanovich.

Girouard, Mark. 1985. *Cities and People.* New Haven, Conn.: Yale University Press.

Gove, Walter, Michael Hughes, and Omer Galle. 1979. "Overcrowding in the Home." *American Sociological Review* 44:59–80.

Pirenne, Henri. 1925. *Medieval Cities.* Princeton, N.J.: Princeton University Press.

Sjoberg, Gideon. 1960. *The Preindustrial City.* New York: Free Press.

Sociology Online

www.socstark10.com

GO TO THE INTERNET AND TYPE: www.socstark10.com.

⬈ **CLICK ON:** 2000 General Social Survey

✓ **SELECT:** MOVERS: When you were 16 years old, were you living in this same (city/town/country)?

⬈ **CLICK ON:** Analyze Now

TRUE OR FALSE: Younger people are significantly more likely than older people to have moved.

How might this finding be explained?

TRUE OR FALSE: Higher income people are significantly more likely to have been movers.

How might this finding be explained?

TRUE OR FALSE: The western region has the highest percentage of movers.

⬈ **CLICK ON:** Select New Data Set

⬈ **CLICK ON:** The 50 States

✓ **SELECT:** % URBAN: Percent who reside in an urban area.

⬈ **CLICK ON:** Analyze Now

TRUE OR FALSE: The most urban states are concentrated in the Northeast.

⬈ **CLICK ON:** Select New Variable

✓ **SELECT:** PUB.TRANSP: Percentage of workers who commute by public transportation (bus, streetcar, subway, elevated railroad, ferryboat, taxicab).

⬈ **CLICK ON:** Analyze Now

TRUE OR FALSE: The use of public transportation by commuters is concentrated in the Northeast.

⬈ **CLICK ON:** Select New Variable

✓ **SELECT:** CARPOOL: Percentage of workers who carpool.

⬈ **CLICK ON:** Analyze Now

TRUE OR FALSE: The use of carpools by commuters is concentrated in the Northeast.

⬈ **CLICK ON:** Select New Data Set

⬈ **CLICK ON:** Nations of the Globe

✓ **SELECT:** % WORK AG: Percentage of the work force employed in agriculture.

⬈ **CLICK ON:** Analyze Now

Which nation is highest? Where does the United States rank?

TRUE OR FALSE: Based on the discussion in the chapter, we can predict that nations with a higher percentage of their people working on farms will be able to produce more food than those with relatively few agricultural workers.

USING INFOTRAC COLLEGE EDITION

GO TO THE INTERNET AND TYPE:
www.InfoTrac-college.com

⬈ **CLICK ON:** Register New Account
You will be asked to enter your Access Code and to create and enter a User Name and Password. Having done so,

⬈ **CLICK ON:** InfoTrac College Edition or Log On.

How do current articles on suburban life, suburban growth, and traffic problems square with the discussions in the chapter?

SEARCH TERMS:
Suburbs
Traffic Congestion

C H A P T E R

20

The first railroad in the United States began operations in 1826. Here, only five years later, these Americans are boarding one of the many trains already in operation. In those days people already knew a lot about building and operating trains, but they were ill-prepared to manage the huge, rapidly growing, new organizations—like railroads—that suddenly were developing. In this chapter we will examine this recent aspect of social change.

The Organizational Age

NAPOLEON BONAPARTE WAS the last Great Captain to exercise direct command of his army, but by the end of his career, even he failed at the task. Armies had simply become too big and the area of the battlefield too vast. Even by standing on a hill and using a telescope, Napoleon could not keep track of everything that was going on. His orders to various units began to arrive too late, and often, they were wrong (Chandler, 1966).

Napoleon's problems were neither unique nor limited to the commands of armies. During the nineteenth century, many human activities grew in size and complexity to the point that no single leader could orchestrate them. When he was president of the United States, George Washington personally evaluated every government employee: There were fewer than 700 of them! Today, the U.S. government has more than

3 million employees, not including members of the armed forces. In fact, General Motors now employs more people than the total able-bodied adult population of the United States in 1776.

When organized human activities reached the scale of Napoleon's Grand Army of 1812 (about 600,000 troops), a crisis developed. Traditional principles of leadership and organization failed. When one of the greatest leaders in history could no longer master affairs on this scale, it was obvious that new methods were needed for directing and coordinating large-scale activities. If no one individual could manage large organizations, somehow several persons had to share the management. But how? How could leadership and decision making be shared among people without a breakdown in coordination? When Napoleon relied on his subordinates to act on their own, the result was often chaos, as different units marched off in different directions and to defeat.

CHAPTER OUTLINE

THE CRISIS OF GROWTH: INVENTING FORMAL ORGANIZATIONS
The Case of the Prussian General Staff
The Cases of Daniel McCallum and
Gustavus Swift
The Case of Civil Service

WEBER'S RATIONAL BUREAUCRACY

RATIONAL VERSUS NATURAL SYSTEMS

Goal Displacement
Goal Conflict
Informal Relations

THE CRISIS OF DIVERSIFICATION
Functional Divisions
Autonomous Divisions

A CLOSER VIEW
Peter M. Blau: Theory of
Administrative Growth

RATIONAL AND NATURAL FACTORS IN DECENTRALIZATION

BUREAUCRACY AND THE BOTTOM LINE

CONCLUSION

REVIEW GLOSSARY

SUGGESTED READINGS

SOCIOLOGY ONLINE

The answer was the creation of a new kind of group, the **formal organization.** Because such an organization was produced by applying reason to problems of management, and because the key to its success lies in operations based on logical rules, it is often called a *rational organization.*

A formal organization has several characteristics that distinguish it from older forms of organization. First of all, it depends on a clear statement of goals. What is it meant to do? Second, a formal organization requires suitable operating principles and procedures for pursuing these goals. Third, leaders must be selected and trained in the use of these operating principles. Fourth, clear lines of authority and communication must be established for issuing instructions, transmitting information, and coordinating the activities of different groups. Finally, to avoid misunderstanding and error, communications must be written, and written records must be kept and organized.

These five elements of the formal organization seem familiar and obvious to us today, for we live in an age dominated by such organizations. Yet formal organizations are very new human creations—so new that we still do not fully understand how to make them work effectively. Moreover, our understanding of formal organizations has been painfully gained by trial and error and in the face of urgent necessity.

When you complain about the government, your employer, or your college or university, chances are your complaint is not really about shortcomings of individuals. More likely, your complaints are about typical features of large formal organizations. Everywhere you turn, you find that some aspect of your life is governed by and occurs within such organizations. That makes it important to know something about the operations of these organizations and the theoretical principles behind their structure and performance. In this chapter you will see that although organizations are created by and for humans, we are not free to create them in any way we like.

Because large formal organizations are so new—very large organizations have existed for little more than 100 years—and because how they are created and operated has such far-reaching effects, they have been the object of intense study. From the beginning, theories of organization have influenced the structure and operations of organizations, and changes in theory and in organizations have gone hand in hand. We can therefore trace the development

© Bettmann/Corbis

Confronting an Austrian army at the village of Wagram on the outskirts of Vienna (July 5–6, 1809), Napoleon Bonaparte used a small telescope to try to see what was going on across a far-flung battlefield on which nearly a half million soldiers were engaged. Communication with his distant units was by mounted messengers such as the one depicted rushing to Napoleon's side. Although the French won, it was becoming evident by this time that armies of such size could no longer be effectively controlled by a single commander—not even by a genius such as Napoleon.

of both organizations and theories of organizations at the same time.

THE CRISIS OF GROWTH: INVENTING FORMAL ORGANIZATIONS

During the nineteenth century, the first large formal organizations were created, and the first social scientific attempts were made to figure out how to control them. These developments went forward in three somewhat independent sectors of industrializing societies: the military, business, and government. Let's examine developments in each of these sectors before we focus on the work of the first great sociologist of formal organizations, Max Weber. And as we shall see, a major limitation of organizations stems from the fact that they are made up of people, and we often put our own interests first.

THE CASE OF THE PRUSSIAN GENERAL STAFF

After Napoleon's defeat at Waterloo in 1815, Europe entered a long period of peace. Armies were cut back to small professional corps, and

interest in military science waned in most nations. Only in Prussia (later to become the major part of Germany) did people study the crises of command that emerged during the last stages of the Napoleonic Wars, when mass armies took to the battlefields. In Prussia they addressed the question head-on: What would happen if war broke out again, and huge armies—made possible by the recent European population explosion and the mass production of arms—once again engaged in battle? Napoleon's failure had shown that such armies could not be led in the traditional way. The Prussians concluded that military command and organization had to be completely revised (Ropp, 1959).

If Napoleon was the last Great Captain of history, then Helmuth von Moltke was the first Great Manager of the modern military era. Moltke took command of the Prussian army in 1857 and rapidly built a new system based on the principle of using *highly trained* and *interchangeable* staff officers. These elite officers were trained in a war academy. Each year 120 young officers were selected from the whole officer corps on the basis of competitive examinations. Of these, only about forty finished the intensive scholastic course of the academy. And of these graduates, Moltke selected only the best twelve

© Bettmann/Corbis

Field Marshal Helmuth von Moltke.

to be trained for the General Staff (Howard, 1962).

In peacetime officers cannot get real experience in their profession, so Moltke arranged for the academy to provide the next best thing: making battle plans for a great variety of hypothetical campaigns and analyzing past battles. By fighting battles on paper, young officers were trained in Prussian strategic and tactical theories. After their academic studies, officers chosen for the General Staff spent several years with Moltke at his headquarters and rode with him through a series of field maneuvers in which real troops participated. Then these officers were assigned a period of duty with a regiment. After that they rotated between assignments on Moltke's staff and regimental duty (Ropp, 1959).

The point of all this training was to overcome the inability of a single commander to direct a war fought with mass armies. Because the supreme commander could not be everywhere at once, the Prussians tried to create many "duplicates" trained to act as he would act. The decision of one leader could then be carried out

through the reflexes which he had already inculcated in his subordinates through previous training: so that, even when deprived of his guidance, they should react to

unexpected situations as he would wish. . . . Thus the Prussian General Staff acted as a nervous system animating the lumbering body of the army, making possible the articulation and flexibility which alone rendered it an effective military force. (Howard, 1962)

Between wars the Prussian General Staff spent time looking ahead, planning in minute detail for future wars, and agreeing on proper tactics and strategies for various circumstances. When war came, each military problem was solved according to the overall plan and the approved methods. Thus, Moltke dealt with the overwhelming scale of modern warfare by training corps of subordinate managers he could count on to be not only his eyes and ears on the battlefield but also his brain.

While Moltke was creating an interchangeable set of military managers, he also perfected another military system that gave these managers standardized units to work with: the **divisional system.** Before the Napoleonic Wars, European armies were organized by armament and function. The cavalry, the infantry, and the artillery were separate branches of service and appeared on the battlefield as separate units under separate commanders. Coordinating these units was the task of the supreme commander of the army, who arrayed these forces into a battle formation and then told infantry units where to march, cavalry units where to charge, and artillery units where to fire.

However, as armies grew, this system proved cumbersome. Under Napoleon the French army began to break up into smaller units, each of which was an independent miniarmy consisting of infantry, cavalry, and artillery and capable of doing battle on its own.

These French formations were of varying size, and their makeup was never standardized. However, Napoleon's British archrival, the Duke of Wellington (who in his long career never lost a battle), adopted this idea of miniarmies and created a standardized unit called the division. British divisions, being complete units, could be detached to fight as self-sufficient units, combined to form larger units, and interchanged. For example, a rested reserve division could replace a fatigued division in combat.

Moltke carried the standardization of Prussian divisions to the point that commanders could easily move from unit to unit. Each division was similar to the others in makeup, training, size, and structure. Indeed, Moltke's divisional system was so detailed that each division had a specified number of spoons and cooking pots.

In 1871 Moltke tested his new military managers and his divisional structure in the Franco-Prussian War. During a lightning campaign, the Prussians utterly routed the much more experienced French army. The Prussians did not win because they were better armed, had more soldiers, or were braver in battle. The French army was their equal in all of these ways. But the French General Staff was only a group of messengers and clerks serving the commander, and the French commander could not control his far-flung armies.

Noting Moltke's success over the French, all major nations soon copied his methods. Later, with the advent of telephones and radios, commanders could better guide their subordinates in the field. But the principle of delegating command to officers on the spot, who are highly trained in a common military theory and in the command of standardized military units, has remained the only workable solution to the problem that overwhelmed Napoleon.

THE CASES OF DANIEL MCCALLUM AND GUSTAVUS SWIFT

It seems fitting that while the key to managing huge military organizations originated in Prussia, the key to managing huge business organizations was first discovered in the United States. For business the rapid growth of railroads in the 1850s was the equivalent of Napoleon's Grand Army—the railroads revealed the inability of traditional organizational principles to cope with large-scale enterprises. The crisis appeared in dramatic fashion: Small railroads made profits while the big railroads lost money.

McCallum In 1855 Daniel C. McCallum, general superintendent of the Erie Railroad, pointed out that the reason his line and other large lines such as the New York Central, the Pennsylvania, and the Baltimore & Ohio were in financial distress was a problem of management. He wrote:

A Superintendent of a road fifty miles in length can give its business his professional attention and may be constantly on the line engaged in the direction of its de-

tails; each person is personally known to him, and all questions in relation to its business are at once presented and acted upon; and any system however imperfect may under such circumstances prove comparatively successful. (in Chandler, 1962)

These comments recall the spectacular ease with which Napoleon dealt with grave military disadvantages when he had only 70,000 or 80,000 troops to maneuver on a single compact battlefield. But McCallum continued, when one attempts to manage a railroad "five hundred miles in length a very different state exists. Any system which might be applicable to the business and extent of a short road would be found entirely inadequate to the wants of a long one." For want of an adequate organizational system, McCallum argued, the large railroads faced financial failure.

McCallum quickly moved to install a management system to replace the overloaded manager. He broke his railroad into **geographical divisions** of manageable size. Each was headed by a superintendent responsible for the operations within his division. Each divisional superintendent was required to submit detailed reports to central headquarters, from where McCallum and his aides coordinated and gave general direction to the operations of the separate divisions. Lines of authority between each superintendent and his subordinates and between each superintendent and headquarters were clearly laid out. In sketching these lines of authority on paper, McCallum created what might have been the first organizational chart for an American business (Chandler, 1962). Soon the other great railroads copied the Erie system, enabling the big railroads to function as effectively as small ones. As a result railroads rapidly became the largest industrial companies of that era.

The railroads had two direct effects on other industrial firms. First, they made it possible for other firms to grow by using rail shipments to reach national rather than just local markets. Rail shipments could carry goods across the nation and bring needed supplies from far away. Second, the railroads provided a first crude organizational model for operating large firms. As other kinds of firms grew, they adopted the idea of divisions, but as we shall see, these were based on function rather than geography. As they grew, new industrial firms

The huge stockyards founded by Gustavus Swift in Chicago, the first of many operated by his company. In Swift's new organizational system, stockyards not only made up a functional division of the company but also were the initial level in the vertical integration of the company: control of each step in the process of bringing beef from the range to the meat counter.

Chicago Historical Society

created **functional divisions** that controlled each step in production through a process called **vertical integration.** These two features of industrial firms came to dominate organizational theory for many decades.

SWIFT The story of Gustavus Swift, who built a huge meat-packing firm in the 1870s and 1880s, reveals how the new industrial organizations came into being.

Swift was a wholesale butcher in New England who moved west to Chicago in the 1870s. The population was concentrated in the East, while the herds of livestock were concentrated on the Great Plains, and getting the meat to market was a cumbersome and inefficient process that depended on the uncoordinated services of small, specialized, local firms. Swift was determined to bring order and efficiency to the process by controlling each step from ranch to retail store. In 1878 he made an experimental shipment of meat from Chicago to the East, using the newly invented refrigerator car. The success of this experiment encouraged Swift and his brother Edwin to found Swift & Co. But they

still faced vast problems. Shipping refrigerated meat east required refrigerated storage facilities at the other end; so, Swift built them. Then the meat had to be sold; so, Swift hired a sales crew and set up a distribution system in each major city. Local butchers tried to prevent the sale of his western meat in eastern markets, even claiming that it was unhealthy to eat "meat killed more than a thousand miles away and many weeks earlier" (Chandler, 1962). Massive advertising was required to convince consumers that Swift meat was safe. Soon Swift built additional packing plants in St. Louis, Omaha, St. Joseph, St. Paul, and Fort Worth.

Then Swift turned his attention to making supplies of meat dependable. He organized stockyards to purchase large numbers of animals on a regular and orderly basis. Finally, he branched out to make use of animal by-products by entering the leather, glue, fertilizer, and soap businesses.

Swift & Co. became a vertically integrated company: It controlled each step in the process of bringing meat products to the consumer. Although Swift did not raise cattle, the company took over at the point of sale and conducted each step thereafter: buying, packing, shipping, and marketing. Furthermore, each of these steps was the province of a *different division* of the company. That is, rather than creating geographical divisions, as the giant railroads had done, Swift *based its divisions on different functions.* In fact, Swift broke up its organizational divisions in the same way that the Industrial Revolution had divided the labor of workers into a few specific production steps. Just as each worker on an assembly line performed only one or a few specialized functions, each division of large industrial firms handled only one aspect of the industry.

Swift had a marketing division, a meatpacking division, a purchasing or stockyards division, a shipping division, a sales division, and an advertising division. Each of these divisions was headed by a manager to whom subordinate managers reported; each manager reported to and received directions from corporate headquarters. As Moltke's General Staff mastered large armies, vertical integration and functional divisions under centralized command made it possible to create and operate huge business firms.

THE CASE OF CIVIL SERVICE

Armies and corporations were not the only organizations that grew to immense size in the modern world. Governments also became very large because of the rapid expansion in size and complexity of the societies they governed. As governments got big, they, too, found that they could no longer function with outdated practices.

In traditional agrarian societies, the government was nothing more than the king's household and court. Such needed functionaries as clerks, accountants, and tax collectors were servants of the king, equal in status to his cooks, grooms, and butlers. When the king needed a general, an advisor, a chief justice, or an administrator of the treasury, he asked one of the noblemen in his court to do the job. These noblemen did not regard a government post as an occupation or even as a full-time activity. Often, they had no special training and little aptitude for their government duties beyond their noble birth and their social graces.

Such a system worked because governments did little governing. Beyond extracting taxes from the populace, maintaining some semblance

National Gallery of Art, Washington, Alisa Mellon Bruce Fund, 1967

A Dutch tax collector in about 1500. His filing system consisted of several spikes on the wall and a ledger. If asked who he was, he would not have said a government employee or a tax collector. Instead, no doubt he would have said "a gentleman." He would find modern civil service practices as strange as computerized accounting systems.

of public order, and defending the realm against invaders, there was little for governments to do. After all, more than 90 percent of the population were peasants leading quiet lives of rural toil. No complex laws, no large regulatory agencies, and no swarms of government experts were needed. Indeed, if the central government had disappeared, people in outlying districts would not have noticed for a long while.

With the growth of population and cities, the complex divisions of labor, and the development of technology, agrarian governments found it increasingly difficult to control their societies. Indeed, modern societies require more control than agrarian societies do.

Governments adopted much the same solutions as did armies and industries. Organizations were created specifically to perform government functions in an orderly and efficient manner, and these organizations were staffed by persons specially trained to carry out their duties. In fact, staffing government positions caused the greatest conflict.

Kings were accustomed to rewarding their loyal and valued friends with government positions. Early democratic governments continued this practice: The party or political faction in control of the government appointed its favorites to office. When the government changed hands, government officeholders were also changed.

Thus, when Thomas Jefferson became president of the United States in 1801, he dismissed hundreds of Federalists appointed by Presidents Washington and Adams and replaced them with his supporters. This practice is known as the **spoils system:** The spoils, or benefits, of public office go to the supporters of winning politicians. The spoils system probably reached its height in the United States during the presidency of Andrew Jackson in 1829, when thousands of officeholders were replaced.

When government is based on the spoils system, disorganization arises from so much turnover, and people are prevented from making a career of government service. The administration of government organizations is forever left in the hands of untrained novices. To combat this problem, governments adopted merit systems for government employment—a practice often referred to as civil service. Civil service systems base government hiring and promotion practices on merit. People are recruited on the basis of their educational and occupational qualifications and by successfully competing with others on a written examination. When a government wants accountants, it hires trained accountants with the highest scores on the civil service examination rather than the brother-in-law of some politician. What Moltke learned was needed for modern war and what Gustavus Swift had discovered about business, modern government also put into practice: a carefully designed organizational system operated by specially selected and trained people.

WEBER'S RATIONAL BUREAUCRACY

At the turn of the twentieth century, Max Weber began to study the new forms of organization being developed for managing large numbers of people in far-flung and complex activities. As a German, he was very familiar with Moltke's development of the General Staff. Furthermore, Germany had been an early leader in developing a civil service. And in Weber's day German industry was rapidly adopting the organizational methods developed in the United States. Surveying this scene, Weber attempted to isolate the elements common to all of these new organizations.

Weber concluded that each of these new large-scale organizations was a **bureaucracy.**

Today, many of us regard *bureaucracy* as a dirty word, suggesting red tape, inefficiency, and officiousness. As we shall see, bureaucracies can develop these features, especially if authority is highly centralized. Weber's purpose, however, was to define the essential features of new organizations and to indicate why these organizations worked so much better than traditional ones. Let us examine the features that Weber found in bureaucracies.

Above all, Weber emphasized that bureaucratic organizations were an attempt to subdue human affairs to the rule of reason—to make it possible to conduct the business of the organization "according to calculable rules." For people who developed modern organizations, the purpose was to find rational solutions to the new problems of size. Weber saw bureaucracy as the rational product of social engineering, just as the machines of the Industrial Revolution were the rational products of mechanical engineering. He wrote:

The decisive reason for the advance of bureaucratic organization has always been its purely technical superiority over any former organization. The fully developed bureaucratic mechanism compares with other organizations exactly as does the machine with nonmechanical modes of production. (Weber, [1921] 1946)

For Weber the term *bureaucracy* was inseparable from the term *rationality*. And we may speak of his concept as a "rational bureaucracy."

But what were the features developed to make bureaucracies rational? We have already met them: (1) functional specialization, (2) clear lines of hierarchical authority, (3) expert training of managers, and (4) decision making based on rules and tactics developed to guarantee consistent and effective pursuit of organizational goals. Weber noted additional features of rational bureaucracies that are simple extensions of the four just outlined. To ensure expert management, appointment and promotion are based on merit rather than favoritism, and those appointed treat their positions as full-time, primary careers. To ensure order in decision making, business is conducted primarily through written rules, records, and communications.

Weber's idea of functional specialization applies both to persons within an organization and to relations between larger units or divisions of the organization. We have already seen how this applied to Swift & Co. Within a Swift pack-

Records are the basis of bureaucratic organizations. For organizations to run on the basis of rational procedures, exact records must be kept of each transaction; only then can the operation of the system be reviewed and improved.

© Burk Uzzle/Woodfin Camp

ing plant, work was broken down into many specialized tasks, including the tasks involved in coordinating the work of others. (This coordination is called administration, or **management.**) Furthermore, Swift was separated into a number of divisions, each specializing in one of the tasks in the elaborate process of bringing meat from the ranch to the consumer. Weber argued that such specialization is essential to a rational bureaucracy and that the specific boundaries separating one functional division from another must be fixed by explicit rules, regulations, and procedures.

For Weber it was self-evident that coordinating the divisions of large organizations requires clear lines of authority organized in a hierarchy. That means there are clear "levels of graded authority." All employees in the organization must know who their boss is, and each person should always respect the chain of command; that is, people should give orders only to their own subordinates and receive orders only through their own immediate superior. In this way the people at the top can be sure that directives arrive where they are meant to go and know where responsibilities lie.

Furthermore, hierarchical authority is required in bureaucracies so that highly trained

experts can be properly used as managers. It does little good to train someone to operate a stockyard, for example, and then have that manager receive orders from someone whose training is in advertising. Rational bureaucracies can be operated, Weber argued, only by deploying managers at all levels who have been selected and trained for their specific jobs. Persons ticketed for top positions in bureaucracies are often rotated through many divisions of an organization to gain firsthand experience of the many problems that their future subordinates must face. (Recall how Moltke rotated his General Staff officers through various regiments.)

Finally, Weber stressed that rational bureaucracies must be managed in accordance with carefully developed rules and principles that can be learned and applied and that transactions and decisions must be recorded so that rules can be reviewed. Only with such rules and principles can the activities of hundreds of managers at different levels in the organization be predicted and coordinated. If we cannot predict what others will do, then we cannot count on them.

Moltke had to be sure that staff officers faced with an unexpected crisis would solve it as he would. To ensure that, officers had to be trained in Moltke's tactical principles and rules. Similarly,

Gustavus Swift had to know that his stockyards would not buy meat faster than his packing plants could process it or that more meat would not be shipped than his eastern refrigerators could accommodate. Of course, it is impossible to spell out detailed rules to fit all contingencies. Therefore, decision makers must be highly trained and must report their decisions promptly and accurately to their superiors.

For a long time, Weber's rational bureaucracy model dominated social science thinking about large, modern organizations. If organizations did not operate quite as Weber had said a bureaucracy should, then the solution was to bring them in line with the ideal bureaucratic procedures. However, by World War II, sharp criticism of Weber's ideas began to surface. Social scientists began to argue that Weber had ignored much of what really went on in organizations—the conflicts, the cliques, and the sidestepping of rules and the chain of command. The problem, according to Philip Selznick (1948, 1957), lay in the fact that bureaucracies were not and could not be like machines because they consisted of human beings. In the final analysis, people simply will not imitate machines.

RATIONAL VERSUS NATURAL SYSTEMS

Weber stressed the rationality of bureaucratic organizations; that is, organizations are created and maintained to pursue clearly defined goals, and the structure and operation of organizations are the result of reasoned, conscious efforts to attain these goals. This approach to studying organizations is called the **rational system approach.**

Alvin Gouldner (1959) described the rational system viewpoint as one in which

the organization is conceived as an "instrument"—that is, as a rationally conceived means to the realization of expressly announced group goals. Its structures are understood as tools deliberately established for the efficient realization of these group purposes. . . . Changes in organizational patterns are viewed as planned devices to improve the level of efficiency. . . . The focus is, therefore, on . . . the formally "blueprinted" patterns.

From this viewpoint organizations that make poor choices and fail to achieve stated standards are guilty of ignorance and miscalculation.

However, many sociologists have criticized the rational approach as too limited. They argue that many important goals of organizations are not the "announced goals" and that often not all members of an organization pursue the same goals. Critics further argue that the real lines of communication and authority in organizations are not always the same as those laid out on the organizational chart. In fact, the real lines may violate the formal structure.

Persons approaching organizations in this way focus on the natural system. They argue that the general principles of the natural behavior of people and groups apply to the behavior of bureaucratic organizations; thus, these principles, not the rational system, reveal what is really going on in the system. A fundamental principle of the **natural system approach** is that the rarely stated but overriding goal of organizations is simply to survive. The living, breathing human beings who staff an organization develop a personal stake in the life of that organization, regardless of the stated goals of the organization. For its members the existence of an organization means the continuation of jobs, careers, and friendships. When an organization folds, its members are cast adrift. People can be expected to resist such a fate even at the expense of the formal goals of the organization. Therefore, features of an organization that appear as miscalculations from the rational perspective may reflect people's efforts to keep the organization alive.

GOAL DISPLACEMENT

The importance of survival over other goals is well illustrated by an organization created during the 1930s to combat the dreaded disease of polio. The National Foundation for the March of Dimes created a huge network of volunteers in each American community who, guided by the professional staff, conducted an annual fund drive. The funds were used to support the treatment of polio victims and to underwrite research for a way to cure or prevent the disease. The March of Dimes was an extremely successful and well-run organization. Then, one day, it was too successful. In the 1950s research supported by the March of Dimes funds led to the discovery of effective vaccines. A massive vaccination drive soon resulted in the virtual elimination of polio. The March of Dimes had achieved its goals.

Yet this achievement was not met with an office party of gleeful people toasting victory and going off happily to pursue new careers. Instead, there was something closer to panic as the March of Dimes staff searched for a new goal to sustain the organization. Quickly, they declared war on birth defects and, as the National Foundation, continue even today to raise funds and to conduct business as usual (Sills, 1957).

Organizations often change, or displace, their goals in the pursuit of survival, but not often as dramatically as did the National Foundation. Philip Selznick (1949) has documented substantial **goal displacement** by the Tennessee Valley Authority (TVA). When created during the 1930s, TVA announced broad goals to transform the whole social structure of the rural farming region served by its hydroelectric power and flood control projects. Many found these goals too radical, and opposition to TVA grew. In reaction to threats to disband the agency, these far-reaching social goals were replaced by the more limited goals of rural electrification and resource management.

These examples illustrate the truth of a principle: When the formal goals of an organization threaten its existence, the goals will be changed.

GOAL CONFLICT

Natural system theorists also argue that different groups within an organization tend to pursue different goals, which often have nothing to do with (and may even conflict with) the goals of the larger organization. This situation is called **goal conflict.**

The official goal of corporations is to earn profits for stockholders, but the managerial revolution—whereby owners were separated from the control of corporations—resulted in the displacement of profit as the overriding goal of modern corporations. Managers as a group have goals that may *conflict* with the profit goal of a company. For example, managers may approve lavish expenses—sales meetings in posh resorts, elegant business entertainment, and company-owned jets and limousines—that cut into potential profits. They do this to reward themselves rather than a mass of anonymous, disorganized owners (stockholders).

A second reason managers may deemphasize profit seeking is because other groups in the corporation effectively pursue goals that also conflict with the profit goal. Consider the case of labor. Maximum profits depend on getting the

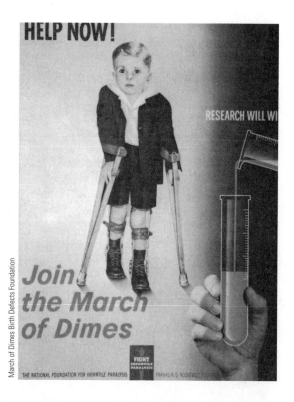

March of Dimes Birth Defects Foundation

Posters like this were on display everywhere when I was growing up. Some of my playmates died of polio (often called infantile paralysis); many more were badly disabled. The line "Research will win" turned out not to be an empty slogan. Research paid for by March of Dimes contributions led to the discovery of the polio vaccine.

greatest amount of production from workers for the lowest possible wage. While all managers seek to raise worker productivity and to hold down labor costs, modern corporate executives do not do so without regard for their own welfare. Unions often resist changes in work procedures and the introduction of new technology and press for the highest possible wages. Therefore, management often does not seek the most profitable labor contracts, settling instead for a compromise that involves the least disruption and stress.

Management and labor are the most obvious examples of groups that strike bargains that affect the structure and operations of an organization. Many such compromises are not designed to make the organization more effective in pursuing its stated goals. Organizations often develop many competing interest groups, and various divisions of a company may strike bargains with one another over who is responsible for what. Such bargains may reflect desires for enhanced status or reduced workloads, but they may lower efficiency.

INFORMAL RELATIONS

Formal organizations are based on clear lines of authority and responsibility. Each member is supposed to know to whom to give orders and from whom to take them. However, natural

A clear organizational chart could easily be created for this busy staff. Everyone has a distinct job title with specific responsibilities, and clear lines of authority state who is supposed to report to whom. But observation would reveal great differences between what the chart shows and what really goes on.

© Frank Herrmann/Masterfile

system theorists point out that people often do not adhere to the formal chain of command. They constantly construct social relations that serve as bases for authority, influence, and communication which often bear little resemblance to relations set out in the formal blueprint.

Suppose Jack in division A needs assistance from division B. He could go to his supervisor, who would then request the assistance from the supervisor of division B. He could also call his friend Sally in division B and get what he needs without either supervisor knowing what is going on. Indeed, Jack and Sally could call on each other all the time. They could thus get around the formal procedures for interdivision cooperation and override the formal system. It is well known in many organizations that you must see certain people to get certain things done, despite the fact that these people do not hold the formal position that is supposed to control these activities. Friendships and enmities may also lead to overriding formal operations.

But if the rational system approach overemphasizes the organizational blueprint, the natural system approach tends to forget that there is one. Although organizations are filled with informal channels of authority, influence, and communication—informal networks of social relations and groups pursuing goals other than

the corporate goals—we can predict from the corporate blueprint most of what people do most of the time. Persons who are hired to work on an automobile assembly line, for example, may cooperate to keep production at the level that they think is best for them and get around their supervisors to alter procedures, but they do not work in the accounting office sorting invoices, in the design division drawing new models, or in the showroom selling cars. When we know the official duties and responsibilities of a particular job, we can predict a great deal about what people hired to fill that job will spend their time doing.

Consequently, neither the rational nor the natural system approach can fully explain formal organizational activities. They are complementary views. The long conflict over which approach better explains organizations is probably nothing more than an academic dispute between college departments. Those who have given the most time to study of the rational system are usually employed in the department of business administration. Those who opt for the natural system are mainly employed in sociology and psychology departments. In the competition for customers (in this case, students), each firm (or department) stresses the merits of its product. But the wise customer knows that making a choice is unnecessary; one need not choose between cereal or fruit for breakfast but may benefit from mixing both in the same bowl.

The important point is that when people design or evaluate an organization, they must pay attention not only to what people are supposed to do but also to what they are apt to do.

THE CRISIS OF DIVERSIFICATION

In the case studies examined at the beginning of this chapter, we saw some solutions to problems in managing large organizations: Gustavus Swift coordinated a large business firm on the basis of vertical integration of functional divisions; Helmuth von Moltke recognized the need for highly trained, interchangeable commanding officers holding a common military doctrine; and civil service developed to ensure experienced and competent managers in government.

As time passed, however, the solutions of Swift, Moltke, and others to the problems of operating large organizations began to fail. As organizations continued to grow in size and complexity, a need arose for even better organizing and managing principles. In part this new crisis occurred because the rational system approach created some organizational problems that exceeded the natural human capacities of managers. There are limits to how much any one person can know and do. These new problems, and the principles they spawned, can best be understood by examining one of the first organizations in which such problems became evident and were solved.

DuPont is one of the oldest and most successful firms in the world. It began in 1802 when Eleuthère Irénée DuPont, an immigrant from France, built a gunpowder factory on the banks of Brandywine Creek near Wilmington, Delaware. The factory grew and prospered. Throughout the nineteenth century, DuPont sons attended West Point or M.I.T., the best engineering schools in the country at that time, and then went into the family business. The firm grew larger and larger, expanding its product line to include blasting powder and dynamite.

FUNCTIONAL DIVISIONS

By the turn of the century, DuPont displayed the same problems of bigness that many other organizations then faced. With many factories

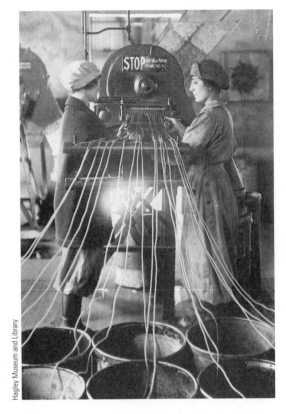

Hagley Museum and Library

Women at work in a DuPont factory making smokeless gunpowder during World War I. Frantic pleas from England and France for powder prompted DuPont to undertake one of the most rapid industrial expansions in history.

requiring massive amounts of raw materials, supplies had to come in regularly and at predictable prices, and the production flowing out of these plants had to be directed to customers in an orderly and efficient manner. When the operations of DuPont became bogged down, the directors responded, as Gustavus Swift had done, by reorganizing the firm into *functional divisions.*

All personnel and facilities involved in manufacturing explosives were grouped into a single administrative unit under a general manager. All purchasing was centralized into a single unit, rather than having factories buy on their own. Sales were coordinated within a single unit, as were engineering, research and development, finance, and legal services. An effective reporting system was instituted so that a steady flow of information kept managers aware of the operations for which each was responsible. This information flowed upward to provide an accurate picture of operations on which the president and directors could base major policy decisions.

Once reorganized, the firm functioned extremely well. In 1914 World War I broke out in Europe and created a sudden, nearly inexhaustible demand for gunpowder and high explosives. To meet this demand, DuPont undertook one of the largest expansions the world of business had ever seen. At the start of the war, DuPont plants had a maximum capacity of slightly more than 8 million pounds of smokeless gunpowder a year. In little more than twelve months, they had expanded their factories by nearly twenty-five times to produce 200 million pounds a year. By 1917 they had increased production by more than fifty times, to 455 million pounds a year. Expansion of production capacities for high explosives grew almost as dramatically. Similarly, the company's payroll expanded enormously. In the fall of 1914, DuPont employed 5,300 people; in 1918 they employed more than 85,000. And the company's investment in plants and equipment grew from $83 million to $309 million.

The new organization based on functional divisions coped very well with these terrific demands. Efficiency did not suffer, and the firm earned excellent profits on its investment. But the enormous growth presented the company with both an opportunity and a challenge. What would they do when the war ended? The demand for gunpowder and explosives would then return to low prewar levels. Should they plan to close down the huge new plants they had built and lay off most of the employees? Or should they try to find a way to convert these new assets into peacetime production?

The DuPonts decided to diversify by entering the growing market for chemical products, a market for which their mastery of explosives manufacture was ideally suited. Demand for early synthetics such as patent leather, synthetic silks, and plastics was growing, and DuPont prepared during the war to enter these new markets when peace returned.

In 1919 these plans were put into operation on a huge scale. The company became a large manufacturer of paint, dyes, plastics, chemicals, fertilizers, and the host of products associated with the chemical industry today. DuPont had immense amounts of money from its wartime profits to invest in these new activities. It had the plants, skilled labor, and highly skilled managers. The quality of its products was excellent, and sales rose rapidly. But the company almost went broke.

What happened? The principle of functional divisions, a hugely successful arrangement for organizations engaged in a relatively narrow set of activities, turned out to be a disastrous arrangement for organizations engaged in a broad range of activities.

AUTONOMOUS DIVISIONS

Before diversifying into its many new activities, DuPont resembled Swift & Co. The company was organized to govern each step in the manufacture of explosives from the acquisition of the needed raw materials to their delivery to customers. But now they were engaged in many parallel operations. For example, before diversification the sales division had dealt with only a few very large customers—primarily governments and manufacturing firms (for example, ammunition manufacturers). To sell paint, however, they had to deal with thousands of small retail merchants and create a demand for DuPont paint among consumers. The same was true of their soap, glue, and finished plastic products. The sales department found itself overwhelmed by an immense array of products that had to be sold in different ways to many different kinds of customers.

The same thing was happening in the manufacturing divisions. Instead of having a number of similar factories, each engaged in similar production processes and thus facing similar difficulties, DuPont now had factories with different concerns. The paint factories bore little resemblance to the plastic factories, the fertilizer factories, and so on. Similarly, the purchasing department no longer searched for a few raw materials but for a huge variety from many different sources.

With diversification of the firm into many businesses, executives faced demands that exceeded their capabilities. Executives in the manufacturing division could not grasp the major technical problems of so many different manufacturing processes. Nor could the sales force master knowledge about so many different products or the appropriate sales techniques for every market. In addition, as the firm diversified, upper management had too many people to supervise.

Considerable research has demonstrated that there are limits on the number of people a given person can supervise effectively. This limit is called the **span of control.** Research suggests that no executive should have more than seven subordinates who report directly to him or her. Beyond that number, confusion begins to set in, and the executive has neither the time nor the memory to serve each subordinate adequately (Drucker, 1967). But at DuPont the span of control imposed on senior executives was far more than seven, with many having direct responsibility for thirty to forty subordinate managers.

Faced with operations of such magnitude, the organizational system began to crack and split open. Just as the large railroads had once lost money while the small railroads were highly profitable, DuPont began to lose money while small specialized competitors flourished. Worse yet, the more DuPont sold of its new products, the more money it lost! Thus, when sales of paint and varnish rose from $1 million to $4 million over three years, annual losses on these products rose from $100,000 to $500,000. Even this giant company with its huge wartime cash reserves couldn't endure such losses, which came to $2.5 million in the first six months of 1921 alone.

Business firms get very nervous when their profits fall because, unlike government agencies, they can go broke. The top managers at DuPont sat down and carefully rethought their entire organization. They knew that they were engaged in the right business because their products sold well. But they also knew that they were not conducting their business properly. As one of their directors put it in a memo to his fellow board members: "The trouble with the Company is right here in Wilmington, and the failure is the failure of administration for which we, as directors, are responsible" (in Chandler, 1962).

After careful consideration, DuPont management realized that the failure of their administration lay in current theories of organization: The functional division system was wrecking them. This system is fine for a firm that produces a narrow line of products or performs a few services. It is an excellent system for an explosives company or a meat company. But it is an unworkable system for a **diversified organization** that manufactures many different kinds of things.

Indeed, what the DuPont managers suddenly realized was that they should no longer think of themselves as a single firm. They had grown so large and were in so many different businesses that they needed to structure themselves into *divisions organized around each business*, not around each function. Functional divisions could be retained but at a lower level of management: only within **autonomous divisions** constituted as independent firms. The larger company of DuPont would consist of a number of divisions, each fully organized to conduct its own business.

DuPont discovered the business version of the military divisions the Duke of Wellington created and Moltke perfected. As a military division included all essential components of an army, a DuPont division included all components of a manufacturing company. It would purchase its raw materials (and pay the market price, even if they were purchased from another DuPont division), supervise its own manufacturing, conduct its own research and development, and operate its own marketing and sales organization.

Each division would be run by a general manager who had the authority to make business decisions for the division and who would answer to top management, who were primarily interested only in the success or failure of the overall operation. The top management at DuPont stepped back from the details of the operations of its divisions and appointed or fired top division managers, decided which divisions to keep and which to dispose of, and made plans for creating and acquiring new divisions. Like Moltke, the DuPont management planned and communicated the grand strategy of the firm but delegated the tactical decisions to commanders on the spot.

Soon most major business firms faced problems similar to DuPont's, and they, too, solved them by establishing autonomous divisions (Fligstein, 1985). In this way Alfred Sloan remodeled sprawling General Motors into the Chevrolet, Pontiac, Oldsmobile, Buick, and Cadillac divisions. With their separate assembly plants, their own systems of exclusive dealerships, their own advertising budgets, and their own financial resources, they competed with one another almost as fully as they did with Ford, Chrysler, and other automobile makers.

These autonomous divisions transformed single firms that had become too big and too complex to manage into a cluster of smaller coordinated firms. This was the second step taken in a new approach to managing big organizations: **decentralization.**

A CLOSER VIEW
PETER M. BLAU: THEORY OF ADMINISTRATIVE GROWTH

We have seen that the solution for managing large corporations lies in creating a division of labor among executives—parceling out administrative responsibilities so that several people perform duties too numerous for any one person to handle. A direct consequence of the growth of organizations has therefore been an even more rapid growth in the number of persons required to manage them. Peter M. Blau (1970, 1972) considered the rapid expansion of managerial positions and formulated a theory of organization. Two of his major propositions were:

1. *Organizational growth causes differentiation.* This is precisely what we have seen throughout this chapter. As organizations get larger, they must be broken down into units so that their activities can be controlled.

2. *As organizations become more differentiated, the size of the administrative component increases relative to the size of other components.* Blau argued that the difficulties of coordinating and communicating are much greater across differentiated units than *within* units. The greater the number of units, the greater the effort required to coordinate them and the greater the number of supervisors and managers needed.

From these two propositions it follows logically that *the larger the organization, the greater the proportion of total resources that must be devoted to management functions.*

By now there is considerable empirical support for **Blau's administrative theory.** Its most obvious implication is that organizations are less costly to run when they are kept smaller and that efficiency may be lost rather than gained when several organizations are combined into one huge organization. The reason for creating larger organizations has always been to achieve savings. For example, a grocery chain can undersell an independent grocer because of the great savings made possible by large-scale purchasing and marketing. However, at some point such savings must be weighed against the greater resources needed to manage the organization. These accelerating management costs are an overhead: They add to the costs of the goods and services offered by an organization. Some organizations are too small and some are too large to be efficient.

It is impossible to give a general answer to the question, How big is too big? For any given organization, the answer comes from weighing administrative costs against savings and then determining when bigness offsets efficiency. But organizational theorists now believe that any organization can get too big.

Blau's theory, therefore, once again leads to the principle of decentralization. When organizations got too big for one person to control, means had to be found to place the control in the hands of many people—that is, to decentralize authority. The various ways to cope with great size that we have examined were all meant to decentralize organizations without letting them become uncoordinated. It turns out, however, that efforts to decentralize organizations tend to run into resistance from the natural system within organizations.

RATIONAL AND NATURAL FACTORS IN DECENTRALIZATION

The goal of top executives in most business organizations has long been to pass decision making on to subordinates. Thus, McCallum ceased making operating decisions for the Erie Railroad and instead asked his division superintendents to make them. Ever since, business has sought ways to push this process of decentralization further. Indeed, maximum decentralization is the main principle of current management science.

Business theorists such as Peter F. Drucker (1946, 1967, 1974) preach that companies ought to create the *smallest possible operating units to give maximum flexibility to the person on the spot.* Drucker argues that it is impossible for higher management to have firsthand experience of the specific conditions in a remote department. Let the people in charge of those departments make the decisions because they will usually make the right ones. As long as the overall performance of a department is satisfactory, leave it alone. If performance fails, then appoint new people. But never try to run it from upstairs.

This doctrine is often called **management by objectives** because managers and their subordinates negotiate the objectives the subordinates should reach. Then managers give subordinates maximum freedom to decide how to achieve the objectives and later judge the subordinates by how well they succeed. In this way decision making is delegated to those in the best position to make particular decisions quickly and correctly.

The key element in the decentralization of organizations is *discretion*, the freedom to make choices and decisions. Decentralization consists of giving the maximum number of subordinates in an organization discretion in running their part of the operation. But discretion involves two closely linked elements: (1) the *responsibility* for making decisions and (2) the *authority* to carry them out. Giving people responsibility is futile unless they are also given the power to meet their responsibilities. This is more than a truism, however. Often, it is very difficult to give members of organizations enough authority to allow the decentralization of decision making to be effective.

Limits on the power to implement decisions come from both within and outside of organizations. Internal limits come from the necessary interdependence of the many subunits of an organization. Limits can also come from having to deal with external factors, such as suppliers, markets, competitors, and regulatory agencies, which organizational members cannot control. For example, a decision to increase efforts to market a particular item may be thwarted when suppliers cannot increase their production, when competitors cut their prices or introduce an improved model, when the market fails to respond, or when the government imposes new requirements.

Reflecting on these aspects of decentralization, James D. Thompson (1967) developed a number of theoretical propositions. Several of these pertain to the conditions under which members of an organization will accept discretion. Thompson proposed that when individuals in an organization believe they cannot adequately control conditions affecting decisions, they will try to evade discretion—they will try to pass responsibility on to someone else (usually someone higher in the organization). Thus, *the more a particular position in an organization depends on other positions in the organization, the less willing people in that position will be to exercise discretion.* Similarly, the more that a decision involves forces outside the organization, the less willing people will be to make that decision.

Second, *the more serious the potential consequences of an error are perceived to be, the less willing people will be to assume discretion.* People want to share responsibility for a decision with all who will be affected by the decision if it turns out to be wrong. A variant of this is that the more discretion assigned to a position in an organization, the more a person holding that position will seek power over those affected by his or her decisions. This approach also minimizes the negative consequences of poor decisions by limiting the power of others to retaliate.

So, we see that despite efforts to impose decentralization on the formal system, forces generated within the natural system tend toward recentralization. People often seek to regain dispersed power to control the conditions affecting their decisions more fully. Or people seek to disperse responsibility for decisions that the organization meant to place in their hands alone.

From the viewpoint of formal organizational models, decentralization has many benefits. It places decisions in the hands of those closest to the scene. Although decentralization disperses power widely, thus diluting the power of top management, it limits an organization's dependence on any given decision maker. That is, decentralized organizations are like ships with many separate watertight compartments. If such a ship strikes an iceberg, flooding is limited to a few compartments, and the ship remains afloat. Similarly, decentralized organizations suffer only limited damage from bad choices made by any given person because each makes only a few decisions. However, Thompson's theory suggests that the natural system within an organization often thwarts these formal arrangements and cuts holes in the watertight compartments.

We have seen why Thompson stated that the natural system of organizations tends to be less decentralized than the formal system. Moreover, coalitions tend to form within an organization (see Chapter 1). Thompson postulated that whenever an individual given discretion has insufficient power to control conditions governing his or her decisions, that individual will seek added power by forming a coalition with other decision makers. For instance, consider someone empowered to decide how much of a product to produce and someone else empowered to decide how to market that product. The marketing decisions depend on supply, and production decisions depend on sales. No matter how the company is organized, the production and marketing managers are likely to get together and act jointly—to form a coalition.

Perhaps the most interesting of Thompson's propositions about coalitions pertains to attempts to control external threats to decisions. When these threats are great, people given substantial discretion will seek to form coalitions with people outside the organization. For example, marketing managers for competing firms may secretly agree to split up the potential market and thereby limit their vulnerability. Or a production manager may make secret agreements with suppliers to reduce uncertainties.

Indeed, newspaper editors have noted that reporters assigned to cover the police form coalitions with police commanders, trading favorable coverage for special access to information. A reporter assigned to the police beat risks his or her job if the police favor competing reporters with inside information, and police commanders fear unsympathetic reports will provoke public criticism. Coalitions therefore serve the interests of both commanders and reporters but undercut the interests of both organizations.

Although the natural system of organizations often circumvents decentralization, many organizations, especially business organizations, are much more decentralized than they used to be. This decentralization has been the major trend for formal organizations generally. The Napoleonic problem of managing large armies was solved by delegating authority. The problem of managing big railroads was corrected by breaking them up into functional units, each the size of a small railroad. And the problems of managing a huge, diversified company like DuPont were solved by treating parts of the organization as independent companies. The adoption of management by objectives, in which authority is delegated to ever smaller internal organizational units, is simply an extension of this same trend. But it would be misleading to end this chapter without noting that not all organizations have followed this decentralizing trend. Thus, while decentralization has dominated organizations in the private sector, government organizations have tended toward ever greater centralization.

Governments grew rapidly in the twentieth century. In most modern societies, the government is by far the largest employer, and government spending makes up a very substantial proportion of all monetary transactions. But as governments have grown huge, they have become ever more centralized. Many government activities that were left wholly to local jurisdictions even twenty years ago are centralized today.

In Europe government is even more centralized than in Canada or the United States. In most nations local government is nearly nonexistent—the central government appoints and controls local mayors. Many colleges and universities are run by a ministry of education in the nation's capital. The ministry not only determines how money is spent but also makes all faculty appointments, chooses textbooks, and even controls student admissions.

Dictatorships are even more centralized. In fact, the government in some undemocratic nations attempts to administer the whole economy—industry, retailing, farming, and so on—from a central bureaucracy. This was, of course, the determining principle of Marxist political ideas, and for a long time it was claimed that Marxist regimes were far more effective than democracies because they could control everything. Since the breakup of the Soviet Union and the liberation of Eastern Europe, we now know that Marxist attempts at central planning were incredibly ineffective, turning economies into a shambles of inefficiency and corruption. This is no surprise. If DuPont was too big for central planning, how could anyone hope to run an entire nation in that style? Yet even democratic governments are reluctant to learn these lessons. Proposals to let "the government" do this or that remain a staple of political life, everywhere.

BUREAUCRACY AND THE BOTTOM LINE

In everyday speech we often use the words *bureaucracy* and *bureaucrat* negatively. They suggest meddlesome people and muddled organizations. Yet as Weber noted, all modern organizations, public or private, are bureaucracies. We usually think of government agencies and personnel when we hear the terms *bureaucracy* and *bureaucrat* because government agencies generally do not work as well as private organizations. They are less efficient because they do not face the same pressures to reexamine and improve their organizational methods.

The strength of private bureaucracies is their vulnerability. They can and often do go broke. This means that they are under constant pressure to earn the right to exist, to achieve their goals, and to adapt to changing circumstances. The DuPont executives stayed late at night in 1921 trying to figure out how to put their organization right not because they were more conscientious than government executives but because time was rapidly running out. If they had not found immediate solutions to their organizational problems, then they soon would have had to inform their creditors that DuPont was bankrupt.

After spending weeks in city offices trying to get a building permit in Brooklyn, artist George Tooker painted *Government Bureau*, which now hangs in New York's Metropolitan Museum of Art. We catch but glimpses of the bureaucrats hiding behind their rules and regulations.

Business organizations therefore have a very clear standard by which to judge their performance: profit and loss. While profit is not the only goal of corporation executives, it is the ultimate one. Day in and day out, profit and loss provide a gauge indicating which operations are working better and which are working worse, which managers are doing better and which are doing worse. The bottom line is always evident. Thus, DuPont executives could not take comfort in the popularity of their new paint products or in their high quality, for to do so meant bankruptcy.

Government bureaucracies do not have such a clear gauge of their success or failure. They do not go broke, they do not count profits, and they can claim great credit when their actions are popular without asking if their actions are effective or efficient. This does not mean that the effectiveness of government bureaucracies cannot be measured. But often, it simply isn't measured.

Recently, many political scientists have proposed that measures of government effectiveness be required. Of course, they do not suggest that government agencies should operate on the basis of profit and loss. They do suggest that

government agencies should measure their results objectively so that their performance can be evaluated. Such measures are not hard to devise if an organization's goals are clearly identified.

Imagine that we are going to reorganize a welfare agency. First, we would need to set its goals. One goal might be to provide jobs for the maximum number of welfare workers. While some people suspect that this is often the goal of public agencies, it would not be acceptable as an official goal. Suppose that we set the goal as getting the maximum amount of welfare money into the hands of poor people. Then we might want to close the welfare agency and send all of its administrative funds to poor people. That is precisely what proposals such as guaranteed income or negative income tax intend to accomplish.

However, we might choose as our goal to minimize the time people spend on welfare by returning a maximum number to economic self-sufficiency as soon as possible. Then we would evaluate the agency by keeping track of the proportion of persons leaving the welfare rolls and the average length of time people were on welfare. With these measures one program could be compared with another. Then we could see if a cheaper program worked as well as or better than a more expensive one, whether some welfare workers were more effective than others, and whether different programs were needed in different places. If decentralization theory is correct, then we would expect a program to work better when local welfare offices were given maximum authority and responsibility to achieve the stated goal.

CONCLUSION

This chapter's fundamental theme is that organizations are human inventions intended to serve human needs. The rational, or formal, organization based on bureaucratic principles was a major social invention necessitated by the great increase in the scale and complexity of human activities in modern times. Perhaps the primary lesson we have learned from our brief experience with formal organizations is that there is no such thing as a perfect organization. Instead, organizational forms and principles that serve extremely well under some conditions can be inadequate or even harmful under others.

Effective organizations are therefore the momentary results of constant reassessment and redesign. When the effectiveness of organizations is not tested, they rapidly tend to become unresponsive and inefficient, taking on the negative features associated with the term *bureaucracy*. In the final analysis, we get the kinds of organizations we deserve. We create them; we run them. If they make our lives unpleasant, we can change them. The fundamental truth about organizations is this: Organizations never do anything; only people do things. While much of what people do is the result of their positions in formal organizations, the fact remains that organizations never make decisions, pursue goals, assess means, or adopt new policies. Only people, acting in the name of organizations, do such things.

By the same token, societies don't cause changes. Society didn't invent the steam engine; James Watt did. Societies don't have babies; women do. And societies become urban only as people move into town. Thus, it is time to shift the focus of this introduction to sociology away from large social structures and back once more to human behavior. To conclude this book, we shall examine how people cause and resist change.

Review Glossary

Terms are listed in the order in which they appear in the chapter.

formal organization Synonymous with rational organization, a group created to pursue definite goals wherein tactics and procedures are designed and evaluated in terms of effectiveness in achieving goals, members are selected and trained to fulfill their roles, and overall operations are based on written records and rules.

divisional system As initially used in armies, the organization of troops into small, identical units, each containing all military elements (infantry, artillery, and cavalry).

geographical divisions Divisions resulting from breaking an organization into smaller units on the basis of geography and making each division relatively independent.

functional divisions Divisions resulting from breaking an organization into smaller units on the basis of specialized activities or functions, such as when a corporation has separate divisions for manufacturing, purchasing, and marketing.

vertical integration The inclusion within an organization of the divisions that control every step in the production and distribution of some product or service.

spoils system System in which the winners take over government jobs after each election.

bureaucracy A formal organization that, according to Weber, is based on (1) functional specialization, (2) clear, hierarchical lines of authority, (3) expert training of managers, (4) decision making depending on rational rules aimed at effective pursuit of goals, (5) appointment and promotion of managers on their merit, and (6) activities conducted by written communications and records.

management Coordination of the work of others.

rational system approach Emphasis on the official and intended characteristics of organizations.

natural system approach Emphasis on the informal and unintended characteristics of organizations.

goal displacement What occurs when the official goals of an organization are ignored or changed.

goal conflict Situation in which one goal of an organization limits the ability of that organization to achieve other goals; for example, the desire to avoid losses due to strikes will conflict with an organization's goal to minimize labor costs.

span of control The number of subordinates one manager can adequately supervise, often estimated as seven.

diversified organization An organization that is not very specialized but instead pursues a wide range of goals.

autonomous divisions Parts of an organization, each of which includes a full set of functional divisions.

decentralization Dispersing of authority from a few central administrators to persons directly engaged in activities.

Blau's administrative theory Theory stating that the larger the organization, the greater the proportion of total resources that must be devoted to management functions.

management by objectives Situation in which managers and subordinates agree on goals that subordinates will try to achieve; subordinates then have maximum freedom in how they will try to reach their objectives.

Suggested Readings

Chandler, Alfred D., Jr. 1962. *Strategy and Structure: Chapters in the History of the American Industrial Revolution.* Cambridge, Mass.: MIT Press.

Drucker, Peter F. 1974. *Management: Tasks—Responsibilities—Practices.* New York: Harper & Row.

Fligstein, Neil. 1985. "The Spread of the Multi-divisional Form among Large Firms, 1919–1979." *American Sociological Review* 50:377–391.

Perrow, Charles. 1979. *Complex Organizations.* 2nd ed. Glenview, Ill.: Scott, Foresman.

Peters, Thomas J., and Robert H. Waterman Jr. 1982. *In Search of Excellence: Lessons from America's Best-Run Companies.* New York: Harper & Row.

Sociology Online

www.socstark10.com

NONE OF THE THREE DATA SETS ONLINE INCLUDES VARIABLES RELEVANT TO THIS CHAPTER.

USING INFOTRAC COLLEGE EDITION

GO TO THE INTERNET AND TYPE:
www.InfoTrac-college.com

↗ **CLICK ON: Register New Account**
You will be asked to enter your Access Code and to create and enter a User Name and Password. Having done so,

↗ **CLICK ON: InfoTrac College Edition or Log On.**

The natural system view of organizations recently has stressed that organizations develop a distinctive "culture." There has been some very interesting writing on this topic. Many other fine discussions of organizations are classified under the subject of organizational behavior.

SEARCH TERMS:
Corporate Culture
Organizational Behavior

REX
THEATRE
FOR COLORED PEOPLE

CHAPTER

21

Until the 1960s southern states had many laws requiring racial segregation. Thus, in Leland, Mississippi, where this picture was taken in 1939, African Americans and whites were required to attend separate theaters. These men would have to pay 11¢ to attend in the afternoon or 15¢ in the evening to see Jack Randall star in *Trigger Smith* and also to see a film of Joe Louis defending his heavyweight championship against Bob Pastor (Louis knocked him out). When a town could support but one theater, African Americans were required to use a separate entrance and sit in the balcony while whites sat on the main floor. Then in the 1950s African Americans began to hold organized protests against segregation. In this chapter we shall see how this came about.

Social Change and Social Movements

O N DECEMBER 1, 1955, MRS. ROSA PARKS, a seamstress in Montgomery, Alabama, boarded a city bus at the end of a long day of work. After paying her fare, she walked down the aisle, passing through the "white section" of the bus as required by law, and took the last seat open in the "colored section." Two stops later, the white section filled up, leaving a white man standing in the aisle. At that point the bus driver, J. P. Blake, asked Parks and three other African American passengers to vacate their seats, which made up the front row of the colored section. No one moved. Blake said, "You all make it light on yourselves and let me have those seats." The other three African American passengers, a man and two women, rose and stood in the aisle. But Parks didn't. Blake told her that if she didn't move, he would call the police and have her arrested for violating Montgomery's bus

segregation law. She told him to go ahead because she was not going to move. Thereupon Blake left the bus and phoned the police. When two officers confronted Parks, she asked them, "Why do you push us around?" One of them answered, "I don't know, but the law is the law and you're under arrest."

Rosa Parks was taken to the police station, booked, and fingerprinted. She was then permitted to call her mother, whose first response was, "Did they beat you?" After her daughter had reassured her and explained her situation, her mother called E. D. Nixon, a local African American leader. Nixon soon arrived at the jail to try to arrange bail for Parks. Meanwhile, something of immense importance was taking place in another part of the city.

After talking with Parks, Nixon had phoned Jo Ann Robinson, a professor of English at Alabama State University, then an all-African

CHAPTER OUTLINE

SOCIOLOGICAL APPROACHES TO SOCIAL MOVEMENTS

SHARED GRIEVANCES
  Economic Domination
  Political Domination
  Personal Domination

HOPE

A PRECIPITATING EVENT

NETWORK TIES

MOBILIZING PEOPLE AND RESOURCES
  Internal Factors: Building the Movement
  External Factors: Opponents and Allies
  The Proliferation of Civil Rights Organizations

FREEDOM SUMMER: MISSISSIPPI, 1964

  A CLOSER VIEW
  Doug McAdam: The Freedom Summer Study

  Becoming a Volunteer
  Biographical Availability
  Attitudes and Values
  Social Networks
  Participants Versus No-Shows
  The Volunteers Twenty Years Later

CONCLUSION

REVIEW GLOSSARY

SUGGESTED READINGS

SOCIOLOGY ONLINE

American school. Robinson was a leader of the newly organized political affairs committee at the Dexter Avenue Baptist Church. She in turn called several other women on the Alabama State faculty who also were on the committee. At midnight they gathered in Robinson's office, under the pretext of grading exams, and began drafting a leaflet:

Another Negro woman has been arrested and thrown into jail because she refused to get up out of her seat on the bus and give it to a white person. . . . Until we do something to stop these arrests, they will continue. . . . The next time it may be you, or your daughter, or mother. This woman's case will come up Monday. We are, therefore, asking every Negro to stay off the buses on Monday in protest of the arrest and trial.

As they worked on the leaflet, the professors realized that there was nowhere they could get it printed—whites owned all of the printing shops. So they stayed through the night and ran off copies on the mimeograph machines at Alabama State. Around 3 A.M. Robinson called Nixon and told him what she and her associates were doing. He was enthusiastic about their plan and agreed to organize a meeting that evening at her church. So, at 5 A.M. Nixon phoned the new pastor of the Dexter Avenue Baptist Church, a twenty-six-year-old from Atlanta by the name of Martin Luther King Jr. When asked to endorse a one-day boycott of the buses, King said, "Brother Nixon, let me think about it and you call me back." An hour later, Nixon called King again. The young minister said he would be willing to take part. Nixon replied, "I'm glad you agreed because I already set the meeting up to meet at your church."

The rest is history.[1]

At the end of the previous chapter, I pointed out that societies never do anything. Only people do things. So, in this closing chapter of the book, I would like to clarify the ways in which *individuals can get together to cause or to prevent social change.* Whenever people organize to cause or prevent social change, we identify them as a **social movement.**

To give substance to an analysis of social movements, we shall study particular groups and events involved in the Civil Rights Movement during the 1950s and early 1960s. We will

1. This account is based on Morris (1984), Branch (1988), Hampton and Fayer (1990), and Weisbrot (1991).

examine why and how African Americans formed groups and mounted organized actions to end segregation and how many whites formed groups and took actions to try to stop them, while other whites joined forces with African Americans to secure an end to racism.

A primary issue to be addressed is *the degree to which people are able to shape history.* Some sociologists argue that we are almost entirely the creations of our society, that our history determines our destiny. Other sociologists disagree. They believe that we have considerable freedom to make our own history—that the course of human events often is decided by what a few people decide to do or not to do. As you read the chapter, you may want to examine the actions of the people involved in a particular event and ask how much impact they had on history or to what extent they simply were puppets dancing to social forces beyond their control.

SOCIOLOGICAL APPROACHES TO SOCIAL MOVEMENTS

In recent years sociological studies of social movements have tended to emphasize one of two basic approaches. The older of these can be described as the **collective behavior approach.** Sociologists committed to this approach emphasize social movements as outbursts of group activity in response to deeply felt grievances (Smelser, 1963; Turner and Killian, 1987). The collective behavior approach also places great significance on the role of ideology in fixing the goals and tactics of a social movement and stresses the importance of emotions and feelings, as opposed to rational decision making, on the part of participants. According to Bruce Fireman and William A. Gamson (1979), the key question asked by sociologists using the collective behavior perspective is, Why do these people want social change so badly and believe that it is possible?

In contrast, the **resource mobilization approach** all but dismisses grievances as the basis for social movements. Sociologists who embrace this approach stress the importance of resources, both human and material, and rational planning both as the source of social movements and as the basis for their success. John D. McCarthy and Mayer N. Zald (1977), who coined the term *resource mobilization,* claim that "there is always enough discontent in any society to supply the

© AP/ Wide World Photos

grass-roots support [for a social movement] if the movement is effectively organized and has at its disposal the power and resources of some established elite group." Indeed, McCarthy and Zald stress the ways in which those seeking to organize a social movement can define, create, and manipulate a sense of grievance among potential followers.

The resource mobilization perspective places its greatest emphasis on leadership. Indeed, it is assumed that social movements sometimes are "stirred up" by leaders having little or no real concern about the issues and problems they identify. This occurs because social movements provide activists with the opportunity for good jobs, power, and even fame, and therefore, there is always sufficient motivation for people to attempt to form social movements.

The key question asked about social movements by sociologists working from the resource mobilization perspective is, How can these people organize, pool resources, and wield them effectively? (Fireman and Gamson, 1979.)

In the judgment of most sociologists who work in this area, neither the collective behavior nor the resource mobilization approach is adequate by itself. Rather, each seems deficient precisely where the other is strongest. Resource mobilization offers a very important corrective

At the Montgomery jail, Rosa Parks is finger-printed after her arrest for refusing to give up her seat on the bus to a white man—a clear violation of the Montgomery bus segregation law.

to the tendency of social scientists to focus too exclusively on the grievances voiced by social movements and to pay too little attention to the organizational realities involved in all social movements. In doing so, however, sociologists committed to the resource mobilization approach seem to pay too little attention to the realities of grievances expressed by movements. Granted that sometimes grievances are largely illusory—and granted, too, that often people must be educated by activists to recognize their grievances—nevertheless, social movements seem to be more durable and generate greater member commitment to the extent that they represent genuine and intensely felt grievances as opposed to dubious claims raised by cynical spokespersons.

The following eight propositions synthesize both approaches to explain how social movements arise and then how they succeed or fail in their quest to change society. Subsequently, each proposition is developed and illustrated at length.

For a social movement to *occur:*

1. Some members of the society must share a *grievance* which they want to correct, either by changing society or by preventing a change they oppose.

2. These people must have *hope*—they must think there is some possibility of success.

3. Often, but not always, a *precipitating event* will ignite pent-up grievances and convince people that the time for action has arrived.

4. People are recruited by social movements through networks of attachments—a point that has been stressed throughout the book. Not only are *individuals* recruited through their network ties, but social movements often *originate within a network,* such as a church congregation, a civic-improvement club, or a trade union. Moreover, once a movement is under way, sometimes whole networks, including those constituting formal organizations, will join at one time.

For a social movement to *succeed:*

5. It must achieve an effective *mobilization of people and resources.* That is, a social movement will tend to be more successful to the degree that it enjoys effective leadership, attracts committed and disciplined members, and is able to secure the necessary finances and facilities. These are classified as *internal* factors influencing a social movement.

6. It must withstand or overcome *external* opposition. Usually, a social movement will generate one or more **countermovements.** That is, if a group organizes to cause social change, it is likely to be opposed by a group (or groups) opposed to this change.

7. The fate of a social movement also depends on enlisting *external allies* from other major groups and powerful institutions in the society—or at least it must be able to keep them neutral.

8. Whenever social movements arise in response to a widely shared grievance, and when substantial resources are available, the movement will tend to be embodied in a number of *separate organizations.* These social movement organizations may cooperate, but often, they compete rather vigorously.

Now let's examine these principles in detail as we attempt to understand the history of the Montgomery bus boycott.

SHARED GRIEVANCES

People don't try to change their society if they are satisfied with things as they are. For a social movement to form, some set of people must share some significant discontent—a grievance that they want to have corrected (Smelser, 1963; Gurr, 1970; Turner and Killian, 1987).

It would be hard to exaggerate the extent and depth of the shared grievances experienced by African Americans in the South prior to the Civil Rights Movement. Aldon D. Morris (1984) describes the domination of southern whites over African Americans as "a comprehensive system" made up of three elements: "Blacks were controlled economically, politically, and personally."

ECONOMIC DOMINATION

Economic control of African Americans rested on discrimination in hiring. Many jobs were simply closed to them; help-wanted ads in the newspapers commonly noted that a job was "For Whites Only." So, most African American men in the South held only the lowest-paying, dirtiest, least desirable jobs. As for African American women, not only were they far more likely than white women to be employed, but half of them worked as domestics—cleaning and cooking for white families. In 1950 African American families nationally earned only 54 percent of the median income enjoyed by white families. In the South the income gap between African Americans and whites was even greater. But it wasn't only income differences that separated the races in the workplace. It was whites who told African Americans what to do. Morris (1984) explains this difference vividly:

Whites had the jobs that required white shirts and neckties. Whites decided who would be promoted, fired, and made to work the hardest. While Black men in greasy work clothes labored in these conditions, their mothers, wives, and sisters cleaned the houses of white women and prepared their meals. Blacks entered into these exploitative economic relationships because the alternative was starvation or at least unemployment, which was usually much higher than average in the Black community.

POLITICAL DOMINATION

Southern African Americans were virtually without any political power. A variety of methods were used to prevent African Americans from registering to vote. A favorite was to require everyone wishing to register to vote to pass a literacy test. White voting registrars saw to it that African Americans always failed. As a result no African Americans held elective office in any southern state despite the fact that they made up a very substantial part of the population. For example, in 1950 African Americans amounted to 45 percent of the population in Mississippi and 39 percent in South Carolina. Yet in many southern counties, not a single African American was registered to vote. In 1955 the Rev. George Lee of Belzoni, Mississippi, was the first African American to register to vote in his county. Shortly thereafter, he was shot to death (despite his gunshot wounds, local authorities ruled that he had died in a traffic accident). A few weeks later in Brookhaven, Mississippi, the only African American to have voted in the recent primary was gunned down in broad daylight on the steps of the courthouse. There were no arrests.

So, whites decided how to spend all the tax dollars, even though African Americans were not exempt from paying their share. And although white leaders in the South long defended segregation on the grounds of separate but equal treatment and facilities, the truth was more often as shown in the photo of the white and colored drinking fountains. In addition to controlling local government, whites also controlled the courts and the police. There were no African American cops.

PERSONAL DOMINATION

African American men of any age were called "boy" or addressed by their first name by whites of any age. However, all African Americans were expected to address all whites in a formal and deferential manner; thus, elderly African Americans would address white children as "Miss" or "Mister." Not only were African Americans required to use different toilets, drinking fountains, and waiting rooms and to sit in their own part of the bus, they also were excluded from white parks, schools, hospitals, hotels, and restaurants. As recently as 1961, when the New York Yankees went to Florida for spring training, their only African American player, Elston Howard, had to board with a local family because no hotel in St. Petersburg would accept African American guests (Houk and Creamer, 1988).

Finally, an African American man had to be extremely cautious when in the presence of white women. The slightest familiarity might cause him to be charged with rape—a crime that, although it carried the death penalty,

Throughout the South stores and public buildings were required to provide separate bathrooms and drinking fountains for whites and African Americans. This practice was defended on the grounds that the facilities, while separate, were equal. But the truth was far more often as shown here. Notice that the water for whites passes through an electric cooler. That for African Americans does not.

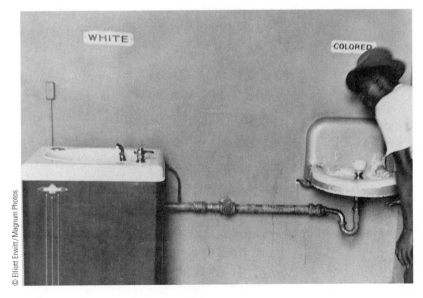

© Elliott Erwitt / Magnum Photos

As the Civil Rights Movement spread, new tactics evolved. Here, on February 2, 1960, four young African American men began a "sit-in" at the Woolworth's lunch counter in Greensboro, North Carolina. They had decided to sit in a seat every day until they were given service. Ironically, African Americans could work behind the counter, but they couldn't be served. After four months, Woolworth's began to serve African Americans.

Reprinted with permission of the *News & Record*

typically triggered white lynch mobs long before charges could be filed or any trial held. In 1955 a fourteen-year-old African American youth from Chicago, visiting relatives in Mississippi, said "Bye, baby" to a white woman as he left a rural store. Several days later, a group of white men armed with shotguns came calling and took young Emmett Till away. Later, his body was discovered in the Tallahatchie River. He had been severely beaten and one eye had been gouged out before he had been shot in the head.

HOPE

People will endure the most terrible situations with considerable stoicism if they believe they must. Thus, Turner and Killian (1987) noted:

Even a state of slavery will be accepted by the majority of its victims without challenge when the slaves are taken out of reach of their homes, denied the skills necessary for independent existence, insulated from the practice of making significant decisions for themselves, and in other respects rendered totally dependent.

By the 1950s African Americans in the South were a long way from the dependent state they had once been in. Indeed, despite segregation, the average African American was far better educated than had been true even a generation before. Moreover, African Americans had been entering professions such as medicine and law, finding their patients and clients within the African American community. In addition segregated schools meant that large numbers of the teachers and principals in the South were African American, and the South also sustained many African American college and university professors, employed at its all-African American schools. Clearly, things were getting better. And with these changes came the ability to hope—to think that something could be done to make things *much* better. These hopes received a very strong boost in 1954 when the Supreme Court ruled against segregated schools.

When people come to believe that things can be made better rapidly, it can lead to what social scientists identify as a **revolution of rising expectations.** The probability that people will take action to change their circumstances increases to the extent that their hopes for change far outstrip the actual progress taking place. The greater the gap between expectations and reality, the more impatient and frustrated people become. James C. Davies (1962) formulated a version of this principle that he called the **J-curve theory of social crisis.** This is depicted in Figure 21-1. As long as what people hope to get and what they actually are getting remain relatively close, no crisis occurs. This remains true even if expectations for change rise rapidly, as long as the actual rate of change keeps pace. However, should the rate of change suddenly slow, the growing gap between what people hope to get and what they actually get begins to cause increasing anger and dissatisfaction. Moreover, should there be an actual decline in what people are getting, there is apt to be a severe crisis. This latter situation resembles a J lying face down, hence the name J-curve.

It is impossible to know whether things suddenly got worse for African Americans in Montgomery. But clearly, the rate of change was falling far behind their rising expectations. For people who had come to believe that there could be an end to white domination over them, the day-to-day humiliations of segregation became increasingly intolerable. Someone was bound to take action sooner or later. And then Mrs. Rosa Parks did.

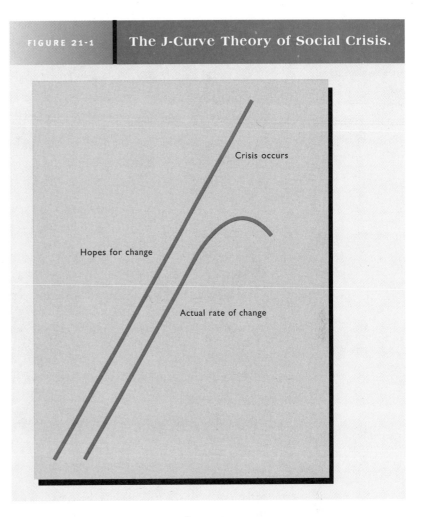

FIGURE 21-1 **The J-Curve Theory of Social Crisis.**

Crisis occurs

Hopes for change

Actual rate of change

A PRECIPITATING EVENT

The birth of a social movement often is prompted by a dramatic event that galvanizes people into action—what Neil J. Smelser (1963) calls a **precipitating event** that serves as "a symbol of cumulated grievances." That is, something specific happens that communicates to people with a shared grievance and increasing hopes for change that *now is the time for action.*

It has been widely believed, on the basis of press reports at the time, that Rosa Parks acted on sudden impulse when she refused to give up her seat on the bus. In fact, she had been doing this on and off for years. So had others. As she tells it:

I happened to be the secretary of the Montgomery branch of the NAACP [the National Association for the Advancement of Colored People, the leading Civil Rights organization of the time] and many cases did come to my attention that nothing came out of 'cause the person abused would be too intimidated to sign an affidavit, or to make a statement. Over the years, I had my own

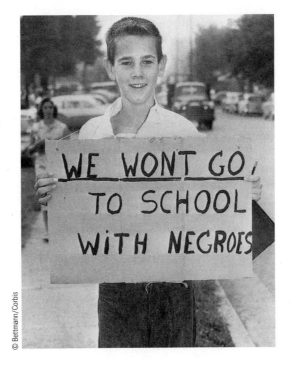

On August 27, 1956, Clinton High School in Clinton, Tennessee, responded to federal court orders and opened its doors to both African American and white students. Some white students, including John Carter, protested outside the school and refused to attend.

© Bettmann/Corbis

After a federal court voided Alabama's bus segregation law, Mrs. Rosa Parks could ride in any seat that was available, and never again was she ordered to give up her seat to a white person.

© Flip Schulke / Black Star

problems with the bus drivers. In fact, some did tell me not to ride their buses if I felt I was too important to go to the back door to get on. One had evicted me from the bus in 1943, which did not cause anything more than just a passing glance. (quoted in Hampton and Fayer, 1990)

In fact, the driver who had forced her off the bus twelve years before was the same one who had her arrested in 1955!

Several things should be noted in Rosa Parks's remarks. First, her many previous actions had not precipitated an arrest. That the police were called on this occasion indicated that the white community had become far more concerned about preventing changes in southern racial norms—a concern prompted by threats of change posed by recent rulings by the Supreme Court, President Eisenhower's termination of legal segregation in the District of Columbia, the institution of strictly enforced integration in the armed forces, and the breaking of the color barrier in professional sports. Second, when this same driver had ordered her off his bus twelve years before, Parks had complied. This time the driver realized she was not going to budge. Her resolve also reflected major changes in the mood of the Montgomery

African American community. Parks was not acting in a vacuum; her decision to risk arrest reflected her belief that many members of the African American community shared her perception that the time for resistance had come. She seems not to have been surprised by the actions taken by Jo Ann Robinson and her colleagues that same night. Third, Parks wasn't the humble seamstress of legend, although that is how she earned her living. She was a very prominent figure in the African American community and a leader in previous efforts to resist white domination. Indeed, she stood at the center of a *dense network of attachments* that could serve as the basis for recruiting people to form a social movement.

NETWORK TIES

Typically, people do not run out and join a social movement because they read about it in the newspaper or heard about it on TV. Instead, they join social movements because of their interpersonal attachments to others who have become involved. Moreover, social movements typically are not founded by a lone individual, who then manages to induce his or her friends and neighbors to join. Instead, social movements often are created by a group, and then group members reach out to others to whom they are attached. It often happens, too, that a movement spreads from group to group, rather than from individual to individual.

All of these forms of building the movement took place in Montgomery. Jo Ann Robinson and her associates, who launched the initial call for a bus boycott, not only were friends and members of the same faculty but also belonged to the same church. All served on the same political affairs committee of that church—and their pastor was none other than Martin Luther King Jr. When E. D. Nixon organized a meeting of African American clergy at King's church, not only did he know all of them, but they all knew one another well. And every one of those mentioned above was well acquainted with the secretary of the Montgomery Chapter of the National Association for the Advancement of Colored People—Rosa Parks. Moreover, for most African Americans in Montgomery, the decision to join the boycott was not an individual act so much as it was a collective action by members of closely knit church groups.

MOBILIZING PEOPLE AND RESOURCES

Social movements don't just happen. They require a number of crucial ingredients, which can be separated into internal and external factors. The primary *internal* factor is a group of people with sufficient commitment and motivation to engage in activities designed to cause change. These actions often will involve danger and sacrifice, so social movements must overcome the "free rider" problem discussed in Chapter 15. Second, social movements require effective leadership: The group must include members with sufficient experience and skill to formulate and coordinate effective plans of action. Third, social movements need material resources of various kinds. They must have means to communicate with one another—which is why Jo Ann Robinson and her friends wrote a leaflet and then used Alabama State University paper and mimeograph machines to run off copies. Movements also benefit from having places to hold meetings and having funds to meet various costs; to bail Rosa Parks out of jail was but one example.

The primary *external* factor is the degree to which the movement can be suppressed by those opposed to its aims. As we shall see, massive pressure was brought to bear on the African Americans in Montgomery to force them to ride the buses. But their leaders were sufficiently independent from white economic pressure, sufficient support for the movement was soon forthcoming from elite groups outside the South, and the federal courts imposed sufficient restraint on white resistance to sustain the movement.

INTERNAL FACTORS: BUILDING THE MOVEMENT

The evolution of Jo Ann Robinson's efforts into a social movement was based on the only institutions over which southern African Americans had total control—their churches. It was no accident that E. D. Nixon phoned Martin Luther King Jr. and asked for the use of his church. The Dexter Avenue Baptist Church already possessed a political action committee, headed by Jo Ann Robinson. No white needed to give permission for Montgomery's African American leaders to gather there. Moreover, these leaders were, nearly without exception, local ministers—including Rev. Ralph D. Abernathy

of the "Brick-a-Day" First Baptist Church, destined to become a leader in the Civil Rights Movement second in prominence only to Martin Luther King Jr.

Not only were African American ministers skilled and experienced organizers, as Aldon D. Morris (1984) pointed out they did not depend on whites for their incomes. They could not be silenced for fear of being fired. Therefore, when the African American ministers gathered at the Dexter Avenue Baptist Church in response to E. D. Nixon's call for a meeting, each represented all the crucial internal ingredients of a social movement: The tightly knit groups of African Americans who made up their congregations would supply the committed and motivated membership, church buildings within which to hold meetings, and the ability to raise money.

Friday evening, the day after Rosa Parks's arrest, about fifty African American leaders gathered in the basement of the Dexter Avenue Church and adopted a resolution worded very similarly to the leaflet produced by Jo Ann Robinson. It urged: "Don't ride the bus to work, to town, to school, or any place Monday, December 5. . . . If you work, take a cab, or share a ride, or walk." The final sentence invited African Americans in Montgomery to "come to a mass meeting Monday at 7:00 P.M. at the Holt Street Baptist Church for further instruction" (Branch, 1988). This resolution was then run off on the Dexter Avenue Church mimeograph machine.

On Monday African Americans did not ride the buses. At midmorning Rosa Parks was convicted as expected, but something previously unheard of took place at the courthouse. As E. D. Nixon left the courtroom to arrange for Parks's release on bail, he confronted a huge, angry crowd of African Americans jamming the corridors and the steps leading back into the street. Nixon tried to calm them and assured them he would return with Parks, unharmed, in a few minutes. Voices in the crowd shouted that if not, they would storm the courthouse. African American militancy such as this had never been seen before in Montgomery.

Monday afternoon the leaders met again and took major steps to channel community anger into effective actions. First, the group created a new organization—the Montgomery Improvement Association (MIA)—and then they elected Martin Luther King Jr. as its president on the grounds that as a newcomer he had no obligations to fulfill to the white power structure and had no enemies among African American leaders.

Then they gave the MIA a new mission. The bus boycott would be continued until segregated seating was discontinued.

Arriving at the Holt Street Church Monday night, King was forced to walk the last several blocks because the crowd gathered for the mass meeting filled the streets. Loudspeakers were being set up so that the large crowd unable to enter the church could hear, too.

When he addressed the crowd, King reminded his audience of their long suffering and humiliation. And when he said, "And you know, my friends, there comes a time when people get tired of being trampled over by the iron feet of oppression," the crowd exploded with cheering that went on for minutes, actually causing the church itself to shake. When he was able to be heard again, King reminded everyone that the MIA was absolutely committed to nonviolence:

I want it to be known throughout Montgomery that we are a Christian people. The only weapon we have in our hands this evening is the weapon of protest. If we were incarcerated behind the iron curtains of a communistic nation—we couldn't do this. If we were trapped in the dungeon of a totalitarian regime—we couldn't do this. But the great glory of American democracy is the right to protest for right.

I want it to be known—that we are going to work with grim and bold determination—to gain justice on the buses in this city. And we are not wrong. We are not wrong in what we are doing. If we are wrong—the Supreme Court of this nation is wrong. If we are wrong—God Almighty is wrong! (quoted in Branch, 1988)

The importance of the bus boycott as a protest tactic lay in the fact that African Americans, not whites, made up the vast majority of passengers. Denied their fares, the bus company began to lose money rapidly. After three weeks, company managers informed the city commissioners that the company faced imminent bankruptcy. The commissioners responded by approving a substantial fare increase.

To get people to their jobs, Montgomery's African Americans organized a massive carpool. Many middle-class African Americans donated their cars to the MIA for the duration of the boycott, knowing they would be returned worn from such heavy use—about 20,000 people a day relied on these carpools. And thousands of people simply walked.

Listen to Gussie Nesbitt, a fifty-three-year-old domestic worker and a member of the Holt Street Baptist Church:

I walked because I wanted everything to be better for us. Before the boycott, we were stuffed in the back of the bus like cattle. . . . And [often] the bus driver would say, "Let me have that seat, nigger." And you'd have to get up. A lot of times that we'd go to the front, he wouldn't let us in the front. He'd take our money at the front, and then before we could come on through the back door he'd drive off and leave us standing there. He done took our money and gone. That's how it was and that's why I walked. I wanted to cooperate with the majority of people on that boycott. I wanted to be one of them that tried to make it better. I didn't want somebody else to make it better for me. I walked. I never attempted to take the bus. Never. I was tired, but I didn't have no desire to get on the bus. (quoted in Hampton and Fayer, 1990)

EXTERNAL FACTORS: OPPONENTS AND ALLIES

Social movements never occur in a vacuum. Whenever people organize to cause social change, they are likely to be perceived as a threat by some others who prefer things to stay as they are. For example, social movements seeking to restrict access to firearms have stirred up angry reactions from various groups who wish to maintain an unrestricted right to own and bear firearms, as guaranteed by the U.S. Constitution.

To the degree that movements proposing change seem to be making headway, opponents are likely to organize social movements of their own to prevent social change. Moreover, both sides will seek to enlist allies—and especially to enlist economic, political, and legal support on behalf of their cause. So, let's look more closely at what was going on outside the Montgomery Improvement Association.

THE WHITE COUNTERMOVEMENT The Supreme Court decision against segregated schools served as a precipitating event that caused organizations to spring up in southern communities with the objective of preventing all federal efforts to end segregation. Many of these groups called themselves White Citizens' Councils. These, too, were social movements.

In the beginning the councils mainly attracted low-status whites, many of them also active in the Ku Klux Klan—the masked and robed men who used violence and murder for nearly a century to keep African Americans "in their place." However, as the boycott continued in Montgomery, prominent whites began to enlist in the local White Citizens' Council.

At a rally of some 1,200 council members held in the Montgomery City Auditorium in early January, Clyde Sellers, the city commissioner in charge of the police, marched to the podium and declared himself a member. Shortly thereafter, Mayor W. A. "Tacky" Gayle and the remaining city commissioners joined the council.

Meanwhile, the white community began to strike back at the boycott. The first blow ended up inflicting more harm on whites than on African Americans. The mayor ordered the town's white women to lay off their African American maids and cooks until or unless these women rode the bus. But he greatly misread the willingness of white housewives to begin, for perhaps the first time in their lives, to clean their own houses. Not only did virtually all of them refuse to comply, but many began to transport their domestics in their own cars. The mayor took to the newspaper to condemn this: "The Negroes are laughing at white people behind their backs. They think it's very funny and amusing that whites who are opposed to the Negro boycott will act as chauffeurs to Negroes who are boycotting the buses."

But the next measures were far uglier. The police began to stop the carpool drivers and ticket them for imaginary violations. The drivers responded by driving very slowly and giving exaggerated turn signals. But the blizzard of tickets continued. Jo Ann Robinson picked up seventeen tickets in two months. Drivers began to have their licenses suspended for too many tickets, and many had their insurance canceled for the same reason. In addition the police began to arrest people waiting at carpool pickup points, charging them with loitering. But the boycott held fast.

Then on January 26, Martin Luther King Jr. noticed two motorcycle policemen pull in behind him as he left a pickup point with a load of African American commuters. King drove far below the speed limit, hoping the officers would depart. Instead, they pulled him over and informed him he was guilty of driving 30 miles an hour in a 25-mile-an-hour zone. But they didn't ticket him for speeding. They arrested him and took him to jail. He feared the worst.

However, King was not in jail long before a huge crowd of African Americans surrounded the jail. Suddenly, a very nervous jailer reappeared and hustled King out the front door of the jail, free on his own recognizance. That night King spoke at a series of mass meetings held in churches across the city. A few days later, while King was again addressing a mass meeting, Ralph Abernathy interrupted him with the news that his home had been bombed. Rushing home, King found that his wife and little daughter had escaped harm, although the front of the house was badly damaged.

From that day on, volunteers guarded the King home around the clock. But there was nothing they could do to prevent King from being arrested once again when the county grand jury brought in dozens of indictments of African American leaders on the grounds of leading an illegal boycott. The MIA had to scramble to raise the needed bail money. However, the arrests turned into an immense public relations victory for the MIA because it gained the attention of the national press. The boycott no longer was simply a local affair.

EXTERNAL SUPPORT As we have noted, social movements operate in a social environment, and characteristics of the environment can play a decisive role in the fate of a movement. Thus far, our focus has been primarily on the hostile aspects of the environment of the MIA—on white opposition and efforts to force an end to the boycott. However, the social environment also imposed clear limits on this opposition. As Martin Luther King Jr. stressed in his speeches, the MIA existed within a democratic society wherein there were legal and normative restraints on white opposition. The police might arrest African American leaders, but they couldn't just line them up and shoot them.

Moreover, the MIA was not confronted by universal white opposition. Even in Montgomery some whites stood up for African American rights, and once the boycott received national press coverage, substantial outside support began to build up.

Almost at once the MIA began to receive donations from church groups in both the North and the South. While these primarily were from African American churches, some white churches sent money, too. Soon other groups began to contribute; for example, the United Auto Workers Union in Detroit donated $35,000.

These contributions proved vital to sustaining the boycott. Much of the money was spent

Rev. Martin Luther King Jr. is dragged into the Montgomery, Alabama, police station and booked for loitering outside the courthouse. To the right is his wife, Coretta Scott King.

to support the carpool system by which African Americans were able to get around the city without using the buses. Other funds were used to pay traffic tickets and to provide bail.

In addition to sending money, prominent African Americans began to visit Montgomery to display their support. And King and Abernathy began to accept speaking engagements in other cities, which gave them access not only to the audience gathered to hear them but to local press coverage in each city as well. The

Montgomery bus boycott was becoming a national cause.

There was another supportive force in the environment, and in the end it ensured victory for the MIA. On November 13, 1956, nearly a year after the boycott began, the U.S. Supreme Court ruled that Alabama's state and local laws requiring segregated buses were unconstitutional.

Thus, the boycott came to a close. But that was the only thing that ended. For that was the only form of segregation covered by the Court's verdict. The drinking fountains, parks, hospitals, hotels, restaurants, theaters, and beaches still were segregated. Much remained for the MIA to do.

Moreover, the Montgomery boycott was prompting African American social movements to organize in many other communities across the South. Indeed, students at Florida A&M (then an all-African American school) had begun a bus boycott in Tallahassee in May of 1956.

King and his associates realized the need for an organization that transcended local communities. Therefore, in January 1957 African American ministers from many southern communities met in Atlanta and formed the Southern Christian Leadership Conference. The Civil Rights Movement was launched.

THE PROLIFERATION OF CIVIL RIGHTS ORGANIZATIONS

Whenever a substantial number of people share a grievance and possess substantial human and material resources, a variety of organizations typically will arise and compete for these resources (McCarthy and Zald, 1977). And so it was in this case.

The MIA, for example, was but one short-lived movement among hundreds that took part in the struggle for civil rights. In fact, there never existed an organization named The Civil Rights Movement. That is merely a collective noun applied to a great variety of different social movement organizations that had in common only the desire to cause social change on behalf of racial equality. Often, these organizations cooperated. But often, they expressed very different aims, employed very different tactics, and produced quite different responses from the general public.

One of these groups was the Student Non-Violent Coordinating Committee (SNCC). To conclude this chapter, let's retrace the high point of SNCC's efforts: its Freedom Summer Project that brought 1,000 white college students to Mississippi during the summer of 1964 where they literally risked their lives in an effort to register African American voters. Then we shall watch sociologist Doug McAdam conduct a study of these student volunteers. Who were they? Why did they go? How were they recruited? And how did this experience influence their subsequent lives?

FREEDOM SUMMER: MISSISSIPPI, 1964

By the fall of 1963, African American student volunteers belonging to SNCC (pronounced "snick") had spent nearly three years attempting to help African Americans in Mississippi organize to overcome the savage racism confronting them. Despite their sacrifices and their willingness to keep going in the face of a succession of unpunished attacks (including murders) by militant whites, they had achieved virtually nothing. The whole structure of white supremacy was as solid in Mississippi as it had ever been.

Only 6.7 percent of African American adults were registered to vote, and these mostly lived in cities where their votes were of limited importance (in 1960 two-thirds of all African Americans in Mississippi lived in rural areas where most were sharecroppers who grew cotton). In five rural Mississippi counties where African Americans made up a majority of residents, not a single African American was registered to vote.

Half of the houses occupied by African Americans in Mississippi lacked running water, and two-thirds did not have toilets. While 42 percent of whites in Mississippi had graduated from high school, only 7 percent of African Americans had done so; the average Mississippi African American had attended school for only six years. Although Mississippi defended its segregated school system on the grounds that it was separate *but equal*, in fact per capita school spending was four times as high for white students as for African American students. During 1964 North Pike County, in the heart of the state's cotton belt, spent more than $30 to educate each of its white students, while spending 76¢ on each of its African American students.

And nowhere else in America were white countermovements so popular, so militant, so well armed, so violent, and so immune from prosecution. Early in 1964 the highly respected syndicated columnist Joseph Alsop reported, "Southern Mississippi is now known to contain no fewer than sixty-thousand armed men organized for what amounts to terrorism." He added that acts of terrorism against local African American residents were "an everyday occurrence." Local law enforcement officers usually were unable to solve such crimes. And even when the culprits were well known, juries refused to convict. When a member of the state legislature gunned down an African American

Freedom Summer volunteers join hands and sing beside the bus bringing them to Mississippi in the summer of 1964.

© 1964 Steve Shapiro/Black Star

man who was trying to register to vote (in response to the SNCC campaign), a coroner's jury determined that the white man had merely acted in "self-defense."

Faced with increasingly bitter disappointments, the near exhaustion of their funds, and little hope for progress, SNCC leaders decided to try a dramatic move. Maybe it wasn't of interest to the national news media when African American students got beaten up or murdered in Mississippi. But what would happen if the victims were white students from Harvard, Yale, Princeton, Stanford, Oberlin, Vassar, Radcliffe, and Berkeley? What if these were white students with rich parents and powerful connections? Would the FBI treat terrorism as a local crime if the press really covered what was going on in Mississippi?

Thus was born the plan for Freedom Summer during which 1,000 white volunteers, recruited from the elite colleges and universities, descended on Mississippi with one goal: to register African American voters.

The volunteers began to arrive in June, as soon as their classes had ended. Sixty days later, four civil rights workers in Mississippi had been murdered, scores had been severely beaten up, and thirty-seven churches and thirty African American homes and businesses had been bombed or burned by white terrorists. Although local law officers discovered grounds for making about 1,000 arrests of SNCC volunteers, they

claimed to be utterly baffled by the crimes committed against the volunteers. Indeed, when the FBI solved the murder of three volunteers in Philadelphia, Mississippi, it was revealed that several local law enforcement officers had been involved. Seven men were convicted on federal charges of violating the victims' civil rights (none served more than six years), but no one was brought to trial for more serious charges in Mississippi state courts until forty-one years later. In June 2005, former Ku Klux Klan leader Edgar Ray Killen was convicted of three counts of manslaughter and sentenced to sixty years in prison. Killen was then eighty years old.

Meanwhile, SNCC expectations about press coverage were fully justified. Daily reports of idealistic white students from good families risking death on behalf of African Americans in Mississippi dominated the national news media. Freedom Summer was a public relations triumph that not only had an immense and lasting impact on American opinion but also generated substantial political and financial support for all civil rights groups.

Indeed, lack of progress brought as much or more approval to Freedom Summer as success could have done. Thus, despite all their efforts, the SNCC volunteers managed to persuade only a small proportion of Mississippi's African American adults to go to their local courthouse and try to register, and of these only 1,600 were allowed to do so—the others having their appli-

Holding aloft posters with pictures of three of the SNCC volunteers murdered in Mississippi, these African American students picket in front of the hall in Atlantic City where the Democratic party was holding its 1964 presidential convention. Left to right, the murdered young men were Michael Schwerner, James Chaney, and Andrew Goodman.

cations rejected by white registrars. But no one has ever suggested that this showed that Freedom Summer hadn't been very effective. Rather, that so little progress could be made despite so immense an effort was taken as certain proof that voting rights needed new federal regulations and enforcement. Thus, on August 6, 1965, President Lyndon Johnson signed the Voting Rights Act. Among the guests present for this historic occasion was Rosa Parks.

In the end Freedom Summer had more impact on civil rights than even its most optimistic advocates could have hoped.[2]

A CLOSER VIEW
DOUG MCADAM: THE FREEDOM SUMMER STUDY

In the early 1980s, Doug McAdam, a sociologist at the University of Arizona, noticed how often prominent participants in various American social movements had begun their careers of activism as participants in the Civil Rights Movement during the late 1950s and early 1960s. In fact, McAdam became convinced that an unusually large number had been among the 1,000 white volunteers who took part in Freedom Summer. To pursue this observation, McAdam tried to locate a list of those who had taken part in Freedom Summer. Were he to find such a list, McAdam planned to locate some of the volunteers and find out what they had been doing since 1964.

McAdam tried a number of archives where documents from the Civil Rights Movement are stored, but he couldn't find the list he needed. But then, at the Martin Luther King Jr. Center in Atlanta, McAdam found something far more important and exciting than a simple list of participants. What he found were the original five-page applications filled out by 959 volunteers before they were accepted to take part in Freedom Summer. In fact, he found not only the applications of those who actually took part but also the applications of an additional 300 persons who had applied, been accepted, but had not gone to Mississippi.

These applications would provide McAdam with data on each of these students from when

2. This account is based on McAdam (1988), Hampton and Fayer (1990), and Weisbrot (1991).

Rita Schwerner, shortly after her husband was murdered in Philadelphia, Mississippi, told the Freedom Summer volunteers that it was more important than ever that the project continue.

© 1964 Steve Shapiro/Black Star

they were seeking to take part in Freedom Summer, data not subject to selective recall and reinterpretation as would be the case with retrospective interviews conducted twenty years after the fact. Moreover, the data on the 300 "no-shows" might provide a useful contrast with those who went. However, McAdam was not content to settle for these data. He still hoped to interview volunteers to see if and how participation in Freedom Summer had influenced their lives. Were most of them active in radical causes and movements?

But first, he needed to obtain current addresses for all these people. Since he knew in what school they had been enrolled when they applied, McAdam was able to write to alumni associations, which supplied him with current addresses of a large number of volunteers. He also discovered that approximately 20 percent of the parents of volunteers still lived at the same address in 1984 as in 1964. He got 101 current addresses this way. Many additional addresses were supplied by volunteers who were contacted by McAdam—volunteers tended to stay in touch.

Once he had current addresses of most volunteers and no-shows, McAdam sent each a lengthy questionnaire. Then McAdam traveled around the country conducting long, personal interviews with forty volunteers and forty no-shows, selected at random from the whole group. Finally, it was time to begin analyzing the data.

BECOMING A VOLUNTEER

The data allowed McAdam to formulate a clear portrait of how people became volunteers. First of all, they were free of obligations or limitations that might have prevented them from taking part; that is, they were what McAdam called **biographically available.** Second, the Freedom Summer Project strongly appealed to their attitudes and values. In fact, many of them had already participated in organizations involved in the Civil Rights Movement—many belonged to their campus chapter of SNCC. Finally, volunteers were linked to the project by preexisting social relationships.

BIOGRAPHICAL AVAILABILITY

When John Lofland and I studied how people came to join the Unification Church (see Chapter 3), we discovered that some people were simply too busy to join. They were genuinely impressed by the teachings of the group, but they had jobs, families, and other responsibilities that prevented them from embracing a full-time religious commitment. This is a primary reason most converts to new religious movements are in their late teens and early twenties. At that age many people have the freedom to make dramatic changes in their lives.

Having the freedom to act influences recruitment to all social movements, be they religious, political, or otherwise. Thus, while it is true that the SNCC leaders designed their recruitment efforts to attract well-to-do students from expensive, private schools, the fact is that such students also were those most likely to be available. A summer spent in Mississippi did not interfere with classes. Students from affluent homes did not need to earn money during their summer vacation; indeed, they could afford to pay their own way south and support themselves while there. Moreover, only 10 percent of applicants were married (and when they were, their spouse usually applied as well). The rest lacked any family responsibilities.

McAdam noted that

for all the social-psychological interpretations that have been proposed to account for the conspicuous role of students in protest, there may be a far more mundane explanation. Students, especially those drawn from privileged classes, are simply free, to a unique degree, of constraints that make activism too time consuming or risky for other groups to engage in.

ATTITUDES AND VALUES

McAdam also pointed out that freedom to act does not lead to action unless it "is joined with particular attitudes and values." Here he found three crucial elements. First of all, volunteers strongly believed in racial equality. Their personal statements explaining why they wished to volunteer all stressed their desire to bring an end to the injustices suffered by African Americans. Second, the volunteers were optimistic. They believed that victory over racism and oppression would be achieved soon. Third, their optimism was closely connected with a strong sense of personal power or potency. They thought that their actions as individuals would make a substantial difference in the world. Notions of the inevitable course of history were not for them. They expected history to yield to their demands.

McAdam traced these attitudes and values of the volunteers to their parents: "Far from using Freedom Summer as a vehicle for rebellion against parents, the applicants simply seem to be acting in accord with values learned at home." When asked later how her parents had responded to her decision to apply to the Freedom Summer Project, one volunteer said, "My father was thrilled. . . . And I'm sure that approval had a lot to do with my applying." Only nine volunteers who went to Mississippi indicated any parental disapproval.

Indeed, the volunteers primarily were the sons and daughters of parents who were themselves active in various liberal and radical causes. This same pattern has turned up so often in studies of student activists during the 1960s and 1970s that it is known as the "red diaper phenomenon"—the notion being that their parents, as committed leftists (or reds), had dressed them in diapers of revolutionary hue (Keniston, 1968; Flacks, 1971; McAdam, 1988).

Background also was responsible for the optimism and sense of personal power expressed by the volunteers. As McAdam put it:

Freedom Summer was an audacious undertaking demanding courage and confidence on the part of planners and participants alike. No doubt much of the self-assurance displayed by the volunteers owed to their generally privileged backgrounds. Class, as we are reminded each day, has its privileges. The roots of these privileges may be material, but the specific advantages enjoyed by those in the upper classes transcend their material base. Among the most important byproducts of class advantage is the psychological heritage that normally accompanies it. Of special interest here is the sense of personal efficacy and felt mastery over one's environment that often characterizes those who are economically well off. . . . Persons in the upper classes do tend to have more control over their environments than those in the lower classes. . . . Thus they are more apt to experience the world as malleable and themselves as master of their fate, than are those who are less well off.

One of the great ironies is that our beliefs about our ability to influence the world tend to be self-fulfilling. Those who are convinced that their actions are futile are unlikely to try to change things and therefore give up any chance of doing so. Those who are convinced they can cause changes will engage in actions that at least *could* result in changes.

SOCIAL NETWORKS

In Chapter 3 we discovered that religious conversion is more a matter of aligning one's behavior to that of one's family and friends than it is the result of a personal search for truth. Most people who ended up joining the Unification Church had not been looking for something to believe in when they encountered the group. Rather, people joined because of their attachment to members.

The same principle applies to political activism. The volunteers were not individuals acting on their own, each seeking to fulfill his or her ideals. People volunteered on the basis of personal ties to others, and their opportunity to volunteer typically arose through their previous activity in civil rights organizations. In fact,

90 percent of the volunteers "had already participated in various forms of activism." Half belonged to a campus civil rights organization such as the Congress of Racial Equality (CORE) or Friends of SNCC. Given that all of the organizers of Freedom Summer were provided by CORE and SNCC, many volunteers already had ties to these activists. And during his interviews McAdam found that nearly two-thirds had a friend who also volunteered.

PARTICIPANTS VERSUS NO-SHOWS

Comparisons between those who actually went to Mississippi and those who volunteered but did not go help illuminate the points covered earlier. McAdam found it impossible to distinguish participants from no-shows on the basis of attitudes and values. All were optimistic idealists. What did distinguish between the groups was their availability and their network ties.

The no-shows were an average of two years younger than participants. Since SNCC required that all volunteers under twenty-one obtain written permission from their parents, this proved an insurmountable barrier for some people. Almost half of no-shows under age twenty cited parental opposition as their primary reason for not going to Mississippi. In contrast, older volunteers were less subject to parental control. As one of them put it: "They weren't particularly thrilled by my decision to go, but, hell, I was 23 at the time. What were they going to do: Ground me? Cut my allowance?"

A higher proportion of female than of male volunteers ended up not going. This, too, seems to have reflected parental opposition: Parents were far more fearful of letting a daughter take part in a dangerous activity.

Participants also were far more likely to be linked to the Freedom Summer Project than were no-shows. For example, participants were three times as likely as no-shows to have friends who also went to Mississippi. McAdam summed up these findings:

The volunteers were already linked to the civil rights community. Whether these links took the form of organizational memberships, prior activism, or ties to other applicants, the volunteers benefited from greater "social proximity" to the project than did the no-shows. In fact, nothing distinguishes the two groups more clearly than this contrast. Biographical availability and attitudinal affinity may have been necessary prerequisites for

applying, but it was the strength of one's links to the project that seem to have finally determined whether one got to Mississippi or not.

THE VOLUNTEERS TWENTY YEARS LATER

Whatever the impact of Freedom Summer on civil rights, the question remains, What was the impact of Freedom Summer on the student volunteers? Did it change them? How? What were their lives like during the next twenty years?

When Doug McAdam surveyed the volunteers during 1984, he discovered that the major impact of participation on the volunteers had been to make them far more radical politically. Most of those who had gone to Mississippi with liberal attitudes and values came back with strong commitments to what was soon to be called the New Left. Moreover, the volunteers took an active role in the many New Left organizations and campaigns of the late 1960s and 1970s—protesting against the draft, the Vietnam War, and nuclear power, and in favor of feminism, abortion, gay rights, and disarmament. Symptomatic of their activism, by 1970 nearly 40 percent of the volunteers were living in the San Francisco Bay Area—which remains the capital of the contemporary American Left.

McAdam's initial observation that volunteers were unusually likely to turn up as leaders in subsequent New Left social movements was fully confirmed by the data. Even in 1984 the overwhelming majority of the veterans of Freedom Summer remained active in social movements, and there was no noticeable tendency for them to have become less committed to the political Left. Some were no longer as confident of their ability to change things as once they had been, but for the most part, their outlooks were much the same. As one volunteer told McAdam: "I don't think my political thinking has changed a hellova lot since then." Another, who was currently devoting his time to a group engaged in civil disobedience to halt nuclear research at the Lawrence Livermore Laboratory of the University of California, claimed to be reexperiencing the same electric excitement he had first felt in Mississippi in 1964 and once again during the student strike at Columbia University in 1968.

Not only were the volunteers far more likely than other Americans of the same age and educational background to be involved in political activism, but they were very different in other ways as well. For one thing, they were far less likely to be married. Thus, in 1984, while 74 percent of American women of equivalent age and education were married, only 42 percent of the women who took part in Freedom Summer were married. Among men, 54 percent of the volunteers were married, as compared with 82 percent of men of similar age and education nationally. Part of the difference was a very high divorce rate among volunteers: More than half of those who had ever married had gotten divorced. But in part, too, it was because volunteers were still biographically available, which may play a role in their high levels of continued activism. However, their activism may also play a role in keeping them available. McAdam suggested that failure to marry may reflect the difficulty people with very strong ideological commitments may have in finding a mate who is politically suitable.

Volunteers also departed from the norm by tending to change jobs very often and by simply being out of the labor market for relatively long periods of time. Moreover, when they did work, volunteers were unusually likely to be employed by social movements, and they were almost six times as likely as others of the same age and education to be college professors.

To conclude his study, Doug McAdam wrote that two qualities set the volunteers apart from the rest of Americans their age:

First, to a remarkable extent, they have remained faithful to the political vision that drew them to Mississippi nearly a quarter of a century ago. Second, they have paid for this lifelong commitment with a degree of alienation and social isolation that has only increased with time. The political and cultural wave that once carried them forward so prominently continues to recede, putting more and more distance between them and mainstream society with each passing day. . . . In their view, it is they who have kept the faith while America has lost it.

CONCLUSION

So, let's return to the question underlying this chapter on social movements as the agents of social change: Are humans able to take substantial control over their affairs, or is the individual pretty much the victim of impersonal historical forces? In my view this is a false dichotomy. Of course our history shapes us. It was history, not defects of personality, that created and sustained racism in the American South. But the history that today shapes the outlooks of southerners—African American and white—is very different from the history that shaped the attitudes of their parents and grandparents. And this history is different, in no small measure, because specific individuals took specific actions at specific times. It may be true that one day segregation of the buses in Alabama would have yielded to change had Rosa Parks never lived—but not until or unless somebody else did something to bring such a change about. Maybe history would not have been so different had Rosa Parks not refused to give up her seat, had Jo Ann Robinson not proposed a boycott, and had E. D. Nixon not organized a meeting of ministers and persuaded a reluctant Martin Luther King Jr. to participate. But that history would not resemble the history we know unless some other people had acted in much the same way at the same time. History doesn't boycott a bus. Only people can walk rather than ride.

By the same token, it did matter that a bunch of students from privileged homes went to Mississippi to join African Americans in the struggle for freedom—and that some of them died there. The news media couldn't have galvanized public support for civil rights without *news to cover.* So, somebody had to do something, somewhere, if the situation of African Americans in America was to be changed.

Indeed, what is history if not the story of human action? Thus, only to the extent that each of us acts does any history get made. To the question "Do we make our history or does it make us?" the answer clearly is "Yes!"

Review Glossary

Terms are listed in the order in which they appear in the chapter.

social movement Any organization created to cause or prevent social change.

collective behavior approach to social movements Emphasizes that social movements are outbursts of group activity in response to deeply felt grievances. Stresses the importance of ideology and of emotions and feelings in group decision making.

resource mobilization approach to social movements Stresses the importance of human and material resources in the development of social movements and places particular emphasis on the role of leaders and of rational planning.

countermovement A social movement that arises in opposition to another social movement.

revolution of rising expectations The probability that people will take action to change their society is maximized during periods when things have been improving but when the hope for change outpaces actual change.

J-curve theory of social crisis One form of the revolution of rising expectations, the J-curve theory emphasizes that an acute crisis occurs when, after a period of rapid improvement, conditions suddenly worsen.

precipitating event A specific occurrence that provokes people to take action.

biographical availability McAdam's term describing people who are free of commitments and involvements (such as jobs and families) that would interfere with their participating in a social movement.

Suggested Reading

Branch, Taylor. 1988. *Parting the Waters: America in the King Years, 1953–63.* New York: Simon & Schuster.

McAdam, Doug. 1988. *Freedom Summer.* New York: Oxford University Press.

McCarthy, John D., and Mayer N. Zald. 1977. "Resource Mobilization and Social Movements: A Partial Theory." *American Journal of Sociology* 82:1212–1241.

Morris, Aldon D. 1984. *The Origins of the Civil Rights Movement: Black Communities Organizing for Change.* New York: Free Press.

Turner, Ralph H., and Lewis M. Killian. 1987. *Collective Behavior.* Englewood Cliffs, N.J.: Prentice Hall.

Sociology Online

www.socstark10.com

GO TO THE INTERNET AND TYPE: www.socstark10.com.

↗**CLICK ON:** Nations of the Globe

✓ **SELECT:** DEMONSTRAT: Percentage who have taken part in a lawful demonstration in support of a cause.

↗**CLICK ON:** Analyze Now

In which nations have at least a third demonstrated? What percentage of Americans have done so?

USING INFOTRAC COLLEGE EDITION

GO TO THE INTERNET AND TYPE:
www.InfoTrac-college.com

↗**CLICK ON:** Register New Account
You will be asked to enter your Access Code and to create and enter a User Name and Password. Having done so,

↗**CLICK ON:** InfoTrac College Edition or Log On.

Demonstrations are a constant occurrence around the world. Some of the articles these search terms will turn up are merely accounts of particular demonstrations. Others attempt serious analyses of why and how they occurred. And some of the articles chart the rise of social movements, including those supporting demonstrations.

SEARCH TERMS:
Demonstrations
Social Movements

Becoming a Sociologist

EPILOGUE OUTLINE

MAJORING IN SOCIOLOGY

GOING TO GRADUATE SCHOOL
Selection
Applying
Succeeding in Graduate School

ON LEARNING TO WRITE

SOCIOLOGICAL CAREERS

IN CHAPTER 1 I confessed that one purpose of this book was to invite readers to become sociologists. Because I assume that you are now considering that option, it seemed useful to add a brief essay explaining how people actually do become sociologists. I have learned from my students that they often have little information about such matters as going to graduate school. For example, many students have worried about how they could afford to go on to graduate school. Clearly, these students didn't know that most graduate students receive full financial support. Nor do undergraduates usually know about how to select a graduate school.

Because it is difficult to be employed as a sociologist without a graduate degree (in fact, a doctorate is needed for many jobs), I shall devote much of this epilogue to graduate education. Then I shall describe the kinds of employment available to sociologists. First, however, something must be said about being an undergraduate major in sociology.

MAJORING IN SOCIOLOGY

Perhaps the most important thing for a would-be sociologist to know is that graduate schools do not require applicants to have an undergraduate major in sociology. That means you have considerable latitude to select courses that interest you and to take advantage of the strengths of the curriculum offered at your school. If another social science department at your school has a stronger program than the sociology department, you could major in that department and still go on in sociology in graduate school.

You should also know that leading graduate schools in sociology place little importance on the quality of one's undergraduate college or university. You will not be at a disadvantage if you are not now enrolled in a well-known institution. And the quality of teaching available to undergraduates is often higher at little-known schools than at famous institutions, where much of the teaching is delegated to graduate students lacking training and experience.

What you major in is not very important in becoming a sociologist; what you learn is. Here are my suggestions about what to take (depending, of course, on what is offered at your particular school).

Take at least one course in statistics. If possible, do not take this in the math department or in the statistics department. Take it in a social science department because then it will emphasize the statistical applications that social scientists actually use. If possible, take a second statistics course.

Take introductory economics. If micro- and macroeconomics are separated into two courses, take micro. It is far more theoretical and more pertinent to sociology.

Take a basic course in logic. You may find a good one in the philosophy department or in the speech communications department. A course that emphasizes the rules of deductive logic will be more useful than a course devoted to modern symbolic logic.

Also, take courses that require you to write. If a good writing course is available, perhaps in the English department, take it. Later in this epilogue, I shall discuss how to learn to write because sociologists make their living by the written word.

Take some history, especially if you can find a course or two that emphasize social history.

Take at least one course in research methods if it is offered at your school. Research methods are not simply statistics, although it may be useful to take a statistics course first.

You will notice I have said nothing about courses in sociology. That is because your sociology department will have a set of courses it recommends or requires for a major. One of these will probably be a course in the history of social thought, and you ought to take it even if it is not required.

But whatever you do, don't specialize in one area of sociology while still an undergraduate. This is the time to sample many different parts of the field to discover what you really like and to gain a broad background that can serve as an adequate base for later specialization.

Finally, you may be able to gain some useful experience in doing sociology while still an undergraduate. Check around and see if some faculty members could use a volunteer assistant. Even if it only involves finding things for them in the library and on the Internet, you can learn from doing it, and you can include this experience in your application to graduate school.

GOING TO GRADUATE SCHOOL

Most graduate schools expect students to apply by the end of December for admission in September. Details of admissions procedures can be obtained by writing to departments you are considering. So by early fall, you ought to have a list of schools you think you would like to attend.

SELECTION

How do you pick graduate schools? The first thing is to know what degree you plan to earn. If you think you want only a master's degree, which will qualify you for some kinds of research jobs (especially in government) and for teaching at some community colleges, then you have a choice of many universities. However, if you think you will aim for a doctorate, which is needed to qualify for most college and university faculty appointments and for most senior research positions, you should enroll in a school that awards a doctorate and that has a "major league" reputation.

Although the reputation of an undergraduate school has little effect on getting into a good graduate school, the reputation of a graduate school has an immense effect on a subsequent professional career. The general rule is that you will never get a job at a school with a substantially better reputation than the one from which you received your doctorate. So, the better the graduate school you attend, the more opportunities you will have.

Fortunately, a number of universities have good reputations. However, the reputation of a university as a whole may not reflect the reputation of any specific department. Thus, some famous schools are not as highly ranked in sociology as some less famous places. Every few years, a national study is conducted by the Council on Higher Education to rank departments in various academic fields. You may want to consult the sociology rankings in the latest of these reports (you can find it in most libraries). You can also get plenty of advice from members of your local sociology department. Moreover, you can request information on the recent placement of graduates from various departments you are considering; this will tell you how well their students are regarded.

Each year, the American Sociological Association publishes a Guide to Graduate Departments, which lists all faculty and various specialties of each sociology department in the United States and Canada. This may help you determine who would be available to work with in any given department.

Be sure to apply to a number of schools, even if your grades and test scores are exceptional. That way you will have a choice. Having been accepted by several schools, you can then telephone the head of admissions at each school and seek further information to help you make your final choice. How much financial support do they offer? Will you have to pay tuition out of it? (Some schools require relatively high tuitions for out-of-state students; however, this tuition may be waived for graduate students.) You may also want to find out how long it takes the average graduate student to earn a degree at this school.

If you are married and your spouse plans to work while you are in graduate school, then you will want to find out the employment situations near the schools you are considering. Usually, it will be easier for a spouse to find proper employment in or very near a large city than in an isolated college town. You will also want to find out about housing costs—some cities are very expensive, but that may be offset by the availability of university-owned housing.

APPLYING

Most departments use their own application forms and have different requirements regarding letters of recommendation and qualifying exams. You should write to each department early in the fall (or even the previous spring) to obtain these materials.

Virtually all good graduate departments require that students take the Graduate Record Examination (GRE) administered by the Educational Testing Service. Your current school will have information on when and where you can take these exams. It is best to take the GRE as early in the fall as possible so that your scores can be sent to graduate schools in time. Many students have found it helpful to prepare for this test by working through one of the practice guides available. Because undergraduate schools differ so greatly in quality and grading practices, graduate schools place strong emphasis on GRE scores.

Letters of recommendation can also greatly influence graduate admissions committees. The more contact you have with several faculty members while you are an undergraduate, the more able they will be to write you an effective recommendation. That's another reason to volunteer to help some faculty with their research, especially if you are attending a very large school. It is not necessary to restrict letters of recommendation to sociology faculty—any faculty member who thinks well of you will be effective. Moreover, you can sometimes find people not in a college or university whose recommendation will be influential. Be careful, however, to ensure that nonacademics can stress your intellectual ability and motivation, not just your good character. Admissions committees tend to be a bit snobbish and do not react well to students who solicit recommendations from their minister, coach, or manager of the fast-food restaurant in which they were employed.

SUCCEEDING IN GRADUATE SCHOOL

There is a great irony about graduate education. People get into graduate schools by having been good students. They succeed in graduate school by learning to cease being students. Until graduate school one is a consumer of knowledge—and one succeeds by learning what other people think about various matters. In graduate school a person must become a producer of knowledge and succeeds by having his or her own thoughts about these same matters.

This is reflected in a whole new style of reading, for example. Rather than reading to understand and to recall what someone has to say on a topic, one now must read to see what the underlying issues are, what remains to be said on that topic, or what is being said that is inconsistent or inadequate. One reads not just to learn but to find opportunities to contribute.

Nevertheless, the first several years of graduate school will revolve around classwork. A number of courses are required of all graduate students, and it is important to do as well as you can in each. But it is equally important to use this opportunity to get to know faculty members and discover with whom you want to work. For after the coursework is completed, graduate school turns into an apprenticeship. Finding the most suitable faculty member to work with as an apprentice will greatly shape your subsequent career.

In negotiating who will be your faculty sponsor in graduate school, let several criteria guide you. Ideally, it ought to be someone who does the kind of work you want to do. But that may be less important than finding someone who will train you well, who is interested in working with you, and who is able to place his or her students in good first jobs. Just as departments differ in reputation and ability to place students in top jobs, so do individual faculty members. Young faculty may be easier to approach and even more pleasant to work under, but they may lack the reputation to place students as effectively as senior faculty members. Sometimes, of course, a junior faculty member will be closely linked with a senior member, and thus, his or her students will enjoy the sponsorship of the senior member.

In the final analysis, of course, it matters less who works with you than how well you do your work. If you begin to produce fine work, anyone will want to sponsor you, and you will be rewarded with quality job opportunities. What does it mean to produce fine work? Many things. But no matter how original and insightful your ideas, no matter how clever or creative your research, you will produce nothing until you write it down.

ON LEARNING TO WRITE

Perhaps the most disabling myth about intellectual activity is that writing is an art that is prompted by inspiration. Some writing can be classified as art, no doubt, but the act of writing is a trade in the same sense that plumbing and automotive repair are trades. Just as plumbers and mechanics would rarely accomplish anything if they waited for inspiration to impel them to action, so writers would rarely write if they relied on inspiration.

You learn to write by writing, just as you learn to plumb by plumbing. And just as any ordinary person can learn to plumb well, so can any ordinary person learn to write well. If you want to become a good writer, you must write. Regularly! Ideally, you should write every day.

In my teens I began to work for newspapers. Nobody taught me to write. I just began to try to do it every day. At first I was slow, and my prose was not very clear, let alone elegant. But just as one learns to make professional pipe joints as one gains experience, so one learns to write more clearly, cleanly, and easily by writing.

Never wait for inspiration; it seldom comes. Approach the job of writing as you would approach household chores—as something you do regularly and routinely. For many years, I have been in the habit of getting up at the same time every morning; as soon as I have had coffee and read the paper, I settle down to write for about five hours. I never have to ask myself if I feel like writing any more than I have to ask whether I feel like brushing my teeth or not. It's just what I do at that time of day.

If you write regularly, you not only get better and better at it but you also get a great deal written. Students often fall into the habit of writing under pressure—of putting on a huge last-minute sprint to get a term paper completed. That's a bad way to write. It mixes writing with anxiety. When you write, you should be able to give your undivided attention to what you are saying, not to impending deadlines. Moreover, when you write regularly, you will find how easy it is to write a lot.

It would make me a nervous wreck had I tried to write this textbook in a series of crash sessions, trying to avert impending deadlines. The experience would have been so terrible I would have probably tried to avoid writing anything again. But that's not the way I did it. I sat down every morning and calmly knocked out 4 or 5 manuscript pages and then quit. That doesn't sound like much, but it is. In just 100 days, that adds up to from 400 to 500 pages, or about half the manuscript of this book. Clearly, then, I would have been content to average only 2 pages a day (and I write more only because I have practiced so long that I am very fast now).

So if you want to write, you should think of it as a routine task, to be approached regularly and calmly. You will be amazed at how rapidly you improve.

When you write, don't agonize over finding the best word or the best phrasing. Get the ideas down no matter how poor your prose. After you have your ideas on paper, then worry about improving the style. When you have a draft, no matter how crude, you can work on improving the writing without getting sidetracked. You do not risk forgetting where you are going as you seek a word or wrestle with a sentence. Indeed, what you are doing now is not writing but editing.

If you are considering graduate school, no matter in what department, keep in mind that you are essentially choosing to be a writer. Great ideas do not become great sociology or great chemistry until they are written and published.

SOCIOLOGICAL CAREERS

Colleges and universities are not the only places that employ people to do sociology. Much sociological research is conducted by people working for local, state, and national governments. Many other sociological researchers are employed by private firms. Thus, of the people who went to graduate school at Berkeley when I did, one ended up doing research on aging for the Social Security Administration, one plans new residential communities for a consortium of banks, one does research on drug abuse for the Department of Justice, another studied parole and prison policies for New York State, two conduct market research for major advertising firms, one studies TV viewership for a rating company, and one went from research on magazine readership to being founder and publisher of a successful magazine. However, most of them ended up on college or university faculties.

College and university positions involve two very different career lines. The one most visible to students consists primarily of teaching. While virtually all college and university faculty members teach, in most schools faculty are expected to devote their major efforts to teaching. In contrast, faculty at the major research-oriented universities are asked to teach only half or even a third as much as faculty in other schools. The remainder of their time is supposed to be spent on research and writing. That is, these people are hired and promoted primarily to do sociology, not to teach it, and it is to them that the rule "publish or perish" primarily applies. In major universities promotion and tenure are awarded almost exclusively on the basis of scholarly publications, with only modest concern given to teaching skills. Many faculty not at the major

research universities also do research and publish, but less emphasis is placed on those activities (they are given less time to devote to them, for one thing) and much more importance is placed on their teaching. Most of these faculty members also did their graduate work at major universities, but they primarily chose the career of college teaching.

For people entering graduate schools today, there will be more opportunities for careers in industry and government than when I got my degree. There probably will be somewhat fewer openings in universities and colleges, since higher education was still expanding when I graduated. If you are thinking about being a sociologist, you will have each of these career lines as possibilities. Moreover, during the course of your career, you will probably have opportunities to switch from one line to another.

Before I became a sociologist, I tried some other careers. I enjoyed being a newspaper reporter. I enjoyed being an advertising writer. For several years, I even enjoyed being a soldier. But there is a considerable difference between enjoying an occupation and being dedicated to it. Although newspaper writing was an interesting job, I was able to leave it without regret. But for me, sociology is different. It is not a job but a way of life. Being a sociologist is not merely what I do but what I am.

People can do a job very well and take considerable pride in it without its being essential to their self-image. But if an occupation is based on self-motivation, dedication is essential. Sociology, like any science, is fundamentally a solitary trade. Even when you are part of a research team, the most important work is not done collectively but in private. The basis of all scientific work is thought. It is very hard to force yourself to sit alone and think about things unless you enjoy thinking about them. I am sure I lack the self-discipline needed to force myself to write every morning. Fortunately, the problem never comes up. It never does when you are doing what you want to do. Dedication, then, means doing a job because you love it.

The first and most important thing I look for in graduate students is dedication. How much talent they have is of much less interest to me. The question I truly want a student to answer is, If you were so rich you didn't need a job, would you still be a sociologist? I would. If you would, too, then we need you.

Aberele, David F., Albert K. Cohen, Arthur K. Davis, Marion J. Levy Jr., and Francis X. Sutton. 1950. "The Functional Prerequisites of a Society." *Ethics* 60:100–111.

Ackerman, Nathan W., and Marie Jahoda. 1950. *Anti-Semitism and Emotional Disorder*. New York: Harper & Row.

Adachi, Ken. 1978. *The Enemy That Never Was: A History of Japanese Canadians*. Toronto: McClelland.

Adorno, Theodore, et al. 1950. *The Authoritarian Personality*. New York: Norton.

Ages, Arnold. 1981. "Antisemitism: The Uneasy Calm." In *The Canadian Jewish Mosaic*, edited by M. Weinfeld, W. Shaffir, and I. Cotler. Toronto: Wiley.

Aguirre, B. E., Rogelio Saenz, and Sean-Shong Hwang. 1989. "Discrimination and the Assimilation and Ethnic Competition Perspectives." *Social Science Quarterly* 70:594–605.

Alba, Richard D. 1977. "Social Assimilation among American Catholic National-Origin Groups." *American Sociological Review* 41:1030–1046.

Alba, Richard D. 1985. "The Twilight of Ethnicity among Americans of European Ancestry: The Case of Italians." *Ethnic and Racial Studies* 8:134–158.

Alba, Richard. 1990. *Ethnic Identity: The Transformation of White America*. New Haven, Conn.: Yale University Press.

Albrecht, S. L., B. A. Chadwick, and D. S. Alcorn. 1977. "Religiosity and Deviance: Application of an Attitude-Behavior Contingent Consistency Model." *Journal for the Scientific Study of Religion* 16:263–274.

Alexander, Karl L., Gary Natriello, and Aaron M. Pallas. 1985. "For Whom the School Bell Tolls: The Impact of Dropping Out on Cognitive Performance." *American Sociological Review* 50:409–420.

Allen, Leslie. 1985. *Liberty: The Statue and the American Dream*. New York: Statue of Liberty–Ellis Island Foundation.

Allport, Gordon. 1958. *The Nature of Prejudice*. New York: Doubleday.

Allport, G. W. 1968. "The Historical Background of Modern Social Psychology." In *The Handbook of Social Psychology*, edited by G. Lindzey and E. Aronson. Reading, Mass.: Addison-Wesley.

Amato, Paul R. 1993. "Children's Adjustment to Divorce: Theories, Hypotheses, and Empirical Support." *Journal of Marriage and the Family* 55:23–38.

Angell, Robert Cooley. 1942. "The Social Integration of Selected American Cities." *American Journal of Sociology* 47:575–592.

Angell, Robert Cooley. 1947. "The Social Integration of Cities of More Than 100,000 Population." *American Sociological Review* 12:335–342.

Angell, Robert Cooley. 1949. "Moral Integration and Interpersonal Integration in American Cities." *American Sociological Review* 14:245–251.

Ardrey, Robert. 1961. *African Genesis*. New York: Dell.

Ardrey, Robert. 1966. *The Territorial Imperative*. New York: Dell.

Ardrey, Robert. 1970. *The Social Contract*. New York: Dell.

Argyle, Michael, and Benjamin Beit-Hallahmi. 1975. *The Social Psychology of Religion*. London: Routledge & Kegan Paul.

Asch, Solomon. 1952. "Effects of Group Pressure upon the Modification and Distortion of Judgements." In *Readings in Social Psychology*, edited by Guy Swanson, Theodore M. Newcomb, and Eugene L. Hartley. New York: Holt, Rinehart & Winston.

Ashton, Basil, et al. 1984. "Famine in China, 1958–61." *Population and Development Review* 10:24–37.

Austen, Ralph. 1979. "The Trans-Saharan Slave Trade: A Tentative Census." In *The Uncommon Market: Essays in the Economic History of the Atlantic Slave Trade*, edited by Henry A. Gemery and Jan S. Hogendorn. New York: Academic Press.

Ausubel, Jesse H., and Cesare Marchetti. 2001. "The Evolution of Transport." *The Industrial Physicist* (Apr./May):20–24.

Azzi, Corry, and Ronald Ehrenberg. 1975. "Household Allocation of Time and Church Attendance." *Journal of Political Economy* 83:27–56.

Bailey, William C., and Ruth D. Peterson. 1989. "Murder and Capital Punishment: A Monthly Time-Series Analysis of Execution Publicity." *American Sociological Review* 54:722–743.

Bain, Reed. 1936. "The Self-and-Other Words of a Child." *American Journal of Sociology* 41:767–775.

Bainbridge, William Sims. 1997. *The Sociology of Religious Movements*. New York: Routledge.

Bainbridge, W. S., and Rodney Stark. 1980. "Client and Audience Cults in America." *Sociological Analysis* 41:199–214.

Bainbridge, W. S., and Rodney Stark. 1981. "Friendship, Religion, and the Occult." *Review of Religious Research* 22:313–327.

Balch, Robert. 1980. "Looking Behind the Scenes in a Religious Cult: Implications for the Study of Conversion." *Sociological Analysis* 41:137–143.

Balch, Robert. 1982. "Bo and Peep: A Case of the Origins of Messianic Leadership," In *Charisma and the Millennium*, edited by Roy Wallis. Belfast, Northern Ireland: Queen's University Press.

Balch, Robert. 1985. "'When the Light Goes Out, Darkness Comes': A Study of Defection from a Totalistic Cult." In *Religious Movements: Genesis, Exodus, and Numbers*, edited by Rodney Stark. New York: Rose of Sharon Press.

Balch, Robert. 1994. "Waiting for the Ships." *Syzygy* 3:95–116.

Balch, Robert, and David Taylor. 1976. "Salvation in a UFO." *Psychology Today* (Oct.):58–66.

Balch, Robert, and David Taylor. 1977. "Seekers and Saucers: The Role of the Cultic Milieu in Joining a UFO Cult." *American Behavioral Scientist* 20:839–860.

Baltzell, E. Digby. 1964. *The Protestant Establishment*. New York: Random House.

Bamforth, Douglas B. 1994. "Indigenous People, Indigenous Violence: Precontact Warfare on the North American Great Plains." *Journal of the Royal Anthropological Institute* (formerly *Man*) 29:95–115.

Bandura, Albert. 1974. "Behavior Theory and the Models of Man." *American Psychologist* 29:859–869.

Barker, Eileen. 1984. *The Making of a Moonie—Brainwashing or Choice?* Oxford: Basil Blackwell.

Barrett, David B. 1982. *World Christian Encyclopedia*. Oxford: Oxford University Press.

Barrett, Richard K., and Martin King Whyte. 1982. "Dependency Theory and Taiwan: Analysis of a Deviant Case." *American Journal of Sociology* 87:1064–1089.

Bart, Pauline. 1972. "Depression in Middle-Age Women." In *Women in Sexist Society*, edited by V. Gornick and B. K. Moran. New York: New American Library.

Barton, Paul E., and Archie Lapointe. 1995. *Learning by Degrees.* Princeton, N.J.: Educational Testing Service.

Batson, C. Daniel, Patricia Schoenrade, and W. Larry Ventis. 1993. *Religion and the Individual: A Social Psychological Perspective.* New York: Oxford University Press.

Baxandall, Rosalyn, and Elizabeth Ewen. 2000. *Picture Windows: How the Suburbs Happened.* New York: Basic Books.

Bean, Frank D., and Marta Tienda. 1987. *The Hispanic Population of the United States.* New York: Russell Sage Foundation.

Becker, Howard S. 1963. *The Outsiders: Studies in the Sociology of Deviance.* New York: Free Press.

Befu, Harumi. 1990. "Four Models of Japanese Society and Their Relevance to Conflict." In *Japanese Models of Conflict Resolution*, edited by S. N. Eisenstadt and Eyal Ben-Ari. London: Kegan Paul.

Beirne, Piers. 1993. *Inventing Criminology.* Albany: State University of New York Press.

Bell, Daniel. 1980. *The Winding Passage.* Cambridge: ABT.

Benedict, Ruth. 1934. *Patterns of Culture.* New York: Houghton Mifflin.

Benitez, J. 1973. "The Effect of Elite Recruitment and Training on Diffuse Socialization Outcomes." Ph.D. dissertation, Stanford University.

Bennett, Neil G., Ann Klimas Blanc, and David E. Bloom. 1988. "Commitment and the Modern Union: Assessing the Link between Premarital Cohabitation and Subsequent Marital Stability." *American Sociological Review* 53:127–138.

Berelson, Bernard. 1978. "Prospects and Programs for Fertility Reduction: What? Where?" *Population and Development Review* 4:579–616.

Berg, E. J. 1966. "Backward-sloping Labor Supply Functions in Dual Economies—the Africa Case." In *Social Change: The Colonial Situation*, edited by Immanuel Wallerstein. New York: Wiley.

Berger, Bennett M. 1981. *The Survival of a Counterculture.* Berkeley: University of California Press.

Berger, Peter L. 1963. *Invitation to Sociology.* New York: Anchor Books.

Berger, Peter L. 1967. *The Sacred Canopy.* Garden City, N.Y.: Doubleday.

Berk, R. A., K. J. Lenihan, and P. H. Rossi. 1980. "Crime and Poverty: Some Experimental Evidence from Ex-Offenders." *American Sociological Review* 45:766–786.

Bestor, Theodore C. 1989. *Neighborhood Tokyo.* Palo Alto, Calif.: Stanford University Press.

Beyer, Peter, 1994. *Religion and Globalization.* London: Sage.

Bian, Yanjie. 1997. "Bringing Strong Ties Back In: Indirect Ties, Network Bridges, and Job Searches in China." *American Sociological Review* 62:366–385.

Bibby, Reginald W. 2002. *Restless Gods: The Renaissance of Religion in Canada.* Toronto: Stoddart.

Biblartz, Timothy J., Adrian E. Raftery, and Alexander Bucur. 1997. "Family Structure and Social Mobility." *Social Forces* 75:1319–1341.

Bickerton, Derek. 1981. *Roots of Language.* Ann Arbor, Mich.: Karoma.

Bickerton, Derek. 1999. "How to Acquire Language without Positive Evidence: What Acquisitionists Can Learn from Creole." In *Language Creation and Language Change*, edited by Michel DeGraff. Cambridge, Mass.: MIT Press.

Bird, Caroline. 1975. *The Case against College.* New York: David McKay.

Blalock, Hubert M., Jr. 1967. *Toward a Theory of Minority-Group Relations.* New York: Capricorn Books.

Blau, Peter M. 1970. "A Formal Theory of Differentiation in Organizations." *American Sociological Review* 35:201–218.

Blau, Peter M. 1972. "Size and the Structure of Organizations: A Causal Analysis." *American Sociological Review* 37:434–440.

Blau, Peter M. 1986. *Exchange and Power in Social Life*, rev. ed. New Brunswick, N.J.: Transaction Books.

Blau, Peter M., and Otis Dudley Duncan. 1967. *The American Occupational Structure.* New York: Wiley.

Blick, Jeffrey P. 1988. "Genocidal Warfare in Tribal Societies as a Result of European-Induced Culture Conflict." *Journal of the Royal Anthropological Institute* (formerly *Man*) 23:654–670.

Bloch, Marc. 1962. *Feudal Society.* Chicago: University of Chicago Press.

Blumenfeld, H. 1971. "Transportation in the Modern Metropolis." In *Internal Structure of the City*, edited by L. S. Bourne. New York: Oxford University Press.

Blumer, Herbert. 1969. *Symbolic Interactionism: Perspective and Method.* Englewood Cliffs, N.J.: Prentice Hall.

Boas, Franz. 1897. "The Social Organization and Secret Societies of the Kwakiutl." *Report of the United States National Museum for 1895.* Washington, D.C.: U.S. Government Printing Office.

Bobrovnikov, Vladimir. 1996. "The Islamic Revival and the National Question in Post-Soviet Dagestan." *Religion, State and Society* 24:233–238.

Bolce, Louis, and Gerald De Maio. 1999. "Religious Outlook, Culture War Politics, and Antipathy toward Christian Fundamentalists." *Public Opinion Quarterly* 63:29–61.

Bonacich, Edna. 1972. "A Theory of Ethnic Antagonism: The Split Labor Market." *American Sociological Review* 37:547–559.

Bonacich, Edna. 1973. "A Theory of Middleman Minorities." *American Sociological Review* 38:583–594.

Bonacich, Edna. 1975. "Abolition, the Extension of Slavery, and the Position of Free Blacks." *American Journal of Sociology* 81:601–628.

Bonacich, Edna. 1976. "Advanced Capitalism and Black/White Race Relations in the United States: A Split Labor Market Interpretation." *American Sociological Review* 41:34–51.

Bongaarts, John. 1990. "The Measurement of Unwanted Fertility." *Population and Development Review* 16:487–506.

Booth, Alan, and James Dabbs Jr. 1993. "Testosterone and Men's Marriages." *Social Forces* 72:463–477.

Bouchard, Thomas J., Jr., David T. Lykken, Matthew McGue, Nancy L. Segal, and Auke Tellegen. 1990. "Sources of Human Psychological Differences: The Minnesota Study of Twins Reared Apart." *Science* 250:223–228.

Boudon, Raymond. 1974. *Education, Opportunity, and Social Inequality: Changing Prospects in Western Society.* New York: Wiley.

Bourdieu, Pierre. 1984. *Distinction: A Social Critique of the Judgement of Taste.* Cambridge, Mass.: Harvard University Press.

Bowker, L. H., ed. 1981. *Women and Crime in America.* New York: Macmillan.

Bowles, Samuel, and Herbert Gintis. 1976. *Schooling in Capitalist America.* New York: Basic Books.

Boyd, Monica, et al. 1981. "Status Attainment in Canada: Findings of the Canadian Mobility Study." *Canadian Review of Sociology and Anthropology* 18:657–673.

Bradshaw, York W., and Michael Wallace. 1996. *Global Inequalities.* Thousand Oaks, Calif.: Pine Forge Press.

Branch, Taylor. 1988. *Parting the Waters: America in the King Years, 1953–63.* New York: Simon & Schuster.

Brantingham, Paul J., and Patricia L. Brantingham. 1982. "Mobility, Notoriety and Crime: A Study in Crime Patterns of Urban Nodal Points." *Journal of Environmental Systems* 11:98–99.

Brantingham, Paul, and Patricia Brantingham. 1984. *Patterns in Crime.* New York: Macmillan.

Braudel, Fernand. 1981. *The Structures of Everyday Life.* New York: Harper & Row.

Brierley, Peter. 1993. "Europe: Where Christianity Matters—and Is in Decline." *MARC Newsletter* 93 (Sept.):2.

Broderick, Francis L. 1974. "W. E. B. Du Bois: History of an Intellectual." In *Black Sociologists,* edited by James E. Blackwell and Morris Janowitz. Chicago: University of Chicago Press.

Brodie, Mollyann, Annie Steffenson, Jaime Valdez, Rebecca Levin, and Robert Suro. 2002. *2002 National Survey of Latinos.* Washington, D.C.: Pew Hispanic Center.

Brophy, I. N. 1945. "The Luxury of Anti-Negro Prejudice." *Public Opinion Quarterly* 9:456–466.

Brough, R. Clayton. 1992. *The Lost Tribes: History, Doctrines, Prophesies and Theories about Israel's Lost Ten Tribes.* New York: Horizon.

Brown, Roger, and Ursula Bellugi. 1964. "Three Processes in the Child's Acquisition of Syntax." *Harvard Educational Review* 34:133–151.

Bruer, John T. 1999. *The Myth of the First Three Years.* New York: Free Press.

Brym, Robert, Michael W. Gillespie, and A. R. Gillis. 1985. "Anomie, Opportunity, and the Density of Ethnic Ties: Another View of Jewish Outmarriage in Canada." *Review of Canadian Sociology and Anthropology* 22:102–112.

Bureau of Justice Statistics. 1988. *Survey of Youth in Custody, 1987.* Washington, D.C.: U.S. Department of Justice.

Bureau of Justice Statistics. 1991. *Profile of Jail Inmates, 1989.* Washington, D.C.: U.S. Department of Justice.

Bureau of Justice Statistics. 1992. *Drugs, Crime, and the Justice System: A National Report.* Washington D.C.: U.S. Government Printing Office.

Burgess, Ernest W., and Charles Newcomb. 1933. *Census Data of the City of Chicago, 1930.* Chicago: University of Chicago Press.

Burgess, Robert L., and Ronald L. Akers. 1966. "A Differential Association-Reinforcement Theory of Criminal Behavior." *Social Problems* 14:128–147.

Burkett, S. R., and M. White. 1974. "Hellfire and Delinquency: Another Look." *Journal for the Scientific Study of Religion* 13:455–462.

Bursik, Robert J., Jr. 1988. "Social Disorganization and Theories of Crime and Delinquency: Problems and Prospects." *Criminology* 26:519–551.

Bursik, Robert J., Jr., and Jim Webb. 1982. "Community Change and Patterns of Delinquency." *American Journal of Sociology* 88:24–42.

Burt, Ronald. 1995. *Structural Holes: The Social Structure of Competition.* Cambridge, Mass.: Belknap Press.

Busch, Lisa. 1993. "Designer Families, Ethical Knots." *U.S. News & World Report* (May 31):73.

Butler, R. E. 1985. *On Creating a Hispanic America: A Nation within a Nation.* Washington, D.C.: Council for Inter-American Security.

Caldwell, John C., I. O. Orubuloye, and Pat Caldwell. 1992. "Fertility Decline in Africa: A New Type of Transition?" *Population and Development Review* 18:221–239.

Calhoun, J. B. 1962. "Population Density and Social Pathology." *Scientific American* 206:139–148.

Campbell, Angus. 1975. "The American Way of Mating: Marriage, Sí; Children, Maybe." *Psychology Today* 8:37–43.

Canadian Centre for Justice Statistics. 1984. *Crime and Traffic Enforcement Statistics.* Ottawa: Statistics Canada.

Cann, Arnie, and William D. Siegfried Jr. 1987. "Sex Stereotypes and the Leadership Role." *Sex Roles* 17:401–408.

Cantor, David, and Kenneth C. Land. 1985. "Unemployment and Crime Rates in the Post-World War II United States: A Theoretical and Empirical Analysis." *American Sociological Review* 50:317–332.

Carmichael, Stokely, and Charles V. Hamilton. 1967. *Black Power: The Politics of Liberation in America.* New York: Random House.

Carnap, Rudolf. 1953. "Testability and Meaning." In *Readings in the Philosophy of Science,* edited by Herbert Feigl and May Brodbeck. New York: Appleton-Century-Crofts.

Centers for Disease Control (CDC). 2002. *Cohabitation, Marriage, Divorce, and Remarriage in the United States.* Washington, D.C.: U.S. Government Printing Office.

Chabris, Christopher F. 1999. "Prelude of Requiem for the 'Mozart Effect'?" *Nature* 400:826–827.

Chagnon, Napoleon A. 1988. "Life Histories, Blood Revenge, and Warfare in a Tribal Population." *Science* 239:985–992.

Chagnon, Napoleon A. 1992. *Yanomamö: The Last Days of Eden.* New York: Harcourt Brace.

Chandler, Alfred D., Jr. 1962. *Strategy and Structure: Chapters in the History of the American Revolution.* Cambridge, Mass.: MIT Press.

Chandler, David. 1966. *The Campaigns of Napoleon.* New York: Macmillan.

Charles, Enid. 1936. *The Menace of Under-Population.* London: Watts.

Chase-Dunn, Christopher. 1975. "The Effects of International Economic Dependence on Development and Inequality: A Cross-National Study." *American Sociological Review* 40:720–738.

Chavez, Linda. 1989. "Tequila Sunrise: The Slow but Steady Progress of Hispanic Immigrants." *Policy Review* (Spring):64–67.

Chesnut, R. Andrew. 2003. *Competitive Spirits: Latin America's New Religious Economy.* New York: Oxford University Press.

Chirot, Daniel. 1976. *Social Change in a Peripheral Society.* New York: Academic Press.

Chirot, Daniel. 1977. *Social Change in the Twentieth Century.* New York: Harcourt Brace Jovanovich.

Chirot, Daniel. 1985. "The Rise of the West." *American Sociological Review* 50:181–195.

Chirot, Daniel. 1994. *How Societies Change.* Thousand Oaks, Calif.: Pine Forge Press.

Chomsky, Noam. 1975. *Reflections on Language.* New York: Pantheon.

Christiansen, K. O. 1977. "A Preliminary Study of Criminality among Twins." In *Biosocial Bases of Criminal Behavior,* edited by S. A. Mednick and K. O. Christiansen. New York: Wiley.

Cloud, John, and Jodie Morse. 2001. "Home Sweet School." *Time* (Aug. 27):46–54.

Cohen, Lawrence E., and Marcus Felson. 1979. "Social Change and Crime Rate Trends: A Routine Activity Approach." *American Sociological Review* 44:588–607.

Cohen, Mark. 1977. *The Food Crisis in Prehistory: Overpopulation and the Origins of Agriculture.* New Haven, Conn.: Yale University Press.

Cohen, Yinon, and Andrea Tyree. 1986. "Escape from Poverty: Determinants of Intergenerational Mobility of Sons and Daughters of the Poor." *Social Science Quarterly* 67:803–813.

Cohn, Werner. 1958. "The Politics of American Jews." In *The Jew: Social Patterns of an American Group*, edited by Marshall Sklare. Glencoe, Ill.: Free Press.

Cohn, Werner. 1976. "Jewish Outmarriage and Anomie." *Canadian Review of Sociology and Anthropology* 13:90–105.

Cole, George F. 1983. *The American System of Criminal Justice.* Monterey, Calif.: Brooks/Cole.

Cole, Sonia. 1975. *Leakey's Luck: The Life of Louis Seymour Bazett Leakey, 1903–1972.* New York: Harcourt Brace Jovanovich.

Coleman, James S. 1990. *Foundations of Social Theory.* Cambridge, Mass.: Harvard University Press.

Coleman, James S., Thomas Hoffer, and Sally Kilgore. 1982. *High School Achievement: Public, Catholic, and Other Private Schools Compared.* New York: Basic Books.

Coleman, James S., et al. 1966. *Equality of Educational Opportunity.* Washington, D.C.: U.S. Government Printing Office.

Coleman, James William. 1987. "Toward an Integrated Theory of White-Collar Crime." *American Journal of Sociology* 93:406–439.

Collins, Randall. 1979. *The Credential Society: An Historical Sociology of Education.* New York: Academic Press.

Collins, Randall. 1994. *Four Sociological Traditions.* New York: Oxford University Press.

Collins, Randall, Janet Salztman Chafetz, Rae Lesser Blumberg, Scott Coltrane, and Jonathan H. Turner. 1993. "Toward an Integrated Theory of Gender Stratification." *Sociological Perspectives* 36:185–216.

Converse, Philip E. 1964. "The Nature of Belief Systems in Mass Publics." In *Ideology and Discontent*, edited by David Apter. New York: Free Press.

Cooley, Charles H. 1909. *Social Organization.* New York: Scribners.

Cooley, Charles Horton. 1922. *Human Nature and the Social Order.* New York: Scribners.

Corman, Hope, and H. Naci Mocan. 2000. "A Time-Series Analysis of Crime, Deterrence, and Drug Abuse in New York City." *The American Economic Review* 90:584–604.

Cornwall, Marie, 1988. "The Influence of Three Agents of Religious Socialization." In *The Religion and Family Connection: Social Science Perspectives*, edited by Darwin Thomas. Provo, Ut.: Brigham Young University Religious Studies Center.

Cortés, J. B. 1982. "Delinquency and Crime: A Biopsychological Theory." In *The Fundamental Connection between Nature and Nurture*, edited by Walter Gove and G. R. Carpenter. Lexington, Mass.: Lexington Books.

Cortés, J. B., and F. M. Gatti. 1972. *Delinquency and Crime: A Biopsychological Approach.* New York: Seminar Press.

Cott, Nancy F. 1977. *The Bonds of Womanhood: "Woman's Sphere" in New England, 1780–1835.* New Haven, Conn.: Yale University Press.

Cott, Nancy F. 1987. *The Grounding of Modern Feminism.* New Haven, Conn.: Yale University Press.

Counts, George Sylvester. 1922. "The Selective Character of American Secondary Education." *Supplementary Educational Monographs*, no. 19. Chicago: University of Chicago Press.

Covello, Leonard. 1967. *The Social Background of the Italo-American School Child.* Leiden, Netherlands: E. J. Brill.

Criminology. 1987. Special issue on theory, 25:783–989.

Croll, Elisabeth. 2001. *Endangered Daughters: Discrimination and Development in Asia.* London: Routledge.

Crutchfield, Robert, Michael Geerken, and Walter R. Gove. 1983. "Crime Rates and Social Integration." *Criminology* 20:467–478.

Curtin, Philip D. 1969. *The Atlantic Slave Trade: A Census.* Madison: University of Wisconsin Press.

Cutright, Phillips, and Lowell Hargens. 1984. "The Threshold Hypothesis: Latin America 1950–1980." *Demography* 21:435–458.

Dabbs, James M., Jr. 1992. "Testosterone and Occupational Achievement." *Social Forces* 70:813–824.

Dabbs, James M., Jr., Robert L. Frady, T. S. Carr, and N. F. Besch. 1987. "Saliva Testosterone and Criminal Violence in Young Adult Prison Inmates." *Psychosomatic Medicine* 49:174–182.

Dabbs, James M., Jr., and Robin Morris. 1990. "Testosterone, Social Class, and Antisocial Behavior in a Sample of 4,462 Men." *Psychological Science* 1:209–211.

Dabbs, James M., Jr., R. Barry Ruback, Robert L. Frady, Charles H. Hopper, and Demetrios S. Sgoutas. 1988. "Saliva Testosterone and Criminal Violence among Women." *Personality and Individual Differences* 9:269–275.

Dahl, Robert. 1956. *A Preface to Democratic Theory.* Chicago: University of Chicago Press.

Dahrendorf, Ralf. 1959. *Class and Class Conflict in Industrial Society.* Palo Alto, Calif.: Stanford University Press.

Daitzman, Reid, and Marvin Zuckerman. 1980. "Disinhibitory Sensation Seeking, Personality and Gonadal Hormones." *Personality and Individual Differences* 1:103–110.

Darcy, R., and Sarah Slavin Schramm. 1977. "When Women Run against Men." *Public Opinion Quarterly* 41:1–12.

Darley, John M., and Bibb Latané. 1968. "Bystander Intervention in Emergencies: Diffusion of Responsibility." *Journal of Personality and Social Psychology* 8:377–383.

Darroch, A. Gordon, and Wilfred G. Marston. 1971. "The Social Class Bias of Ethnic Residential Segregation: The Canadian Case." *American Journal of Sociology* 77:491–510.

Davies, James C. 1962. "Toward a Theory of Revolution." *American Sociological Review* 27:5–19.

Davis, Kingsley. 1940. "Extreme Social Isolation of a Child." *American Journal of Sociology* 45:523–535.

Davis, Kingsley. 1945. "The World Demographic Transition." *Annals of the American Academy of Political and Social Sciences* 237:1–11.

Davis, Kingsley. 1947. "Final Note of a Case of Extreme Isolation." *American Journal of Sociology* 50:432–437.

Davis, Kingsley. 1949. *Human Society.* New York: Macmillan.

Davis, Kingsley. 1951. *The Population of India and Pakistan.* Princeton, N.J.: Princeton University Press.

Davis, Kingsley. 1965. "The Population Impact of Children in the World's Agrarian Countries." *Population Review* 9:17–31.

Davis, Kingsley. 1971. "The World's Population Crisis." In *Contemporary Social Problems*, 2nd ed., edited by Robert K. Merton and Robert Nisbet. New York: Harcourt Brace Jovanovich.

Davis, Kingsley. 1976. "The World's Population Crisis." In *Contemporary Social Problems*, 3rd ed., edited by Robert K. Merton and Robert Nisbet. New York: Harcourt Brace Jovanovich.

Davis, Kingsley, and Wilbert E. Moore. 1945. "Some Principles of Stratification." *American Sociological Review* 10:242–249.

Davis, Kingsley, and Wilbert E. Moore. 1953. "Replies to Tumin." *American Sociological Review* 18:394–396.

Deck, Leland. 1971. "Short Workers of the World, Unite!" *Psychology Today* 5:102.

Delacroix, Jacques. 1977. "The Export of Raw Materials and Economic Growth: A Cross-National Study." *American Sociological Review* 42:795–808.

Delacroix, Jacques. 1992. "A Critical Empirical Test of the 'Common Interpretation' of *The Protestant Ethic and the Spirit of Capitalism.*" Paper presented at the International Association for Business and Society, Leuven, Belgium.

de la Garza, Rodolpho O., Louis DiSipio, F. Chris Garcia, John Garcia, and Angelo Falcon. 1992. *Latino Voices: Mexican, Puerto Rican, and Cuban Perspectives on American Politics.* Boulder, Colo.: Westview Press.

DeLoache, Judy S., Deborah J. Cassidy, and C. Jan Carpenter. 1987. "The Three Bears Are All Boys: Mother's Gender Labelling of Neutral Picture Book Characters." *Sex Roles* 17:163–178.

del Pinal, Jorge H., and Carmen De Navas. 1990. *The Hispanic Population of the United States, March 1989.* Washington, D.C.: U.S. Bureau of the Census.

Demerath, N. J., and Phillip E. Hammond. 1969. *Religion in Social Context.* New York: Random House.

de Vaus, David A. 1984. "Workforce Participation and Sex Differences in Church Attendance." *Review of Religious Research* 25:247–258.

Diamond, Jared. 1998. *Gunds, Germs, and Steel.* New York: Norton.

Dickeman, Mildred. 1975. "Demographic Consequence of Infanticide in Man." *Annual Review of Ecology and Systematics* 6:107–137.

DiMaggio, Paul, and Hugh Louch. 1998. "Socially Embedded Consumer Transactions: For What Kinds of Purchases Do People Most Often Use Networks?" *American Sociological Review* 63:619–637.

Dohrenwend, Bruce P. 1966. "Social Status and Psychological Disorder: An Issue of Substance and an Issue of Method." *American Sociological Review* 31:14–34.

Douglas, Mary. 1975. *Implicit Meanings: Essays in Anthropology.* London: Routledge & Kegan Paul.

Drucker, Peter F. 1946. *Concept of the Corporation.* New York: John Day.

Drucker, Peter F. 1967. *The Effective Executive.* New York: Harper & Row.

Drucker, Peter F. 1969. *The Age of Discontinuity: Guidelines to Our Changing Society.* New York: Harper & Row.

Drucker, Peter F. 1974. *Management: Tasks—Responsibilities—Practices.* New York: Harper & Row.

Drummond, J. C., and Anne Wilbraham. 1957. *The Englishman's Food: A History of Five Centuries of English Diet*, rev. ed. London: Cape.

Du Bois, W. E. B. 1904. "The Atlanta Conferences." *Voice of the Negro* 1:85–89.

Duesenberry, J. S. 1960. "Comment." In Universities-National Bureau Committee for Economic Research, *Demographic and Economic Change in Developed Countries.* Princeton, N.J.: Princeton University Press.

Durkheim, Émile. [1886] 1994. "Review of Part VI of the *Principles of Sociology* by Herbert Spencer." *Revue philosophique de la France et de l'étranger* 21:61–69, translated and published by W. S. F. Pickering, *Durkheim on Religion.* Atlanta, Ga.: Scholars Press.

Durkheim, Émile. [1895] 1982. *The Rules of Sociological Method.* New York: Macmillan.

Durkheim, Émile. [1897] 1966. *Suicide.* New York: Free Press.

Eberstadt, Nicholas. 1994. "Demographic Shocks in Eastern Germany, 1989–93." *Europe-Asia Studies* 46:519–533.

Edwards, Harry. 1982. "Race in Contemporary American Sports." *National Forum* 62:19–22.

Eibl-Eibesfeldt, Irenaus. 1970. *Ethology: The Biology of Behavior.* New York: Holt, Rinehart & Winston.

Ekstrand, Laurie E., and William Eckert. 1981. "The Impact of Candidate's Sex on Voter Choice." *Western Political Quarterly* 34:78–87.

El-Badry, M. A. 1969. "Higher Female Than Male Mortality in Some Countries of South Asia: A Digest." *Journal of the American Statistical Association* 64:1234–1244.

Eltis, David. 1993. "Europeans and the Rise and Fall of African Slavery." *American Historical Review* (Dec.).

Emerson, Richard M. 1962. "Power-Dependence Relations." *American Sociological Review* 27:31–41.

England, Robert Stowe. 2005. *Aging China.* Westport, Conn.: Praeger.

Epstein, Cynthia Fuchs. 1976. "Sex Roles." In *Contemporary Social Problems*, 4th ed., edited by Robert K. Merton and Robert Nisbet. New York: Harcourt Brace Jovanovich.

Erickson, Bonnie H. 1996. "Culture, Class, and Connections." *American Journal of Sociology* 102:217–251.

Erickson, Maynard L. 1971. "The Group Context of Delinquent Behavior." *Social Problems* 19:114–129.

Evans-Pritchard, Sir Edward. 1981. *A History of Anthropological Thought.* New York: Basic Books.

Exter, Thomas. 1987. "How Many Hispanics?" *American Demographics* 9:36–39, 67.

Fair, Charles. 1971. *From the Jaws of Victory.* New York: Simon & Schuster.

Faris, Robert E. L. 1967. *Chicago Sociology: 1920–32.* San Francisco: Chandler.

Faris, Robert E. L., and Warren Dunham. 1939. *Mental Disorders in Urban Areas.* Chicago: University of Chicago Press.

Featherman, David L., and Robert M. Hauser. 1976. "Sexual Inequalities and Socioeconomic Achievement in the U.S., 1962–1973." *American Sociological Review* 41:462–483.

Feeney, Floyd. 1986. "Robbers as Decision-Makers." In *The Reasoning Criminal*, edited by D. B. Cornish and R. V. Clarke. New York: Springer-Verlag.

Felson, Marcus. 2002. *Crime and Everyday Life.* 3rd ed. Thousand Oaks, Calif.: Sage.

Ferguson, R. Brian. 1984. *Warfare, Culture, and Environment.* New York: Academic Press.

Ferguson, R. Brian. 1992. "Tribal Warfare." *Scientific American* 266(1):108–113.

Filiatov, Sergei Borisovich, and Dmitri Efimovich Furman. 1993. *Sociological Research* (formerly *Soviet Sociology*) (Jul.-Aug.):38–53.

Finke, Roger, and Rodney Stark. 1988. "Religious Economies and Sacred Canopies: Religious Mobilization in American Cities, 1906." *American Sociological Review* 53:41–49.

Finley, M. I. 1982. *Economy and Society in Ancient Greece.* New York: Viking Press.

Firebaugh, Glenn. 2003. *The New Geography of Global Income Inequality.* Cambridge, Mass.: Harvard University Press.

Firebaugh, Glenn, and Frank D. Beck. 1994. "Does Economic Growth Benefit the Masses? Growth, Dependence, and Welfare in the Third World." *American Sociological Review* 59:631–653.

Firebaugh, Glenn, and Kenneth E. Davis. 1988. "Trends in Antiblack Prejudice, 1972–1984: Region and Cohort Effects." *American Journal of Sociology* 92:251–272.

Fireman, Bruce, and William A. Gamson. 1979. "Utilitarian Logic in the Resource Mobilization Perspective." In *The Dynamics of Social Movements*, edited by M. N. Zald and J. D. McCarthy. Cambridge, Mass.: Winthrop.

Fischer, Claude S. 1975. "Toward a Subcultural Theory of Urbanism." *American Journal of Sociology* 80:1319–1341.

Fischer, Claude S. 1976. *The Urban Experience.* New York: Harcourt Brace Jovanovich.

Fishberg, Maurice. 1911. *The Jews.* New York: Scribners.

Fishman, Robert. 1987. *Bourgeois Utopias: The Rise and Fall of Suburbia.* New York: Basic Books.

Flacks, Richard. 1971. *Youth and Social Change.* Chicago: Markam.

Flavell, J. H., et al. 1968. *The Development of Role-Taking and Communication Skills in Children.* New York: Wiley.

Fleming, Joyce Dudney. 1974. "The State of the Apes." *Psychology Today* 7:31–38.

Fligstein, Neil. 1985. "The Spread of the Multidivisional Form among Large Firms, 1919–1979." *American Sociological Review* 50:377–391.

Fogel, John K., and Stanley L. Engerman. 1974. *Time on the Cross: The Economics of American Negro Slavery.* Boston: Little, Brown.

Ford, Clellan S. 1952. "Control of Conception in Crosscultural Perspective." *World Population Problems and Birth Control. Annals of the New York Academy of Sciences* 54:763–768.

Fossett, Mark A., and K. Jill Kiecolt. 1993. "Mate Availability and Family Structure among African-Americans in U.S. Metropolitan Areas." *Journal of Marriage and the Family* 55:288–302.

Frank, Andre Gunder. 1969. *Latin America: Underdevelopment or Revolution.* New York: Monthly Review Press.

Freeman, Derek. 1983. *Margaret Mead and Samoa: The Making and Unmaking of an Anthropological Myth.* Cambridge, Mass.: Harvard University Press.

Freidson, Eliot. 1973. "Professions and the Occupational Principle." In *The Professions and Their Prospects*, edited by Eliot Freidson. Beverly Hills, Calif.: Sage.

Freud, Sigmund. 1927. *The Future of an Illusion.* Garden City, N.Y.: Doubleday.

Fried, Morton H. 1967. *The Evolution of Political Society: An Essay in Political Anthropology.* New York: Random House.

Fuguitt, G. V., and J. J. Zuiches. 1973. "Residential Preferences and Population Distribution: Results of a National Survey." Paper presented to Rural Sociological Society, College Park, Md.

Fuller, J. L., and W. R. Thompson. 1960. *Behavior Genetics.* New York: Wiley.

Furstenberg, Frank F., Jr., and Graham B. Spanier. 1984. *Recycling the Family: Remarriage after Divorce.* Beverly Hills, Calif.: Sage.

Gains, Donna. 1991. *Teenage Wasteland: Suburbia's Dead End Kids.* New York: Pantheon.

Galle, O., and W. Gove. 1978. "Overcrowding, Isolation and Human Behavior: Exploring the Extremes in Population Distribution." In *Social Demography*, edited by Karl Taeuber and James Sweet. New York: Academic Press.

Gallup, George H. 1972. *The Gallup Poll: Public Opinion 1935–1971*, 3 vols. New York: Random House.

Gallup, George H. 1978. *The Gallup Poll: Public Opinion 1972–1977*, 2 vols. Wilmington, Del.: Scholarly Resources.

Gallup, George H. 1985. *The Gallup Poll: Public Opinion 1972–1984.* New York: Random House.

The Gallup Poll. 1967–1984. *The Gallup Report* (formerly *The Gallup Opinion Index*), published monthly.

Galtung, J. 1971. "A Structural Theory of Imperialism." *Journal of Peace Research* 8:81–117.

Gans, Herbert. 1962. *The Urban Villagers: Group and Class in the Life of Italian-Americans.* New York: Free Press.

Gardiner, R. Allen, and Beatrice T. Gardiner. 1969. "Teaching Sign Language to a Chimpanzee." *Science* 165:664–672.

Gibbons, Don C., and Gerald F. Blake. 1976. "Evaluating the Impact of Juvenile Diversion Programs." *Crime and Delinquency* 22:411–420.

Gibbs, Jack P. 1975. *Crime, Punishment, and Deterrence.* New York: Elsevier.

Giordano, Peggy C., Stephen A. Cernkovich, and Jennifer L. Rudolph. 2002. "Gender, Crime, and Desistance: Toward a Theory of Cognitive Transformation." *American Journal of Sociology* 107: 990–1064.

Glazer, Nathan. 1955. "Social Characteristics of American Jews, 1654–1954." *American Jewish Yearbook* 56:3–41.

Glazer, Nathan. 1971. "Blacks and Ethnic Groups: The Difference, and the Political Difference It Makes." *Social Problems* 18:444–461.

Glazer, Nathan, and Daniel P. Moynihan. 1970. *Beyond the Melting Pot*, 2nd ed. Cambridge, Mass.: MIT Press.

Glock, Charles Y., and Rodney Stark. 1966. *Christian Beliefs and Anti-Semitism.* New York: Harper & Row.

Goffman, Erving. 1959. *The Presentation of Self in Everyday Life.* New York: Doubleday.

Goffman, Erving. 1961. *Asylums: Essays on the Social Situation of Mental Patients and Other Inmates.* Chicago: Aldine.

Goffman, Erving. 1963. *Behavior in Public Places.* New York: Free Press.

Goffman, Erving. 1971. *Relations in Public.* New York: Basic Books.

Goldstein, David B. 1999. Paper on the cohanim genetic signature among the Lemba. Read at a conference on human evolution, Cold Spring Harbor Laboratory, Long Island, N.Y.

Goldstein, Sidney. 1971. "American Jewry, 1970: A Demographic Profile." In *American Jewish Yearbook*. New York: American Jewish Committee.

Goldstone, Jack A. 2000. "The State." In *Encyclopedia of Sociology*, edited by Edgar F. Borgatta and Rhonda J. V. Montgomery. New York: Macmillan.

Goodall, Jane. 1971. *In the Shadow of Man*. Boston: Houghton Mifflin.

Goodall, Jane. 1990. *Through a Window: My Thirty Years with the Chimpanzees of Gombe*. Boston: Houghton Mifflin.

Goring, Charles. 1913. *The English Convict*. London: His Majesty's Stationery Office.

Gottesman, Irving I. 1991. *Schizophrenia Genesis: Origins of Madness*. New York: W. H. Freeman.

Gottesman, Irving I. 1993. "Origins of Schizophrenia: Past as Prologue." In *Nature, Nature, and Psychology*, edited by Robert R. Plomin and Gerald E. McLearn. Washington, D.C.: American Psychological Association.

Gottfredson, Michael R., and Travis Hirschi. 1990. *A General Theory of Crime*. Palo Alto, Calif.: Stanford University Press.

Gough, E. Kathleen. 1974. "Nayar: Central Kerala." In *Matrilineal Kinship*, edited by David Schneider and E. Kathleen Gough. Berkeley: University of California Press.

Gouldner, Alvin W. 1959. "Organizational Analysis." In *Sociology Today*, edited by Robert K. Merton, Leonard Broom, and Leonard S. Cottrell Jr. New York: Basic Books.

Gove, Walter. 1975. *The Labelling of Deviance*. New York: Halsted.

Gove, Walter R. 1985. "The Effect of Age and Gender on Deviant Behavior: A Biopsychosocial Perspective." In *Gender and the Life Course*, edited by Alice S. Rossi. New York: Aldine.

Gove, Walter R., Michael Hughes, and Omer R. Galle. 1979. "Overcrowding in the Home." *American Sociological Review* 44:59–80.

Granovetter, Mark S. 1973. "The Strength of Weak Ties." *American Journal of Sociology* 78:1360–1380.

Grant, Madison. 1916. *The Passing of the Great Race*. New York: Scribners.

Grant, Michael. 1978. *History of Rome*. London: Faber and Faber.

Gray, Diana. 1973. "Turning Out: A Study of Teenage Prostitution." *Urban Life and Culture* 1:401–425.

Greeley, Andrew M. 1974. *Ethnicity in the United States*. New York: Wiley.

Greeley, Andrew. 1991. *The Persistence of the Primordial: The Italian Family in Europe and America*. Chicago: NORC.

Greeley, Andrew. 1994. "A Religious Revival in Russia?" *Journal for the Scientific Study of Religion* 33:253–272.

Green, Dan S., and Edwin D. Driver, eds. 1978. *W. E. B. Du Bois on Sociology and the Black Community*. Chicago: University of Chicago Press.

Gualtieri, Francesco M. 1929. *We Italians: A Study on Italian Immigration in Canada*. Toronto: Italian World War Veterans' Association.

Guerry, A. M. 1860. *Statistique morale de l'Angleterre comparée avec la statistique morale de la France*. Paris: Baillière.

Guest, Avery M., and James A. Weed. 1976. "Ethnic Residential Segregation: Patterns of Change." *American Journal of Sociology* 81:1088–1111.

Gurr, Ted Robert. 1970. *Why Men Rebel*. Princeton, N.J.: Princeton University Press.

Guttentag, Marcia, and Paul F. Secord. 1983. *Too Many Women? The Sex Ratio Question*. Beverly Hills, Calif.: Sage.

Habermas, Jurgen. 1975. *The Legitimation Crisis*. Boston: Beacon Press.

Hage, Jerald, Maurice A. Garnier, and Bruce Fuller. 1988. "The Active State, Investment in Human Capital, and Economic Growth: France 1825–1975." *American Sociological Review* 53:824–837.

Halberstam, David. 1981. *The Breaks of the Game*. New York: Ballantine.

Hall, Peter. 1966. *The World Cities*. New York: McGraw-Hill.

Hallam, Elizabeth M. 1986. *Domesday Book through Nine Centuries*. London: Her Majesty's Stationery Office.

Hampton, Henry, and Steve Fayer. 1990. *Voices of Freedom: An Oral History of the Civil Rights Movement from the 1950s through the 1980s*. New York: Bantam Books.

Handlin, Oscar. 1957. *Race and Nationality in American Life*. Garden City, N.Y.: Anchor Press/Doubleday.

Hanssen, F. Andrew, and Torben Andersen. 1999. "Has Discrimination Lessened Over Time? A Test Using Baseball's All-Star Vote." *Economic Inquiry* (Jun.):326–352.

Hardin, Garrett, ed. 1964. *Population, Evolution and Birth Control*. San Francisco: W. H. Freeman.

Hare, E. H. 1956. "Mental Conditions and Social Conditions in Bristol." *Journal of Mental Science* 102:349–357.

Harlow, Harry F., and Margaret K. Harlow. 1965. "The Affectional Systems." In *Behavior in Non-Human Primates: Modern Research Trends*, vol. 2, edited by Allan Schrier, Harry Barlow, and Fred Stollnitz. New York: Academic Press.

Harris, Marvin. 1979. *Cultural Materialism: The Struggle for a Science of Culture*. New York: Random House.

Harry, Bruce, and Henry Steadman. 1988. "Arrest Rates of Patients Treated in a Community Mental Health Center." *Hospital and Community Psychiatry* 39:862–866.

Hatt, Paul K., and Cecil C. North. 1947. "Jobs and Occupations: A Popular Evaluation." *Opinion News* 9:1–13.

Haub, Carl. 1995. "Population Change in the Former Soviet Republics." *Population Bulletin* 49:1–52.

Hay, David A. 1985. *Essentials of Behaviour Genetics*. Melbourne: Blackwell Scientific Publications.

Hayes, K. J., and C. Hayes. 1951. "The Intellectual Development of a Home-Raised Chimpanzee." *Proceedings of the American Philosophical Society* 95:105–109.

Haynie, Dana L. 2001. "Delinquent Peers Revisited: Does Network Structure Matter?" *American Journal of Sociology* 106:1013–1057.

Heath, Antony. 1976. *Rational Choice and Exchange Theory*. Cambridge: Cambridge University Press.

Heaton, Tim B. 1986. "Sociodemographic Characteristics of Religious Groups in Canada." *Sociological Analysis* 47:54–65.

Hechter, Michael. 1974. *Internal Colonialism: The Celtic Fringe in British National Development*. Berkeley: University of California Press.

Hechter, Michael. 1978. "Group Formation and the Cultural Division of Labor." *American Journal of Sociology* 84:293–318.

Hechter, Michael. 1987. *Principles of Group Solidarity*. Berkeley: University of California Press.

Hechter, Michael, and Satoshi Kanazawa. 1993. "Group Solidarity and Social Order in Japan." *Journal of Theoretical Politics* 5:455–493.

Heer, David M. 1980. "Intermarriage." In *Harvard Encyclopedia of American Ethnic Groups*. Cambridge, Mass.: Belknap Press.

Hexham, Irvin, Raymond F. Currie, and Joan B. Townsend. 1985. "New Religious Movements." In *The Canadian Encyclopedia*. Edmonton, Alberta: Hurtig.

Heyneman, Stephen P., and William A. Loxley. 1983. "The Effect of a Primary-School Quality on Academic Achievement across Twenty-Nine High- and Low-Income Countries." *American Journal of Sociology* 88:1162–1194.

Heyns, Barbara. 1978. *Summer Learning and the Effects of Schooling*. New York: Academic Press.

Higgins, P. C., and G. L. Albrecht. 1977. "Hellfire and Delinquency Revisited." *Social Forces* 55:952–958.

Hill, Charles T., Zick Rubin, and Letitia Anne Peplau. 1976. "Breakups before Marriage: The End of 103 Affairs." *Journal of Social Issues* 32:147–168.

Hiller, Harry W. 1978. "Continentalism and the Third Force in Religion." *Canadian Journal of Sociology* 3:183–207.

Hindelang, Michael, Travis Hirschi, and Joseph G. Weis. 1981. *Measuring Delinquency*. Beverly Hills, Calif.: Sage.

Hintze, Otto. 1975. *The Historical Essays of Otto Hintze*. New York: Oxford University Press.

Hirschi, Travis. 1969. *Causes of Delinquency*. Berkeley: University of California Press.

Hirschi, Travis, and Michael Gottfredson. 1983. "Age and the Explanation of Crime." *American Journal of Sociology* 89:551–575.

Hirschi, Travis, and Michael Gottfredson. 1987. "Causes of White-Collar Crime." *Criminology* 25:949–974.

Hirschi, Travis, and Rodney Stark. 1969. "Hellfire and Delinquency." *Social Problems* 17:202–213.

Hirschman, Charles, and C. Matthew Snipp. 1999. "The State of the American Dream: Race and Ethnic Socioeconomic Inequality in the United States, 1970–90." In *A Nation Divided*, edited by Phyllis Moen, Donna Dempster-McClain, and Henry A. Walker. Ithaca, N.Y.: Cornell University Press.

Hirschman, Charles, and Morrison G. Wong. 1984. "Socioeconomic Gains of Asian Americans, Blacks, and Hispanics: 1960–1976." *American Journal of Sociology* 90:584–607.

Hirschman, Charles, and Morrison G. Wong. 1986. "The Extraordinary Educational Attainment of Asian-Americans: A Search for Historical Evidence and Explanations." *Social Forces* 65:1–27.

Hite, Shere. 1987. *Women and Love: A Cultural Revolution in Progress*. New York: Knopf.

Ho, Ping-Ti. 1959. *Studies on the Population of China, 1368–1953*. Cambridge, Mass.: Harvard University Press.

Hobbes, Thomas. [1651] 1956. *Leviathan*. Chicago: Henry Regnery.

Hodge, Robert W., P. M. Siegal, and Peter Rossi. 1964. "Occupational Prestige in the United States, 1925–1963." *American Journal of Sociology* 70:286–302.

Hodge, Robert W., Donald J. Treiman, and Peter H. Rossi. 1966. "A Comparative Study of Occupational Prestige." In *Class, Status, and Power*, 2nd ed., edited by Reinhard Bendix and Seymour Martin Lipset. New York: Free Press.

Holcomb, William R., and Paul R. Ahr. 1988. "Arrest Rates among Young Adult Psychiatric Patients Treated in Inpatient and Outpatient Settings." *Hospital and Community Psychiatry* 39:52–57.

Hollander, E. P. 1985. "Leadership and Power." In *The Handbook of Social Psychology*, 3rd ed., edited by Gardner Lindzey and Elliott Aronson. New York: Random House.

Hollingshead, August B., and Frederic C. Redlich. 1958. *Social Class and Mental Illness*. New York: Wiley.

Homans, George C. 1974. *Social Behavior: Its Elementary Forms*. New York: Harcourt Brace Jovanovich.

Houk, Ralph, and Robert W. Creamer. 1988. *Season of Glory: The Amazing Saga of the 1961 New York Yankees*. New York: Putnam.

Hout, Michael. 1982. "The Association between Husbands' and Wives' Occupations in Two-Earner Families." *American Journal of Sociology* 88:397–409.

Howard, Michael. 1962. *The Franco-Prussian War*. New York: Macmillan.

Howe, Irving. 1976. *World of Our Fathers*. New York: Harcourt Brace Jovanovich.

Hudson, Valerie M., and Andrea M. Den Boer. 2004. *Bare Branches: The Security Implications of Asia's Surplus Male Population*. Boston: MIT Press.

Hunter, Alfred A., and Margaret A. Denton. 1984. "Do Female Candidates 'Lose Votes'?: The Experience of Female Candidates in the 1979 and 1980 Canadian General Elections." *Canadian Review of Sociology and Anthropology* 21:395–406.

Huntington, Samuel P. 1996. *The Class of Civilizations and the Remaking of World Order*. New York: Simon & Schuster.

Hyland, Ann. 1994. *The Medieval Warhorse: From Byzantium to the Crusades*. London: Grange Books.

Illich, Ivan. 1970. *Deschooling Society*. New York: Harper & Row.

Interpol. 1992. *International Crime Statistics, 1990*. Lyons, France: General Secretariat, Interpol.

Introvigne, Massimo, PierLuigi Zoccatelli, Nelly Ippolito Macrina, and Verónica Roldán. 2001. *Enciclopedia delle religioni in Italia*. Turin, Italy: Elledici.

Irwin, John. 1970. *The Felon*. Englewood Cliffs, N.J.: Prentice Hall.

Isajiw, Wsevolod W. 1980. "Definitions of Ethnicity." In *Ethnicity and Ethnic Relations in Canada*, edited by Jay E. Goldstein and Rita M. Bienvenue. Toronto: Butterworths.

Jacobs, Jerry A., and Frank F. Furstenberg Jr. 1986. "Changing Places: Conjugal Careers and Women's Marital Mobility." *Social Forces* 64:714–732.

Jackson, Kenneth T. 1985. *Crabgrass Frontier: The Suburbanization of the United States*. New York: Oxford University Press.

Jaffe, Dennis. 1975. "Couples in Communes." Ph.D. dissertation, Yale University.

Jencks, Christopher, et al. 1972. *Inequality: A Reassessment of the Effects of Family and Schooling in America*. New York: Basic Books.

Jensen, Gary F. 1969. "Crime Doesn't Pay: Correlates of Shared Misunderstanding." *Social Problems* 17:189–201.

Jensen, Gary F. 1972. "Parents, Peers and Delinquency Action: A Test of the Differential Association Perspective." *American Journal of Sociology* 78:562–575.

Jewkes, John, David Sawers, and Richard Stillerman. 1969. *The Sources of Invention*. New York: Norton.

Jiobu, Robert M. 1988. "Ethnic Hegemony and the Japanese of California." *American Sociological Review* 53:353–367.

Johnson, Benton. 1963. "On Church and Sect." *American Sociological Review* 28:539–549.

Johnson, Paul. 1976. *A History of Christianity*. New York: Atheneum.

Jones, Landon Y. 1980. *Great Expectations: America and the Baby Boom Generation*. New York: Coward, McCann & Geohegan.

Julian, Teresa, and Patrick C. McKenry. 1989. "Relationship of Testosterone to Men's Family Functioning at Mid-Life." *Aggressive Behavior* 15:281–289.

Kage, Joseph. 1981. "Able and Willing to Work: Jewish Immigration and Occupational Patterns in Canada." In *The Canadian Jewish Mosaic*, edited by M. Weinfeld, W. Shaffir, and I. Cotler. Toronto: Wiley.

Kalmijn, Matthijs. 1993. "Trends in Black/White Intermarriage." *Social Forces* 72:119–146.

Kanter, Rosabeth Moss. 1972. *Commitment and Community*. Cambridge, Mass.: Harvard University Press.

Kantrowitz, Nathan. 1973. *Ethnic and Racial Segregation in the New York Metropolis*. New York: Praeger.

Kaplan, Steven, Tudor Pafitt, and Emanuela Trevisan Semi, eds. 1995. *Between Africa and Zion*. Jerusalem: Ben-Zvi Institute.

Keeley, Lawrence H. 1996. *War before Civilization*. New York: Oxford University Press.

Kegl, Judy, Ann Senghas, and Marie Coppola. 1999. "Creation through Contact: Sign Language Emergence and Sign Language Change in Nicaragua." In *Language Creation and Language Change*, edited by Michel DeGraff. Cambridge, Mass.: MIT Press.

Kellogg, W. N., and L. A. Kellogg. 1933. *The Ape and the Child*. New York: McGraw-Hill.

Kelly, Barbara. 1993. *Expanding the American Dream*. Albany: State University of New York Press.

Keniston, Kenneth. 1968. *Young Radicals*. New York: Harcourt, Brace & World.

Kennedy, Robert E., Jr. 1972. "The Social Status of the Sexes and Their Relative Mortality in Ireland." In *Readings in Population*, edited by William Peterson. New York: Macmillan.

Kephart, William. 1957. *Racial Factors and Urban Law Enforcement*. Philadelphia: University of Pennsylvania Press.

Keuchler, Manfred. 2000. "Voting Behavior." In *Encyclopedia of Sociology*, edited by Edgar F. Borgatta and Rhonda J. V. Montgomery. New York: Macmillan.

Kihumura, Akemi, and Harry H. L. Kitano. 1973. "Interracial Marriage: A Picture of Japanese Americans." *Journal of Social Issues* 29:69–73.

Kitano, Harry H. L. 1969. *Japanese Americans*. Englewood Cliffs, N.J.: Prentice Hall.

Klein, Malcolm W. 1976. "Issues and Realities in Police Diversion Programs." *Crime and Delinquency* 22:421–427.

Kluckhohn, Clyde. 1939. "On Recent Applications of Association Coefficients to Ethnological Data." *American Anthropologist* 41:345–377.

Kluckhohn, Clyde. 1962. "Values and Value-Orientations in the Theory of Action." In *Toward a General Theory of Action*, edited by Talcott Parsons and Edward Shils. New York: Harper Torchbooks.

Kluckhohn, Clyde, and Henry A. Murray. 1956. *Personality: In Nature, Society, and Culture*. New York: Knopf.

Kobbervig, Wayne, James Inverarity, and Pat Lauderdale. 1982. "Deterrence and the Death Penalty: A Comment on Phillips." *American Journal of Sociology* 88:161–164.

Kohlberg, Lawrence, and Carol Gilligan. 1971. "The Adolescent as a Philosopher: The Discovery of the Self in a Postconventional World." *Daedalus* 100:1051–1086.

Kohlstedt, Sally, and Helen Longino. 1997. "The Women, Gender, and Science Question: What Do Research on Women in Science and Research on Gender and Science Have to Do with Each Other?" *Osiris* 12:3–15.

Kohn, M. L. 1959. "Social Class and Parental Values." *American Journal of Sociology* 64:337–351.

Kohn, M. L., and Carmi Schooler. 1969. "Class, Occupation, and Orientation." *American Sociological Review* 34:659–678.

Kohn, Melvin L., and Carmi Schooler. 1982. "Job Conditions and Personality: A Longitudinal Assessment of Their Reciprocal Effects." *American Journal of Sociology* 87:1257–1286.

Kohn, Melvin L., and Carmi Schooler. 1983. *Work and Personality*. Norwood, N.J.: Ablex.

Kornhauser, Ruth. 1978. *Social Sources of Delinquency: An Appraisal of Analytic Models*. Chicago: University of Chicago Press.

Kox, Willem, Wim Meeus, and Harm 't Hart. 1991. "Religious Conversion of Adolescents: Testing the Lofland and Stark Model of Religious Conversion." *Sociological Analysis* 52:227–240.

Krakowski, Menachem, Jan Valovka, and David Brizer. 1986. "Psychopathology and Violence: A Review of the Literature." *Comprehensive Psychiatry* 27:131–148.

Kristof, Nicholas D. 1993. "Chinese Peasants Using Ultrasound to Have Sons." *The New York Times* Service in the *Seattle Post-Intelligencer* (Jul. 21):A10.

Kroeber, Alfred L. 1925. *Handbook of American Indians of California*. Bulletin 78. Washington, D.C.: Smithsonian Institution, Bureau of American Ethnology.

Kunstler, James Howard. 1993. *The Geography of Nowhere: The Rise and Decline of America's Man-Made Landscape*. New York: Simon & Schuster.

Lacey, W. K. 1968. *The Family in Classical Greece*. London: Thames & Hudson.

Lamm, Richard, and Gary Imhoff. 1985. *The Immigration Time Bomb: The Fragmenting of America*. New York: Truman Tally Books.

Landes, David S. 1998. *The Wealth and Poverty of Nations*. New York: Norton.

Langer, William L. 1972. "Checks on Population Growth, 1750–1850." *Scientific American* 226:92–99.

Laslett, Peter. 1965. *The World We Have Lost*. London: Kegan Paul.

Laslett, Peter. 1977. *Family Life and Illicit Love in Earlier Generations*. London: Cambridge University Press.

Laub, John H., Daniel S. Nagin, and Robert J. Sampson. 1998. "Trajectories of Change in Criminal Offending: Good Marriages and the Desistance Process." *American Sociological Review* 63:225–238.

Lee, James Daniel. 2002. "More Than Ability: Gender and Personal Relationships Influence Science and Technology Involvement." *Sociology of Education* 75:349–373.

Lemert, Edwin M. 1951. *Social Pathology.* New York: McGraw-Hill.

Lemert, Edwin M. 1967. *Human Deviance, Social Problems, and Social Control.* Englewood Cliffs, N.J.: Prentice Hall.

Lenski, Gerhard. 1954. "Status Crystallization: A Nonvertical Dimension of Social Status." *American Sociological Review* 19:405–413.

Lenski, Gerhard. 1956. "Social Participation and Status Crystallization." *American Sociological Review* 21:458–464.

Lenski, Gerhard. 1966. *Power and Privilege.* New York: McGraw-Hill.

Lenski, Gerhard. 1976. "History and Social Change." *American Journal of Sociology* 82:548–564.

Levine, Gene N., and Darrel M. Montero. 1973. "Socioeconomic Mobility among Three Generations of Japanese Americans." *Journal of Social Issues* 29:40–45.

Lewis, Bernard. 1990. *Race and Slavery in the Middle East.* New York: Oxford University Press.

Li, Bobai, and Andrew G. Walder. 2001. "Career Advancement as Part Patronage: Sponsored Mobility in the Chinese Administrative Elite, 1949–1996." *American Journal of Sociology* 106:1371–1408.

Lieberson, Stanley. 1961. "The Impact of Residential Segregation on Ethnic Assimilation." *Social Forces* 40:52–57.

Lieberson, Stanley. 1963. *Ethnic Patterns in American Cities.* New York: Free Press.

Lieberson, Stanley. 1973. "Generational Differences among Blacks in the North." *American Journal of Sociology* 79:550–565.

Lieberson, Stanley. 1980. *A Piece of the Pie: Blacks and White Immigrants since 1880.* Berkeley: University of California Press.

Lieberson, Stanley, and Mary C. Waters. 1993. "The Ethnic Responses of Whites: What Causes Their Instability, Simplification, and Inconsistency?" *Social Forces* 72:421–450.

Light, Ivan H. 1972. *Ethnic Enterprise in America: Business and Welfare among Chinese, Japanese and Blacks.* Berkeley: University of California Press.

Linden, Rick, and Cathy Fillmore. 1981. "A Comparative Study of Delinquency Involvement." *Canadian Review of Sociology and Anthropology* 18:343–359.

Lindesmith, Alfred R. 1965. *The Addict and the Law.* Bloomington: Indiana University Press.

Lindsay, Jack. 1968. *The Ancient World: Manners and Morals.* New York: Putnam's Sons.

Link, Bruce G., Howard Andrews, and Francis T. Cullen. 1992. "The Violent and Illegal Behavior of Mental Patients Reconsidered." *American Sociological Review* 57:275–292.

Linton, Ralph. [1936] 1964. *The Study of Man: An Introduction.* Englewood Cliffs, N.J.: Prentice Hall.

Lipset, Seymour Martin, and Reinhard Bendix. 1959. *Social Mobility in Industrial Society.* Berkeley: University of California Press.

Liska, Allen E. 1987. *Perspectives on Deviance.* Englewood Cliffs, N.J.: Prentice Hall.

Litwak, Eugene, and Peter Messeri. 1989. "Organizational Theory, Social Supports, and Mortality Rates." *American Sociological Review* 54:49–66.

Lofland, John, and Rodney Stark. 1965. "Becoming a World-Saver: A Theory of Conversion to a Deviant Perspective." *American Sociological Review* 30:862–875.

Lombroso-Ferrero, Gina. 1911. *Criminal Man.* Montclair, N.J.: Patterson-Smith.

Lorenz, Konrad. 1966. *On Aggression.* New York: Harcourt Brace Jovanovich.

MacMullen, Ramsay. 1981. *Paganism in the Roman Empire.* New Haven, Conn.: Yale University Press.

Malinowski, Bronislaw. 1935. *The Foundations of Faith and Morals.* Oxford: Oxford University Press.

Mallory, Walter H. 1926. *China: Land of Famine.* New York: American Geographical Society.

Malson, Lucien. 1972. *Wolf Children and the Problem of Human Nature.* New York: Monthly Review Press.

Martin, David. 1981. "Disorientations to Mainstream Religion: The Context in New Religious Movements." In *The Social Impact of New Religious Movements,* edited by Bryan Wilson. New York: Rose of Sharon Press.

Martin, David. 1989. "Speaking in Latin Tongues." *National Review* (Sept. 29):30–35.

Martin, David. 1990. *Tongues of Fire: The Explosion of Protestantism in Latin America.* Oxford: Basil Blackwell.

Marx, Gary T. 1967. *Protest and Prejudice.* New York: Harper & Row.

Marx, Karl, and Friedrich Engels. [1848] 1967. *Communist Manifesto.* New York: Pantheon.

Mason, Robert, and Lyle D. Calvin. 1978. "A Study of Admitted Tax Evasion." *Law & Society* (Fall):73–89.

Matras, Judah. 1973. *Populations and Societies.* Englewood Cliffs, N.J.: Prentice Hall.

Mayhew, Pat. 1984. "Target-Hardening: How Much of an Answer?" In *Coping with Burglary,* edited by R. V. G. Clarke and T. Hope. Boston: Kluwer-Nijhoff.

McAdam, Doug. 1988. *Freedom Summer.* New York: Oxford University Press.

McBurnett, Keith, Benjamin B. Lahey, Paul J. Rathouz, and Rolf Loeber. 2000. "Low Salivary Cortisol and Persistent Aggression in Boys Referred for Disruptive Behavior." *Archives of General Psychiatry* 57:38–43.

McCarthy, John D., and Mayer N. Zald. 1977. "Resource Mobilization and Social Movements: A Partial Theory." *American Journal of Sociology* 82:1212–1241.

McClean, Charles. 1978. *The Wolf Children.* New York: Hill & Wang.

McClosky, Herbert. 1964. "Consensus and Ideology in American Politics." *American Political Science Review* 58:361–382.

McCord, William, and Joan McCord. 1959. *Origins of Crime: A New Evaluation of the Cambridge-Somerville Youth Study.* New York: Columbia University Press.

McCusker, Claire. 2002. "Home Schoolers Arrive on Campus." *Insight* (Sept. 9):47.

McDougall, W. 1908. *An Introduction to Social Psychology.* Boston: Luce.

McDougall, William. 1932. *The Energies of Men: A Study of the Fundamentals of Dynamic Psychology.* London: Methuen.

McEachern, A. W. 1968. "The Juvenile Probation System." *American Behavioral Scientist* 11:1–10.

McFarland, Bentson H., Larry R. Faulkner, Joseph E. Bloom, Roxy Hallaux, and J. Donald Bray. 1989. "Chronic Mental Illness and the Criminal Justice System." *Hospital and Community Psychiatry* 27:131–148.

McFarland, H. Neill. 1967. *The Rush Hour of the Gods: A Study of New Religious Movements in Japan.* New York: Macmillan.

McGahey, Richard M. 1980. "Dr. Ehrlich's Magic Bullet: Econometric Theory, Econometrics, and the Death Penalty." *Crime and Delinquency* 17:485–502.

McGovern, Thomas. 1981. "The Economics of Extinction in Norse Greenland." In *Climate and History: Studies in Past Climates and Their Impact on Man*, edited by T. M. L. Wigley, M. J. Ingram, and G. Farmer. Cambridge: Cambridge University Press.

McKenzie, R. F. 1933. *The Metropolitan Community.* New York: McGraw-Hill.

McKinney, William, and Wade Clark Roof. 1982. "A Social Profile of American Religious Groups." In *Yearbook of American and Canadian Churches: 1982*, edited by Constant H. Jacquet Jr. Nashville, Tenn.: Abingdon Press.

McLuhan, Marshall. 1964. *Understanding Media: The Extensions of Man.* New York: McGraw-Hill.

McNeill, William H. 1976. *Plagues and Peoples.* New York: Basic Books.

McNeill, William. 1982. *The Pursuit of Power.* Chicago: University of Chicago Press.

McWilliams, Carey. 1945. *Prejudice: Japanese-Americans.* Boston: Little, Brown.

Mead, George Herbert. 1925. "The Genesis of the Self and Social Control." *International Journal of Ethics* 35:251–273.

Mead, George Herbert. 1934. *Mind, Self, and Society: From the Standpoint of a Social Behaviorist*, edited by Charles W. Morris. Chicago: University of Chicago Press.

Mead, Margaret. 1935. *Sex and Temperament in Three Primitive Societies.* New York: Morrow.

Mead, Margaret. 1950. *Sex and Temperament in Three Primitive Societies*, 2nd ed. New York: Dell.

Meadows, Donella, et al. 1972. *The Limits to Growth: A Report for the Club of Rome's Projection on the Predicament of Mankind.* New York: Universe Books.

Mednick, S. A., et al. 1984. "Genetic Influences in Criminal Convictions: Evidence from an Adoption Cohort." *Science* 224: 891–894.

Meeks, Wayne A. 1983. *The First Urban Christians: The Social World of the Apostle Paul.* New Haven, Conn.: Yale University Press.

Meeks, Wayne. 1993. *The Origins of Christian Morality: The First Two Centuries.* New Haven, Conn.: Yale University Press.

Melton, J. Gordon. 1978. *Encyclopedia of American Religions.* Wilmington, N.C.: McGrath.

Melton, J. Gordon. 1993. *Encyclopedia of American Religions.* 4th ed. Detroit: Gale Research.

Melton, J. Gordon. 2002. *The Encyclopedia of American Religions.* 7th ed. Detroit: Thomson/Gale.

Melton, J. Gordon, and Jolen Marya Gedridge. 1998. *The Encyclopedia of American Religions.* Detroit: Gale Research.

Meltzoff, Andrew N., and M. Keith Moore. 1977. "Imitation of Facial and Manual Gestures by Human Neonates." *Science* 198:75–78.

Meltzoff, Andrew N., and M. Keith Moore. 1983a. "Newborn Infants Imitate Adult Facial Gestures." *Child Development* 54:702–709.

Meltzoff, Andrew N., and M. Keith Moore. 1983b. "The Origins of Imitation in Infancy: Paradigm, Phenomena, and Theories." *Advances in Infancy Research* 2:266–301.

Meltzoff, Andrew N., and M. Keith Moore. 1994. "Imitation, Memory and the Representation of Persons." *Infant Behavior and Development* 17:83–99.

Mendelssohn, Kurt. 1976. *The Secret of Western Domination.* New York: Praeger.

Merton, Robert K. 1938. "Social Structure and Anomie." *American Sociological Review* 3:672–682.

Messick, David M., et al. 1983. "Individual Adaptations and Structural Change as Solutions to Social Dilemmas." *Journal of Personality and Social Psychology* 44:294–309.

Meyer, John W. 1977. "The Effects of Education as an Institution." *American Journal of Sociology* 83:55–77.

Miller, Alan S., and John P. Hoffmann. 1995. "Risk and Religion: An Explanation of Gender Differences in Religiosity." *Journal for the Scientific Study of Religion* 34:63–75.

Miller, Alan S., and Rodney Stark. 2002. "Gender and Religiousness: Can Socialization Explanations Be Saved?" *American Journal of Sociology* 107:1399–1423.

Miller, G. Tyler. 1998. *Living in the Environment*, 10th ed. Belmont, Calif.: Wadsworth.

Miller, Walter B. 1962. "The Impact of a 'Total Community' Delinquency Control Project." *Social Problems* 10:168–191.

Millis, H. A. 1915. *The Japanese Problem in the United States.* New York: Macmillan.

Mills, C. Wright. 1956. *The Power Elite.* New York: Oxford University Press.

Mills, C. Wright. 1959. *The Sociological Imagination.* New York: Oxford University Press.

Mills, J. P. 1922. *The Lhota Nagas.* London: Macmillan.

Minor, W. William, and Joseph Harry. 1982. "Deterrent and Experiential Effects in Perceptual Deterrence Research: A Replication and Extension." *Journal of Research in Crime and Delinquency* 19:190–203.

Miyamoto, Frank. 1939. "Social Solidarity among the Japanese in Seattle." *University of Washington Publications in the Social Sciences* 2:57–130.

Miyamoto, Frank S., and Sanford M. Dornbusch. 1956. "A Test of Interaction Hypothesis of Self-Conception." *American Journal of Sociology* 61:399–403.

Mizruchi, Ephraim H. 1964. *Success and Opportunity: A Study of Anomie.* Glencoe, Ill.: Free Press.

Montgomery, Field Marshal Viscount. 1968. *A History of Warfare.* New York: World.

Moody, James. 2001. "Race, School Integration, and Friendship Segregation in America." *American Journal of Sociology* 107:679–716.

Mooney, James. 1896. *The Ghost Dance Religion and the Sioux Outbreak of 1890.* Fourth Annual Report of the Bureau of Ethnology to the Secretary of the Smithsonian Institution. Washington, D.C.: U.S. Government Printing Office.

Morioka, Kiyomi. 1975. *Religion in Changing Japanese Society.* Tokyo: University of Tokyo Press.

Morris, Aldon D. 1984. *The Origins of the Civil Rights Movement: Black Communities Organizing for Change.* New York: Free Press.

Morselli, Henry. 1882. *Suicide: An Essay on Comparative Moral Statistics.* New York: Appleton.

Mosca, Gaetano. [1896] 1939. *The Ruling Class.* New York: McGraw-Hill.

Mowat, Farley. 1965. *Westviking.* Boston: Little, Brown.

Murdock, George P. 1949. *Social Structure.* New York: Macmillan.

Murdock, George P., and Douglas R. White. 1969. "Standard Cross-Cultural Sample." *Ethnology* 8:329–369.

Musmanno, Michael A. 1965. *The Story of the Italians in America.* New York: Doubleday.

Myrdal, Gunnar. 1944. *An American Dilemma: The Negro Problem and Modern Democracy.* New York: Harper & Row.

National Center for Education Statistics. 2001. *Homeschooling in the United States: 1999.* Washington, D.C.: U.S. Department of Education.

National Institute of Justice. 1991. *Drug Use Forecasting.* Washington, D.C.: U.S. Department of Justice.

Newman, Jody. 1994. *Perception and Reality: A Study Comparing the Success of Men & Women Candidates.* Washington, D.C.: National Women's Political Caucus.

Niebuhr, H. Richard. 1929. *The Social Sources of Denominationalism.* New York: Henry Holt.

Nisbet, Robert. 1980. *History of the Idea of Progress.* New York: Basic Books.

Nordhoff, Charles. [1875] 1966. *The Communistic Societies of the United States.* New York: Dover.

North, Douglass C., and Robert Paul Thomas. 1973. *The Rise of the Western World: A New Economic History.* Cambridge: Cambridge University Press.

Notestein, Frank W. 1950. "The Population of the World in the Year 2000." *Journal of the American Statistical Association* 42:335–349.

Noyes, John Humphrey. 1870. *The History of American Socialisms.* Philadelphia: Lippincott.

Nozick, Robert. 1974. *Anarchy, State and Utopia.* New York: Basic Books.

Ogburn, William F. 1932. *Social Change.* New York: Viking Press.

Olsen, Mancur. 1965. *The Logic of Collective Action.* Cambridge, Mass.: Harvard University Press.

Ornstein, Michael D. 1981. "The Occupational Mobility of Men in Ontario." *Review of Canadian Sociology and Anthropology* 18:183–214.

Ortiz, Vilma. 1989. "Language Background and Literacy among Hispanic Young Adults." *Social Problems* 36:149–164.

Ossowski, Stanislaw. 1963. *Class Structure in the Social Consciousness.* New York: Free Press.

Ostrom, Carol M. 2002. "New Technique Lets Parents Pick Baby's Gender." *Seattle Times* (Oct. 16).

Paige, Jeffery M. 1974. "Kinship and Polity in Stateless Societies." *American Journal of Sociology* 80:301–320.

Parfitt, Tudor, and Emanuela Trevisan Semi, eds. 1999. *The Beta Israel in Enthiopia and Israel.* Surrey: Curzon Press.

Park, Robert E., Ernest W. Burgess, and Roderick McKenzie. 1925. *The City.* Chicago: University of Chicago Press.

Parker, J., and H. G. Grasmick. 1979. "Linking Actual and Perceived Certainty of Punishment: An Exploratory Study of an Untested Proposition in Deterrence Theory." *Criminology* 17:336–379.

Parker, Robert Nash. 1989. "Poverty, Subculture of Violence, and Type of Homicide." *Social Forces* 67:983–1007.

Parsons, Talcott. 1937. *The Structure of Social Action.* New York: McGraw-Hill.

Parsons, Talcott. 1951. *The Social System.* Glencoe, Ill.: Free Press.

Patterson, G. R. 1980. "Children Who Steal." In *Understanding Crime: Current Theory and Research,* edited by Travis Hirschi and Michael Gottfredson. Beverly Hills, Calif.: Sage.

Patterson, Orlando. 1982. *Slavery and Social Death: A Comparative Study.* Cambridge, Mass.: Harvard University Press.

Pearlin, L. I., and M. L. Kohn. 1966. "Social Class, Occupation, and Parental Values: A Cross-National Study." *American Sociological Review* 31:466–479.

Perlmann, Joel. 1988. *Ethnic Differences: Schooling and Social Structure among the Irish, Italians, Jews, and Blacks in an American City, 1880–1935.* Cambridge: Cambridge University Press.

Petersen, William. 1971. *Japanese Americans: Oppression and Success.* New York: Random House.

Petersen, William. 1975. *Population.* New York: Macmillan.

Petersen, William. 1978. "Chinese Americans and Japanese Americans." In *Essays and Data on American Ethnic Groups,* edited by Thomas Sowell. Washington, D.C.: Urban Institute.

Peterson, Iver. 1999. "Some Perched in Ivory Tower Gain Rosier View of Suburbs." *The New York Times* (Dec. 5):1, 43.

Pfeiffer, John E. 1977. *The Emergence of Society.* New York: McGraw-Hill.

Phillips, David P. 1980. "The Deterrent Effect of Capital Punishment: New Evidence on an Old Controversy." *American Journal of Sociology* 86:139–148.

Piaget, Jean. 1926. *The Language and Thoughts of the Child.* New York: Harcourt.

Piaget, Jean. 1970. "Piaget's Theory." In *Carmichael's Manual of Child Psychology,* 3rd ed., edited by Paul Mussen. New York: Wiley.

Piaget, Jean, and Barbel Inhelder. 1969. *The Psychology of the Child.* New York: Basic Books.

Piliavin, Irving, and Scott Briar. 1964. "Police Encounters with Juveniles." *American Journal of Sociology* 70:206–214.

Pineo, Peter C. 1961. "Disenchantment in the Later Years of Marriage." *Marriage and Family Living* 23:4.

Pineo, Peter C. 1976. "Social Mobility in Canada: The Current Picture." *Sociological Focus* 9:1091–1123.

Pineo, Peter C. 1977. "The Social Standing of Ethnic and Racial Groupings." *Canadian Review of Sociology and Anthropology* 14:147–157.

Pineo, Peter C. 1981. "Prestige and Mobility: The Two National Surveys." *Canadian Review of Sociology and Anthropology* 18:615–626.

Pineo, Peter C., and John Porter. 1967. "Occupational Prestige in Canada." *Canadian Review of Sociology and Anthropology* 4:24–40.

Pinker, Steven. 1994. *The Language Instinct.* New York: Morrow.

Pirenne, Henri. 1925. *Medieval Cities.* Princeton, N.J.: Princeton University Press.

Pomeroy, Sarah B. 1975. *Goddesses, Whores, Wives, Slaves: Women in Classical Antiquity.* New York: Schocken Books.

Porter, John. 1965. *The Vertical Mosaic: An Analysis of Social Class and Power in Canada.* Toronto: University of Toronto Press.

Portes, Alejandro. 1981. "Modes of Structural Incorporation and Present Theories of Immigration." In *Global Trends in Migration*, edited by Mary M. Kritz, Charles B. Keely, and Sylvano M. Tomasi. Staten Island, N.Y.: CMS Press.

Portes, Alejandro. 1987. "The Social Origins of the Cuban Enclave Economy of Miami." *Sociological Perspectives* 30:340–372.

Portes, Alejandro, and Robert L. Bach. 1985. *Latin Journey: Cuban and Mexican Immigrants in the United States.* Berkeley: University of California Press.

Portes, Alejandro, and Leif Jensen. 1989. "The Enclave and the Entrants: Patterns of Ethnic Enterprise in Miami before and after Mariel." *American Sociological Review* 54:929–949.

Portes, Alejandro, and Robert D. Manning. 1986. "The Immigrant Enclave: Theory and Empirical Examples." In *Competitive Ethnic Relations*, edited by Susan Olzak and Joane Nagel. New York: Academic Press.

Powers, Edwin, and Helen Witmer. 1951. *An Experiment in the Prevention of Delinquency: The Cambridge-Somerville Youth Study.* New York: Columbia University Press.

Preston, Samuel E. 1975. "Estimating the Proportion of American Marriages That End in Divorce." *Sociological Methods and Research* 3:435–460.

Pullum, Thomas W. 1982. "The Eventual Frequencies of Kin in a Stable Population." *Demography* 19:549–565.

Quetelet, Adolphe. 1835. *Sur l'homme et le développement de ses facultés.* Paris: Bachelier.

Quinley, Harold E., and Charles Y. Glock. 1979. *Anti-Semitism in America.* New York: Free Press.

Rabkin, Judith. 1979. "Criminal Behavior of Discharged Mental Patients." *Psychological Bulletin* 86:1–27.

Ragin, Charles C. 1987. *The Comparative Method.* Berkeley: University of California Press.

Raine, Adrian, Todd Lencz, Susan Bihrle, Lori LaCasse, and Patrick Colletti. 2000. "Reduced Prefrontal Gray Matter Volume and Reduced Autonomic Activity in Antisocial Personality Disorder." *Archives of General Psychiatry* 57:119–127.

Rapp, Rayna, and Ellen Ross. 1983. "The Twenties' Backlash: Compulsory Heterosexuality, the Consumer Family, and the Waning of Feminism." In *Class, Race and Sex: The Dynamics of Control*, edited by Amy Swerdlow and Hanna Messinger. Boston: Hall.

Rauscher, Frances, Gordon Shaw, and Katherine Ky. 1993. "Music and Spatial Task-Performance." *Nature* 365:611.

Rawson, Beryl, ed. 1986. *The Family in Ancient Rome.* Ithaca, N.Y.: Cornell University Press.

Ray, Brian. 1997. *Strengths of Their Own: Home Schoolers across America.* Salem, Oreg.: National Home Education Research Institute Publications.

Reiss, Albert J., Jr. 1961. *Occupations and Social Status.* New York: Free Press.

Reiss, Ira L. 1988. *Family Systems in America.* New York: Holt, Rinehart & Winston.

Rhodes, A. L., and A. J. Reiss, Jr. 1970. "The Religious Factor and Delinquent Behavior." *Journal of Research in Crime and Delinquency* 7:83–98.

Richer, Stephen. 1984. "Sexual Inequality and Children's Play." *Review of Canadian Sociology and Anthropology* 21:166–180.

Riddle, M., and A. H. Roberts. 1977. "Delinquency, Delay of Gratification, Recidivism, and the Porteus Maze Tests." *Psychological Bulletin* 84:417–425.

Riesman, David. 1961. *The Lonely Crowd.* New Haven, Conn.: Yale University Press.

Rodriguez, Gregory. 1999. *From Newcomers to New Americans: The Successful Integration of Immigrants into American Society.* Washington, D.C.: National Immigration Forum.

Ro'i, Yaacov. 1996. "Islam in the Soviet Union after the Second World War." *Religion, State and Society* 24:158–164.

Roncek, Dennis W., and Antoinette Lobosco. 1983. "The Effect of High Schools on Crime in Their Neighborhoods." *Social Science Quarterly* 64:598–613.

Ropp, Theodore. 1959. *War in the Modern World.* Durham, N.C.: Duke University Press.

Rose, Richard J., Markku Koskenvuo, Jaako Kaprio, Seppo Sarna, and Heimo Langinvainio. 1988. "Shared Genes, Shared Experiences, and Similarity of Personality: Data from 14,288 Adult Finnish Co-Twins." *Journal of Personality and Social Psychology* 54:161–171.

Rosenberg, Charles E. 1975. "Introduction: History and Experience." In *The Family in History*, edited by Charles E. Rosenberg. Philadelphia: University of Pennsylvania Press.

Rosenberg, Stuart E. 1970. *The Jewish Community in Canada.* Toronto: McClelland & Stewart.

Rosenfeld, Richard, and Scott Decker. 1993. "Discrepant Values, Correlated Measures: Cross-City and Longitudinal Comparisons of Self Reports and Urine Tests of Cocaine Use among Arrestees." *Journal of Criminal Justice.*

Rosenfeld, Richard, Matthew Perkins, and Eric Baumer. 1994. "'Any Decent Thing': The Manifest and Latent Functions of Gun Buy-Back Programs." Read at the annual meetings of the American Society of Criminology.

Rosenthal, David. 1970. *Genetic Theory and Abnormal Behavior.* New York: McGraw-Hill.

Ross, E. A. 1914. *The Old World in the New.* New York: Century.

Ross, Marc Howard. 1983. "Political Decision Making and Conflict: Additional Cross-Cultural Codes and Scales." *Ethnology* 22:169–192.

Rossi, Peter H., Richard A. Berk, and Kenneth J. Lenihan. 1982. "Saying It Wrong with Figures: A Comment on Zeisel." *American Journal of Sociology* 88:390–393.

Rothman, Sheila. 1978. *Woman's Proper Place.* New York: Basic Books.

Rubinow, Israel. 1907. "The Economic Condition of Jews in Russia." *Bulletin of the Bureau of Labor*, no. 72. Washington, D.C.: U.S. Government Printing Office.

Russell, J. C. 1958. "Late Ancient and Medieval Population." *Transactions of the American Philosophical Society* 48:1–152.

Sack, Benjamin G. 1965. *History of the Jews in Canada.* Montreal: Harvest House.

Sampson, Robert J., and John H. Laub. 1990. "Crime and Deviance over the Life Course: The Salience of Adult Social Bonds." *American Sociological Review* 55:609–627.

Sampson, Samuel F. 1969. "Crisis in a Cloister." Ph.D. dissertation, Cornell University.

Samuelsson, Kurt. 1957. *Religion and Economic Action.* New York: Basic Books.

Sanders, Jimmy M., and Victor Nee. 1987. "Limits of Ethnic Solidarity in the Enclave Economy." *American Sociological Review* 52:745–773.

Sanders, Ronald. 1969. *Downtown Jews.* New York: Harper & Row.

Sapir, Edward. 1921. *Language.* New York: Harcourt, Brace, and World.

Sawyer, P. H. 1982. *Kings and Vikings.* London: Methuen.

Scheff, Thomas J. 1984. *Being Mentally Ill: A Sociological Theory.* 2nd ed. Chicago: Aldine.

Schmidt, Constance R., and Scott G. Paris. 1984. "The Development of Verbal Communicative Skills in Children." *Advances in Child Development and Behavior* 18:2–47.

Schmookler, Jacob. 1966. *Invention and Economic Growth.* Cambridge, Mass.: Harvard University Press.

Schoefield, J. W. 1995. "Improving Intergroup Relations among Students." In *Handbook of Research on Multicultural Education,* edited by J. A. Banks and C. A. M. Banks. New York: Macmillan.

Schollaert, Paul T., and Donald Hugh Smith. 1987. "Team Racial Composition and Sports Attendance." *The Sociological Quarterly* 28:71–87.

Schuckit, Marc, et al. 1972a. "The Half-Sibling Approach in a Genetic Study of Alcoholism." *Life History Research in Psychopathology* 2:120–127.

Schuckit, Marc, et al. 1972b. "A Study of Alcoholism in Half-Siblings." *American Journal of Psychiatry* 128:1132–1136.

Schuckit, Marc, et al. 1979. *The Genetic Aspects of Psychiatric Syndrome Relating to Antisocial Problems in Youth.* Seattle, Wash.: Center for Law and Justice.

Schur, Edwin. 1971. *Labeling Deviant Behavior.* New York: Harper & Row.

Schwartz, Richard, and Jerome H. Skolnick. 1962. "A Study of Legal Stigma." *Social Problems* 10:133–138.

Seielstad, Mark T., Eric Minch, and L. Luca Cavalli-Sforza. 1998. "Genetic Evidence for a Higher Female Migration Rate in Humans." *Nature Genetics* 20:278–280.

Selznick, Gertrude J., and Stephen Steinberg. 1969. *The Tenacity of Prejudice.* New York: Harper & Row.

Selznick, Philip. 1948. "Foundations of the Theory of Organization." *American Sociological Review* 13:25–35.

Selznick, Philip. 1949. *TVA and the Grass Roots.* Berkeley: University of California Press.

Selznick, Philip. 1957. *Leadership in Administration.* New York: Harper & Row.

Shapiro, Susan P. 1990. "Collaring the Crime, Not the Criminal: Reconsidering the Concept of White-Collar Crime." *American Sociological Review* 55:346–365.

Shattuck, Roger. 1980. *The Forbidden Experiment.* New York: Farrar, Straus & Giroux.

Shaw, Clifford R., and Henry D. McKay. 1929. *Delinquency Areas.* Chicago: University of Chicago Press.

Shaw, Clifford R., and Henry D. McKay. 1931. *Report on the Causes of Crime,* vol. 12, no. 13. Washington, D.C.: National Commission on Law Observance and Enforcement.

Shaw, Clifford R., and Henry D. McKay. 1942. *Juvenile Delinquency and Urban Areas.* Chicago: University of Chicago Press.

Sheldon, W. H. 1940. *The Varieties of Human Physique.* New York: Harpers.

Sherif, Muzafer, and Carolyn W. Sherif. 1953. *Groups in Harmony and Tension: An Integration of Studies on Intergroup Relations.* New York: Harper & Row.

Sherman, Lawrence W., Patrick R. Gartin, and Michael E. Buerger. 1989. "Hot Spots of Predatory Crime: Routine Activities and the Criminology of Place." *Criminology* 27:27–55.

Sherman, Stephanie L., John C. DeFries, Irving I. Gottsman, John C. Loehlin, Joanne M. Meyer, May Z. Pelias, John Rice, and Irwin Waldman. 1997. "Recent Developments in Human Behavioral Genetics: Past Accomplishments and Future Directions." *American Journal of Human Genetics* 60:1265–1275.

Shioji, Hiroki. 1980. "The Japanese Family: Economic Pressures Affecting Cultural Values within the Home." Ph.D. dissertation, University of Arizona.

Shirer, William L. 1969. *The Collapse of the Third Republic.* New York: Simon & Schuster.

Shorter, Edward. 1975. *The Making of the Modern Family.* New York: Basic Books.

Sigelman, Lee, and Susan Welch. 1993. "The Contact Hypothesis Revisited: Black-White Interaction and Positive Racial Attitudes." *Social Forces* 71:781–795.

Silberman, Charles E. 1971. *Crisis in the Classroom: The Remaking of American Education.* New York: Vintage.

Sills, David L. 1957. *The Volunteers.* Glencoe, Ill.: Free Press.

Simcha-Fagan, Ora, and Joseph E. Schwartz. 1986. "Neighborhood and Delinquency: An Assessment of Contextual Effects." *Criminology* 24:667–699.

Simmel, Georg. 1905. "A Contribution to the Sociology of Religion." *American Journal of Sociology* 11:359–376.

Simon, Julian L. 1981. *The Ultimate Resource.* Princeton, N.J.: Princeton University Press.

Simpson, John H. 1996. "'The Great Reversal': Selves, Communities, and the Global System." *Sociology of Religion* 57:115–125.

Sjoberg, Gideon. 1960. *The Preindustrial City.* New York: Free Press.

Sjoberg, Gideon. 1965. "Cities in Developing and in Industrialized Societies: A Cross-Cultural Analysis." In *The Study of Urbanization,* edited by P. H. Hauser and L. F. Schnore. New York: Wiley.

Skeels, H. M. 1966. *Adult Status of Children with Contrasting Early Life Experiences.* Monographs of the Society for Research in Child Development.

Skeels, H. M., and H. A. Dye. 1939. "A Study of the Effects of Differential Stimulation in Mentally Retarded Children." *Proceedings of the American Association for Mental Deficiency* 44:114–136.

Skorecki, Karl, Sara Selig, Shraga Blazer, Robert Bradman, Neil Bradman, P. J. Warburton, Monica Ismajlowicz, and Michael F. Hammer. 1997. "Y Chromosomes of Jewish Priests." *Nature* 385 (Jun. 2):32.

Slater, Miriam K. 1969. "My Son the Doctor: Aspects of Mobility among American Jews." *American Journal of Sociology* 34:359–373.

Smelser, Neil J. 1963. *Theory of Collective Behavior*. New York: Free Press.

Smith, Adam. [1776] 1937. *An Inquiry into the Nature and Causes of the Wealth of Nations*. New York: Modern Library.

Smith, Daniel S., and M. S. Hindus. 1975. "Premarital Pregnancy in America, 1640–1971: An Overview and Interpretation." *Journal of Interdisciplinary History* 4:537–570.

Smith, Robert J. 1983. *Japanese Society: Individual, Self and the Social Order*. Cambridge: Cambridge University Press.

Smith, W. Robertson. 1889. *The Religion of the Semites: Fundamental Institutions*. Edinburgh: Adam and Charles Black.

Sorokin, Pitirim A. 1937. *Social and Cultural Dynamics*. New York: American Books.

South, Scott J. 1993. "Racial and Ethnic Differences in the Desire to Marry." *Journal of Marriage and the Family* 55:357–370.

South, Scott J., and Steven F. Messner. 1987. "The Sex Ratio and Women's Involvement in Crime: A Cross-National Analysis." *Sociological Quarterly* 28:171–188.

South, Scott J., and Katherine Trent. 1988. "Sex Ratios and Women's Roles: A Cross-National Analysis." *American Journal of Sociology* 93:1096–1115.

Sowell, Thomas, ed. 1978. *Essays and Data on American Ethnic Groups*. Washington, D.C.: Urban Institute.

Sowell, Thomas. 1981. *Ethnic America: A History*. New York: Basic Books.

Sowell, Thomas. 1994. *Race and Culture: A World View*. New York: Basic Books.

Spada, A. V. 1969. *The Italians in Canada*. Ottawa: Canada Ethnica VI.

Spiro, Melford E. 1966. "Religion: Problems of Definition and Explanation." In *Anthropological Approaches to the Study of Religion*, edited by Michael Banton. New York: Praeger.

Srole, Leo, Thomas S. Langer, Stanley T. Michael, Marvin K. Opler, and Thomas A. C. Rennie. 1962. *Mental Health in the Metropolis*. New York: McGraw-Hill.

Stack, Steven. 1987. "Publicized Executions and Homicide, 1950–1980." *American Sociological Review* 52:532–540.

Stark, Rodney. 1981. "Must All Religions Be Supernatural?" In *The Social Impact of New Religious Movements*, edited by Bryan Wilson. New York: Rose of Sharon Press.

Stark, Rodney. 1984. "The Rise of a New World Faith." *Review of Religious Research* 26:18–27.

Stark, Rodney. 1987. "Estimating Church Membership Rates: 1971–1980." *Review of Religious Research* 29:69–77.

Stark, Rodney. 1992a. "Epidemics, Networks, and the Rise of Christianity." *Semeia* 56:159–175.

Stark, Rodney. 1992b. "Do Catholic Societies Really Exist?" *Rationality and Society* 4:261–271.

Stark, Rodney. 1993. "Modernization and Mormon Growth: The Secularization Thesis Revisited." In *A Sociological Analysis of Mormonism*, edited by Marie Cornwall, Tim B. Heaton, and Lawrence Young. Champaign: University of Illinois Press.

Stark, Rodney. 1995. "The Role of Women in the Rise of Christianity." *Sociology of Religion* 56:242–244.

Stark, Rodney. 1996a. "Why Religious Movements Succeed or Fail: A Revised General Model." *Journal of Contemporary Religion* 11:133–146.

Stark, Rodney. 1996b. *The Rise of Christianity: A Sociologist Reconsiders History*. Princeton, N.J.: Princeton University Press.

Stark, Rodney. 1996c. "Religion as Context: Hellfire and Delinquency One More Time." *Sociology of Religion* 57:163–173.

Stark, Rodney. 1997. "Catholic Contexts: Competition, Commitment, and Innovation." *Review of Religious Research* 38.

Stark, Rodney. 1999. "Atheism, Faith and the Social Scientific Study of Religion." *Journal of Contemporary Religion* 14:41–62.

Stark, Rodney. 2001a. *One True God: Historical Consequences of Monotheism*. Princeton, N.J.: Princeton University Press.

Stark, Rodney. 2001b. "Gods, Rituals, and the Moral Order." *Journal for the Scientific Study of Religion*. 40:619–636.

Stark, Rodney. 2002. "Physiology and Faith: Addressing the 'Universal' Gender Difference in Religiousness." *Journal for the Scientific Study of Religion* 41:495–507.

Stark, Rodney, and W. S. Bainbridge. 1980. "Secularizations, Revival, and Cult Formation." *Annual Review of the Social Sciences of Religion* 4:85–119.

Stark, Rodney, and W. S. Bainbridge. 1981. "American-Born Sects: Initial Findings." *Journal for the Scientific Study of Religion* 20:130–149.

Stark, Rodney, and W. S. Bainbridge. 1985. *The Future of Religion: Secularization, Revival and Cult Formation*. Berkeley: University of California Press.

Stark, Rodney, and William Sims Bainbridge. [1987] 1996. *A Theory of Religion*, new ed. New Brunswick, N.J.: Rutgers University Press.

Stark, Rodney, and William Sims Bainbridge. 1997. *Religion, Deviance, and Social Control*. New York and London: Routledge.

Stark, Rodney, and Roger Finke. 2000. *Acts of Faith: Explaining the Human Side of Religion*. Berkeley: University of California Press.

Stark, Rodney, and Laurence R. Iannaccone. 1997. "Why the Jehovah's Witnesses Grow So Rapidly: A Theoretical Application." *Journal of Contemporary Religion* 12:133–157.

Stark, Rodney, Lori Kent, and Daniel P. Doyle. 1982. "Religion and Delinquency: The Ecology of a 'Lost' Relationship." *Journal of Research in Crime and Delinquency* 19:4–24.

Stark, Rodney, and James C. McCann. 1993. "Market Forces and Catholic Commitment: Exploring the New Paradigm." *Journal for the Scientific Study of Religion* 32:111–124.

Stark, Rodney, et al. 1971. *Wayward Shepherds: Prejudice and the Protestant Clergy*. New York: Harper & Row.

Steffensmeier, Darrell J., Emilie Anderson Allan, Miles D. Harer, and Cathy Streifel. 1989. "Age and the Distribution of Crime." *American Journal of Sociology* 94:803–831.

Steinberg, Stephen. 1974. *The Academic Melting Pot*. New York: McGraw-Hill.

Steinberg, Stephen. 1981. *The Ethnic Myth: Race, Ethnicity, and Class in America*. Boston: Beacon Press.

Stephan, Karen H., and G. Edward Stephan. 1973. "Religion and the Survival of Utopian Communities." *Journal for the Scientific Study of Religion* 12:89–100.

Stinchcombe, Arthur L. 1968. *Constructing Social Theory*. New York: Harcourt Brace Jovanovich.

Stoll, David. 1990. *Is Latin America Turning Protestant? The Politics of Evangelical Growth.* Berkeley: University of California Press.

Stolzenberg, Ross M. 1990. "Ethnicity, Geography, and Occupational Achievement of Hispanic Men in the United States." *American Sociological Review* 55:143–154.

Stuller, Jay. 1991. "Cleanliness Has Only Recently Become a Virtue." *Smithsonian* 21:126–135.

Sunahara, Ann. 1981. *The Politics of Racism: The Uprooting of Japanese Canadians during the Second World War.* Toronto: Lorimer.

Sutherland, Edwin. 1983. *White-Collar Crime: The Uncut Version.* New Haven, Conn.: Yale University Press.

Swanson, Guy E. 1968. "To Live in Concord with Society: Two Empirical Studies of Primary Relations." In *Cooley and Sociological Analysis*, edited by A. J. Reiss. Ann Arbor: University of Michigan Press.

Swanson, Guy E. 1969. *Rules of Descent: Studies in the Sociology of Parentage.* Ann Arbor: Museum of Anthropology, University of Michigan.

Taeuber, Cynthia M., and Victor Valdisera. 1986. *Women in the American Economy.* Bureau of the Census, Current Population Reports, Special Studies Series P-23, no. 146. Washington, D.C.: U.S. Government Printing Office.

Taeuber, Karl E., and Alma F. Taeuber. 1964. "The Negro as an Immigrant Group: Recent Trends in Racial and Ethnic Segregation in Chicago." *American Journal of Sociology* 69:347–382.

Taeuber, Karl E., and Alma F. Taeuber. 1969. *Negroes in Cities.* New York: Atheneum.

Tainter, Joseph A. 1988. *The Collapse of Complex Societies.* Cambridge: Cambridge University Press.

Tanner, James M. 1970. "Physical Growth." In *Carmichael's Manual of Child Psychology*, 3rd ed., edited by Paul Mussen. New York: Wiley.

Tardola, H. 1970. "The Needle Scene." In *The Participant Observer: Encounters with Social Reality*, edited by G. Jacobs. New York: Braziller.

Taylor, Marylee C. 1983. "The Black-White Model of Attitude Stability: A Latent Class Examination of Opinion and Nonopinion in the American Public." *American Journal of Sociology* 89:373–401.

Taylor, Pamela, and John Gunn. 1984. "Violence and Psychosis. I, Risk of Violence among Psychotic Men." *British Medical Journal of Clinical Research* 288:1945–1949.

Taylor, Ralph B., and Jeanette Covington. 1988. "Neighborhood Changes in Ecology and Violence." *Criminology* 26:553–589.

Taylor, Ralph B., et al. 1980. "The Defensibility of Defensible Space." In *Understanding Crime: Current Theory and Research*, edited by Travis Hirschi and Michael Gottfredson. Beverly Hills, Calif.: Sage.

Tellegen, Auke, David T. Lykken, Thomas J. Bouchard, Jr., Kimberly J. Wilcox, Nancy Segal, and Stephen Rich. 1988. "Personality Similarity in Twins Reared Apart and Together." *Journal of Personality and Social Psychology* 54:1031–1039.

Tepperman, Lorne. 1976. "A Simulation of Social Mobility in Industrial Societies." *Canadian Review of Sociology and Anthropology* 13:26–42.

Thomas, Hugh. 1979. *A History of the World.* New York: Harper & Row.

Thomas, Mark G., Karl Skorecki, Haim Ben-Ami, Tudor Parfitt, Neil Bradman, and David B. Goldstein. 1998. "Origins of Old Testament Priests." *Nature* 394 (Jul. 9):138–140.

Thompson, Edward H. 1991. "Beneath the Status Characteristic: Gender Variations in Religiousness." *Journal for the Scientific Study of Religion* 30:381–394.

Thompson, James D. 1967. *Organizations in Action.* New York: McGraw-Hill.

Thornberry, Terrance P. 1973. "Race, Socio-Economic Status and Sentencing in the Juvenile Justice System." *Journal of Criminal Law, Criminology, and Police Science* 64:90–98.

Tittle, Charles R., and Robert F. Meier. 1990. "Specifying the SES/Delinquency Relationship." *Criminology* 28:271–299.

Tittle, C. R., et al. 1978. "The Myth of Social Class and Criminality: An Empirical Assessment of the Empirical Evidence." *American Sociological Review* 43:643–656.

Toby, Jackson. 1957. "Social Disorganization and Stake in Conformity: Complementary Factors in the Predatory Behavior of Hoodlums." *Journal of Criminal Law, Criminology, and Police Science* 48:12–17.

Toby, Jackson. 1965. "Early Identification and Intensive Treatment of Pre-delinquents: A Negative View." *Social Work* 6:3–13.

Torrey, E. Fuller. 1988. *Nowhere to Go.* New York: Harper & Row.

Treiman, Donald, and Kermit Terrell. 1975. "Sex and the Process of Status Attainment: A Comparison of Working Men and Women." *American Sociological Review* 40:174–200.

Trent, Katherine, and Scott J. South. 1989. "Structural Determinants of the Divorce Rate: A Cross-Societal Analysis." *Journal of Marriage and the Family* 51:391–404.

Trowbridge, Gordon. 2002. "Locale Links Segregated Cities." *The Detroit News* (Jan. 14).

Turk, Austin T. 1969. *Criminality and the Legal Order.* Chicago: Rand McNally.

Turnbull, Colin. 1965. "The Mbuti Pygmies of the Congo." In *Peoples of Africa*, edited by James L. Gibbs. New York: Holt, Rinehart & Winston.

Turner, R. Jay, and Morton O. Wagenfeld. 1967. "Occupational Mobility and Schizophrenia: An Assessment of the Social Causation and Social Selection Hypotheses." *American Sociological Review* 32:104–113.

Turner, Ralph H., and Lewis M. Killian. 1987. *Collective Behavior.* Englewood Cliffs, N.J.: Prentice Hall.

Tylor, Edward Burnett. [1871] 1958. *Religion in Primitive Culture.* New York: Harper and Brothers.

Udry, J. Richard. 1988. "Biological Predispositions and Social Control in Adolescent Sexual Behavior." *American Sociological Review* 53:709–722.

Uggen, Christopher. 2000. "Work as a Turning Point in the Life Course of Criminals: A Duration Model of Age, Employment, and Recidivism." *American Sociological Review* 67:529–546.

Ujimoto, K. Victor. 1976. "Contrasts in the Prewar and Postwar Japanese Community in British Columbia: Conflict and Change." *Canadian Review of Sociology and Anthropology* 13:81–89.

United Nations. 2001. *World Population Prospects: The 2000 Revision.* New York: Population Division, Department of Economic and Social Affairs.

U.S. Department of Justice. *Uniform Crime Reports.* Washington, D.C.: U.S. Government Printing Office. Published annually.

Vallance, Theodore R. 1993. *Prohibition's Second Failure: The Quest for a Rational and Humane Drug Policy.* Westport, Conn.: Praeger.

Vallee, Frank G. 1981. "The Sociology of John Porter: Ethnicity as Anachronism." *Canadian Review of Sociology and Anthropology* 18:639–648.

van de Walle, Etienne. 1992. "Fertility Transition, Conscious Choice, and Numeracy." *Demography* 29:487–502.

Veblen, Thorstein. 1899. *The Theory of the Leisure Class.* New York: Macmillan.

Vigod, Bernard L. 1984. *The Jews in Canada.* Ottawa: Canadian Historical Association.

Vogt, William. 1948. *Road to Survival.* New York: Sloane.

von Bertalanffy, Ludwig. 1967. "General System Theory." In *System, Change and Conflict*, edited by N. J. Demerath III and Richard A. Peterson. New York: Free Press.

Vorontsova, Lyudmila, and Sergei Filiatov. 1994. "The Changing Pattern of Religious Belief: *Perestroika* and Beyond." *Religion, State and Society* 22:89–96.

Wade, Nicholas. 1999. "DNA Backs Tribe's Tradition of Early Descent from the Jews." *New York Times* (May 9):1, 20.

Waldo, Gordon P., and Simon Dinitz. 1967. "Personality Attributes of the Criminal: An Analysis of Research Studies, 1950–1965." *Journal of Research in Crime and Delinquency* 4:185–202.

Wallace, Anthony F. C. 1956. "Revitalization Movements." *American Anthropologist* 58:264–281.

Wallace, Anthony F. C. 1966. *Religion: An Anthropological View.* New York: Random House.

Wallace, W. L. 1966. *Student Culture: Social Structure and Continuity in a Liberal Arts College.* Chicago: Aldine.

Wallerstein, Immanuel. 1974. *The Modern World System.* New York: Academic Press.

Wallis, Roy. 1975. *Sectarianism.* New York: Wiley.

Wallis, Roy. 1982. *Millennialism and Charisma.* Belfast, Northern Ireland: The Queen's University.

Walzer, Michael. 1963. *The Revolution of the Saints.* Cambridge, Mass.: Harvard University Press.

Ward, David. 1971. *Cities and Immigrants.* New York: Oxford University Press.

Warren, John Robert, Robert M. Hauser, Jennifer T. Sheridan. 2002. "Occupational Stratification across the Life Course: Evidence from the Wisconsin Longitudinal Study." *American Sociological Review* 67:432–455.

Washington, M. H., ed. 1975. *The Black Woman and the Disappointment of Romantic Love.* Garden City, N.Y.: Anchor Books.

Waters, Mary C. 1990. *Ethnic Options: Choosing Identities in America.* Berkeley: University of California Press.

Weber, Max. [1921] 1946. "Politics as a Vocation." In *From Max Weber*, edited by Hans Gerth and C. Wright Mills. New York: Oxford University Press.

Weed, James A. 1980. "National Estimates of Marriage Dissolution and Survivorship." *Vital and Health Statistics*, series 3, no. 19.

Weir, Alex A. S., Jackie Chappell, and Alex Kacelnik. 2002. "Shaping of Hooks in New Caledonian Crows." *Science* (Aug. 9) 297:981.

Weis, J. G. 1977. "Comparative Analysis of Social Control Theories of Delinquency: The Breakdown of Adequate Social Controls."

In *Preventing Delinquency: A Comparative Analysis of Delinquency Prevention Theory.* Washington, D.C.: National Institute for Juvenile Justice and Delinquency Prevention.

Weisbrot, Robert. 1991. *Freedom Bound: A History of America's Civil Rights Movement.* New York: Penguin.

Welch, Kevin. 1983. "Community Development and Metropolitan Religious Commitment: A Test of Two Competing Models." *Journal for the Scientific Study of Religion* 22:167–180.

West, Donald, and David Farrington. 1977. *The Delinquent Way of Life.* New York: Crane Russak.

Westhues, Kenneth. 1976. "Religious Organization in Canada and the United States." *International Journal of Comparative Sociology* 17:245–261.

White, L. A. 1949. *The Science of Culture.* New York: Farrar, Straus & Giroux.

White, Lynn, Jr. 1962. *Medieval Technology and Social Change.* London: Oxford University Press.

White, Lynn K., and Alan Booth. 1985. "The Quality and Stability of Remarriages: The Role of Stepchildren." *American Sociological Review* 50:689–698.

White, Lynn, and John N. Edwards. 1990. "Emptying the Nest and Parental Well-Being: An Analysis on National Panel Data." *American Sociological Review* 55:235–242.

Whiten, A., J. Goodall, W. C. McGrew, T. Nishida, V. Reynolds, Y. Sugiyama, C. E. G. Tutin, R. W. Wrandham, and C. Boesch. 1999. "Cultures in Chimpanzees." *Nature* 399 (Jun. 17):682–685.

Wilkinson, Karen. 1980. "The Broken Home and Delinquent Behavior." In *Understanding Crime: Current Theory and Research*, edited by Travis Hirschi and Michael Gottfredson. Beverly Hills, Calif.: Sage.

Williams, R. M., Jr. 1947. *The Reduction of Intergroup Tensions.* New York: Social Science Research Council.

Wilson, Bryan. 1959. "An Analysis of Sect Development." *American Sociological Review* 24:2–15.

Wilson, Bryan. 1961. *Sects and Society.* Berkeley: University of California Press.

Wilson, Bryan. 1970. *Religious Sects.* New York: McGraw-Hill.

Wilson, James Q. 2002. *The Marriage Problem: How Our Culture Has Weakened Families.* New York: HarperCollins.

Wilson, James Q., and Richard J. Herrnstein. 1985. *Crime and Human Nature.* New York: Simon & Schuster.

Wimberly, Dale W., and Rosaria Bello. 1992. "Effects of Foreign Investment, Exports, and Economic Growth on Third World Food Consumption." *Social Forces* 70:895–921.

Wirth, Louis. 1928. *The Ghetto.* Chicago: University of Chicago Press.

Wirth, Louis. 1938. "Urbanism as a Way of Life." *American Journal of Sociology* 44:8–20.

Wolfgang, Marvin, Robert Figlio, and Thorsten Sellin. 1972. *Delinquency in a Birth Cohort.* Chicago: University of Chicago Press.

Woodsworth, James S. [1909] 1972. *Strangers within Our Gates: Or Coming Canadians.* Toronto: University of Toronto Press.

World Bank. *World Development Report, 1990, 1991, 1992, 1993, 1994.* Oxford: Oxford University Press.

Wrigley, E. A. 1969. *Population and History.* New York: McGraw-Hill.

Wrong, Dennis H. 1961. "The Oversocialized Conception of Man in Modern Sociology." *American Sociological Review* 26:183–193.

Wuthnow, Robert. 1976. *The Consciousness Reformation.* Berkeley: University of California Press.

Yearbook of Jehovah's Witnesses. 1996. Published annually.

Yunker, James A. 1982. "The Relevance of the Identification Problem to Statistical Research on Capital Punishment." *Crime and Delinquency* 28:96–124.

Zablocki, Benjamin. 1980. *Alienation and Charisma.* New York: Free Press.

Zahavi, Yacov, and James M. Ryan. 1980. "Stability of Travel Components over Time." *Transportation Research Record* 750:19–26.

Zahavi, Yacov, and Antli Talvitie. 1980. "Regularities in Travel Time and Money Expenditures." *Transportation Research Record* 750:13–19.

Zborowski, Mark, and Elizabeth Herzog. 1962. *Life Is with People: The Culture of the Shtetl.* New York: Schocken Books.

Zeisel, Hans. 1982a. "Comment on the Deterrent Effect of Capital Punishment." *American Journal of Sociology* 88:167–169.

Zeisel, Hans. 1982b. "Disagreement over the Evaluation of a Controlled Experiment." *American Journal of Sociology* 88:378–389.

Zeisel, Hans. 1982c. *The Limits of Law Enforcement.* Chicago: University of Chicago Press.

Zuckerman, Michael. 1975. "Dr. Spock: The Confidence Man." In *The Family in History,* edited by Charles E. Rosenberg. Philadelphia: University of Pennsylvania Press.

vi: © Stone/The Hulton Getty Picture Collection; **vii:** Library of Congress; **vii:** © Brown Brothers; **viii:** Library of Congress; **ix:** © Culver Pictures; **ix:** Library of Congress; **x:** © Leonard Freed/ Magnum Photos; **xi:** American Correctional Association; **xi:** Copyright the Dorothea Lange Collection, The Oakland Museum of California, City of Oakland. Gift of Paul S. Taylor. Reproduced by permission; **xii:** © David Phillips; **xiii:** From the Benny Goodman Papers in the Music Library of Yale University; **xiv:** Library of Congress; **xiv:** Courtesy of Rodney Stark; **xix:** Marion Post Wolcott for the Farm Security Administration, 1939; **xv:** Copyright the Dorothea Lange Collection, The Oakland Museum of California, City of Oakland. Gift of Paul S. Taylor. Reproduced by permission; **xvi:** Freer Gallery of Art, Smithsonian Institution, Washington, D.C.; **xvi:** © David Phillips; **xvii:** Library of Congress; **xviii:** U.S. Census Bureau, Suitland, MD; **xviii:** © David Phillips

Chapter 1 © The Hulton Getty Picture Collection/Stone/Getty Images; **(left to right):** © Harald Sund; © Bibliotheque Royale Albert, Bruxelles (Cabinet des Estampes) Jean-Baptiste Madou, artist; © Elizabeth Crews; **4:** © Harald Sund; **5:** © Bibliotheque Royale Albert, Bruxelles (Cabinet des Estampes) Jean-Baptiste Madou, artist; **7:** © Bettmann/ Corbis; **8:** University of Chicago Library Photo, Ann Colley; **8:** Photographs and Prints Division Schomburg Center For Research in Black Culture New York Public Library Astor, Lenox and Tilden Foundations; **23:** © Elizabeth Crews.

Chapter 2 31: © Brown Brothers; **(left to right):** all, Library of Congress; **41:** © Brown Brothers; **42:** © Tim Fitzharris/Masterfile; **45:** Library of Congress; **49:** © Alter Kacyzne/Raphael Abramovich Collection. Reproduced by permission; **50:** © Brown Brothers; **51:** Library of Congress; **52:** Library of Congress **54:** Eaton's of Canada Archives; **58:** Library of Congress; **59:** Immigrants Aboard the S.S. Westerland. Photograph ca. 1890. Museum of the City of New York Print Archives. Reprinted by permission; **60:** Bank of America; **60:** Bank of America.

Chapter 3 67: Library of Congress; **(left to right):** Library of Congress; Courtesy of Lynne Roberts; © Corbis/Sygma; **69:** © Jose L. Pelaez/

Corbis **71:** © Mike Valeri/Taxi/Getty Images; **72:** © Juan Silva/The Image Bank/Getty Images; **72:** University of Missouri Archives; **73:** Bentley Historical Library/University of Michigan; **74:** Library of Congress; **75:** Courtesy of Lynne Roberts; **88:** © AP/Wide World Photos; **89:** © Corbis Sygma.

Chapter 4 95: Library of Congress; **(left to right):** © Ellis Herwig/Stock Boston; © 1997, James Marvy/The Stock Market; AP/Wide World Photos; **98:** © Arthur Tilley; **99:** © Jeff Zaruba/Stone/Getty Images; **101:** © Ellis Herwig/Stock Boston; **105:** © James Marvy/Corbis; **113:** AP/Wide World Photos; **118:** The Film Study Center, Harvard University.

Chapter 5 124: © Culver Pictures; **(left to right):** Courtesy of Gordon and Gary Shepherd; Courtesy of Alex A. S. Weir, Jackie Chappell, Alex Kacelnik, Oxford University Department of Zoology. Reprinted by permission; Courtesy of Lynne Roberts; © Culver Pictures; **127:** A. Meltzoff & M. K. Moore, "Imitations of facial and manual gestures by human neonates." Science, 1977, 198, 75–78. American Association for the Advancement of Science and A. Meltzoff; **128:** Courtesy of Gordon and Gary Shepherd; **130:** Carnegie Museum of Art, Pittsburgh; **133:** Manuscripts, Special Collections, University Archives, University of Washington Libraries Neg. # #UW1880; **134:** © Charles Kennard/ Stock Boston; **137:** Courtesy of Dr. Tudor Parfitt; **139:** © Ian Berry/Magnum Photos; **140:** Baron Hugo van Lawick/© National Geographic Society; **141:** Courtesy of Alex A. S. Weir, Jackie Chappell, Alex Kacelnik, Oxford University Department of Zoology. Reprinted by permission; **142:** Harlow Primate Laboratory, University of Wisconsin; **143:** Courtesy of Lynne Roberts.

Chapter 6 146: Library of Congress; **(left to right):** Courtesy of Judy Shepard-Kegl, Nicaraguan Sign Language Project; Neg. #2A 5161, American Museum of Natural History; Courtesy of Lynne Roberts; **147:** Library of Congress; **149:** © Thomas Hoepker/Woodfin Camp; **152:** © Yves de Braine/ Black Star; **156:** Courtesy of Judy Shepard-Kegl, Nicaraguan Sign Language Project; **158:** Neg. #2A 5161, American Museum of Natural History; **159:** © Barbara Kirk/Peter Arnold, Inc.; **163:** Courtesy of Lynne Roberts; **165:** © David Wells/ The Image Works; **168:** © Jacques Langevin/ Corbis Sygma.

Chapter 7 174: © Leonard Freed/Magnum Photos; **(left to right):** © Corbis Sygma; St. Duroy/ Rapho; Scala/Art Resource, NY; **179:** © Corbis Sygma; **182:** © Bettmann/Corbis; **190:** St. Duroy/ Rapho; **191:** © Richard Kalvar/Magnum; **195:** Scala/Art Resource, NY.

Chapter 8 208: American Correctional Association; **(left to right):** © Sepp Seitz/Woodfin Camp; Library of Congress; © Alex Webb/Magnum Photos; **209:** American Correctional Association; **216:** © Sepp Seitz/Woodfin Camp; **217:** Library of Congress; **218:** © Eddie Adams/TimePix/Getty Images; **225:** © Alex Webb/Magnum Photos.

Chapter 9 230: Copyright the Dorothea Lange Collection, The Oakland Museum of California, City of Oakland. Gift of Paul S. Taylor. Reproduced by permission; **(left to right):** all, Library of Congress; **231:** Copyright the Dorothea Lange Collection, The Oakland Museum of California, City of Oakland. Gift of Paul S. Taylor. Reproduced by permission; **233:** Library of Congress; **234:** Library of Congress; **234:** Archives of Labor & Urban Affairs, University Archives, Wayne State University; **236:** © German Information Center; **237:** © Bettmann/Corbis; **239:** © Ken Heyman; **243:** Library of Congress.

Chapter 10 254: © David Phillips; **(left to right):** © Bettmann/Corbis; French Cultural Services; © Jean Gaumy/Magnum Photos; **255:** © David Phillips; **257:** Neg. #17023. photo R. M. Anderson, American Museum of Natural History; **258:** © Lorne Resnick/Stone/Getty Images; **260:** Art Resource, NY; **262:** Giraudon, Art Resource, NY; **263:** © Bettmann/Corbis; **264:** Scala/Art Resource, NY; **265:** Art Resource, NY; **268:** © Jean Gaumy/ Magnum Photos; **269:** French Cultural Services.

Chapter 11 282: From the Benny Goodman Papers in the Music Library of Yale University **(left to right):** Special Collections Division, University of Washington Libraries, Neg. #UW526; © Margaret Bourke-White Estate/ TimePix/Getty Images; Library of Congress; **283:** From the Benny Goodman Papers in the Music Library of Yale University; **287:** Laurie Platt Winfrey, Inc.; **292:** © L. Delevigne Photo; **295:** New York Public Library; **298:** Library of Congress; **300:** Library of Congress; **303:** © Brown Brothers; **305:** Special Collections Civision, University of Washington Libraries, Neg. #UW526; **306:** Library and Archives Canada/ Department of National Defence collection/ PA-037468; **309:** California Historical Society, San Francisco/Arnold Genthe, FN#23115; **313:** Library of Congress; **316:** © Margaret Bourke-White Estate/TimePix/Getty Images; **318:** AP/Wide World Photos; **321:** © Rick Reinhard.

Chapter 12 324: Library of Congress; **(left to right):** Library of Congress; Manuscripts, Special Collections, University Archives, University of Washington Libraries photographer Clifford, neg. #11; National Maritime Museum, San Francisco; **325:** Library of Congress; **327:** © David Powers Photography, 1988; **(top to bottom):** Courtesy of Judy Shepard-Kegl, Nicaraguan Sign Language

A

Aberele, David F., 70
Abernathy, Ralph D., 595, 597–598
Ackerman, Nathan W., 290
Acton, John, 1st Baron Acton of
    Aldenham, 424
Adachi, Ken, 306
Adorno, Theodore, 290
Ages, Arnold, 44
Ahr, Paul R., 186
Akers, Ronald L., 188
Alba, Richard D., 47, 61, 286
Albrecht, G. L., 102
Albrecht, S. L., 102
Alcorn, D. S., 102
Alexander, Karl L., 458
Allen, Leslie, 47
Allport, Gordon W., 125, 291–292,
    296, 307
Alsop, Joseph, 599
Amato, Paul R., 379
Andersen, Torben, 317
Andrews, Howard, 186
Angell, Robert Cooley, 200
Applewhite, Herff, 88–89
Araiza, Francisco, 327
Ardrey, Robert, 138
Argyle, Michael, 393
Aristotle, 3, 392
Armstrong, Louis, 131
Asch, Solomon, 81–82
Ashton, Basil, 507, 528
Augustus (Caesar), 500
Austen, Ralph, 293
Ausubel, Jesse H., 550
Azzi, Corry, 393

B

Bach, Robert L., 308, 309
Bader, Chris, 17–19
Bailey, William C., 221
Bain, Reed, 157
Bainbridge, William Sims, 5, 87, 104,
    106–107, 112, 195, 200, 388, 393,
    399–400, 402, 404, 405–406, 410, 437
Balch, Robert, 88
Baltzell, E. Digby, 238
Bamforth, Douglas B., 259
Bandura, Albert, 153
Barker, Eileen, 95

Barrett, David B., 407
Barrett, Richard K., 489
Bart, Pauline, 382
Barton, Paul E., 456
Batson, C. Daniel, 393
Baumer, Eric, 200
Baxandall, Rosalyn, 550
Bear, Chief, 479
Beck, Frank D., 491, 493
Becker, Howard S., 176
Befu, Harumi, 211
Beirne, Piers, 4, 5, 183
Beit-Hallahmi, Benjamin, 393
Bell, Daniel, 399
Bello, Rosaria, 491
Bellugi, Ursula, 154
Bendix, Reinhard, 266, 267, 269–270,
    274, 276
Benedict, Ruth, 158
Benitez, J., 465
Bennett, Neil G., 379
Berelson, Bernard, 514, 521–522
Berg, E. J., 300
Berger, Bennett M., 104, 106
Berger, Peter L., 7, 388
Bestor, Theodore C., 212
Beyer, Peter, 43
Bian, Yanjie, 36, 278
Bibby, Reginald W., 405
Biblartz, Timothy J., 273
Bickerton, Derek, 155
Bird, Caroline, 462
Blake, Gerald F., 202
Blake, J. P., 587
Blalock, Hubert M., Jr., 131, 301
Blanc, Ann Klimas, 379
Blau, Peter M., 76, 270–272, 274,
    448, 578
Blick, Jeffrey P., 259
Bloch, Marc, 260, 261
Bloom, David E., 379
Blumberg, Rae Lesser, 354–355
Blumenfeld, H., 537
Blumer, Herbert, 72, 73–74
Boas, Franz, 158, 161, 259
Bobrovnikov, Vladimir, 409
Bolce, Louis, 291
Bonacich, Edna, 297–301
Bonaparte, Napoleon, 563, 565, 567
Bongaarts, John, 523

Booth, Alan, 136, 381–382
Borgatta, Edgar F., 307
Bosch, Hieronymus, 195
Bouchard, Thomas J., Jr., 131–132
Boudon, Raymond, 463, 466
Bourdieu, Pierre, 241, 276, 277
Bowers, W. J., 220
Bowker, L. H., 184
Bowles, Samuel, 464
Boyd, Monica, 273, 274, 275
Bradburn, Norman, 98
Bradshaw, York W., 43, 493
Branch, Taylor, 595, 596
Brantingham, Patricia L., 179, 213
Brantingham, Paul J., 179, 213
Braudel, Fernand, 134, 264–265, 266, 481
Breughel, Pieter, 368, 535
Breuglel, Pieter, 368, 535
Briar, Scott, 222
Brierley, Peter, 17
Brizer, David, 186
Broderick, Francis L., 8
Brodie, Mollyann, 311
Brophy, I. N., 291
Brough, R. Clayton, 137
Brown, Roger, 154
Bruer, John T., 151
Brunk, Robert, 89
Brym, Robert, 48
Bucur, Alexander, 273
Buerger, Michael E., 192
Burgess, Ernest W., 551–552, 555
Burgess, Robert L., 188
Burkett, S. R., 102
Burroughs, Edgar Rice, 472
Bursik, Robert J., Jr., 192
Burt, Ronald, 36
Busch, Lisa, 527
Bush, George W., 10

C

Cabot, John, 46
Cabot, Richard Clarke, 214
Caldwell, John C., 523
Calhoun, J. B., 556
Calvin, John, 483–484
Calvin, Lyle D., 193
Campbell, Angus, 377
Cann, Arnie, 349
Cantor, David, 192, 213
Carmichael, Stokely, 319

Carnap, Rudolf, 79
Carpenter, C. Jan, 168–169
Carter, John, 593
Cassidy, Deborah J., 168–169
Catherine the Great, 265
Chabris, Christopher F., 151
Chadwick, B. A., 102
Chafetz, Janet Saltzman, 354–355
Chagnon, Napoleon A., 117
Champneuf, Jacques Guerry de, 4
Chandler, David, 563, 567, 568, 577
Charlemagne, 263
Charles, Enid, 524
Chase-Dunn, Christopher, 488
Chestnut, R. Andrew, 407
Chirot, Daniel, 39, 112–113, 256,
    485, 488
Chomsky, Noam, 154, 156
Christiansen, K. O., 183
Cloud, John, 461
Cohen, Lawrence E., 213
Cohen, Mark, 258
Cohen, Yinon, 272
Cohn, Werner, 48, 238
Cole, George F., 224
Cole, Sonia, 138
Coleman, James S., 68, 75, 76, 241, 421,
    456–457, 461
Coleman, James William, 194
Collins, Randall, 3, 69, 73, 354–355, 464
Coltrane, Scott, 354–355
Comte, Auguste, 8, 398
Converse, Philip E., 438, 439–440
Cooley, Charles Horton, 14, 73–74
Cooper, Bill, 419
Coppola, Marie, 156
Corman, Hope, 223
Cornwall, Marie, 393
Cortés, J. B., 183, 185
Cott, Nancy F., 343, 344–345, 351
Counts, George Sylvester, 50
Covello, Leonard, 50–51, 57, 61
Covington, Jeanette, 192
Creamer, Robert W., 591
Crick, Francis, 136
Crockett, Davy, 517
Croll, Elizabeth, 526
Crutchfield, Robert, 200
Cullen, Francis T., 186
Currie, Raymond F., 406
Cutright, Phillips, 514

D
Dabbs, James M., Jr., 135, 136, 185
Dahl, Robert, 427

Dahrendorf, Ralf, 235, 244–245, 249
Daitzman, Reid, 135, 185
Darcy, R., 435
Darley, John M., 97
Darroch, A. Gordon, 551, 552
Darwin, Charles, 285, 508
Davies, James C., 593
Davis, Kenneth E., 317
Davis, Kingsley, 148, 245–247, 248, 257,
    506–507, 511, 512–514, 525, 535, 536
de la Garza, Rodolpho O., 311
De Maio, Gerald, 291
de Mondeville, Henri, 508
de Paul, Vincent, 335
de Peyronnet, Comte (Count), 4
de Tocqueville, Alexis, 268–269
de Vaus, David A., 393
Deck, Leland, 349
Delacroix, Jacques, 484, 489–491
DeLoache, Judy S., 168–169
Demerath, N. J., 483
Den Boer, Andrea M., 527
Denton, Margaret A., 434–435
Dewey, John, 125
DeWitt, J. L., 304
Diamond, Jared, 480, 508
Dickeman, Mildred, 333
Dickens, Charles, 346
DiMaggio, Paul, 37
Dinitz, Simon, 187
Dohrenwend, Bruce P., 555
Dornbusch, Sanford M., 74
Douglas, Kirk, 320
Douglas, Mary, 392
Doyle, Daniel P., 102
Driver, Edwin D., 8
Drucker, Peter F., 266, 448, 450, 576, 578
Drummond, J. C., 511
Du Bois, W. E. B., 8, 24
Dublin, Louis I., 523
Duesenberry, J. S., 68
DuMont, Allen B., 477
Duncan, Otis Dudley, 270–272, 274, 448
Dunham, Warren, 555, 556
DuPont, Eleuthère Irénée, 575
Durkheim, Émile, 6–7, 7, 8, 14, 34, 67,
    183, 194, 196, 199–200, 391, 554
Dye, H. A., 149

E
Earle, Pliny, 6
Eckert, William, 437
Edwards, Harry, 318
Edwards, John N., 382–383
Ehrenberg, Ronald, 393

Ehrlich, Isaac, 218, 220–221
Ehrlich, Paul, 527
Eibl-Eibesfeldt, Irenaus, 141
Einstein, Albert, 464
Eisenhower, Dwight D., 439, 594
Ekstrand, Laurie E., 437
El-Badry, M. A., 335
Elizabeth II, Queen of England, 426
Eltis, David, 293
Emerson, Richard M., 338
Engels, Friedrich, 232
Engerman, Stanley L., 312, 353
England, Robert Stowe, 525, 527
Epstein, Cynthia Fuchs, 344
Erickson, Bonnie H., 277
Erickson, Maynard L., 189
Evans-Pritchard, Sir Edward, 115
Ewen, Elizabeth, 550
Exter, Thomas, 310

F
Fair, Charles, 245
Falwell, Jerry, 437
Faris, Robert E. L., 8, 555, 556
Farnsworth, Philo T., 477
Farrington, David, 189
Fayer, Steve, 594, 596
Featherman, David L., 275
Feeney, Floyd, 178
Felson, Marcus, 213
Ferguson, R. Brian, 259
Ferraro, Geraldine, 432–433
Figlio, Robert, 189
Filiatov, Sergei Borisovich, 409
Fillmore, Cathy, 198–199
Finke, Roger, 87, 557
Finley, M. I., 293, 336, 337
Firebaugh, Glenn, 317, 491, 493
Fireman, Bruce, 589
Fischer, Claude S., 539, 549, 556
Fishberg, Maurice, 320
Fishman, Robert, 550
Flacks, Richard, 603
Flavell, J. H., 157
Fleming, Joyce Dudney, 143
Fligstein, Neil, 577
Fogel, John K., 312, 353
Ford, Clellan S., 513
Fossett, Mark, 353
Fouts, Roger, 143
Frank, Andre Gunder, 488
Franklin, Benjamin, 306, 484
Freeman, Derek, 160
Freidson, Eliot, 250
Freud, Sigmund, 125, 398

Fried, Morton H., 257
Friedan, Betty, 344
Frohock, Jane, 343
Fuguitt, G. V., 549
Fuller, Bruce, 491
Fuller, J. L., 129
Furman, Dmitri Efimovich, 409
Furstenberg, Frank F., Jr., 275, 380–381

G
Gains, Donna, 550
Galle, Omer R., 556, 557
Gallup, George H., 429–431
Galtung, J., 488
Gamson, William A., 589
Gandhi, Indira, 433
Gannett, Henry, 44, 46–99
Gans, Herbert, 37
Gardiner, Beatrice T., 142–143
Gardiner, R. Allen, 142–143
Garnier, Maurice A., 491
Gartin, Patrick R., 192
Gatti, F. M., 183, 185
Gayle, W. A. (Tacky), 597
Geerken, Michael, 200
Giannini, Amadeo P., 59–60
Gibbons, Don C., 202
Gibbs, Jack, 218–219, 221
Gillespie, Michael W., 48
Gilligan, Carol, 153
Gillis, A. R., 48
Gintis, Herbert, 464
Giordano, Peggy C., 196
Glazer, Nathan, 57, 288, 307, 308
Glock, Charles Y., 47, 290, 291
Goffman, Erving, 165
Goldstein, David B., 137
Goldstein, Sidney, 48
Goldstone, Jack A., 420
Goodall, Jane, 138–140
Goodman, Benny, 283
Goring, Charles, 183
Gottesman, Irving I., 132
Gottfredson, Michael R., 176–178, 180–181, 187, 194, 202, 203, 380, 394
Gough, E. Kathleen, 361
Gouldner, Alvin, 572
Gove, Walter R., 184–185, 196, 200, 202, 556, 557, 558
Goyder, John, 273
Granovetter, Mark, 35–36, 277–278
Grant, Madison, 46
Grant, Michael, 293, 300
Grasmick, H. G., 219
Graunt, John, 538

Gray, Diana, 189–190
Greeley, Andrew M., 61, 296, 307, 310, 408
Greeley, Horace, 332
Green, Dan S., 8
Gualtieri, Francesco M., 46
Guerry, André Michel, 3, 4–5, 13, 33, 200
Guest, Avery M., 551–553
Gunn, John, 186
Gurr, Ted Robert, 590
Guttentag, Marcia, 330–331, 332–333, 335–336, 338–340, 344–345, 351, 353, 354, 355, 378

H
Habermas, Jurgen, 114
Hage, Jerald, 491
Halberstam, David, 318
Hall, Peter, 556
Hallam, Elizabeth M., 499
Hamilton, Charles V., 319
Hammond, Phillip E., 483
Hampton, Henry, 594, 596
Hampton, Lionel, 283
Handlin, Oscar, 320
Hanssen, F. Andrew, 317
Hardin, Garrett, 416
Hare, E. H., 555
Hargens, Lowell, 514
Harlow, Harry F., 141–142, 149
Harlow, Margaret K., 141–142, 149
Harold Godwin, King of England, 499
Harris, Marvin, 160, 249, 257, 258, 260, 513
Harry, Bruce, 186
Harry, Joseph, 219
Hatt, Paul K., 446
Haub, Carl, 524
Hauser, Robert M., 275
Hay, David, 132
Hayes, C., 142
Hayes, K. J., 142
Haynie, Dana L., 189
Heath, Anthony, 70–71, 421
Heaton, Tim B., 310
Hechter, Michael, 76, 211, 286, 301, 308
Heer, David M., 305
Herrnstein, Richard J., 183, 187, 189, 202, 204, 380
Herzog, Elizabeth, 48–49, 50
Hexham, Irving, 406, 484
Heyneman, Stephen P., 462
Heyns, Barbara, 457–458
Higgins, P. C., 102

Hill, Charles T., 197
Hiller, Harry W., 405
Hindelang, Michael, 192
Hindus, M. S., 349
Hintze, Otto, 425
Hirschi, Travis, 100–102, 176–178, 180–181, 187, 189, 191–192, 194–195, 196, 197, 198, 202, 203, 380, 394
Hirschman, Charles, 304
Hite, Shere, 377
Ho, Ping-Ti, 333, 508
Hobbes, Thomas, 1, 422
Hobson, J. A., 486
Hodge, Robert W., 446, 447, 448
Hoffer, Thomas, 461
Hoffman, John P., 393–394
Holcomb, William R., 186
Hollander, E. P., 349
Hollerith, Herman, 499
Hollingshead, August B., 555
Homans, George C., 2, 13, 21–22, 76–77, 82, 195, 242, 247
Houk, Ralph, 591
Hout, Michael, 275, 276, 381
Howard, Elston, 591
Howard, Michael, 566
Howe, Irving, 299
Hudson, Valerie M., 527
Hughes, Michael, 557
Hunter, Alfred A., 434–435
Huntington, Samuel P., 494
Hyland, Ann, 262

I
Illich, Ivan, 464
Introvigne, Massimo, 406
Inverarity, James, 221
Irwin, John, 222, 465
Isajiw, Wsevolod, 286

J
Jackson, Andrew, 570
Jackson, Kenneth T., 550
Jacobs, Jerry A., 275, 381
Jaffe, Dennis, 106
Jahoda, Marie, 290
Jefferson, Thomas, 294, 570
Jencks, Christopher, 272, 456
Jensen, Gary F., 189, 219, 222
Jensen, Leif, 309
Jewkes, John, 486
Jiobu, Robert M., 303
John, King of England, 425
Johnson, Benton, 395, 397

Johnson, Lyndon B., 601
Jones, Frank E., 273, 515
Jordan, Michael, 131, 318
Julian, Teresa, 135

K
Kage, Joseph, 44, 53
Kalmijn, Matthijs, 317, 352
Kanazawa, Satoshi, 211
Kanter, Rosabeth Moss, 104, 244
Kantrowitz, Nathan, 551
Kaplan, Steven, 137
Keeley, Lawrence H., 259
Kegl, Judy, 155–156
Kellogg, L. A., 142
Kellogg, W. N., 142
Kelly, Barbara, 550
Keniston, Kenneth, 603
Kennedy, John F., 308
Kennedy, Joseph, 308
Kennedy, Robert E., Jr., 335
Kent, Lori, 102
Kephart, William, 292
Kiecolt, K. Jill, 353
Kihumura, Akemi, 305
Kilgore, Sally, 461
Killen, Edgar Ray, 600
Killian, Lewis M., 589, 590, 592
Kim, Young Oon, 85–86, 88
King, Cameron H., Jr., 301
King, Coretta Scott, 598
King, Martin Luther, Jr., 588, 594–596,
    597–599, 605
Kirsch, Irwin, 455
Kitano, Harry H. L., 303, 305
Klein, Malcolm W., 202
Kluckhohn, Clyde, 116, 157, 437
Kobbervig, Wayne, 221
Kohlberg, Lawrence, 153
Kohlstedt, Sally, 169
Kohn, Melvin L., 162–164
Kornhauser, Ruth, 196
Kox, Willem, 87
Krakowski, Menachem, 186
Kristof, Nicholas D., 526
Kroeber, Alfred L., 256
Krupa, Gene, 283
Kuechler, Manfred, 432
Kunstler, James Howard, 550
Ky, Katherine, 150

L
Lacey, W. K., 336
Land, Kenneth C., 192, 213
Landes, David S., 480, 485
Landon, Alfred M, 429

Lange, Dorothea, 299, 387
Langer, William L., 335
Lapointe, Archie, 456
Laslett, Peter, 365
Latané, Bibb, 97
Laub, John H., 184, 196
Lauderdale, Pat, 221
Leakey, Louis, 138–140
Lee, George, 591
Lee, James Daniel, 169
Lemert, Edwin M., 201
Lenin, Vladimir I., 486, 487
Lenski, Gerhard, 112, 237–238, 248, 249,
    256, 257, 260, 261, 265
Levine, Gene N., 305
Lewis, Bernard, 293, 294
Li, Bobai, 278
Lieberson, Stanley, 286, 297, 308, 312,
    315, 320, 321, 452, 551
Light, Ivan H., 303
Lincoln, Abraham, 438
Linden, Rick, 198–199
Lindesmith, Alfred R., 176
Lindsay, Jack, 336
Link, Bruce G., 186
Linton, Ralph, 477
Lipset, Seymour Martin, 266, 267,
    269–270, 274, 276
Liska, Allen E., 187, 189, 201–202
Litwak, Eugene, 14
Lloyd, William Forster, 416
Lobosco, Antoinette, 213
Lofland, John, 84–87, 95, 602
Lombroso, Cesare, 182–183, 200
Lombroso-Ferrero, Gina, 182
Longino, Helen, 169
Lorenz, Konrad, 138
Louch, Hugh, 37
Louis, Joe, 131, 587
Loxley, William A., 462
Luther, Martin, 483

M
Macedonicus, Quintas Caecilius Metellus,
    371
MacMullen, Ramsay, 392
Macphail, Agnes Campbell, 435
Madison, James, 426
Magee, Charles M., 419
Malinowski, Bronislaw, 391
Mallory, Walter H., 507
Malson, Lucian, 147
Malthus, Thomas Robert, 508–514, 527
Manning, Robert D., 308
Marchetti, Cesare, 550
Marston, Wilfred G., 552

Martel, Charles, 262
Martin, David, 399, 407
Marx, Gary T., 238
Marx, Karl, 8, 107, 113–115, 232–235,
    236, 238, 242–245, 266, 424–425,
    481–482, 485, 486, 494, 580
Mason, Robert, 193
Matras, Judah, 333
Mayhew, Pat, 179
McAdam, Doug, 599, 601–605
McBurnett, Keith, 184
McCallum, Daniel C., 567, 578
McCann, James C., 408
McCarthy, John D., 589, 599
McClean, Charles, 147
McClosky, Herbert, 439
McCord, Joan, 215–216
McCord, William, 215–216
McCusker, Claire, 461
McDougall, William, 125
McEachern, A. W., 202
McFarland, Bentson H., 186, 400
McGahey, Richard M., 221
McGovern, Thomas, 480
McKay, Henry D., 192, 200
McKenry, Patrick C., 135
McKenzie, Roderick F., 544, 551
McKinney, William, 396
McLuhan, Marshall, 493
McNeill, William H., 481, 483, 507–508,
    508, 536
McRoberts, Hugh A., 273
McWilliams, Carey, 301
Mead, George Herbert, 74–75, 156–157
Mead, Margaret, 126, 157–161
Meadows, Donella, 518
Mednick, S. A., 183
Meeks, Wayne A., 392, 395
Meeus, Wim, 87
Meier, Robert F., 191
Meir, Golda, 433
Melton, J. Gordon, 401, 406
Meltzoff, Andrew N., 127
Mendelssohn, Kurt, 481
Merton, Robert K., 191–192
Messeri, Peter, 14
Messick, David M., 417–419
Messner, Stephen F., 184, 354
Meyer, Henry J., 307
Meyer, John W., 465–467
Miller, Allan S., 393–394, 395
Miller, Walter B., 216
Millis, H. A., 300
Mills, C. Wright, 7, 25, 428
Mills, J. P., 392
Ming, Yao, 133

Minor, W. William, 219
Miyamoto, Frank S., 74, 303
Mizruchi, Ephraim H., 200
Mocan, H. Naci, 223
Moise, Chief of the Salish tribe, 255
Mole, Arthur S., 95
Moltke, Helmuth von, 565–567, 569, 570, 571, 575, 577
Mondale, Walter, 433
Montaigne, Michel Eyquem, 522
Montero, Darrel M., 305
Montgomery, Field Marshal Viscount, 262–263, 508
Moody, James, 292
Moon, Sun M., 85
Mooney, James, 400
Moore, M. Keith, 127
Moore, Wilbert E., 245–247, 248
Morioka, Kiyomi, 400
Morris, Aldon D., 590, 591, 595
Morris, Robin, 136
Morse, Jodie, 461
Morselli, Henry, 6, 7, 14, 28, 43, 200
Mosca, Gaetano, 245
Moses, 500
Mott, Lucretia, 343
Mowat, Farley, 479
Moynihan, Daniel Patrick, 288, 307
Murdock, George Peter, 116, 160, 256, 361
Murray, Henry A., 157
Musmanno, Michael A., 44, 46
Myrdal, Gunnar, 294

N
Natriello, Gary, 458
Nee, Victor, 310
Nesbitt, Gussie, 596
Nettles, Bonnie, 88–89
Newcomb, Charles, 555
Newman, Jody, 436
Niebuhr, H. Richard, 395–397, 409
Nisbet, Robert, 473
Nixon, E. D., 587–588, 594–595, 605
Nordhoff, Charles, 244, 360
North, Cecil C., 446
North, Douglass C., 485, 486
Notestein, Frank W., 524
Noyes, John Humphrey, 244
Nozick, Robert, 420, 485

O
Ogburn, William F., 476
Okun, John P., 419
Olsen, Mancur, 419–420
Ornstein, Michael D., 274

Ossowski, Stanislaw, 249
Ostrom, Carol M., 527

P
Paige, Jeffery M., 118–120, 488
Pallas, Aaron M., 458
Pankhurst, Christabel, 349
Parent-Duchatelet, Alexandre, 5
Parfitt, Tudor, 137
Paris, Scott G., 157
Park, Robert E., 551–552
Parker, Fess, 517
Parker, J., 219
Parker, Robert Nash, 190
Parks, Rosa, 587, 589, 593–595, 605
Parsons, Talcott, 67, 107
Pascal, Blaise, 394
Pastor, Bob, 587
Patterson, Orlando, 261, 295, 312
Pearlin, L. I., 164
Pearson, Karl, 183
Penn, William, 224
Peplau, Letitia Anne, 197
Perkins, Matthew, 200
Perlmann, Joel, 55–57, 62
Perry, Matthew, 287
Petersen, William, 303, 333, 506–507, 508
Peterson, Iver, 550
Peterson, Ruth D., 221
Pfeiffer, John E., 259
Phillips, David P., 221
Phillips, John, 353
Piaget, Jean, 151–154, 157–158, 241
Pierce, Glenn L., 220
Piliavin, Irving, 222
Pineo, Peter C., 273, 274, 377, 446, 448
Pinker, Steven, 153–154, 154
Pirenne, Henri, 535
Plato, 217, 232, 424, 427
Polo, Marco, 477
Pomeroy, Sarah B., 336, 337
Porter, John, 273, 274, 446, 448
Portes, Alejandro, 307, 308, 309
Powell, Colin, 321
Powers, Edwin, 214–215
Preston, Samuel E., 376
Pu Yi, 415
Pullum, Thomas W., 373

Q
Quetelet, Adolphe, 5, 23
Quinley, Harold E., 290

R
Rabkin, Judith, 186
Raftery, Adrian E., 273

Ragin, Charles C., 115
Raine, Adrian, 184
Randall, Jack, 587
Rapp, Rayna, 344
Rasmussen, Knud, 257
Rauscher, Frances, 150
Rawson, Beryl, 371
Redlich, Frederic C., 555
Rehnquist, William, 432
Reiss, Albert J., Jr., 102, 446
Reiss, Ira L., 361, 373
Rhodes, A. L., 102
Riddle, M., 187
Riesman, David, 428
Roberts, A. H., 187
Robinson, Bill (Bojangles), 131
Robinson, Jackie, 130
Robinson, Jo Ann, 587–588, 594–595, 597, 605
Rodman, Henrietta, 344
Rodriguez, Gregory, 305
Ro'i, Yaacov, 408
Roncek, Dennis W., 213
Roof, Wade Clark, 396
Roosevelt, Franklin Delano, 304, 429
Ropp, Theodore, 565, 566
Rose, Richard J., 131–132
Rosenberg, Charles E., 365
Rosenberg, Stuart E., 44
Rosenfeld, Richard, 200
Rosenthal, David, 129
Ross, E. A., 307
Ross, Ellen, 344
Ross, Marc Howard, 260
Rossi, Peter H., 446, 447, 448
Rothman, Sheila, 344
Rousseau, Jean-Jacques, 138
Rubin, Zick, 197
Rubinow, Israel, 53
Russell, J. C., 535
Ryan, James M., 550

S
Sack, Benjamin G., 44
Sampson, Robert J., 184
Sampson, Samuel F., 16
Samuelsson, Kurt, 484
Sanders, Jimmy M., 310
Sanders, Ronald, 50
Sandow, Eugene, 125
Sapir, Edward, 155
Sawers, David, 486
Sawyer, P. H., 479
Scheff, Thomas J., 186, 201
Schmidt, Constance R., 157
Schmidt (laborer), 449

Schmookler, Jacob, 485
Schoenrade, Patricia, 393
Schofield, J. W., 292
Schollaert, Paul T., 318–319
Schooler, Carmi, 164
Schramm, Sarah, 435
Schuckit, Marc, 132
Schur, Edwin, 201
Schwartz, Joseph E., 192
Schwartz, Richard, 201
Secord, Paul F., 332–333, 335–336, 338–340, 340–342, 344–345, 351, 353, 354, 378
Sellers, Clyde, 597
Sellin, Thorsten, 189
Selznick, Gertrude J., 290
Selznick, Philip, 572, 573
Semi, Emanuela Trevisan, 137
Senghas, Ann, 156
Shakespeare, William, 165
Shapiro, Susan P., 193
Shattuck, Roger, 147
Shaw, Clifford R., 192, 200
Shaw, Gordon, 150
Sheldon, W. H., 185
Shepherd, Gary, 128
Shepherd, Gordon, 128
Sherif, Carolyn W., 292
Sherif, Muzafer, 292
Sherman, Lawrence W., 192
Sherman, Stephanie L., 129, 132
Shioji, Hiroki, 212
Shirer, William L., 503
Shorey, William T., 352
Shorter, Edward, 366, 369, 370, 371, 372
Siegal, P. M., 446, 447, 448
Siegfried, William D., Jr., 349
Sigelman, Lee, 292, 317
Silberman, Charles E., 457
Sills, David L., 573
Simcha-Fagan, Ora, 192
Simmel, Georg, 388
Simon, Julian L., 513, 518, 519, 528, 554
Simpson, John H., 43, 494
Singley, B. L., 67
Sjoberg, Gideon, 259, 539
Skeels, H. M., 149
Skolnick, Jerome H., 201
Skorecki, Karl, 137
Slater, Miriam K., 48
Sloan, Alfred, 577
Small, Albion, 8
Smelser, Neil J., 589, 590, 593
Smith, Adam, 3, 75, 294, 408, 508, 509
Smith, Bessie, 351

Smith, Daniel S., 349
Smith, Donald Hugh, 318–319
Smith, Robert J., 211
Smith, W. Robertson, 391
Snipp, C. Matthew, 305
Snow, C. P., 527
Sorokin, Pitirim, 260
South, Scott J., 184, 353, 354, 378
Sowell, Thomas, 267, 293, 294, 303, 307, 308, 312, 315–316
Spada, A. V., 44, 47, 58
Spanier, Graham B., 381
Spencer, Herbert, 8, 391
Spiro, Melford E., 388
Spock, Benjamin, 515
Srole, Leo, 555
Stack, Steven, 221
Stanton, Elizabeth Cady, 343
Stark, Andrew, 359
Stark, Rodney, 5, 17–19, 47, 84–87, 95, 100–102, 103, 106–107, 112, 195, 200, 291, 354, 388, 392, 393, 395, 398–399, 400, 402, 404–405, 406, 408, 410, 437, 518, 527, 557, 602
Steadman, Henry, 186
Steinberg, Stephen, 48, 49, 50, 52–53, 54–55, 290
Stephan, G. Edward, 107
Stephan, Karen H., 107
Stevenson, Adlai, 439
Stillerman, Richard, 486
Stinchcombe, Arthur L., 110–111, 247
Stoll, David, 407
Stolzenberg, Ross M., 311
Stuller, Jay, 285
Sunahara, Ann, 306
Sutherland, Edwin H., 188, 192, 194
Sutton, Willie, 179
Swift, Edwin, 568
Swift, Gustavus, 568, 570, 572

T
Taeuber, Alma F., 551, 552, 553
Taeuber, Cynthia M., 348–349
Taeuber, Karl E., 551, 552, 553
Tainter, Joseph A., 256
Talvitie, Antli, 550
Tanner, James M., 134
Tarde, Gabriel, 183, 200
Taylor, David, 88
Taylor, Frederick W., 449–450
Taylor, Marylee C., 440
Taylor, Pamela, 186
Taylor, Ralph B., 192, 213
Tellegen, Auke, 131–132

Tepperman, Lorne, 276
Terrell, Kermit, 275
't Hart, Harm, 87
Thatcher, Margaret, 433
Thomas, Hugh, 481
Thomas, Mark G., 137
Thomas, Robert Paul, 485, 486
Thompson, Edward H., 394
Thompson, James D., 579
Thompson, W. R., 129
Thornberry, Terrance P., 202
Thorndike, E. L., 125
Till, Emmett, 592
Tittle, Charles R., 191
Toby, Jackson, 194, 216
Todd, E. Lillian, 325
Tönnes, Ferdinand, 8, 554
Tooker, George, 581
Torrey, E. Fuller, 186
Townsend, Joan B., 406
Treiman, Donald J., 275, 448
Trent, Katherine, 378
Turk, Austin T., 176
Turnbull, Colin, 257
Turner, Jonathan H., 354–355
Turner, R. Jay, 555
Turner, Ralph H., 589, 590, 592
Tylor, Edward Burnett, 391–392
Tyree, Andrea, 272
Tz'u-hsi, dowager empress of China, 415

U
Udry, J. Richard, 135
Uggen, Christopher, 197
Ujimoto, K. Victor, 306

V
Valdisera, Victor, 348–349
Vallance, Theodore R., 176
Vallee, Frank G., 273
Valovka, Jan, 186
van de Walle, Etienne, 522
Veblen, Thorstein, 264
Ventis, W. Larry, 393
Vigod, Bernard L., 44
Vogt, William, 519
von Bertalanffy, Ludwig, 108
Vorontsova, Lyudmila, 409

W
Wade, Nicholas, 137
Wagenfeld, Morton O., 555
Wagner, Adolf, 6
Walder, Andrew G., 278
Waldo, Gordon P., 187

Wallace, Anthony F. C., 398, 400
Wallace, Michael, 43, 493
Wallace, W. L., 465
Waller, Thomas (Fats), 131, 351
Wallerstein, Immanuel, 266, 486–488
Wallis, Roy, 397, 400
Walzer, Michael, 486
Ward, David, 546, 548
Warren, John Robert, 272
Washington, George, 563
Washington, M. H., 353
Washoe (chimp), 143–144
Waters, Mary C., 61, 286
Watson, James D., 136
Watt, James, 582
Webb, Jim, 192
Weber, Max, 114, 236–238, 244–245, 395, 400–401, 420, 483–485, 486, 494, 565, 570–572, 580
Weed, James A., 376, 551–553
Weir, Alex A. S., 140
Weis, Joseph G., 192, 216
Welch, Susan, 292, 317, 404
Wellington, Arthur Wellesley, Duke of, 566, 577

West, Donald, 189
Westhues, Kenneth, 405
White, Douglas R., 116, 160
White, L. A., 126
White, Lynn, Jr., 262–263
White, Lynn K., 381–383
White, M., 102
Whiten, A., 141
Whyte, Martin King, 489
Wilbraham, Anne, 511
Wilke, Henk, 417–419
Wilkinson, Karen, 380
William the Conqueror, 499–500
Williams, R. M., Jr., 283
Wilson, Bryan, 397
Wilson, James Q., 183, 187, 189, 202, 204, 353, 380
Wilson, Teddy, 283
Wimberly, Dale W., 491
Wirth, Louis, 307, 554–555, 556
Witmer, Helen, 214–215
Wolfgang, Marvin, 189
Wong, Morrison G., 304
Woodsworth, James S., 46, 57, 58–59
Wrigley, E. A., 540

Wrong, Dennis H., 68–69
Wundt, Wilhelm, 8
Wuthnow, Robert, 436–437
Wyatt-Edgell, Edgell, 5

Y
Yoshimura, Irwin, 305
Yunker, James A., 221

Z
Zablocki, Benjamin, 104–107
Zahavi, Yacov, 550
Zald, Mayer N., 589, 599
Zborowski, Mark, 48–49, 50
Zeisel, Hans, 221, 224
Ziegfeld, Flo, 125
Zuckerman, Marvin, 135, 185
Zuckerman, Michael, 373
Zuiches, J. J., 549
Zworykin, Vladimir K., 477

A

Abolitionist movement, 438

Abortion, 9–10, 11, 12, 526

Abstractions Ideas or mental formulations that are apart from the concrete, material world. "Beauty" is an abstract mental formulation that exists only in our mental responses to the world rather than as a quality of the things observed, a fact that is acknowledged in the common saying "Beauty is in the eye of the beholder.", 78

Accommodation An agreement between two groups to ignore some cultural differences between them and emphasize common interests instead, 43, 49, 288

Achieved status. *See* Status, achieved

Adoptees, 132, 183

African Americans. *See also* Civil Rights Movement

  barriers to progress of, 318–321

  capital punishment and, 220

  domination of, 590–592, 599–600

  education, 267, 313–314, 315, 452, 457, 464

  family composition, 315–316, 351–353

  history in U. S., 312–321

  income of, 315–316, 590–591

  integration and, 317

  migration north, 297–298, 312–315

  as musicians, 131, 283

  population statistics, 310

  poverty, 313–315, 590–591, 599

  prejudice, 239, 284, 291–292, 295–296, 299, 317–318

  progress of, 310, 315–317, 592–593

  segregation and, 552, 553, 590–591

  sex ratios and, 341, 350, 351–353

  slavery and, 284, 293–296, 318–319, 592

  in sports, 130–131, 318–319

  status of, 238, 239

  unemployment among, 452, 590–591

Age

  crime and, 179, 184–185, 196, 197

  divorce and, 376

  marriage and, 340–341

  sex ratios and, 340–341

Age structure The proportion of persons of various age groups making up a total population, 504–506, 525–526

Aggregate A collection of people lacking social relations; for example, pedestrians waiting for a walk light, 14

Agrarian societies Societies that live by farming. Although these were the first societies able to support cities, they usually require that about 95 percent of the population be engaged in agriculture, 259–265

  food supply in, 259, 260, 506–507

  freedom of the commons, 415–417, 425

  government in, 265, 569

  Marx on peasants, 233

  ruling elites in, 261, 264–265, 485

  stratification in, 259–265, 423–424

  warfare in, 260, 261–263, 423

Agreement, law of The more the members of a group like one another, the more apt they are to agree with each other, 76–77, 242

Agricultural revolution, 540–543

Agriculture. *See also* Agrarian societies

  child labor, 513

  common grazing, 415–417, 425

  Marx on peasants, 233

  population growth and, 508–511, 519

  specialization and urbanization, 259

  technology/modernization, 449, 511, 519

Alcoholism/alcohol use, 132, 202, 203

Allocation theories Theories that argue that the primary function of schools is to allocate status, to place students in the stratification system, rather than to train them, 464–466

Allport's theory of contact Theory holding that contact between groups will improve relations only if the groups are of equal status and do not compete with one another, 291–292

Altruism Unselfish actions done entirely for the benefit of others, 69, 70–71. *See also* Social movements

American dilemma Term used by Gunnar Myrdal to describe the contradiction of a society committed to democratic ideals but sustaining racial segregation, 293–296

American Institute of Public Opinion (AIPO), 430

American Revolution, 426, 473

American Sign Language, 142–144

Anarchists Followers of a political philosophy that regards the state as inevitably repressive and unjust and who, therefore, propose to destroy the state and live without laws or government, 244, 422

Animal behavior

  chimpanzees, 138–141, 142–144

  vs. human behavior, 137–138

  language and, 142–144

  monkeys and isolation, 141–142, 149

Anomie A condition of normlessness in a group or even a whole society when people either no longer know what the norms are or have lost their belief in them, 199–200, 554–556

Anti-Semitism Prejudice and discrimination against Jews, 45, 46, 50

Arithmetic increase A constant rate of growth (or decline) that speeds up as an increasingly larger number of units is added (or subtracted) each cycle, as in 1-2-3-4-5, 509

Asch experiment, 81–83

Ascribed status. *See* Status, ascribed

Asian Americans, 300, 302–306, 310, 552

Assimilation The process by which an individual or a group reacts to a new social environment by adopting the culture prevalent in that environment, 42–43

  of European Muslim immigrants, 526

  of Hispanic Americans, 311, 312

  intergroup conflict and, 286–287

  of Jews and Italians, 47–61

  language and, 308, 310, 311

Astrologers, 17–19

Atheists, 390, 409

Athens, 336

Attachment(s) A stable and persistent pattern of exchange between two people when positive sentiments are among the "commodities" exchanged. That is, attachments involve bonds of liking or affection between two people, 76. *See also* Friendships

  conversion and, 86–89

  criminal behavior and, 184–185, 187–190, 195–196, 198, 199, 201, 214, 225

  emotional, 76, 364–365, 370–373

  norms and, 77

  parent-child, 369–370, 371, 374

  religion and, 400, 404

  romantic love, 370–371, 372–373

  self-conception and, 74

  social movements and, 594, 603–604

  urbanization and, 554–556

Australian aborigines, 283
Autonomous divisions Parts of an organization, each of which includes a full set of functional divisions, 576–577

B
Baby boom A brief period of high fertility in many Western industrial nations immediately following World War II, 515–517, 518
Bank of America, 60
Behavior. *See also* Criminal behavior; Heredity *vs.* environment; Sexual relations
  of animals, 137–144, 149
  hormones and, 135–136
  instinctual theories of, 125–127
  learning theory, 188–189
  predictability/causes of, 1, 5, 23–24
  primary sources of, 67–69
  thrill-seeking/risk-taking, 185, 187, 394–395, 406
Behavioral genetics A scientific field that attempts to link behavior, especially human behavior, with genetics, 129–133, 183–185
Beliefs Our notions about how we ought to act, 197. *See also* Ideology; Religion
  agreement and conformity, 77
  deviance and conformity, 197–198
  political, 437–438, 439–440
  self-fulfilling, 603
Biographical availability McAdam's term describing people who are free of commitments and involvements that would interfere with their participating in a social movement, 602–603
Biology. *See* Heredity *vs.* environment
Birth cohort All persons born within a given time period, usually one year, 503–504
Birth control, 522, 523
Birthrate, crude The total number of births for a year divided by the total population that year, 502
Black Death, 507
Blau's administrative theory Theory stating that the larger the organization, the greater the proportion of total resources that must be devoted to management functions, 577–578
Born criminals Lombroso's term for people whose deviance he attributed to their more primitive biology, 182–184
Bourgeoisie Marx's name for the class made up of those who own the means of production; the employer or owner class, 232–233, 236, 244, 482, 483
Bribes, 192–193

Bridge position One having bridge ties, 36
Bridge ties Links across holes between groups, 36
Bureaucracy A formal organization that, according to Weber, is based on functional specialization; clear, hierarchical lines of authority; expert training of managers; decision making depending on rational rules aimed at effective pursuit of goals; appointment and promotion of managers on their merit; and activities conducted by written communications and records, 570–572, 580–582
Burglary, 177, 179, 213, 222
Business organizations, 567–569, 573–575, 575–578, 580–581

C
Calvinists, 483–484
Canada
  baby boom in, 515–517
  birthrates in, 502
  capital punishment and, 217
  delinquency study in, 198–199
  educational attainment in, 452, 454, 455
  fertility rates, 512, 523
  French- *vs.* English-speaking, 284, 307
  immigrants from, 284
  immigration into, 44, 45, 46, 47, 53, 55
  Japanese in, 305–306
  networks in, 38, 276–277
  occupational prestige in, 448
  politics in, 433–435, 440–441
  religious economy in, 405–406
  sex ratios in, 341, 350
  status attainment in, 273–275
Capitalism An economic system based on free-market exchanges and individual property rights with the result that any individual can benefit from becoming more productive, 482
  Marx on, 232–235, 481–483, 485
  modernization theories, 481–491
  Protestant Ethic and, 483–485
  spirit of, defined, 484
Capital punishment The death penalty, 217–218, 219, 220–221
Caste system A stratification system wherein cultural or racial differences are used as the basis for ascribing status, 35, 239, 288, 301
Catholics
  in Canada, 405
  divorce and, 378
  in Latin America, 407
  monastery network, 16–17
  as a monopoly, 395
  politics and, 238

  Protestants and, 296–297
  school system, 459, 460, 461
Cause Something that makes something else happen. A cause is anything producing a result, an effect, or a consequence. To demonstrate causation, three criteria, or tests, must be met: time order, correlation, and spurious relationships, 78–79, 83, 87
Census A population count, often recorded in terms of such categories as age, occupation, material status, and the like. The U.S. Census is conducted during the first year of each decade, 98, 499–501
Census Metropolitan Area (CMA) (Canada) Includes an urban core of 100,000 and surrounding communities if 40 percent of the labor force work in the urban core or 25 percent of those employed in the outlying area commute from the urban core, 545
Change. *See* Social change
Charisma The unusual ability of some religious leaders to influence others, 400
Child development. *See also* Socialization
  accelerating/suppressing, 149–151, 153
  cognitive, 151–156, 157
  culture and personality, 158–161
  emotions, 156–158
  height, 133–135, 265
  isolation and, 147–148
  language acquisition, 153–156, 157
  stages of, 75, 151–154, 157
Child labor, 366, 369, 513
Child-rearing practices
  deviant behavior and, 213–216, 380
  in premodern societies, 159–161, 167–170, 366–367, 369–370
  self-control and, 188–189
  sex-role socialization and, 159–161, 167–169, 337
  social class and, 163–164
  socialization and, 361
Children. *See also* Infants
  bonds with parents, 369–370, 371, 374
  desired family size, 522–523
  emergence of self in, 156–157
  feral, 147–148
  mortality rates, 366, 369, 513
  in one-parent families, 379–380
  remarriage and, 367, 380, 381–383
Chimpanzees, 138–141, 142–144
China
  famine in, 507
  first contact with Europeans, 284
  government power, 415, 420, 422, 507
  industrialization in, 483

infanticide in, 526
networks in, 277–278
population, 520, 521–522, 525,
526–527
religion in, 390–391, 392
sex ratios, 333
Chinese Americans, 134, 304–305, 307,
309. *See also* Asian Americans
Christianity, 292, 354, 393, 397, 401,
483–485. *See also specific sects*
Chromosomes Complex genetic structures inside the nucleus of a cell, each
containing some of the basic genetic
units (genes) of the cell. Chromosomes combine in pairs; thus, in humans twenty-three chromosomes from
the father combine with twenty-three
from the mother, 127–128, 129
Churches Religious bodies in a relatively
low state of tension with their environment, 395–397. *See also* Religion
Church of Jesus Christ of Latter-day
Saints, 400
Church-sect theory The proposition
that, in time, successful sects will be
transformed into churches, thereby
creating the conditions for the eruption of new sects, 395–397, 399
Cities According to the U.S. Census,
communities having at least 50,000
inhabitants, 554. *See also* Urbanization
anomie theories and, 554–556
crime rates, 192, 213, 537–538, 539
crowding in, 537, 556–558
effects of living in, 535–538, 554–558
ethnic neighborhoods, 550–553
immigration and, 44
population of, 259, 450, 533, 537
preindustrial, 534–540
suburbs, 548, 549, 550
suicide rates, 6
Civilization, rise of, 264
Civil Rights Movement
building of, 83–84, 592–599
bus segregation/boycott, 587–588, 589,
593–599
discontent/grievances, 590–592
effects of, 474
external support, 597–599
Freedom Summer Project, 599–605
precipitating event, 593–594, 596–597
white countermovement, 596–597, 599
Civil rights organizations, 594, 599, 604
Civil service, 569–570
Class (social classes) Groups of people
who share a similar position in the
stratification system, 35, 231–232. *See
also* Ethnic and racial mobility; Social
mobility; Status; Stratification

in ancient Greece and Rome, 232
child-rearing practices and, 163–164
church network structure and, 395–397
crime/deviance and, 191, 196
economic dimension of, 235
education and, 464–466
Marx's concept of, 232–235, 236, 237, 244
middle (*See* Middle class)
personal sense of control and, 603
religion and, 397
ruling class, 113–114, 261, 422,
424–425 (*See also* Power elite)
segregation and, 552
"status" *vs.* "class," 236, 237–238
Weber's dimensions of, 236–237
working, 163–164, 242
Class conflict, 107, 114, 233–234
Class consciousness The concept Marx
used to identify the awareness of members of a class of their class interests
and enemies, 233–235
Classless societies, 242–245
Class, subjective The class to which a
person thinks he or she belongs,
234–235
Coercion, 67, 249–250, 395, 419, 420, 422,
424, 427, 522
Cognitive development, 151–156, 157
Cognitive structures General rules or
principles that govern reasoning,
151–152
Cohorts, 503–504
**Collective behavior approach to social
movements** Emphasizes that social
movements are outbursts of group activity in response to deeply felt grievances. Stresses the importance of
ideology and of emotions and feelings
in group, 589–590
Collective goods. *See* Public or collective
goods
Command economies Economic systems
wherein property rights are not secure
and much productive activity is based on
coerced labor, 482–483, 485, 486–487
Commons, 415–419
Communal societies, 117
Commune A group of people who organize to live together, often choosing to
equally share duties, resources, and
finances. Typically, communes also
attempt to live a distinctive lifestyle in
accord with an ideology that sets them
apart from the surrounding society,
104–107, 244, 361
Communication. *See also* Language; News
media
global, 493
television, 477–478, 516, 517

Communism, 244, 408–409, 482
Commuting, 549–550
Comparative research Comparisons of
large social units, typically whole nations or societies but sometimes states,
cities, or counties, 115–120
global perspectives and, 11–13
social structures and, 117–120
stratification systems, 255–256
Concepts Names used to identify some
set or class of things that are said to be
alike. Concepts are the building blocks
of theories, 13, 21
Concrete operational stage According to
Piaget, the period from seven until
about twelve during which humans
develop a number of cognitive structures, including the rule of conservation, 152–153
Conflict. *See* Social conflict; Warfare
Conflict theory An explanation of social
structures and cultural patterns based
on conflicts between classes and status
groups, each seeking to gain the most
benefits, 113–115, 339
Conflict theory of stratification A theory
that holds that individuals and groups
will always exploit their positions in an
effort to gain a larger share of the rewards in a society, and therefore societies will often be much more stratified
than functionalism can explain. Put
another way, this theory holds that the
stratification system of any society is
the result of conflicts and compromises
between contending groups,
249–251, 267
Conformity. *See also* Control theory;
Norms; Social control
agreement and, 77
class and, 163–164
crime and, 100–102, 184, 194–195, 196,
198, 199
group solidarity and, 81–83
public good and, 421
religion and, 85–87, 100–103,
392–393, 395
subcultural deviance *vs.*, 190
Conformity, law of The more intense the
group solidarity is, the more intense
will be the demand for conformity, 77
Conformity, stake in Those things a person risks losing by being detected
committing deviant behavior; what a
person protects by conforming to the
norms, 194, 198, 199. *See also* Social
bonds
Conjugal careers The histories of individuals in terms of marriages, 380

Conscience. *See* Norms, internalization of

Constant Something that never changes (or varies), 77–78

Constitution, U.S., 422, 426

Contextual effect The dependence of a relationship found among individuals in social contexts, when different results occur in different social surroundings. Contextual effects mark the borderline between micro- and macro- sociology, 102–103

Contraception, 522, 523

Control group, 214

Control theory A theory that stresses how weak bonds between the individual and society free people to deviate, whereas strong bonds make deviance costly, 194–199

Core nations According to Wallerstein, those most modernized nations, having diversified economies and stable internal politics, that dominate the world system, 487

Correlation Variation in unison, another test of causation. For something to be the cause of something else, the two must be correlated, or vary in unison. Causes must produce changes in their supposed effects, 78–79, 80, 83, 87. *See also* Spurious relationships

Correlation coefficient A measure of the degree of correlation between two variables; it varies from 0.0 (no correlation) to 1.0 (perfect correlation), 80

Cosmopolitan growth principle The opportunities for the growth of a movement are maximized when movements can establish ties to cosmopolitan networks having many nonredundant ties. This is another example of the strength of weak ties, 88

Cosmopolitan networks. *See* Networks, cosmopolitan

Countermovements Social movements that arise in opposition to another social movement, 590, 596–597, 599

Court system, 223–224, 598

Covert observers Field researchers who are operating without the knowledge or permission of those who are being studied, 85

Creole A complex and grammatical language evolved from a pidgin, 155, 156

Crime Acts of force and fraud undertaken in pursuit of self-interest, 175–207. *See also* Delinquency; Deviance
anomie and, 199–200
burglary, 177, 179–180, 213, 222
drugs and, 202–203

vs. legality, 176–177
murder (*See* Homicide)
ordinary, 177–181
prevention, 213–216 (*See also* Deterrence)
reporting of, 221–222
theft, 4, 5, 178, 222
white-collar, 192–194

Crime rates
capital punishment and, 217–218, 219, 220–221
in cities, 213, 537–538, 539
climate, season and, 200–201
in Japan, 212
social control and, 211
stability *vs.* variation in, 4–6
statistics on, 3–6, 177–181, 184, 185, 198, 199, 220, 221–222, 223
studies launching sociology, 3–6

Criminal behavior. *See also* Deviance
age and, 179, 184–186, 196, 197
biological theories of, 182–185
"born criminals," 182–184
characterized, 181–182, 187
control theories and, 194–199
gender and, 4, 5, 101, 184–185, 199
labeling theory and, 201–202
mental illness and, 186–187
personality theories and, 187
self-control and, 187
socialization and, 213–216
structural strain theories and, 190–192, 194
testosterone and, 135, 136, 185
versatility of, 180–181

Criminal justice system, 221–226, 591, 599–600. *See also* Police

Criminologists, 8, 175–176, 183, 395

Crowding, 537, 556–558

Cuban Americans, 54, 353

Cult formation The process by which a person or persons with new revelations succeed in gathering a group of followers, 399–400

Cults Religious movements that represent faiths that are new and unconventional in a society, 399–400
in Canada, 406
conversion/recruitment, 84–89, 410
in Europe, 406
in United States, 404–405

Cultural capital Assets based on knowledge, style, speech, tastes, and the like, which can be used to "purchase" privileges and power, 241

Cultural determinism The claim that an almost infinite array of cultural and social patterns is possible and that human nature can be shaped into almost any form by cultural forces, 158–161

Cultural division of labor A situation in which racial or ethnic groups tend to specialize in a limited number of occupations, 301

Cultural lag Period of delay following a change in one part of a society when other parts of the society have not yet readjusted, 475–476
population growth and, 518–519

Cultural pluralism, 286–287

Culture The sum total of human creations—intellectual, technical, artistic, physical, and moral. Culture is the complex pattern of living that directs human social life, the things each new generation must learn and to which they eventually may add, 39–43. *See also* Socialization
animals and, 137, 140–141, 143
assimilation, 42–43
class and, 241–242, 276
cross-cultural samples, 115–116
deviance and, 189–190
diffusion of, 479
ethnicity and race, 285–286
ethnic mobility and, 48–51, 61–62
European Muslim immigrants and, 526
globalization and, 493–494
heredity and biology, 125–126
highbrow, 276
internalization of, 68–69
multiculturalism, 41–42
personality and, 157, 158–161
preferences and tastes, 70
social change and, 473–474, 475–476
society and, 43, 57

D

Data collection Systematic fact gathering, 96, 100

Deafness and sign language, 155–156

Death. *See also* Infanticide; Infant mortality
causes of (1632), 538
of family members, 367, 371

Death rates (mortality rates). *See also* Infant mortality; Life expectancy
age-specific, 502
disease and, 507–508, 509, 511
famine and, 506–507, 509
in less developed nations, 518–519
modernization and, 512, 513
in preindustrial societies, 367, 369, 506–508, 538
technology and, 518–519
war and, 508

Death rates, age-specific The number of deaths per year of persons within a given age range divided by the total number of persons within that age range, 502, 503

Death rates, crude The total number of deaths for a year divided by the total population that year, 501–502

Decentralization Dispersing of authority from a few central administrators to persons directly engaged in activities, 577–580

Delinquency. *See also* Crime; Deviance control and differential association, 198–199
  deviant friends and, 189
  effect of religion on, 100–103, 200
  in one-parent families, 380
  personality theories and, 187
  prevention efforts, 213–216
  self-report data, 19, 199

Democracy
  "American dilemma," 293–296
  China and, 420, 422
  direct *vs.* representative, 428, 429
  industrialization and, 267, 278
  main threats to, 426
  pluralism and, 427–428
  "power elites" in, 428–429
  power in, 427, 597
  public interest in, 429–432
  slavery and racism, 295
  tyranny of the majority/tyranny of the minority, 426
  voter choices in, 441

Demographic transition A shift in population trends from high fertility, controlled by high mortality, to one of low mortality and low fertility, 512–513

Demographic transition theory Theory stating that the demographic transition was caused by modernization, which reduced the need for and the value of larger numbers of children, 512–514, 519, 521, 525

Demography Literally, written description of the people; the field of sociology devoted to the study of human populations with regard to how they grow, decline, or migrate, 500–506

Dense origins, principle of Social movements tend to originate within dense local networks, 88

Dependence, principle of The more dependent that members are upon a group, the more they conform to group norms. By dependence is meant the extent to which a group is the only available source of important rewards, 211

Dependency theory Theory stating that some nations become modernized by exploiting other nations and that their continuing exploitation prevents less developed nations from becoming fully modernized, 486–491

Dependents Family members unable to support themselves, 364

Dependent variable The consequence (or the thing that is being caused). To help you remember the difference, think that variables being caused are dependent on the causal variable, while causal variables are not dependent but are independent, 78, 80

Deterrence The use of punishment (or the threat of punishment) in order to make people unwilling to risk deviance, 216–221, 222–223

Deterrence theory The proposition that the more rapid, the more certain, and the more severe the punishment for a crime, the lower the rate at which that crime will occur, 219

Deviance Behavior that violates norms, 175–207. *See also* Crime
  anomie and, 199–200
  attachment theories and, 187–190, 195–196, 198, 199, 201, 225
  biological theories of, 182–185
  control theories and, 194–199
  differential association and, 188–189, 198, 199, 219
  labeling approach to, 201–202, 225
  as a matter of definition, 190
  mental illness and, 186–187
  personality theories and, 187
  religious cults as, 400
  self-control and, 187
  structural strain theory and, 190–192, 194
  subcultural, 189–190, 225

Deviance, primary In labeling theory actions that cause others to label an individual deviant. More generally, any deviant acts that result in the commission of other deviant acts, 201, 202

Deviance, secondary In labeling theory actions carried out in response to having been labeled as deviant. More generally, any deviant acts committed as a result of committing other deviant acts; for example—burglaries committed to support a drug habit, 201–202

Deviant role A set of norms attached to a position that in turn violates the norms adhered to by the larger society. For example, a proper performance of the role of burglar will deviate from other people's norms, 166

Diet, 370, 492, 506–507, 511

Differential association theory A theory that traces deviant behavior to association with other persons who also engage in this behavior, 188–189, 198, 199, 219

Diffusion The process by which innovations spread from one society to another, 477–478

Discrimination Actions taken against a group to deny its members rights and privileges available to others, 42. *See also* Prejudice; Segregation
  breaking barriers, 130–131
  economic/political/personal, 590–592
  housing, 47, 307, 551–553
  against immigrants, 44–48, 300–301, 305–306
  intergroup conflict and, 285–288
  religion and, 296–297, 408–409
  status and, 239

Disease
  mortality rate and, 507–508, 511, 518–519, 535–536
  population trends and, 507–508, 509, 518–519
  in preindustrial cities, 535–537

Diversified organization An organization that is not very specialized but instead pursues a wide range of goals, 575–578

Divisional system As initially used in armies, the organization of troops into small, identical units, each containing all military elements, 566–567

Division of labor
  among executives, 577–578
  cultural, 301
  gender and, 167, 347
  within households, 382–383
  stratification and, 248

Divorce
  causes of, 377
  children of, 379, 381–383
  international comparisons in, 377–378
  modernization and, 374–377
  in premodern societies, 336, 362
  rates of, 374–376, 378
  remarriage and, 380–383
  stepchildren and, 380, 381–383

DNA and culture, 136–137

Domesday Book An outstanding medieval census conducted by William the Conqueror following his takeover of England in 1066, 499, 500

Dominance, dependency and, 488

Downward mobility. *See* Mobility, upward and downward

Drug use, 104, 135, 185, 202–203

DuPont, 575–577

Dyadic power The capacity of each member of a dyad to improve his or her will on the other member, 338

Dysfunctions Social arrangements that harm or distort a social system, 111

E

Ecological sources of change, 478–480

Economic development. *See also* Industrialization; Modernization
among ethnic and racial minorities, 308–310
education and, 453, 454, 489–491
fertility rate and, 521–522
life expectancy and, 80
property rights and, 485–486
racism and, 474

Economics (field of), 8, 68, 75, 508

Economic status/inequality. *See also* Income; Poverty; Wealth
of African Americans, 299, 313–315, 315–316, 452, 590–591
class and, 235, 236, 237
enclave economic theory, 308–309
gender and, 347–349
of immigrants, 31, 32–33, 48, 51, 52, 53, 296–300, 306, 308–310
international statistics on, 492–493
of minorities, 321
prejudice and, 296, 297–300
remarriage and, 381
segregation and, 551–553

Economic systems
Economies, command, 485, 486–487
and education, 453
in "global village," 493, 494
in precapitalist societies, 482–483, 485, 486–487 (*See also* Capitalism; Communism)

Economic unit, family as, 364

Education What a person has learned, 445–469, 454. *See also* Schools; Universities
economic benefits of, 462–464
homework and, 459, 460
industrialization and, 266–267, 454, 489–491
international data, 454–455, 462
of military officers, 565–566
occupational status and, 272, 273, 274, 275
occupation and, 445–446, 448, 462–464
political ideology and, 439
prejudice and, 50, 290
prestige and, 448
proficiency test scores, 455–456
vs. schooling, 454
social control and, 211, 212
socializing function of, 464–466
of sociologists, 608–611
summer vacation and, 457–458
teaching credentials and, 464–465
transformation of, 453–456

upward mobility of minorities, 52–56, 303–304, 306–310, 311
of women, 336, 337
women with professional degrees, 451–452

Educational achievement
of African Americans, 267, 313–314, 315, 452, 464
of Asian Americans, 303–304, 306
ethnic mobility and, 48–50, 52–56, 303–304, 306
of Hispanic Americans, 464
of Italian immigrants, 48, 50–51, 61–62
of Jews, 48–50, 52–56, 310
reading and, 458

Educational Testing Service (ETS), 455–456

Elites. *See* Bourgeoisie; Power elite

Elitist state A society ruled by a single elite group; such states repress and exploit nonelite members, 426–428

Emotional development, 156–158

Emotional support, 361, 364–365, 370–373, 374, 377. *See also* Attachments

Empirical implications, 2

"Empty nest syndrome," 382–383

Enclave economic theory Theory that proposes that the spatial concentration of an ethnic group permits it to create its own business enterprises, thus speeding the economic progress of the group, 308–309

England
census of, 499, 500
death rate in, 535
pasture use in, 415–417
pluralism in, 424–425
population of, 511, 512, 526–527
urbanization of, 533

English language proficiency, 55, 57

Environment (ecological)
city pollution, 536–537
social change and, 478–480

Environment (social). *See* Heredity *vs.* environment; Socialization

Equilibrium A state of balance among interdependent parts of a system, 110

Equity, 418

Eskimos, 480

Estrogen The female sex hormone, 135

Ethnic and racial groups. *See also* Immigrants; Prejudice; *specific ethnic groups*
ancestry of Americans, 286–288
ascribed status, 285
divorce and, 376
economic status of, 31, 32–33, 315, 452, 590–591

education, 459, 460
identifiability, 301–302
intergroup conflict, 285–288
intermarriage, 47
labor market conflict and, 296–301
names and, 320
political power, 299, 307, 321, 591
remarriage and, 380
slavery/"American dilemma," 293–296
in sports, 130–131, 318–319
status attainment, 274
as status groups, 114, 237
urban neighborhoods, 550–553
visibility, 320–321

Ethnic and racial mobility
African Americans, 308, 310, 315–317, 592–594
cultural theory of, 48–51, 61–62
Hispanic Americans, 54, 310–312
Italian immigrants, 48, 50–51, 56, 57–62
Japanese Americans, 303–305, 306
Jewish immigrants, 48–50, 52–56, 310
mechanisms of, 306–310
reference groups and traditionalism, 57–62
social theory of, 51–57
sports and entertainment, 130–131

Ethnic groups Groups that think of themselves as sharing special bonds of history and culture that set them apart from others, 286. *See also specific ethnic groups*

Ethnic succession, theory of Theory stating that ethnic and racial groups will be the targets of neighborhood segregation only until they achieve economic parity and that slum neighborhoods will therefore house a succession of ethnic and racial groups, 551–553

Europe
cities in, 534, 535–537
crime statistics, 179, 180
exploitation by, 487
politics in, 440–441
population of, 506, 507, 508, 525, 526, 534
Protestantism and, 483–485
rapid social change, 480–481
religion in, 406
suicide rates (1870), 6
technological progress/modernization, 474, 480–481, 488
traditional families in, 360, 365–372
warfare in, 563, 565–567

Evolutionary theory of stratification A theory that holds that because culture accumulates in human societies, eventually it happens that no one can master the whole of a group's culture. At that point cultural specialization will be more valued and inequality, or stratification, will exist, 248–249, 251

Evolution, biological
sociological theory and, 111–113
supremacist ideas and, 285

Evolution, social. *See* Social evolutionary theories

Exchange mobility Mobility that occurs because some people fall, thereby making room for others to rise in the stratification system, 240–241, 275–276

Exchange theory A microtheoretical perspective with its central concern to explain the key processes by which people seek to maximize by exchanging rewards with one another, 72, 75–77, 338, 421

Experimental group, 214

Experiment A method wherein the researchers are able to (1) manipulate the independent variable (make it vary as much as they wish, whenever they wish) and (2) randomly assign persons to groups exposed to different levels of the independent variable, 82–83. *See also* Research

Exploitation All profit in an exchange in excess of the minimum amount needed to cause an exchange to occur, 249
in capitalist societies, 482
dependency and, 486–491
within nations, 492–493
by the ruling class, 424–425
by the state, 415–419
stratification and, 245, 249, 260
of women, 351

Exponential increase A rate of growth (or decline) that speeds up as an increasingly larger number of units is added (or subtracted) each cycle, as in 1-2-4-8-16, 509

Extended family. *See* Family, extended

Extensiveness, principle of The greater the scope and extent of norms upheld by the group are, the greater the contribution to overall social order will be, 211, 212

F

Facial expressions, 127
Factional societies, 117–120
Facts, simple and proportional, 96, 98, 98
Facts, social, 3–6, 67
False consciousness A term that Marx applied to members of one class who think they have common interests with members of another class, 234

Family A small kinship-structured group with the key function of nurturant socialization of the newborn, 359–385, 361. *See also* Child-rearing practices; Divorce; Marriage
child care, 361, 369–370
composition of, 351–354, 361–362, 366–367, 368–369, 373–374
crowding, 368
decline of, 359, 360
definition of, 359–363
of delinquents/criminals, 188–189, 196, 380
as economic unit, 364
functions of, 361, 364–372
household chores, 382–383
Italian, 61–62
modernization and, 372–377
in premodern societies, 362
as primary group, 14
quality of emotional bonds, 373–374
remarriage, 367, 380–383
single-parent, 315, 352–353, 367, 379–380
size of, 366–367, 373–374, 513, 522–523, 525
stepchildren in, 367, 380, 381–383
traditional, 360, 365–372
types of, 111, 362

Family, extended Families made up of a least two adult couples, 111, 362–363, 366, 367, 373

Family, female-headed
African Americans, 315, 352–353
poverty and, 379–380
in preindustrial societies, 367
statistics on, 379

Family, nuclear A family group containing one adult couple and their children, 111, 362, 366

Famine, 506–507, 509, 527–528

Farming. *See* Agriculture

Female Empowerment, 326, 328

Females. *See also* Family, female-headed; Gender; Sex ratio; Sex-roles
as criminals, 184, 185
education of, 336, 337
geographical mobility of, 136–137
in Gold Rush, 333, 334
in Greece, 336–338
infanticide of, 117, 333, 335, 336, 337, 371
in labor force, 274, 275, 345–349, 451–452
as political candidates, 432–436
with professional degrees, 451–452
remarriage, 380–383

in Rome, 371
in science careers, 169–170
status attainment and, 272, 275
subordination of, 166, 371
women's movement, 340, 342–345

Feminism An ideology having three essential features: (1) opposition to all forms of stratification based on gender, (2) belief that biology does not consign females to inferior status, and (3) a sense of common experience and purpose among women to direct their efforts to bring about change, 344–345

Feminist movement, 84, 344–345, 351

Feral children The name often applied to children who, because of severe neglect, act as if they were raised in the wild, 147–148

Fertility
baby boom, 515–517
control over, 510, 513, 522–523, 525, 526–527
demographic transition, 512–514, 519, 521, 525
economic development and, 80, 521–522
employment of women and, 451
factors affecting, 502
gender bias and, 526–527
international statistics, 519–527

Fertility rate The total number of births for a year divided by the total number of women in their childbearing years, 502
crude birthrate *vs.*, 502
cultural lag and, 518–519
decline in, 512, 517, 523–527
of European Muslim immigrants, 526
in less developed nations, 518–519
modernization and, 373, 512, 513
in premodern societies, 366

Fertility, replacement-level (also called zero population growth) Point at which the number of births each year equals the number of deaths, 512, 520, 524

Fertility, wanted The number of children a couple wishes to have, 523

Feudalism, 262–263, 425

Field research Going out to observe people as they engage in the activities that the social scientist wants to understand: so called because the research is conducted in the field—in the natural settings in which the people and activities of interest are normally to be found, 83–88, 115–120

Fixed-rail metropolis A city whose form and size are determined by the routes of rail systems, 545

Foreigners, prejudice against, 288, 289–291

Formal operational stage According to Piaget, the time after about age twelve when some humans develop the capacity for abstract thought—that is, for using theories rather than only empirical observations, 153

Formal social control. *See* Social control, formal

France, 503–504, 566, 567

Frankish society, 262–263

Freedom of the commons, 415–419

Freedom Summer (1964), 599–605

Free market, 482, 485

Free riding, 421, 595

Freeway metropolis A city developed after the widespread use of autos and trucks freed city structures from rail dependence, 546, 547–548

Free will The philosophical and theological doctrine that humans possess the capacity for choosing among alternatives and, therefore, can be held responsible for the choices they make, 23–24, 68–69, 157

Friendships. *See also* Attachments; Networks
  beliefs and, 77
  civil rights volunteers, 603–604, 604
  deviant attachments and, 189, 198–199
  interracial, 317
  rank and, 76
  religious conversion, 85–86, 87

Functional alternative The existence of more than one system structure that satisfies the same system need, 111

Functional divisions Divisions resulting from breaking an organization into smaller units on the basis of specialized activities or functions, such as when a corporation has separate divisions for manufacturing, purchasing, and marketing, 568–569, 575–576

Functionalist theories Theories that attempt to explain some part of a system by showing its consequences for some other part of the system. These consequences are called functions; for example, the function of the sweat gland is to keep organisms from overheating, 110–111
  vs. conflict theory, 114
  family and, 360–363
  vs. social evolution, 112
  of stratification, 249, 251

Functionalist theory of stratification A theory that holds that inequality is built into the roles of any society because some roles are more important and harder to fill, and to ensure that the most qualified people will seek to fill the most important positions, it is necessary to reward these positions more highly than others, 245–248, 249, 251, 267

G

Gallup Poll, 98, 429–432

Gemeinschaft A German word meaning community, used to describe the intimacy of life in small villages, 554

Gender, 325–357. *See also* Females; Sex ratio; Sex-roles
  background/summary, 329
  crime and, 4, 5, 101, 184–185, 199
  extramarital sex and, 377–378
  hormones and behavior, 135–136, 185
  household chores and, 382–383, 450, 451
  income and, 347–349
  inequality and, 166, 325–326, 336–338
  labor force participation, 274, 275, 345–349, 353, 378, 451–452
  religious commitment and, 393–395
  sexual "revolution," 349–351
  social movements and, 340, 342–345
  status and, 114, 275
  voting rights, 343, 344, 345, 435

Gender bias
  female infanticide, 117, 333, 335, 336, 337, 371, 526–527
  fertility rates and, 526–527

Gender roles. *See* Sex-roles

General Social Survey, 26–28, 317, 438

Genes The basic units of heredity within which specific genetic instructions are encoded in complex chemical chains, 127–128. *See also* Heredity

Genotype The sum total of genetic instructions contained in an organism's genes, 128–129, 137

Geographical divisions Divisions resulting from breaking an organization into smaller units on the basis of geography and making each division relatively independent, 567, 569

Geographical mobility. *See also* Migration
  economic opportunity and, 38–39
  families and, 373, 374
  of females, 136
  sex ratios and, 333, 334, 341

Germany, 503–504

Gesellschaft A German word meaning society or association, used to describe the impersonality of life in cities, 554

Ghetto, 49

Globalization The development of global communications, a global economy, and a global culture, 43
  comparison methods, 11–13
  culture and, 493–494
  dependency and, 486–491
  inequality and, 492–493
  main elements of, 493–494

Goal conflict Situation in which one goal of an organization limits the ability of that organization to achieve other goals, 573

Goal displacement What occurs when the official goals of an organization are ignored or changed, 572–573

God(s), 389–393

Gold Rush, 333, 334

Government. *See* State

Greece, ancient, 232, 262, 293, 336–338, 392

Greenland, 479–480

Gross domestic product (GDP) The final value of all economic activities within a society. When GDP is divided by the total population of a society, the result is per capita GDP, 489–490

Group solidarity. *See* Social solidarity

Group(s) Two or more persons who maintain a stable pattern of social relations over a significant period of time, 1–29. *See also* Network
  agreement and conformity in, 76–77
  defined/types of, 13, 14–15
  inequality and segregation, 551–554
  networks within, 15–17
  peers, as primary, 371–372
  properties of, 33–34, 104–107
  in research, 81–83
  size of, 97, 98, 99
  as units of analysis, 9–10, 11, 12, 13–15

Groups, primary Groups whose members have close and intimate emotional attachments to one another, 14, 371–372

Groups, secondary Groups whose members have only limited emotional attachments to one another, 14–15

Growth rate Population gains or losses computed by dividing the net gain or loss for a particular period by the population total at the start of that period, 501

H

Health. *See also* Death rates; Disease; Infant mortality; Life expectancy; Mental illness
  criminal activity and, 186–187
  diet and, 370, 492, 506–507, 511
  growth patterns and, 134–135, 265

Industrial Revolution and, 511
networks and, 37
population growth and, 518–519
in preindustrial cities, 535–537
in preindustrial societies, 506–508
sex ratios and, 335
urbanization and, 554
Heaven's Gate, 88–89
Height, 133–135, 265
Heredity
genes, 127–129
genetic "signatures," 137
mental illness and, 129, 132
Heredity vs. environment
behavioral genetics, 129–133, 183–185
cultural determinism, 158–161
deviance and crime, 182–185
height/growth, 133–135
human vs. animal behavior, 137–144
isolation and behavior, 141–142
language acquisition, 153–156
theories of, 125–129
tracing human relatedness, 136–137
Highbrow culture Culture that consists primarily of acquired tastes—the appreciation of art, literature, music, furnishings, and food and wine that requires experience and instruction, 276
Hispanic Americans
economic conditions of, 312
education and, 311, 464
English fluency, 311
ethnic diversity of, 310–311
ethnic mobility of, 54, 310–312
population statistics, 310, 311
sex ratios and, 353–354
History, shaping of, 589, 603, 605
Homeschooling, 461
Homicide. See also Infanticide
of African Americans, 591, 592, 599–600
among the Yanomamö, 117
capital punishment and, 219–221
causes of, 190, 202
civil rights workers, 600, 607
evolution and, 138
seasonal variation, 201
solving of, 222
statistics on, 180, 185
Hormones "Chemical messengers" excreted into the bloodstream by the glands comprising the endocrine system, 135–136, 184, 185
Horses, 262–263
Housework, 382–383, 450, 451
Hunting and gathering societies The most primitive human societies; their rather small numbers of members (often fewer than fifty) live by wandering in pursuit of food from animals and plants, 167, 256–257, 480
Hypothesis A statement about the expected relationship between (or among) observable measures of concepts, 21–22, 55, 79, 153

I

Identity. See Self
Ideology A connected set of strongly held beliefs based on a few very abstract ideas, used to guide one's reactions to external events; for example, a political ideology is used to decide how societies ought to be run, 84, 437
communes and, 104, 107
conformity and, 100–103
elites and politicians, 440
organizations and, 440–441
political party elites, 439
public opinion and politics, 436–441
religious conversion and, 84, 87
of women's movement, 343, 344
Illegitimacy ratio The proportion of all births that occur out of wedlock, 353
Immigrants. See also Assimilation; Ethnic and racial groups; Italian immigrants; Jewish immigrants
African Americans studied as, 315, 319–320
assimilation, 47–61, 286–287, 308, 310, 311, 312, 526
Catholics, 44, 296–297
class origins and occupations, 52–55, 308–310
countries of origins, 43–48
Cuban, 54
economic exploitation of, 299, 301
economic status of, 31, 32–33, 48, 51, 52, 53, 296–300, 306
European Muslims, 526
Japanese, 300, 302–304
population, effect on, 524, 526
prejudice against, 31–33, 43–48, 296–300, 302–307
returning to homeland, 300, 320
social mobility (See Ethnic and racial mobility)
in sports, 130–131
statistics, 31–33, 44–47, 524, 526
success, cultural theory of, 48–51, 55, 61
success, social theory of, 51–57
success, traditionalism and, 57–62
Vietnamese, 54
Immigration. See also Migration
from Canada, 284
into Canada, 44, 45, 46, 47, 53, 55
economic opportunity and, 320
impact of, 31–33
quotas on, 46–47, 300–301, 306
sex ratios and, 340–342, 353–354
Immigration Commission, 31–33, 43, 53, 57
Impression management Conscious manipulation of role performance, 165
Incapacitation The inability of offenders to commit new offenses while they are in jail or prison, 222–223
Incest taboo Prohibition against sexual relations between certain members of the same family, 363
Income. See also Economic status/inequality; Poverty
of African Americans, 267, 315, 590–591
of city residents, 538
divorce and, 376
education and, 462–464
of immigrants, 33
of Japanese Canadians, 306
language fluency and, 57
marriage and, 272, 364
of minorities, 297–300
occupational prestige and, 448
prejudice and, 290
stratification and, 266, 270, 271
supply and demand, 482
of two-wage families, 272, 451
of women, 347–349
Independent variable Something we think might be the cause of something else, 78, 80, 82–83
Index of dissimilarity A measure of the degree to which a given ethnic or racial group lives in integrated or segregated neighborhoods; it compares the ethnic makeup of city blocks with the ethnic makeup of the city as a whole, 552, 553
India, 35, 239, 507
Industrialization The process by which technology is substituted for manual labor as the basis of production, 266–267. See also Economic development; Modernization; Technology
agricultural revolution and, 540–543
crime rates and, 6
education and, 266–267, 454
specialization and, 543
standard of living and, 243
urbanization and, 540–543
Industrial Revolution
agriculture and, 511
causes of, 267, 481, 484, 486–487
education and, 454
health impact of, 134–135
Industrial societies Societies with economies based on manufacturing in which machines perform most of the heavy labor, 265–278

Industrial societies, *continued*
  networks in, 276–278
  social mobility in, 268–278
  status attainment in, 271–278
  stratification in, 266–268
Inequality. *See also* Class; Economic status/inequality; Gender; Status; Stratification
  in capitalist societies, 482
  global dimensions of, 492–493
  segregation and, 551–554
Inequality, Homans' law of Emotional attachments among members of a group will be weaker among members of different ranks than among those of similar rank, 2, 3, 21, 76
Infanticide The practice of killing infants soon after birth, often done by simply abandoning them out of doors, 333
  female, 117, 333, 335, 336, 337, 371, 526–527
Infant mortality
  African Americans, 335
  in America, 502
  cultural lag and, 476
  Mexican Americans, 354
  parent-child attachments and, 369–370
  in premodern societies, 366
  sex ratios and, 335, 352, 354
Infants. *See also* Children
  facial expressions, 127
  in preindustrial societies, 370
Influence and ties, 36, 278
Informal social control. *See* Social control, informal
Instinct Any behavior that occurs in all normal members of a species without having been learned, 125–126, 141, 153–156
Institutions. *See* Social institutions
Integration. *See* Moral integration; Segregation; Social integration; Vertical integration
Intelligence, 131, 132, 137–138, 142, 149–153, 183, 187
Interdependence A relationship among parts of a system such that if one part changes, at least one other part is affected, 109–110
Intergroup conflict Conflict between groups that are racially or culturally different, 285. *See also* Prejudice
  cultural pluralism and, 286–287
  ethnicity and race, 285–288, 321
  prejudice and, 288–293
Intermarriage, 47
Internalization of norms. *See* Norms, internalization of

Internalization The process by which people accept the norms, values, roles, beliefs, and other primary aspects of their culture to the point that these are a fundamental basis for all of their decision making, 68–69
International relations. *See* Globalization; World system theory
International Social Survey Program (ISSP), 61
Internet: instructions for using this text's site, 26–29
Inventions, 473, 477, 485–486. *See also* Technology
Investments The costs expended to construct a satisfactory life and the current and potential plan of rewards expected, 197, 198
Involution A retrograde or degenerative change in the survival capacity of societies, 112
Involvements The amount of time and energy expended in nondeviant activities, 197, 198
Irish Americans, 130, 307
Islam. *See* Muslims
Isolation, 141–142, 147–148, 160, 242, 555
Issue publics, 439–440
Italian immigrants/Italian Americans
  assimilation and accommodation of, 47–48, 61–62
  economic status of, 32, 33, 300
  educational achievement, 50–51, 56
  education and upward mobility, 48, 50–51, 56, 57–62
  prejudice/discrimination, 43–48
  in sports, 130
  traditionalism, reference groups and social mobility, 57–62
  values of, 48, 51, 58, 61

J
Japan, 134, 211–212, 284, 287, 365
Japanese Americans, 302–306, 308
Japanese Canadians, 305–306
Japanese immigrants, 300, 302–304, 307
J-curve theory of social crisis One form of the revolution of rising expectations, the J-curve theory emphasizes that an acute crisis occurs when, after a period of rapid improvement, conditions suddenly worsen, 593
Jealousy, 106, 107
Jehovah's Witnesses, 190
Jewish immigrants
  assimilation and accommodation of, 47–48
  economic status of, 32–33, 52, 53
  education and upward mobility, 48–50, 52–56, 310

  prejudice/discrimination, 43–48, 50
  social class and occupational skills, 52–55
Jews
  intermarriage rate among, 47–48
  labor unions and, 299–300
  the Lemba of Africa, 137
  prejudice against, 45, 46, 50, 53, 288, 289–290, 301
  sex ratios among, 335
  in sports, 130
  stratification and, 236, 238, 301
  values of, 49–50

K
Kibbutz, 361, 364
Kinship, 118–120, 361–362, 373–374. *See also* Family
Knights, 262–263
Ku Klux Klan, 596–597, 600

L
Labeling theory A theory that explains deviant behavior as a reaction to having been socially identified as a deviant, 201–202, 225
Labor. *See also* Occupations
  in agriculture, 540–543
  attitude/motivation, 481, 482, 483
  child labor, 366, 369, 513
  class consciousness and, 234
  conflict theory, 250–251
  criminal behavior and, 201
  discrimination/exploitation, 260, 299, 590–591
  division of (*See* Division of labor)
  functionalist theory, 246–247
  in industrialized societies, 267–268
  inequality and prejudice, 296, 297–301
  job satisfaction, 453
  productivity and, 485–486
  Protestant Reformation and, 483
  shortages, 517
  transformation of, 448–450, 453
Labor force Those persons who are employed or seeking employment, 450. *See also* Occupations
  African American females in, 353, 591
  African Americans in, 590–591
  females in, 169–170, 274, 275, 345–349, 353, 378, 451–452
  in Japan, 211, 212
  job-seeking and networks, 35–36, 278
  population change and, 526
  transformation of, 450–453
  unemployment, 452–453
Labor unions. *See* Unions
Language. *See also* Literacy
  acquisition of, 151, 153–156, 157
  in Canada, 55

immigrant success and, 55, 57, 308, 310, 311
nonhuman, 142–144
pidgin and creole, 155, 156
sign language, 142–144, 155–156
social class and, 241–242
symbols and, 73
television and, 478
Language instinct An inborn, elementary Universal Grammar (as Chomsky calls it) that enables normal infants to rapidly acquire complex languages and even to create languages, 153–156
Latin America, 407–408, 520, 523
Latinos. *See* Hispanic Americans; *specific groups*
Law of agreement. *See* Agreement, law of
Law of conformity. *See* Conformity, law of
Law of inequality. *See* Inequality, Homans' law of
Law of liking. *See* Liking, law of
Leaders/leadership. *See also* Power elite
charismatic religious, 400
exploitation and repression by, 262–263, 416–419
of large organizations, 563
philosopher-kings, 424
political parties, 435
of social movements, 594, 595, 604
Learning, models of, 151–156
Learning theory, 188–189
Lemba tribe, 137
Libraries, 458
Life expectancy, 80, 492, 519, 554. *See also* Death rates
Liking, law of Participation in common activities (exchanges) causes people to like one another, 76, 195
Literacy, 54, 455–456, 457–458, 514
Living together, 378–379
Local networks. *See* Networks, local
Long-distance mobility Mobility that occurs when an individual or group rises from the bottom to the top of the stratification system, 270
Longitudinal study Research in which observations are made of the same people at different times, 164
Looking glass self The process by which our sense of self develops through interaction as we come to see ourselves as others see us, 74–75
Los Angeles, 547–548
Love, 71, 104–107, 370–371, 372–373. *See also* Attachments
Lumpenproletariat Literally, the "ragamuffin proletariat"; the people on the very bottom of society, whom Marx labeled "social scum", 233

M
Macrosociology The study of large groups and even of whole societies, 11, 95–123
comparative research, 115–120
contextual effects and, 102–103
group properties and, 104–107
populations, describing, 96–102
social systems, 108–115
societies as units of analysis, 107
units of analysis, 21
Magna Carta, 425
Malthusian theory of population Theory stating that populations will always rise to, and then go somewhat above, the limits of subsistence and then will be reduced by the positive checks, only to rise again and be checked again, 508–511, 511–512, 513
Management by objectives Situation in which managers and subordinates agree on goals that subordinates will try to achieve; subordinates then have maximum freedom in how they will try to reach their objectives, 578
Management Coordination of the work of others, 571–572, 577–578
Manipulate In experimental designs controlling the value or level of the independent variable to which subjects are exposed, 83
March of Dimes, 572–573
Markers (cultural or racial) Noticeable differences between two or more groups that became associated with status conflicts between the groups, 297
Marriage A formal commitment between a couple to maintain a long-term relationship involving specific rights and duties toward one another and toward their children, 362. *See also* Divorce; Family
after living together, 379
age and, 340–341
in ancient Greece, 336, 337, 338
crime and, 196
extramarital sex and, 336, 338, 377–378
happiness in, 102, 103
household chores and, 382–383
income and, 272, 364
intermarriage, 47, 305, 306, 317, 352
political activists and, 605
polygamy/polygyny, 363
in premodern societies, 362, 363, 370–373
remarriage, 367, 374, 380–383
residence norms, 119–120
romantic love in, 370–371, 372–373
stepchildren and, 367, 380, 381–383
Marxism, 424–425
Mass media. *See* News media; Television

Mass society theory, 556
Matrilocal rule of residence A situation in which married couples live with or near the woman's family, 119–120
Meaning, social construction of, 73–74
"Meaning systems," 437. *See also* Ideology
Meaning, ultimate, questions about Questions about the very meaning of life, the universe, reality—for example, Does life have a purpose? Is death the end? Why do we suffer?, 388, 392
Means of production Everything, except human labor, that is used to produce wealth, 233, 235, 244–245, 260, 261, 482
Mental illness
attachments and, 77
crime and, 186–187
hereditary basis of, 129, 132
urbanization and, 555–556, 556–557
Metropolis A city and its sphere of influence, 544–550. *See also* Urbanization
Mexican Americans, 353–354. *See also* Hispanic Americans
Microsociology The study of small groups and of face-to-face interaction among humans, 67–93
basic approaches to, 71–72
defined, 11
exchange theory and, 71–72, 75–77
field research, 83–88
group solidarity and, 81–83
vs. macrosociology, 71, 102, 107
rational choice proposition, 69–71
symbolic interaction, 71, 72–75
theory testing/research, 77–79, 80
units of analysis, 21
Middle class
capitalism and, 233
child-rearing and values of, 163–164
family life, 373
immigrant success and, 54–55, 57, 310
industrialization and, 266
Middleman minorities Racial or ethnic groups restricted to a limited range of occupations in the middle, rather than lower, level of the stratification system, 301
Migration. *See also* Geographical mobility
of African Americans, 297–298, 312–315
to cities, 450
Gold Rush, 334
for jobs, 38–39, 297–298, 300
sex ratios and, 333, 334, 341
temporary, 57–58, 300, 320
workers and wages, 297–298, 300
Military domination, 261–263, 423, 478. *See also* Warfare
Military organization, 563, 565–567
Military technology, 249

Mind Our understanding of symbols. According to Mead, the mind arises entirely through repeated interaction with others—a learning process that in infants is largely a matter of trial and error, 74–75

Minorities. *See* Ethnic and racial groups; Immigrants

Mobility. *See* Ethnic and racial mobility; Geographical mobility; Long-distance mobility; Social mobility; Structural mobility

Mobility, upward and downward A change of position within the stratification system, 35, 239, 269–270

Modernization theories
  capitalism (Marx), 481–483
  dependency theory, 486–491
  Protestant ethic (Weber), 483–485
  state theory, 485–486

Modernization The process of industrialization, economic development, and technical innovation by which a culture achieves high standards of living and maximizes control over the physical environment, 43, 472. *See also* Industrialization
  divorce and, 374–377
  employment of women and, 451
  fertility and, 514, 521–522
  industrialization and, 266, 494
  inventions/science and, 473, 474, 475
  kinship and, 373–374
  mortality rates and, 512, 513
  population and, 494, 511
  rise of the West, 480–481
  romance and, 372–377
  social relationships and, 14
  urbanization and, 533–534

Monarchy, 262–263, 424, 425, 569
Monopolies, 249, 395, 420, 422, 483
Montgomery Improvement Association (MIA), 595–596, 597, 599

Moral communities Groups within which there is a very high agreement on the norms and strong bonds of attachment among members, 200

Moral integration The degree to which members of a group are united by shared beliefs, 200

Morality, 367, 391–393
"Moral statistics," 4–6, 8
Mormons, 400
Mortality rates. *See* Death rates
"Mozart Effect," 150–151

Multiculturalism The presence of several significant cultures, 41–42

Muslims
  European immigrants, 526
  polygyny, 363

  prejudice against, 288, 289–290
  religion and morality, 393
  slave trade, 293
  in Soviet Union, 408
  women, 354

N

NAACP (National Association for the Advancement of Colored People), 594
National pride, 108
Nation, defined, 33
Native Americans, 255, 283–284, 293, 310, 479, 480

Natural system approach Emphasis on the informal and unintended characteristics or organizations, 572, 578–580

Neanderthals, 387
Neighborhoods, urban, 550–553

Network A pattern of ties or connections among some set of units, as a computer network links many computers, allowing them to communicate and exchange, 15, 35–39

Network analysis A research method that has been developed specifically to test hypotheses about properties of social networks. It involves constructing measures of network variables and using networks as the units of analysis, 104–107

Network recruitment principle Books, articles, advertisements, and speeches may serve many important functions, such as boosting the solidarity and confidence of members, but recruitment is primarily a person-to-person phenomenon, 88

Networks, cosmopolitan Networks that are relatively full of holes, consisting of weaker, nonredundant ties. The word cosmopolitan means worldly, at home throughout the world, or widely distributed. Hence, members of a cosmopolitan network seldom engage in face-to-face interaction and tend to be scattered geographically, 37–38, 76

Networks, local Dense networks with strong, redundant ties. They are called local because members often engage in the direct, person-to-person interaction that is necessary to form and sustain strong ties. Hence, members tend to be clustered geographically, 37–38, 76, 85–86, 88, 242

Network, social A pattern of social relationships or links among some set of social units-usually people, but sometimes groups, 15–17. *See also* Attachment

  in Canada, 276–277
  in China, 278
  churches/religion, 85, 87, 395–397, 594, 595
  class cultures and, 241–242, 276
  in communes, 104–107
  deviance and, 189
  global village and, 493, 494
  in organizations, 574
  power and, 35–37, 276–277
  social construction of self and, 494
  social movements and, 594, 603–604
  social order and, 211
  social solidarity and, 15–16, 76
  societies as, 38–39
  stratification and, 241–242

Network variety The number of classes, status groups, occupations, and cultures included in one's social network, 276–277

New Guinea, 159–163
New Left, 604
News media
  civil rights reporting, 597, 600, 605
  crime reporting, 177–178, 580
  education reporting, 459
  occult topics in, 18
  opinion polls, 429–432
  police/reporter coalitions, 580
  race relations in, 317
  stereotypes, 45
  TV viewing habits, 432
  unemployment reporting, 452

Norms Rules that define the behavior that is expected, required, or acceptable in particular circumstances, 40. *See also* Conformity
  anomie, 199–200, 554–556
  deviance and, 190, 191
  education and, 49, 51
  origins of, 77
  personality and, 157–158
  public good and, 421
  religion and, 86, 103
  sexual relations, 104–105, 336, 339, 349–351 (*See also* Sex-roles)
  violating (*See* Deviance)

Norms, internalization of The sociological synonym for conscience; refers to the tendency of people not simply to learn what the norms are but also to come to believe the norms are right, 68–69, 197

Nuclear family. *See* Family, nuclear
Numeracy, 522–523

O

Observer effects Changes that people make in their behavior when they know they are being studied, 85

Occupational prestige The respect people receive on the basis of their job; often refers to a score on a standard system for rating occupations, 446–448

Occupations, 445–469. *See also* Labor; Labor force
child-rearing practices and, 163–164
creating of, 466
credentialing and, 464–465
education and, 445–446, 448
gender and, 169–170, 451–452
immigrant success and, 52–55, 308
of marriage partners, 381
middleman minorities, 301
minority specialization, 308–309
women with professional degrees, 451–452

Open-minded principle It will be very hard to recruit someone who is strongly tied to a network wherein most people are opposed to the ideology of the recruiting movement, but recruitment does not depend on people having any prior attitudes favorable to the ideology of the movement, 87–88

Open system A system that is open to external influences. A society is open to such forces as weather, topography, seismic activity, natural resources, and other societies, 110

Operationalize To select measures of concepts in order to make it possible to perform observational operations on it, 21

Opinion polls, 98–99, 429–432

Opportunity theory, 213

Organizations, 563–585. *See also* Social movements
administrative theory, 577–578
bureaucracy and, 570–572, 580–582
creation of, 565–570
decentralization of, 577–580
diversification of, 575–578
emergence of large-scale, 563–570
goals: rational *vs.* natural systems, 572–575
ideologies and, 440–441

Organizations, formal Groups created to pursue definite goals wherein tactics and procedures are designed and evaluated in terms of effectiveness in achieving goals; members are selected and trained to fulfill their roles, and overall operations are based on written records and rules, 564. *See also* Bureaucracy
business, 567–569, 573–575, 575–578, 580–581

government, 564, 569–570, 580–581
military, 563, 565–567

Overt observers Field researchers who are operating with the knowledge and permission of those who are being studied, 85

P

Panel studies Public opinion surveys that interview the same respondents several times, 439–440

Parliament, Canadian, 433–435

Parliament, U.K., 425

Pascal's wager, 394

Patent laws, 485–486

Patrilocal rule of residence A situation in which married couples live with or near the man's family, 119–120

Peer groups, 371–372

Peripheral nations According to Wallerstein, those nations in the world system that are forced to specialize in the export of unprocessed raw materials and food to the core nations and that must import manufactured goods. This makes them dependent on the core nations, which in turn forces them to adopt economic and social policies that prevent them from modernizing, 487, 488

Personality Consistent patterns of thoughts, feelings, and actions, 157. *See also* Socialization
authoritarian, 290
criminal behavior and, 187
culture and, 158–161
formation of, 157–158, 466

Phenotype The observable organism as it has developed out of the interplay between the genotype and the environment, 128–129, 134

Philosopher-kings, 424, 427

Philosophies ("world views"), 437. *See also* Ideology

Pidgin A jargon of made-up nouns and verbs, often including some borrowed from the language of the dominant group, highly variable in terms of word order and with little in the way of grammar, 155

Pluralism, cultural, 286–287

Pluralism/pluralist state A system or a society in which power is dispersed among many competing elites who act to limit one another's power and minimize the repression and exploitation of members, 425
evolution of, 424–426
power elites *vs.*, 428–429, 441
religious, 395, 486

Police, 221–223, 580, 591, 600

Political candidates, 432–436

Political participation
civil rights volunteers, 604–605
personal beliefs and, 436–441, 603–604
petitions, 429, 431
voter characteristics, 432
voting patterns, 432–436, 438

Political parties, 434, 435, 439, 440–441

Political power. *See also* State
gender and, 325
of minorities, 299, 307, 321, 591
voting rights, 343, 344, 345, 427, 435, 591, 599, 600–601

Politics
abolitionist movement, 438
indifference to, 429, 432
public opinion on, 436–438
status inconsistency and, 238

Pollution, 536–537

Polygamy All plural marriages, wherein one person (of either sex) has multiple partners, 363

Polygyny The proper term for marriages involving one man and multiple wives, 363

Population, 499–531. *See also* Death rates; Fertility
African Americans, 310, 321
birthrate decline, 475–476, 519–527
of cities, 259, 450, 533, 537
cultural lag and, 475–476
demographic transition, 512–514
density of, 33–34, 537, 556–558
"depopulation," 523–527
growth of, 415–416, 474–475, 494, 509, 514–519, 519–523
immigration and, 31, 44–47, 524, 526
in less developed nations, 518–519
Malthusian theory of, 508–511, 511–512, 513
measurement techniques, 501–506
of minorities, 310, 311
modernization and, 494, 511, 519
predictions, 527–528
in preindustrial societies, 506–508
in research design, 96–99
second explosion in, 514, 518–519
social change and, 474–475, 494
in the United States, 504, 506, 525

Population structure, constrictive An age structure in which younger cohorts are smaller than the ones before them; such a population is shrinking, 505, 506

Population structure, expansive An age structure in which each younger cohort is larger than the one before it; such a population is growing, 504, 505

Population structure, stationary An age structure in which younger birth cohorts are the same size as older ones were before mortality reduced them; such a population neither grows nor declines, 505, 506

Positive checks According to Malthus, famine, disease, and war—the primary factors that check or stop population growth, 509

Poverty. *See also* Economic status/inequality
of African Americans, 313–315, 590–591, 599
capitalism and, 482
crime and, 191–192
education and, 272, 462
"equality of," 481
female-headed households and, 315, 379–380
social pathology and, 555
unemployment and, 452
in U. S. South, 313–314, 316

Power The ability to get one's way despite the opposition of others, 237. *See also* Political power
conflict theory and, 114
gender and inequality, 325–331, 336–338
military, 261–263, 423 (*See also* Warfare)
networks and, 35–37, 276–278
state, limiting, 424–429, 486
state, repression/exploitation by, 416–419
stratification and wealth, 236, 245
structural, 339–340

Power dependence The dependence of one member of a dyad on the other is equal to their inability to achieve their goals outside the dyad, 338

Power elite A term C. Wright Mills used to identify an inner circle of military, government, and business leaders he believed controls the United States, 428–429. *See also* Bourgeoisie
emergence of, 261, 264–265
expropriation of wealth, 485
warfare and, 261–264

Precipitating event Specific occurrence that provokes people to take action, 593–594, 596–597

Prejudice Negative or hostile attitudes toward and beliefs about a group, 42. *See also* Discrimination; Segregation; *specific minority groups*
decline in, 302–306, 317–318
against foreigners, 288–293
housing segregation and, 551

identifiability and, 301–302
against immigrants, 31–33, 43–48, 297–301, 302–306, 306–307
intergroup conflict and, 285–288
labor market conflict and, 297–301
slavery and, 293–296, 318–319
status inequality and, 296–301, 322
theories regarding, 288–293, 296–297, 301–302

Preoperational stage According to Piaget, the period from age two to about seven during which a child learns to take the role of the other, 152

Prestige Social honor or respect; synonymous with Weber's term status, 236–238, 264, 446–448

Prevention As a form of social control, all efforts to remove the opportunity for deviance or to deactivate its causes, 213–216

Primary groups. *See* Groups, primary

Primitive societies, 117–120

Prisons, 223–226

Privacy, 212, 373, 557–558

Professions Occupational organizations that can prevent their functions from being performed by those not certified as adequately trained and qualified in an extensive body of knowledge and technique, 250

Progress, idea of The philosophical doctrine that technological and social progress is inevitable, 473

Proletariat The name that Marx applied to the class made up of those who do not own the means of production; the employee or working class, 232–233, 234, 236, 244

Property The term many sociologists use to identity what Weber called class. Property includes all economic resources and opportunities owned or controlled by an individual or a group, 236, 244
means of production as, 260
private, and democracy, 424–426
private, and free market, 482

Proportional fact The distribution of something, or even the joint distribution of several things, among a number of cases, 97, 98

Prostitution, 371, 377

Protestant ethic According to Weber, doctrines holding that economic success reflects God's grace, 483–485

Protestant Reformation The separation of Protestant Christians from the Roman Catholic Church during the sixteenth century; usually associated

with the founding of the Lutheran Church in Germany, 483–484, 486

Protestants
capitalism and, 486
Catholics and, 296–297
conflicts between denominations, 474
in Latin America, 407–408
networks of, 395–397

Psychology, 7, 8, 67

Public opinion
ideology and, 436–438
Lincoln's election and, 438
opinion polls, 98–99, 429–432
on political issues, 439–440
voting patterns and, 438

Public or collective goods Things necessary for group life that individual members of a society cannot provide for themselves and that require cooperative actions by many members, 419–420, 422

Puerto Rican Americans, 353, 354

Punishment, 216–221, 223–226

Q

Quakers, 224, 473

R

Race A human group having some biological features that set it off from other human groups, 285–286. *See also* Ethnic and racial groups

Racism, 295–296, 316, 474, 554, 599–600. *See also* Discrimination; Prejudice; Segregation

Railroads, 546, 567–568

Randomization The use of chance procedures to determine who will be exposed to which value of the independent variable. Sometimes experimenters do what Asch did and just flip a coin to assign subjects to a group; more typically, they use a table of random numbers, 83

Random selection A method of selection whereby all cases have an equal (or at least a known) probability of being included in the sample, 98–99

Rape, 201, 202, 219, 222, 336, 338, 591–592

Rational bureaucracy, 570–572

Rational choice proposition The proposition that within the limits of information and available choices, guided by their preferences and tastes, humans will tend to maximize, 69–71, 75

Rational choice theory. *See* Exchange theory

Rational organizations. *See* Organizations, formal

Rational system approach Emphasis on the official and intended characteristics of organizations, 572–575

Reading, 458, 459. *See also* Literacy

Reason, 23–24, 151–153

Recidivism rate The proportion of persons convicted for a criminal offense who are later convicted for committing another crime. Sometimes this rate is computed as the proportion of those freed from prison who are sentenced to prison again, 224, 225, 226, 227

Recruitment, 83–89

Redundant tie One that duplicates links among members of the same network, 36

Reference group A group a person uses as a standard for self-evaluation, 57–60

Religion Any socially organized pattern of beliefs and practices concerning ultimate meaning that assumes the existence of the supernatural, 387–413. *See also specific religions*
accommodation and, 43, 48
charisma and, 400
church affiliation statistics, 401, 402, 403, 404, 405, 406
church-sect theory, 395–397
communes, 106–107
conversion, 84–89, 602
cult formation, 84–89, 399–400, 404–405, 406, 410
delinquency and, 100–103, 197, 198, 200
diversity and statistics, 390, 401–405, 405–409
divorce and, 376
in Europe, 390, 406
free will and, 23–24
gender and, 393–395, 437
industrialization and, 483–485
instinct and, 125
intermarriage and, 47–48
in Latin America, 390, 407–408
nature and function of, 388–395
Neanderthals and, 387
prejudice and, 288, 289–291, 296–297
remarriage and, 380
revival, 388, 398–399, 401–402, 409
in Russia, 390, 399, 408–409
sects, 395–397, 402
secularization and, 387–388, 398–399, 401–405
social change and, 478
subcultures, 42
universal appeal of, 409–410
values and, 39, 49

Religious economy The set of competing faiths, and their adherents, within a given society or geographical area of a society, 387, 395
in Canada, 405–406
church-sect theory and, 399
cult formation and, 400
in Latin America, 408
in United States, 400–405

Religious innovation The appearance of new religions in a society either by founding of a new faith or by importing a new faith from another society, 388. *See also* Cult formation

Religious pluralism The existence of several religions in the same society, 395

Replaceability A measure of the functional importance of a role based on the extent to which other roles can substitute for or take on the duties of that particular role. For example, a doctor can easily substitute for an orderly, but the reverse is not so, 246–247, 249–251, 464

Replication Repetition of research to guard against incorrect findings or findings that occurred merely by chance, 86–87, 102, 150–151

Research The process of making appropriate empirical observations or measurements—this usually is referred to as collecting data or facts. The purpose of research is to test theories or to gain sufficient knowledge about some portion of reality so that it becomes possible to theorize about it, 2–3. *See also* Experiment; Science
bias in, 20–21
comparative, 11, 13, 115–120, 255–256
field research, 83–89
longitudinal studies, 164
methodology, 17–21, 77–87, 96–102, 214, 215
populations, 96–99
on self-aware subjects, 17–21
statistics, 107, 150–151
survey research, 100–102
units of analysis, 9–10, 11, 12

Resocialization Efforts to change a person's socialization, that is, to socialize a person over again in hopes of getting him or her to conform to the norms, 225–226

Resource mobilization approach to social movements Stresses the importance of human and material resources in the development of social movements and places particular emphasis on the role of leaders and of rational planning, 589–590

Responsibility, 97

Revival Movements within religious organizations, or the breaking away of new organizations, to reaffirm less secularized versions of a faith, 388, 398–399, 401–402, 409

Revolution of rising expectations The probability that people will take action to change their society is maximized during periods when things have been improving but when the hope for change outpaces actual change, 592–593

Rewards
as basis of social interaction, 71–72
crime/deviance and, 182, 188, 189, 194
exchange theory and, 72, 75–77, 338, 421
rational choices and, 70–71
religion and, 391, 397
sex ratios and, 338–340
social control and, 211
stratification and, 246, 247, 248, 249

Rights. *See also* Civil Rights Movement
in democracies, 596
property, 424–426
voting rights, 343, 344, 345, 427, 435, 591, 599, 600–601

Robbery, 178, 185

Role of the other, taking the, 75, 157

Role performance The actual behavior of people in a particular role, in contrast to how they are supposed to behave, 165–166

Role(s) A set of expectations governing the behavior of persons holding a particular position in society; a set of norms that defines how persons in a particular position should behave, 41, 161–162. *See also* Sex-roles; Socialization
education and, 49, 465
impression management, 165
inequality and, 245–248
political ideology and, 439

Rome, ancient, 232, 263, 371, 392, 500

Ruling elite. *See* Power elite

Russia, 390, 399, 408–409, 525

S

Sample A set of cases (or units) randomly selected from the entire set of cases or units to be described, 98–99

Sanitation, 536–537

Schooling Amount of time spent in an institution dedicated to educating and which often confers degrees and diplomas on those who complete a period of enrollment, 454

Schools. *See also* Education; Educational achievement; Universities
baby boom and, 516–517
class/status and, 464–466
compulsory school law, 453
effectiveness of, 456–458
high schools, 458–461
homeschooling, 461
population changes and, 475–476
public dissatisfaction with, 459
public *vs.* private, 459, 460, 461
segregation of, 53, 592, 596, 599
socializing function of, 464–466

Science A method for describing and explaining why and how things work (human as well as material "things"). It consists of two components: theory and research, 1–3. *See also* Research; Social sciences; Technology; Theory
attitudes towards advances, 473–474, 475
experiments, 82–83
purpose of, 158
religion and, 398
scientific method, 20
women in, 169–170

Scientific management The application of scientific techniques to increase efficiency, 449

Secondary groups. *See* Groups, secondary

Sect formation The breaking off of a group from a conventional religion in order to move into a higher degree of tension with the environment, 397, 402

Sects Religious bodies in a relatively high state of tension with their environment but which remain within the conventional religious traditions of their society, 395–397

Secularization The process by which particular religious organizations become more worldly and offer a less vivid and less active conception of the supernatural, 387–388, 398–399, 401–405

Segregation. *See also* Discrimination; Prejudice
of African Americans, 317, 552, 553, 587–588, 589, 591
of cities, 551–554
legal, termination of, 593, 598–599
of minorities, 307, 309, 551–554
of schools, 592, 593, 596, 599

theories of, 551–553
world perspective, 553–554

Selective reinforcement, 188

Self Our learned understanding of the responses of others to our conduct. Through long experience in seeing others react to what we do, we not only gain a central notion of who we are but also are able to put ourselves in another's place, 71, 74–75, 156–157, 494

Self-conceptions, 201
Self-control, 187, 188–189, 394
Self-esteem, 71
Self-interest, 107, 114, 249, 415–419, 420, 481, 565

Semiperipheral nations Wallerstein's term for nations that fall in between core and peripheral nations, being more industrialized than the latter and less industrialized than the former, 487

Sensorimotor stage According to Piaget, the period from birth to about age two years during which the infant develops perceptual abilities and body control and discovers the rule of object permanence: that things still exist even when they are out of sight, 152

Sex ratio The number of persons of one gender relative to the number of persons of the other gender, unusually expressed as the number of males per 100 females, 327, 330–331
African American family and, 341, 350, 351–353
among early Christians, 354
in Athens and Sparta, 336–338
causes of unbalanced, 333–336, 526–527
Hispanic American family and, 353–354
in medieval Europe, 340
power dependency and, 338–340
social pathology and, 556
as a social structure, 34, 331
statistics, 330–332, 333, 340–342
supply *vs.* demand study of gender relations, 327, 330–331, 332–333
theory of Guttentag and Secord, 332–340

Sex-roles
culture and, 158–162, 168–169
dehumanization and, 351
employment and, 345–349
socialization into, 162–164, 167–170, 184, 339, 348–349
traditional, 326, 327, 328, 333

Sex structure The proportions of males and females in a population, 504

Sexual attraction, 377
Sexual relations
in communes, 104–106
extramarital, 336, 338, 377–378
within families, 361, 363–364
norms and, 363–364
sex ratios and, 335

Sexual "revolution," 349–351

Significance test, 150–151

Significance, test of A calculation of the odds that a given correlation (or a difference between groups in an experiment) was caused by random variations (or flukes), 83, 150–151

Sign language, 142–144, 155–156

Simple fact An assertion about a concrete, quite specific, and limited state of affairs, often merely the claim that something happened or exists and usually having to do with only one or very few causes, 96

Slavery
abolitionist movement, 438
dependence and, 592
emergence of, 261, 264
in history, 293–296
and pidgin development, 155
in United States, 284, 293–296, 318–319

SLOPS (self-selected listener opinion polls), 98–99, 429–430

Smallpox, 507–508

Social bonds Bonds that, as used in control theory, consist of attachments, investments, involvements, and beliefs, 194–196

Social change, 471–497. *See also* Industrialization; Modernization; Social movements
cultural lag and, 475–476
in Europe, 480–481
external sources of, 476–480
internal sources of, 472–475, 477, 490–491
scope and pace of, 471–472

Social class. *See* Class

Social conflict Unfriendly interactions between groups, ranging in degree of seriousness from disagreements to violent encounters. Notice that conflicts between two individuals are excluded from this definition. For a conflict to be social, at least three people must be involved, 15
class consciousness and, 233–235
kinship and, 118–120
social change and, 474, 478
social networks and, 16–17, 38

Social construction of reality, 73–75

Social control All collective efforts to ensure conformity to the norms, 209–229. *See also* Norms
crime and, 192, 221–224, 226–227
deterrence as, 216–221
prevention as, 213–216
reform and resocialization, 224–226
Social control, formal Actions by organizations and groups that exist to uphold the norms, 209, 210, 212–213, 227
Social control, informal Direct social pressure from those around us, 209, 210–212, 227
Social drift, 555
Social evolutionary theories Theories that account for the existence of a social structure on the basis of its survival benefits for societies. For example, technologically superior societies will be better able to withstand environmental challenges; hence, societies will evolve toward increased technological capacity, 111–113, 248–249, 255–256, 477
Social facts, 3–6, 67
Social institutions Relatively permanent patterns or clusters of specialized roles, groups, organizations, customs, and activities devoted to meeting fundamental social needs. Five major social institutions are the family, economy, religion, political order, and education, 109
Social integration The degree to which persons in a group have many strong attachments to one another, 200
Social interaction The process by which humans seek to influence one another, 72–77
Socialization The process by which culture is learned and internalized by each normal member of a society—much of which occurs during childhood, 68–69, 147–173. *See also* Child development
crime and deviance, 213–216
culture and personality, 158–161
education and, 464–466
vs. free will, 68–69
human development, 149–158
occupation and, 169–170, 448
sex-roles, 167–170 (*See also* Sex-roles)
Socialization, adult Processes by which adults are enabled to perform new roles, 161, 164, 466
Socialization, differential The process by which different members of the same society or even the same family are raised differently because of varying expectations about the roles that

each will need to fill as an adult, 161–166, 167, 169, 448
Social mobility A change of position within the stratification system, 238–242. *See also* Status attainment
class cultures and networks, 241–242
exchange mobility, 240–241, 275–276
in industrialized societies, 268–275
international comparisons, 270
long-distance, 270
of minorities (*See* Ethnic and racial mobility)
parents and, 272, 273, 274, 275
structural, 239–240, 269, 275–276
upward and downward, 35, 239, 269–270
Social movements Organizations created to cause or prevent social change, 83–89, 587–607. *See also* Civil Rights Movement; Feminist movement
Anti-Slavery Society, 438
approaches to, 589–590
building/supressing, 595–599
countermovements, 590, 596–597, 599
gender and, 340, 342–345
hope/expectations, 592–593, 603
network ties and, 594, 603–604
precipitating events, 593–594
reasons for, 590–592
recruitment for, 83–89, 594–596, 600–601, 602–608
volunteer characteristics, 601–605
Social network. *See* Network, social
Social order The extent to which citizens comply with important norms, 211–212
Social relationship Repeated actions between social units, or the persistence of stable, shared features among units, 15
Social sciences Those scientific fields devoted to the study of human behavior, including sociology, psychology, economics, political science, anthropology, criminology, and some branches of history, 7–9. *See also* Sociology
emergence of, 3–9, 508
free will and, 23–24
goal of, 24
Social solidarity The density and emotional intensity of attachments within a group; put another way, the capacity of group members to generate a sort of glue that enables them to stick together, to "belong," to be loyal, 15. *See also* Social order
conformity and, 81–83, 103
crime/deviance and, 195
exchange theory and, 72, 76, 77
social networks and, 15–16, 37
social theories and, 114–115

Social structure A characteristic of a group rather than of an individual, 33–34. *See also* Organizations
conflict and, 117–120
group size experiment, 97
sex ratios as, 34, 331
social change and, 474
Society A relatively self-sufficient and self-sustaining group of people who are united by social relationships and who live in a particular territory, 33
basic elements of, 107
components of, 108–110
culture and, 43, 57
as networks, 38–39
simple, 256–259
as systems, 108–115
Sociological imagination A term coined by American sociologist C. Wright Mills to describe the ability to see the link between incidents in the lives of individuals and large social forces, 6–7
Sociologists
early achievements, 8
education of, 608–611
perspective of, 6–9, 11, 13
writing skills, 611–612
Sociology The scientific study of the patterns and processes of human social relations, 7–9. *See also* Research
careers in, 608, 612
comparative, 115–120
emergence of, 3–6, 8
global perspectives, 11, 13
instructions for using this text's Internet site, 26–29
scientific process in, 21–22
at universities, 8, 608–610
Solidarity. *See* Social solidarity
Soviet Union, former, 408–409
Span of control The number of subordinates one manager can adequately supervise, often estimated as seven, 576–577
Sparta, 336–338
Specialization
culture and stratification, 248
economic development and, 308–309
industrialization and, 543
military, 261
in organizations, 570–571
political/military, 427
rise of cities, 538
urbanization and, 259, 543
Sphere of influence The area whose inhabitants depend on a city for jobs, recreation, newspapers, television, and a sense of community, 554–555

Spirit of capitalism According to Weber, a nonreligious version of the Protestant ethic; values favoring hard work, thrift, and the importance of economic success, 484

Spoils system System in which the winners take over government jobs after each election, 570

Sports, 130–131, 133, 135, 277, 318–319

Spurious relationships A false correlation between a proposed cause and its effect-for instance, both the "cause" and its "effect" may be the result of some third factor, 78–79, 80, 83, 87, 101–102, 220–221

Sri Lanka, 518–519

Stage/backstage, 165

Stake in conformity. *See* Conformity, stake in

Standard Cross-Cultural Sample (Murdock & White), 116, 160, 261, 362, 423

Standard Metropolitan Statistical Area (SMSA) According to the U.S. Census, a central city (with at least 50,000 residents) and all surrounding counties where 75 percent of the labor force is not in agriculture and where either 15 percent of the workers commute to the city or 25 percent of the workers commute from the city, 545

"Stateless societies," 423

State The organized monopoly on the use of force (or coercion) within a society; synonymous with government, 415–443

   centralization of, 580
   definition of, 420, 422
   emergence of, 261
   employees, number of, 563
   functions of, 420, 422–424
   individual freedom *vs.* regulation, 415–419
   limiting power of, 424–429, 486
   Marxist, 424–425, 580
   organizational system, 564, 569–570, 580–581
   pluralism and, 424–426, 441
   public goods and, 419–420
   repressive, 423–424
   in simple societies, 422–423
   types of, 426–428

State theory of modernization Theory that wherever the power of the state to seize private property is curtailed, free markets will appear, capitalism will develop, and modernization will occur as a result of mass efforts to become more productive, 485–486

Statistical significance, 150–151

Statistical Society of London, 5

Status The position or rank of a person or group within the stratification structure, 35. *See also* Economic status/inequality
   allocation theories, 464–466
   deviance and, 202
   education and, 464–466
   of immigrants, 52–55
   prejudice and, 291, 292, 296–301

Status, achieved A position gained on the basis of merit (in other words by achievement), 35
   in industrialized societies, 239, 268, 269, 270–271
   minorities and, 301
   rules governing, 238–239

Status, ascribed A position assigned to individuals or groups without regard for merit but because of certain traits beyond their control, such as race, sex, or parental social standing, 35
   in agrarian societies, 261, 264–265
   prejudice and discrimination, 42
   racial and ethnic differences, 285
   stratification and, 238–239

Status attainment. *See also* Ethnic and racial mobility; Social mobility
   in Canada, 273–275, 276–277
   in China, 277–278
   factors influencing, 271–273
   trends in, 275–276
   women, 272, 275

Status attainment model The process by which individuals achieve their positions in the stratification system, 271–273

Status group A subculture having a rather specific rank (or status) within the stratification system, 114

Status inconsistency A condition in which a person holds a higher position (or status) on one dimension of stratification than on another. For example, an uneducated millionaire displays status inconsistency, 237–238

Status inconsistency theories Theories built on the proposition that persons who experience status inconsistency will be frustrated and will therefore support political movements aimed at changing the stratification system, 237–238

Stereotypes
   gender and, 167–170
   of immigrants, 45, 46, 47
   prejudice and, 292
   racial/ethnic, 320–321

Steroids, 135

Stimulus-response (S-R) learning theory The theory in which behavior (responses) of organisms is said to be the result of external stimuli; that is, organisms only repeat behavior that has been reinforced by the environment, 151–156

Strangers, 283, 285

Stratification The unequal distribution of rewards (or of things perceived as valuable) among members of a society; the class structure, 35, 231–253, 255–281. *See also* Class; Status
   in agrarian societies, 259–261, 264–265, 423–424
   among nations, 486–491
   in capitalist societies, 232–233
   conflict theory of, 249–251, 267
   deviance and, 191–192
   evolutionary theory of, 248–249
   in feudal societies, 262–263
   functionalist theory of, 245–248, 249, 251, 267
   in industrial societies, 265–278
   roots of, debate on, 242–245
   in simple societies, 255, 256–259
   social evolution theory of, 248–249
   social mobility and, 238–242
   state political structure and, 423–424
   status inconsistency and, 237–238
   Weber's dimensions of, 236–237

Strength of weak ties, the Mark Granovetter's proposition that, for purposes of spreading information, weak ties are stronger, or more effective, than strong ones, 35–37, 277–278

Structural holes Unlinked pairs that show up on the network diagrams as blank space, 36

Structural mobility Mobility that occurs because of changes in the relative distribution of upper and lower statuses in a society, 239–240, 269, 275–276

Structural power Power based on statuses within social structures, 339–340

Structural strain Frustration or discontent caused by being in a disadvantaged position in the social structure, 190–192

Structural strain theories Theories that blame deviance on the stress of structural strain; for example, one such theory claims that people commit crimes because of their poverty, 191–192, 194

Student Non-Violent Coordinating Committee (SNCC), 599–605

Studied nonobservance The way in which people pretend not to notice minor lapses in one another's role performance, 165–166

Subcultural deviance Behavior through which a person deviates from the norms of the surrounding society by conforming to the norms of a subculture, 189–190, 225

Subculture A culture within a culture; a group that maintains or develops its own set of beliefs, morals, values, and norms, which usually are at variance with those of the dominant culture, 41–42, 104, 114, 242

Subjects Persons on whom an experiment is based; they are subjected to the experiment, 82

Suburb An urban place in the immediate vicinity of a city, 548, 549, 550, 554

Suicide rates, 4, 5, 6–7, 26–29, 200

Supernatural That which is beyond natural laws and limits, 389, 392–393, 405, 409

Survey research A method of sociological research that has two identifying elements. First, it is based on personal interviews or on having each individual fill out a questionnaire. Second, it is based on a random sample of the relevant population. All studies having these two features are correctly identified as survey research, 100–102, 219

Swift & Co., 567–569, 570–571

Symbolic interaction A microsociological theoretical approach that regards interaction among human beings as the fundamental social process but places far more emphasis on how people influence one another and communicate than it does on the exchange process as such, 72–75, 127

Symbol Something that stands for or indicates something else. A cross worn on a chain around a person's neck is often a symbol indicating that person is a Christian, 72, 73

Syphilis, 555–556

System(s) Anything with these three features: (1) a set of components (or parts) that are (2) interdependent and (3) maintain some degree of stability or equilibrium, 108–115, 572–575

T

Taboos, 362, 363

Taoism, 392

TARP experiment, 225–226, 227

Taxation, 263, 424, 425, 426, 483, 485, 499

Technology. *See also* Industrialization; Modernization

animal use of tools, 139–141

capitalism and, 481

computer prevalence, 267

diffusion of, 477–478, 481

fertility and, 526–527

industrialization and, 266, 476

inventions/progress, 473, 474

military, 249, 262–263

mortality rate and, 518–519

progress and, 485–486

social change and, 473

transformation of work and, 448–450

Television, 432, 477–478, 516, 517. *See also* News media

Tennessee Valley Authority, 573

Terrorism, Southern racism as, 599–600

Testosterone The male sex hormone, 135–136, 185

Theft, 4, 5, 177, 179, 213, 222

Theory An abstract statement that explains why and how certain things take place, whether these things be eclipses of the moon, chemical reactions, or outbursts of racism. In addition, scientific theories must have empirical implications. That is, theories make definite predictions and prohibitions; they say some things will happen under certain circumstances and that other things will not happen, 2, 13, 21, 77–83

Tiananmen Square demonstration, 420, 422

Tie Another word for a link or a relationship among persons or groups, 35–37, 88, 277–278. *See also* Attachments; Networks; Strength of weak ties

Time order The sequence in which variables occur and a vital test of causation. A cause must occur before its effect. This is simply to recognize that the process of cause and effect cannot work backward in time, 78, 82–83, 87, 100

Titanic (ship), 1, 75

Tool An object altered or adapted from natural materials in order to increase the ability of an organism to achieve some goal, 139–141

Trade relations, 488

Tragedy of the commons, 415–419, 425

Transportation

bus segregation/boycott, 587–588, 589, 593–599

railroads, 546, 567–568

urbanization and, 535, 545, 546, 547–550, 558

Treatment group, 214

Twins, studies of, 128, 129–132, 183

Tyranny of the majority/tyranny of the minority, 426

U

UFOs, 18, 19–20, 88–89

Unemployed Persons of legal working age who are not enrolled in school, who do not have a job, and who are actively looking for one, 452

Unification Church, 85–87, 404, 602

Unions Occupational organizations that can prevent their functions from being performed by others on the basis of contractual rights, 234, 250–251, 299–300, 573

Units of analysis The "things" on which a set of research observations are based. Sociologists use many different units of analysis, among them are individuals, small groups, large organizations, countries, cities, states, and nations, 9–10, 11, 12, 13–14, 21, 116

Universal Grammar, 154–155

Universities. *See also* Educational achievement

benefits of degrees, 462–463, 466

creating occupations, 466

departments of sociology in, 8, 608–610

enrollment in, 454

ethnic origins of faculty in, 50, 52

increased enrollment, 517

Jewish enrollment in, 50

women with professional degrees, 451–452

Unobtrusive measures Techniques used to measure behavior without disturbing the behavior of the subjects, 17–19

Upward mobility. *See* Mobility, upward and downward

Urbanization, 533–561. *See also* Cities

food supply and, 259

industrialization and, 540–543

theories of impact of, 554–558

transportation and, 547–550

Urban neighborhoods, 550–553

Urban place According to the U.S. Census, a community having at least 2,500 inhabitants, the majority of whom do not farm, 554

Urban society A society in which the majority of people do not live in rural areas, 533

Utopian One who tries to design a perfect society, 243–244

V

Validation research Studies conducted to determine whether particular measures used in research are accurate, 19–20

Values Ideals or ultimate aims; general evaluative standards about what is desirable, 39–40

family size, 523

social class and, 163–164

social movement participation and, 603, 604, 605

Variable Something that changes (varies) in that it can take more than one value, 77–78, 80, 82–83

Vertical integration The inclusion within an organization of the divisions that control every step in the production and distribution of some product or service, 568–569

Vietnamese immigrants, 54

Vietnam veterans study, 136

Vikings, 479–480

Violence. *See also* Homicide; Warfare
among the Yanomamö, 117
by chimpanzees, 140
factionalism and, 117–120
in families, 370–371
mental illness and, 186
nonviolence, 596
testosterone and, 135, 136

Violent crimes, 180–181, 184, 186–187

Visibility The degree to which a racial or an ethnic group can be recognized-how easily those in such a group can pass as members of the majority, 320–321

Visibility, principle of To the extent that behavior of group members is easily observed (or otherwise monitored) by other members, their degree of conformity to group norms will be greater, 211, 212

Voting patterns, 432–436, 437–438

Voting rights, 343, 344, 345, 427, 435, 591, 599, 600–601

W

Wages. *See* Income

Warfare
in agrarian societies, 260, 261, 262–263, 423, 424
child rearing and, 160–161
Civil War, 284
death rates, 503–504, 508
fertility and, 503–504
formal organizations and, 565–567
in hunting and gathering societies, 259
military service, 419
Napoleonic Wars, 563, 564
population trends and, 503–504, 508
in primitive societies, 118
revolutions, 233
sex ratios and, 335
state political structure and, 423
World War II, 303, 304, 306

Wealth
in agrarian societies, 260, 265
attitude shift, 484–485
in capitalist societies, 482, 484–485
of core nations, 488
expropriation of, 485
property, power, prestige, 236–237, 238
unequal distribution of, 492–493

Weather, 200–201, 479–480

Wheel, invention of, 473, 474

White-collar crime According to Sutherland, crimes committed by "a person of respectability and high social status in the course of his [her] occupation", 192–194

Women. *See* Females; Gender

Women's movement, 340, 342–345

Work. *See* Labor; Labor force; Occupations

Working class, 163–164, 242

World system theory Theory stating that some nations become modernized by exploiting other nations and that their continuing exploitation prevents less developed nations from becoming fully modernized, 486–491

World War II, 303, 304, 306, 347

Y

Yanomamö, 117

Z

Zero population growth Point at which the number of births each year equals the number of deaths, 512